Talent Identification and Development in Youth Sports

Talent Identification and Development in Youth Sports

Editors

Adam Leigh Kelly
Sergio L. Jiménez Sáiz
Sara Santos
Alberto Lorenzo Calvo

MDPI • Basel • Beijing • Wuhan • Barcelona • Belgrade • Manchester • Tokyo • Cluj • Tianjin

Editors
Adam Leigh Kelly
Centre for Life and Sport Sciences (CLaSS)
Birmingham City University
Birmingham
United Kingdom

Sergio L. Jiménez Sáiz
Centre for Sport Studies
Universidad Rey Juan Carlos
Madrid
Spain

Sara Santos
Research Center in Sports Sciences
University of Trás-os-Montes and Alto Douro
Vila Real
Portugal

Alberto Lorenzo Calvo
Faculty of Sport Sciences
Universidad Politécnica de Madrid
Madrid
Spain

Editorial Office
MDPI
St. Alban-Anlage 66
4052 Basel, Switzerland

This is a reprint of articles from the Special Issue published online in the open access journal *Sports* (ISSN 2075-4663) (available at: www.mdpi.com/journal/sports/special_issues/Talent_Sports).

For citation purposes, cite each article independently as indicated on the article page online and as indicated below:

LastName, A.A.; LastName, B.B.; LastName, C.C. Article Title. *Journal Name* **Year**, *Volume Number*, Page Range.

ISBN 978-3-0365-7693-0 (Hbk)
ISBN 978-3-0365-7692-3 (PDF)

Cover image courtesy of August Phlieger

Contents

About the Editors

Adam Leigh Kelly

Adam L. Kelly, PhD, is a Senior Lecturer and Course Leader for Sports Coaching and Physical Education at Birmingham City University, United Kingdom. Alongside attaining his PhD from the University of Exeter, United Kingdom, Adam is a Senior Fellow of the HEA, BASES Sport and Exercise Scientist, and FA UEFA A and AYA Licenced Coach. Broadly, his research interests explore organisational structures in youth sports to better understand the athlete development process and create more appropriate settings. He is currently collaborating with a number of regional, national, and international organisations across a range of sports including cricket, rugby, soccer, squash, and swimming.

Sergio L. Jiménez Sáiz

Sergio L. Jiménez-Sáiz, PhD, is a Senior Lecturer at Universidad Rey Juan Carlos in Madrid, Spain and professional basketball coach in ACB League (1st Spanish League). His research considers the long-term talent development in sports and physical activity in different areas of sports science, such as strength training, monitoring training load, nutrition, sleep, supplementation and psychosocial factors. He works with high performance sports teams and athletes around the world to help achieve the goals they have set for themselves.

Sara Santos

Sara Santos, PhD, holds a position as researcher at the University of Trás-os-Montes and Alto Douro (UTAD), Portugal. She is an integrated member of the Research Center in Sports Sciences, Health Sciences and Human Development (CIDESD) - CreativeLab Research Community. Her main research interests are focused on exploring the influence of enrichment environments to nurture creative behaviour in youths. Additionally, she is the mentor of the Skills4Genius program, which was awarded by the Calouste Gulbenkian Foundation and the Olympic Committee of Portugal. Finally, she is the 2018 recipient of the Ruth B. Noller Award. This award supports paradigm shifting research with high potential for impact communities of practice and was offered by the Creative Education Foundation, USA.

Alberto Lorenzo Calvo

Alberto Lorenzo Calvo, PhD, is a Senior Lecturer at Universidad Politécnica in Madrid (Spain) and a professional basketball coach in Spain (ACB League, Leb Gold) or at Great Britain (Head Coach of GB basketball). His research interests explore talent identification and long-term talent development, usually in team-sports, in different areas of sports science, such as monitoring training load and psychosocial factors. He is currently working with high performance sports teams and athletes around the world to help achieve the goals they have set for themselves.

Preface to "Talent Identification and Development in Youth Sports"

How do talented children achieve expertise at adulthood? The identification of young athletes with the potential to develop and then subsequently excel as senior professionals in their specialist sport remains one of the major contemporary challenges for national governing bodies, sports clubs, coaches, and practitioners. As such, talent identification and development strategies have become pivotal to sporting organisations. The complexity of the process of nurturing youth through development trajectories towards adulthood has enhanced over recent years, with the implementation of modern multidisciplinary paradigms becoming increasingly prevalent in youth sport settings.

As part of this Special Issue, we sought to collaborate with researchers within the disciplines of talent identification and development in youth sports. More specifically, the overarching aim was to explore how youth's personal engagement in activities (i.e., the what), quality social dynamics (i.e., the who), and appropriate settings and organisational structures (i.e., the where) can foster immediate, short- and long-term developmental outcomes (see the Personal Assets Framework by Côté et al., 2014, 2016). In doing so, it was hoped that contributing research can inform evidence-based youth sport policies and athlete development programmes. Submissions were encouraged from a diverse range of qualitative and quantitative data collection procedures to explore the current context of talent identification and development in youth sports. In addition, commentaries, conceptual papers, and reviews were welcome to synthesise expert knowledge within this discipline.

In total, 34 articles (31 empirical studies and three reviews) were included from 128 authors, many of whom are internationally recognised scholars and emerging researchers in this field. These authors represent universities or sport institutions (i.e., professional sport teams or national governing bodies) from across 20 countries. Owing to our approach, it is not surprising that 17 sports are studied within this Special Issue, which includes 13 studies comprising female participants. Overall, we believe our initial aims of progressing the talent identification and development literature have been achieved, and now hope that the research presented can be utilised by key stakeholders (e.g., administrators, coaches, parents, practitioners) and organisational structures (e.g., national governing bodies, professional clubs, recreational teams, youth sport associations) to create more appropriate youth sport settings.

Adam Leigh Kelly, Sergio L. Jiménez Sáiz, Sara Santos, and Alberto Lorenzo Calvo
Editors

sports MDPI

Editorial

Special Issue "Talent Identification and Development in Youth Sports"

Adam Leigh Kelly [1,*], Alberto Lorenzo Calvo [2], Sara Diana Leal dos Santos [3] and Sergio Lorenzo Jiménez Sáiz [4]

1 Research Centre for Life and Sport Sciences (CLaSS), Department of Sport and Exercise, School of Health Sciences, Birmingham City University, Birmingham B15 3TN, West Midlands, UK
2 Departamento de Deportes, Facultad de Ciencias de la Actividad Física y del Deporte—INEF, Universidad Politécnica de Madrid, 28040 Madrid, Spain
3 Research Center in Sports Sciences, Health Sciences and Human Development (CIDESD), University of Trás-os-Montes and Alto Douro, 5000-801 Vila Real, Portugal
4 Centre for Sport Studies, Universidad Rey Juan Carlos, Fuenlabrada, 28942 Madrid, Spain
* Correspondence: adam.kelly@bcu.ac.uk

Citation: Kelly, A.L.; Calvo, A.L.; dos Santos, S.D.L.; Jiménez Sáiz, S.L. Special Issue "Talent Identification and Development in Youth Sports". *Sports* **2022**, *10*, 189. https://doi.org/10.3390/sports10120189

Received: 28 October 2022
Accepted: 21 November 2022
Published: 24 November 2022

1. Introduction

We are delighted to share our Special Issue on Talent Identification and Development in Youth Sports. In 2020, the editorial team had several informal discussions about the growing interest from researchers and practitioners in these disciplines and subsequently wanted to create a platform to help advance this field of literature. Following these conversations, we decided to use the Personal Assets Framework [1–3] to outline our objectives and the potential research topics for our Special Issue. The overarching aim was to explore how youth's personal engagement in activities (i.e., the what), quality social dynamics (i.e., the who), and appropriate settings and organisational structures (i.e., the where) can foster immediate, short-term, and long-term developmental outcomes in sport. In doing so, it was hoped that the studies included can inform evidence-based youth sport policies and athlete development programmes. Submissions were encouraged from a diverse range of quantitative and qualitative research methods to examine the current context of talent identification and development in youth sport, as well as reviews to synthesise knowledge within these disciplines.

In light of the articles that have been included within our special issue, we believe our initial aims of progressing the talent identification and development literature have been achieved. We now hope that the research presented can be utilised by key stakeholders (e.g., administrators, coaches, parents, practitioners) and organisational structures (e.g., national governing bodies, professional clubs, recreational teams, youth sport associations) to create more appropriate youth sport settings. To summarise the key messages of the studies in our special issue, this editorial focusses on two fundamental considerations: (a) contextual, and (b) methodological.

2. Contextual Considerations

Our Special Issue has supplemented the existing literature to show that contextual factors (e.g., age, gender, nationality, sport popularity, sport type) play an important role in talent identification and development. In particular, the role of birthday and birthplace have been strongly associated with a greater likelihood of being selected into talent pathways and achieving professional status. In relation to 'relative age effects', relatively older athletes (i.e., those born near the start of the selection cut-off date) are generally considered to have greater potential and likelihood of being selected. This provides them with more exposure to organised activities (e.g., practice, competition) and resources (e.g., facilities, specialist support), which can facilitate their long-term performance towards adulthood [4].

Conversely, relatively younger athletes (i.e., those born near the end of the selection cut-off date) are less likely to access the same opportunities, which could lead to a negative impact on participation and personal development in the long-term [5,6]. Relative age effects appeared to be the most popular topic in our Special Issue, encompassing twelve studies [7]. For instance, Kelly and colleagues [8] showed relatively older athletes (i.e., those born in the first three months of the year) were up to ten times more likely to be selected into the Basketball England Talent Pathway compared to relatively younger athletes (i.e., those born in the last three months of the selection year); however, findings were more significant in males compared to females. Moreover, McCarthy and colleagues [9] explained the possible mechanisms of relative age effects in rugby union, proposing how challenge was an ever-present feature of all players journeys, especially at the point of transition to senior rugby, whilst psycho-behavioural factors seemed to be a primary mediator of the response to challenge. In addition, Romann and colleagues [10] suggested relative age effects led to inefficient talent selection and an accompanying waste of money when exploring male professional football players' market values.

Another contextual factor studied that can influence the probability of an athlete participating in high-performance sport is their birthplace (i.e., where the athlete is born and raised). A range of quantitative studies have concluded that athletes born in small cities (<500,000 inhabitants) are more likely to play in professional leagues compared with athletes born in larger cities (>500,000 inhabitants). These 'birthplace effects' are likely due to smaller cities being associated with greater development opportunities, effortlessness in the mobility of the athlete, and safety conditions for practice and competition. Within our Special Issue, Maayan and colleagues [11] used a qualitative approach to explore athlete and coach perceptions of birthplace effects from a range of sports, revealing that growing up in cities of small and medium sizes was more beneficial than growing up in towns or cities of other sizes. Most of the coaches they interviewed believed that certain characteristics of the place or city where the athlete grew up (e.g., proximity to sport facilities, access to organised activities) is a significant contributing factor towards talent development.

It is important that the impact of relative age and/or birthplace is not considered to be homogenous. Moving forward, it will be worthwhile to increase the studies of relative age and birthplace to better understand the appropriate climate to develop young athletes in a variety of contexts. Practitioners should be cautious of these influences during talent identification and development, whilst researchers should focus on advancing our understanding of the potential barriers and relationships. This should be conducted both across and within different countries, as the interaction between contextual factors can help explain the trajectory and performance of athletes [12]. Moreover, understanding the connection between relative age and birthplace effects will support researchers and policy makers to design sports systems and policies that help nurture talent more accurately and equitably. However, we already know that producing favourable environments for talent identification and development are highly complex tasks due to the multidimensional nature of development coupled with the continued evolution of sport performance. Thus, researchers are encouraged to design more longitudinal, multidimensional, and prospective studies in order to capture the trajectories of youth athletes in diverse youth sports settings.

Another approach that has received growing attention, as highlighted by Sæther and colleagues [13], is the 'talent transfer'. This refers to the intention of a talented athlete choosing to invest in other sports, such as transferring athletes from donor sports into target sports or the transition from summer (e.g., kayaking or rowing) to winter (e.g., cross-country skiing) sports. Thus, it is important to deepen our knowledge to understand: (a) the range of elements (e.g., technical, tactical, physical, psychological, social) that act as facilitators to help athletes successfully transfer from one sport to another, (b) whether there are donor sports that are more suitable for recruitment into other sports, and (c) the effectiveness of existing talent transfer programmes. Considering this, practitioners should be aware that there are many critical determinants to talent identification and development; therefore, the incorporation of multi-method research approaches across and within different sporting and sociocultural contexts could be an avenue for future research.

3. Methodological Considerations

The editorial team made a concerted effort to recruit diverse authors who could support this research topic, and we are incredibly grateful to all the researchers for their strong contributions. We also aimed to capture varied methodological approaches to ensure the Special Issue offered a unique contribution to the field of talent identification and development in youth sport. The 34 articles (31 empirical studies and 3 reviews) that appear were penned by 128 authors (several whom appear more than once), many of which are internationally recognised scholars in this field. These authors represent universities or sport institutions (i.e., professional sport teams or national governing bodies) from across 20 countries. Owing to our approach, it is not surprising that 17 sports are studied within this Special Issue, which includes 13 studies comprising female participants. While soccer remains at the fore of the evidence-base with twelve articles, it is pleasing to see more under-researched talent development contexts also being considered (e.g., [14]). Ultimately, we believe that the diverse authorship, contexts, and samples have resulted in a unique Special Issue, which has significantly advanced the field through its various approaches that should facilitate thoughtful discussion about talent identification and development.

A range of quantitative (e.g., Bayesian machine learning, chi-square, coding, discriminant analyses, mixed multilevel logistic models) and qualitative (e.g., abductive hierarchical content analysis, ethnography, inductive approach, realist evaluation approach, reflexive thematic analysis) data collection and analysis methods were used throughout this Special Issue. Not only do these approaches provide novel insights into talent identification and development, but they also offer researchers the opportunity to replicate studies in different settings. One rapidly emerging quantitative analysis approach that was used twice in this Special Issue [15,16] is machine learning. For instance, Owen and colleagues [16] used Bayesian machine learning to create predictive models for selected and non-selected Welsh male U16 and U18 rugby players. Whilst they showed their physiological and psychosocial models correctly classified 67.5% and 62.3% of all players, respectively, they also provided a unique method to explore selection into talent pathways that may be replicable to other researchers in the future.

From a qualitative perspective, Lara-Bercial and McKenna [17,18] produced a two-paper series using a season long ethnography of a youth performance sport club based on a novel realist evaluation approach. In the first study [17], the authors detailed the perceptions of club stakeholders to build a set of programme theories, with the resulting network of outcomes (i.e., self, emotional, social, moral, and cognitive) and generative mechanisms (i.e., the attention factory, the greenhouse for growth, the personal boost, and the real-life simulator) providing a nuanced understanding of stakeholders' views and experiences. In the second study [18], the lead author spent a full season in the club, whereby the collection of context–mechanism–outcome networks (CMONs) described in the first study was used to guide the researcher during their immersive period. Such qualitative approaches offer a different lens to those typically incorporated in talent identification and development, which may shed light on findings that may not be captured during traditional approaches.

It is also important to consider how articles can be grounded or discussed through relevant models or theories (e.g., [19,20]). As a novel example, Kelly and colleagues [21] used the Personal Assets Framework to explain the immediate, short-term, and long-term developmental outcomes due to relative age effects in English male cricket. Indeed, using such an approach helps give the study a well-defined and proven basis of argument or phenomena, offers an explanation of the study's significance and validity, and shows where the researcher intends to fill in gaps of knowledge and practice. Moving forward, we lend our guidance using Barraclough and colleagues' [22] narrative review, which summarised methodological approaches to talent identification in team sports. The authors highlight the benefits of longitudinal, multidisciplinary, and ecologically valid research designs. Specifically, they outline three key areas for consideration for future research: (a) the timespan of the research design, (b) the use of monodisciplinary or multidisciplinary variables, and (c) the fidelity of the methodological approaches to the assessment of talent. One final methodological consideration, as highlighted by Mosher and colleagues [23], is the growing issue of the language that is used in sport science literature. Following their investigation into early specialisers, the authors underscored that the main rationale related to their study was the lack of terminological and conceptual concreteness related to this topic. As such, it is necessary to improve and develop new methodologies and concrete definitions that allow us to access much more reliable and externally valid data.

4. Future Directions

Grounded on contemporary talent identification and development knowledge, this Special Issue provides useful insights to drive future advances in research and practice. Currently, gaps arise when considering the limitations of unidimensional assessment models (i.e., static and isolated variables) to identify talent and capture the dynamic nature of sports. Thus, it is necessary that future studies conduct more valid, reliable, and multidisciplinary assessment procedures in an effort to more accurately and resourcefully identify and develop talented youth athletes (e.g., [24,25]). Furthermore, several studies included in our Special Issue suggest that future directions should consider longitudinal tracking of interacting factors [26], such as the quantity, quality, and type of practice during sport participation [27], as well as monitoring biological maturation [28,29], physical performance [30,31], technical skills [32], and psychological profiles [33,34]. Evaluating how stakeholders communicate and implement these theoretical findings into applied settings will also be important for future research to ensure they are adequately deployed (e.g., [35–37]). Finally, it is vital for future research to consider the contextual and methodological implications of COVID-19 in youth sport (see [38] for an overview), since the immediate, short-term, and the long-term impact of a global pandemic on the identification and development of young athletes remains relatively unknown and could have enduring consequences [39].

5. Summary

The main purpose of research is to enhance real-life settings by advancing knowledge through the development of scientific theories, concepts, and ideas. In relation to this Special Issue, it is hoped that the articles presented can be utilised by key stakeholders and organisational structures to create more appropriate youth sport settings. It was a pleasure to assemble this resource and is hoped the contextual and methodological considerations presented throughout this editorial provide researchers and practitioners with a range of thought-provoking concepts for their respective agendas. Thank you to all the contributing authors and reviewers, without whom, this research topic would not be possible.

Author Contributions: Conceptualization, A.L.K., A.L.C., S.D.L.d.S. and S.L.J.S.; writing—original draft preparation, A.L.K.; writing—review and editing, A.L.K., A.L.C., S.D.L.d.S. and S.L.J.S. All authors have read and agreed to the published version of the manuscript.

Funding: This research received no external funding.

Acknowledgments: The Guest Editors express sincere gratitude to the valuable contributions by all authors, reviewers, and the editorial team of Sports.

Conflicts of Interest: The authors declare no conflict of interest.

References

1. Côté, J.; Turnnidge, J.; Evans, M.B. The Dynamic Process of Development Through Sport. *Kinesiol. Slov. Sci. J. Sport* **2014**, *20*, 14–26.
2. Côté, J.; Turnnidge, J.; Vierimaa, M. A Personal Assets Approach to Youth Sport. In *Handbook of Youth Sport*; Smith, A., Green, K., Eds.; Routledge: Abingdon, UK, 2016; pp. 243–256.

3. Côté, J.; Turnnidge, J.; Murata, A.; McGuire, C.; Martin, L. Youth Sport Research: Describing the Integrated Dynamics Elements of the Personal Assets Framework. *Int. J. Sport Psychol.* **2020**, *51*, 562–578. [CrossRef]
4. Huard Pelletier, V.; Lemoyne, J. Early Sport Specialization and Relative Age Effect: Prevalence and Influence on Perceived Competence in Ice Hockey Players. *Sports* **2022**, *10*, 62. [CrossRef] [PubMed]
5. Smith, K.L.; Weir, P.L. An Examination of Relative Age and Athlete Dropout in Female Developmental Soccer. *Sports* **2022**, *10*, 79. [CrossRef] [PubMed]
6. Smith, K.L.; Bélanger, M.; Chittle, L.; Dixon, J.C.; Horton, S.; Weir, P.L. Does Relative Age Influence Organized Sport and Unorganized Physical Activity Participation in a Cohort of Adolescents? *Sports* **2022**, *10*, 97. [CrossRef]
7. Peterson, T.; Ek, S.; Thorsson, O.; Karlsson, M.K.; Dencker, M.; Wollmer, P. Changes in Athletic Performance in Children Attending a Secondary School with a Physical Activity Profile. *Sports* **2022**, *10*, 71. [CrossRef]
8. Kelly, A.L.; Jiménez Sáiz, S.L.; Lorenzo Calvo, A.; de la Rubia, A.; Jackson, D.T.; Jeffreys, M.A.; Ford, C.; Owen, D.; Santos, S.D.L.D. Relative Age Effects in Basketball: Exploring the Selection into and Successful Transition Out of a National Talent Pathway. *Sports* **2021**, *9*, 101. [CrossRef] [PubMed]
9. McCarthy, N.; Taylor, J.; Cruickshank, A.; Collins, D. Happy Birthday? Relative Age Benefits and Decrements on the Rocky Road. *Sports* **2022**, *10*, 82. [CrossRef]
10. Romann, M.; Javet, M.; Cobley, S.; Born, D.-P. How Relative Age Effects Associate with Football Players' Market Values: Indicators of Losing Talent and Wasting Money. *Sports* **2021**, *9*, 99. [CrossRef]
11. Maayan, Z.; Lidor, R.; Arnon, M. The Birthplace Effect in 14–18-Year-Old Athletes Participating in Competitive Individual and Team Sports. *Sports* **2022**, *10*, 59. [CrossRef] [PubMed]
12. Leite, N.; Arede, J.; Shang, X.; Calleja-González, J.; Lorenzo, A. The Influence of Contextual Aspects in Talent Development: Interaction Between Relative Age and Birthplace Effects in NBA-Drafted Players. *Front. Sport. Act. Living* **2021**, *3*, 642707. [CrossRef] [PubMed]
13. Sæther, S.A.; Iversen, M.; Talsnes, R.K.; Sandbakk, Ø. Comparison of High and Low Responders to a Cross-Country Skiing Talent Transfer Program: A Coach's Perspective. *Sports* **2021**, *9*, 138. [CrossRef] [PubMed]
14. Storli, L.; Aune, M.A.; Lorås, H. Aspects of Developmental Pathways toward World-Class Parasport. *Sports* **2022**, *10*, 123. [CrossRef]
15. Kelly, A.L.; Williams, C.A.; Cook, R.; Sáiz, S.L.J.; Wilson, M.R. A Multidisciplinary Investigation into the Talent Development Processes at an English Football Academy: A Machine Learning Approach. *Sports* **2022**, *10*, 159. [CrossRef]
16. Owen, J.; Owen, R.; Hughes, J.; Leach, J.; Anderson, D.; Jones, E. Psychosocial and Physiological Factors Affecting Selection to Regional Age-Grade Rugby Union Squads: A Machine Learning Approach. *Sports* **2022**, *10*, 35. [CrossRef] [PubMed]
17. Lara-Bercial, S.; McKenna, J. Looking Back and Looking Around: How Athletes, Parents and Coaches See Psychosocial Development in Adolescent Performance Sport. *Sports* **2022**, *10*, 47. [CrossRef]
18. Lara-Bercial, S.; McKenna, J. Roots to Grow and Wings to Fly: An Ethnography of Psychosocial Development in Adolescent Performance Sport. *Sports* **2022**, *10*, 48. [CrossRef]
19. Dimundo, F.; Cole, M.; Blagrove, R.C.; Till, K.; Kelly, A.L. A Multidisciplinary Investigation into the Talent Development Processes in an English Premiership Rugby Union Academy: A Preliminary Study through an Ecological Lens. *Sports* **2022**, *10*, 13. [CrossRef]
20. Duncan, M.J.; Martins, R.; Noon, M.; Eyre, E.L.J. Perception of Affordances for Dribbling in Soccer: Exploring Children as Architects of Skill Development Opportunity. *Sports* **2022**, *10*, 99. [CrossRef]
21. Kelly, A.L.; Brown, T.; Reed, R.; Côté, J.; Turnnidge, J. Relative Age Effects in Male Cricket: A Personal Assets Approach to Explain Immediate, Short-Term, and Long-Term Developmental Outcomes. *Sports* **2022**, *10*, 39. [CrossRef]
22. Barraclough, S.; Till, K.; Kerr, A.; Emmonds, S. Methodological Approaches to Talent Identification in Team Sports: A Narrative Review. *Sports* **2022**, *10*, 81. [CrossRef]
23. Mosher, A.; Fraser-Thomas, J.; Wilson, M.J.; Baker, J. Proportions of Early Specializers Varies According to Methods and Skill Level. *Sports* **2022**, *10*, 34. [CrossRef] [PubMed]
24. Banjevic, B.; Zarkovic, B.; Katanic, B.; Jabucanin, B.; Popovic, S.; Masanovic, B. Morphological Characteristics and Situational Precision of U15 and U16 Elite Male Players from Al-Ahli Handball Club (Bahrein). *Sports* **2022**, *10*, 108. [CrossRef]
25. Lemoyne, J.; Brunelle, J.-F.; Huard Pelletier, V.; Glaude-Roy, J.; Martini, G. Talent Identification in Elite Adolescent Ice Hockey Players: The Discriminant Capacity of Fitness Tests, Skating Performance and Psychological Characteristics. *Sports* **2022**, *10*, 58. [CrossRef]
26. Fuhre, J.; Øygard, A.; Sæther, S.A. Coaches' Criteria for Talent Identification of Youth Male Soccer Players. *Sports* **2022**, *10*, 14. [CrossRef] [PubMed]
27. Aalberg, M.; Roaas, T.V.; Aune, M.A.; Bjerke, Ø.; Aune, T.K. In Pursuit of a Comprehensive Understanding of Expertise Development: A Comparison between Paths to World-Class Performance in Complex Technical vs. Endurance Demanding Sports. *Sports* **2022**, *10*, 16. [CrossRef]
28. Radnor, J.M.; Staines, J.; Bevan, J.; Cumming, S.P.; Kelly, A.L.; Lloyd, R.S.; Oliver, J.L. Maturity Has a Greater Association than Relative Age with Physical Performance in English Male Academy Soccer Players. *Sports* **2021**, *9*, 171. [CrossRef] [PubMed]
29. Larkin, P.; Woods, C.T.; Haycraft, J.; Pyne, D.B. Physical and Anthropometric Characteristics Do Not Differ According to Birth Year Quartile in High-Level Junior Australian Football Players. *Sports* **2021**, *9*, 111. [CrossRef]

30. Abarghoueinejad, M.; Baxter-Jones, A.D.G.; Gomes, T.N.; Barreira, D.; Maia, J. Motor Performance in Male Youth Soccer Players: A Systematic Review of Longitudinal Studies. *Sports* **2021**, *9*, 53. [CrossRef]
31. Keiner, M.; Kapsecker, A.; Stefer, T.; Kadlubowski, B.; Wirth, K. Differences in Squat Jump, Linear Sprint, and Change-of-Direction Performance among Youth Soccer Players According to Competitive Level. *Sports* **2021**, *9*, 149. [CrossRef]
32. Sørensen, A.; Haugen, E.C.; van den Tillaar, R. Is There a Sex Difference in Technical Skills among Youth Soccer Players in Norway? *Sports* **2022**, *10*, 50. [CrossRef] [PubMed]
33. Consoni, C.; Pesce, C.; Cherubini, D. Early Drop-Out from Sports and Strategic Learning Skills: A Cross-Country Study in Italian and Spanish Students. *Sports* **2021**, *9*, 96. [CrossRef]
34. Greenlees, I.; Parr, A.; Murray, S.; Burkitt, E. Elite Youth Soccer Players' Sources and Types of Soccer Confidence. *Sports* **2021**, *9*, 146. [CrossRef] [PubMed]
35. Dean, F.; Kavanagh, E.; Wilding, A.; Rees, T. An Examination of the Experiences of Practitioners Delivering Sport Psychology Services within English Premier League Soccer Academies. *Sports* **2022**, *10*, 60. [CrossRef] [PubMed]
36. McBurnie, A.J.; Dos'Santos, T.; Johnson, D.; Leng, E. Training Management of the Elite Adolescent Soccer Player throughout Maturation. *Sports* **2021**, *9*, 170. [CrossRef]
37. Taylor, J.; Ashford, M.; Collins, D. Tough Love—Impactful, Caring Coaching in Psychologically Unsafe Environments. *Sports* **2022**, *10*, 83. [CrossRef]
38. Kelly, A.L.; Erickson, K.; Turnnidge, J. Youth sport in the time of COVID-19: Considerations for researchers and practitioners. *Manag. Sport Leis.* **2022**, *27*, 62–72. [CrossRef]
39. Costello, I.; Belton, S.; MacNamara, Á. Stand up and Fight: A Case Study of a Professional Rugby Club Negotiating a COVID-19 Crisis, a Talent Development Perspective. *Sports* **2022**, *10*, 124. [CrossRef]

sports

Article

A Multidisciplinary Investigation into the Talent Development Processes at an English Football Academy: A Machine Learning Approach

Adam L. Kelly [1,*], Craig A. Williams [2], Rob Cook [1], Sergio Lorenzo Jiménez Sáiz [3] and Mark R. Wilson [2]

1 Research Centre for Life and Sport Sciences (CLaSS), Faculty of Health, Education and Life Sciences, Birmingham City University, Birmingham B15 3TN, West Midlands, UK
2 Children's Health and Exercise, Research Centre and Sport and Health Sciences, College of Life & Environmental Sciences, University of Exeter, Exeter EX1 2LU, Devon, UK
3 Centre for Sport Studies, Universidad Rey Juan Carlos, Fuenlabrada, del Molino, 5, 28942 Madrid, Spain
* Correspondence: adam.kelly@bcu.ac.uk

Abstract: The talent development processes in youth football are both complex and multidimensional. The purpose of this two-fold study was to apply a multidisciplinary, machine learning approach to examine: (a) the developmental characteristics of under-9 to under-16 academy players (n = 98; Study 1), and (b) the characteristics of selected and deselected under-18 academy players (n = 18; Study 2). A combined total of 53 factors cumulated from eight data collection methods across two seasons were analysed. A cross-validated Lasso regression was implemented, using the glmnet package in R, to analyse the factors that contributed to: (a) player review ratings (Study 1), and (b) achieving a professional contract (Study 2). Results showed non-zero coefficients for improvement in subjective performance in 15 out of the 53 analysed features, with key findings revealing advanced percentage of predicted adult height (0.196), greater lob pass (0.160) and average dribble completion percentage (0.124), more total match-play hours (0.145), and an older relative age (BQ1 vs. BQ2: −0.133; BQ1 vs. BQ4: −0.060) were the most important features that contributed towards player review ratings. Moreover, PCDEQ Factor 3 and an ability to organise and engage in quality practice (PCDEQ Factor 4) were important contributing factors towards achieving a professional contract. Overall, it appears the key factors associated with positive developmental outcomes are not always technical and tactical in nature, where coaches often have their expertise. Indeed, the relative importance of these factors is likely to change over time, and with age, although psychological attributes appear to be key to reaching potential across the academy journey. The methodological techniques used here also serve as an impetus for researchers to adopt a machine learning approach when analysing multidimensional databases.

Keywords: talent identification; expertise; psychological characteristics; physical characteristics; technical and tactical; elite youth soccer

Citation: Kelly, A.L.; Williams, C.A.; Cook, R.; Sáiz, S.L.J.; Wilson, M.R. A Multidisciplinary Investigation into the Talent Development Processes at an English Football Academy: A Machine Learning Approach. *Sports* **2022**, *10*, 159. https://doi.org/10.3390/sports10100159

Academic Editors: Adam Baxter-Jones and Michael Chia

Received: 19 July 2022
Accepted: 13 October 2022
Published: 19 October 2022

1. Introduction

It is widely acknowledged that the talent development processes in youth football are both complex and multidimensional [1]. Although various factors have been identified to influence the talent development processes in youth football, only a few multidisciplinary studies exist. As an example, Huijgen and colleagues [2] applied a battery of objective field tests and questionnaires within the four domains of technical, tactical, physiological, and psychological characteristics to players aged 16 to 18 years. It was revealed that selected players outperformed their deselected counterparts, whereby performance in the technical skill of dribbling, the tactical characteristics of positioning and deciding, and the physiological attribute of sprinting correctly classified 69% of talented players. Moreover, Forsman and colleagues [3] examined multiple factors of youth football players at aged 15 years that eventually contributed to successful football performance at aged 19 years. Performance at aged 19 years was associated with technical (i.e., passing), tactical (i.e., centering), physiological (i.e., agility), and psychological (i.e., motivation) attributes that were displayed at aged 15 years. In addition, Zuber and colleagues [4] observed holistic patterns as an instrument for predicting the performance in promising young football players over a three-year period. They revealed that highly skilled players scored above average on all technical, physiological, and psychological factors. Collectively, this research reinforces the importance of providing a multidimensional research methodology in youth football when exploring the talent development processes.

The multidisciplinary nature of the talent development process is also reflected in several theoretical (e.g., Personal Assets Framework [5,6]) and practical (e.g., Locking Wheel Nut Model [7]) frameworks. One practically based model that is particularly relevant to the talent development processes in youth football is The Football Association's Four Corner Model (FCM) [8]. The FCM is often adopted in professional football clubs and organisations in England (amongst other countries), which advocates the assessment and development of players according to: (a) *technical/tactical*, (b) *physical*, (c) *psychological*, and (d) *social* attributes. Towlson and colleagues [9] applied the FCM to their qualitative methodology whilst examining the perceived importance that practitioners placed on the four sub-components during player selection in academy football. It was discovered that the psychological sub-component was rated significantly higher than the other three sub-components. Likewise, Kelly and colleagues [10] adopted the FCM in a quantitative analysis of factors differentiating those who 'play-up' an age-group compared to those who do not based on age phase (i.e., Foundation Development Phase [FDP]: under-9 to under-11; Youth Development Phase [YDP]: under-12 to under-16). Technical/tactical and social characteristics appeared to differentiate those who play-up compared to those who do not at ages 8 to 11 years, whereas there were measures representing all four sub-components from the FCM for those aged 11 to 16 years. Since the FCM is a tool that is perceived to be relevant and useful for football coaches and practitioners [11,12], it provides a salient framework for understanding the factors associated with talent development and thus may facilitate important knowledge translation.

With talent development being inherently multifactorial, explorative studies must employ analysis techniques that can handle multiple competing, and possible correlated, features. Traditional regression techniques inherently struggle to estimate model coefficients when the number of independent variables (IVs) is comparable to the number of observations, though the emerging family of feature selection algorithms from the machine learning discipline offer possible solutions. In the case of regression, feature selection is often achieved by including a penalization term during the model fit, such as the Lasso first proposed by Tibshirani [13], creating a scenario whereby the optimal model is that which explains the most of the data with the fewest parameters. These penalized regression routines present themselves as a potential tool for rapidly summarising observational retrospective data, as well as generating new insights and testable hypotheses. The advantage of such approaches is that they can effectively process large amounts of data for key features in a cost effective and timely manner. Within the remit of talent development, the machine learning approaches here do not aim to answer the deep questions of what leads to optimal performance, but instead seek to demonstrate a method to leverage some of the quantities of available data to generate new hypotheses and insights [14].

Although the use of machine learning as a statistical analysis method in sport science research is very much in its infancy, there is an increasing amount of literature that has applied such methods, including competition outcome predictions [15], human movement [16], practice history [17], and injury risk [18]. From a talent development perspective, preliminary studies in cricket have used non-linear machine learning (pattern recognition) techniques to examine various factors that contribute to 'super-elite' status [19,20]. As an example, Jones and colleagues [19] showed how a subset of 18 features (from 658 collected) differentiated 'super-elite' (i.e., high-profile international) and 'elite' (domestic professional) senior batsmen with excellent classification accuracy (96%). Moreover, Musa and colleagues [21] classified and predicted 'high-potential' archers from a set of variables trained on a variation of k-NN algorithms and logistic regression. Weighted k-NN outperformed all the tested models with reasonably good accuracy (83%) for the prediction of 'high-potential' (i.e., top of group) and 'low-potential' (i.e., bottom of group) developmental level (aged 13 to 20 years) archers. Most recently, Owen and colleagues [22] used a Bayesian machine learning approach to identify the physiological and psychosocial models that predict selection to a regional age-grade rugby union team. They showed their physiological models correctly classified 67.55% of all players, whereas their psychosocial models correctly classified 62.26% of all players.

It is also important to consider the intra- and inter-contextual factors when designing and evaluating talent development processes, since predictive features (i.e., technical/tactical, physical, psychological, social) in youth football can differ compared to other sports (e.g., cricket, archery, rugby union) and within football-specific environments (e.g., soccer, beach soccer, sepak takraw [23]), respectively. Thus, this current study aims to add to the growing body of literature that has applied machine learning techniques in sport to better understand talent development processes in youth football. The purpose of this two-fold study was to use machine learning algorithms to: (a) explore the multidimensional developmental characteristics of under-9 to under-16 football academy players based on coaches review ratings, and (b) examine the multidimensional characteristics that differentiated selected and deselected under-18 football academy players based on achieving a professional contract.

2. Exploring the Developmental Characteristics of under-9 to under-16 Football Academy Players

Professional football academies and governing bodies aim to foster player development pathways towards expertise through adopting evidence-based philosophies (see the Elite Player Performance Plan via The Premier League [24]). In England, young players join an academy on schoolboy terms between the ages of 8 and 16 years (i.e., part time attendance). At aged 16 years, those players who show continued progress are selected to undertake a two year, full-time youth training scheme known as an academy scholarship. Upon completion of their scholarship, players either sign a professional contract or are released. These developmental stages have been divided into three phases to capture the possible age-specific requirements: (a) FDP (under-9 to under-11), (b) YDP (under-12 to under-16), and (c) Professional Development Phase (PDP; under-17 to under-21). In the pursuit of developing male players towards their respective senior team, professional clubs and organisations continue to invest a significant monetary outlay towards human (e.g., coaches, specialist support staff) and physical (e.g., facilities, specialist equipment) resources [25]. In order to better understand the talent development processes in youth football, it is important to identify factors that may influence the progression of schoolboy players (i.e., FDP and YDP). By doing so, it will also help inform key stakeholders (e.g., coaches, practitioners, policy makers) to create evidence-based policies that will offer each individual the most suitable opportunity to achieve a scholarship and professional contract (i.e., PDP).

The purpose of this study was to examine a range of factors based on the FCM (i.e., technical/tactical, physical, psychological, social) that may have contributed to under-9 to under-16 academy football players' review ratings across two seasons using a machine learning approach.

2.1. Methods

2.1.1. Sample

Ninety-eight male participants were recruited from under-9 to under-16 age groups. All the participants were from the same tier four English professional football club and their category three academy. The average weekly training and match-play time was

9–10.5 training hours/week with one match-play hour/week for the FDP players, and 10–14.5 training hours/week with one match-play hour/week for the YDP players. Goalkeepers were not included in this study due to their contrasting position-specific requirements [26]. Parental consent and player assent were collected prior to the study commencing. The study was approved by the Ethics Committee of Sport and Health Sciences at the University of Exeter.

2.1.2. Measures and Procedures

The dataset comprised of eight data collection methods that were collected twice during two football seasons (2014–2015 and 2015–2016) to test year-on-year developmental outcomes (see Figure 1 for a timeline of the data collection). These measures were then allocated into the four sub-components in-line with the FCM: (1) *Technical/Tactical*; (a) technical tests [27,28], (b) match analysis statistics [28], and (c) perceptual-cognitive expertise (PCE) video simulation tests [29]. (2) *Physical*; (a) anthropometric measures, and (b) fitness tests [29]. (3) *Psychological*; (a) the Psychological Characteristics for Developing Excellence Questionnaire (PCDEQ) [30–32]. And, (4) *Social*; (a) Participation History Questionnaire (PHQ) [33], and (b) postcode data [30].

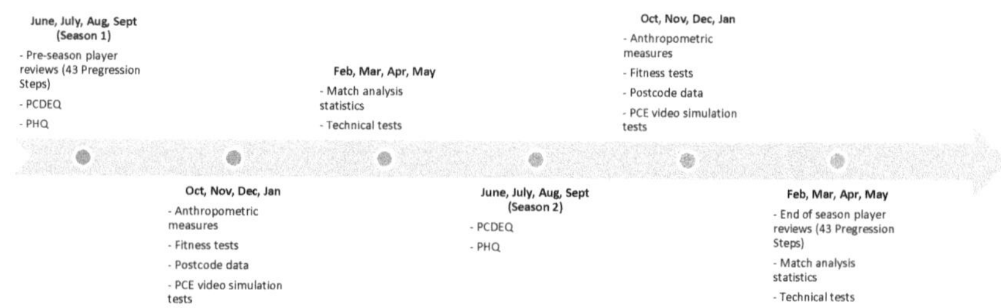

Figure 1. Timeline of the data collection.

A combined total of 53 characteristics were cumulated from the eight methods [34]: (1) *Four football-specific technical tests*; (a) ball juggling, (b) slalom dribble, (c) shooting accuracy, and (d) lob pass. (2) *Eight match analysis statistics from across the entire season*; (a) reliability in possession, (b) pass completion, (c) tackle completion, (d) block completion, (e) loose balls retrieved, (f) dribble completion, (g) total touches, and (h) goals scored. (3) *Three PCE video simulation tests*; (a) 'pre' execution occlusion, (b) 'during' execution occlusion, and (c) 'post' execution occlusion. (4) *Eight anthropometric measures*; (a) height, (b) body mass, (c) body mass index, (d) body fat percentage, (e) estimated adult height, (f) percentage of estimated adult height attained, (g) maturity status, and (h) birth quartile. (5) *Eight fitness tests*; (a) 0–10 m sprint test, (b) 0–30 m sprint test, (c) 10–30 m sprint test, (d) L-agility left test, (e) L-agility right test, (f) L-agility test combined, (g) countermovement jump height, and (h) countermovement jump flight time. (6) *Six factors from the 59-item PCDEQ*; (a) Factor 1 (support for long-term success), (b) Factor 2 (imagery use during practice and competition), (c) Factor 3 (coping with performance and developmental pressures), (d) Factor 4 (ability to organise and engage in quality), (e) Factor 5 (evaluating performances and working on weaknesses), and (f) Factor 6 (support from others to compete to my potential). (7) *Ten items from the PHQ*; (a) age started playing football, (b) age started playing academy football, (c) total match-play hours, (d) total coach-led practice hours, (e) total peer-led play hours, (f) total individual practice hours, (g) total football hours, (h) total sports played, (i) total multisport hours, and (j) total football and multisport hours. Finally, (8) *six measures from postcode data*; (a) home area code, (b) home financial risk, (c) home social classification, (d) school area code, (e) school financial risk, and (f) school social classification. The procedures for each of these methods have been outlined in our previously published work [27–34], which have been added below (Section 2.1.2) for the convenience of the reader.

Four Football-Specific Technical Tests

Four football-specific technical tests were used to measure technical ability: (a) ball juggling, (b) slalom dribble, (c) shooting accuracy, and (d) lob pass. First, the slalom dribble test required the player to control the ball through nine cones (2 m apart) from the start to the end line and return. The timings were recorded using timing gates (Brower TC Timing System, Draper, Utah, USA), with each player completing two trials and the quicker of the two recorded for analysis. Second, the lob pass test required the player to kick the football from a distance of 20 m into a target area divided into three concentric circles (3 m, 6 m, and 9.15 m in diameter). Each kick was scored by the circle in which the ball initially landed (3, 2, and 1 point, respectively). Ten attempts (five with each foot) were executed with a maximum of 30 points available. Third, the shooting accuracy test required the player to kick the ball at a 16 m wide goal target from a shooting distance of 20 m and central to the goal. The goal was divided into five parallel zones, whereby the centre was, 2 m wide (3 points), with two areas 3 m on each side of the centre (2 points), and two areas 4 m wide at each extreme (1 point). Ten attempts (five with each foot) were executed with a maximum of 30 points available. Fourth, the ball juggling test required the player to keep a football off the ground with the total number of touches recorded. Two trials were completed, with a maximum of 100 touches per attempt permitted, allowing a maximum number of 200 touches. Each player completed these tests in an indoor sports hall with a hard-wood floor, with generic training kit being worn. In addition, age group-specific balls were used for the tests in-line with the Football Association regulations, with size three for under-9, size four for under-10 to under-13, and size five for under-14 to under-16 [27,28].

Eight Match Analysis Statistics from across the Entire Season

Video footage examined each player during competitive match-play as they performed each skill behaviour. An average score of each skill behaviour was computed from across an entire football season, including: (a) reliability in possession percentage, (b) pass completion percentage, (c) number of tackles, (d) number of blocks, (e) number of loose balls retrieved, (f) successful dribble completion, (g) total touches, and (h) goals scored. As a standard pro-forma of match analysis statistics within each academy varies based on its philosophy, this current study applied the academy's existing protocol for its data collection. The specialist software Gamebreaker© was used to perform participant analysis for each game and trained, club-appointed Performance Analysts (who were not part of the research team and were blind to the grouping of the study participants) adopted technical expert definitions to code behaviours (n = 10). Twenty matches (25% of the data) of the matches that were included in the current study were used to calculate the Performance Analysts' reliability (15-day test–retest analysis). One match per team was randomly selected to carry out the intra- and inter-reliability analysis. An intra-class correlation coefficient test was executed to analyse the reliability levels (poor, <0.50; moderate, 0.50 to 0.75; good, 0.76 to 0.90; excellent, 0.91 to 1.00). Results showed the intra-observer reliability ranged from 0.76 to 1.00 and the inter-observer reliability ranged from 0.71 to 1.00 [28].

Three Perceptual-Cognitive Expertise Video Simulation Tests

Film-based simulation tests were applied to examine the players' decision-making skill. Action sequences were selected from live football match footage of academy players aged 18 to 19 years engaging in a competitive game, filmed from an elevated angle above and behind the goal. Following general build-up play of five to ten seconds in duration, the clips unexpectedly occlude immediately prior to a critical decision moment. At this point, an occlusion display appears that shows the pitch lines (i.e., boundaries, eighteen yard box, and half way line) and the location of the ball on a white screen. This screen was frozen for 7-s whereby the participant had to select their answer on the response sheet before the next clip automatically begins. Forty-five clips were created for three different phases that were used for analysis: (a) 'pre', (b) 'at', and (c) 'post' execution. Thus, 135 clips were viewed by the players in total. 'Pre' clips are considered more difficult as the occlusion happens 0.5 s prior to the action that is executed, whereas the 'at' clips occlude during the moment the action is executed, as opposed to the 'post' clips that are considered the easiest as they are occluded after the execution with a duration 0.5 s longer. Consequently, clips are viewed in this order, with a response sheet completed separately and collected before the next batch of clips begin, to prevent players changing their answer when they see the longer clips. The 45 film-based simulations are distributed into three decision-making skills, including 'select action', 'select direction', and 'select pass recipient', thus creating 15 clips for each [29].

Eight Anthropometric Measures

The physiological measures included: (a) height, (b) body mass, (c) body mass index (BMI), and (d) body fat percentage. Height measures were recorded to the nearest 0.1 cm (Seca 213 Leicester Height Measure). Body mass measures were recorded to the nearest 0.1 kg (Tanita BF-350 Body Composition Monitor). Body mass index was calculated through dividing weight (kg) by height (m) and dividing that number by height (kg/m^2). Body fat percentage was also estimated (Tanita BF-350 Body Composition Monitor). Players completed these procedures bare footed with their training shorts and t-shirt on. Moreover, the Khamis-Roche method was used to analyse: (a) predicted adult height, (b) percentage of predicted adult height attained, and (c) PHV status. The Khamis-Roche method is based on a mathematical calculation using the child's gender, current height and body mass, and the height of both parents. The formula applied to predicted height in inches is: =((age factor) * (age in years)) + ((height factor) * (height in inches)) + ((body mass factor) * (body mass in pounds)) + ((parental height factor) * (parental height in inches)) + (beta coefficient) [35]. The participants predicted adult height then identifies the percentage of predicted adult height attained. Additionally, the growth curve attained from this data identifies the participants PHV status: (a) pre-, (b) circa-, and (c) post-PHV. Lastly, birth quartile was measured by dividing the twelve months of the year into four quarters, conforming to the strategy applied to distribute chronological age groups. Due to the start of the section year beginning in September in England, this is recognised as 'month 1' while August is 'month 12'.

Eight Fitness Tests

Fitness tests were conducted with the participants to measure specific physical parameters, including acceleration, sprint, agility, and jump abilities. These tests were executed by the first author and have been proved valid and reliable measures for talent development research in youth football. Players were already familiarised with these testing procedures since they were already part of the academy fitness testing protocol. The 0–30 m sprint test started 1 m behind the first set of timing gates (Brower TC Timing System, Draper, UT, USA). Participants sprinted until passing the final set of timing gates. Timings for 0–10 m, 0–30 m, and 10–30 m were taken to observe acceleration and sprint speed, respectively. The L-agility test required the participants to start 1 m behind the first set of timing gates (Brower TC Timing System, Draper, UT, USA), then run forwards 5 m around the tall centre cone, run 5 m to the left hand cones and place one foot between the two marker cones, and then turn and follow the same path back to the start. In the second trial, players performed the same test, but this instance running 5 m to the cones on the right-hand side. Timings were recorded for the right, left, and combined. During the CMJ test (Just Jump system, Probotics Inc. 8602 Esslinger CT, Huntsville, AL, USA), players were instructed on the importance of using a countermovement and the need to take-off and land with straight legs, with the jump height (cm) and time (s) recorded for analysis. Three trials were completed for each test with the best result taken for investigation. Players conducted these fitness tests in a sports hall, whilst players completed a familiarity session prior to the data collection to counteract any earning effects.

Six Factors from the 59-Item Psychological Characteristics for Developing Excellence Questionnaire

The 59-item PCDEQ was used to assess psychological characteristics across six dimensions. Each of the questionnaire's items is placed on a six-point Likert scale with a similarity response method from '1' (very unlike me) to '6' (very like me). This ensured participants were not allowed to remain neutral and therefore encouraged them to think more carefully about whether they agree or disagree with the statement leading to greater accuracy. Additionally, a mixture of positively and negatively worded items is included to minimise the danger of acquiescent bias. The PCDEQ is designed for youth athletes, thus offers user-friendly language that is applicable to this cohort (see MacNamara and Collins [31] for the psychometric properties of the PCDEQ). The participants completed the PCDEQ in a classroom setting. They were allocated 45-min to complete it and the researcher was available to help answer any questions if the participants were unsure.

Ten Items from the Participation History Questionnaire

The PHQ is a retrospective recall questionnaire, which is used to elicit information regarding the activities in which players have engaged in during their development. The test–retest reliability and the concurrent validity of the PHQ have been previously established by Ford and colleagues [33]. The PHQ contains three sections including milestones within football, engagement within football activities, and engagement in other sport activities. Initially, the football-specific milestones include both: (a) the age at which the player first engaged in football, and (b) the age they began participation in a professional football academy. The second section of the PHQ is designed to elicit information from four football-specific activities: (a) match-play, (b) coach-led practice, (c) individual practice, and (d) peer-led play. The hours per week and months per year in each of these football activities, as well as the accumulation of time spent engaged in all of these activities, were recorded in the PHQ for each year from the current season back to the year the participant stared playing football. Finally, the third section of the PHQ is designed to produce information concerning engagement in other sport activities, including: (a) total sports played, and (b) total multi-sports hours. It contains a list of sports from which players were required to indicate those in which they have participated in regularly for at least a total minimum period of three months. Players were not required to record other sport activities engaged in during Physical Education (PE) classes in school. Total football and multi-sport hours were also included as a measure. The participants completed the PHQ in a classroom setting. Each participant was given one hour to complete the PHQ under supervision from the lead author, while allowing questions to facilitate individual understanding.

Six Measures from Postcode Data

Social classification and credit score are proxy indicators of socioeconomic status. In the UK, postcodes are associated with data pertaining to the locale to which they correspond. These data include income, employment, education, health, and crime levels, which can be accessed in multiple ways. For this study, the UK General Registrar Classification system was adopted that uses the average credit rating applying the Cameo™ geodemographic database. This provided a social classification (A, B, C1, C2, D, and E) determined by the UK's Office for National Statistics and an average credit score (out of 999) for where each participant lives and goes to school. The social classification was scored numerically, with a higher score relating to a lower social classification (i.e., A = 1, B = 2, C1 = 3, C2 = 3, D = 4, and E = 5). The credit score denotes those with a higher score to have lower financial risk from '0' (low) to '999' (high). The participants area code was also included to test outline whether they are from urban and rural settings.

2.1.3. Player Review Ratings

Player profiling is a widely used tool that is utilised within professional football academies [36]. Indeed, coach opinion is central to the subjective nature of youth football, with modern objective information readily available to professional coaches to support their judgement [37,38]. This study applied a unique progress assessment to measure each individual's development. This tool, named the *43 Progression Steps*, applies a holistic approach during the player review process. This includes capturing the club's pre-existing philosophy of developing core skills within *mental, physical, technical,* and *tactical* variables. These four sub-components grade specific characteristics that are considered necessary for development and progress towards senior professional status within this particular football club. The scoring system for the player profiling reports has a continual and progressive pattern rather than identical Likert scales. For example, the under-9 rating scale ranges from 1 (below average) to 4 (excellent), while the under-16 rating scale ranges from 26 (significantly below the required standard) to 33 (pushing towards the under-18 s). Throughout the development process, these specific grades are not prescribed within age groups, with players able to move through the tool seamlessly if they are developing or playing in certain areas above or below their chronological age.

The player review ratings were initially completed by the players who give their perception of themselves, and then the coaches subsequently provided their ratings alongside specific individual learning objectives. These reports were completed three times (i.e., pre-season, mid-season, and end of season), with each coach having completed the participants' review ratings throughout the two seasons included in this study across the under-9 to under-16 age groups. Only the accumulated scores for all the components within each participant's 43 Progression Steps rating were recorded at the start of season one and the end of season two in order to create two time points and analyse year-on-year developmental outcomes. Comparing the differences between the overall scores from the two player review ratings illustrated each player's total development over two years, which was the score used for the data analysis in this current study. Two coaches from each age group (n = 16), who were deemed suitably qualified assessors (UEFA Pro, 'A', or 'B' Licenced alongside either the FA Advanced Youth Award or the FA Youth Award), graded each participant's player review ratings for each of the specific characteristics. See Kelly's doctoral thesis [34] for a comprehensive overview of the 43 Progression Steps player review tool.

2.1.4. Data Analysis

The dataset was analysed via Lasso linear regression using cross-validated Lasso regression as implemented in the glmnet package in R [39]. Analysis of the improvement in player score across the two seasons used a coach assessed outcome measure, with the scores standardized at an age group aggregate:

$$y_{i,\,t} = \frac{x_{i,t} - \mu_t}{\sigma_t}$$

where $y_{i,t}$ is the corrected scores for the ith member of age group t, $x_{i,t}$ is the uncorrected scores for the ith member of age group t, μ_t is the mean of the $x_{i,t}$ scores and σ_t is the standard deviation of the $x_{i,t}$ scores. The independent variables (IVs) were divided into categorical ("Home Postcode Social Grade", "School Postcode Social Grade", "PHV Status", and "Birth Quarter") and numeric (see SI for full list). Each numeric IV was standardized for mean at standard deviation at an age group aggregate, and the categorical IVs underwent a one-hot vector encoding [40]. Hence, coefficient estimates for numeric IVs reflect the change in DV per standard deviation from the average, while categorical IVs reflect the change in DV where the variable possess the relative value. The cross-validation technique first learned a model penalisation parameter, λ, by optimising the model performance characteristic (mean squared error) under 10-fold cross validation. The results for the optimal value of λ were then extracted to identify key contributing factors towards player review ratings.

2.2. Results

The summary of the Lasso regression techniques is outlined in Table 1. Results showed non-zero coefficients for improvement in subjective performance in 15 out of the 53 analysed features. Key findings revealed advanced percentage of predicted adult height (0.196), greater lob pass (0.160) and average dribble completion percentage (0.124), more total match-play hours (0.145), and an older relative age (BQ1 vs. BQ2: −0.133; BQ1 vs. BQ4: −0.060) were the most important features that contributed towards player review ratings.

Table 1. Summary of non-zero coefficients for improvement in subjective performance.

Feature	Coefficient/SD of Feature
Ball juggling	0.083
Lob pass	0.160
Average dribble competition percentage	0.124
PCE 'at'	0.091
PCE 'post'	0.062
PCDEQ Factor 3	0.062
Total match-play hours	0.145
Total individual practice hours	−0.027
0–30 m sprint	−0.041
CMJ height	0.053
Percentage of predicted adult height attained	0.196
Birth quarter 2 (reduced relative to birth quarter 1)	−0.133
Birth quarter 4 (reduced relative to birth quarter 1)	−0.060
Home postcode social grade 2 (reduced in comparison to social grade 1 or 4)	−0.082
School postcode social grade 3 (reduced relative to social grade 1 or 4)	−0.045

2.3. Discussion

The purpose of this exploratory study was to examine the multidimensional factors that contributed to player review ratings across two seasons by applying a machine learning approach. Results showed a total of 15 of the 53 analysed features were important contributors towards player review ratings, which were representative of all four subcomponents from the FCM (i.e., technical/tactical, physical, psychological, and social). Most notably, advanced percentage of predicted adult height, greater lob pass and average dribble completion percentage, more total match-play hours, and an older relative age were the largest features. Taken together, these findings underscore the holistic nature of the talent development processes in youth football.

Advanced percentage of predicted adult height had the greatest influence on player review ratings. The variation of maturation status (i.e., early, on-time, and late) between players within a single chronological age group can lead to up to 5-years difference in biological age [41]. The trainability and performance of physical competencies are closely aligned with maturity status [42]. Male players who experience their adolescent growth spurt mature earlier than their peers are invariably taller and heavier from late childhood and possess greater absolute and relative lean mass [43–45]. As a result of their advanced maturity, early maturing players also tend to outperform their less mature peers on tests of strength, power, speed, agility, and endurance [45,46]. However, from a psychological perspective, Cumming and colleagues [47] showed how later maturing players are more likely to possess and/or develop more adaptive self-regulation skills in the long-term, in particular self-evaluation and reflection. Moving forward, coaches should reflect on how an advanced maturity status can influence football-specific developmental outcomes from a holistic perspective (e.g., technical/tactical, physical, psychological, social). Since maturity status can significantly influence football-specific skills (e.g., physical competencies, self-regulation), coaches should observe and/or review players based on their maturity status (e.g., bio-banding [48]), rather than just their chronological age. This would support the long-term development of a wider pool of potential talent and focus on retaining later maturing players, whilst move the focus on short-term performance results that largely benefit early maturing players [25].

Greater lob pass and average dribble completion percentage (technical), as well as PCE 'at' and PCE 'post' (tactical), were important contributing features towards player review ratings. Coaches are the decision-makers in the player review rating process and often have a greater understanding of technical/tactical features compared to the other subcomponents of the FCM. Thus, it is not surprising that technical/tactical skills featured within these current results, as it is possible that greater value may be placed on these characteristics compared to the other subcomponents. This is emphasised by the traditional coach education and sport-specific qualifications that often focus on athlete competence compared to other developmental factors (e.g., confidence, connection, and character [6]) [49]. As such, although further evidence is needed, it is suggested coaches and organisations involve other stakeholders (e.g., Sport Scientists, Sport Psychologists, Strength and Conditioning Coaches) as part of a broader, holistic decision-making strategy when reviewing young players development [50].

There has been an ongoing search for the most appropriate activities that facilitate long-term player development in youth football (see Ford & Williams [51] for an overview). Findings from this current study found more total match-play hours had the largest contribution towards player review ratings. This may be explained through the coaches who are providing the players with their review rating being the same coaches who are selecting the players for the matches, and thus may be offering the players they perceive as better with more game time. However, it's important to note that these coaches would have only been responsible for selecting these players in the recent years, whereas the *total* match-play hours accumulate the numbers the player has engaged in since they began playing football. Thus, another possible explanation is the benefits of engaging in match-play that may have contributed to player development. For instance, small-sided games have been shown to develop and refine young players' skills and movements [52–54]. As such, coaches should consider how to offer a rich games programme to their players, through both competitive match-play and small-sided games, which could contribute to the holistic development of young players.

Birth quarter played an important role in influencing player review ratings, which favoured those born in the first three month of the year. This aligns with a wealth of relative age literature in youth soccer. As an example, initial research from Barnsley and colleagues [55] showed 45% of players selected for the 1989 U17 World Cup were born in the first three months of the annual selection year, whereas only 7.7% were born in the last three months of the annual selection year, with similar results shown across the U20 team squads. Since this preliminary research, the last three decades has generated various studies that shows how those born earlier in the selection year are overrepresented in talent pathways [56], accrue more league points [57], and win more games [58]. However, these benefits at youth level do not necessarily translate into success at adulthood in professional football [59–61]. As such, it is plausible to suggest that coaches perceive greater development in those who are relatively older largely due to their advanced age. As such, it is important for future research to explore the mechanisms of relative age effects and how they impact coaches perceived potential. Practitioners and researchers should also work collaboratively to design, implement, and evaluate a range of relative age solutions to help mitigate against these effects.

3. The Junior-to-Senior Transition from Youth Academy to Professional Level: Exploring the Characteristics of Selected and Deselected under-18 Players

Becoming a professional footballer is the aspiration of many academy prospects. However, it is well documented that only a small proportion of young players successfully graduate into senior professional levels. As an example, Dugdale and colleagues [62] showed how only 10% of 537 male players made the successful transition to professional level across a twelve-year period at a Scottish professional football club. Similarly, spanning an eleven-year period at an English professional football club, Kelly and colleagues [63] revealed how only 7.4% of 364 male players who entered the academy from under-9 to under-18 achieved a professional contract at aged 18 years. To better understand the junior-to-senior level transition, it is important to consider the characteristics that differentiate those academy players who achieve professional status and those who do not. By doing so, it will enable key stakeholders (e.g., coaches, practitioners, policy makers) employed in talent development programmes to allocate resources more efficiently, as well as facilitate a science-based support system [25].

The purpose of this study was to examine a range of factors based on the FCM (i.e., technical/tactical, physical, psychological, social) that may have contributed to under-18 academy football players achieving a professional contract.

3.1. Methods

3.1.1. Sample

Eighteen under-18 male participants were recruited from the same tier four English professional football club and their category three academy. Their average weekly training and match-play time was 15 training hours/week and 1.5 match-play hours/week. Goalkeepers were not included in this study due to their contrasting position-specific requirements [26]. Parental consent and player assent were collected prior to the study commencing. The study was approved by the Ethics Committee of Sport and Health Sciences at the University of Exeter.

3.1.2. Measures and Procedures

The same 53 factors from the eight measures outlined in Study 1 were collected for this study across two seasons (2014–2015 and 2015–2016). Player review ratings were also added since they were not used as an outcome measure. This dataset was then used to compare selected (i.e., offered a professional contract; $n = 8$) and deselected (i.e., not offered a professional contract; $n = 10$) players as they reached the end of their academy scholarship.

3.1.3. Data Analysis

The dataset was analysed via Lasso regression techniques using cross-validated Lasso regression as implemented in the glmnet package in R. Analysis of the 'Selection' for professional play was performed using binomial Lasso regression, coding the outcome as 1 for 'Selected' and 0 for 'Deselected'. The cross-validation technique first learned a model penalisation parameter, λ, by optimising the model performance characteristic (binomial deviance) under 10-fold cross validation. The results for the optimal value of λ were then extracted to identify key contributing factors. In reporting the results of the binomial Lasso regression, the exponential of the coefficients was included. In the case of a logistic binomial mode, the exponential of the coefficients is equivalent to the change in odds ratio for each increase of the dependent variable by 1, one standard deviation in this case [13,40].

3.2. Results

The summary of the Lasso regression techniques are outlined in Table 2. The relatively small parameter space of importance in Table 2 is not indicative that few features matter, but instead due to the limited quantity of data available. The size of the effect of the psychological factor, while arising from a limited quantity of data, should be noted. Having included the possibility for multiple confounding factors, the strongest marker for signing was the psychological outcomes of the player. What is not clear, given the observational nature of the study, is if improvements in psychological factors would lead to a greater chance of signing a professional contract or if it is in fact a proxy variable marking out players with a range of sought after factors.

Table 2. Summary of non-zero coefficients for likelihood of signing a professional contract.

Feature	Coefficient/SD of Feature	Odds Ratio/SD of Feature
43 progression steps rating	0.64	1.89
Slalom dribble	0.01	1.01
PCDEQ Factor 3	0.44	1.55
Home postcode social grade 2	−0.12	0.89

The prominence of Factor 3 (coping with performance and developmental pressures) with regard to both end points posed a key question: is this factor more prominent than the other five factors, or are all the six closely correlated in the dataset and the Lasso is selecting the most informative? To quantify the relative associations of the six PCDEQ factors, the correlation matrix was calculated for the progression data set. Of the PCDEQ factors, Factor 3 only shows a reasonably strong correlation with Factor 4 (ability to organise and engage in quality practice), with only weak links to the other terms. Hence, we conclude that PCDEQ Factor 3 and Factor 4 are the strongest discriminatory variables relating to signing a professional contract. Hence, it appears valid to suggest Factors 1, 2, 5, and 6 may pose no contribution to signing a professional contract.

3.3. Discussion

The junior-to-senior transition is arguably the most defining moment in a promising young player's career. Indeed, by achieving their first professional contract, a player moves one-step closer to fulfilling their aspirations of competing for their respective senior first team. To the author's knowledge, this was the first study to explore the characteristics of selected and deselected under-18 academy players using machine learning techniques. Key findings revealed how PCDEQ Factor 3 and Factor 4 were important contributing factors towards achieving a professional contract. Moreover, player review ratings (i.e., higher coach scores), slalom dribble (i.e., quicker dribble times), and a lower home social classification (i.e., derived from more deprived areas) also provided a small contribution.

PCDEQ Factor 3 and Factor 4 were important contributing factors within this current study. Indeed, psychological factors have been previously identified as important attributes that are required during the junior-to-senior transition. As an example, the current findings are consistent with previous studies that found 'good developers' within team sports had a significantly greater perceived ability to cope with performance and developmental pressures (e.g., such as overcoming struggles, set-backs, injury, or a decline in performance) compared to 'poor developers' [32]. These current findings also compliment the opinions of coaches as derived from qualitative studies. First, Mills and colleagues' [64] analysis of ten expert coaches revealed six factors, including resilience, that were perceived to either positively or negatively influence player development. Second, Cook and colleagues [65] reported four general dimensions of mental toughness, including competitiveness with self and others, mind-set, resilience, and personal responsibility, that are associated with the ability to cope with the pressures inherent in the academy environment. Similarly, Holt and Mitchell [66] identified a deficiency in coping behaviours of professional football players near to being released, whereas Holt and Dunn [67] revealed how discipline, commitment, resilience, and social support were associated with becoming a professional football player. While it is plausible to suggest that these psychological characteristics are generally accepted as crucial factors for positive developmental outcomes, further investigation is required to design, implement, and evaluate effective psychological development strategies within academy environments [68]

When compared to other specialist support staff in youth soccer environments (i.e., Sport Scientists, Strength and Conditioning Coaches, Performance Analysts), Sport Psychologists appear to be less common (particularly in a full-time capacity) [68]. Since the development of psychological characteristics appears to be an important contributing factor towards both coaches perceived development outcomes (i.e., Study 1) as well as achieving a professional contract (i.e., Study 2), professional clubs and governing bodies should consider how they can formalise their psychological support and invest in qualified practitioners. By doing so, it will enable young players the opportunity to access psychological support when required, as well as help with coach development to ensure effective strategies are consistently implemented throughout coaching provision [69].

Unsurprisingly, player review ratings contributed to selection. This is likely due to the fact that those coaches who are rating the players are the same stakeholders who are part of the professional contract decisions. More unexpectedly, dribbling ability and social classification also made a small contribution towards selection. First, dribbling has been previously identified as an important technical attribute as part of a multidisciplinary study when comparing selected and deselected players [2]. Thus, these current findings further support the significance of possessing ball dribbling skills, which could be incorporated into developmental programmes to ensure players are adequately prepared as they navigate their ways towards senior levels. Second, the results of a lower home social classification contributing to achieving a professional contract reflect the stereotype of football being a sport participated by individuals with a lower socioeconomic status [70]. This might imply that football retains a traditional divide between socioeconomic status and participation [71], which may have implications on opportunities to achieve a professional contract [72]. However, it is important to consider the exploratory nature of these findings, as well as the limited number of participants included in this current study. Therefore, future research is encouraged to further explore the significance of possessing ball dribbling skills and the role of socioeconomic status in developing expertise in football.

4. Limitations

The key limitations of this study are the role of retrospective analysis and predictive models. The techniques used here are best viewed as exploratory and hypothesis generating, rather than confirmatory, as they do not seek to provide evidence for or against any pre-existing mechanisms but generate new insight and an optimal predictive model given the available data. In the case of the 'Selection' analysis, the inherently small dataset does provide limitations to the analysis, whereby machine learning approaches with small datasets inherently run the risk of memorising the sample rather than generating transferable lessons [73]. In general, the subset selection algorithms demonstrated here could provide stakeholders within the football development community with insight into the operational data currently held. Data collection and storage mechanisms have increasingly become cheaper over the past two decades (e.g., the rise of wearable technology and cloud storage). Moving forward, an important question is how to leverage such data to aid decision making. Techniques such as those shown here are invaluable in being able to quickly and easily reduce data to interpretable models and highlight key signals.

5. Conclusions

It appears the key factors associated with positive developmental outcomes in youth soccer are not always technical and tactical in nature, which is where youth coaches often have their expertise and/or focus their attention on talent development. Indeed, the relative importance of these factors is likely to change over time, and with age, although psychological attributes appear to be influential to reaching potential across the academy journey. Therefore, coaches are encouraged to focus on long-term potential as opposed to short-term performance. The techniques used here also serves as an impetus for researchers to adopt machine learning approaches when analysing multidimensional databases for talent development purposes.

Author Contributions: Conceptualization, A.L.K.; Data curation, A.L.K.; Formal analysis, R.C.; Methodology, R.C.; Software, R.C.; Supervision, C.A.W. and M.R.W.; Writing—original draft, A.L.K.; Writing—review and editing, C.A.W., S.L.J.S. and M.R.W. All authors have read and agreed to the published version of the manuscript.

Funding: This research was co-funded by the University of Exeter, College of Life & Environmental Sciences, the Open Innovation Platform at the University of Exeter, and Exeter City Football Club Academy. Thanks to the players, parents, and staff at Exeter City Football Club Academy for their participation and support in this project.

Institutional Review Board Statement: The study was approved by the Ethics Committee of Sport and Health Sciences at the University of Exeter. Reference number: 141203/B/15.

Informed Consent Statement: Written parental consent and player assent were collected prior to the study commencing.

Data Availability Statement: Data can be obtained via the lead author.

Conflicts of Interest: The authors declare no conflict of interest.

References

1. Sarmento, H.; Anguera, M.T.; Pereira, A.; Araújo, D. Talent Identification and Development in Male Football: A Systematic Review. *Sports Med.* **2018**, *48*, 907–931. [CrossRef] [PubMed]
2. Huijgen, B.C.H.; Elferink-Gemser, M.T.; Lemmink, K.A.P.M.; Visscher, C. Multidimensional performance char-acteristics in selected and deselected talented soccer players. *Eur. J. Sport Sci.* **2014**, *14*, 2–10. [CrossRef] [PubMed]
3. Forsman, H.; Gråstén, A.; Blomqvist, M.; Davids, K.; Liukkonen, J.; Konttinen, N. Development of perceived competence, tactical skills, motivation, technical skills, and speed and agility in young soccer players. *J. Sports Sci.* **2016**, *34*, 1311–1318. [CrossRef] [PubMed]
4. Zuber, C.; Zibung, M.; Conzelmann, A. Holistic patterns as an instrument for predicting the performance of prom-ising young soccer players—A 3-years longitudinal study. *Front. Psychol.* **2016**, *7*, 1088. [CrossRef] [PubMed]
5. Côté, J.; Murata, A.; Turnnidge, J.; Hancock, D.J. Situating birth advantages within the youth sport system. In *Birth Advantages and Relative Age Effects in Sport: Exploring Organizational Structures and Creating Appropriate Settings 2*; Kelly, A., Côté, J., Jeffreys, M., Turnnidge, J., Eds.; Routledge: London, UK, 2021; pp. 15–29.
6. Côté, J.; Turnnidge, J.; Evans, M.B. The dynamic process of development through sport. *Kinesiol. Slov. Sci.-Entific J. Sport* **2014**, *20*, 14–26.
7. Kelly, A.L.; Wilson, M.R.; Williams, C.A. Developing a football-specific talent identification and development profiling concept—The Locking Wheel Nut Model. *Appl. Coach. Res. J.* **2018**, *2*, 32–41.
8. The Football Association. The FA's 4 Corner Model [Online]. 2020. Available online: https://thebootroom.thefa.com/resources/coaching/the-fas-4-corner-model (accessed on 2 February 2022).
9. Towlson, C.; Cope, E.; Perry, J.L.; Court, D.; Levett, N. Practitioners' multi-disciplinary perspectives of soccer talent according to phase of development and playing position. *Int. J. Sports Sci. Coach.* **2019**, *14*, 528–540. [CrossRef]
10. Kelly, A.L.; Wilson, M.R.; Jackson, D.T.; Goldman, D.E.; Turnnidge, J.; Côté, J.; Williams, C.A. A multidisciplinary investigation into "playing-up" a chronological age group in an English football academy. *J. Sports Sci.* **2021**, *39*, 854–864. [CrossRef]
11. Goldman, D.E.; Turnnidge, J.; Côté, J.; Kelly, A.L. "Playing-up" in youth soccer. In *Birth Advantages and Relative Age Effects in Sport: Exploring Organizational Structures and Creating Appropriate Settings*; Kelly, A.L., Côté, J., Jeffreys, M., Turnnidge, J., Eds.; Routledge: London, UK, 2021; pp. 77–94.
12. Turnnidge, J.; Kelly, A.L. Organizational structures: Looking back and looking ahead. In *Birth Advantages and Relative Age Effects in Sport: Exploring Organizational Structures and Creating Appropriate Settings*; Kelly, A., Côté, J., Jeffreys, M., Turnnidge, J., Eds.; Routledge: London, UK, 2021; pp. 239–246.
13. Tibshirani, R. Regression shrinkage and selection via the Lasso. *J. R. Stat. Soc. Ser. B (Method-Ological)* **1996**, *58*, 267–288. [CrossRef]
14. Oquendo, M.A.; Baca-García, E.; Artés, A.; PerezCruz, F.; Galfalvy, H.; Blascofontecilla, H.; Madigan, D.; Duan, N. Machine learning and data mining: Strategies for hypothesis generation. *Mol. Psychiatry* **2012**, *17*, 956–959. [CrossRef]
15. Bunker, R.; Thabtah, F. A machine learning framework for sport result prediction. *Appl. Comput. Inform.* **2019**, *15*, 27–33. [CrossRef]
16. Unni, M.P.; Menon, P.P.; Livi, L.; Wilson, M.R.; Young, W.R.; Bronte-Stewart, H.M.; Tsaneva-Atanasova, K. Data-Driven prediction of freezing of gait events from stepping data. *Front. Med. Technol.* **2020**, *2*, 581264. [CrossRef] [PubMed]
17. Barth, M.; Emrich, E.; Güllich, A. A machine learning approach to "revisit" specialization and sampling in institu-tionalized practice. *SAGE Open* **2019**, *9*, 1–18. [CrossRef]
18. Jauhiainen, S.; Kauppi, J.-P.; Leppänen, M.; Pasanen, K.; Parkkari, J.; Vasankari, T.; Kannus, P.; Äyrämö, S. New machine learning approach for detection of injury risk factors in young team sport athletes. *Laryngo-Rhino-Otologie* **2020**, *42*, 175–182. [CrossRef]
19. Jones, B.D.; Hardy, L.; Lawrence, G.; Kuncheva, L.I.; Brandon, R.; Bobat, M.; Thorpe, G. It ain't what you do—It's the way that you do it: Is optimizing challenge key in the development of super-elite batsmen? *J. Expert.* **2020**, *3*, 144–168.
20. Jones, B.D.; Hardy, L.; Lawrence, G.; Kuncheva, L.I.; Du Preez, T.; Brandon, R.; Such, P.; Bobat, M. The identification of "game changers" in England cricket's developmental pathway for elite spin bowling: A machine learning approach. *J. Expert.* **2019**, *2*, 92–120.

21. Musa, R.M.; Majeed, A.P.P.A.; Taha, Z.; Chang, S.W.; Nasir, A.F.A.; Abdullah, M.R. A machine learning ap-proach of predicting high potential archers by means of physical fitness indicators. *PLoS ONE* **2019**, *14*, e0209638.
22. Owen, J.; Owen, R.; Hughes, J.; Leach, J.; Anderson, D.; Jones, E. Psychosocial and physiological factors affecting selection to regional age-grade rugby union squads: A machine learning approach. *Sports* **2022**, *10*, 35. [CrossRef]
23. Musa, R.M.; Majeed, A.P.P.A.; Kosni, N.A.; Abdullah, M.R. *Machine Learning in Team Sports: Performance Analysis and Talent Identification in Beach Soccer & Sepak-Takraw*; Springer: New York, NY, USA, 2020.
24. The Premier League. Elite Player Performance Plan [Online]. 2011. Available online: https://www.premierleague.com/youth/EPPP (accessed on 2 February 2022).
25. Kelly, A.L.; Williams, C.A. Physical characteristics and the talent identification and development processes in male youth soccer: A narrative review. *Strength Cond. J.* **2020**, *42*, 15–34. [CrossRef]
26. Gil, S.M.; Zabala-Lili, J.; Bidaurrazaga-Letona, I.; Aduna, B.; Lekue, A.B.; Santos-Concejero, J.; Granados, C. Talent identification and selection process of outfield players and goalkeepers in a professional soccer club. *J. Sports Sci.* **2014**, *32*, 1931–1939. [CrossRef]
27. Vaeyens, R.; Malina, R.M.; Janssens, M.; Van Reterghem, B.; Bourgois, J.; Vrijens, J.; Philippaerts, R.M. A multidis-ciplinary selection model for youth soccer: The Ghent Youth Soccer Project. *Br. J. Sports Med.* **2006**, *40*, 928–934. [CrossRef] [PubMed]
28. Kelly, A.; Wilson, M.R.; Jackson, D.T.; Williams, C.A. Technical testing and match analysis statistics as part of the talent development process in an English football academy. *Int. J. Perform. Anal. Sport* **2020**, *20*, 1035–1051. [CrossRef]
29. Kelly, A.L.; Wilson, M.R.; Jackson, D.T.; Turnnidge, J.; Williams, C.A. Speed of thought and speed of feet: Examining perceptual-cognitive expertise and physical performance in an english football academy. *J. Sci. Sport Exerc.* **2020**, *3*, 88–97. [CrossRef]
30. Kelly, A.L.; Williams, C.A.; Jackson, D.T.; Turnnidge, J.; Reeves, M.J.; Dugdale, J.H.; Wilson, M.R. Considering the role of socioeconomic status and psychological characteristics in talent development: An exploratory study in an English soccer academy. *Sci. Med. Footb. submitted.*
31. MacNamara, A.; Collins, D. Development and initial validation of the Psychological Characteristics of Developing Excellence Questionnaire. *J. Sport Sci.* **2011**, *29*, 1273–1286. [CrossRef]
32. MacNamara, A.; Collins, D. Do mental skills make champions? Examining the discriminant function of the psy-chological characteristics of developing excellence questionnaire. *J. Sport Sci.* **2013**, *31*, 736–744. [CrossRef]
33. Ford, P.; Ward, P.; Hodges, N.J.; Williams, A.M. The role of deliberate practice and play in career progression in sport: The early engagement hypothesis. *High Abil. Stud.* **2009**, *20*, 65–75. [CrossRef]
34. Kelly, A.L. A Multidisciplinary Investigation into the Talent Identification and Development Process in an English Football Academy. Ph.D. Thesis, University of Exeter, Exeter, UK, 2018. Available online: https://ore.exeter.ac.uk/repository/handle/10871/34621 (accessed on 2 February 2022).
35. Khamis, H.J.; Roche, A.F. Predicting adult stature without using skeletal age: The Khamis-Roche method. *Pediatrics* **1994**, *94*, 504–507.
36. Morley, D.; Morgan, G.; McKenna, J.; Nicholls, A.R. Developmental contexts and features of elite academy football players: Coach and player perspectives. *Int. J. Sports Sci. Coach.* **2014**, *9*, 217–232. [CrossRef]
37. Sieghartsleitner, R.; Zuber, C.; Zibung, M.; Conzelmann, A. science or coaches' eye?—Both! beneficial collaboration of multi-dimensional measurements and coach assessments for efficient talent selection in elite youth football. *J. Sports Sci. Med.* **2019**, *18*, 32–43.
38. Tangalos, C.; Robertson, S.J.; Spittle, M.; Gastin, P.B. Predictors of individual player match performance in junior australian football. *Int. J. Sports Physiol. Perform.* **2015**, *10*, 853–859. [CrossRef] [PubMed]
39. Friedman, J.H.; Hastie, T.; Tibshirani, R. Regularization paths for generalized linear models via coordinate descent. *J. Stat. Softw.* **2010**, *33*, 1–22. [CrossRef] [PubMed]
40. Dunn, P.K.; Smyth, G.K. *Generalized Linear Models with Examples in R*; Springer: Berlin/Heidelberg, Germany, 2018.
41. Malina, R.M.; Bouchard, C.; Bar-Or, O. *Growth, Maturation and Physical Activity*, 2nd ed.; Human Kinetics: Champaign, IL, USA, 2004.
42. Lloyd, R.S.; Oliver, J.L. The Youth Physical Development Model: A new approach to long-term athletic development. *Strength Cond. J.* **2012**, *34*, 61–72. [CrossRef]
43. Clemente, F.M.; Clark, C.C.T.; Leão, C.; Silva, A.F.; Lima, R.; Sarmento, H.; Figueiredo, A.J.; Rosemann, T.; Knechtle, B. Exploring relationships between anthropometry, body composition, maturation, and selection for competition: A study in youth soccer players. *Front. Physiol.* **2021**, *12*, 651735. [CrossRef] [PubMed]
44. Malina, R.M.; Eisenmann, J.C.; Cumming, S.P.; Ribeiro, B.; Aroso, J. Maturity-associated variation in the growth and functional capacities of youth football (soccer) players 13–15 years. *Eur. J. Appl. Physiol.* **2004**, *91*, 555–562. [CrossRef]
45. Meylan, C.; Cronin, J.; Oliver, J.; Hughes, M. Talent identification in soccer: The role of maturity status on physical, physiological and technical characteristics. *Int. J. Sports Sci. Coach.* **2010**, *5*, 571–592. [CrossRef]
46. Guimarães, E.; Ramos, A.; Janeira, M.A.; Baxter-Jones, A.D.G.; Maia, J. How does biological maturation and training experience impact the physical and technical performance of 11–14-year-old male basketball players? *Sports* **2019**, *7*, 243. [CrossRef]
47. Cumming, S.P.; Searle, C.; Hemsley, J.K.; Haswell, F.; Edwards, H.; Ryan, D.; Scott, S.; Gross, A.; Lewis, J.; White, P.; et al. Biological maturation, relative age and self-regulation in male professional academy soccer players: A test of the underdog hypothesis. *Psychol. Sport Exerc.* **2018**, *39*, 147–153. [CrossRef]
48. Cumming, S.P.; Lloyd, R.S.; Oliver, J.L.; Eisenmann, J.C.; Malina, R.M. Bio-banding in sport: Applications to competition, talent identification, and strength and conditioning of youth athletes. *Strength Cond. J.* **2017**, *39*, 34–47. [CrossRef]

49. Lefebvre, J.S.; Evans, M.B.; Turnnidge, J.; Gainforth, H.L.; Côté, J. Describing and classifying coach development programmes: A synthesis of empirical research and applied practice. *Int. J. Sports Sci. Coach.* **2016**, *11*, 887–899. [CrossRef]
50. Piggott, B.; Müller, S.; Chivers, P.; Papaluca, C.; Hoyne, G. Is sports science answering the call for interdisciplinary research? A systematic review. *Eur. J. Sport Sci.* **2019**, *19*, 267–286. [CrossRef] [PubMed]
51. Ford, P.R.; Williams, M.A. Sport Activity in Childhood: Early Specialization and Diversification. In *Routledge Handbook of Talent Identification and Development in Sport*; Baker, J., Cobley, S., Schorer, J., Wattie, N., Eds.; Routledge: London, UK, 2017; pp. 117–132.
52. Coutinho, D.; Gonçalves, B.; Santos, S.; Travassos, B.; Wong, D.P.; Sampaio, J. Effects of the pitch configuration design on players' physical performance and movement behaviour during soccer small-sided games. *Res. Sport. Med.* **2019**, *27*, 298–313. [CrossRef] [PubMed]
53. Santos, S.; Coutinho, D.; Gonçalves, B.; Schöllhorn, W.; Sampaio, J.; Leite, N. Differential Learning as a Key Training Approach to Improve Creative and Tactical Behavior in Soccer. *Res. Q. Exerc. Sport* **2018**, *89*, 11–24. [CrossRef] [PubMed]
54. Travassos, B.; Duarte, R.; Vilar, L.; Davids, K.; Araújo, D. Practice task design in team sports: Representativeness enhanced by increasing opportunities for action. *J. Sport. Sci.* **2012**, *30*, 1447–1454. [CrossRef] [PubMed]
55. Barnsley, R.H.; Thompson, A.H.; Legault, P. Family planning: Football style. *Relat. Age Eff. Football. Int. Rev. Sociol. Sport* **1992**, *27*, 77–87. [CrossRef]
56. Finnegan, L.; Richardson, D.; Littlewood, M.; McArdle, J. The influence of date and place of birth on youth player selection to a National Football Association elite development programme. *Sci. Med. Footb.* **2017**, *1*, 30–39. [CrossRef]
57. Verbeek, J.; Lawrence, S.; van der Breggen, J.; Kelly, A.L.; Jonker, L. The average team age method and its potential to reduce relative age effects. In *Birth Advantages and Relative Age Effects in Sport: Exploring Organizational Structures and Creating Appropriate Settings*; Kelly, A., Côté, J., Jeffreys, M., Turnnidge, J., Eds.; Routledge: London, UK, 2021; pp. 107–124.
58. Söderström, T.; Brusvik, P.; Lund, S. Factors underlying competitive success in youth football. A study of the Swedish national U15 football talent system. *Scand. Sport Stud. Forum* **2019**, *10*, 139–162.
59. Ashworth, J.; Heyndels, B. Selection bias and peer effects in team sports: The effect of age grouping on earnings of German soccer players. *J. Sports Econ.* **2007**, *8*, 355–377. [CrossRef]
60. Grossmann, B.; Lames, M. Relative Age Effect (RAE) in Football Talents–the Role of Youth Academies in Transition to Professional Status in Germany. *Int. J. Perform. Anal. Sport* **2013**, *13*, 120–134. [CrossRef]
61. Ramos-Filho, L.; Ferreira, M.P. The reverse relative age effect in professional soccer: An analysis of the Brazilian National League of 2015. *Eur. Sport Manag. Q.* **2021**, *21*, 78–93. [CrossRef]
62. Dugdale, J.H.; Sanders, D.; Myers, T.; Williams, A.M.; Hunter, A.M. Progression from youth to professional soccer: A longitudinal study of successful and unsuccessful academy graduates. *Scand. J. Med. Sci. Sport* **2021**, *31*, 73–84. [CrossRef] [PubMed]
63. Kelly, A.L.; Wilson, M.R.; Gough, L.A.; Knapman, H.; Morgan, P.; Cole, M.; Jackson, D.T.; Williams, C.A. A longitudinal investigation into the relative age effect in an English professional football club: The 'underdog hypothesis'. *Sci. Med. Footb.* **2020**, *4*, 111–118. [CrossRef]
64. Mills, A.; Butt, J.; Maynard, I.; Harwood, C. Identifying factors perceived to influence the development of elite youth football academy players. *J. Sport Sci.* **2012**, *30*, 1593–1604. [CrossRef] [PubMed]
65. Cook, C.; Crust, L.; Littlewood, M.; Nesti, M.; Allen-Collinson, J. 'What it takes': Perceptions of mental toughness and its development in an English Premier League Soccer Academy. *Qual. Res. Sport Exerc. Health* **2014**, *6*, 329–347. [CrossRef]
66. Holt, N.L.; Mitchell, T. Talent development in English professional football. *Int. J. Sport Psychol.* **2006**, *37*, 77–98.
67. Holt, N.L.; Dunn, J.G.H. Towards a grounded theory of the psychosocial competencies and environmental condi-tions associated with soccer success. *J. Appl. Sport Psychol.* **2004**, *16*, 199–219. [CrossRef]
68. Pain, M.A.; Harwood, C.G. Knowledge and perceptions of sport psychology within English soccer. *J. Sport Sci.* **2004**, *22*, 813–826. [CrossRef]
69. Harwood, C.; Barker, J.B.; Anderson, R.J. Psychosocial development in youth soccer players: Assessing the effec-tiveness of the 5C's intervention program. *Sport Psychol.* **2015**, *29*, 319–334. [CrossRef]
70. Bourke, A. The dream of becoming a professional soccer player. *J. Sport Soc. Issues* **2003**, *27*, 399–419.
71. Winn, C.O.N.; Ford, P.R.; McNarry, M.A.; Lewis, J.; Stratton, G. The effect of deprivation on the developmental activities of adolescent rugby union players in Wales. *J. Sport Sci.* **2016**, *35*, 2390–2396. [CrossRef]
72. Lawrence, D. Sociodemographic profile of an Olympic team. *Public Health* **2017**, *148*, 149–158. [CrossRef] [PubMed]
73. Zhang, Y.; Ling, C. A strategy to apply machine learning to small datasets in materials science. *NPJ Comput. Mater.* **2018**, *4*, 25. [CrossRef]

sports MDPI

Article

Stand up and Fight: A Case Study of a Professional Rugby Club Negotiating a COVID-19 Crisis, a Talent Development Perspective

Ian Costello [1,2], Sarahjane Belton [2] and Áine MacNamara [2,*]

[1] Munster Rugby, V94 X9VK Limerick, Ireland
[2] School of Health and Human Performance, Dublin City University, D09 W6Y4 Dublin, Ireland
* Correspondence: aine.macnamara@dcu.ie

Abstract: A wealth of literature examines the role of challenge from an individual psychological perspective, but research investigating how a talent development system can proactively support athletes to successfully meet the ever-increasing demands of top-level professional sport is less prevalent. This study takes advantage of a naturally occurring but highly atypical developmental challenge as a result of COVID-19 to examine factors influencing the efficacy and effectiveness of the talent development pathway at Munster Rugby. Players and staff (n = 12) took part in semi-structured interviews exploring their experiences of the build-up to the event, the game itself, and the impact post-event. The data were subsequently analysed using Reflexive Thematic Analysis. Players and coaches highlight the groundwork undertaken to establish alignment and coherence, both horizontally and vertically across the talent development environment, and how this contributed to navigating the challenge successfully. The findings support the necessity of both the player and the talent development system being prepared to enable players to perform at the highest level. The findings point to an overlap between the development and performance phases of a player's journey and the need to integrate short- and long-term objectives within a talent development system.

Keywords: challenge; development; coherence; talent development

Citation: Costello, I.; Belton, S.; MacNamara, Á. Stand up and Fight: A Case Study of a Professional Rugby Club Negotiating a COVID-19 Crisis, a Talent Development Perspective. *Sports* **2022**, *10*, 124. https://doi.org/10.3390/sports10080124

Received: 27 June 2022
Accepted: 4 August 2022
Published: 16 August 2022

1. Introduction

Over the last 30 years, considerable attention has been paid to the broad area of talent development in sport, with research focused on the principles underpinning effective talent development environments [1] incorporating a holistic environmental approach [2], factors impacting the development of high potential young players [3], the talent development system [4], talent development coaching [5], and talent development coaches [6]. Much of the research has been positioned against a pragmatic framework [7], motivated by the desire to generate effective answers to talent development problems directly applicable in the 'real world'. Collins and Kamin [8] refer to this as science *for* sport as opposed to science *of* or *through* sport; an approach underpinned by a motivation to generate practically meaningful knowledge.

Reflecting the need for coherence as an integral part of the talent development pathway [9], there is increasing consideration of how best to design and operationalise the talent development system. Although there is a well-established positive correlation between financial investment and performance outcomes in sport [10], some successful sports systems benefit from significant financial investment (e.g., in excess of GBP 100 million investment by UK Sport, 2015), while others still succeed with significantly less resources. In the latter case, appropriate talent development pathways are critical to this success, and must be carefully resourced and structured to optimise the efficacy and output of their pathway. In professional sports such as rugby, the talent development pathway is frequently structured as an academy that is structurally and systematically linked to the club's performance pathway. The objective of an academy in these settings is primarily the development and nurturing of high-potential players to facilitate their progression to first-team performance status [1]. To achieve this outcome, it is important that there is a coherence between what is happening at different stages of the talent development pathway [5], along with a shared understanding of processes and behaviours [11], at different phases to facilitate this movement.

Coherence in Talent Development Systems

It is widely accepted that talent development is a complex, dynamic, and non-linear process [9], with effective talent development environments characterised as offering individualised development, deploying long-term aims and methods, and focusing on a breadth of psycho-behavioural and psycho-social skills with coherence across stakeholders [12]. As such, effective talent development systems are characterised by inputs that are complementary, framed, and structured against long-term agendas. A strategic strength of an effective talent development system is a cohesive philosophy between stakeholders operating both horizontally (i.e., within a particular stage of development; e.g., a professional sports academy) and vertically (i.e., between stages of development; e.g., across the development system) [13], where shared values, expectations, and behaviours support the development of high-potential players [14]. In reality, however, there is often a diverse range of philosophies, beliefs, and methodologies being applied to many aspects of talent development, with what is best described as a lack of joined-up thinking. A lack of horizontal and vertical coherence, along with a lack of shared mental models [15], has been identified as a barrier to optimising the TD experience [5], especially where relationships are suboptimal between stakeholders, [16].

In this regard, a lack of organisational proximity and communication within a talent development system has been suggested to hinder player development [9]. Navigating coherence becomes even more complicated when players co-inhabit multiple pathways (e.g., club, academy, senior team), unless there is clarity in the message and support received at each level. Of course, coaches and players operating within a performance setting and a development environment will, by necessity, have different aims and objectives; the former focused primarily on outcome success, and the latter on longer-term development. Although the talent development system is conceptualised as a pathway, in reality, academy, club, and performance coaching and development often happen independently [17], with coaches and players operating in a siloed manner without shared goals and visions [4]. This is further complicated by the fact that high-potential young players rarely reside exclusively in either the performance or development worlds, but instead regularly move between these environments. Reflecting this, there have been calls to consider how the talent development system is designed to optimise the methods, structures, and opportunities afforded to high-potential athletes to allow them to navigate in-career transitions [1].

The coherence and effectiveness of the talent development system may be especially tested when young players are presented with non-normative transitions that stretch their skills and capabilities. In fact, the role of challenge in talent development has drawn considerable attention (and challenge!) since seminal work in 2012, which identified that, 'talent needs trauma' [18]. Traditional talent development approaches were often based on optimising the support offered to high-potential young athletes and minimising challenges for an athlete in order to smooth their development pathway. However, despite the face validity of such an approach, there is a growing research base suggesting that experiencing challenges on the development pathway, and then debriefing and learning from that experience, supports the development of talent, and therefore, young athletes benefit from a variety of challenges/bumps that they encounter on the 'rocky road'. Although some transitions are well flagged e.g., normative transition from youth to senior teams, transition into and out of academy settings [19]; high-potential young players often encounter a range of non-normative and unexpected transitions as part of their development journey. Wylleman & Lavelle [20] proposed a developmental model on transitions faced by athletes that takes a linear perspective from beginning to end, reflecting the nature of normative transitions at athletic, psychosocial, academic, and vocational levels. Stambulova's [21] Athletic Career Transition model focuses on the importance of an effective coping response for an athlete to make a successful transition from one developmental stage to another. These transitions can be challenging if the athlete does not have a well-developed skillset (physical, technical, tactical, and psycho-behavioural) and support system. In this respect, the importance of psycho-behavioural skills as the mechanism that supports the navigation of the challenge, and as the foundation for further learning and development as a result of the challenge, has been well explored [22,23]. Critically, the importance of identifying developmental challenges that can be optimally deployed to develop this skillset and maximise growth is recognised [5]. Although some attention has focused on the purposeful operationalisation of artificial 'speedbumps' on the pathway [23] as a means of ensuring the skills are tested, taught, practised and embedded, it is also apparent that high-potential young players will encounter a range of naturally occurring challenges as part of their trajectory. Athletes' and coaches' experiences of navigating and learning from such naturally occurring challenges represent a naturalistic lab to test the efficacy of the talent development system, and the ability of young athletes to navigate and grow from challenges. Once such, a naturalistic lab opportunity arose when Munster Rugby lost 48 players and staff from their senior team for a European Cup match in 2021, with young academy and National Talent Squad (NTS) players from the Munster pathway required to 'step into the breech'. The purpose of this current study, therefore, was to provide an in-depth and rich understanding of key stakeholders' experiences of navigating this naturally occurring challenge on the talent development pathway.

2. Materials and Methods

In this paper, we present a case study of a talent development system within a professional rugby club (Munster Rugby), with a particular focus on how one particular challenge was navigated; namely, the loss of 48 players and staff from the senior team for a European Cup match in 2021. The case study approach allowed us to explore this phenomenon in context, using a variety of data sources [24]. An ethnographic approach enabled us to capture the phenomenon from the viewpoint of the coaches and players involved in the match by using fieldwork, observation, and informal interviews to gain a rich understanding of the experiences within Munster Rugby and examine the complexities of the interactions and experiences in the group. The prolonged and in-depth engagement required by an ethnographic approach was facilitated by the first author, who was employed as the academy manager at the club throughout this period.

2.1. Context

Munster Rugby was founded in 1879 and is one of four provincial professional rugby teams in Ireland. Munster has a proud history in the European cup, reaching or surpassing the semi-finals 14 times since the tournament's inception in 1995. The most successful era in the club's history resulted in the club lifting the trophy on two occasions, in 2006 and 2008. It has a strong tradition of competing against international touring sides, with the most famous victory being against the touring All Blacks in 1978. In December 2021, during the worldwide COVID-19 pandemic, Munster Rugby travelled to South Africa on a two-game tour as part of the United Rugby Championship. Immediately on their return, they were scheduled to play Wasps in the first round of the Heineken European Champions Cup. Following Covid -19 testing, 14 players and staff were forced to remain in quarantine in South Africa, while a further 34 members of the squad were required to be isolated in Ireland for ten days upon their return. This resulted in all members of the squad who travelled to South Africa, both players and staff, being unavailable for preparation (and selection) for the game against Wasps. A small number of senior contracted players ($n = 10$) had not travelled with the squad to South Africa (seven staying due to participation in an Irish Rugby series of matches, while three were returning from injury). To fulfil the

Wasps fixture, Munster were forced to select a squad comprising of those ten senior players, along with young academy and NTS players chosen from the Munster pathway. Fourteen players were selected to make their European debut in this game, with twelve of those representing Munster at senior level for the first time. In addition, the senior coaching staff were unavailable due to COVID-19 protocols meaning that the academy staff led preparations for the game, and coached and managed the team on game day itself.

2.2. An Insider Perspective

In July 2021, six months prior to the game in question, I (first author) had been appointed as Munster Academy Manager, after returning from coaching professionally in the UK for five seasons. I had previously held positions in Munster Rugby including as a senior team assistant coach from 2010–2016 and as academy coach from 2008–2010. Prior to this, I had coached at each stage of the Munster pathway from youths and schools level through to All-Ireland League club level. With all Munster senior coaches forced to isolate as described above, I became the defacto head coach for the Wasps game in December 2021. Coincidentally, my last three years coaching in the UK, immediately prior to my return to Munster, were spent working as an Assistant coach with Wasps Rugby, at which time I had also enrolled on a Professional Doctorate exploring talent development practices in rugby. Without doubt this added extra spice to what was already an intriguing mixture.

The challenges of this unprecedented situation provided a test of Munster Rugby's talent development pathway, and an opportunity to explore key stakeholders' experiences and perceptions of a unique developmental challenge. It also provided a real-life case-study to examine the efficacy of the principles underpinning the Munster pathway; principles that the academy staff had spent the previous six months aiming to instil and embed across Munster's talent environment. These principles are outlined in the following section.

2.3. Principles Underpinning the Munster Pathway

Starting in my role as Munster Rugby Academy Manager, I prioritised a number of principles that I felt were essential to developing an effective TDE at Munster Rugby. These principles were based on a combination of my experiences in my coaching career to date, intimate knowledge of the Munster Rugby landscape, and knowledge acquired from TD research I had carried out up to that point on my Professional Doctorate journey. The build-up to, and the Munster versus Wasps match itself, also provided a unique opportunity for a 'temperature check' on the efficacy and impact of these principles in terms of preparing staff and players to embrace (and successfully navigate) many of the challenges encountered during this period. These principles were:

- Understanding and aligning the multiple stakeholders [4,25];
- Vertical and horizontal coherence, and shared mental models [5];
- Supportive but challenging environment, with players moving quickly through different stages of the pathway [18,26];
- Emphasis on developing skills, players, and staff [12,14].

Munster, a Complex Landscape

Munster Rugby is a particularly complex landscape with many challenges to creating an effective, coherent TD pathway. Taylor & Collins [5] described the high-pressure milieu of professional sport as being a significant barrier to coherence, with cultural and socio-political realities often having a negative impact on the long-term success of a Talent Development Environment (TDE). In Munster there are long standing historical and political considerations that impact decision making at all levels of the organisation. The province consists of a vast geographical region and incorporates multiple stakeholders across the entire TDE.

Vertically, the academy is completely aligned with the senior team, follows the senior programme, and operates out of the same training base in the Munster High Performance Centre under the remit of the Academy Manager. Below this level, the National Talent

Squad (NTS) comprises of a group of players that have been identified as potential academy entrants, with a different training base, a separate training schedule, and an entirely different group of coaches (partly due to Covid -19 'bubbles'). The management of this programme also falls under the remit of the Academy Manager.

Horizontally, the clubs and schools are key stakeholders in the TDE and are the primary feeders of the Munster pathway. The majority of academy and NTS players play club rugby in the amateur All-Ireland League each week. The European Champions Cup represents the highest standard of club rugby in Europe, and it is incredibly rare that a player who is not a contracted senior team player would be selected to play in the competition. On the occasion examined in this paper, 13 non-senior players (i.e., players from the Munster pathway) were selected in the match day 23 for the European Champions Cup match against Wasps, presenting a challenge of unprecedented magnitude.

2.4. Data Sources

Data for this study was collected by the first author using a combination of field notes and informal interviews. Ethical approval for the study was granted by the authors' institutional Research Ethics Committee DCUREC/2022/015.

2.4.1. Field Notes

The first author recorded daily observations and reflections of coaching and strategic decisions within his environment from July 2021 to February 2022. These consisted of oral and written records of incidents, events, documents, unusual occurrences, meetings, decisions, and observations. Reflecting the importance of prolonged engagement in ethnographic research, this study was completed over thirty-two weeks of daily contact.

2.4.2. Informal Interviews

The first author carried out a series of internal group and individual interviews with players (n = 11) and coaching staff (n = 4) involved in the Wasps-Munster match preparation and game day. These interviews were conducted informally at the Munster Rugby training base at times that were convenient for the participants. All participants were involved in the preparation for the Wasps-Munster match either as NTS players, academy players, first team players or coaching staff. The interviews lasted between forty-five and sixty minutes, with the first author also following up with individuals post-hoc for clarifications and further probing as part of a member reflection process [27]. During the interview, participants were asked questions about the preparation for the game, the environment during the two weeks' preparation, their experience of the game, how groundwork in the previous six months supported them during the event, and if they had experienced any post-event impact.

2.5. Data Analysis

Firstly, it must be acknowledged that data analysis in ethnographic research is complicated, and therefore no single method is universally accepted as a strategy for data analysis [28]. As such, an iterative process of theme building was undertaken [29] using the dataset as a whole. As a first step, text from field notes, observations and interview transcripts were labelled and then categorised and sorted according to a theme using a constant comparative method of analysis [30,31]. This step involved multiple reviews of the data and then coding for similarities, differences, groupings, patterns, and items of particular significance [32].

2.6. Trustworthiness and Integrity

This study was underpinned by a pragmatic approach to research, and therefore decisions about methodology and data analysis were chosen with these philosophical underpinnings in mind. Of course, given the nature of this study, it was important to consider and implement steps to ensure the trustworthiness and integrity of the data collection and data analysis. Reflecting Bradshaw et al.'s [33] suggestion, a number of quality markers were implemented:

1. Firstly, reflecting the importance of in-depth and prolonged engagement and to ensure the credibility of the findings, data were collected over 32 weeks by the first author, who held a thorough and rich understanding of the context;
2. Triangulation of data was employed using a range of data sources, including observations, interviews, focus groups, and reflections, in order to increase the rigour of the data;
3. Throughout the study, from conception through to write-up, critical friends were employed to address the trustworthiness and credibility of the findings;
4. Rich and in-depth accounts of each theme are offered using contextualised data to offer the reader an opportunity to consider the transferability of the findings to other contexts.

As described earlier, the first author was appointed academy manager at the start of the data collection process, thus bringing an insider perspective, deeper knowledge, and interactions to the research [34]. Of course, this also presented some potential challenges to the rigour of the research. To counter this, reflectivity was employed to enhance rigour [35], with the first author keeping a reflective diary to record their thoughts, behaviours, and actions during the data collection. This process supported the development of self-awareness, criticality, and flexible thinking during the data collection and data analysis phases of the research.

3. Results and Discussion

The results derived from the multiple sources in this ethnographic study provide an in-depth and rich understanding of key stakeholders' experiences of navigating a naturally occurring challenge on the talent development pathway at Munster Rugby. The three key themes, as highlighted in Table 1, were:

- Groundwork to establish alignment and coherence on the pathway;
- Preparation for the Munster-Wasps game;
- Impact post challenge.

Table 1. Thematic Analysis.

Higher Order Themes	Lower Order Themes	Raw Data Themes
Groundwork to establish alignment and coherence on the pathway	Integration/alignment of academy and senior team (players and staff)	Training programme alignment
		Senior and academy coaches alignment
		Social integration between academy and first team
		Connections and communications with senior coaches
	Connecting the pathway	Coaching up and down the pathway
		Shared mental models/common coaching framework
		Relationships between academy coaches and pathway players
	Creating a high performing environment	High challenge support environment
		Clear purpose and vision guiding behaviours/practices
	Horizontal alignment with key stakeholders in the TDE	Links to clubs and mutual support
		Practices to inform player selection
		Increased visibility of Munster coaches at club games

Table 1. *Cont.*

Higher Order Themes	Lower Order Themes	Raw Data Themes
Preparation for the Munster-Wasps game	Dual management—Player empowerment and responsibility	Empowering senior players
		Building trust between coaches and players
		Shared ownership of game preparation
		Player led sessions and meetings
	Squad integration, building trust and connections	Getting on the same page quickly
		Connecting the squad
	Creating a positive task focused environment	Creating balance between work and enjoyment
		Managing information flow
		Managing pressure and challenge
		Excitement to embrace the challenge
	Role models and mentoring	Buddy system
		Inspired by role models
		Support from senior players
	Playing with a cause and a deep sense of purpose	Creating a legacy, a Munster moment
		Clear sense of identity and history
		Representing the whole club
Impact post challenge	Inspired and highly motivated players	Motivation from first cap
		Inspired and hungry for more
	Increased vertical alignment and interpersonal relationships between staff and players	Acceptance as a Munster senior player
		Established trust between senior and academy staff
		Increased engagement and inclusivity in the HPC
		Stronger relationships across the club
		Connection through shared experiences
	Newfound confidence and belief to play at this level	Confidence in own ability
		Belief that opportunity is within touching distance
	Post event anti-climax	Disappointment with limited playing opportunities post Wasps game
		Academy staff experienced an anti-climax slump
	Confidence that the future is bright	Confidence in players coming through pathway
		Belief in young players ability

Each theme is supported using vignettes to capture the participants' voices in portraying events.

3.1. Groundwork to Establish Alignment and Coherence on the Pathway

As outlined previously, prior to the Munster-Wasps game, a number of key principles were prioritised, many of which were based around establishing alignment and coherence on the pathway. The groundwork undertaken in the six months prior to the Munster vs. Wasps games certainly appears to have been an influential factor in how players and coaches successfully navigated the challenge. The complete integration and alignment of the senior and academy training programmes meant that academy players trained on a full-time basis with the senior team, and in addition, the academy coaches had full access to all senior meetings and sessions. An academy player who played in the match talked about how seamless the transition was as a result:

"We train the exact same as the seniors. If we hadn't been training week in week out with the seniors we wouldn't have been aligned, we wouldn't have known all the roles and every single call, all as one group the whole time, it was seamless really".

This vertical alignment within the club [9] enabled the academy players and coaches to get on the same page quickly when preparing for the game. The groundwork that had been carried out in relation to establishing shared mental models [5] (e.g., game plan, playing details, coaching practices) proved to be invaluable when presented with so many potential challenges during the build-up to the game.

Recent improvements to connections and communications that had been made between senior and academy coaches were considered a positive factor in alignment between the groups [13], and proved important when it came to making informed selection decisions in assembling a squad that included so many players outside the traditional senior squad. An academy coach highlighted the following:

"We started the practice of having regular pathway meetings with the senior coaches (every 6 weeks) so that they were familiar with the players coming through but also so we have a shared understanding of what we were looking for in players".

Another academy player referenced the fact that, unlike in the past, academy players were now included in senior social functions, which *"really helped us all to work together"* in challenging circumstances during the build-up to the game when a group of players who had never played together needed to come together as a team, connect and build relationships quickly.

Connecting the pathway was a critical part of the groundwork in establishing alignment and coherence from top to bottom [36]. The academy coaches consistently work at all stages up and down the pathway, and had spent eight weeks the previous summer coaching the NTS players. This particular group of players were pivotal during the two-week preparation leading into the game, and in fact, two NTS players (sub-academy) were named in the match-day squad. This regular contact between academy coaches and pathway players enabled coaches to develop strong relationships with these players and exposed these pathway players to the training methods and practises employed at academy and senior level. An academy coach talked about the value of this:

"The sessions we did in the summer (with the NTS players) was very similar to what they experienced during the two weeks. They had trained that type of format and they were familiar with the academy coaching staff. Coaching up and down the pathway stood to us in the prep".

The deliberate practice of academy coaches and pathway coaches regularly working together meant that, in relation to both coaching philosophies and coaching practices, there was a shared mental model and a common coaching framework [6], which allowed everyone to get on the same page quickly.

"Academy coaches and the NTS and Talent coaches worked closely together for the 6 months leading into game, so when all the players and staff came together to prepare for the Wasps game we were on the same page".

Striving to create a high-performing environment in the academy, based on high challenge and support, was recognised as an important piece of the groundwork undertaken in the previous six months that enabled the coaches to operate effectively and cohesively under unusually demanding circumstances in the build-up to the Wasps game [37]:

"Our coaching environment (academy) is very open and honest and based on challenge and support, we have each other's backs but we challenge each other to be operating at the highest standards".

A process facilitated by the academy's performance psychologist, began six months prior to establish a clear purpose for the academy with an agreed set of values to guide behaviours. At the core of the purpose was a focus on raising standards across the pathway and a commitment to long-term development versus a short-term performance focus.

> *"As a newly formed academy team (Multi-disciplinary Team, MDT) we had a focus on raising the standards right across the board in the whole Pathway. This began with establishing a clear purpose and an agreed set of values and this was brought to life through our behaviours. We aligned to a key message of the whole Munster organisation 'we rise by lifting each other'"*.

As described earlier, Munster is a complex landscape, and the local schools and All-Ireland League (AIL) clubs are key stakeholders in the TDE. Horizontal alignment between Munster and these important stakeholders was a specific focus upon coming into the Academy Manager role six months earlier, with the aim of establishing strong relationships and building on them throughout the season [38]. The majority of players that are in the senior squad have come through the academy and from local clubs and schools. Most academy players will play their rugby each week in the AIL league, and the importance of aligning that experience to the academy was stressed by one of the coaches:

> *"We communicated with each AIL Head coach twice per week regarding Academy players and we offered the coaching support to all the clubs. This was vital when we needed club players to prepare for the Wasps game, we had incredible support with releasing players to train etc."*.

One academy coach spoke about the process of mapping out more regular visits to AIL matches to scout academy and pathway players and to build relationships with club coaches; *"Each month we mapped out a schedule so that coaches rotated regularly to watch different club games"*. This process proved invaluable when it came to making informed selections for the Wasps game and also getting cooperation from club coaches to release their players to train for the game, with another academy coach noting: *"Increased visibility of all the academy coaches at club games helped to get to know coaches and players which helped with selection for Wasps"*.

3.2. Preparation for the Wasps Versus Munster Game

As the COVID-19 situation unfolded and it became apparent that 45 senior players and staff would be unavailable for the Munster-Wasps game, a number of alignment meetings with coaching staff and players were organised to navigate the unique challenges that were presented. The coaching staff made a number of decisions to employ specific strategies based on bringing an unfamiliar squad together, building connections and integrating the group to get them on the same page as quickly as possible, to optimise the two-week preparation for and performance in the Wasps game.

One of the first decisions made by the coaching staff was to organise a meeting with the senior players that had not travelled to South Africa, with the purpose of outlining a 'dual management' model (shared ownership) between the coaches and senior players on all aspects of game preparation [39]. This involved empowering senior players and giving them the responsibility to lead the group and provide input across both the on-pitch and off-pitch components of preparation. An academy coach recalled that *"Seniors players had input into game plan design and session content and they presented throughout the two weeks in training reviews and opposition previews"*. The senior players bought in to this approach and invested heavily in preparing the squad for the game. An academy player observed the impact this had on the wider group:

> *"The coaches gave the senior players a licence to help take the sessions as well, there was an emphasis that they would lead from the front. It gave us confidence that they trusted us and the coaches trusted us. There was real sense of trust around and that brought a closeness"*.

This approach is based on an approach employed by the All Blacks (New Zealand rugby team) as documented by Hodge et al. [39], and is consistent with the coaching philosophy of the academy coaches at Munster. Many of the principles associated with dual-management are consistent with autonomy-supportive coaching [40], and emotional leadership coaching [41]. One of the senior players who was part of this leadership group captured the impact that it made during the preparation:

> *"If you empower players it creates a different type of trust and bond in the group, coaching each other, not you (coaches) deciding what we are doing but coming up with it together".*

Following on from the theme of player empowerment, the participants noted the impact of a buddy system [42], between each senior player and a younger academy or pathway player that was implemented during the preparation phase. This provided a mechanism for connecting players that were either new to the group, or had never played at European Cup level before, with senior and established players. An academy coach noted that:

> *"Every senior player was partnered with a young player, took him under their wing and supported him throughout the two weeks right up to presenting the jersey to their 'mentee' the night before the game".*

The influence of role models is well documented [43,44], with proximity and access to role models documented as one of the key Environmental Success Factors (ESF) in talent development environments.

The buddy system concept was deployed to optimise the resources available to help all the new players get up to speed with Munster game principles and specific details in the game plan. This was particularly important as key coaching staff were also absent from the preparation phase. This dual-management approach complemented the role of the senior players by giving them more responsibility in preparation for the Wasps game while also serving to build connections between players that had never played together. In addition to the technical and tactical benefits in relation to match preparation, an academy player talked about how he was inspired by working so closely with his role models [12]:

> *"It was kind of inspiring as well because you're having intimate conversations about your game with players who you grew up watching that you really respect and admire".*

It is worth noting that, in addition to the short-term benefits observed from this mentoring approach, the relationships forged between the players remained long after the Wasps game; this is described in a later section as part of the third higher order theme.

From the outset, there was a general acceptance from all participants that there were going to be aspects of preparation that would be far from perfect (in fact, it became a bit of an inside joke), and to this end, both players and coaches accepted that they were in a highly untypical situation and, as described by one coach, embraced a theme of '*improvise, adapt, overcome*' for the two weeks prior to and during the match. For example, the coaches observed the mentoring role the senior players played in relation to creating an environment that accepted that mistakes would be made in these circumstances; *"the senior lads were great at helping the younger players move past mistakes, it happens, learn, move on"*. This approach to the game appeared to be supported by the steps taken to integrate the squad and build trust and connections as quickly as possible, which was a primary focus of initial preparation. An academy coach pointed out the importance of this; *"Every meeting and session always included an element of connecting players, building relationships, we had to bring the squad together, get on the same page quickly"*. The team psychologist played a major role in connecting the squad early in the preparation phase. For both the inexperienced players and senior internationals alike, this match represented one of the biggest challenges of their rugby careers to date, but steps were taken to ensure it was presented as an opportunity to write their own history rather than as an insurmountable challenge. As one example, the team psychologist organised formal team meetings centred on developing team culture to augment the technical and tactical preparation. Developing this vision through formal meetings and facilitation by team staff was identified as a critical moment in the team's preparation, as highlighted by a senior player; *"The psychologist's meeting had a big role to play in that, the connection and vulnerability piece settled everyone as a group. We left that meeting feeling like a real team"*.

Central to preparation for the Wasps game was creating a positive, task-focused environment [45]. Due to the unusual circumstances surrounding the game, there was plenty of distraction, media, etc., that had the potential to affect the preparation of the young squad. There was ongoing discussion within the group to stay in the present, concentrate on the current task at hand, and continuously reinforce the importance of 'winning the next moment'. As stated above, there was a genuine acceptance that preparation would be disrupted and that mistakes would be expected. In what was potentially a high-pressured situation, there was a deliberate focus on creating a balance between working hard but also embracing and enjoying the novelty of the situation. A senior player remarked that, "*there was a good balance of work and enjoyment, the days just flowed better. It was enjoyable to work*".

As is well-documented, team cohesion is a trait associated with successful teams in sport [46]. Lack of cohesion represented a potential significant challenge with this particular group of players, most of whom had never played together, and many of whom had never played a professional game of rugby. To counterbalance a potential lack of cohesion, the coach's and senior players' focus (reflecting the implementation of a dual-management approach) was on creating a task-focused environment by managing the information flow and stripping back the level of detail in the game plan to give (especially the younger) players a freedom to play. One of the coaches mentioned how the game plan had been simplified, "*We gave the players a little bit less detail, were very concise, gave them freedom to play*". As part of creating a positive, task-focused environment, the coaches' intention was to manage the level of challenge and the pressure on the players, which less experienced players in particular responded positively to:

> *"You had freedom to show what you could do, show your skills, what you had learned up until now, don't get bogged down on errors and go out and enjoy it. An example of that was a coach saying 'I just can't wait to see you play' gave me great confidence anyway".*

There was a tangible sense of playing for a cause and playing with a deep sense of purpose among the playing group and the staff. There are numerous examples of successful teams in many different sports that try to connect to a deep sense of purpose or a cause that allows them to elevate their performance by being part of something greater than themselves [47]. It was clear that these steps were built around the identity of Munster, and the coaches deliberately tapped into this fact by positioning the game, "*as a legacy, something special to be a part of, another special Munster moment*". A senior player who had been involved in many of what would be considered famous games from Munster's past, felt that it was "*a sense of the old Munster when we were in a corner fighting our way out. It was another chance to add to the Munster history, which it has*".

As outlined earlier, throughout the entire preparation, there were players and staff that remained in South Africa, others in quarantine at home in Ireland, and the staff and players fortunate enough to be involved in preparing for the game embraced the privilege and responsibility of representing the entire club and its proud history. An academy player captured the sense of responsibility by saying that everyone was, "*representing the players that weren't there, you were doing a job for them*".

3.3. Impact Post Challenge

In the period following the Munster vs. Wasps game, it was clear that there were both positive and negative impacts on the team. Following the match, there seemed to be an increased vertical alignment along the pathway and an articulation of how interpersonal relationships were significantly enhanced since the game. Some players reported that they were inspired and highly motivated by playing in the game, with others sensing that they were within touching distance of playing in top-level games. Some of the academy and pathway players described how their involvement in the match instilled confidence and belief in their ability to play at senior level. Senior players reported more confidence in the young players coming through the pathway and their ability to make the step into a senior performance environment. Academy staff reported an enhanced level of trust from the senior players and staff, which all contributed to increased vertical alignment between both groups:

> *"Positive things that have come from the situation are that we* [academy staff] *are trusted a lot more now by senior staff and by players due to the feedback from the senior players that were here".*

Both players and staff reported improved interpersonal relationships in the High Performance Centre creating a sense of being one big squad:

> *"The dynamic between the younger and senior players is different now, it's like they are one squad now. The senior players are far more inclusive and engaged with the academy players, any previous divisions are gone".*

Previous authors [18,48] cited a lack of alignment on the talent development pathway as a barrier to creating an effective, coherent environment. Enhanced trust and improved interpersonal relationships between the academy and senior squad at Munster potentially eliminate, or at least reduce, that barrier and should facilitate greater alignment and coherence on the clubs' pathway in the future if maintained.

For several academy players, the Munster-Wasps match acted as a catalyst for their development, with many describing how their experience left them inspired, hungry for more and with a renewed confidence in their ability to progress. One academy player reflected that, *"It was my first taste, my first cap, it will probably be my motivation for quite a while"*. While another academy player's account was that, *"It left you wanting more. It was so much more enjoyable to be centrally involved in the team. I wish it was the same every week"*. An academy coach, in relation to the players that were involved in the game, felt that,

> *"Some of them definitely have grown a lot, the confidence that experience has given them, being able play at a level above what they thought they were capable".*

For those players who helped the team prepare but didn't make match day 23 there was also a benefit, with the realisation that they were within touching distance,

> *"Looking at player A, player B and player C playing, who you train with all the time, you're thinking that if they are able to do that there's no reason why I can't do that as well, it gave you belief, that it's there for you".*

In addition, these findings point to the importance of proximal role models on the talent pathway, along with the need to support players prior to and during challenges and transitions to optimise progression [18].

In contrast to the post-event bounce experienced by many players, a small number of players and staff experienced a post-event slump, or anti-climax, after the success of the Wasps game. This is not uncommon with sportspeople who have been involved in what they consider to be major sports events [49]. In the two weeks following the game, the majority of the regular squad members returned from quarantine and were selected in subsequent matchday squads. Many of the young players that had experienced the high of a European Cup game now found themselves with limited opportunities and a return to lower levels of competition, *"There was definitely an anti-climax, I feel like I'm at this level and you're not playing for the next few weeks, the disappointment because you have a taste of it"*. This feeling wasn't reserved solely for the players, however, as academy staff also reported the same sense of anti-climax after being centrally involved in preparation for such a big game, *"People were almost disappointed to go back into normal weeks, a come down, after an occasion as special as this. It was different energy"*.

As mentioned earlier, the benefits of the mentoring system employed during the two-week preparation were evident as the season continued, suggesting that there was longevity in the impact of the match, and the match preparations, on systems and behaviours. Many inter-player relationships that have been forged during the build-up to and during the Wasps game have continued to develop since then. One of the senior players talked about, *"texting player A & player B during the U20 six nations, you wouldn't normally have a relationship with them but now I do and I've a genuine interest in how they are getting on"*. Another senior player talked about academy players *"coming up to me looking for advice and if we didn't go through that experience together that wouldn't be happening"*. Providing developing players with access to role models in their environment is a factor strongly associated with successful TD environments [2], and the ongoing interpersonal relationships between academy and senior players, arguably a fortuitous spin off from the buddy system employed during game preparation, is a very welcome and positive one.

Finally, senior players noted a renewed confidence in the quality of players that are coming through the Munster system. One senior player stated that we, *"have a different calibre of player coming through now"* and a second senior player addressed the fact that he was previously concerned with the throughput of players from within our pathway, but after this experience he believes the future is bright, *"My biggest thing now is that I'm not worried what's coming through, I had a big worry but not anymore"*. While the perceived 'security' of the club is without doubt of benefit, there are also unseen, and perhaps unexpected, additional advantages to this, in that senior players will be aware of very genuine competition for places in the starting team. It has been well documented that this genuine competition for places amongst a squad is an essential part of a successful TD ecosystem, which can have a direct impact on enhanced levels of performance within teams [50].

3.4. General Discussion

The goal of a talent development pathway in sport is to efficiently develop athletes with the ability to perform at the senior level. With this in mind, it is important that talent development environments such as professional sports academies are fit for purpose and promote talent through the pipeline. This study took advantage of a naturally occurring (but highly atypical) challenge (i.e., Munster Rugby's loss of 48 players and staff due to COVID-19 restrictions for a vital European Rugby Cup fixture) to pressure test the efficacy and effectiveness of a professional rugby club's talent pathway.

This scenario provided a real-life case study from which to examine the groundwork, principles, and practices that underpinned the talent development environment and how navigating a significant challenge provided opportunities for growth and development of both players and staff following the experience [51] Findings from this study clearly demonstrate the importance of an effective talent pathway in supporting performance objectives at senior level [9] and the need for coherent and aligned systems on the pathway that support players to be ready for the next step in their journey. Arguably, this may be even more important in (relatively) less resourced sports such as rugby union clubs, where squad size and depth is constrained by budget limitations [1].

At a system's level, the results point to the importance of vertical and horizontal coherence in the TDE, with development and performance clearly overlapping on the talent pathway [5]. In these particular circumstances, the academy (development) players performed and the senior (performance) players developed. This raises an interesting question in terms of 'where does performance begin and development end?'. Given the dynamic environment of elite sport, and the need for high potential young players to optimise development opportunities when they arise, it would seem that the performance-development dichotomy is more an academic delineation rather than an applied reality. Given the performance pressures and competitive characteristics that are inevitable features of high performance, the need for a TD pathway to be both efficient (getting players there as quickly as possible) and effective (getting players there as 'ready' as possible), points to the importance of a TD pathway having both short-term and longterm objectives [12].

Preparing for and learning from challenges is central to the talent systems approach [23]. The role of challenge as a development catalyst is well established [52], provided the TD system supports its players to proactively prepare for challenge and, just as importantly, players are managed post challenge to ensure learning and growth are facilitated moving forward [51]. There exists an abundance of literature examining the role of challenge in

TD from an individual psychological perspective [22], but the findings of this current study highlight the importance of both the player being prepared and the TD system being prepared to proactively support its athletes. Our findings support that the steps taken to connect the pathway by (a) aligning the key stakeholders from top to bottom, and (b) increasing coherence between different groups at various stages, were influential in navigating the significant challenges that were encountered. The importance of empowering players [39], creating a high-performing, positive, task-focused environment [37], utilising role models [53], and establishing trust, connections, and a deep purpose in a playing group, were all considered influential in achieving a successful outcome in the Wasps-Munster match at the centre of this study.

Given the considerable discussion about the role of TD in sport, from a system perspective, examining the effectiveness and efficiency of the pathway holds both academic and applied interest moving forward. Effective TDEs are characterised as deploying long-term aims and methods and being structured against long-term agendas [4]. In his study involving the Ajax soccer academy, Larsen [54], highlighted several environmental success factors, including the focus on long-term development rather than winning the next match. Ajax provided opportunities for individual long-term development in preference to short-term success. To reinforce this philosophy, the academy complex is very aptly named 'De Toekomst', which translates to 'The Future'.

In this current study, the findings point to an obvious overlap in the pathway between development and performance phases. This creates potential conflict between a 'win later' development approach and a 'win now' performance philosophy, usually associated with a much shorter-term agenda. Often, development coaching is not clearly conceptualised, and again, as explored above, it can be difficult to identify where development coaching ends and performance coaching begins. There is a lack of research into the nuances of development coaching versus performance coaching, and more research is needed to investigate the in situ contextual practice of coaches working in successful TDE's. Such research would serve to inform and develop coaching practice *for* the coach, rather than the coach, in the TD space. With increasing pressure in professional sport to get more players ready to perform at the top level, even earlier in their development, continuing to explore how we optimise the design and operation of effective TD pathways is critical, along with further research investigating, informing, and developing practices of (development) coaches.

Finally, although the insider perspective provided by the first author's position added richness to the data, it must be acknowledged that there are some limitations to this approach. Such a scenario can lead to cultural bias, and due to the first author's position as academy manager, there may also be a degree of 'impression management' among the participants during the interviews. In addition, this was one event, in one season, on one specific pathway, so it is important that practitioners question why this practice works in this context with this particular group, rather than assuming to generalise. As such, caution needs to be exercised when interpreting findings and applying them to other contexts and environments.

Author Contributions: Conceptualization, I.C., S.B. and Á.M.; methodology, I.C., S.B. and Á.M.; formal analysis, I.C. and Á.M.; investigation, I.C.; writing— I.C., S.B. and Á.M.; draft preparation, I.C., S.B. and Á.M.; writing—review and editing, I.C., S.B. and Á.M.; supervision, S.B. and Á.M. All authors have read and agreed to the published version of the manuscript.

Funding: This research received no external funding.

Institutional Review Board Statement: The study was conducted in accordance with the Declaration of Helsinki, and approved by the Institutional Review Board (or Ethics Committee) of Dublin City University (February 2022)". for studies involving humans.

Informed Consent Statement: Informed consent was obtained from all subjects involved in the study, including information about the possibility to withdraw from the study at any time.

Data Availability Statement: Not applicable, as the study includes data protected by the Norwegian Personal Data Act and The General Data Protection Regulation from the EU (GDPR).

Conflicts of Interest: The authors declare no conflict of interest.

References

1. Till, K.; Baker, J. Challenges and [Possible] Solutions to Optimizing Talent Identification and Development in Sport. *Front. Psychol.* **2020**, *11*, 664. [CrossRef]
2. Henriksen, K.; Stambulova, N.; Roessler, K.K. Holistic approach to athletic talent development environments: A successful sailing milieu. *Psychol. Sport Exerc.* **2010**, *11*, 212–222. [CrossRef]
3. Gledhill, A.; Harwood, C.; Forsdyke, D. Psychosocial factors associated with talent development in football: A systematic review. *Psychol. Sport Exerc.* **2017**, *31*, 93–112. [CrossRef]
4. Martindale, R.J.; Collins, D.; Daubney, J. Talent Development: A Guide for Practice and Research Within Sport. *Quest* **2005**, *57*, 353–375. [CrossRef]
5. Taylor, J.; Collins, D. Getting in the Way: Investigating Barriers to Optimizing Talent Development Experience. *J. Expertise* **2021**, *4*, 315–332. Available online: https://www.journalofexpertise.org (accessed on 1 March 2022).
6. Williams, G.G.; MacNamara, Á. Coaching on the Talent Pathway: Understanding the Influence of Developmental Experiences on Coaching Philosophy. *Int. Sport Coach. J.* **2020**, *8*, 141–152. [CrossRef]
7. Giacobbi, P.R.; Poczwardowski, A.; Hager, P.F. A Pragmatic Research Philosophy for Applied Sport Psychology. *Sport Psychol.* **2005**, *19*, 18–31. Available online: http://hdl.handle.net/20.500.12648/2434 (accessed on 1 March 2022). [CrossRef]
8. Collins, D.; Kamin, S. The performance coach. In *The Oxford Handbook of Sport and Performance Psychology*; Murphy, S.M., Ed.; Oxford University Press: Oxford, UK, 2012; pp. 692–706. [CrossRef]
9. Webb, V.; Collins, D.; Cruickshank, A. Aligning the talent pathway: Exploring the role and mechanisms of coherence in development. *J. Sports Sci.* **2016**, *34*, 1799–1807. [CrossRef]
10. Johnson, D.K.N.; Ali, A. A Tale of Two Seasons: Participation and Medal Counts at the Summer and Winter Olympic Games. *Soc. Sci. Q.* **2004**, *85*, 974–993. [CrossRef]
11. Bjørndal, C.T.; Ronglan, L.T. Orchestrating talent development: Youth players' developmental experiences in Scandinavian team sports. *Sports Coach. Rev.* **2018**, *7*, 1–22. [CrossRef]
12. Martindale, R.J.J.; Collins, D.; Abraham, A. Effective Talent Development: The Elite Coach Perspective in UK Sport. *J. Appl. Sport Psychol.* **2007**, *19*, 187–206. [CrossRef]
13. Mills, A.; Butt, J.; Maynard, I.; Harwood, C. Examining the Development Environments of Elite English Football Academies: The Players' Perspective. *Int. J. Sports Sci. Coach.* **2014**, *9*, 1457–1472. [CrossRef]
14. Larsen, C.H.; Alfermann, D.; Henriksen, K.; Christensen, M.K. Successful talent development in soccer: The characteristics of the environment. *Sport Exerc. Perform. Psychol.* **2013**, *2*, 190–206. [CrossRef]
15. McComb, S.; Simpson, V. The concept of shared mental models in healthcare collaboration. *J. Adv. Nurs.* **2014**, *70*, 1479–1488. [CrossRef]
16. Ivarsson, A.; Stenling, A.; Fallby, J.; Johnson, U.; Borg, E.; Johansson, G. The predictive ability of the talent development environment on youth elite football players' well-being: A person-centered approach. *Psychol. Sport Exerc.* **2015**, *16 Pt 1*, 15–23. [CrossRef]
17. Curran, O.; MacNamara, Á.; Passmore, D. Singing off the same hymn sheet? Examining coherence in a talent development pathway (part 1). *J. Sports Sci.* **2021**, *39*, 1709–1716. [CrossRef]
18. Collins, D.; MacNamara, Á. The Rocky Road to the Top. *Sports Med.* **2012**, *42*, 907–914. [CrossRef]
19. Wylleman, P.; Reints, A. A Lifetime perspective on the career of talented an elite athletes: Perspectives of high-intensity sports. *Scand. J. Sci. Med. Sports* **2010**, *20*, 88–94. [CrossRef]
20. Wylleman, P.; Lavallee, D. A developmental perspective on transitions faced by athletes. In *Developmental Sport and Exercise Psychology: A Lifespan Perspective*; Fitness Information Technology Publishers: Morgantown, WV, USA, 2004; pp. 507–527.
21. Stambulova, N.B. Athlete's crises: A developmental perspective. *Int. J. Sport Psychol.* **2000**, *31*, 584–601.
22. MacNamara, A.; Button, A.; Collins, D. The Role of Psychological Characteristics in Facilitating the Pathway to Elite Performance Part 1: Identifying Mental Skills and Behaviors. *Sport Psychol.* **2010**, *24*, 52–73. [CrossRef]
23. Savage, J.; Collins, D.; Cruickshank, A. Exploring Traumas in the Development of Talent: What Are They, What Do They Do, and What Do They Require? *J. Appl. Sport Psychol.* **2017**, *29*, 101–117. [CrossRef]
24. Hafiz, K.; Baxter, P.; Jack, S. Case study example Qualitative Case Study Methodology: Study Design and Implementation for Novice Researchers. In *The Qualitative Report*; 2008; Volume 13, No. 4. Available online: http://www.nova.edu/ssss/QR/QR13-4/baxter.pdf (accessed on 1 March 2022).
25. Gesbert, V.; von Roten, F.C.; Hauw, D. Reviewing the role of the environment in the talent development of a professional soccer club. *PLoS ONE* **2021**, *16*, e0246823. [CrossRef]
26. Nash, C.; Taylor, J. 'Just Let Them Play': Complex Dynamics in Youth Sport, Why It Isn't So Simple. *Front. Psychol.* **2021**, *12*, 700750. [CrossRef]

27. Smith, B.; McGannon, K. Developing rigor in qualitative research: Problems and opportunities within sport and exercise psychology. *Int. Rev. Sport Exerc. Psychol.* **2018**, *11*, 101–121. [CrossRef]
28. Gibbs, G. *Analyzing Qualitative Data*; Sage Publications Ltd.: London, UK, 2007. [CrossRef]
29. Fetterman, D.M. *Ethnography: Step by Step*, 3rd ed.; Sage Publications: Thousand Oaks, CA, USA, 2010.
30. Glaser, B.; Srauss, A. *The Discovery of Grounded Theory*; Aldine Publishing Co: Chicago, IL, USA, 1967.
31. Lincoln, Y.S.; Guba, E.G. *Naturalistic Inquiry*; Sage Publications: Beverley Hills, CA, USA, 1985.
32. Mason, J. *Qualitative Researching*; Sage Publications: London, UK, 1996.
33. Bradshaw, C.; Atkinson, S.; Doody, O. Employing a Qualitative Description Approach in Health Care Research. *Glob. Qual. Nurs. Res.* **2017**, *4*, 1–8. [CrossRef]
34. Greene, M. On the Inside Looking In: Methodological Insights and Challenges in Conducting Qualitative Insider Research. *Qual. Rep.* **2014**, *19*, 1–13. [CrossRef]
35. Berger, R. Now I see it, now I don't: Researcher's position and reflexivity in qualitative research. *Qual. Res.* **2015**, *15*, 219–234. [CrossRef]
36. Mills, A.; Butt, J.; Maynard, I.; Harwood, C. Identifying factors perceived to influence the development of elite youth football academy players. *J. Sports Sci.* **2012**, *30*, 1593–1604. [CrossRef]
37. Lara-Bercial, S.; Mallett, C.J. The Practices and Developmental Pathways of Professional and Olympic Serial Winning Coaches. *Int. Sport Coach. J.* **2016**, *3*, 221–239. [CrossRef]
38. Hall, A.J.; Jones, L.; Martindale, R.J. The Talent Development Environment Questionnaire as a Tool to Drive Excellence in Elite Sport Environments. *Int. Sport Coach. J.* **2019**, *6*, 187–198. [CrossRef]
39. Hodge, K.; Henry, G.; Smith, W. A Case Study of Excellence in Elite Sport: Motivational Climate in a World Champion Team. *Sport Psychol.* **2014**, *28*, 60–74. [CrossRef]
40. Lyons, M.; Rynne, S.B.; Mallett, C.J. Reflective Practice International and Multidisciplinary Perspectives. Reflection and the art of coaching: Fostering high-performance in Olympic Ski Cross. *Reflective Pract.* **2012**, *13*, 359–372. [CrossRef]
41. Chan, J.T.; Mallett, C.J. The Value of Emotional Intelligence for High Performance Coaching. *Int. J. Sports Sci. Coach.* **2011**, *6*, 315–328. [CrossRef]
42. Bushmaker, T.; Lund, M.; Meyer, J.; Mcnulty, E.; Wait, K.; Braun, S. Effectiveness of a Buddy System on Strength Training, Adherence, Confidence, and Perceived Health in College Females. *Int. J. Res. Ex. Phys.* **2019**, *15*, 23–38.
43. Henriksen, K.; Stambulova, N.; Roessler, K.K. Successful talent development in track and field: Considering the role of environment. *Scand. J. Med. Sci. Sports* **2010**, *20* (Suppl. 2), 122–132. [CrossRef]
44. Adronikos, G.; Westbury, T.; Brazo-Sayavera, J.; Olivares, P.; Martindale, R. Factors contributing to the quality of the junior-senior transition in Greek athletes. *Int. J. Sport Exerc. Psychol.* 2021, 1–18, *Ahead-of-print*. [CrossRef]
45. Jackson, S.A.; Roberts, G.C. Positive Perfonnance States of Athletes: Toward a Conceptual Understanding of Peak Performance. *Sporl Psychol.* **1992**, *6*, 156–171.
46. Bruner, M.W.; Eys, M.A.; Wilson, K.S.; Côté, J. Group cohesion and positive youth development in team sport athletes. *Sport Exerc. Perform. Psychol.* **2014**, *3*, 219–227. [CrossRef]
47. Johnson, T.; Martin, A.J.; Palmer, F.R.; Watson, G.; Ramsey, P. A Core Value of Pride in Winning: The All Blacks' Team Culture and Legacy. *Int. J. Sport Soc.* **2013**, *4*, 1–14. [CrossRef]
48. Taylor, J.; Collins, D. The Highs and the Lows—Exploring the Nature of Optimally Impactful Development Experiences on the Talent Pathway. *Sport Psychol.* **2020**, *34*, 319–328. [CrossRef]
49. Bradshaw, H.; Howells, K.; Lucassen, M. Abandoned to manage the post-Olympic blues: Olympians reflect on their experiences and the need for a change. *Qual. Res. Sport Exerc. Health* **2021**, *13*, 1–18. [CrossRef]
50. Passos, P.; Araújo, D.; Davids, K. Competitiveness and the Process of Co-adaptation in Team Sport Performance. *Front. Psychol.* **2016**, *7*, 1562. [CrossRef]
51. Taylor, J.; Collins, D. Navigating the winds of change on the smooth sea—The interaction of feedback and emotional disruption on the talent pathway. *J. Appl. Sport Psychol.* **2021**, *34*, 886–912. [CrossRef]
52. Taylor, J.; Collins, D. Shoulda, Coulda, Didnae—Why Don't High-Potential Players Make it? *Sport Psychol.* **2019**, *33*, 85–96. [CrossRef]
53. Henriksen, K.; Larsen, C.H.; Christensen, M.K. Looking at success from its opposite pole: The case of a talent development golf environment in Denmark. *Int. J. Sport Exerc. Psychol.* **2014**, *12*, 134–149. [CrossRef]
54. Larsen, C.H.; Storm, L.K.; Sæther, S.A.; Pyrdol, N.; Henriksen, K. A world class academy in professional football: The Case of Ajax Amsterdam. *Scand. J. Sport Exerc. Psychol.* **2020**, *2*, 33–43. [CrossRef]

sports

Article

Aspects of Developmental Pathways toward World-Class Parasport

Lise Storli [1], Morten Andreas Aune [1] and Håvard Lorås [2,*]

[1] Department of Sport Science, Sport and Human Movement Science Research Group (SaHMS), Nord University, 7600 Bodø, Norway
[2] Department of Teacher Education, Faculty of Social and Educational Sciences, NTNU—Norwegian University of Science and Technology, 7491 Trondheim, Norway
* Correspondence: havard.loras@ntnu.no

Abstract: The developmental pathways of athletes with a physical disability into world-class parasport are much less researched and understood compared to able-bodied athletes' participation histories. The purpose of this study was to investigate the developmental pathways of para-athletes toward elite performance. Data from eight athletes with physical disabilities ranked among the top performers in Paralympics, World Championships, and/or European Championships were gathered. Thematic analysis of retrospective semi-structured interviews led to the identification of four themes. The findings showed the importance of early childhood sport-related encounters in a family environment followed by sampling of various organized and coach-led sports throughout the childhood period. The youth sport period was highly heterogenous, albeit with important transitions towards elite-level practice environments, competition, coaching, equipment adaptability and the first intentions of becoming an elite-level athlete. Throughout, significant contributions are attributed towards parents, friends, coaches, athletes, and others, towards fostering a sustained motivational climate focused on improvement and further progress. These findings could provide useful information to tailor developmental models towards elite-level performance in parasport.

Keywords: para-athletes development; sport expertise; deliberate play; deliberate practice; developmental model of sport participation (DMSP)

Citation: Storli, L.; Aune, M.A.; Lorås, H. Aspects of Developmental Pathways toward World-Class Parasport. *Sports* **2022**, *10*, 123. https://doi.org/10.3390/sports10080123

Academic Editor: Kevin Till

Received: 2 June 2022
Accepted: 10 August 2022
Published: 15 August 2022

1. Introduction

The developmental pathways towards expertise and peak performance in sport has been investigated in abundance the past three decades, highlighting the individual and sport-specific interplay between factors such as deliberate practice or play, specialization or diversification, participation in organized vs. unorganized activities, and coach-led vs. self-initiated practice in training and competition [1,2]. In contrast, the developmental pathways of elite-level athletes with a disability have not been subjected to similar empirical investigations in order to disentangle important aspects of their pathways towards elite level performance. Parasport is increasingly becoming recognized as a legitimate component of international sport, although in some instances the elite-level athletes with disabilities are still not taken as seriously as elite-level able-bodied athletes [3].

Parasport is simultaneously used as a synonym for Paralympic sport, which addresses athletes who compete in the Paralympic Games [4]. The term captures practice and competition for persons with a physical or intellectual disability, in which the type of disability determines types of competitive categories in order to ensure fair competition amongst athletes with similar disabilities [5]. To be eligible for parasport, a person must have an impairment type according to the minimum disability criteria outlined in the World Para Athletics Classification Rules and Regulations [6]. Illustrating the substantial individuality and diversification amongst athletes with a disability in general, World Para Athletics describes eight eligible physical impairment types which apply to various competitive events, creating around ~100 different competitive classes. The main aim of the current study was to investigate the developmental pathways towards elite performance (as defined by top-level performance in Paralympics and/or World/European championships) among athletes with a physical disability. As parasport clearly contains a substantial heterogenous sport population due to individual differences related to (amongst other factors) type and degree of disability, the developmental pathways towards elite-level performance might potentially be more diversified in athletes with a disability compared to able-bodied athletes who typically demonstrate at least some degree of homogeneity in physical characteristics according to sport-specific demands. Thus, data on developmental pathways towards elite-level performance obtained from able-bodied athletes might not necessarily be applicable toward elite-level parasport.

Apart from the potential differences in physical characteristics between able-bodied and athletes with a disability that might implicate differences in developmental pathways, young participants in parasport typically have far less opportunities in terms of organized club-sports, local, regional and national developmental/talent programmes/teams, highly experienced and educated coaches, overall funding, as well as other organizational factors that exist for able-bodied athletes [4]. Many nations worldwide still lack appropriate systems from local up to national level that address parasport in general, and particularly when it comes to high-level parasport [7]. Thus, athletes with a disability might already experience substantial barriers towards sport participation (in general) at a local level, which might impact upon their developmental pathways towards elite performance levels.

Existing research on key determinants in development of elite sporting performance amongst athletes with a physical disability is relatively scarce. In a systematic review from 2017, Denghansai et al. [8] noted that few studies had examined issues relating to trajectories, training modifications, competition, or recovery processes throughout the sporting careers of athletes with a disability, from their introduction to sport-related activities through all phases on the path to competing at the highest level of international competition in their particular sport. Some authors have claimed that athletes with a disability pass through the same sport-related developmental stages as able-bodied athletes, only with some modest differences in rate of progress depending on the degree of disability [9]. Results from two studies on elite Canadian wheelchair basketball players, however, indicate that factors such as athletes' disability classification and acquired vs. congenital disabilities introduce inter-individual variability in terms of reaching performance-related milestones, and impacts upon developmental pathways in training and competition history [10–12]. Further, Denghansai et al. [13] compared developmental trajectories between elite able bodies, and para-athletes with both congenital disability and acquired disability, and the results showed both similarities and differences between the three groups, but more interestingly, it was shown that congenital disability Para-athletes had more developmental similarities to able-bodied athletes.

Developmental pathways toward elite-level sport performance, from the first experiences with sport up to top performances in major international tournaments, are typically examined by referring to athlete development models. In the current study, we examine developmental pathways towards elite-level parasport through the lens of the Developmental Model of Sport Participation (DMSP) as a theoretical framework [14]. Albeit caution needs to be exercised in applying a model of athlete development initially designed for able-bodied athletes, the DMSP postulates three broad stages in which high-performance athletes move through on their way to expertise through childhood and adolescence, involving sampling, specialization, and investment. Sport specialization is defined as year-round training greater than 8 months, choosing and focusing on a single primary sport [4], while sampling involves sustained participation in several sports. The key element in sport sampling is that children experience and develop a number of different motor skills and cognitive skills that are crucial to optimize skills in the main sport, and sports specialization is aimed to develop sport-specific motor coordination patterns through a high amount of specific training [1,15].

Thus, the DMSP outlines two possible developmental trajectories towards elite-level in sport: (1) elite performance might emerge from a combination of primary sampling in the childhood years and later specialization in adolescence, or (2) elite-level performance emerges through early and sustained specialization in childhood and adolescence. The DMSP predicts that elite performance through sampling consists of specializing years around the ages of 13–15 succeeded by primary sampling years in childhood. In this pathway, high amounts of deliberate play are suggested as a significant factor towards staying motivated with sustained enjoyment. In comparison, the second developmental trajectory toward elite-level performance in sport contains early specialization and focus on one sport and performing high amounts of sport-specific and individualized coach-led deliberate practice [1]. The past two decades of work associated with the DMSP framework on elite athletes has generated seven postulates that have received sufficient support from research [16,17]. Despite the general consistency of trends leading up to postulates in elite able-bodied athletes, their relevance for understanding the development pathways of elite-level athletes with disabilities is relatively unknown. The first three postulates address early diversification (sampling of various sport-related experiences) as favorable for long-term sport involvement, positive youth development, and development of high-level performance. Regarding athletes with disabilities, they might in principle not have the opportunity to 'sample' several sports in childhood due to a lack of organized practice and competitions from local clubs and sport organizations. Taking part in several sports has been suggested to optimize the development of motor skills to athletes' main sport [15,18], although in regard to parasport some studies could not draw a similar conclusion [19].

Two other postulates originating from the DMSP highlight the importance of high amounts of deliberate play in the sampling years to build strong intrinsic motivation and establish a range of motor and cognitive competencies that children can ultimately bring to a main sport [16]. Deliberate play concerns the involvement in youth-led informal sports-play with peers (such as playing, leisure time, backyard soccer, street hockey, basketball in the driveway, ice-hockey on a frozen lake, etc.) without adult/coach supervision [15,20]. These activities are for the purpose of maximizing enjoyment, freely chosen, and are regulated intrinsically by the children/adolescents themselves where rules are typically adapted from original formal sports. However, athletes with a disability might experience similar barriers for participating in such informal physical activities, as for participation in organized sport. One typical recurring theme when it comes to sport participation for athletes with a disability concerns the important role of individualized equipment [21]. Thus, lack of resources and/or availability of equipment might be just as consequential for athletes with a disability towards participating in deliberate play-related activities as for their entrance and participation in organized sport.

The last two postulates from DMSP address the adolescence period and stress the opportunity for athletes to choose to specialize in their favorite sport in this developmental period [16]. Especially in late adolescence, the physical, cognitive, social, emotional, and motor skills are considered developed at necessary levels for youth athletes to invest effort in highly specialized training in one sport [1,22]. However, this might not necessarily apply to all parasport athletes, because selecting a sport implies that there is a pool of sports to select from. Furthermore, the necessary levels of elite-sport readiness in terms of biological and psychological characteristics might not be similar in athletes with a disability compared to able-bodied athletes.

Based on the presented considerations, the main aim of the current study was to further examine the developmental pathways towards elite-level performance, through childhood and adolescence, amongst athletes with a physical disability that can be defined as experts in their fields. Thus, athletes included in the current study are competing at the highest level and are ranked amongst the top performers in Paralympics, World Championships, and/or European Championships. By conducting retrospective qualitative interviews, the purpose was to examine the developmental histories and experiences of elite athletes with a physical disability, including exploring their perspectives and reflections in terms of the proposed pathways in the DMSP. The goal was to further understand aspects of the various developmental periods and milestones involved in pathways toward world-class performance in parasport, in an attempt to shed light on and address some of the proposed problems with applying non-disability athlete development models such as the DMSP towards understanding athletes with a physical disability. Further understanding of the developmental process by recognizing the experiences of these athletes could potentially inform the training and development of elite parasport.

2. Materials and Methods

A strategic sample of eight elite-level world-class athletes with a physical disability were recruited by contacting sport federations and through personal contacts. To capture some of the variations in elite parasport, five male and three female athletes were included, and they were recruited from both winter and summer sports as well as individual or team sports. Furthermore, the sample contained five athletes with acquired disabilities in the lower extremities (progressive neurological disease, amputation, and/or paralysis) and three athletes with congenital disabilities in the lower extremities (paralysis and/or spinal hernia). To be included, they had to compete and perform at the top level in Paralympics, The World Championships, The European Champions, or several of these. In the recruited sample, the average age of the participants at the time of data collection was 27 years (SD = 7.5 years). As a testimony of the elite level of the athletes with a physical disability in the current study, six of them had won medals (1st, 2nd or 3rd place) in one or several of the mentioned competitions, and the two remaining participants had competed in the Paralympics and/or World Championship with results among the top 5. All participants provided written informed consent before completing the interview, and the project was approved by The Norwegian Centre for Research Data. Because of the Norwegian law regulated through the Norwegian Personal Data Act and The General Data Protection Regulation from the EU (GDPR), we are prohibited from providing more personal information about the participants in order to preserve their anonymity. Truly elite-level world-class athletes with a physical disability are not a substantial population, and connecting information through, e.g., reporting gender, type of sport, type of disability, age, etc., together could easily lead to readers potentially identifying a specific participant.

2.1. Philosophical Assumptions

The study adopted a qualitative approach as it provides an analytical framework that potentially generates deep insight into parasport athletes lives during their developmental pathway toward world-class performance. Informed by a post-positivistic epistemological and ontological position that stresses the potential bias in considering the researcher as an independent and completely objective observer [23], the authors acknowledged the close connection between researchers and the research participants. The research is thus not conducted on the elite parasport athletes, rather, the research unfolds to learn from the athletes. This calls for a strive to engage in social construction of a narrative regarding the developmental pathway of individual participants, in an attempt to activate the respondent's knowledge and experiences.

To gain insight into athlete's developmental pathways toward elite-level performance, there needs to be reflections made by the research team toward how one's values, beliefs and experiences influence especially the analytical process. The researchers involved in the project had a mixture of experiences from various practical fields such as physical activity, sport, and physical education, with scientific approaches derived from the fields of psychology, human movement science and sport science and with a diversity of research methodology. All members of the team also had experience in working within sport and physical recreation with able-bodied children, adolescents and adults, and the first author has experience with coaching athletes with a disability. When the study was conducted, however, none of the researchers were directly involved with parasport. Thus, the diversification of the research group helped strike a balance between taking a distanced view and being sensitive to contextual features. By acknowledging that valid knowledge claims might emerge through differences in understanding [24], interpretational possibilities of interwoven ideas from issues raised during the interviews, the participants' reactions, and our interpretations of these, were discussed and negotiated among the members of the research group. Research in this mode thus required an ability to 'see the whole picture' while avoiding simply aggregating data in order to arrive at an overall conceptualization [25].

2.2. Retrospective Interviews

The athletes participated in an interview based on retrospective recall methodology, designed to examine their sport practice and developmental histories. All interviews were collected individually by telephone and the audio was recorded by a SONY IC recorder ICD-PX370 (Tokyo, Japan). The interview was conducted in the mother tongue of both interviewer and participants and lasted approximately 40 min, and all were conducted by the same interviewer (the first author). The final part of each interview was a member-checking procedure [26], in which participants were provided a summary from the interviewer's notes and asked to provide input on whether it accurately reflected their developmental experiences.

Retrospective semi-structured interviews are considered as one of the superior approaches to collect historical practice data in sport as they allow for additional probing to obtain as rich information as possible [27]. They provide an opportunity for minimizing misunderstandings, and openings for additional clarification [28]. Thus, they have been used extensively in previous research on athletes' development toward elite (expert) levels [1,27,29]. In the current study, structured and habitual sport activities were addressed in athletes that are currently engaged in sport participation, which provides for shorter recall periods with salient events, that has been shown to provide accuracy in retrospective recall data [30]. Furthermore, elite athletes also typically produce training dairies at an early age and follow individually tailored training schedules developed by their coaches which enhance memory of previous practice schedules. It should be noted however, that elements of bias necessarily apply to all types of retrospective recall data [30], albeit studies have demonstrated that athletes can reliably report details of physical activity patterns up to 25 years retrospectively [31–33].

2.3. Structure and Topics of the Semi-Structured Interview

According to the developmental model of sport participation (DMSP), the semi-structured interview was portioned into four main developmental periods: childhood (up to 12 years of age), youth (from age 13 up to 16), junior (age 17–19), and senior level (age 19+). Across all periods, they were asked to reflect upon participation in organized competitive sports (practice and competition), transitions between periods, time spent in organized practice (with a coach) and unorganized self-monitored practice. The organized training was characterized by coach-led training sessions with instructions and feedback while unorganized training was defined as individual or self-led activities [28]. Furthermore, the athletes were also asked to reflect upon the development of relevant equipment (as this is a recurring theme in parasport), training schedules, and their contact with local, regional and national sport federations.

Explicitly for the developmental period defined as childhood, the athletes were asked to recall their first sporting experiences, i.e., when they first tried a sport-related activity (organized or unorganized). In youth, transitions from childhood were considered in terms of changes in type of sport, unorganized/organized activities, competitions, and whether they began reflecting upon becoming a high-performance athlete. In junior age, the transition from primary school to high school was investigated, and how sport participation affected everyday life. According to DMSP, many athletes develop a strong motivation for becoming a world-class athlete in their sport, and they were thus asked to reflect upon important factors for becoming elite performers especially in this period.

2.4. Thematic Analysis

Recorded audio was transcribed and analyzed by a six-step thematic analysis [34]. This included a reflective approach in which we systematically and iteratively worked forward and backward through the steps of (I) familiarization with the data, including transcribing and noting down initial reflections, (II) generating initial codes for each participant's transcripts (across the entire dataset), (III) development and searching for themes across participant's individual codes, (IV) reviewing themes according to initial codes and the entire dataset, (V) defining and naming themes and (VI) writing up results. As a thematic analysis emerges in the interaction between research data in relation to the researcher's theoretical assumptions, knowledge, skills, and experience [34], the approach in the current study can be defined as a more explicitly analyst-driven theoretical thematic analysis driven by the current study's approach rooted in the developmental model of sport participation (DMSP). Albeit this form of thematic analysis might risk a somewhat reduced description of the overall data in exchange for a detailed deductive analysis of some aspect of the data, the data analysis was still data-driven and guided by the above systematic thematic analysis phases advocated by Braun et al. [34,35].

2.5. Study Rigor

All three authors were involved in the data analysis and contributed to the analytical rigor. Data were managed using NVivo 11 (QSR International, Chadstone, Australia). Throughout the analysis, actions were undertaken to ensure rigor, including archiving individual data analysis files and decisions from critical discussions [26]. First, the researchers operated in isolation from each other to independently code data without negotiation, before further critical discussions between three authors were undertaken and maintained throughout the six-step thematic analysis. These discussions focused on the importance of exploring how athlete development towards elite levels was understood and described by elite athletes with a physical disability, as opposed to previous studies that focused on the developmental pathways of able-bodied athletes. The discussions were also vital in ensuring and maintaining reflexive examination of assumptions and the authors' own prior views throughout each stage of data analysis.

3. Results

The overarching theme of the current study was the developmental pathways of elite athletes with a physical disability, viewed through the lens of the developmental model of sport participation (DMSP). Although each athlete's developmental pathway into elite parasport was unique, strong commonalities across their experiences were also apparent and were captured in the following four themes: (a) initiation of sport-related activity, (b) sport sampling in childhood, (c) youth sport transitions, and (d) significant others. The theme names were selected in keeping with the athlete's descriptions and the DMSP terminology.

3.1. Initiation of Sport-Related Activity

A recurrent theme among the interviewed elite athletes with a physical disability, was the relatively early introduction to sport-related activities in their childhood, in which several described experiences already in the preschool age (2–4 years old). This activity was typically in leisure time with their families, and many had vivid memories of these events:

> I was 4 years old when I got a ski-sledge ... I wanted to ski on my own, so this was one of the best winter seasons [at this age] as I could move around on snow on my own without any help ...

Around the age of entering the first grade at school (5–7 years old), all athletes were introduced to what might be defined as organized sport (typically by a local sport club).

Many athletes in the current sample report this as an important part of participating on the same level as their peers and looking for sport opportunities rather than seeing the activity limitations of their disability:

> It has always been very important for me to do the same as my friends. So, I played multiple sports. It has always been important to see opportunities instead of limitations . . .

3.2. Sport 'Sampling' in Childhood

There are several instances in the transcripts that can be defined as sampling according to the developmental model of sport participation. Altogether the athletes report a total of 16 different team- and individual sports, in which none reported sustained activity in less than 3 of these throughout their childhood years (6–12 years old). As one athlete noted:

> I tried and practiced handball, athletics, swimming, football, alpine skiing, table tennis, tennis, badminton and sit-skiing. I have tried most sports, although in some it was initially difficult to participate, but with a little adjustment, it went very well! . . .

Besides the overall multi-sport approach as an indication of sampling, the participation rates also indicate substantial time and involvement in the various sports. All participated in organized sport training several times a week and reported an average rate of 118 yearly practice sessions in childhood, as well as some local club-initiated competitions, in addition to unorganized sports-related training/activity in leisure time typically with their friends. Still, the playful approach is clearly visible across all their activities:

> It was mostly training for fun when we were little. It was some organized training with the team, but since we were so small it was not serious . . .

3.3. Youth Sport Transitions

The youth sport period seemed to overall represent the developmental period of greatest diversity among the athletes in terms of their sport practice and competition. This is captured by the many transitional experiences laid out by the participants. First, albeit all report a mixture of organized and unorganized (self-initiated) sport activities and an increase in overall sustained activity levels, some increase their involvement in organized sport while yet others spend more time in self-initiated sport activities before eventually changing back to organized sport:

> I practiced with coaches [organized] in soccer and handball, but unorganized on skateboards and snowboards. I also practiced some organized floorball . . .

All of the elite athletes in the current study also reflected upon their transitions towards more organized and professional training and competition environments. This shift in environment was due to change of schools (sometimes involving moving to another place in country) and/or being approached (or seeking out on their own) organized parasport typically in bigger sport clubs in terms of memberships. Amongst other things, these transitions involved the introduction of training schedules, albeit most of the athletes did not get a professional schedule developed in collaboration with a coach until late adolescence. The initial schedules were typically described as very general and contained very little individual adaptions:

> It was a very general training schedule, some physical things such as strength and endurance, but very different training methods I should try out. Getting a feel of the different heart rate zones . . .

The youth sport period also seems to represent a significant shift towards participation in competition:

> Initially, I participated mostly in regional competitions. I had not yet been introduced to parasports, so I competed mostly against the able-bodied. From the

> time I was 14–15 years old, I was classified as a para-athlete and got to participate in the National Championships and joined the national team for young upcoming athletes.

Thus, many athletes experienced their first transition towards both practicing with, and competing against, other athletes with a disability at a relatively high level. As the transcript indicates, some had their first experiences altogether with parasport competitions in the youth sport period.

Individual adaptations in equipment represent a reoccurring theme in relation to parasport at all levels. In the current study, the participants also mention the role of equipment as a significant developmental factor. As they typically start out with limited access due to expensive (and sometimes non-existent) equipment and may have to rely on custom-made equipment made by their families, the youth sport period is where most experience transitions towards lighter and more individualized equipment. This is especially important as the athletes also pass through puberty with a possibility for substantial variations in growth and maturation that might amplify the need for specially adapted equipment:

> The [available] equipment has been fine, however too expensive. ... I know several athletes who have had problems getting their own equipment. I went to buy my own equipment because they had not brought that type to my country yet ...

A fifth transitional factor occurring in the youth sport period amongst the athletes in the current study concerned a shift towards intentions of becoming an elite performer:

> My first international competition was at the age of 14, and I did not perform at a high level. It was around that time I realized that this was something for me if I put in a lot of training ...

Thus, the competitive spirit and desire for further improvement typically emerges around their first experiences with high-level sport or around other transitional factors such as changing team/club or sport:

> When I moved to a bigger city, there was an opportunity to practice with the very best players. In fact, there were several players so you could play and practice as a team ...

3.4. Significant Others in a Sport Context

A recurring theme amongst the athletes in the current study, was the role of significant others throughout their developmental pathways, and their contribution toward initiation and sustained involvement and investment in sport. In childhood, parents play a particularly important role:

> Parents are important supporters. They do not hold back, instead they push forward for further development ...

Parents, coaches, peers, and family are all assigned the crucial role of helping to see beyond the functional limitations associated with various disabilities and focusing on possibilities. In the youth sport transitions especially, training and practicing with qualified and experienced coaches across an extended period of time are highlighted as a key factor towards further development:

> I had a good coach who was very experienced and who knew how to push me to get even better ... He has followed me all the way from early years and is the national team's head coach today ...

However, coaches had also been situated in clubs with a community and culture of sharing, where athletes show interest in each other's development:

> To get into an environment and training group have been extremely good for me. It is an enormous sharing culture, which has given me access to the world's best athletes training journals . . .

Many of the elite athletes in the current study also address the possibility for training and practice with athletes without disabilities:

> I have gained a lot in terms of technique and training methods when I have pushed myself with able-bodied athletes. To participate with them, you get into a culture that is very professional, and it does not get much better . . .

National-level sport federations working with the development of elite athletes both with and without disabilities have an important role in this regard, and are often mentioned as a key for individual development from sub-elite towards elite sport performance:

> [the national federation] . . . has been there from an early age, and we received training schedules and got feedback . . . when you have such resources it is important to use them . . .

4. Discussion

The findings of the present study offer support for some of the propositions in the developmental model of sport participation (DMSP) applied to elite athletes with a physical disability. Specifically, there seems to be an important role for sampling organized sport in childhood, i.e., trying out and participating in several sports. Also consistent to a degree with the DMSP, is the opportunity for athletes with a disability to choose to specialize in their favorite sport during the period of adolescence [16]. This includes sustained investment and effort and increased specialized training in one sport. The reflections of elite-level athletes with a physical disability on the various developmental periods, however, offer important nuances of the DMSP framework applied to elite parasport and highlight important factors involved in determining the opportunities for pursuing high-performance and elite-level parasport. It needs to be acknowledged that any developmental history of sport participation is in essence highly individualized. As so in the current sample of elite-level athletes with a physical disability, the youth sport period was especially heterogenous, in which the common factor was in principle that it was highly transitional in many aspects.

4.1. Initiation of Sport-Related Activity

The DMSP framework initiates with 'entry into sport', and further describes either sampling or specialization through sport participation in childhood [14]. There is, however, no substantiation of what constitutes the entry-phase and what might be crucial determinants in terms of getting children into sport or whether some factors might be important in terms of long-term participation towards elite-level performance.

As our study shows, all para-athletes started with sport-related activities at relatively young ages. However, this is in contrast to studies on Brazilian para-athletes with both congenital disabilities and acquired disabilities, where it was shown that para-athletes tend to start with sport in older ages compared with able-bodied athletes, and indicate that early starting age in parasport is not crucial to achieve expert level [12]. Respectively, this underpins that para-athletes can still be successful and maintain long-lasting careers in sport, even though they started with sport in older ages [12]. These, to some extent, conflicting results might be explained since both studies include different respondents and subsequently indicate that starting age is highly individual in parasport. However, a novel aspect of the current study was when the elite athletes were asked to reflect upon their entry into sport, they did not initially talk about their first introduction to organized and coach-led sport in local club, rather, they reflected about their early childhood experiences in leisure time with their families trying out different sport-related activities typically before the age of 5–6 years old. In this period, they had important experiences with physical activities that might have initially 'shaped' their personal understanding as a possible sport participant, including first experiences with personalized equipment that allowed them to move around freely during all seasons.

These early family experiences thus seem to be an integrated part of the entry into sport for the elite athletes in the current sample. Indeed, an emerging body of research strongly suggests that socialization in the family and 'family culture' is one of the strongest influences on children's propensities to take up sport [20,36] and can have profound and lasting effects on individuals' sports participation across the life course [37]. Thus, parents are thought to be the most influential socializing agents for children's early sport-related learning experiences and certainly the first point of socialization into sport and other leisure activities [38]. The family culture might therefore be especially important for athletes with a disability, as parents in particular can facilitate the introduction of many different sport-related activities and broaden their horizon towards seeing physical activity possibilities, rather than negotiating their disability. This also includes addressing the need for, and application of, individualized equipment as a part of the demonstration of sport opportunities and thus perhaps 'setting the stage' for identifying themselves as a possible athlete and sport participant.

As an integrated part of the entry into sport, the current sample of elite athletes with a physical disability also stressed the importance of the desire to do the same as their friends and being allowed to participate with their able-bodied peers. As peer environments may vary considerably because of demographic and socioeconomic factors, athletes with disabilities might be introduced to a vast continuum of social environments. These might range from dedicated peer groups where peers share similar disabilities and even experiences, reverse-integrated or inclusive groups where peers vary regarding ability and may include able-bodied participants, up to being the single minority in an otherwise able-bodied peer group [39,40]. At the time of entry into sport, all elite athletes in the current study experienced the latter scenario, and thus were integrated into an otherwise able-bodied peer group in the local sport club situated in a relatively small community. This clearly illustrates that early on in the developmental pathway towards elite parasport, it might be just as (or perhaps more) important to participate with able-bodied friends and peers in sport, as there is a need for dedicated peer groups/clubs exclusively for young people with disabilities. Participating with able-bodied peers might be empowering, and important for building the long-term motivation needed for pursuing a pathway toward elite parasport [41]. The potential role of integrated sport environments in long-term development towards elite performance in athletes with a disability seems thus to be an important avenue for further research. Furthermore, these empirical findings lend to the premise that an age-related athlete development model may not be an optimal approach in parasports, but need rather to focus on allowed people with disabilities participating in activity together with peers and able-bodied friends.

4.2. Sport 'Sampling' in Childhood

Multiple studies have shown that elite able-bodied athletes have participated in several sports through childhood [22,29,42,43]. The principle of sport sampling in accordance with the DMPS framework is built on the concept that children experience several different sports with a focus on deliberate play and with a low focus on deliberate practice [44]. Sport sampling allows children to experience and develop different motor skills and cognitive skills that are required when specializing in one sport [1,45]. The elite-level para-athletes in the current sample did indeed reflect upon participating and trying out multiple sports throughout childhood, thus adhering to the notion of sport sampling as an important aspect of the developmental trajectory towards world-class performance.

A distinction emerges, however, between the transcripts in the current study and that of the DMSP, when considering the relative contribution of deliberate practice and deliberate play in sport sampling. The sport participation in childhood appeared to be highly coach-led, organized, and focused on sport-specific skills, which are typical indicators of deliberate practice [1,45]. Thus, albeit the athletes in the current study participated in various sports several times a week (a multi-sport approach), there seemed to be less opportunity for sport-related activities that can be less adult-organized and more self-regulated/peer-regulated, in which the athletes themselves are discovering and exploring activities in deliberate play [1,15,45].

Typically, deliberate play with an unorganized approach to skill acquisition is often designed to maximize enjoyment in contrast to deliberate practice where the specific purpose of the organized training sessions is to increase performance (e.g., not for enjoyment or external rewards) [45]. There are, however, examples of studies supporting that deliberate practice activities can result in enjoyable effects [46], and the athletes in the current study indeed reported to experience enjoyment when participating in coach-led and organized sport. Importantly, this sport participation facilitated social contact with their typically developing and able-bodied peers. This illustrates that some postulates generated from the DMPS are not directly transferable to developmental pathways towards elite-level parasport. Participating in organized sport activities with peers might lead athletes with a disability to discover sport possibilities and become a part of a group where teammates can contribute to the development of sport expertise by sharing experiences and knowledge about the sport and help match their training exercises towards their current skill level [47].

4.3. Youth Sport Transitions

The youth developmental period seemed to contain the greatest inter-individual differences between the athletes in the current study. Indeed, there are examples of athletes who have reached the high-performance levels without any specific organized training during adolescence, but rather have similarities with the developmental trajectory in the DMSP which contains a high amount of deliberate play and activities that focus on fitness and health [14]. This clearly demonstrates that any developmental history of sport is highly individualized, and that athletes report a mixture of the pathways indicated in the DMSP. Indeed, multiple studies [14,22,43,44,48] indicate that expert able-bodied athletes did not follow a particular trajectory in the youth sport period, but rather changed between different trajectories when developing toward expert performance.

The findings in the current sample of expert athletes with a physical disability are that they initiated their focus on a main sport relatively late in adolescence and are in line with other studies, where Huxley et al. [49] showed that the most para-athletes did not focus on their main sport in both training and competition before at least the age of 16. As captured by the DMPS model, para-athletes are not physically or psychosocially ready to invest in specialized training before the late adolescent period. While able-bodied athletes often engage considerably in their primary sport also during childhood [50], it seems as though athletes in the current sample had an extended sampling period before they followed a more specializing trajectory. As a consequence of these findings, it might be that para-athletes must be recommended to sample a variety of sport before specialization in late adolescence. Furthermore, these transitions in youth sport seem to be highly influenced by sometimes random encounters with, e.g., elite sport federations, teams and/or coaches that can assist in developing individualized training schedules tailored for athletes with disabilities. It thus appears that across sport for both able-bodied athletes and athletes with a disability, that in order to reach elite-level sport at the international level, young athletes need to initiate highly structured and organized training no later than during their mid-adolescence years [49,51,52]. Similarly, transitions also need to take place in terms of competition levels during youth sport. As a part of the specializing pathway, most of the athletes in the current study first entered international level competitions at this age, and four of the athletes also reported achieving their first international medal at a junior age (17–19).

The important role of individualized equipment is a recurring theme in sport for athletes with a disability and analyses of transcripts from the current study also clearly indicated that in order to transit to specialized and high-level practice and competition, participation often requires costly equipment that might be difficult to find altogether.

In Kean et al. [21] investigation of wheelchair basketball players, financial barriers were also reported due to expensive equipment. These barriers have in some instances led the athletes in the current sample to buy their own equipment, which follows from the transitions towards developing strong intentions and motivations towards becoming elite-level performers. These motivations and intentions seem to emerge alongside the increased involvement in a particular organized sport and shift towards more organized and professional training and competition environments. The need for better equipment thus follows from these changes, and the transitional youth sport period might thus contain several biopsychosocial challenges to overcome if one is to pursue a pathway towards elite-level parasport.

4.4. Significant Others in a Sport Context

As mentioned previously, parents, peers and family are influential socializing agents [38] for early sport-related learning experiences for the current sample of elite athletes. Similarly, coaches, sport organizations and other athletes are deemed especially important in youth sport transitions towards the development of elite-level performance. It is beyond the scope of models such as the DMSP to pinpoint such key aspects of elite-level pathways, albeit it was a recuring theme amongst the individual retrospective reflections provided in the current study. Altogether, the significant others as jointly termed in the current study (e.g., parents, family, peers, friends, coaches, sport organizations, etc.) seemed to have created a sport environment around the young developing athletes with disabilities that shared two interlocked traits: (i) Focus on further development, not necessarily performance and (ii) seeing beyond the limitations that might reside with the disabilities. The various person's influence on the athlete might have different contributions at different times, but need to focus on what can be done instead of what cannot be done in training and practice. It seems to be of vital importance that people are knowledgeable of the interaction between technical and tactical sport-specific skills and the different individual disabilities. Coaches, especially, need to go beyond the knowledge required for able-bodied practice to provide athletes with disability-relevant and sport-specific input [53,54]. Altogether, it seems highly important for athletes with a disability that influential persons in a sport context foster a motivational climate that focuses on skill improvement and individual progress.

4.5. Limitations and Future Directions

Several limitations exist with this study, that will motivate further examination. Typical for an exploratory and qualitative study, the focus was on a modest sample of athletes with a physical disability. The athletes had sustained performance at the highest levels, however, and such top-ranked athletes do not typically exist in high numbers. Hence, they were not just competing at the highest levels, they were also among the very best in the world in their specific sport. Still, the study obtained insights from a small group of participants, and their responses may have differed from a quantitative study gathering data from larger samples of athletes with a physical disability. Furthermore, although no systematic differences seemed to emerge in the current study's thematic analysis between those with congenital and acquired disabilities, important insights might still be gained if developmental stories are collected from these specific sub-samples. The current study was also explicitly viewed through the lens of the developmental model of sport participation, and it seems clear that additional in-depth interviews should focus on specific developmental periods in order to further elucidate facilitating factors and barriers. Cultural differences might also have substantial impact upon the perceptions of disability, thus limiting the current study findings to European and industrial societies with high levels of welfare. As the understanding of elite sport development in parasport is further increased, it should be possible to develop specific models that can be addressed by various methods, e.g., developing a survey instrument based upon the qualitative research.

5. Conclusions

The study examined developmental pathways towards elite-level performance, through childhood and adolescence, amongst athletes with a physical disability that can be defined as top-performers in their sports. The retrospective interviews and thematic analysis were viewed through the lens of the developmental model of sport participation and explored their perspectives and reflections in terms of the proposed pathways in the DMSP. The findings of the present study offer support for some of the propositions in the model, although also addressing some of the problems with applying non-disability athlete development models towards understanding athletes with a disability. It appears that early sport-related encounters typically in a family environment have been important for the athletes in the current sample, shaping their further interest in sport, and beginning the journey of seeing past their disability. This is typically followed by childhood experiences with a substantial amount of sport sampling through participating in a number of different organized and coach-led sports. This sampling facilitates contact with their able-bodied peers and seems in some instances to work as a substitute for the substantial amount of peer-led unorganized activities typically found in the developmental histories of able-bodied high-level athletes. The youth sport period seems to be particular transitional, in which further adaptation of equipment follows from the transition to more organized and structured practice and higher levels of competition in their main sport. The intention of becoming a high-level athlete emerges in the youth period, and family, coaches, friends and athletes all seem to play a role in sustaining motivation towards improvement and further progress. These findings, and those of others, might be applied in creating developmental models tailored towards parasport, in which they can inform policymakers and organizations, on how to best tailor the parasport experience.

Author Contributions: Conceptualization, L.S., M.A.A. and H.L.; methodology, L.S., M.A.A. and H.L.; formal analysis L.S., M.A.A. and H.L.; investigation, L.S. and H.L.; data curation, L.S., M.A.A. and H.L.; writing—original draft preparation, L.S., M.A.A. and H.L.; supervision, H.L. All authors have read and agreed to the published version of the manuscript.

Funding: This research received no external funding.

Institutional Review Board Statement: The study was conducted in accordance with the Declaration of Helsinki, and approved by the Norwegian center for research data (NSD) with reference number 215429.

Informed Consent Statement: Informed consent was obtained from all subjects involved in the study, including information about the possibility to withdraw from the study at any time.

Data Availability Statement: Not applicable, as the study includes data protected by the Norwegian Personal Data Act and The General Data Protection Regulation from the EU (GDPR).

Conflicts of Interest: The authors declare no conflict of interest.

References

1. Ericsson, K.A.; Krampe, R.T.; Tesch-Römer, C. The role of deliberate practice in the acquisition of expert performance. *Psychol. Rev.* **1993**, *100*, 363–406. [CrossRef]
2. Coutinho, P.; Mesquita, I.; Fonseca, A.M. Talent development in sport: A critical review of pathways to expert performance. *Int. J. Sports Sci. Coach.* **2016**, *11*, 279–293. [CrossRef]
3. DePauw, P.K.; Gavron, J.S. *Disability Sport*, 2nd ed.; Human Kinetics: Champaign, IL, USA, 2005.
4. Patatas, J.M.; De Bosscher, V.; Legg, D. Understanding parasport: An analysis of the differences between able-bodied and parasport from a sport policy perspective. *Int. J. Sport Policy Politics* **2017**, *10*, 235–254. [CrossRef]
5. Dieffenbach, K.D.; Statler, T.A. More Similar than Different: The Psychological Environment of Paralympic Sport. *J. Sport Psychol. Action* **2012**, *3*, 109–118. [CrossRef]
6. Tweedy, S.M.; Bourke, J. IPC Athletics Classification Project for Physical Impairments: Final Report-Stage 1. 2009, Bonn. IPC Athletics. pp. 1–107. Available online: https://www.paralympic.org/sites/default/files/document/120725114512622_2010_07_16_Stage_1-Classification_Project_Final_report__for_2012_forward.pdf (accessed on 9 August 2022).

7. Kohe, G.Z.; Purdy, L.G.; Litou, A. Athlete welfare, stakeholder responsibility, and ethics of care in elite sport. An examination of para-sport organization approaches in France. In *Developing and supporting Athlete Wellbeing. Person First, Athlete Second*, 1st ed.; Campbell, N., Brady, A., Tincknell-Smith, A., Eds.; Routledge: London, UK, 2021.
8. Dehghansai, N.; Lemez, S.; Wattie, N.; Baker, J. A Systematic Review of Influences on Development of Athletes with Disabilities. *Adapt. Phys. Act. Q.* **2017**, *34*, 72–90. [CrossRef] [PubMed]
9. Higgs, C.; Bluechardt, M.; Balyi, I.; Way, R.; Jurbala, P.; Legg, D. No Accidental Champions. Long-Term Athlete Development for Athletes with Disabilities, 2nd ed.; 2016; pp. 1–36. Available online: https://sportforlife.ca/wp-content/uploads/2016/06/NAC_ENGLISH_SCREEN_rev2013.pdf (accessed on 9 August 2022).
10. Dehghansai, N.; Lemez, S.; Wattie, N.; Baker, J. Training and development of Canadian wheelchair basketball players. *Eur. J. Sport Sci.* **2017**, *17*, 511–518. [CrossRef] [PubMed]
11. Lemez, S.; Wattie, N.; Dehghansai, N.; Baker, J. Developmental pathways of Para athletes: Examining the sporting backgrounds of elite Canadian wheelchair basketball players. *Curr. Issues Sport Sci.* **2020**, *5*, 002. [CrossRef]
12. Patatas, J.M.; De Bosscher, V.; Derom, I.; Winckler, C. Stakeholders' perceptions of athletic career pathways in Paralympic sport: From participation to excellence. *Sport Soc.* **2020**, *25*, 299–320. [CrossRef]
13. Dehghansai, N.; Spedale, D.; Wilson, M.J.; Baker, J. Comparing Developmental Trajectories of Elite Able-Bodied and Wheelchair Basketball Players. *Adapt. Phys. Act. Q.* **2020**, *37*, 338–348. [CrossRef]
14. Côte, J.; Fraser-Thomas, J. The health and developmental benefits of youth sport participation. In *Sport Psychology: A Canadian Perspective*; Crocker, P., Ed.; Pearson: Toronto, ON, Canada, 2007; pp. 266–294.
15. Côté, J.; Baker, J.; Abernethy, B. Practice and Play in the Development of Sport Expertise. In *Handbook of Sport Psychology*; Tenenbaum, G., Eklund, R.C., Eds.; John Wiley & Sons, Inc.: Hoboken, NJ, USA, 2007; pp. 184–202. [CrossRef]
16. Côté, J.; Vierimaa, M. The developmental model of sport participation: 15 years after its first conceptualization. *Sci. Sports* **2014**, *29*, S63–S69. [CrossRef]
17. Huxley, D.J.; O'Connor, D.; Larkin, P. The pathway to the top: Key factors and influences in the development of Australian Olympic and World Championship Track and Field athletes. *Int. J. Sports Sci. Coach.* **2017**, *12*, 264–275. [CrossRef]
18. Abbott, D.; Townsley, R.; Watson, D. Multi-agency working in services for disabled children: What impact does it have on professionals? *Health Soc. Care Community* **2005**, *13*, 155–163. [CrossRef]
19. Patatas, J.M.; De Rycke, J.; De Bosscher, V.; Kons, R.L. It's a Long Way to the Top: Determinants of Developmental Pathways in Paralympic Sport. *Adapt. Phys. Act. Q.* **2021**, *38*, 605–625. [CrossRef]
20. Birchwood, D.; Roberts, K.; Pollock, G. Explaining differences in sport participation rates among young adults: Evidence from the South Caucasus. *Eur. Phys. Educ. Rev.* **2008**, *14*, 283–298. [CrossRef]
21. Kean, B.; Gray, M.; Verdonck, M.; Burkett, B.; Oprescu, F. The impact of the environment on elite wheelchair basketball athletes: A cross-case comparison. *Qual. Res. Sport, Exerc. Health* **2017**, *9*, 485–498. [CrossRef]
22. Baker, J.; Cote, J.; Abernethy, B. Sport-Specific Practice and the Development of Expert Decision-Making in Team Ball Sports. *J. Appl. Sport Psychol.* **2003**, *15*, 12–25. [CrossRef]
23. Kastenholz, E.; Weiler, B.; Black, R.; Hong, J.; Zizzo, N.; Bell, E.; Lafontaine, A.L.; Racine, E.; Zuckerman, D.; Duncan, V.; et al. Achieving Research Impact Through Co-Creation in Community-Based Health Services: Literature Review and Case Study. *Milbank Quart.* **2016**, *94*, 392–429. [CrossRef]
24. Phillips, D.C.; Burbules, N.C. *Postpositivism and Educational Research*; Rowman & Littlefield: Oxford, MI, USA, 2000.
25. Ryan, A.B. Post-Positivist Approaches to Research. In *Researching and Writing your Thesis: A Guide for Postgraduate Students*; MACE: Maynooth, Ireland, 2006; pp. 12–26.
26. Smith, B.; McGannon, K. Developing rigor in qualitative research: Problems and opportunities within sport and exercise psychology. *Int. Rev. Sport Exerc. Psychol.* **2016**, *11*, 101–121. [CrossRef]
27. Côté, J.; Ericsson, K.A.; Law, M.P. Tracing the Development of Athletes Using Retrospective Interview Methods: A Proposed Interview and Validation Procedure for Reported Information. *J. Appl. Sport Psychol.* **2005**, *17*, 1–19. [CrossRef]
28. Ericsson, K.A.; Hoffman, R.R.; Kozbelt, A.; Williams, A.M. *The Cambridge Handbook of Expertise and Expert Performance*, 2nd ed.; Cambridge University Press: New York, NY, USA, 2018.
29. Baker, J.; Côté, J.; Deakin, J. Expertise in Ultra-Endurance Triathletes Early Sport Involvement, Training Structure, and the Theory of Deliberate Practice. *J. Appl. Sport Psychol.* **2005**, *17*, 64–78. [CrossRef]
30. Brewer, B.W.; Van Raalte, J.L.; Linder, D.E.; Van Raalte, N.S. Peak Performance and the Perils of Retrospective Introspection. *J. Sport Exerc. Psychol.* **1991**, *13*, 227–238. [CrossRef]
31. Dex, S. The Reliability of Recall Data: A Literature Review. *Bull. Sociol. Methodol.* **1995**, *49*, 58–89. [CrossRef]
32. Macdonald, D.J.; Horton, S.; Kraemer, K.; Weir, P.; Deakin, J.M.; Côté, J. Application and Reliability of the Retrospective Interview Procedure to Trace Physical Activity Patterns in Master Athletes and Nonactive Older Adults. *Educ. Gerontol.* **2009**, *35*, 1107–1122. [CrossRef]
33. Hayman, R.; Polman, R.; Taylor, J. The validity of retrospective recall in assessing practice regimes in gold. *Int. J. Sport Exerc. Psychol.* **2012**, *10*, 329–337. [CrossRef]
34. Braun, V.; Clarke, V. Using thematic analysis in psychology. *Qual. Res. Psychol.* **2006**, *3*, 77–101. [CrossRef]
35. Braun, V.; Clarke, V.; Weate, P. Using thematic analysis in sport and exercise research. In *Routledge Handbook of Qualitative Research in Sport and Exercise*; Smith, B., Sparkes, A.C., Eds.; Routledge: London, UK, 2016; pp. 191–205.

36. Wheeler, S. The significance of family culture for sports participation. *Int. Rev. Sociol. Sport* **2011**, *47*, 235–252. [CrossRef]
37. Wheeler, S.; Green, K. Parenting in relation to children's sports participation: Generational changes and potential implications. *Leis. Stud.* **2012**, *33*, 267–284. [CrossRef]
38. Day, J. Negotiating fatherhood: Sport and family practices. *Leis. Stud.* **2020**, *39*, 463–464. [CrossRef]
39. Hutzler, Y.; Chacham-Guber, A.; Reiter, S. Psychosocial effects of reverse-integrated basketball activity compared to separate and no physical activity in young people with physical disability. *Res. Dev. Disabil.* **2013**, *34*, 579–587. [CrossRef]
40. Evans, M.B.; Shirazipour, C.H.; Allan, V.; Zanhour, M.; Sweet, S.N.; Ginis, K.M.; Latimer, A. Integrating insights from the parasport community to understand optimal Experiences: The Quality Parasport Participation Framework. *Psychol. Sport Exerc.* **2018**, *37*, 79–90. [CrossRef]
41. Allan, V.; Smith, B.; Côté, J.; Ginis, K.A.M.; Latimer, A. Narratives of participation among individuals with physical disabilities: A life-course analysis of athletes' experiences and development in parasport. *Psychol. Sport Exerc.* **2018**, *37*, 170–178. [CrossRef]
42. Vaeyens, R.; Güllich, A.; Warr, C.R.; Philippaerts, R. Talent identification and promotion programmes of Olympic athletes. *J. Sports Sci.* **2009**, *27*, 1367–1380. [CrossRef] [PubMed]
43. Güllich, A. Many roads lead to Rome—Developmental paths to Olympic gold in men's field hockey. *Eur. J. Sport Sci.* **2014**, *14*, 763–771. [CrossRef] [PubMed]
44. Côte, J.; Horton, S.; MacDonald, D.; Wilkes, S. The benefits of sampling sports during childhood. *Phys. Health Educ. J.* **2009**, *74*, 6–11.
45. Côté, J.; Abernethy, B. A Developmental Approach to Sport Expertise. In *The Oxford Handbook of Sport and Performance Psychology*; Murphy, S.M., Ed.; Oxford University Press: New York, NY, USA, 2012. [CrossRef]
46. Starkes, J. The road to expertise: Is practice the only determinant? *Int. J. Sport Psychol.* **2000**, *31*, 431–451.
47. Kragen, J.L. *Developing Talent in Young People*; Ballantine Books: New York, NY, USA, 2004; pp. 1–5.
48. Güllich, A. International medallists' and non-medallists' developmental sport activities—A matched-pairs analysis. *J. Sports Sci.* **2016**, *35*, 2281–2288. [CrossRef]
49. Huxley, D.J.; O'Connor, D.; Bennie, A. Olympic and World Championship track and field athletes' experiences during the specialising and investment stages of development: A qualitative study with Australian male and female representatives. *Qual. Res. Sport Exerc. Health* **2017**, *10*, 256–272. [CrossRef]
50. Leite, N.; Baker, J.; Sampaio, J. Paths to expertise in portuguese national team athletes. *J. Sports Sci. Med.* **2009**, *8*, 560–566.
51. Moesch, K.; Elbe, A.-M.; Hauge, M.-L.T.; Wikman, J.M. Late specialization: The key to success in centimeters, grams, or seconds (cgs) sports. *Scand. J. Med. Sci. Sports* **2011**, *21*, 282–290. [CrossRef]
52. Güllich, A.; Emrich, E. Considering long-term sustainability in the development of world class success. *Eur. J. Sport Sci.* **2011**, *14*, S383–S397. [CrossRef]
53. Culver, D.M.; Werthner, P. Voices: Para athletes speak. *Qual. Res. Sport, Exerc. Health* **2017**, *10*, 167–175. [CrossRef]
54. Bentzen, M.; Alexander, D.; Bloom, G.A.; Kenttä, G. What Do We Know About Research on Parasport Coaches? A Scoping Review. *Adapt. Phys. Act. Q.* **2021**, *38*, 109–137. [CrossRef]

sports MDPI

Article

Morphological Characteristics and Situational Precision of U15 and U16 Elite Male Players from Al-Ahli Handball Club (Bahrein)

Boris Banjevic [1], Boris Zarkovic [2], Borko Katanic [3], Blazo Jabucanin [4], Stevo Popovic [1,4,*] and Bojan Masanovic [1,4]

1 Faculty for Sport and Physical Education, University of Montenegro, 81400 Niksic, Montenegro; boris.banjevic@gmail.com (B.B.); bojanma@ucg.ac.me (B.M.)
2 Faculty of Sport and Physical Education, University of Novi Sad, 21000 Novi Sad, Serbia; zarkovicboris@gmail.com
3 Faculty of Sports and Physical Education, University of Nis, 18000 Nis, Serbia; borkokatanic@gmail.com
4 Western Balkan Sport Innovation Lab, 81000 Podgorica, Montenegro; blazojabucanin@yahoo.com
* Correspondence: stevop@ucg.ac.me; Tel.: +382-(40)-235-207

Abstract: The aim of this cross-sectional study was to determine the differences in the morphological characteristics and situational precision among younger and older groups of handball players. The sample of participants consisted of 30 handball players, members of the younger category of the Al-Ahli handball club (Bahrein), divided into two groups: older (U16, $n = 18$) and younger (U15, $n = 12$). To evaluate their morphological characteristics, eight variables were measured, while two standardized tests were used to evaluate their situational precision. The results indicate that a statistically significant difference between the groups was noticeable for nine variables in total, seven in morphology (body height, $p = 0.010$; body mass index, $p = 0.049$; arm length, $p = 0.009$; upper arm length, $p = 0.016$; lower arm length, $p = 0.040$; the planimetric parameter of the hand, $p = 0.005$; hand length $p = 0.004$) and two in situational precision (the standing shot, $p = 0.003$; the jump shot, $p = 0.17$), and that the achieved difference ranges from a medium to a large effect. For only one variable (body mass, $p = 0.734$), significant difference was not determined between the groups. It was also determined (by Cohen's criterion) that handball players with higher longitudinal dimensionality achieve better results for specific precision. Therefore, when selecting young handball players, the aforementioned dimensions should be taken into consideration as predictors of success.

Keywords: handball; specific precision; anthropometry; longitudinal dimensionality

Citation: Banjevic, B.; Zarkovic, B.; Katanic, B.; Jabucanin, B.; Popovic, S.; Masanovic, B. Morphological Characteristics and Situational Precision of U15 and U16 Elite Male Players from Al-Ahli Handball Club (Bahrein). *Sports* **2022**, *10*, 108. https://doi.org/10.3390/sports10070108

Academic Editor: Jon Oliver

Received: 11 May 2022
Accepted: 4 July 2022
Published: 6 July 2022

1. Introduction

Current research indicates that the morphological characteristics of athletes are significantly associated with success in sport [1–5]. In relation to the demands of a certain sport, every athlete should possess optimal morphological characteristics [4,6–8], and therefore Milanovic et al. [9] cite that their determination represents one of the three most frequently tested dimensions of athletes. Insight into the morphological characteristics of athletes has practical significance when adapting the training process to the individual characteristics of an athlete, as well as during the evaluation of their ultimate abilities [10]. When it comes to young athletes, Masanovic and Vukasevic [5] point out that they too must have optimal morphological characteristics and motor skills so as to meet the requirements of a certain sport. It should be added that morphological characteristics of young athletes are susceptible to change and can to a significant extent be improved [11].

Handball is a demanding contact sport which in its structure contains elements of movement such as running, jumping, sprinting, swinging, hitting, blocking and pushing an opponent [12]. Consequently, the performance of handball players is affected by morphological characteristics, technical and tactical skills, and physical abilities which develop with age [13]. Among young handball players of various ages, the determination of morphological measurements represents an important part of the effective modelling of the training process and selection [14].

Current research carried out on handball players of various age categories and levels of competition indicates that handball requires higher values of body height and body mass [15]. Even though authors have dealt with the morphological differences between young handball players and other athletes such as soccer players, basketball players, and volleyball players [8], there are still not enough studies which have analyzed differences in the morphological features among young handball players in relation to age categories.

Moreover, one of the most important skills for success in handball is the ability to throw [16] and for that reason numerous studies have focused on the throwing technique [16–19]. It was determined that success in throwing is affected by precision [17–19] and speed of the ball [16,19]. Precision and the speed of the throw are increasingly gaining in importance in the end results of a game and represent two basic factors, both of which are important for the effectiveness of the throw in handball [20]. Despite this, certain authors point out that precision is more important, and that if athletes focus on precision, speed decreases [21]. Still, insufficient data are available regarding the precision of handball players, especially handball players in younger categories.

Therefore, this study aim to determine the differences in the morphological characteristics and situational precision between older and younger male groups of handball players.

2. Methods

2.1. The Sample of Participants

The sample of participants consisted of 30 handball players, all members of younger categories of the Al-Ahli club (Bahrein), divided into two groups: an older group of boys born in 2002 (n = 18, 15.60 ± 0.30 yrs.) and a younger group of boys born in 2003 (n = 12, 14.69 ± 0.32 yrs.). The sample included participants who were psychologically and physically healthy, who had no pronounced aberrations concerning the locomotor apparatus. In addition, the inclusion criteria for the participants were that they have actively been taking part in handball for at least a period of 3 years, that they train at least 4 times a week, and that are competing in the appropriate age category. All of the participants, as well as their parents, were well acquainted with the procedure and purpose of the testing, and the parents gave their consent on the condition that they could withdraw their children from the study at any point (which none of them did). All procedures performed in studies involving human participants were in accordance with the 1964 Helsinki Declaration and its later amendments or comparable ethical standards.

Skinfold thicknesses (mm) were measured at six sites: triceps skinfold, forearm skinfold, thigh skinfold, calf skinfold, chest skinfold, and abdominal skinfold thickness (using a skinfold caliper).

2.2. The Sample of Measuring Instruments

Standardized anthropometric instruments were used to measure the morphological characteristics. The measurements were taken according to a predetermined International Biological Program (IBP) [22]. To evaluate the morphological characteristics, eight variables were measured: body height, body mass, body mass index, arm length, upper arm length, lower arm length, the planimetric parameter of the hand, and hand length [23]. Height was measured to the nearest 0.1 cm using a fixed stadiometer (Seca, Leicester, UK) and weight was measured to the nearest 0.1 kg with an electronic weighing machine (HD-351, Tanita, Illinois, USA). The body mass index was calculated using the standard formula: BMI = body mass (kg) ÷ body height2 (meters). Length measures were measured to the nearest 0.1 cm, and taken with the Martin anthropometer (GPM, Bachenbülach, Switzerland). At the same time, to evaluate the situational precision of the handball players, two standardized situational-motor tasks of a composite type were used, selected on the basis of existing studies: precision of the throw from a standing position on both legs from 7 m, and precision of the throw from a jump at 9 m [24–27]. The protocol is the same for both tests, and the target is a unilateral triangle whose sides are 50 cm in length, while the hypotenuse is an elastic band 5 cm wide (Figure 1). Each test consists of seven attempts of four shots each into every corner of the goal (the participant makes the shots in the following sequence: the upper left-hand corner, the upper right-hand corner, the lower left-hand corner, the lower right-hand corner). Every shot that landed within the triangle carried two points, the ones that landed on the frame of the triangle one point, and a miss carried zero points. The result represents the sum of all the points from all seven attempts.

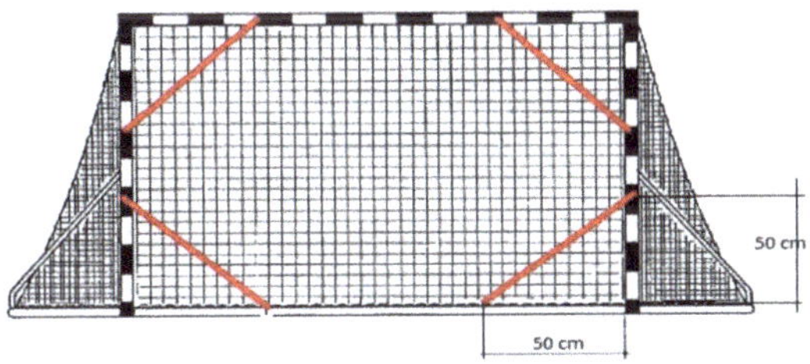

Figure 1. Marked targets on the goal.

2.3. *Statistical Analysis*

The data analysis was carried out using the IBM SPSS Statistics 26 software (Chicago, IL, USA). The means and standard deviation were calculated for each variable. To evaluate the normality of the distribution, the Shapiro-Wilk's test was used. Differences in the morphological characteristics and situational precision between the groups were determined using a univariate analysis of variance (ANOVA), while the level of significance was set at $p < 0.05$. What was then calculated was Eta squared (η^2), as an indicator of effect size. Cohen's classification indicates the size of the effect whereby 0.01 is a small effect, 0.06 is a medium effect and 0.14 is a large effect [28].

3. Results

The normal distribution of the data was confirmed by the Shapiro-Wilk test (Table 1). One variable indicated a deviation from normal distribution (Lower arm length, $p = 0.021$), while the remaining variables have normal distribution ($p > 0.05$) and, along with a visual overview of the histogram and Q-Q plot, it was determined that the data were normal, and that a parametric technique could be used.

Table 1. The Shapiro-Wilk test of normality.

	Shapiro-Wilk	
	Statistic	**Sig.**
Body height (cm)	0.961	0.326
Body mass (kg)	0.969	0.512
BMI (score)	0.974	0.650
Arm length (cm)	0.954	0.216
Upper arm length (cm)	0.953	0.203
Lower arm length (cm)	0.916	0.021
Planimetry of the hand(cm)	0.950	0.166
Hand length (cm)	0.948	0.154
The standing shot (score)	0.965	0.420
The jump shot (score)	0.951	0.184

Legend: BMI = body mass index.

When we view the average values (Mean) between the younger and older group of handball players (Table 2), it can be noted that the older group achieved greater values for all the variables which pertain to morphological measurements and situational precision, except for the body mass index for which higher values were noted for the younger group.

Table 2. Descriptive data and the differences between younger and older groups handball players (ANOVA).

	Older (n = 18)	Younger (n = 12)	ANOVA		
Variables	Mean ± SD		F	p	η^2
Body height (cm)	174.28 ± 4.57	168.00 ± 7.90	7.628	0.010 *	0.211
Body mass (kg)	56.67 ± 4.98	56.03 ± 5.05	0.118	0.734 ^	0.004
BMI (score)	18.66 ± 1.53	19.88 ± 1.69	4.219	0.049 *	0.131
Arm length (cm)	78.01 ± 1.65	76.06 ± 2.14	7.963	0.009 *	0.221
Upper arm length (cm)	31.31 ± 1.33	29.78 ± 1.92	6.611	0.016 *	0.191
Lower arm length (cm)	29.49 ± 1.00	28.30 ± 2.02	4.637	0.040 *	0.142
Planimetry of the hand (cm)	22.23 ± 1.39	20.87 ± 0.88	9.120	0.005 *	0.246
Hand length (cm)	19.13 ± 0.95	17.88 ± 1.25	9.652	0.004 *	0.256
The standing shot (score)	30.44 ± 3.59	26.17 ± 3.43	10.598	0.003 *	0.275
The jump shot (score)	22.06 ± 5.39	16.58 ± 6.30	6.492	0.017 *	0.188

Legend: Mean = Arithmetic mean; SD = Standard deviation; F = ANOVA test value; p = Statistical significance; η^2 = Eta square; ^ = Non-significant; * = Significant difference.

The ANOVA determined that only the values of body mass did not show a significant difference ($p > 0.05$) between the older and younger group of handball players, while a significant difference between the groups was noted for the remaining nine variables (p = 0.003–0.049), which also goes for the seven morphological variables (p = 0.004–0.49), and two variables of situational precision, the standing shot (p = 0.003) and the jump shot (p = 0.17).

Based on Eta squared and Cohen's criterion [28], we can note that the actual difference between the means for the given nine variables has a medium to large effect, of which the variable for the body mass index has medium effect size (η^2 = 0.131), while all the other variables have a large effect size (η^2 = 0.142–0.275).

4. Discussion

This study showed that there are differences between the morphological characteristics of the two studied groups of young handball players, members of the Al-Ahli handball club, and also between their results for situational precision. Statistically significant differences with greater values for the older boys were noted for all the variables which refer to the longitudinal dimensionality of the skeleton, which is congruent with the results of Vuleta et al. [14], who also noted higher values for the variables of longitudinal dimensionality for older boys. Considering the difference in chronological age and that the participants are in a period of turbulent growth and development, which is still enabling the growth of various dimensions of the body [29], it could be said that these differences were expected. Contrary to this, a statistically significant difference for the BMI variable, with higher values for the young group of boys, was not expected. Even though the intensity and scope of training were identical for both groups of handball players, greater values of BMI were noted for the group of younger players, which can be ascribed to an inappropriate diet, biological age, and genetic conditioning. Based on the descriptive data which refer to body height and body mass, we can see that the data for young handball players from Al-Ahlija do not deviate much from the results of similar studies which were carried out locally [30,31]. If we were to compare the morphological characteristics of Bahraini young handball players to those of their peers from Croatia and Serbia [14,32], who have been at the top of European sports for years, we could say that, despite the decreased population from which young handball players are extracted [33], Bahraini coaches do carry out proper talent identification.

When we look at the results of tests which measured the specific precision of young handball players (the standing shot, and the jump shot) we can see that there were statistically significant differences on both tests with better results scored by the older group of boys. Šibila and Pori [34] point out that handball at an elite level requires certain morphological characteristics which are reflected in the longitudinal dimensionality of the skeleton, while Taborsky [35] states that greater dimensionality of the hand leads to better control of the ball. According to their claims and the analysis of the results of anthropometric measurements which refer to the size of the hand and the results on specific tests of precision, our results suggested that handball players with greater longitudinal dimensionality of the skeleton scored better results. This observation is also congruent with previous research which dealt with the same topic [26,27]. The mentioned differences can also be conditioned by preforming a throw in situational conditions with increased speed. During fast movements, the possibility of visual and kinesthetic monitoring is significantly reduced, which ultimately reflects on the technique and precision of performance [36]. Given that older handball players have many years of practice and greater handball experience, and therefore significantly better and more consistent sports technique, it was to be expected that they would achieve better results in situational precision.

In order to compare the precision of young Bahraini handball players, which along with the strength of the throw is the most important component of an effective throw in handball [19], there is a need for such studies to be carried out on larger samples and on handball players who compete in environments where elite handball is played. Such studies could be used to evaluate the quality of the training process itself, that is, the level of technical skills of young Bahraini handball players compared to European ones. In addition, the players should also be classified based on their position on the team, since satisfactory knowledge of the morphological characteristics of players for each position is of key importance for coaches when it comes to the orientation of players in certain positions. Accordingly, when selecting young handball players, the aforementioned dimensions should be taken into consideration as predictors of success.

Author Contributions: B.K. performed statistical analyses, presented the results and wrote the manuscript. B.M. designed the study, overviewed previous studies and discussed the results. B.B., B.J. and S.P. overviewed previous studies and revised the manuscript. B.Z. collected the data, discussed the results and wrote the manuscript. All authors have read and agreed to the published version of the manuscript.

Funding: This research received no external funding.

Institutional Review Board Statement: The study was conducted according to the guidelines of the Declaration of Helsinki, and approved by the Institutional Review Board (or Ethics Committee) of Bahrain Handball Federation (protocol code: 03-2018; date of approval: 18 March 2018).

Informed Consent Statement: Informed consent was obtained from all subjects involved in the study.

Data Availability Statement: Not applicable.

Conflicts of Interest: The authors declare no conflict of interest.

References

1. Malina, R.M.; Bouchard, C.; Bar-Or, O. *Growth, Maturation, and Physical Activity*; Human Kinetics: Champaign, IL, USA, 2004.
2. López-Plaza, D.; Alacid, F.; Muyor, J.M.; López-Miñarro, P.Á. Sprint kayaking and canoeing performance prediction based on the relationship between maturity status, anthropometry and physical fitness in young elite paddlers. *J. Sports Sci.* **2017**, *35*, 1083–1090. [CrossRef]
3. Garcia-Gil, M.; Torres-Unda, J.; Esain, I.; Duñabeitia, I.; Gil, S.M.; Gil, J.; Irazusta, J. Anthropometric parameters, age, and agility as performance predictors in elite female basketball players. *J. Strength Cond. Res.* **2018**, *32*, 1723–1730. [CrossRef] [PubMed]
4. Slimani, M.; Nikolaidis, P.T. Anthropometric and physiological characteristics of male soccer players according to their competitive level, playing position and age group: A systematic review. *J. Sports Med. Phys. Fit.* **2019**, *59*, 141–163. [CrossRef] [PubMed]
5. Masanovic, B.; Vukasevic, V. Differences in anthropometric characteristics between junior handball and volleyball players. *J. Anthropol. Sport Phys. Educ.* **2020**, *4*, 9–14. [CrossRef]

6. Hatzimanouil, D.; Oxizoglou, N. Evaluation of the morphological characteristics and motor skills in the national junior handball teams of Greece and Yugoslavia. *J. Hum. Mov. Stud.* **2004**, *46*, 125–140.
7. Fieseler, G.; Hermassi, S.; Hoffmeyer, B.; Schulze, S.; Irlenbusch, L.; Bartels, T.; Delank, K.S.; Laudner, K.G.; Schwesig, R. Differences in anthropometric characteristics in relation to throwing velocity and competitive level in professional male team handball: A tool for talent profiling. *J. Sports Med. Phys. Fit.* **2017**, *57*, 985–992. [CrossRef]
8. Masanovic, B. Comparative Study of Morphological Characteristics and Body Composition between Different Team Players from Serbian Junior National League: Soccer, Handball, Basketball and Volleyball. *Int. J. Morphol.* **2019**, *37*, 612–619. [CrossRef]
9. Milanovic, D.; Jukic, I.; Vuleta, D.; Simek, S.; Sentija, D. Measurement and Evaluation of Fitness Characteristics of Croatian Handball Players. In Proceedings of the 9th Sport Kinetics International Conference; Sport Kinetics: Rimini, Italy, 2005; pp. 444–448.
10. Alarcón-Jimenez, J.; Pardo-Ibáñez, A.; Romero, F.J.; Gámez, J.; Soriano, J.M.; Villarón-Casales, C. Kinanthropometric Assessment of Individual, Collective and Fight Sport Players from the Spanish National Sport Technification Program. *Int. J. Morphol.* **2020**, *38*, 888–893. [CrossRef]
11. Zhao, X.; Jia, L. Training and transfer effects of interference control training in children and young adults. *Psychol. Res.* **2019**, *83*, 1519–1530. [CrossRef]
12. Ghobadi, H.; Rajabi, H.; Farzad, B.; Bayati, M.; Jeffreys, I. Anthropometry of world-class elite handball players according to the playing position: Reports from men's handball world championship 2013. *J. Hum. Kinet.* **2013**, *39*, 213. [CrossRef]
13. Chelly, M.S.; Hermassi, S.; Shephard, R.J. Relationships between power and strength of the upper and lower limb muscles and throwing velocity in male handball players. *J. Strength Cond. Res.* **2010**, *24*, 1480–1487. [CrossRef] [PubMed]
14. Vuleta, D.; Nikolic, A.; Krakan, I. Differences in the anthropometric characteristics of cadet and younger handball players. In *Ljetna Skola Kineziologa Republike Hrvatske*; Hrvatski Kinezioloski Savez: Zagreb, Croatia, 2011; pp. 110–116.
15. Vukotic, M.; Corluka, M.; Vasiljevic, I.; Bubanja, M. Differences in the morphological characteristics and body composition of handball players WHC Levalea in Montenegro and WHC Grude in Bosnia and Herzegovina. *J. Anthropol. Sport Phys. Educ.* **2018**, *2*, 49–53. [CrossRef]
16. Gorostiaga, E.M.; Granados, C.; Ibanez, J.; Izquierdo, M. Differences in physical fitness and throwing velocity among elite and amateur male handball players. *Int. J. Sports Med.* **2005**, *26*, 225–232. [CrossRef] [PubMed]
17. Van den Tillaar, R.; Ettema, G. Influence of instruction on velocity and accuracy of overarm throwing. *Percept. Mot. Ski.* **2003**, *96*, 423–434. [CrossRef]
18. Van den Tillaar, R.; Ettema, G. A three-dimensional analysis of overarm throwing in experienced handball players. *J. Appl. Biomech.* **2007**, *23*, 12–19. [CrossRef] [PubMed]
19. Ziv, G.; Lidor, R. Physical characteristics, physiological attributes, and on-court performances of handball players: A review. *Eur. J. Sport Sci.* **2009**, *9*, 375–386. [CrossRef]
20. Karisik, S.; Bozic, D.; Tiric, T. Influence of ball resin to shot accuracy in handball. *Eur. J. Phys. Educ. Sport Sci.* **2018**, *4*, 39–46.
21. Rivilla-Garcia, J.; Grande, I.; Sampedro, J.; Van Den Tillaar, R. Influence of opposition on ball velocity in the handball jump throw. *J. Sports Sci. Med.* **2011**, *10*, 534.
22. Marfell-Jones, M.J.; Stewart, A.D.; De Ridder, J.H. *International Standards for Anthropometric Assessment*; International Society for the Advancement of Kinanthropometry: Wellington, New Zealand, 2012.
23. Eston, R.; Reilly, T. *Kinanthropometry and Exercise Physiology Laboratory Manual: Tests, Procedures and Data, Exercise Physiology*; Routledge: London, UK, 2013; Volume 2.
24. Pavlin, K.; Simenc, Z.; Delija, K. An analysis of the reliability and factor validity of situational motor tests in handball. *Kineziologija* **1982**, *14*, 177–187.
25. Hosek, A.; Pavlin, K. The association between morphological dimensions and efficiency in handball. *Kinesiology* **1983**, *15*, 145–151.
26. Djukic, M.; Djukic, M.; Zaric, D.; Tuzovic, A. Impact of the anthropometric characteristics on specific motor abilities of young handball players. *Sport Sci. Int. Sci. J. Kinesiol.* **2016**, *9*, 49–51.
27. Karisik, S.; Milicevic, L.; Bozic, D. The qualitative indicators of the ball throw in handball. *Sport I Zdr.* **2016**, *11*, 19–27.
28. Cohen, J. *Statistical Power for the Behavioural Sciences*; Lawrence Erlbaum: New York, NY, USA, 1988.
29. Misigoj-Durakovic, M. *Kinantropologija: Bioloski Aspekti Tjelesnog Vjezbanja;* Kinezioloski fakultet Sveucilista u Zagrebu: Zagreb, Croatia, 2008.
30. Ostroski, S.; Milanovic, M.; Metikos, B. The differences in fitness features between soccer and handball school sports club players aged 13 to 14. In Proceedings of the 23rd Summer School of Kinesiology of the Republic of Croatia, Poreč, Croatia, 24–26 June 2022; Hrvatski Kinezioloski Savez: Zagreb, Croatia, 2014; pp. 231–236.
31. Grujic, S.; Peric, D.; Ahmetovic, Z.; Okicic, T.; Isakovic, M. Anthropometric and motor characteristics of young handball players for the purpose of early selection. *Facta Univ. Ser. Phys. Educ. Sport* **2018**, *16*, 033–045. [CrossRef]
32. Grujic, S. The Model Characteristics of Young Handball Players in Relation to Morphological and Motor Features. Ph.D. Thesis, University Edukons, Novi Sad, Serbia, 2016.
33. Hasan, A.A.A.; Rahaman, J.A.; Cable, N.T.; Reilly, T. Anthropometric profile of elite male handball players in Asia. *Biol. Sport* **2007**, *24*, 3–12.
34. Sibila, M.; Pori, P. Position-related differences in selected morphological body characteristics of top-level handball players. *Coll. Antropol.* **2009**, *33*, 1079–1086. [PubMed]

35. Taborsky, F. *The Body Height and Top Team Handball Players*; EHF Web Periodical: Vienna, Austria, 2007.
36. Peric, D. *Introduction to Sports Anthropomotorics*; Sports Academy: Belgrade, Serbia, 1997.

sports

Article

Perception of Affordances for Dribbling in Soccer: Exploring Children as Architects of Skill Development Opportunity

Michael J. Duncan *, Ricardo Martins, Mark Noon and Emma L. J. Eyre

Centre for Sport, Exercise and Life Sciences, Coventry University, Coventry CV1 5FB, UK; marti234@uni.coventry.ac.uk (R.M.); aa5349@coventry.ac.uk (M.N.); ab2223@coventry.ac.uk (E.L.J.E.)
* Correspondence: aa8396@coventry.ac.uk

Abstract: This study examined affordance perception for soccer dribbling using a mixed-methods approach in male grassroots soccer players. We examined how children construct and perceive skills practices for dribbling in soccer. Fourteen boys aged 10–11 years (Mean ± SD = 10.8 ± 0.4 years) who were regularly engaged in grassroots soccer participated in the present study. Children were provided with ten soccer cones and asked to create their own soccer dribbling pattern that would enable them to maximise the number of touches with a football and then dribble the ball in the pattern they had created for a 1 min period. Children were interviewed to explore their perception of affordances for soccer dribbling. The test of gross motor development-3 was used to assess fundamental movement skills (FMS), and the UGent soccer dribbling test was used to assess soccer dribbling skills. Children self-rated their own ability for soccer dribbling, as did their coaches. Pearson's correlations were employed to examine the associations between quantitative variables, and thematic analysis was used to explore qualitative data. Results of the present study suggest that those children who created patterns with less space between cones accrued more touches of the football in their dribbling task ($r = -0.671$, $p = 0.03$). Children with a higher perception of their own dribbling ability had higher scores for FMS ($r = 0.604$, $p = 0.049$). Those children who scored better in actual soccer dribbling had higher scores for FMS ($r = -0.746$, $p = 0.012$) and were rated as better dribblers by their coaches ($r = -0.67$, $p = 0.03$). Interview data suggest a feedback loop between perception of ability and actual ability, which influenced the dribbling patterns that were created. This suggests that dribbling performance is scaled to the (perceived) action capabilities of the children, and children can act as architects in their own skill development.

Keywords: motor competence; motor skill; youth; grassroots; talent development; mixed methods

Citation: Duncan, M.J.; Martins, R.; Noon, M.; Eyre, E.L.J. Perception of Affordances for Dribbling in Soccer: Exploring Children as Architects of Skill Development Opportunity. *Sports* **2022**, *10*, 99. https://doi.org/10.3390/sports10070099

Academic Editor: Michael Chia

Received: 21 May 2022
Accepted: 20 June 2022
Published: 24 June 2022

1. Introduction

The concept of affordances is well established in the sport and exercise science and sports coaching literature [1,2]. Affordances refer to opportunities for action that are available to a given individual in a given environment [2]. There is considerable research, which demonstrates human sensitivity to affordances [1]. Such research is varied, focusing on a range of different movements from sitting on a chair [3], crossing the street [4], using playgrounds [5], scoring in soccer [6], as well as kicking for distance and precision [7]. Through development, a child's varied movement contexts provide different opportunities or affordances for action that are fundamental to promoting motor competence and sport skills [8]. In the context of sport, perception of affordances and opportunities to develop such perceptions are useful for talent development [9,10].

In soccer, there is good evidence that manipulation of constraints can promote perception of affordances, and subsequently, more effective skill development, as a player's ability to adapt their perception and action to continuous changes in spatial and temporal information in a game underpins successful performance [11,12]. Soccer performance and the execution of motor skills within soccer are complex, comprising technical, tactical, physical and psychological aspects, which collectively contribute to decision making to execute motor skills at an opportune time to result in a successful outcome. Dribbling, passing and shooting are recognised as the most important motor skills in soccer [13] and are consequently a focus for coaches working with children at grassroots levels. The ability to dribble the ball and go past the opposing players is a particularly fundamental aspect to the game of soccer [14] and is a complex skill where players must apply cognitive, perceptual and motor skills to rapidly changing situations when dribbling with the ball to maximise their performance during the game [15]. However, while studies have specifically examined affordances for passing in soccer [16,17], shooting [6] and kicking for distance and/or precision [7], few studies have examined affordances for dribbling in soccer. In the context of assessment, the assessment of dribbling skill has been criticised [14], as the majority of the methods available are based on travelling with a ball around cones placed 2–4 m away from each other in a figure-of-eight movement pattern or similar [18–20]. Such a process also translates to how dribbling skill is trained in a blocked and constant manner, where there is repetition of dribbling around cones placed a set distance apart [21,22].

This standardisation of approach does not necessarily translate to an optimal way to coach, assess and improve sport-specific motor skills. Motor skills primarily develop when skills are performed in different ways [23]; perceptual sensitivity to affordances also changes when an individual acquires better motor skills [24]; and standardisation, where distances between cones tend to be equal, does not facilitate variability of practice [5]. Instead, the standardised approach invites the child to continuously replicate a movement pattern. Crucially, affordances exist by virtue of the relationship between the physical properties of the world and the action capabilities of the individual [2]. Whether a space affords the development of soccer dribbling depends on the size of the space, the cones/obstacles/other performers in that space and the rules for use of that space, relative to the skill capabilities of the performer themselves. This means that the same space and constraints can afford different behaviour to children with different action capabilities. Understanding how children may use the same space differently in relation to soccer skill development, and the extent to which other factors, such as perception of their own competence or technical skill, might impact how they use such space, have not been fully examined. Likewise, to date, research has not examined how effective children are as architects of their own soccer skill development. Such information and understanding are useful for coaches in better structuring practices to maximise skill development. The current study sought to explore this issue by examining affordance perception for soccer dribbling using a mixed-methods approach in boys who play grassroots soccer. Using an experimental paradigm, followed by a qualitative contextualisation, we sought to uncover a more in-depth understanding of perception of affordances for dribbling in soccer. We hypothesised that children who perceived themselves as better dribblers would create more challenging dribbling patterns and would also accrue a greater number of touches on the ball in the dribbling pattern they created.

2. Materials and Methods

2.1. Participants

Fourteen boys aged 10–11 years (Mean $\pm$ SD = 10.8 $\pm$ 0.4 years, 145.7 $\pm$ 3.7 cm, 37.0 $\pm$ 4.1 kg) who regularly played organised grassroots soccer for clubs in England participated in the study following institutional ethics approval (protocol code 131207), informed parental consent and child assent. We employed the Fédération Internationale de Football Association (FIFA) [25] definition of grassroots soccer in the present study. To be eligible for participation, children had to be registered/playing with a grassroots soccer club, including participation in training and fixtures against other teams within England's County FA structure. Participants also had to have at least 1 year of playing experience prior to taking part (Mean $\pm$ SD playing experience was 4.2 $\pm$ 1.0 years).

2.2. Design and Procedure

An exploratory mixed-methods approach was employed. All assessments took place over two days and were conducted on an artificial football pitch surface. On the first day, anthropometric assessment was conducted, followed by child self-rating of dribbling skill, as was technical skill and general motor competence assessment. This was followed on the second day by the dribbling affordances task and qualitative interview. All assessments were conducted by trained researchers, and the participants' soccer club coaches were not involved in any way.

2.3. Anthropometry

Stature (cm) and body mass (kg) were assessed to the nearest 0.1 cm and 0.1 kg using a SECA anthropometer and weighing scales (SECA Instruments Ltd., Hamburg, Germany), respectively.

2.4. Child Perception of Dribbling Ability

Children's perception of dribbling ability was determined by asking each child individually to rate, on a visual analogue scale, how good they considered their football dribbling ability to be. Each child was asked to score themselves from 0 to 10, with 0 being not able to dribble at all and 10 being able to dribble confidently in a diversity of situations. Each child completed their perception of dribbling ability on an individual basis, prior to any assessment of movement competence or dribbling ability, and followed processes previously employed in the assessment of perceived soccer competence [18] and affordance in gap-crossing in children [5].

2.5. Motor Competence Assessment

Motor competence was assessed using selected skills from the test of gross motor development-3 (TGMD-3) [26]. To provide a balance of both locomotor and object control skills, children were assessed on the run, jump, hop, overhand throw, underarm throw and catch skills. The kick skill was excluded to avoid confounding the assessment of motor competence and technical soccer skills. Skill mastery on the TGMD-3 requires each component of each skill to be demonstrated, and each skill comprises 3–5 components. Trials of each skill were video recorded (Sony Handicam CX405b, Sony, Tokyo, Japan). Skills were subsequently edited into individual movie clips using the Quintic Biomechanics software v21 (Quintic Consultancy Ltd., Birmingham, UK). As per TGMD-3 guidelines [26], scores from the two trials were summed (scored 0–50) to create an overall score reflecting total FMS.

2.6. Technical Skills

Soccer dribbling skill was assessed using the Ghent University (UGent) dribbling test as previously described by Vandendriessche et al. [27]. All testing was completed with a size 4 football (the official ball size for age band U10–12) as recommended by the Football Association. Testing was completed individually by the participants to minimise any peer pressure to perform.

Participants completed a set circuit with four left and four right turns at different angles, with a distance between cones ranging between 1 and 2.2 m [27]. Following familiarisation and a practice trial, each participant undertook two attempts at the test. Each test was performed as quickly as possible in two steps per test; the first step was performed without the ball and the second step with the ball. The time of each attempt was measured to the nearest 0.01 s with a handheld stopwatch. The time taken to complete the dribbling course without the ball was deducted from the time with the ball to give a skill differential reflecting the dribbling skill. This test has a good reliability, shown by an intra-class correlation coefficient (ICC) of 0.81 [28] and an ICC of 0.82 in a subsample (n = 30) of the current sample [28].

2.7. Affordances for Dribbling Task

The affordances for dribbling task was designed based on procedures previously employed in assessing affordances for playground play in children by Jongneel et al. [5] and Sporrell et al. [29]. The affordances for the dribbling task took part in two phases. In the first phase, each child was asked to create a space/formation to maximise their own soccer dribbling. To this end, each child was provided with 10 cones and a space measuring 9 m length by 5 m width. The cones were standard mini soccer cones that were circular, 12 cm in diameter and 4 cm high. Each child was instructed to create their own space that would allow them to have the most touches on the ball within a 1 min period. The child could place the cones wherever they wanted in the space and could use a maximum of 10 cones. To experience whether the constructed space was in line with their desires, the child could move in and around the space as they wished during this phase. Following this, the children were allowed to adapt the course during this phase. Three of the children adapted their course following the trial, making minor changes to placement of a cone. After the child completed the first phase, the formation of cones, trial and adaption of the cones, the distances between cones were measured with a measuring tape. In the second phase of the task, each child was asked to dribble the soccer ball in the space they had created for a period of 1 min. The child's playing behaviour was video recorded (Sony Handicam CX405b, Sony, UK) and, based upon this, we determined how many touches of the ball were made, with what foot and what patterns of movement occurred in the space.

2.8. Qualitative Interviews

Once each child had completed the affordances task one to one, interviews were held with each child in a comfortable room that the child was familiar with. The interview used a structured guide with open questions, images and videos of their and other players' drills created to use the probe memory. The interview asked 4 main questions and was based on recommendations for conducting qualitative research and validation of interview scheduling [30]: 1. (show their pattern) Why did you create this patten to get the most touches?; 2. How do you feel about the drill you created versus the drill created by your coaches to get the most touches they created versus ones created by their coaches?; 3. (show player's drills). How do you feel about this drill and the number of touches you would be able to complete doing this drill?; 4. Perceptions of their dribbling ability. Prior to all interviews, the interviewer followed the eight-stage interview preparation stage identified by McNamara [30], which included: 1. Choosing a setting to minimise distraction; 2. Explaining the purpose of the interview; 3. Addressing confidentiality; 4. Explaining interview format; 5. Stating the length of interview duration; 6. Providing researcher contact information; 7. Asking if participants had any questions before starting; and 8. Asking for permission to record their views during the interview. The interview was led by one facilitator. Probing was used throughout the interview to gain further understanding.

2.9. Analysis

Descriptive statistics (mean $\pm$ SD) were calculated for touches/min, child perception of dribbling ability, UGent dribbling test score, total FMS score, coach rating of dribbling ability and the average distance between cones in the pattern the child had created. Pearson's product moment correlations were then used to explore relationships between these aforementioned variables. The Statistical Package for Social Sciences (SPSS V25) was used for all analyses.

2.10. Interview Analysis

Interviews were transcribed verbatim during the interviews. Each participant was anonymised and given a code from 1 to 11. Transcripts were analysed using inductive analysis following the steps proposed by Braun and Clarke [31], such as familiarisation of data, reading and re reading the data, code generation, categorisation, search and reviewing themes and defining and naming themes. This process resulted in themes, sub themes and their associated quotes. This enabled a broad flexible approach for the analysis of the data collected to produce an enriched and detailed account of the findings [31]. Analyst triangulation was conducted to increase the quality and credibility of the findings [32,33] using a second independent analyst who conducted a thematic analysis, which was then compared with the primary researcher, assessing the potential selective perception and blind interpretive bias [32,33]. Frequent de-briefing sessions between authors facilitated the discussion, debate and re-definition of the themes. Following completion of the coding process, pen profiles were created to help present the emergent themes via diagrams with the number of times the themes were mentioned [34].

3. Results

Mean ± SD of touches/min, child perception of dribbling ability, time taken on the UGent dribbling test and total FMS score are presented in Table 1.

Table 1. Mean ± SD of touches/min, child perception of dribbling ability, time taken on the UGent dribbling test, total FMS score, coach rating of dribbling ability and the average distance between cones in each participant-created dribbling pattern.

Touches (No/Min)		Perceived Dribbling Ability (0–10)		UGent Dribbling Test (Secs)		Total FMS (0–54)		Coach Rating of Dribbling Ability (0–10)		Average Distance between Cones (cm)	
M	SD	M	SD	M	SD	M	SD	M	SD	M	SD
66.5	26.1	7.1	1.4	25.9	3.3	35.1	5.2	6.0	1.6	99.7	38.7

The number of touches undertaken by children in the 1 min period was spread, ranging from 32 to 118. Pearson's product moment correlations also revealed significant associations between the number of touches/min and average distance between cones ($r = -0.671$, $p = 0.03$), with a greater number of touches being associated with a smaller average distance between cones. There were also significant negative relationships between UGent dribbling test time and coach rating of dribbling ability ($r = -0.67$, $p = 0.03$) and total FMS ($r = -0.746$, $p = 0.012$), as well as between child perception of dribbling ability and total FMS score ($r = 0.604$, $p = 0.049$). All other relationships between the variables were non-significant ($p > 0.05$). Figure 1 presents the different patterns created by children as architects of the dribbling task.

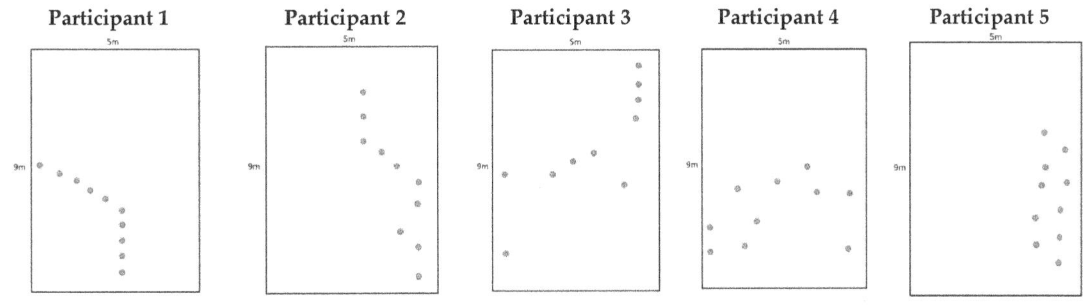

Figure 1. *Cont.*

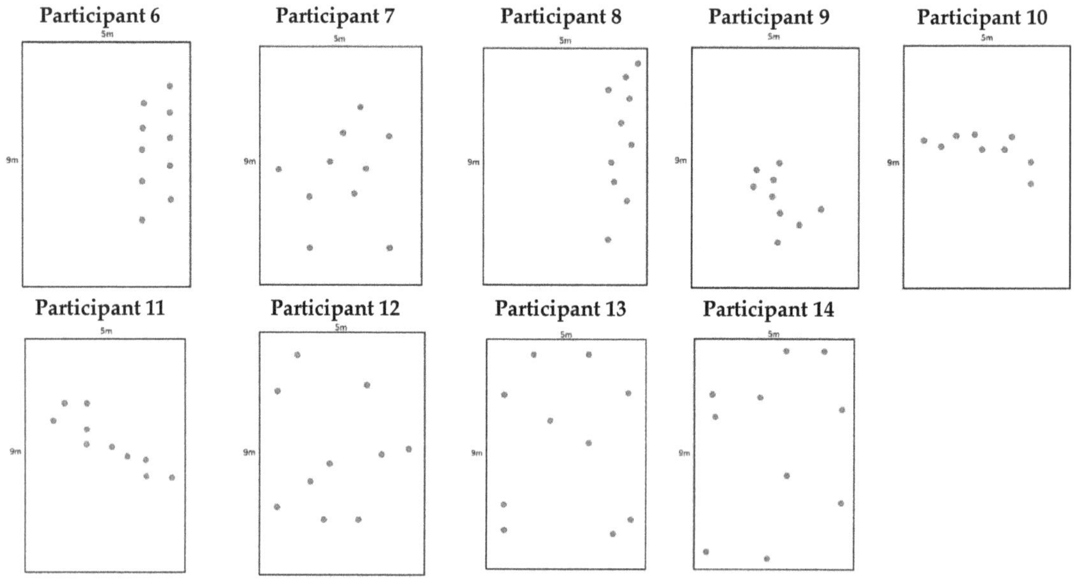

Figure 1. Individual dribbling patterns created by the participants.

With regard to the thematic analysis, three emergent but interconnected themes arose, focusing on: perception of ability and actual ability, and challenge and autonomy. Those children who perceived themselves as 'good dribblers' tended to create patterns where the cones were more closely spaced, while those who perceived themselves as poorer dribblers created patterns where the cones were more widely spaced (See Figure 2). The patterns of cones, which were more closely spaced, were created by the children to get more touches and make it more challenging as a task on the basis that this increased challenge resulted in better progress or development of the dribbling skill itself. Conversely, children who perceived themselves to be poor or average dribblers created patterns of cones with larger gaps to ensure they could 'get a lot more touches' while staying in control of the ball. When children were asked about their preference for developing dribbling skills using their own patterns or those provided by the coaches, there was a relatively even split between those who preferred the coach ($n = 5$) to create drills or those who would prefer autonomy to create the drills themselves ($n = 6$). Those whose preference was for self-created drills seemed to explicitly relate to challenge, for example: 'because coaches have a lot of space between cones usually—these do not put the pressure on my dribbling and I would like to create challenge' (P9) and: *'I feel more comfortable because I designed it so I could do it and it isn't too hard for me'* (P2). Comments from the participants also related to having autonomy in practice, for example: *'If I need to work on something, I can do it as how I need to work on something'* (P10), and: *'Because it allows you to do what you would like to do with the ball, instead of [the coach], saying this is what you're going to do with the ball'* (P11).

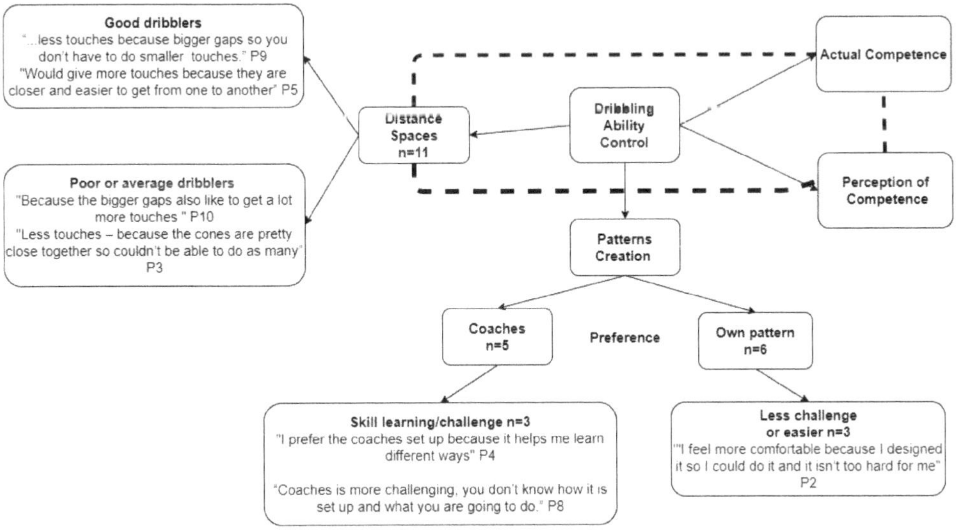

Figure 2. Pen profile diagram of children's perceptions of the affordances for dribbling task.

4. Discussion

The current study examined how children afford spaces for soccer dribbling using a constraints led approach. We uniquely examined children as architects of their soccer skill development in exploring how their perception of affordances for soccer dribbling influenced the number of touches on the ball in a soccer dribbling task. Understanding this process is important in developing practices to maximise soccer skill development for children and to inform coach practice related to player development. The results of the current study are congruent with the perspective that the environment consists of possibilities for action and is perceived as such, and in different ways by different children [2]. The present study suggests that those children who created patterns with less space between the cones accrued more touches of the football in their dribbling task. According to the affordances perspective, when asked to create a task, such as that used in the present study, a child would perceive the distance between two cones not in terms of a set distance but in terms of whether the gap between cones is sufficient for him/her to dribble through [2]. Thus, the current study suggests that affordances are not only primary in the children's perception of the environment but also in the design of learning spaces and, if children can modify their environment, they do so in accordance with their perceived action capabilities for their physical ability and body [1].

The qualitative data suggest there is a feedback loop between a child's perception of ability and their actual ability, which influences the types of dribbling patterns they create, which in turn relates to considerations by the children of challenge and autonomy. Those children who perceived themselves to be better dribblers created dribbling patterns that were more closely spaced, which, in turn, resulted in a greater number of touches on the ball. The creation of such patterns was anchored in the children wanting to challenge themselves in terms of their skill development and having the autonomy to create dribbling patterns to either challenge themselves, support themselves in gaining as many touches of the football as possible or to make the drill easier to ensure success. In this context, greater challenge via less space between cones ensured less time was afforded to move the ball between cones and thus required better performance of the motor skills to successfully do so.

Collectively, the results from this small-scale study suggest that engaging children as active participants in the creation of the dribbling drill appeared to reveal a discrepancy between the types of soccer dribbling typically employed in soccer coaching [18] and the patterns created by the children themselves. Such an observation is congruent with prior research investigating children's active involvement in play space design [5] and is suggestive that historical design processes for movement, typically using standardisation of spaces [35], may be different when a child creates movement opportunities based on their (perceived) action capabilities [5]. Whether children engaging in soccer practice using practices they have wholly designed themselves results in greater skill development compared to practices prescribed by coaches would be an interesting next step arising from this exploratory study. Although we demonstrate that self-organisation of the cones by children increased the number of touches accrued in conditions of greater proximity, we did not examine whether this process might influence game performance. There is evidence, from a study of 15 male youth players, that team composition (e.g., fielding midfielders versus attackers) in small-sided games can influence the players' capability of action that emerges during performance [36]. Whether self-organisation of practice in the way we operationalised in the present study translates into game performance is a natural progression from the findings of the present study. It is also important to note that different types of practice may be more or less important as a focus depending on the stage of development. While a blocked and constant approach to developing motor skill has been criticised [15], it may be useful in the first stages of fundamental development, while using a child-centred approach, such as that demonstrated in the current study, may be more beneficial once children have gone beyond the early development of soccer-specific motor skills.

No study to date has examined how children's perceptions of affordances affect the development of soccer skill, and, although exploratory, the current study's approach, by using a mixed-methods design, represents an original contribution to the field. Using an experimental paradigm followed by a qualitative contextualisation of the affordance activity enables a richer understanding to be uncovered. The combination of quantitative and qualitative approaches in the present study is a novel contribution to the literature examining affordances for skill development in children's soccer and in understanding how children structure space and work with constraints as architects of their own skill development.

We are cognisant that the results of the present study are based on a relatively small sample of grassroots footballers, all of whom were male. These should be considered limitations of the current study. Future work examining this process in girls and across different developmental stages of football skill is also needed to understand if involving children at earlier, or later, stages of skill development changes the way in which children afford dribbling in soccer. In addition, we did not assess biological maturation of the participants. It is possible that some of the sample participants were starting onset of growth spurt, given this is typically at around 12 years of age in boys [37], and our sample were aged 10–11 years. As biological maturation can change performance parameters, future research should consider the measurement of maturation when examining how children interact with the environment in sports-related circumstances. A key focus of the current study was in exploring whether the mixed-methods approach, employing an experimental design followed by qualitative exploration, adds value to scientific understanding of children's soccer skill development. Additional research, using the approach employed in the current study, would therefore be useful in establishing the effectiveness of children as architects of their own skill development. The present study has some practical applications; notably, we demonstrate that coaches could empower children themselves to construct their own skill development practices for soccer, and, if they do, the children themselves are likely to scale their practices to their actual ability. This would suggest the co-creation of soccer practices between children and coaches could be an effective strategy for soccer skill development at grassroots level. Despite this, and to conclude, the current mixed-methods study demonstrates that children can create their own opportunities for skill development and that, when they do, these opportunities are related to their actual dribbling ability, reinforcing the affordances construct in practice. Dribbling performance is therefore scaled to the (perceived) action capabilities of the children, and children can act as architects in their own skill development.

Author Contributions: Conceptualisation, M.J.D. and E.L.J.E.; methodology, M.J.D., R.M., M.N. and E.L.J.E.; validation, M.J.D.; formal analysis, M.J.D., R.M. and E.L.J.E.; investigation, M.J.D., R.M., M.N. and E.L.J.E.; resources, M.J.D.; writing—original draft preparation, M.J.D.; writing—review and editing, R.M., M.N. and E.L.J.E.; supervision, M.J.D. All authors have read and agreed to the published version of the manuscript.

Funding: This research received no external funding.

Institutional Review Board Statement: The study was conducted in accordance with the Declaration of Helsinki and approved by the Institutional Review Board (or Ethics Committee) of Coventry University (protocol code 131207 on 20 February 2022).

Informed Consent Statement: Informed consent was obtained from all subjects involved in the study.

Data Availability Statement: Data supporting the results of the current study are available on request to the corresponding author.

Conflicts of Interest: The authors declare no conflict of interest.

References

1. Withagen, R.; van Wermeskerken, M. The Role of Affordances in the Evolutionary Process Reconsidered: A Niche Construction Perspective. *Theory Psychol.* **2010**, *20*, 489–510. [CrossRef]
2. Gibson, J.J. *The Ecological Approach to Visual Perception: Classic Edition*; Psychology Press: London, UK, 2014.
3. Mark, L.S. Eye height-scaled information about affordances: A study of sitting and stair climbing. *J. Exp. Psych. Hum. Percep. Perf.* **1987**, *13*, 361–370. [CrossRef]
4. Grechkin, T.Y.; Chihak, B.J.; Cremer, J.F.; Kearney, J.K.; Plumert, J.M. Perceiving and acting on complex affordances: How children and adults bicycle across two lanes of opposing traffic. *J. Exp. Psych. Hum. Percep. Perf.* **2013**, *39*, 23–36. [CrossRef] [PubMed]
5. Jongeneel, D.; Withagen, R.; Zaal, F.T.J.M. Do children create standardized playgrounds? A study on the gap-crossing affordances of jumping stones. *J. Env. Psych.* **2015**, *44*, 45–52. [CrossRef]
6. Paterson, G.; van der Kamp, J.; Bressan, E.S.; Savelsbergh, G.J. The effect of perception action coupling on perceptual decision-making in a self-paced far aiming task. *Int. J. Sports Psych.* **2013**, *44*, 179–196.
7. Peker, A.T.; Böge, V.; Bailey, G.; Wagman, J.B.; Stoffregen, T.A. Perception of Affordances in Soccer: Kicking for Power Versus Kicking for Precision. *Res. Q Exerc. Sport* **2022**, *93*, 144–152. [CrossRef]
8. Flôres, F.S.; Rodrigues, L.P.; Copetti, F.; Lopes, F.; Cordovil, R. Affordances for Motor Skill Development in Home, School, and Sport Environments: A Narrative Review. *Percep. Mot. Skills* **2019**, *126*, 366–388. [CrossRef]
9. Williams, A.M.; Reilly, T. Talent identification and development in soccer. *J. Sports Sci.* **2000**, *18*, 657–667. [CrossRef]
10. Baker, J.; Farrow, D. *Routledge Handbook of Sport Expertise*; Routledge: London, UK, 2015.
11. Travassos, B.; Araújo, D.; Davids, K.; Esteves, P.T.; Fernandes, O. Improving Passing Actions in Team Sports by Developing Interpersonal Interactions between Players. *Int. J. Sports Sci. Coach* **2012**, *7*, 677–688. [CrossRef]
12. Corrêa, U.C.; Vilar, L.; Davids, K.; Renshaw, I. Informational constraints on the emergence of passing direction in the team sport of futsal. *Eur. J. Sport Sci.* **2014**, *14*, 169–176. [CrossRef]
13. Rampinini, E.; Impellizzeri, F.M.; Castagna, C.; Coutts, A.J.; Wisløff, U. Technical performance during soccer matches of the Italian Serie A league: Effect of fatigue and competitive level. *J. Sci. Med. Sport* **2009**, *12*, 227–233. [CrossRef] [PubMed]
14. Bekris, E.; Gissis, I.; Ispyrlidis, I.; Mylonis, E.; Axeti, G. Combined visual and dribbling performance in young soccer players of different expertise. *Res. Sports Med.* **2018**, *26*, 43–50. [CrossRef] [PubMed]
15. Ali, A. Measuring soccer skill performance: A review. *Scand. J. Med. Sci. Sports* **2001**, *21*, 170–183. [CrossRef]
16. Oppici, L.; Panchuk, D.; Serpiello, F.R.; Farrow, D. Futsal task constraints promote transfer of passing skill to soccer task constraints. *Eur. J. Sport Sci.* **2018**, *18*, 947–954. [CrossRef]
17. Travassos, B.; Araújo, D.; Davids, K.; Vilar, L.; Esteves, P.; Vanda, C. Informational constraints shape emergent functional behaviors during performance of interceptive actions in team sport. *Psych. Sport Exerc.* **2013**, *14*, 216–223.
18. Forsman, H.; Gråstén, A.; Blomqvist, M.; Davids, K.; Liukkonen, J.; Konttinen, N. Development of perceived competence, tactical skills, motivation, technical skills, and speed and agility in young soccer players. *J. Sports Sci.* **2016**, *34*, 1311–1318. [CrossRef] [PubMed]
19. McGregor, S.; Hulse, M.; Strudwick, A.; Williams, C. The reliability and validity of two tests of soccer skill. *J. Sports Sci.* **1999**, *17*, 815.
20. Vänttinen, T. *Growth-Associated Variation in Body Size, Hormonal Status, Physical Performance Characteristics and Perceptual-Motor Skills in Finnish Young Soccer Players, a Two-Year Follow-Up Study in the U11, U13 and U15 Age Groups*; University of Jyväskylä: Jyväskylä, Finland, 2013.

21. Hoff, J. Training and testing physical capacities for elite soccer players. *J. Sports Sci.* **2005**, *23*, 573–582. [CrossRef]
22. Hosseini-Khezri, S.A.; Zaxarova, A.V. The Underpinning of Technical Training in Soccer Players 10–12 Years Old. *J. Phys. Fit. Treat. Sports* **2018**, *5*, 555663.
23. Chow, Y.Y.; Davids, K.; Hristovski, R.; Araujo, D.; Passos, P. Nonlinear pedagogy: Learning design for self-organizing neurobiological systems. *New Ideas Psych.* **2011**, *29*, 189–200. [CrossRef]
24. Seifert, L.; Button, C.; Davids, K. Key properties of expert movement systems in sport: An ecological dynamics perspective. *Sports Med.* **2013**, *43*, 167–178. [CrossRef] [PubMed]
25. FIFA. *Grassroots Football*; FIFA: Zurich, Switzerland, 2011. Available online: https://www.fifa.com/who-we-are/news/grassroots-goes-live-fifa-com-1462069 (accessed on 20 May 2022).
26. Ulrich, D.A. *Test of Gross Motor Development*, 3rd ed.; PRO-ED: Austin, TX, USA, 2013.
27. Vandendriessche, J.B.; Vaeyens, R.; Vandorpe, B.; Lenoir, M.; Lefevre, J.; Philippaerts, R.M. Biological maturation, morphology, fitness, and motor coordination as part of a selection strategy in the search for international youth soccer players (age 15–16 years). *J. Sports Sci.* **2012**, *30*, 1695–1703. [CrossRef] [PubMed]
28. Duncan, M.; Richardson, D.; Morris, R.; Eyre, E.; Clarke, N. Test-retest reliability of soccer dribbling tests in children. *J. Mot. Learn Dev.* **2021**, *9*, 526–532. [CrossRef]
29. Sporrel, K.; Caljouw, S.R.; Withagen, R. Gap-crossing behavior in a standardized and a nonstandardized jumping stone configuration. *PLoS ONE* **2017**, *12*, e0176165. [CrossRef] [PubMed]
30. McNamara, C. General Guidelines for Conducting Interviews. 2009. Available online: http://managementhelp.org/evaluatn/intrview.htm (accessed on 20 May 2022).
31. Braun, V.; Clarke, V. Reflecting on reflexive thematic analysis. *Qual. Res. Sport Exerc. Health* **2019**, *11*, 589–597. [CrossRef]
32. Cohen, D.; Crabtree, B. Qualitative Research Guidelines Project Retrieved 11, 2021. 2006. Available online: http://www.qualres.org/homesemi-3629.html (accessed on 20 May 2022).
33. Patton, M.Q. Enhancing the quality and credibility of qualitative analysis. *Health Serv. Res.* **1999**, *34*, 1189–1208.
34. Ridgers, N.; Knowles, Z.; Sayers, J. Encouraging play in the natural environment: A child-focused case study of Forest School. *Child. Geogr.* **2012**, *10*, 49–65. [CrossRef]
35. Olwig, K.R. Designs upon children's special places? *Child. Environ. Q* **1990**, *7*, 53.
36. Laakso, T.; Davids, K.; Luhtanen, P.; Liukkonen, J.; Travassos, B. How football team composition constrains emergent individual and collective tactical behaviours: Effects of player roles in creating different landscapes for shared affordances in small-sided and conditioned games. *Int. J. Sports Sci. Coach* **2022**, *17*, 346–354. [CrossRef]
37. Limony, Y.; Kozieł, S.; Friger, M. Age of onset of a normally timed pubertal growth spurt affects the final height of children. *Pediatr. Res.* **2015**, *78*, 351–355. [CrossRef]

sports

Article

Does Relative Age Influence Organized Sport and Unorganized Physical Activity Participation in a Cohort of Adolescents?

Kristy L. Smith [1,*], Mathieu Bélanger [2,3], Laura Chittle [1], Jess C. Dixon [1], Sean Horton [1] and Patricia L. Weir [1]

[1] Department of Kinesiology, University of Windsor, Windsor, ON N9B 3P4, Canada; chittlel@uwindsor.ca (L.C.); jdixon@uwindsor.ca (J.C.D.); hortons@uwindsor.ca (S.H.); weir1@uwindsor.ca (P.L.W.)
[2] Department of Family Medicine, Université de Sherbrooke, Sherbrooke, QC J1H 5H3, Canada; mathieu.belanger@umoncton.ca
[3] Centre de Formation Médicale du Nouveau-Brunswick, Moncton, NB E1A 3E9, Canada
* Correspondence: smith43@uwindsor.ca

Abstract: Despite their prevalence, the longitudinal impacts of relative age effects (RAEs) on sport and other forms of physical activity (PA) are understudied. This study examined longitudinal participation patterns in organized sport (team and individual), unorganized PA, and non-participation with respect to RAEs in a prospective cohort of adolescents. Data from the first 24 cycles of the MATCH study were used for analyses. Elementary students (n = 929) were recruited from 17 schools in Atlantic Canada. Respondents self-reported PA three times/year. Mixed multilevel logistic models compared the likelihood of participating in each context across birth quarter. Chronological age and gender were considered, along with the interaction between chronological and relative age. Individuals born in Quarter 1/Quarter 2 were more likely to report participation in organized team sport but not individual sports. Relatively older participants born in Quarter 2 were more likely to report participation in unorganized PA. Increasing chronological age was associated with decreased participation in organized sport (particularly team-based) and increased non-participation. Gender was not associated with organized sport participation, but girls were under-represented in unorganized PA and more likely to report non-participation. The interaction parameters suggested that RAEs were consistent throughout adolescence in each context. Longitudinal analyses suggest RAEs are context dependent.

Keywords: relative age effects; participation trends; longitudinal; sport context; physical activity

Citation: Smith, K.L.; Bélanger, M.; Chittle, L.; Dixon, J.C.; Horton, S.; Weir, P.L. Does Relative Age Influence Organized Sport and Unorganized Physical Activity Participation in a Cohort of Adolescents?. *Sports* **2022**, *10*, 97. https://doi.org/10.3390/sports10070097

Academic Editors: Adam Leigh Kelly, Sergio L. Jiménez Sáiz, Sara Diana Leal dos Santos and Alberto Lorenzo Calvo

Received: 28 April 2022
Accepted: 21 June 2022
Published: 23 June 2022

1. Introduction

Despite efforts to communicate the importance of regular physical activity, high levels of physical inactivity remain a global concern [1]. Adolescence is a particularly concerning period as it is often marked by significant declines in physical activity participation [2–4]. "Physical activity" encompasses a variety of activities such as competitive sport, leisure walking, and occupational and domestic activities, among others [5,6]. Consequently, context is an important consideration when evaluating physical activity patterns [7]. Likewise, recommendations for increasing physical activity can incorporate different domains such as active transport (e.g., walking to school), participation in unorganized sport or active play [8], and even breaking up sedentary time with light intensity activity [9]. While participation in organized sport has been associated with a diverse range of positive outcomes among youth (e.g., enhanced resilience, character development, social skills [10,11]), the health benefits of physical activity (e.g., cardiorespiratory fitness, improved posture) might also be obtained in an unorganized context [12,13]. Thus, both organized and unorganized activity provide beneficial opportunities to increase participation. While the work to date by Bengoechea et al. [14] and Hardy et al. [15] present equivocal findings with respect to whether participation is higher in organized or unorganized activities, it is clear that through adolescence, both categories of participation are important contributors to overall activity level [16].

One determinant that has been associated with participation in the physical activity context of organized sport is relative age. Relative age refers to the difference in age between individuals in a cohort [17,18], such as those grouped within the same grade in the school system. These age groupings are intended to help promote developmentally appropriate instruction and competition [19,20]. However, a one- or two-year age difference can result in significant physical, psychological, and experiential differences, particularly at young ages when a few months can represent a considerable proportion of a child's lived experience. In addition to experiential differences, age differences within a given age grouping can include considerable variability in biological maturity from childhood through to adolescence [18,21].

The term relative age effects (RAEs) refers to the (dis)advantages commonly associated with the aforementioned age groupings [22], providing a benefit to those who are relatively older (meaning that they were born closer to but following an arbitrary cut-off date) and disadvantaging those who are relatively younger in the same age group or cohort. With respect to sport, the advantage conveyed to those who are relatively older is assumed to be present when an over-representation of relatively older players is observed among sport participants (e.g., on a specific team), particularly at elite levels. While RAEs have predominantly been studied in team sports such as ice hockey, soccer, volleyball, and basketball (see Cobley et al. [23] for a review), individual sports, where success is predicted by strength or endurance capabilities, such as tennis [24], cross-country and alpine skiing [25], and sprinting [26,27], may also be affected. Potential consequences of RAEs include relatively younger players dropping out of organized sport at an earlier age [28–30] or adapting by taking up forms of activity that are less structured [31] or less competitive [32] in nature. To date, RAEs have primarily been found in male sport contexts ([23]), but a meta-analysis of RAEs in female sport shows a pervasive effect there as well (see Smith et al. [33]). Sex is nevertheless believed to be a significant moderator of the effect (Baker et al., 2010), with previous findings suggesting RAEs in females are more variable and weaker in magnitude when compared to males [23,34].

Several explanations and theoretical models of RAEs have been presented over the years. For instance, many relative age studies published to date have focused on maturation and selection (e.g., [20,23,35]). Being relatively older is often associated with greater physical and anthropometric maturity (e.g., height, weight, strength) [36]. During the maturational period, up to and including puberty, relatively older individuals may be viewed as more talented on the basis of their more advanced physical and anthropometric development. As a result, relatively older sport participants may benefit from a higher likelihood of being selected to elite teams, providing enhanced access to coaching and facilities, thereby facilitating greater skill development than their relatively younger peers.

Hancock et al. [37] argue that RAEs are largely tied to social influences. Specifically, RAEs are attributed to early experiences of success and the manner by which expectations of coaches and parents influence athletes' behaviors and self-perceptions, which ultimately impacts their level of achievement. Finally, a model proposed by Wattie et al. [38] suggests that the impact of relative age comes at the intersection of individual (e.g., gender, relative age), task (e.g., team or individual sport, organized vs. unorganized activity), and environmental constraints (e.g., age grouping policies) within developmental systems theory. Relative age can be viewed as one of many factors influencing the probability of sport participation, and there is opportunity for change throughout development. This developmental perspective is adopted within the current longitudinal study.

Relative age research has largely been cross-sectional in nature and focused on performance, examining the athlete population of a specific sport (e.g., basketball [39]) or sporting event (e.g., Olympic Games [40]). While this strategy has provided useful descriptions of the prevalence of RAEs, it does not inform about the potential longer-term effects of relative age advantages and/or disadvantages on sport and/or physical activity participation for children and adolescents. Furthermore, evidence pertaining to RAEs in sport predominantly emerges from studies in organized sport and competitive contexts and is less commonly studied in recreational sport [23]. Observations from the few studies published to date suggest that RAEs may be highly variable within recreational contexts and influenced by a variety of factors and/or constraints [41–43]. Furthermore, published studies concerning the role of RAEs on outcomes associated with participation in unorganized sport and/or physical activity in children and adolescents are virtually non-existent.

Relative age effects have also been observed in a variety of areas within the school system such as physical education [44], high school leadership activities [45], standardized test scores (math and science [46]), university attendance [46], and emotional regulation [47]. These findings suggest that comparisons with peers occur in other contexts and could potentially impact the overall course of development for youth [48]. If relatively younger youth are habitually disadvantaged in organized activities and have fewer opportunities to benefit from sport system development, they could suffer from reduced perceptions of competence and eventually disengage from a physically active lifestyle. For example, perceived competence in a physical education setting has been positively associated with leisure-time physical activity (e.g., [49,50]). The impact of relative age may be evidenced in children's degrees of involvement not only in organized activities but also in unorganized activities or in a lack of participation altogether; these possibilities deserve further consideration.

To address the aforementioned gaps in the literature, the primary purpose of the current study is to examine the longitudinal patterns of participation in organized sport, unorganized physical activity, and non-participation with respect to relative age, in a prospective study of adolescents. The impact of gender and chronological age are considered, along with the interaction between chronological age and relative age within each participation/non-participation context. We hypothesize that a greater proportion of those who are relatively older participate in organized sport (both team and individual), while those who are relatively younger gravitate toward unorganized physical activities (where talent selection activities that favor the relatively oldest are notably absent), or a lack of participation altogether. Furthermore, it is expected that the hypothesized trends will become more pronounced with increasing chronological age for boys as they enter adolescence and become less pronounced for girls in alignment with previous meta-analytical findings [23,33].

2. Materials and Methods

2.1. Respondents and Data Collection

To accomplish the research objectives, the database for the Monitoring Activities of Teenagers to Comprehend their Habits (MATCH) study was queried to examine the same respondents in multiple activities rather than a single sport in isolation, which is a noted limitation of previous work. The MATCH study is an ongoing longitudinal prospective study in the province of New Brunswick, Canada. The complete study protocol is provided in Bélanger et al. [51]. Briefly, a convenience sample of 19 schools were recruited, and it included both French- and English-speaking students from a range of socioeconomic neighborhoods located in both rural and urban areas. Two of the 19 schools were excluded due to a low return rate of consent forms (below 50%). All grade five and six students from the remaining 17 schools were invited to participate in the study. Respondents ranged from 10 to 12 years of age at the time of recruitment, representing a critical timeframe when physical activity levels are often found to attain a peak before typically experiencing a decline as the transition from childhood to adolescence occurs [2,52]. Initial recruitment in the fall of 2011 included 802 of 1545 eligible students (response proportion of 52%), and other students from the participating schools were allowed to join the study in subsequent cycles (total number of respondents = 929). Please note within this study, the term 'respondent' refers to any individual who participated in the MATCH study, 'survey response' indicates participation in a single MATCH data collection cycle, and 'participant' indicates reported participation in sport and/or physical activity (outlined further below).

The same respondents were asked to complete self-report questionnaires three times per year (fall, winter, and spring) at four-month intervals, which were combined to capture seasonal variations in one year of physical activity participation [6,51,53]. The first eight years of the MATCH study were used for the current analysis (n = 24 cycles). The questionnaires were completed under the supervision of a trained research assistant in a classroom setting, which was scheduled at a time that was convenient for the teacher. The initial session was 45–60 min in duration, and subsequent questionnaires were completed in 20–30 min. Respondents answered questionnaires in the first language of instruction at the school.

2.2. Measures

Respondents reported all forms of leisure time physical activity over the preceding four months from a list of 36 activities commonly practiced by youth in Atlantic Canada [54], which was similar to other validated physical activity checklists [55–57]. Respondents could also report additional activities under "other". The students were specifically instructed to disregard physical activity taking place during physical education classes because youth do not have control over the content of these classes [58]. The frequency of participation in each activity was collected using the response options "never", "once per month or less", "two to three times per month", "once per week", "two to three times per week", "four to five times per week", and "almost every day". Respondents were also asked to report "with whom" they had most often engaged in each activity by selecting from "by myself", "organized group or team", "sibling(s)", "friend(s)", or "parent(s)". The answer to this question was considered to be an indicator of the context in which each activity was most often practiced [59]. Respondents were described as 'participants' in the physical activity context being examined if they reported participating in any activity pertaining to each category "once per week" or more, similar to previous studies [14,60,61], thereby creating a binary measure (participant/non-participant).

For each of the cycles in the longitudinal sample, the classification of "relatively older" versus "relatively younger" must be made with reference to the cut-off date used to group individuals within the specific sporting activities. The dates used for activities recognized as "sports", defined by the existence of a recognized, provincial organization registered with Sport New Brunswick (SportNB), were obtained by contacting each respective organization listed in SportNB's online directory. If no response was received from the provincial organization, the national organization was contacted through the Sport Canada website. The majority of sports employed a 31 December cut-off date to group athletes. Other organized activities use cut-off dates as well, including the education system, which utilizes 31 December in the province of New Brunswick [62]. Therefore, the decision was made to code all respondents using this date because it is the most commonly encountered age grouping experienced by the respondents in this study. The coding of birth quarter proceeded as follows: Quarter 1—January to March; Quarter 2—April to June; Quarter 3—July to September; Quarter 4—October to December. Thus, the relatively oldest were identified as 'Quarter 1' subsequently followed by Quarter 2, Quarter 3, and Quarter 4 (i.e., the relatively youngest). This method of coding is consistent with previous studies (e.g., [63–65]).

Similar to other studies employing the MATCH sample, respondents were classified as "organized sport participants" if they reported participation with an "organized group or team" [61] in one of the recognized sports in New Brunswick. Organized sports were further categorized as "organized team sports" if the sport is most often practiced or played with more than one participant competing at the same time or multiple participants are a necessary element for competition. Alternatively, organized sports were classified as "organized individual sports" if participation most often occurs with only a single competitor competing at any given time. These definitions were consistent with previous research [58,61]. Organized team sports that met the criteria outlined above included: ice hockey, ringette, baseball/softball, basketball, soccer, volleyball, and handball. Organized individual sports that met these criteria included: bicycling, track and field, jogging or running, badminton, tennis, downhill skiing or snowboarding, and cross-country skiing. The specific activity identified by the respondent was not important but rather the classification of "participant" in a particular context (i.e., organized team and/or individual sport).

To ensure accurate representation of each activity, sports that used a cut-off date other than 31 December were excluded to prevent bias from the use of an alternative date (i.e., 1 July for ice skating and 1 August for golf). Gymnastics was omitted because a reversal of the classic relative age pattern would be expected in this context. Late maturation is valued in aesthetic sports with peak performance occurring prior to the completion of maturation [38,66]. Accordingly, an over-representation of relatively younger participants and an under-representation of relatively older participants might be expected (i.e., reversal of the typical RAE pattern). This expectation was supported in the current data and therefore, gymnastics was removed from the organized sport analysis for this study to prevent an underestimation of the typical RAE present in this sample. Swimming was excluded because participants are categorized based on their chronological age on the day of competition. Other activities were excluded because the cut-off was unknown (i.e., football, boxing/wrestling), or multiple activities were grouped together (i.e., canoe/kayak and karate/judo/tai chi/taekwondo) in the questionnaire and could not be categorized consistently as an organized sport due to the absence of a provincial governing body for one or more of the activities.

Using the classification system derived by Ward et al. [61] and validated by MacKenzie and colleagues [67], five of the 36 activities were categorized as "unorganized physical activity", regardless of the context in which they were most often practiced, including trampoline, jump rope/skipping, games (e.g., chase, tag, hide and seek), home exercise, and weight training. Indoor chores and outdoor chores were excluded given the non-volitional and low intensity nature of these activities. Participation in the remaining activities was also considered to be unorganized if the individual reported to most often practice it alone, with friends, with siblings, or with a parent. A respondent who did not participate in any activity within a classification category was considered a "non-participant". A summary of the classifications is available in Appendix A.

2.3. Data Analysis

Prior to analyses, missing data for birth date (0.002%) was reviewed and corrected across all 24 cycles. A total of 12,061 survey responses were available for examination. Respondents (n = 929) participated in an average of 13 survey cycles each. Three objectives guided statistical analyses. First, the relative age trends in each of the five participation/non-participation contexts were assessed (i.e., any organized sport, organized team sport, organized individual sport, unorganized physical activity, and non-participation). Second, the impact of gender and chronological age were considered for each of the five contexts. Gender was self-reported by respondents in the MATCH study. In seven survey responses, (<0.01%) the respondent selected "other" as their gender category. These instances could not be included in the analyses due to insufficient sample size. Third, the longitudinal interaction between chronological age and relative age was considered within each participation/non-participation condition. To accomplish these objectives, a mixed multilevel logistic model was conducted for each of the five participation/non-participation contexts. The potential for intra-class correlation because of repeated measures among respondents and school-level clustering was accounted for by including random intercepts for these variables. Birth quarter (representing relative age), chronological age (by year – included as a continuous variable), and gender were included as independent variables, along with the interaction between chronological age and relative age. The fourth quarter (Quarter 4) was used as the comparison group. The interaction did not meet the criteria for statistical significance ($p > 0.05$) in any of the participation/non-participation contexts.

Thus, the models presented in this paper do not include estimates for interaction parameters in the interest of parsimony.

3. Results

Our primary findings with respect to relative age trends, as indicated by the representation of participants in each birth quarter, are outlined below. Results are presented for organized sport (overall, team, and individual), unorganized physical activity, and non-participation, and they include findings for gender, chronological age, and the interaction between chronological age and relative age in each respective context. Of the participation contexts examined, unorganized PA had the most participants, and organized individual sport had the lowest number of participants throughout adolescence (Table 1). In general, there were more boys than girls reporting participation in all types of sport/PA.

Table 1. Average proportion of respondents in each context across the 24 survey cycles included in the analyses.

Context [1]	Girls (n = 511, 55%)	Boy (n = 418, 45%)	Total (n = 929)
Organized sport [2]	37.2%	45.8%	41.9%
Team sport	31.8%	40.1%	35.6%
Individual sport	11.3%	13.6%	12.3%
Unorganized PA	79.0%	82.6%	80.6%
Non-participant	14.3%	12.0%	13.3%

[1] Respondents could be included in organized sport and unorganized physical activity (PA) simultaneously, but the non-participant category is exclusive. [2] Participants included in organized sport overall may have reported team and/or individual sport participation.

3.1. Organized Sport

In comparison to those born in Quarter 4, those born in Quarter 3 were less likely to report participation in any organized sport overall, t (976) = −2.205, $p < 0.05$ (Table 2a). However, when team and individual sport were analyzed separately, Quarter 1- and Quarter 2-born participants (i.e., the relatively oldest) were more likely to report participation in organized team sports when compared to Quarter 4 (i.e., the relatively youngest), t (1093) = 1.995, $p < 0.05$ and t (1003) = 2.138, $p < 0.05$, respectively (Table 2b). No statistical differences based on relative age were noted for organized individual sport participation suggesting equal representation of participants from each birth quarter (Table 2c).

Gender was not observed to be an important variable for any of the organized sport participation contexts (overall, team, or individual). Negative estimates for chronological age indicated that as participants became older, their participation in organized sports overall, t (915) = 1.356, $p < 0.05$, and organized team sports, t (3643) = −2.801, $p < 0.01$, declined but their reported participation in organized individual sport was maintained. The lack of statistical significance for any of the interaction parameters suggest that RAEs observed for organized sport participation overall, as well as for team and individual sport were consistent throughout adolescence (i.e., the same patterns of over-representation by birth quarter were observed across the years examined).

3.2. Unorganized Physical Activity

Participants born in the second quarter (i.e., relatively older) were found to be statistically over-represented compared to the fourth quarter (i.e., relatively younger) in unorganized physical activity participation, t (1544) = 2.731, $p < 0.01$ (Table 3).

Table 2. Odds ratios of participation in organized sport across relative age, (**a**) overall; (**b**) team; (**c**) individual.

Fixed Effects	Odds Ratios [1]	95% Confidence Intervals	
		Lower	Upper
(a) Organized Sport (Overall)			
Relative age (Q1 vs. Q4) [2]	1.03	0.98	1.08
Relative age (Q2 vs. Q4)	1.04	0.99	1.09
Relative age (Q3 vs. Q4)	**0.95**	0.90	0.99
Gender (Girl vs. Boy)	1.02	0.99	1.06
Chronological age	**0.98**	0.96	1.00
(b) Organized Sport (Team)			
Relative age (Q1 vs. Q4)	**1.06**	1.00	1.13
Relative age (Q2 vs. Q4)	**1.06**	1.01	1.12
Relative age (Q3 vs. Q4)	0.95	0.90	1.01
Gender (Girl vs. Boy)	1.00	0.97	1.05
Chronological age	**0.97** *	0.94	0.99
(c) Organized Sport (Individual)			
Relative age (Q1 vs. Q4)	1.02	0.98	1.07
Relative age (Q2 vs. Q4)	1.03	0.99	1.07
Relative age (Q3 vs. Q4)	0.97	0.93	1.01
Gender (Girl vs. Boy)	1.01	0.99	1.04
Chronological age	1.01	0.99	1.03

[1] Statistically significant values are indicated with bold text ($p < 0.05$) and * ($p < 0.01$). [2] Birth quarter is identified by the following: Quarter 1 (Q1), Quarter 2 (Q2), Quarter 3 (Q3), and Quarter 4 (Q4).

Table 3. Odds ratios of participation in unorganized physical activity across relative age.

Fixed Effects	Odds Ratios [1]	95% Confidence Intervals	
		Lower	Upper
Relative age (Q1 vs. Q4) [2]	1.02	0.98	1.05
Relative age (Q2 vs. Q4)	**1.04** *	1.01	1.08
Relative age (Q3 vs. Q4)	1.03	1.00	1.06
Gender (Girl vs. Boy)	**0.97** *	0.95	0.99
Chronological age	0.99	0.97	1.00

[1] Statistically significant values are indicated with bold text ($p < 0.05$) and * ($p < 0.01$). [2] Birth quarter is identified by the following: Quarter 1 (Q1), Quarter 2 (Q2), Quarter 3 (Q3), and Quarter 4 (Q4).

Participation in unorganized physical activity did not change with chronological age, but we noted that girls were less likely than boys to report participation in an unorganized context, t (1473) = −2.850, $p < 0.01$. The test of the interaction suggested that relative age trends (by year) were invariant in unorganized physical activity participation across the years examined.

3.3. Non-Participation

With respect to non-participation (i.e., individuals who did not report participation in an organized and/or unorganized context), Quarter 2-born individuals were found to be significantly under-represented compared to those born in Quarter 4, t (1556) = −3.374, $p < 0.01$ (Table 4).

Table 4. Odds ratios of non-participation across relative age.

Fixed Effects	Odds Ratios [1]	95% Confidence Intervals	
		Lower	Upper
Relative age (Q1 vs. Q4) [2]	0.98	0.95	1.01
Relative age (Q2 vs. Q4)	**0.95** *	0.93	0.98
Relative age (Q3 vs. Q4)	0.98	0.95	1.01
Gender (Girl vs. Boy)	**1.02**	1.00	1.04
Chronological age	**1.02** *	1.00	1.03

[1] Statistically significant values are indicated with bold text ($p < 0.05$) and * ($p < 0.01$). [2] Birth quarter is identified by the following: Quarter 1 (Q1), Quarter 2 (Q2), Quarter 3 (Q3), and Quarter 4 (Q4).

Non-participation increased with chronological age, t (5216) = 2.657, $p < 0.01$, and there was an increased likelihood of non-participation among girls versus boys, t (1482) = 2.012, $p < 0.05$. The longitudinal trends (by year) with respect to relative age were stable in the non-participation category, as indicated by the lack of statistical significance in the test of the interaction.

4. Discussion

The longitudinal analyses undertaken in this study were based on eight years of data spanning late childhood to late adolescence. The primary purpose was to examine participation with respect to relative age in five participation/non-participation contexts. Secondary objectives included consideration of gender and chronological age as well as the interaction between chronological age and relative age. We found that relatively older (Quarter 1 and Quarter 2) respondents were more likely to report participation in organized team sport when compared to the relatively youngest (Quarter 4). This finding is consistent with past research and supports our hypotheses for the organized team sport context. However, in contrast to previous findings (e.g., [24–27,33]) and our hypotheses, no relative age trends were observed for organized individual sport participation. Furthermore, RAEs appeared to be associated with participation in unorganized physical activities, with Quarter 2-born participants over-represented. Notably, gender did not play a role in organized sport participation, but girls were under-represented in unorganized physical activities and more likely to report non-participation. Increasing chronological age was associated with decreased participation in organized sports overall and specifically in team-based contexts (but not individual sports); it was also associated with increased non-participation in this sample. Relative age effects appeared to be consistent throughout the developmental years examined in this study (i.e., the interaction did not meet the criteria for statistical significance ($p > 0.05$) in any of the participation/non-participation contexts).

An over-representation of relatively older participants in team-based sports supports the role of talent selection processes in perpetuating RAEs [23]. "Tryouts", whereby coaches and scouts scrutinize athletes' skills for the purpose of identifying the most talented individuals, are often required to gain membership on an organized team. These activities are believed to favor those who are relatively older, as they are likely to be taller/stronger and have acquired more life experience as a result of being chronologically older [18]. Once membership is gained, these selected individuals continue to benefit from being relatively older by having access to higher levels of training, coaching, and competition [20], resulting in an accumulated advantage from initial (albeit subtle) differences in growth, development, and experience [23,68].

Evidence for RAEs in individual sports has been less consistent in comparison to team sports [25]. Previous reports of RAEs have emerged from studies of individual sports for which physical attributes might provide an advantage such as tennis [24], cross-country and alpine skiing [25], and sprinting [26,27]. However, the examination of sport contexts that rely more predominantly on technical skills (vs. physical prowess) have not produced equivalent findings (e.g., golf [69], shooting sports [70]). It has been theorized that RAEs may be less prominent in individual sports due to comparisons between peers occurring after performance, and they are often informed by more objective scores such as race time or a deductive rating system administered by a group of judges. In contrast, the peer-to-peer comparisons associated with team selection processes are often made in the midst of competition or scrimmages where physical differences might be more apparent on the field of competition or during practice [25]. While "individual sport" in this study included several activities associated with physical demands, the lack of evidence for biased birth distributions suggests this type of sport context may be more likely to provide equitable opportunities for success. However, the compiled sample included a variety of sports, and each respective sport would need to be examined in detail to identify exact trends along with sport-specific factors contributing to the existence of RAEs or lack thereof. It is also important to note that several individual sport contexts were not available for examination, and competition/skill level was not considered in these analyses (i.e., a stronger risk of RAEs would be expected at elite levels of performance and may be somewhat diminished in a heterogenous sample) [23,33].

Gender (or sex, as reported in some relative age studies) has been identified as a significant moderator of RAEs in previous research, with male RAEs reported to be greater in magnitude versus the female effect [23], even when participation numbers are similar [34]. However, reported gender did not appear to modify trends observed in this longitudinal sample. This is not completely unexpected or incongruent with previous findings, given that RAEs were primarily associated with team-based sport in these analyses and likewise have been consistently identified in cross-sectional samples of both boys and girls in other team-based contexts (e.g., [23,33]). Furthermore, while the pattern of risk varies when samples are compared, RAEs are almost always present and expected to be present during the adolescent transition years, when variability in maturational differences is greatest [18,21].

An over-representation of Quarter 2-born participants (versus Quarter 4) was observed for unorganized physical activity participation. This finding may suggest that the impact of RAEs extends beyond the commonly examined context of organized sport, as talent selection processes would not be required to gain membership in these activities and engagement would be based on individual and/or parental volition. While direct comparisons for children and youth are not available in previous research for unorganized physical activity participation (Larouche et al. [71] reported findings for physical activity participation in a sample that included both adolescents and adults), this trend highlights the need to address developmental biases in both organized and unorganized contexts for this demographic given the importance of regular participation on health outcomes [12,72]. Reduced levels of cardiorespiratory fitness have been reported for the relatively youngest among children ages 9–12 years [73], and these findings may be associated with the extent to which these individuals feel confident about their ability to engage in various physical activity contexts.

An increased focus on the development of fundamental movement skills may support participation in unorganized forms of physical activity among the relatively youngest and could also support girls who were less likely to report participation in this context compared to boys. A future consideration is to examine whether basic movement skills (e.g., sprint, jump, throw, catch, balance) differ within a same-age cohort along with consideration of gender. Coincidentally, New Brunswick is one of the few Canadian provinces that require elementary schools to have a dedicated physical education specialist on staff [8], which could increase the likelihood that the development of basic movement skills is being promoted. These skills could potentially promote successful participation in a variety of physical activities across the lifespan and minimize impact of RAEs in an unorganized context [74,75]. It should also be noted that "unorganized" included a large number of activities (total of 36) in a variety of settings. Future studies should consider breaking this category down in order to identify any existing patterns in certain types of unorganized activity (e.g., group versus individual activities, domestic chores versus active transportation versus leisure pursuits).

"Non-participation" was defined as a lack of self-reported participation in an organized sport or unorganized physical activity at a minimum frequency of once per week, which is consistent with previous studies [14,60,61]. The relatively oldest participants born in Quarter 2 were less likely to report non-participation when compared to the relatively youngest born in Quarter 4. This further supports the notion that RAEs in organized contexts may potentially impact other aspects of health behavior during adolescence. Future work could examine whether this finding extends to sedentary behavior (e.g., screen time). An increased risk of non-participation was also observed for girls and with increasing chronological age. These findings are consistent with previous observations that activity levels begin to decline during adolescence (e.g., [76,77]) and should be taken into consideration when designing interventions to address non-participant behavior.

Returning to the theoretical framework of Wattie et al. [38], we have demonstrated that different constraints may be operating that impact adolescent participation in activities across a broad range of contexts. Individual constraints of chronological age and relative age continue to be present within the realm of organized sport overall, and specifically team-based activities, while relative age and gender impact less structured activities. Task-related constraints (e.g., team versus individual, organized versus unorganized) also differentiate participation. From a practical standpoint, the Wattie et al. framework and the findings of this longitudinal analysis (and future work) should be used to develop strategies to support the relatively youngest in various contexts. It may be necessary to address constraints at all levels (individual, task, and environmental) beyond the realm of organized sport. For instance, schools and/or community organizations may want to promote extracurricular opportunities by grouping students according to birth quarter to minimize the impact of age-related disparities. Children and youth could also be grouped according to their level of competency in a given activity. Furthermore, interventions are required to overcome RAEs throughout adolescence, since the relative age trends observed in this study were consistent across the eight years examined, suggesting that the relatively youngest did not overcome participation disadvantages at any point during the adolescent transition years.

The strengths of this study include that it is one of the first to provide a longitudinal examination of RAEs and it does so during the "pivotal" years of youth sport development [78]. Furthermore, the same respondents were evaluated in multiple contexts rather than in a single sport (e.g., team and individual organized sport, unorganized physical activity). The limitations of this study include those inherent in self-report questionnaires such as recall bias and social desirability bias, both of which could have resulted in misclassification and the over- or under-estimation of organized sport or unorganized physical activity participation. No objective measures of participation were utilized, and some organized sport contexts were not available for examination, or the cut-off date was unknown; however, the use of the questionnaire allowed for the investigation of several types of participation in addition to context. Furthermore, the study sample was not designed to be representative of a population, meaning that more research is needed to determine if the results of this study are generalizable to different youth populations [79].

5. Conclusions

Longitudinal analyses suggest RAEs are context dependent and persist through adolescence. The relatively oldest respondents (Quarter 1 and Quarter 2) were over-represented with respect to reported organized team sport participation, while individual sport contexts appeared to be unaffected, as evidenced by the lack of statistically significant differences in the birth distribution. Relative age inequities also appeared to be associated with participation in unorganized physical activities, with Quarter 2-born participants over-represented. Chronological age impacted organized sport participation overall and team-based contexts, while gender was observed to influence participation in unorganized physical activity and non-participation. Detailed research into specific constraints contributing to the observed patterns would be valuable.

Author Contributions: The data were collected by M.B., Principal Investigator of the MATCH study. K.L.S., M.B., L.C., J.C.D., S.H. and P.L.W. contributed to the conceptualization and design of the relative age analyses. K.L.S. prepared the data for statistical analyses, summarized the results, and drafted the manuscript with assistance from P.L.W. All authors assisted with review and editing of the manuscript. All authors have read and agreed to the published version of the manuscript.

Funding: The data used in these analyses were drawn from the Monitoring Activities of Teenagers to Comprehend their Habits (MATCH) project, which was funded by the Social Sciences and Humanities Research Council (SSHRC, 435-2016-0888 and 435-2020-0809), by the SSHRC and Sport Canada joint Sport Participation Research Initiative (862-2010-0001 and 862-2014-0002), and by the New Brunswick Health Research Foundation (20130729). Support was also received through a SSHRC Doctoral Fellowship (K.L.S.). The funding sources had no direct involvement in the study design, in the collection, analysis, or interpretation of data, in writing of the report, or in the decision to submit the article for publication.

Institutional Review Board Statement: The MATCH study was reviewed and approved by the Comité d'Éthique de la Recherche du Centre Hospitalier de l'Université de Sherbrooke.

Informed Consent Statement: All respondents and their parents or guardians provided signed informed assent and consent to participate in the MATCH study, respectively.

Data Availability Statement: The original data analyzed for this study are available through a data-sharing agreement with the MATCH study research team. More information on this may be obtained from the principal investigator of the MATCH study, Mathieu Bélanger.

Acknowledgments: The authors would like to thank Chris McCoy for assistance with the statistical analyses. The authors are also grateful to Joe Baker for his helpful comments in developing the framework for the study.

Conflicts of Interest: The authors declare no conflict of interest.

Appendix A

Table A1. Adapted from Ward et al. [61].

Physical Activities	Recognized Sport in New Brunswick	Sports Using 31 December for Age Groupings [1]		Unorganized Activities	
		Team	Individual	In Any Context	Certain Contexts [2]
Street hockey, floor hockey					X
Ice hockey	X	X			X
Ringette	X	X			X
Ice skating (not for hockey or ringette)	X				X
In-line skating					X
Skateboarding or scooter					X
Bicycling	X		X		X
Walking for exercise					X
Track and field	X		X		X
Jogging or running	X		X		X
Golfing	X				X
Swimming	X				X
Gymnastics	X				X
Aerobics, yoga, exercise class					X
Home exercise (push-ups, sits-ups)				X	
Baseball or softball	X	X			X
Weight training	X			X	
Basketball	X	X			X
Football	X				X
Soccer	X	X			X
Volleyball	X	X			X

Table A1. *Cont.*

Physical Activities	Recognized Sport in New Brunswick	Sports Using 31 December for Age Groupings [1]		Unorganized Activities	
		Team	Individual	In Any Context	Certain Contexts [2]
Badminton	X		X		X
Tennis	X		X		X
Kayak, canoe	X				X
Dance					X
Trampoline				X	
Skipping rope				X	
Handball or mini handball	X	X			X
Ball playing (dodge ball, kickball, catch)					X
Games (chase, tag, hide and seek)				X	
Downhill skiing or snowboarding	X		X		X
Boxing, wrestling	X				X
Karate, judo, tai chi, taekwondo	X				X
Cross-country skiing	X		X		X

[1] Participation reported with an "organized group or team". [2] Participation reported with "sibling(s)", "friend(s)", "parent(s)/grandparent(s)", or "by myself".

References

1. World Health Organization. Physical Activity. Fact Sheet. Available online: https://www.who.int/news-room/fact-sheets/detail/physical-activity (accessed on 30 May 2022).
2. Nader, P.R.; Bradley, R.H.; Houts, R.M.; McRitchie, S.L.; O'Brien, M. Moderate-to-Vigorous Physical Activity from Ages 9 to 15 Years. *J. Am. Med. Assoc.* **2008**, *300*, 295–305. [CrossRef] [PubMed]
3. Statistics Canada. Physical Activity and Screen Time among Canadian Children and Youth, 2016 and 2017. Health Fact Sheet. Available online: https://www150.statcan.gc.ca/n1/pub/82-625-x/2019001/article/00003-eng.htm (accessed on 30 May 2022).
4. Sember, V.; Jurak, G.; Kovač, M.; Đurić, S.; Starc, G. Decline of Physical Activity in Early Adolescence: A 3-Year Cohort Study. *PLoS ONE* **2020**, *15*, e0229305. [CrossRef] [PubMed]
5. Abu-Omar, K.; Rütten, A. Relation of Leisure Time, Occupational, Domestic, and Commuting Physical Activity to Health Indicators in Europe. *Prev. Med.* **2008**, *47*, 319–323. [CrossRef] [PubMed]
6. Bélanger, M.; Townsend, N.; Foster, C. Age-Related Differences in Physical Activity Profiles of English Adults. *Prev. Med.* **2011**, *52*, 247–249. [CrossRef]
7. Eime, R.M.; Harvey, J.T.; Sawyer, N.A.; Craike, M.J.; Symons, C.M.; Polman, R.C.J.; Payne, W.R. Understanding the Contexts of Adolescent Female Participation in Sport and Physical Activity. *Res. Q. Exerc. Sport* **2013**, *84*, 157–166. [CrossRef]
8. Participaction. The Biggest Risk Is Keeping Kids Indoors. The 2015 ParticipACTION Report Card on Physical Activity for Children and Youth. Available online: https://participaction.cdn.prismic.io/participaction%2F61cf55e8-c1c0-42c7-ba6b-1480fd2c29b9_participaction-2015-report-card-full.pdf (accessed on 27 August 2015).
9. Chaput, J.-P.; Carson, V.; Gray, C.; Tremblay, M. Importance of All Movement Behaviors in a 24 Hour Period for Overall Health. *Int. J. Environ. Res. Public. Health* **2014**, *11*, 12575–12581. [CrossRef]
10. Eime, R.M.; Young, J.A.; Harvey, J.T.; Charity, M.J.; Payne, W.R. A Systematic Review of the Psychological and Social Benefits of Participation in Sport for Children and Adolescents: Informing Development of a Conceptual Model of Health through Sport. *Int. J. Behav. Nutr. Phys. Act.* **2013**, *10*, 98. [CrossRef]
11. Fraser-Thomas, J.; Côté, J. Understanding Adolescents' Positive and Negative Developmental Experiences in Sport. *Sport Psychol.* **2009**, *23*, 3–23. [CrossRef]
12. Janssen, I.; LeBlanc, A.G. Systematic Review of the Health Benefits of Physical Activity and Fitness in School-Aged Children and Youth. *Int. J. Behav. Nutr. Phys. Act.* **2010**, *7*, 40–55. [CrossRef]
13. Sidlauskiene, A.; Strukcinskiene, B.; Raistenskis, J.; Stukas, R.; Strukcinskaite, V.; Buckus, R. The Association between the Level of Physical Activity with Spinal Posture and Physical Fitness Parameters in Early Adolescence. *Vojnosanit. Pregl.* **2019**, *76*, 1209–1216. [CrossRef]
14. Bengoechea, E.G.; Sabiston, C.M.; Ahmed, R.; Farnoush, M. Exploring Links to Unorganized and Organized Physical Activity During Adolescence: The Role of Gender, Socioeconomic Status, Weight Status, and Enjoyment of Physical Education. *Res. Q. Exerc. Sport* **2010**, *81*, 7–16. [CrossRef] [PubMed]
15. Hardy, L.L.; O'Hara, B.J.; Rogers, K.; St. George, A.; Bauman, A. Contribution of Organized and Nonorganized Activity to Children's Motor Skills and Fitness. *J. Sch. Health* **2014**, *84*, 690–696. [CrossRef] [PubMed]
16. Kjønniksen, L.; Torsheim, T.; Wold, B. Tracking of Leisure-Time Physical Activity during Adolescence and Young Adulthood: A 10-Year Longitudinal Study. *Int. J. Behav. Nutr. Phys. Act.* **2008**, *5*, 69. [CrossRef]

17. Barnsley, R.H.; Thompson, A.H.; Barnsley, P.E. Hockey Success and Birth Date: The Relative Age Effect. *Can. Assoc. Health Phys. Educ. Recreat.* **1985**, *51*, 23–28.
18. Musch, J.; Grondin, S. Unequal Competition as an Impediment to Personal Development: A Review of the Relative Age Effect in Sport. *Dev. Rev.* **2001**, *21*, 147–167. [CrossRef]
19. Baker, J.; Schorer, J.; Cobley, S. Relative Age Effects: An Inevitable Consequence of Elite Sport? *Sportwissenschaft* **2010**, *40*, 26–30. [CrossRef]
20. Helsen, W.F.; Starkes, J.L.; Van Winckel, J. The Influence of Relative Age on Success and Dropout in Male Soccer Players. *Am. J. Hum. Biol.* **1998**, *10*, 791–798. [CrossRef]
21. Malina, R.M.; Bouchard, C.; Bar-Or, O. Maturity-Associated Variation in Growth and Performance. In *Growth, Maturation, and Physical Activity*; Malina, R.M., Bouchard, C., Bar-Or, O., Eds.; Human Kinetics: Champaign, IL, USA, 2004; pp. 337–365.
22. Wattie, N.; Cobley, S.; Baker, J. Towards a Unified Understanding of Relative Age Effects. *J. Sports Sci.* **2008**, *26*, 1403–1409. [CrossRef]
23. Cobley, S.; Baker, J.; Wattie, N.; McKenna, J. Annual Age-Grouping and Athlete Development: A Meta-Analytical Review of Relative Age Effects in Sport. *Sports Med.* **2009**, *39*, 235–256. [CrossRef]
24. Edgar, S.; O'Donoghue, P. Season of Birth Distribution of Elite Tennis Players. *J. Sport Sci.* **2005**, *23*, 1013–1020. [CrossRef]
25. Baker, J.; Janning, C.; Wong, H.; Cobley, S.; Schorer, J. Variations in Relative Age Effects in Individual Sports: Skiing, Figure Skating and Gymnastics. *Eur. J. Sport Sci.* **2014**, *14*, S183–S190. [CrossRef] [PubMed]
26. Romann, M.; Cobley, S. Relative Age Effects in Athletic Sprinting and Corrective Adjustments as a Solution for Their Removal. *PLoS ONE* **2015**, *10*, e0122988. [CrossRef]
27. Saavedra-García, M.; Gutiérrez-Aguilar, Ó.; Sa-Marques, P.; Fernández-Romero, J.J. Relative Age Effect in Spanish Athletics. *Cuad. Psicol. Deporte* **2016**, *16*, 275–286. (In Spanish)
28. Delorme, N.; Chalabaev, A.; Raspaud, M. Relative Age Is Associated with Sport Dropout: Evidence from Youth Categories of French Basketball. *Scand. J. Med. Sci. Sports* **2011**, *21*, 120–128. [CrossRef]
29. Lemez, S.; Baker, J.; Horton, S.; Wattie, N.; Weir, P. Examining the Relationship between Relative Age, Competition Level, and Dropout Rates in Male Youth Ice-Hockey Players. *Scand. J. Med. Sci. Sports* **2014**, *24*, 935–942. [CrossRef]
30. Cobley, S.P.; Till, K. Participation Trends According to Relative Age across Youth UK Rugby League. *Int. J. Sports Sci. Coach.* **2017**, *12*, 339–343. [CrossRef]
31. Mota, J.; Esculcas, C. Leisure-Time Physical Activity Behavior: Structured and Unstructured Choices According to Sex, Age, and Level of Physical Activity. *Int. J. Behav. Med.* **2002**, *9*, 111–121. [CrossRef]
32. Delorme, N.; Radel, R.; Raspaud, M. Relative Age Effect and Soccer Refereeing: A 'Strategic Adaptation' of Relatively Younger Children? *Eur. J. Sport Sci.* **2013**, *13*, 400–406. [CrossRef]
33. Smith, K.L.; Weir, P.L.; Till, K.; Romann, M.; Cobley, S. Relative Age Effects Across and Within Female Sport Contexts: A Systematic Review and Meta-Analysis. *Sports Med.* **2018**, *48*, 1451–1478. [CrossRef]
34. Baker, J.; Schorer, J.; Cobley, S.; Bräutigam, H.; Büsch, D. Gender, Depth of Competition and Relative Age Effects in Team Sports. *Asian J. Exerc. Sport Sci.* **2009**, *6*, 1–7.
35. Lovell, R.; Towlson, C.; Parkin, G.; Portas, M.; Vaeyens, R.; Cobley, S. Soccer Player Characteristics in English Lower-League Development Programmes: The Relationships between Relative Age, Maturation, Anthropometry and Physical Fitness. *PLoS ONE* **2015**, *10*, e0137238. [CrossRef] [PubMed]
36. Tanner, J.M.; Whitehouse, R.H. Clinical Longitudinal Standards for Height, Weight, Height Velocity, Weight Velocity, and Stages of Puberty. *Arch. Dis. Child.* **1976**, *51*, 170–179. [CrossRef]
37. Hancock, D.J.; Adler, A.L.; Côté, J. A Proposed Theoretical Model to Explain Relative Age Effects in Sport. *Eur. J. Sport Sci.* **2013**, *13*, 630–637. [CrossRef] [PubMed]
38. Wattie, N.; Schorer, J.; Baker, J. The Relative Age Effect in Sport: A Developmental Systems Model. *Sports Med.* **2015**, *45*, 83–94. [CrossRef]
39. Delorme, N.; Raspaud, M. The Relative Age Effect in Young French Basketball Players: A Study on the Whole Population. *Scand. J. Med. Sci. Sports* **2009**, *19*, 235–242. [CrossRef]
40. Baker, J.; Schorer, J.; Cobley, S.; Schimmer, G.; Wattie, N. Circumstantial Development and Athletic Excellence: The Role of Date of Birth and Birthplace. *Eur. J. Sport Sci.* **2009**, *9*, 329–339. [CrossRef]
41. Smith, K.L.; Weir, P.L. An Examination of the Relative Age Effect in Developmental Girls' Hockey in Ontario. *High Abil. Stud.* **2013**, *24*, 171–184. [CrossRef]
42. Wattie, N.; Tietjens, M.; Cobley, S.; Schorer, J.; Baker, J.; Kurz, D. Relative Age-Related Participation and Dropout Trends in German Youth Sports Clubs. *Eur. J. Sport Sci.* **2014**, *14*, S213–S220. [CrossRef]
43. Chittle, L.; Horton, S.; Weir, P.; Dixon, J.C. Investigating the Relationship between the Relative Age Effect and Leadership Behaviors among Male Ice Hockey Players. *Int. Rev. Sociol. Sport* **2017**, *52*, 751–768. [CrossRef]
44. Cobley, S.; Abraham, C.; Baker, J. Relative Age Effects on Physical Education Attainment and School Sport Representation. *Phys. Educ. Sport Pedagogy* **2008**, *13*, 267–276. [CrossRef]
45. Dhuey, E.; Lipscomb, S. What Makes a Leader? Relative Age and High School Leadership. *Econ. Educ. Rev.* **2008**, *27*, 173–183. [CrossRef]

46. Bedard, K.; Dhuey, E. The Persistence of Early Childhood Maturity: International Evidence of Long-Run Age Effects. *Q. J. Econ.* **2006**, *121*, 1437–1472. [CrossRef]
47. Thompson, A.H.; Barnsley, R.H.; Battle, J. The Relative Age Effect and the Development of Self-Esteem. *Educ. Res.* **2004**, *46*, 313–320. [CrossRef]
48. MacDonald, D.J.; Baker, J. Circumstantial Development: Birthdate and Birthplace Effects on Athlete Development. In *Conditions of Children's Talent Development in Sport*; Côté, J., Lidor, R., Eds.; Fitness Information Technology: Morgantown, WV, USA, 2013; pp. 197–208.
49. Cox, A.E.; Smith, A.L.; Williams, L. Change in Physical Education Motivation and Physical Activity Behavior during Middle School. *J. Adolesc. Health* **2008**, *43*, 506–513. [CrossRef]
50. McDavid, L.; Cox, A.E.; McDonough, M.H. Need Fulfillment and Motivation in Physical Education Predict Trajectories of Change in Leisure-Time Physical Activity in Early Adolescence. *Psychol. Sport Exerc.* **2014**, *15*, 471–480. [CrossRef]
51. Bélanger, M.; Caissie, I.; Beauchamp, J.; O'Loughlin, J.; Sabiston, C.; Mancuso, M. Monitoring Activities of Teenagers to Comprehend Their Habits: Study Protocol for a Mixed-Methods Cohort Study. *BMC Public Health* **2013**, *13*, 649. [CrossRef] [PubMed]
52. Trost, S.G.; Pate, R.R.; Sallis, J.F.; Freedson, P.S.; Taylor, W.C.; Dowda, M.; Sirard, J. Age and Gender Differences in Objectively Measured Physical Activity in Youth. *Med. Sci. Sports Exerc.* **2002**, *34*, 350–355. [CrossRef]
53. Bélanger, M.; Gray-Donald, K.; O'Loughlin, J.; Paradis, G.; Hanley, J. Influence of Weather Conditions and Season on Physical Activity in Adolescents. *Ann. Epidemiol.* **2009**, *19*, 180–186. [CrossRef]
54. Craig, C.; Cameron, C.; Russell, S.; Beaulieu, A. Increasing Physical Activity Participation: Supporting Children's Participation. *Can. Fit. Lifestyle Res. Inst.* **2001**.
55. Crocker, P.R.; Bailey, D.A.; Faulkner, R.A.; Kowalski, K.C.; McGrath, R. Measuring General Levels of Physical Activity: Preliminary Evidence for the Physical Activity Questionnaire for Older Children. *Med. Sci. Sports Exerc.* **1997**, *29*, 1344–1349. [CrossRef]
56. Janz, K.F.; Lutuchy, E.M.; Wenthe, P.; Levy, S.M. Measuring Activity in Children and Adolescents Using Self-Report: PAQ-C and PAQ-A. *Med. Sci. Sports Exerc.* **2008**, *40*, 767–772. [CrossRef] [PubMed]
57. Sallis, J.F.; Condon, S.A.; Goggin, K.J.; Roby, J.J.; Kolody, B.; Alcaraz, J.E. The Development of Self-Administered Physical Activity Surveys for 4th Grade Students. *Res. Q. Exerc. Sport* **1993**, *64*, 25–31. [CrossRef] [PubMed]
58. Goguen Carpenter, J.; Bélanger, M.; O'Loughlin, J.; Xhignesse, M.; Ward, S.; Caissie, I.; Sabiston, C. Association between Physical Activity Motives and Type of Physical Activity in Children. *Int. J. Sport Exerc. Psychol.* **2017**, *15*, 306–320. [CrossRef]
59. Pate, R.R.; Sallis, J.F.; Ward, D.S.; Stevens, J.; Dowda, M.; Welk, G.J.; Young, D.R.; Jobe, J.B.; Strikmiller, P.K. Age-Related Changes in Types and Contexts of Physical Activity in Middle School Girls. *Am. J. Prev. Med.* **2010**, *39*, 433–439. [CrossRef]
60. Vella, S.A.; Cliff, D.P.; Okely, A.D. Socio-Ecological Predictors of Participation and Dropout in Organised Sports during Childhood. *Int. J. Behav. Nutr. Phys. Act.* **2014**, *11*, 62. [CrossRef]
61. Ward, S.; Bélanger, M.; Donovan, D.; Caissie, I.; Goguen, J.; Vanasse, A. Association Between School Policies and Built Environment, and Youth's Participation in Various Types of Physical Activities. *J. Sch. Health* **2015**, *85*, 423–432. [CrossRef]
62. Government of New Brunswick. Education and Early Childhood Development: Kindergarten Registration. Available online: https://www2.gnb.ca/content/gnb/en/departments/education/k12/content/kindergarten.html (accessed on 31 October 2016).
63. Baker, J.; Logan, A.J. Developmental Contexts and Sporting Success: Birth Date and Birthplace Effects in National Hockey League Draftees 2000-2005. *Br. J. Sports Med.* **2007**, *41*, 515–517. [CrossRef]
64. Wattie, N.; Baker, J.; Cobley, S.; Montelpare, W.J. A Historical Examination of Relative Age Effects in Canadian Hockey Players. *Int. J. Sport Psychol.* **2007**, *38*, 178–186.
65. Weir, P.L.; Smith, K.L.; Paterson, C.; Horton, S. Canadian Women's Ice Hockey—Evidence of a Relative Age Effect. *Talent Dev. Excell.* **2010**, *2*, 209–217.
66. Ste-Marie, D.M.; Starkes, J.L.; Cronin, A.; Fletcher, H. The Relative Age Effect: A "Flip-Flop" Phenomenon in Gymnasts. In Proceedings of the Canadian Society for Psychomotor Learning and Sport Psychology Conference, Waterloo, ON, Canada, 26–29 October 2000.
67. MacKenzie, J.; Brunet, J.; Boudreau, J.; Iancu, H.-D.; Bélanger, M. Does Proximity to Physical Activity Infrastructures Predict Maintenance of Organized and Unorganized Physical Activities in Youth? *Prev. Med. Rep.* **2015**, *2*, 777–782. [CrossRef]
68. Murray, C. *Human Accomplishment: The Pursuit of Excellence in the Arts and Sciences. 800 B.C. to 1950*; Harper Collins: New York, NY, USA, 2003.
69. Côté, J.; Macdonald, D.J.; Baker, J.; Abernethy, B. When "Where" Is More Important than "When": Birthplace and Birthdate Effects on the Achievement of Sporting Expertise. *J. Sports Sci.* **2006**, *24*, 1065–1073. [CrossRef] [PubMed]
70. Delorme, N.; Raspaud, M. Is There an Influence of Relative Age on Participation in Non-Physical Sports Activities? The Example of Shooting Sports. *J. Sports Sci.* **2009**, *27*, 1035–1042. [CrossRef] [PubMed]
71. Larouche, R.; Laurencelle, L.; Grondin, S.; Trudeau, F. Influence of Birth Quarter on the Rate of Physical Activities and Sports Participation. *J. Sports Sci.* **2010**, *28*, 627–631. [CrossRef] [PubMed]
72. Warburton, D.E.R.; Nicol, C.W.; Bredin, S.S.D. Health Benefits of Physical Activity: The Evidence. *Can. Med. Assoc. J.* **2006**, *174*, 801–809. [CrossRef]
73. Roberts, S.J.; Boddy, L.M.; Fairclough, S.J.; Stratton, G. The Influence of Relative Age Effects on the Cardiorespiratory Fitness Levels of Children Age 9 to 10 and 11 to 12 Years of Age. *Pediatr. Exerc. Sci.* **2012**, *24*, 72–83. [CrossRef]

74. Birch, S.; Cummings, L.; Oxford, S.W.; Duncan, M.J. Examining Relative Age Effects in Fundamental Skill Proficiency in British Children Aged 6–11 Years. *J. Strength Cond. Res.* **2016**, *30*, 2809–2815. [CrossRef]
75. Higgs, C.; Balyi, I.; Way, R.; Cardinal, C.; Norris, S.; Bluechardt, M. *Developing Physical Literacy: A Guide for Parents of Children Ages 0 to 12*; Canadian Sport Centres: Vancouver, BC, Canada, 2008.
76. Allison, K.R.; Adlaf, E.M.; Dwyer, J.J.M.; Lysy, D.C.; Irving, H.M. The Decline in Physical Activity Among Adolescent Students: A Cross-National Comparison. *Can. J. Public Health.* **2007**, *98*, 97–100. [CrossRef]
77. Van Mechelen, W.; Twisk, J.W.R.; Post, G.B.; Snel, J.; Kemper, H.C.G. Physical Activity of Young People: The Amsterdam Longitudinal Growth and Health Study. *Med. Sci. Sports Exerc.* **2000**, *32*, 1610–1616. [CrossRef]
78. Malina, R.M. The Young Athlete: Biological Growth and Maturation in a Biocultural Context. In *Children and Youth in Sport: A Biopsychosocial Perspective*; Smoll, F.L., Smith, R.E., Eds.; Brown & Benchmark: Dubuque, IA, USA, 1996; pp. 161–186.
79. Lidor, R.; Côté, J.; Arnon, M.; Zeev, A.; Cohen-Maoz, S. Relative Age and Birthplace Effects in Division 1 Players—Do They Exist in a Small Country? *Talent Dev. Excell.* **2010**, *2*, 181–192.

sports MDPI

Article

Tough Love—Impactful, Caring Coaching in Psychologically Unsafe Environments

Jamie Taylor [1,2,3,*], Michael Ashford [1,4] and Dave Collins [1,3]

1 Grey Matters Performance Ltd., Stratford upon Avon CV37 9TQ, UK; mike@greymattersuk.com (M.A.); dave@greymattersuk.com (D.C.)
2 Faculty of Science and Health, School of Health and Human Performance, Dublin City University, Glasnevin, D09 NA55 Dublin, Ireland
3 Moray House School of Education and Sport, The University of Edinburgh, Holyrood Campus, Edinburgh EH8 8AQ, UK
4 Faculty of Health and Life Sciences, Coventry University, Coventry CV1 5FB, UK
* Correspondence: jamie@greymattersuk.com

Abstract: (1) Background: The interpersonal dimensions of coaching in high performance sport have been subject to increasing scrutiny but with limited evidence to guide practice. Similarly, there is increasing practical interest in the concept of psychological safety, often portrayed as an implicitly desirable characteristic of all sporting environments but, as yet, still to receive research attention in high performance. As a first step to addressing these deficiencies, the present study addressed two research aims: (a) to examine the extent to which matched groups of international and released professional rugby union players perceived psychological safety to be an adaptive feature of their developmental experience and (b) to understand what elements of the player's coaching experience were perceived to be enabling or disenabling of future progress. (2) Methods: Seven rugby union players who had 'made it' and eight players who had been released from their professional contracts took part in a semi-structured interview exploring their developmental experiences. Data were subsequently analysed using Reflexive Thematic Analysis. (3) Results: Both groups of players found each of their talent development and high performance environments to be psychologically unsafe. Furthermore, players perceived coaches who were the most impactful in their development as offering 'tough love'. This included a range of 'harder' and 'softer' interpersonal approaches that presented the player with clear direction, role clarity and a sense of care. It appeared that this interpersonal approach helped the player to navigate, and benefit from, the psychologically unsafe high performance milieu. (4) Conclusions: There appear to be a number of balances for the coach in the high performance setting to navigate and a need for more nuance in applying constructs such as psychological safety.

Keywords: psychological safety; care; elite performance; talent development; challenge

Citation: Taylor, J.; Ashford, M.; Collins, D. Tough Love—Impactful, Caring Coaching in Psychologically Unsafe Environments. *Sports* **2022**, *10*, 83. https://doi.org/10.3390/sports10060083

Academic Editors: Beat Knechtle, Adam Leigh Kelly, Sergio L. Jiménez Sáiz, Sara Diana Leal dos Santos and Alberto Lorenzo Calvo

Received: 20 April 2022
Accepted: 20 May 2022
Published: 25 May 2022

1. Introduction

In recent times there has been a significant interest in the development of talent towards high performance (HP) sport [1]. This has generated a significant volume of publications from across contexts [2] and enormous sums of money spent across the world to promote athlete development. Despite this interest, however, there remains a paucity of research that is 'for' the coach, rather than 'of' coaching [3,4]. In essence, there is limited research that offers the 'granularity' [5] necessary for coaches and practitioners to make truly evidence-informed decisions [6]. This may be one of the reasons why research appears not to have made significant impact on applied practice, where a wide variety of different, even contradictory, implicitly and explicitly held constructs influence coaching [7]. Specifically in HP sport, there is a limited body of literature that has investigated truly 'elite' performers [8]. Similarly, there has been a failure to distinguish between levels of performance and understanding further development when athletes do reach the elite level [9].

1.1. Psychological Safety

In the applied domain, there is growing interest in a construct with a long history in the organisational literature, psychological safety (PS) [10]. Although there are multiple definitions in use, the concept was initially defined by Edmondson [11] as: 'a shared belief by members of a team that the team is safe for interpersonal risk taking', underpinned by a number of scale items, including: the lack of 'fear of mistakes being held against you', 'it is safe to take a risk on this team' and 'working with members of this team, my unique skills and talents are valued and utilised' [11]. In the most recent definition, Edmondson suggests a dual effect: 'in psychologically safe environments, people believe that if they make a mistake or ask for help, others will not react badly' [12]. Perhaps in recognition that there are issues with the application of Edmondson's definition in sport, a recent paper redefined PS as: 'the perception that one is protected from, or unlikely to be at risk of, psychological harm in sport', with harm including perceptions of threat or fear [13].

Several issues with the application of the construct in HP have been raised. As highlighted in a recent critical review, even by redefining the construct, it is difficult to see how it can be applied to athletes in HP [14]. At the very least, it has been suggested that there is a need to apply PS in a critical manner, understanding potential downsides such as lower overall performance without accountability, or even unethical behaviour [15,16]. Similarly, there is a need to recognise the realities of the HP milieu, one in which there will nearly always be consequences for poor performance [17,18] that may challenge one's implicit safety.

Notably, despite a wave of interest in the applied domain, there remains very limited empirical investigation of the construct in sport. For instance, current examples involve high level university team sport athletes [19] and amateur rugby [20]. There is even less at higher levels of performance, for example NCAA basketball [21] and, notably, given the complex and multidimensional nature of the construct, no qualitative data that we are aware of.

1.2. Challenging Coaching

This lack of empirical investigation is notable, as one of the most robust findings in talent development (TD) research is the central role of challenge dynamics in promoting elite performance [22–26]. An example of these dynamics is the impact of early advantage in talent pathways, with athletes experiencing less challenge earlier in the pathway being more likely to be deselected as the challenge level increases [27,28]. As a result, it has been proposed that athletes can benefit from a range of affective conditions through development, as they experience a range of positive and negative experiences, so long as they can deploy the requisite psycho-behavioural skills to learn from their experiences [29,30]. Therefore, in the HP domain, it appears that a number of situations have the potential to be particularly emotionally disturbing and therefore a useful means to provoke further development [30]. Given that in nearly all cases, the coach will be responsible for a significant number of emotionally disturbing athlete experiences, whether they be positive or negative (e.g., selection), the coach has a central role to play in the wider development of the athlete, which may usefully provoke perceptions of an unsafe environment.

A number of conceptual framings of coaching practice have been suggested to understand the characteristics of effective TD environments, or 'all aspects of the coaching situation' [31]. For the coach seeking to take advantage of these dynamics, several features of TD have the potential to be impactful for the athlete including coherence of experience [32], individualisation [33] and the quality of messages received and sent by the athlete [34]. Yet, despite understanding what characterises effective coaching practice, limited literature has actively considered what the coach actually does. For example, the work of Collins and colleagues [23] suggested that the TD coaching of 'super champion' athletes was characterised by a blend of high challenge, facilitative and relatively non-directive input, in contrast to less successful athletes where the coach drove the agenda.

There is a similar limited volume of empirical work considering the role of the HP coach, particularly as regards the enhancement of performance. As an example, the work of Lara-Bercial and Mallett [35] found the need for coaches to enforce a culture of high expectations and standards, producing an inherently challenging environment. They also introduced the term 'driven benevolence' to describe an approach where coaches cared for others and maintained optimal development through pressure. As such, there appear to be a number of interpersonal dimensions that are particularly pertinent to the coaches of current or future high performers. These include the need for setting high expectations, ensuring that athletes have role clarity [36] and doing this with a level of emotional 'elasticity' to make and communicate tough decisions, albeit in a considerate manner [35].

1.3. Care

Elsewhere there has been a growth in the literature examining the role of care in coaching. Whilst seen as a core feature of coaching practice, it is often portrayed as an addition to performance or development [37]. In contrast, the recent literature has drawn on the work of Noddings' theory of care in pedagogical relationships [38,39] to suggest a dual strand of care, where the coach should care 'for' and 'about' the athlete. To 'care for' an athlete, coaches express 'devotion and desire' to meet the needs and wants of that individual [40,41]. Such devotion is characterised by a coach's engrossment, motivational displacement and reciprocity in the coaching relationship. Engrossment refers to the sustained attention of the coach in relation to the athlete's needs and wants, consistently offering high levels of empathetic concern. Motivational displacement captures the behaviour of a coach who aims to meet the athlete's needs even if it supersedes that of their own. Finally, reciprocity implies that the care offered to a player can only be confirmed when the player receives and acknowledges it. In contrast, caring about an athlete describes a coach's emotional concern for an individual, without the need for engrossment, motivational displacement nor a reciprocal interaction.

Interestingly, others have highlighted limitations to this gentle view [41,42], suggesting that the nature of the HP milieu provides individuals with a social context where coaches may care for athletes through interactions that are harder in nature. For instance, Cronin and colleagues [40] explored the role of care within an English Premiership football environment, where the milieu was characterised as cut-throat, competitive and volatile [43,44]. They offer the conclusion that care is inherently contextual and situated within a social milieu [40]. It is this view that sits in contrast to views of dichotomous perspectives of coaching practice, particularly in HP, that have suggested a separation of person and performer, with an emphasis on the former. This false dichotomy encourages the view that care in coaching is more 'caring about', than 'caring for' an athlete. If the latter is about meeting the wants and needs of the individual, particularly in HP populations, it is likely that it will be expressed differently to a participant with different motivations [45]. Worthy of note is the idea that high performers, or ambitious developing athletes, may want or require a more robust coaching approach, which is already established in the literature [30]. Yet, confusingly, it has recently been suggested that the concept of 'tough love' may be used as a euphemism for abusive coaching practice [46]. This seems at odds with the suggestion that care should be understood within particular social circumstances; that those aspiring to an elite level of performance may require a different approach from their coaches, and the often-used idea that coaching requires holding people to high standards. Reflecting these apparent conflicts, there is a need to understand what athletes perceive to be truly developmental coaching.

As a result of the dynamics presented in the literature and the lack of overall empirical investigation, as practicing coaches and coach developers working in TD, we felt a need to investigate the dynamics involved in development and the role of the coach in this process. Thus, we sought to understand the experiences and perceptions of athletes who had progressed through a TD pathway, reaching or failing to achieve senior elite status. Given the need to understand what was considered effective and ineffective, two matched groups of successful and unsuccessful athletes were considered appropriate as a means of understanding different experiences. This takes account of calls for comparison groups in talent research [2] and the value of using negative case studies, rather than risking survivorship bias in successful samples [47]. Our ultimate purpose being to inform the practice of coaches working with athletes in the TD or HP setting. As a result, the aims of this study were: (a) to examine the extent to which two matched groups of international and released professional rugby union players perceived PS to be an adaptive feature of their developmental experience and (b) to understand what elements of the player's coaching experience were perceived to be enabling or disenabling of future progress.

2. Methods

2.1. Research Philosophy, Design and Methods

Given the aims of this study and our wish to critically explore the practical utility of popular ideas and concepts in HP sport, a pragmatic research philosophy was employed [48]. The pragmatic approach encourages the use of methods that provide findings and implications that are practically meaningful, without need to adhere to a specific epistemological view [49]. Thus, the comparison of player's perceptions of PS and their coaching experiences, between those who 'made it' and those who were released provides a pertinent investigation for anyone interested in the elite sport context.

In line with our pragmatic approach, qualitative research methods allow for a deep examination of the player group's developmental experiences [50]. Qualitative research methods allow authors to collect rich descriptive data, with an aim of producing a useful interpretation of a practical problem, rather than one that is absolute [51]. Furthermore, a pragmatic research philosophy encourages the consideration of biases and preferences to make sense of findings. Reflecting these considerations, it is important to note that this study was aided by our experience as active practitioners within HP sport and specifically in rugby union [52,53].

2.2. Participants

Two purposefully sampled groups of male rugby union players were recruited against distinct inclusion and exclusion criteria. In both cohorts, players were matched by the criteria of: (a) progressing through the English Premiership academy system between 2012 and 2018, (b) had represented their country at junior international level at either U18 and/or U20 level and (c) had signed a professional contract at a senior elite team in the English Premiership (the highest level of performance nationally). As a point of difference, the first group were players whose career status was defined as 'made it'. All had progressed through the domestic game and subsequently been selected to play for their country at senior international level ($n = 7$; Age, $M = 22.14$). Importantly, for perspective on the career status of the player, interviews took place within 6 months of players entering an international camp or playing for their country for the first time. The second group were players whose career status was defined as those who were 'released'. Those who, despite matching the first cohort as having progressed through the academy system, played junior international rugby and signed a professional contract, had been released from their professional contract ($n = 8$; Age $M = 22.75$; See Table 1 for participant career status). The comparison between these two groups allowed us to explore the developmental experiences and perceptions of coaching along similar pathways that resulted in divergent outcomes. Finally, all players were recruited to take part through personal contact and, following the protocol approval by the University Ethics Committee, completed informed consent.

Table 1. Participant career status.

	Career Status at Time of Interview	International Status
Player 1 (S1)	Professional player at Premiership club	U18/U20/Senior
Player 2 (S2)	Professional player at Premiership club	U20/Senior
Player 3 (S3)	Professional player at Premiership club	U20/Senior
Player 4 (S4)	Professional player at Premiership club	U18/U20/Senior
Player 5 (S5)	Professional player at Premiership club	U18/U20/Senior
Player 6 (S6)	Professional player at Premiership club	U18/U20/Senior
Player 7 (S7)	Professional player at Premiership club	U18/U20/Senior
Player 8 (R1)	Released	U18
Player 9 (R2)	Released	U18/U20
Player 10 (R3)	Released	U18
Player 11 (R4)	Released	U18/U20
Player 12 (R5)	Released	U20
Player 13 (R6)	Released	U18/U20
Player 14 (R7)	Released	U18/20
Player 15 (R8)	Released	U20

2.3. Data Collection

Both groups of players were invited to participate in two stages of data collection, both of which took place within a wider semi-structured interview. A semi-structured interview guide was developed and refined through a pilot interview with former professional players (n = 2) with similar profiles to the released population, minus experience of junior international rugby. These pilot studies led to subtle refinements of the semi-structured interview guide, including clearer guidance regarding the use of graphic timeline and the refinement of questions to enhance clarity. The interview consisted of open-ended questions that elicited responses informed by appropriate literature whilst follow-up probes and prompts were planned for and used to allow expansion on key points [54]. Interviews were conducted by both the first (n = 10) and second (n = 5) authors. Rather than for the purpose of data analysis and instead as a means to enhance recall and generate rich dialogue, the first stage of the interview asked players to sketch out their playing journey in rugby union on a timeline, highlighting key moments and transitions for the player along the Y axis [55]. Players were asked to score their relative development in terms of progress (+5) and stagnation (−5) along the X axis. Following this initial stage, the interview then asked players to go through the timeline again, this time focusing more deeply on specific perceived critical periods and, importantly, the time between perceived critical events. The second stage of the interview included questions regarding players' perceptions of PS, as per definition of Edmondson [11] and coaching experiences within each age/stage of their pathway at academy club, junior international, senior club, and for the made it group, senior international levels. The interview guide is available on request from the first author. Abiding by regulations put in place by the University Ethics Committee to mitigate against the risk of COVID-19, interviews were arranged over video-conferencing software (Zoom Video Communications, California, Version 5.7) at a time and date that suited the participant. Prior to this, a pre-briefing allowed them to reflect on the timeline task and interview questions ahead of the interview. The interviews lasted between 60 and 105 min (M = 79 min) and were recorded for subsequent analysis.

2.4. Data Analysis

All interviews were transcribed verbatim and subsequently checked for accuracy against audio recordings. Data were then analysed using a Reflexive Thematic Analysis (TA) approach [56], using QSR NVivo Version 12 software. Against our pragmatic orientation, TA was chosen given the need to examine patterns of shared meaning across the player cohort [57]. Indeed, a core feature of TA is the recognition that the researcher plays a key role in the process of generating themes through engagement with the data. This allowed for deep reflexive engagement between researcher, data and theory (Braun and Clarke, 2019). Analysis was conducted by the first author, utilising each of the six phases initially outlined by Braun and Clarke [58]. Importantly, this took place flexibly, with appropriate non-linear movement between phases [56]. At the first stage, the first author became familiar with the content, highlighting and annotating areas of interest. Second, coding was conducted on a surface (semantic) level, before capturing the assumptions that underpin surface meaning through multiple sweeps of analysis [57]. Third, initial themes were generated, organised and captured from the initial coding process. At the fourth stage, the second author, acting as a critical friend, supported the review and refinement of themes to quality check if they were 'coherent, consistent and distinctive' [59]. The fifth phase involved defining and naming themes based on attribution of shared meaning. The final stage was the write up and report of data [57].

2.5. Trustworthiness

Several measures were taken to ensure trustworthiness in our approach. First, member reflections were solicited by email following completion of the six-phase TA process [60]. This involved all participants being contacted and sent a tabulated form of the final themes through the TA to seek their reflections on generated themes. In addition, participants were asked if the themes reflected their own experiences and if they had any further comments or considerations. Nearly all participants chose to take part in member reflections (n = 14 from 15) and their additional reflections have been incorporated into the results section. Players were universally supportive of the concept of 'tough love' (as defined by this study), and something they perceived as crucially important to their overall development. One way in which member reflections deepened overall analysis was the emphasis that players put on the technical and tactical competence of the coach, in addition to the interpersonal and pedagogic dimensions of their practice. As an example, player 9 commented: 'as regards to accountability, I am all for that and with coaches who really care about my development. But it is so important that they hold you accountable for the right stuff, stuff that would have genuinely improved me'.

In addition, during data collection, the first author kept a reflexive journal reflecting on key differences and similarities between players' perceptions of PS and coaching experiences and key areas of interest in line with the research questions. This journal was also used as an audit trail, to critically consider the methodological approach and support the initial generation of codes [61]. Finally, to ensure resonance in our approach, the second author, who is an experienced qualitative researcher, acted as a critical friend throughout the process [59].

3. Results

The purpose of the study was two-fold. Firstly, to understand the perceptions of both groups of players to understand the extent to which perceptions of PS were an effective and clear feature of their development experience. Secondly, to explore what elements of the player's coaching experience were perceived to be enabling or disenabling of future progress. The first part of the results section addresses the first of these research questions, with two developed themes. The first developed theme concerns the extent to which player's perceived it to be possible that the HP rugby union environment had the potential to offer them a sense of PS. The second generated theme presents the adaptive and maladaptive consequences of these perceptions.

3.1. A Lack of Psychological Safety

Across both groups of players, a lack of PS as a near constant factor in their professional career was reported. Players described this lack of safety as being driven by the judgement inherent to the HP milieu, making them subject to consistent scrutiny and criticism. Yet, for this group, the core feature of their experience that appeared to drive perceptions of a lack of safety was the judgement conferred by selection.

> There's no comfort, no safety, if you don't work hard or you perform poorly, there's eight other guys that are going to happily step in your shoes. It can be

> exhausting, but it definitely helps. It's one of those places where you bring your gumshield to breakfast, you don't want to put a foot wrong. (S6)

Players perceived their coaches to be central in engineering a lack of safety. They expressed that this was due to the coach's role in making selection and contractual decisions:

> It's the most uncomfortable environment I've been in ... if you were making a mistake because you weren't switched on, it was a major issue. I wasn't on it for a session, (coach) picked up on it straight after, he told me that I was going to be in the starting 15, but I'd worked my way out of it and that it would be tough to get back in. It was a good lesson for me, I reacted my usual way, when I've had a kick up the a*se, that's where I've sat down and properly had a think about what I need to do and where I need to be ... when I'm under that intense pressure, I feel a bit on edge, at the time it doesn't feel great, afterwards, you know you've progressed a lot. (S5)

Additionally, where players were part of strong playing squads, this lack of safety was also experienced because of peer comparison, selection pressure and the consequent intra-group competition. In essence, at both the individual and organisational level, players felt that safety was incongruous with the nature of the HP milieu.

3.2. A Double-Edged Sword

Rather than these perceptions being debilitating, the ability to cope with and develop in a psychologically unsafe environment appeared critical to their progress. As a particularly prominent factor amongst the 'made it' group, players appeared to respond adaptively to a lack of safety. Take for example player three's (S3) early experience of international rugby:

> (International coach) pulled me in on the Wednesday morning before we trained. He says: 'we're gonna start you' ... I woke up a day later to a text saying you need to see (coach). He said: 'I'm not picking you, you didn't train well, XXX had the upper hand on you'. I went away and had a think, I knew what (coach) wanted and what he was trying to do ... I had to find gears I never had before in training. I thought I was leaving it all out there, but somehow, I wasn't. It brought far more out of me. The game is mad, you can be on top of the world for 48 h before and then rock bottom. That camp was a perfect example and although it was traumatic at the time, I look back on it and I loved it, it definitely improved me.

As a comparison, consider the response of player eight (S8), reflecting on deselection from an international U20 squad:

> A couple of players got picked ahead of me and I was thinking, what more could I be doing? I'm playing well, starting in (on loan in lower league) ahead of players who are on the bench (at a lower level). Why is this fair? Why is this happening? It especially annoyed me because I felt like they took (player) because of his brother. It was a stitch up.

As such, what appeared to be adaptive or positive was not the perception of safety. Rather, it was how individual players responded to the nature of the milieu. Indeed, maladaptive responses to this lack of safety were a prevalent and indeed, derailing factor for some of those players unable to transition from age group international to senior elite performance. One that seemed to be exacerbated where a player's developmental experience had been overly safe and lacking the demands of the senior game: 'through the academy, life was easy, rugby was easy. Then, you go from being big fish in a small pond to a small fish in a huge pond. I found it so hard' (R6). Notably, players who perceived their development experience to be very safe, expressed that such environments did not adequately prepare them for the demands of the HP milieu.

This does not suggest that players believed that they were consistently able to respond adaptively; their experiences seemed to also change over time. Whilst the nature of the milieu remained the same, it was the player that seemed to change. Take for example the changing experiences of player four (S4) as he began to deploy psycho-behavioural skills to manage the demands placed on him as he reflected on two stages of his career. Firstly, as he began in the senior training environment:

> In my first year, I was terrified of making mistakes. You're new on the scene, you're a young guy and thrown into training with players like XXX [a senior international and club stalwart] and you don't want to f*ck up. I felt I trained within myself for a lot of that period. I've addressed that now but there's still a lot of pressure ... nobody's position is safe.

Players from both groups held the perspective that a lack of safety, rather than being debilitating, was an adaptive feature of their environment. Note the perceived differences for player four (S4), as he began embedding himself into a senior international squad:

> When you don't have expectation, you don't perform at the level you need to. If you feel like a bit safe, when there's no expectation, you think you can coast. When everyone is competing for a spot, that gets the best out of everyone. The squad depth at (international) is ridiculous. Everyone's battling so hard to play, it's so competitive. I thrive in that. I like the intensity but there's a couple of players who don't thrive in those environments.

When players did perceive a level of safety, often reported to be outside of the HP milieu, this was seen subsequently as a negative factor in their development; seemingly prompting stagnation over the longer term. In the case of player 13 (R6), when he was loaned out to a semi-professional team:

> It sounds silly, but I was very confident in my position (at loan club). I wasn't going to dropped for a 19 year old student or a 28 year old plumber. I think that's the problem, especially during that time. I needed to be pushed, I just wasn't. I accepted that I'll be playing for this team every weekend on loan.

It appeared that, rather than seeking safety and the certainty of their position, players were appreciative of being held accountable for their performance and the processes necessary for their continued development. Where this accountability and judgement was absent, it was perceived to be a negative developmental factor. Players also perceived this to be important at the group level, where the collective response to a lack of safety was seen as a key determinant of team performance:

> (Club coach), when you don't get selected, looks at how hard you train. It's almost 10 times harder when you're not getting picked because you got to push yourself even harder. In that sense it's a good environment, when you play the non-selected in training and all of them are all p*ssed off, that preps you for the weekend. It's probably a difference between teams because in some places players don't get picked and sulk. This place is completely different, it fires people up. (S3)

Players also described a lack of safety as prompting increased effort and attention to detail. The pressure conferred by a lack of safety was also significantly fatiguing and long-term exposure left players exhausted:

> I was just so tired, not so much physically, but mentally. I just couldn't get away from all the pressure. The lockdown (COVID) came at just the right time for me, I really enjoyed it. It was just what I needed because those two years, not doing well at (club) were really tough. I took that quite personally and struggled with it. So the lockdown was almost like a restart button. (S1)

> I was straight out of school. I finished the (international U18) stuff, finished my last exam then the next day I was straight into (club). I couldn't get used to the pressure, needing to be on it all the time. I was tired and just didn't cope with it, it was a chore from the beginning. (R7)

The perception of fatigue was most strongly felt where levels of safety were lowest and for extended periods of time. Despite this, however, there was a perception that this was highly developmental, both in the short term, driving them to higher standards and greater effort and, when this pressure was reduced, there was a perception that players were better able to cope with lower pressure in other environments:

> Being at (international), you feel like you have to step up a level, or you will just get sent home. It feels like you are on X Factor! I feel like I have to be on point every session or I am going to be embarrassed. You have (coach) prowling around watching everything as well. Everything that you do is going to get judged, it either pushes you into your shell, or drives you to perform. I found it hard to adjust, but going back into club rugby, I felt so much more relaxed. I felt like I could go at the same intensity, but I wasn't as stressed out, it was like I had adapted to the stress and could get the same results. It helped me massively from a mental point of view. (S7)

Players' reflections were complex; there was a highly individual and contextually mediated response to a lack of PS. In nearly all cases, players felt that high pressure and a lack of safety was very uncomfortable at the time, but through the deployment of different psycho-behavioural skills, negative impacts were moderated and could become a developmental stimulus. Where players lacked the psycho-behavioural skillset to cope, or where the pressures were prolonged and uncontrollable, the inherent lack of safety was a significant risk factor for the player's overall development.

3.3. *Tough Love*

In seeking to understand what approaches players considered to be the most effective in supporting their development, two themes were developed as prominent features of their experience. The first theme, coaches use of 'harder' approaches in their coaching practice, concerned holding players accountable to high standards, giving them role clarity, engaging in robust feedback processes and attention to detail. The second generated theme included the 'softer' approaches used by coaches in offering the player a level of openness and care. Perhaps summarising the needs that players perceived they had throughout the early stages of their careers, player three (S3) noted: 'I've worked with so many good coaches, we always had good conversations. They know when someone needs an arm on the shoulder or know when they need a kick up the a*se'. The reflections of players were complex and multidimensional; players believed that they derived important but differential benefits from different coaching behaviours. This didn't appear to be a balance between hard and soft approaches. Instead, players perceived coaches as offering more of one or the other through to offering both simultaneously. Across the interviews, both groups frequently used the term 'tough love' to describe what they perceived to be the most enabling coaching approach. Players consistently reflected on their desire and the necessity for the coach to adopt 'hard' approaches (tough) and the need for 'softer' approaches (love).

Table 2 presents the data based on player perception, identifying the approaches used by coaches that they perceived to be enabling of their future progress and the opposite pole [62], perceived to be disenabling. The perspectives of the players suggest a non-dichotomous perspective on effective coaching. Players strongly believed that effectiveness was the result of both 'harder' and 'softer' approaches to coaching practice. In short, players strongly desired highly competent coaches who presented them with a clear direction and held them to high standards, whilst also caring for and about them.

Table 2. Player perception of effective coaching.

	Characterised as Enabling	Raw Data Example	Characterised as Disenabling	Raw Data Example
Harder coaching approaches	High challenge and accountability to high standards	It was robust, often negative, tough love if you like, but I needed it. It was whenever I had a coach who was hard on me, but respected me as a person, that's what I always took the most from, seeing me as more than a player (R5)	Lack of accountability or challenge from coach	(1st team coach) came in as head coach and I got along really well with him. It was brilliant, my mate is the head coach kind of thing. I had (1st team coach) who was practically like my rugby Dad, he was looking after me, putting his arm around me. On a personal level it was immense, but my performance wasn't good, it was the time I stagnated the most probably because I felt so comfortable. (S6)
	Offering Role clarity	(International coaches) I learned so much from them, they were really personal with it. They told me exactly what they saw from me, and what I needed to do to improve and it was a really good environment to learn. (S4)	Ambiguity of role	I came off a loan and I felt as though I started to make a bit of progression. Then I got told I was meant to play in the European games by (head coach). He told me before that I was going to get picked and then for whatever reason I didn't get picked for those games (R1)
	Robust two-way and actionable feedback	I had a conversation with (coach) about selection. I asked why another player was being picked over me and (coach) said: 'he's more physical in the carry and it suits the game this week'. I'd say: 'I'll work on being more aggressive than the carry'. He said: 'No, you're not the same carrier as (player). Keep working on what you're good at. Your link play, your tips and playing out the back. You've got good feet you've got good handoff, use that to beat defenders, not just run straight through them like (player) does'. That kind of feedback is massive (S2)	Non-actionable, or inauthentic feedback	I like a coach that tells me what I'm doing well, what I'm not doing well and how I need to improve on it. Instead, at that time, all I was getting was coaches just trying to please me and build relationships. It was like 'oh yeah, you're doing really well' and then realistically I just wouldn't get any feedback. So I felt like that was a real point of stagnation. It made me angry at the time (S5)
	Building understanding through attention to detail	Having the technical knowledge is absolutely massive. With (1st team coach), I had never seen a coach who could say: 'watch this clip from a game in 2010, it would be perfect for you' to help you. I was sitting at home one night and he messaged me at 10 pm, he asked if I was watching the European Cup game and said: 'someone just made a really good read, I thought you would want to see it' (laughter). He was so passionate and wanting to help you. The tiny details have helped so much. (S7)	Lacking attention to detail	No one had ever said: 'you're better off carrying the ball into hands, because then you can pass, kick and run. No one cared about that, which I think, you know, looking back, someone's job probably was to say 'try carrying in two hands'. It's such a fundamental part of the game to be successful, particularly in my position. You need that attention to detail and that accountability for developing. (R2)
Softer coaching approaches	Care and empathic accuracy	I was developing because (school coach) was someone who has been really important to me. I still speak to him quite frequently, he's someone that cares about your wellbeing and cares about you as a person, but he was still pushing me, expecting better performance. (S4)	Lack of care	If coaches called you the wrong name, or your name was being missed off team sheets for training, you don't feel great. It's a big thing for you, but the coaches won't think too much about it. (R3)
	Coach openness to player input	I asked (head coach) for a chat. We sat down and I said: 'I want you to just to have faith in me. Just trust me. I'm going to work as hard as I can. We found some common ground. Usually, players keep it all on their chest. It was just a very good open conversation, now I think he's one of the best coaches I've ever had. It just took time for us to find common ground. (S6)	Coaches unavailable or unapproachable	When the pressure is on, coaches turn to focus on winning every week. At that point, they become less available. I'd go 8 weeks without speaking to them. You'd end up waiting around for hours trying to chase people. It got to the point where there was no dialogue, it was just a loan sheet. It was no one's job to check on us. I remember them saying: 'own your own development'. They never saw it as their job to help you. (R4)

3.3.1. Harder Approaches

Across both groups, there was a near universal perception that players wanted coaches who adopted 'harder' approaches to their coaching. Their descriptions indicated that harder approaches seemed to offer players direction, motivation and robust feedback, guiding the reflective patterns that appeared to be supportive of further performance development. For example, player 13 (R6) suggested:

> I needed a coach, who offered some tough love, who would be on top of me, that would be honest, but also respect me. I only got that in the early years of my career. Other coaches, who I had a really good relationship with, there was no: 'this wasn't good enough'. I needed a coach to say: 'your performance at this level is fine, but I want you to think about your ambitions as a player and if you want to push on, your performance at the weekend was crap'. There was never any of that. I knew if I went out with no preparation, had a beer the night before, hadn't really done much mental prep ... I'd still have been a seven out of ten. I needed coach to tell me: 'seven out of ten at this level is not good enough'.

There was a strong perception across the group that the challenges of the elite game required these harder approaches to drive and promote the extra levels of performance necessary to make the jump to truly elite performance. Players also reflected on the need for role clarity to make sense of where they stood in a challenging environment and mitigate against the ill effects of a lack of PS.

> It was a complete emotional rollercoaster, one week I was captain, the next week I couldn't captain the tiddlywinks team. I had no idea where I stood, it was like the team performance rested on me. I heard from another coach that one week I was being recommended to (international coach), being offered 2 year contract on way more money, the next (club coach) wasn't going to pick me at all. Whether there was an agenda to get rid of me I don't know, but there was a lot going on behind the scenes. (S1)

The absence of role clarity was felt strongest in relation to the frequent experience of emotionally disturbing challenges, such as selection and contractual matters. This lack of clarity was experienced most strongly amongst the group of 'didn't' players but was also a feature of the whole group's experience. Without knowing where they stood within an environment, or what they needed to do in order to obtain opportunities to further their careers, players felt confused and directionless:

> I was told that I'd hit all goals set for me, but the goals had changed. I got told I wasn't going to be contracted, but that I'm doing exactly what they'd asked of me, but they'd moved the goal posts at some point. I hadn't been told. (R8)

Players were also well aware of the scale of the challenge presented by the HP milieu and were accepting of the commitment necessary to reach the elite level. Coaches who guided this process were perceived to engage players in robust feedback processes, that were often emotionally laden but acted to give clear direction to the player and enabled them to take action:

> (Head coach) was known to be a bit of a 'yes man'. He'd always say: 'yeah, yeah, you'll get your chance'. The type of person who avoids the tricky conversations. Other coaches would just tell you something like (Head coach), you'd try and do it, go and do it but it wouldn't change anything. (R4)

Where robust feedback was offered to the player, it was seen as a performance enhancer, helping players to make sense of the difficulties they faced. The opposite of which, players found especially difficult to understand or energise their next steps:

> Wishy washy feedback was a killer. I was being released and it just wasn't clean: 'we really rate you, but we aren't contracting you.' I go away and what the hell am I supposed to do with that? Now, it's like: 'you're not getting picked because

> of this, this and this.' It's just being straight up and honest. I think coaches need to think deeply about their approach to those conversations, they can kick players on, or beat them down. (S3)

Players also put significant weighting on the ability of the coach's competence in guiding their progress. This was seen both in terms of the coach's competence and knowledge base, but also through their individual rather than group-focused attention to detail:

> When you dropped that ball, or when you made that pass, he asked: 'what do you think you could have done?'. (Coach) remembers all the details, and it was a huge attention to detail. It just broke everything down for me, it was so much easier. Their attention to detail, it was eye opening. (S6)

Notably, where coaches did not have the perceived knowledge base, particularly as players reached higher levels of the game, regardless of other facets of support, they were rejected as helpful agents.

3.3.2. Softer Approaches

Regarding the player's perceptions of 'softer' approaches used by coaches, 'softer' approaches in the mind of the players did not seem to mean gentle. Instead, these reflected a genuine concern for their welfare and performance, offering them a voice to express thoughts and concerns:

> (1st team coach) was tough in terms of having high expectations for everybody in the squad and to take responsibility. No one was able to sort of rest on their laurels, there was an expectation for everyone to improve, all the time. But also, this person was a generous, kind and compassionate human being ... From my point of view, I thought it was absolutely fantastic. (R2)

This compassion was reflected by the attention paid to the player as an individual; understanding their broader life, with the perception that the player was able to approach the coach regarding their life outside of the game. However, this perception of care was also multidimensional, and, more prevalently, players referred to care as being a coach's investment of time and effort in their development as a performer. In essence, it seemed that, whilst players valued coaches paying attention to and recognising their wider lives, players cared more about coaches who invested in their careers as athletes.

The final generated theme concerned the extent to which coaches made themselves open to conversations or invited a player's input. Coaches who were perceived as open, allowed players to approach them to discuss issues and to understand the player's point of view: 'Having spent time with a lot of coaches, those that are open enough to invite challenge and debate is such a massive thing for my development' (S7). For others, with coaches where these conversations did not take place, there seemed to be a barrier for coaches in understanding the psycho-emotional state of the player. In some cases, players felt that coaches were deliberately undermining them and had closed themselves off to the input of players. Where players perceived this, it seemed to change the player's interpretation of the coach's input. For example, in one case, a coaching group was perceived to favour a certain type of player based on non-performance relevant characteristics:

> (International age group coaches) didn't want me there, it was basically all private school lads down there and the coaches were quite cliquey because nearly all the players had been in their system for ages. I remember in my first session I made a tackle and I folded a (club) back rower. The coaches were like: 'oh the (club) lads are here'. I was the only player from my club, it was clear that they didn't like me. They had their favourites from certain clubs and the louder ones who were more extroverted. I couldn't approach them. (R3)

In addition to the perceptions of the player regarding the extent to which the coach could offer a 'tough love' approach, it appeared that the key mediating factor was the extent to which the player's perceived the coach to be competent. Often demonstrated through technical understanding of the sport, or of a player's needs, where a coach's competence was questioned, it often led to their input being rejected.

> I respected coaches more if I thought they knew what they were talking about. With (international age group), I was told I wasn't fit enough but I ran a 4.51 bronco (fitness test). It was so annoying, they didn't know what I needed to work on. They were just covering their a*ses. (R5)

At times, players held different perspectives on the same behavioural approach from the same coach. Compare, for example, the perceptions of players one and seven of a 1st team coach:

> I was like, wow, this man who's achieved so much is willing to take a gamble on me. He was really direct with me, but I remember feeling like I had to give everything I can to learn from him. (S1)

> For me, it is finding the coaches who are doing it for the right reason, and you look at their track record, you know they are doing something right. Some coaches will just shout at you for the sake of it, (coach) was just an authoritarian, I stopped listening. (S7)

In essence, amongst this group of players, there was a nuanced necessity for the coach to deploy a range of interpersonal skills as a medium to support their development. This was experienced on a highly individual basis and often moderated, not only by the type of relationship between coach and athlete, but also perceptions of competence, power and the wider network of relationships held by the coach with other players.

4. Discussion

The aims of this study were: (a) to examine the extent to which two matched groups of international and released professional rugby union players perceived PS to be an adaptive feature of their developmental experience and (b) to understand what elements of the player's coaching experience were perceived to be enabling or disenabling of future progress. The findings present two overarching points of discussion including the player's perceptions regarding PS, firstly the near universal lack of PS throughout their career and second, the role of the coach, including their experiences of interpersonal dimensions and soft skills of their coaches throughout the pathway. Subsequent evidence-informed implications are offered to coaches and practitioners within HP and TD settings.

4.1. Psychological Safety

In line with recent critiques of the uncritical application of PS as a conceptual model in HP sport, data in this study suggest a complex and nuanced relationship between perceptions of safety and performance development [14]. Much like data collected in other contexts, it does appear that the construct may not always be a performance enhancer [15]. The experiences of the players suggest that a lack of safety appeared to be a ubiquitous feature of their experience and, indeed, where players had the requisite psycho-behavioural skills to cope with a lack of safety, it appeared to be a performance enhancer [55]. Indeed, this may be a key differentiating feature of the HP milieu and non-elite settings, cf. [20]. In this regard, with growing and significant interest in the construct, there remains a lack of empirical investigation and limited conceptual clarity [13]. In essence, the players in this sample were well aware that if they made mistakes, it would be held against them, that risk taking was not safe and the pressures of selection meant that their unique talents would not always be valued [11,12]. In addition, the response to a lack of safety seemed to differentiate between those that were able to make the next step in the professional game towards senior international status and those who were not. Yet, importantly, the present sample also alluded to a number of potential issues associated with these perceptions of a lack of safety, including the high fatigue associated with the pressure to perform over extended periods [63].

Selection and deselection are clear realities of the HP milieu; in the case of elite rugby union, faced on a weekly basis. Coaches simply cannot offer athletes an environment where mistakes will not be held against you, where risk taking is safe and unique skills will always be valued [11]. Taking a different definition, nor is it possible to offer an experience that is: 'protected from, or unlikely to be at risk of, psychological harm in sport (including fear, threat, and insecurity)' with shared perceptions of comfort [13]. Therefore, in terms of theory, rather than aiming for PS as a universally desirable outcome, we might be a little more realistic and consider safety on a continuum, with differential effects depending on the extent of the lack of safety. Low levels of safety (when coupled with an appropriate psycho-behavioural skillset) potentially drive increased levels of effort and attention to detail, whilst higher levels of safety may allow the athlete the space to experiment and recover from the demands placed on them, cf. [29].

This would suggest that for the athlete, if there are differential effects from more or less safe experiences, through appropriate planning, it may be desirable to periodise the extent of the pressure experienced by the athlete, through an athlete's career [33,64]. Either way, a more individual view of safety is likely appropriate [13], rather than the blanket suggestion that an entire environment can be classified as being 'safe' [65]. In essence, if the concept of PS is to offer a meaningful impact in HP sport, there is a need for further critical investigation, framed by a real world understanding of the HP milieu. In short, and as with so much else in coaching, 'it depends' on the coach using professional judgement and decision making (PJDM) to select the most appropriate approach for each context [66].

4.2. *Role of the Coach*

Coherent with the findings of previous research, athletes put a significant weighting on the interpersonal dimensions of their interactions with coaches [67]. Importantly, the players highlighted multiple coaches, along with other stakeholders, playing an active role in their development [34,68]. The interpersonal dimensions of these interactions were complex and in many ways seemed to contradict a variety of coach education advice [23]. Players did not perceive the role of the coach to be offering unremitting positivity. Instead, players sought genuinely developmental input from their coach [29]. This input required the coach to be knowledgeable across multiple domains [69] and operationalise this through a variety of interpersonal approaches [70,71].

Prominently, the coach's ability to help the player generate appropriate role clarity was supportive of development and protected against maladaptive consequences of a lack of PS. That is, whilst players were well aware of the judgement and the inherent potential for them to be deselected, understanding the criteria they would be judged on and help from the coach to make progress towards these ends appeared to be central to the coach's role. For the most part, the coach appeared to enable this by deploying a variety of 'harder' approaches. Importantly, however, the longer-term vehicle for improvement was also supported by 'softer' approaches [72]. This appeared especially important where players perceived a significant performance gap between themselves and the senior elite level.

4.2.1. Interpersonal Dimensions

This has significant implications for the existing care literature in coaching. Rather than 'tough love' being used as a euphemism to justify abusive coaching practices, 'tough love' was perceived by players to be an essential feature of coaching practice that characterised both the hard and soft approaches that they deployed [46]. The data here present an orthogonal relationship between the role of the coach and players' perceptions of their decisions and interactions. Coaches who were perceived by players to offer them 'tough love' were characterised as both caring for and caring about their players [39,40]. Our data are consistent with the assertions that a consideration of 'care', needs to be ecologically situated [40] and take account of the nature of the HP milieu. Put simply, context is key. Similar to research conducted in TDEs [42], it also suggests that care does not always need to be demonstrated through softer approaches. If we take this view in relation to the findings within this study, players offer numerous accounts where care is expressed by their coaches through interactions that, without reference to social context, may be perceived as uncaring [73,74]. These include instances of stern corrective feedback, the threat of deselection, increasing psychological and social pressures and the impact of contractual decisions. It seems unanimous from the findings that the players perceived effective coaching interactions as tough but caring, as they were often perceived as necessary given the challenges they regularly face in their day-to-day life as a professional rugby union player [42,75]. Additionally, players were clear that overly positive interactions with coaches, aimed at 'pleasing' them, or building relationships, were perceived as detrimental, especially without the developmental input that they needed [42]. Thus, when addressing the HP milieu, 'caring for' athletes whose needs are clearly expressed and self-determined by their own long term success as performers, offers a paradox to coaches who are presented with key conceptual precursors of PS [14].

The findings from our study present clear differences between the two groups' descriptions of their dialogue with their coaches. On the one hand, those players who were able to progress, demonstrated 'reciprocity' between feedback and their choice to accept it and engage with it [40,76]. Clear connections can be drawn here to two relevant concepts. Players perceived role clarity [36], that is where they fit within the wider group and the accepted roles and responsibilities that come with it, and their feedback literacy [34], which addresses how well an athlete can interact with feedback. Progression was supported when players understood where they stood in the social context and the roles and responsibilities that were expected of them. In contrast, stagnation seemed to be coupled with coach decisions about the destination of loan clubs, or (de)selection for the senior team. Similarly, the group who 'made it', described numerous instances where they took time to consider and process critical feedback, often relishing these types of challenges [23]. Interestingly, the coach's capacity to care for their players in an HP environment seemed dependent on the interrelation between the capacity to engage in open dialogue with players, the creation of reciprocity in highly challenging situations, ensuring player role clarity and, finally, the nurturing of a player's feedback literacy over time.

We hope this adds further nuance to the existing base of evidence in the coaching field, which in recent years has emphasised the interpersonal dimensions of the coaching process [77]. To be clear, this is not to suggest that abusive coaching practice is justified in any domain, not only for the obvious ethical reasons, but also for the sake of performance. Such an extreme 'either-or' position is unfortunately commonly expressed but not grounded in any reality of which we are aware. Where players in this sample did feel uncared for, it seemed to have a negative performance consequence. Therefore, whilst there may be a case for performance sport to use the concept of PS under very specific circumstances, we would suggest there is a need for a more mechanistic understanding of the dynamics at play. In addition, a clear need for the field to be conceptually clear, especially in the realm of TD and elite sport, where the demands of the milieu are very different to those experienced in community sport or organisational life. This is especially important if findings are to be used in applied practice, as we cannot ignore the realities of HP. In this regard, we would suggest that conceptions of care seem to present significantly more transferability, based on environmental nuance.

The perception that the coach really cared about the player is important. Care from the perspective of the player seemed to be driven by a level of complementarity; that player and coaches converged towards the same aims [78]. Significantly, for the team sport coach, this convergence seemed to be towards the interests of the individual player rather than the organisation. In essence, there was a perception that the coach understood and cared for that player's developmental needs and career ambitions. For some players, it was important to them that the coach understood their life beyond being a rugby player. Yet, for most players, the latter was not perceived to be an essential feature of an effective coaching experience. This is a notable finding and one that deepens our understanding of the coach in the TD and HP setting.

There are of course nuances presented by the sample given that even those players who 'didn't make it' could be classified as 'competitive elite' [8]. Thus, it may be that the relative 'eliteness' of the sample has implications for desired coaching practice as all participants were, at least for a time, full time professional athletes. Thus, it is likely that their participatory aims were more geared to performance [45]. We might justifiably suggest that this may not be the case in populations with a more participatory focus. Therefore, we suggest that blanket statements designed to cover the realm of 'coaching' from the U5's to the podium are clearly not appropriate but, rather, counterproductive and confusing.

4.2.2. Soft Skills

On the basis that the role of the coach seemed to be so fundamental to the perceived progress of the players and that such a complex picture was presented by their reflections, as has long been identified, interpersonal skills are a core feature of coaching practice [79]. Taking this idea, colloquially, interpersonal skills are often referred to as being 'soft skills'. On the basis of the data presented in this sample and others [30], this label would seem wholly inappropriate. Instead, the label 'interpersonal skills' seems far more appropriate to describe the range of different stances and approaches that were adopted and perceived to be supportive of player progress. In this particular instance, the interpersonal approaches adopted by coaches need to be matched to the nature of the HP milieu [17]. Perhaps challenging the common discourse around coaching that is suggesting that care in coaching may need to rely on 'softer' approaches, both harder and softer approaches were used by coaches [72]. Additionally, given the complexity of both the setting and the athlete's needs, it is also clear that effective interpersonal skills were fundamentally intertwined with the coach's ability to utilise a wide body of knowledge in their practice [80]. For example, understanding the role of feedback on the player and deciding an appropriate way to engage the athlete in a feedback process, requires pedagogic and interpersonal knowledge and skills, along with the technical, tactical, physical and psychological knowledge by which to begin a feedback process [69,80].

Players also saw the competence and power of the coach as a primary mediator of their messaging. That is, unless they believed that the coach was competent enough to offer them input, what they heard was rejected and added to a sense of overall frustration [34]. Much like the literature in pedagogic sciences, such as understanding impactful feedback as a process [34,81], the data presented here suggest a complex interaction between coach and athlete. The interpersonal effectiveness of the coach was underpinned by a range of other coaching knowledge and skills [71]. Indeed, the data presented here suggest that high quality coaching practice was supported by an effective coach–athlete relationship [67]. However, as specifically highlighted by a number of participants, this was not sufficient for high quality coaching.

5. Limitations

Given our attempt to successfully address the research aims, there are of course limitations in our approach. First, the retrospective nature of the methods employed within this study have often been criticised in research within HP and TD environments as they may offer an invalid and untrustworthy representation of athlete experiences [2,82]. Secondly, the analysis may have benefited from the triangulation of coach perspectives, if players had been matched to coaches that had been working with players during critical periods. This option was discounted given the desire for a longer term, career perspective of each individual player, each of whom had worked with large numbers of support personnel and, as such, coaches may only be able to offer a limited perspective on the experience of each individual. Given these limitations it is essential to first consider possible alternate methods of data collection in respect to the research aims and second, address the attempts to mitigate the impact of these limitations in respect to the research findings.

In considering the study design, alternative methods of data collection may have utilised more in situ ethnographic methods over prolonged periods of time within a single TD pathway [2]. Whilst this approach would have resulted in data that explore an environment in a much deeper and broader fashion, it also would result in a much smaller sample size and a failure to capture the similarities, perceptions and experiences of rugby players across different clubs and environments, and at different times, who eventually ended up at the same ultimate destination. Furthermore, to mitigate athlete data, which may be invalid or untrustworthy, the graphic timeline task successfully stimulated the player's memory and supported the recall of particular events, perceptions of the environment and experience of the coach's support [83]. This is a likely reason as to why the member reflection process corroborated the tabulation of our TA.

Finally, to mitigate possible survivorship bias, the inclusion/exclusion criteria was employed to explore two populations who had experienced the same overall TD system, albeit within different contexts, which resulted in different athlete perceptions. To truly address the research aims, the perspective of athletes from both populations were valued, offering an equal investigation into those who had and had not made it [47].

6. Applied Implications

This paper raises several challenges for the coach, coach development and broader coaching practice, perhaps reflecting a level of nuance that is very difficult to capture outside of the complexity of the context and social milieu where coaching happens. To this point, much of the literature that has considered the role of the coach has focused on the softer side of the interpersonal dimensions of the coaching relationship. In practice, this has meant that much coach development support has tended to be unable to offer coaches support on harder interpersonal approaches (e.g., role clarity, provision of challenge). We therefore believe that there is a significant opportunity for practitioners to expand the scope of their work beyond the boundaries of 'soft skills' and begin to consider the extent to which their interpersonal approach deploys a range of both hard and soft approaches [72]. Importantly, this is not an 'either/or', but, rather, should be seen as an 'and/both'.

6.1. Role Clarity

We would suggest that for the individual coach, the suggestion that coaches should spend time explaining to the athlete the 'why' behind their coaching decisions would be a good starting point [84]. In addition, engaging in open dialogue with athletes may be the necessary starting point for effective coaching, especially if we are to ensure that care is a core feature of the coaching environment [39]. Without this open but potentially uncomfortable dialogue [85], there is a risk that athletes see the coach as a nice person who is unable to help them improve, especially in HP populations. Athletes may well need to feel psychologically unsafe, at least in the short term. A core dimension of open dialogue should be the athlete's generation of role clarity (cf. PCDEs)—[86]. The coach has a critical role to play in minimising role ambiguity as a means of driving a player's developmental journey, especially as they work through a variety of challenging experiences [36]. This means players knowing what their role in a group was, what a coach expected of them and what they would be judged on. This may be especially important as players navigate through a psychologically unsafe landscape.

With no exceptions, all players in this sample experienced issues with coaches and practice that they considered to be poor. This suggests important implications for player development, where TD systems may need to consider the extent to which players are prepared to work with coaches who are unable to provide optimal developmental conditions. This may be especially important in settings where player development is more distributed across a wider network of influences (e.g., [68]). How the coach optimally prepares athletes for periods where they may have to be the predominant driver of their development seems to be critical [87]. Importantly, unlike the experiences of a player in this sample, this does not mean leaving players to their own devices and calling it empowerment.

6.2. Safety

Perhaps controversially, we would suggest that, rather than being an implicitly desirable feature of all coaching environments, this study clearly shows that a lack of PS was an ever-present feature of the experience of the players in this sample. Indeed, in contrast to the literature in the organisational domain, a lack of safety was often regarded by individual players as a performance enhancer [88]. As a result, it seems that PS, where present, rather than being universally positive may limit development, cf. [14,15]. In essence, there is the fascinating dimension, clearly warranting further empirical attention, that so long as coupled by a level of role clarity, it may be desirable for coaches to deliberately reduce PS. At the very least, we would suggest that the uncritical attempt to apply the construct across performance sport settings without an evidence base is both unwarranted and potentially harmful to the athlete interests presented as of paramount importance. For the time being, it would at least appear wholly unwarranted for PS to be seen as an unquestionably positive element of athlete development, or in many cases whether it is actually possible for PS to be a feature of the HP milieu [14].

Whilst it appears that HP athletes will rarely experience PS as defined by Edmondson [11] or Vella et al. [13], suggestions that the HP milieu can be made psychologically safe are unlikely, and perhaps, undesirable. However, in specific conditions, a coach may choose to generate a specific set of circumstances that allows for increased perceptions of safety, either to manage the fatigue of an athlete, or where they may be struggling with the demands of the environment. This circles on the centrality of role clarity as a means of supporting performance and player development. If safety is seen on a continuum, rather than the characteristic of an environment, there may be situations where higher levels of safety may be adaptive. For example, players in this sample discussed transitioning out of international squads, back to their club, or where players were exhausted from constant judgement and competitive demands. For this reason, there are potentially transferable dimensions of the concept of the 'safe container', a temporally constrained block of time where increased perceptions of safety can be generated via contracting lower levels of judgment [89].

Thus, as has been the case for coaching practice for a long time, the coach has the chance to deploy strategies such as telling an athlete that they will be selected for a given period of time, regardless of their performance, or an athlete might be removed from competitive games for a block of time to make technical adjustments. By the same token, just as was the case for athletes in this sample, coaches may remove a sense of safety but take advantage of a lack of safety, offering role clarity for the athlete. For example, in a centimetres, grams, seconds (CGS) sport, an athlete will know the performance level they might need to reach to achieve a higher level of funding or selection for competition. Similarly, in team sports, an example is the time taken by the coach on an individual basis to show and articulate to the player exactly what they need to do to be selected. In all of this, it is of course worth noting that if one athlete is guaranteed selection, then another athlete is made unsafe because they are not being selected.

Much of the data presenting in this sample bears similarities to discussions of learning and performance more generally [90]. There is obviously more research to be conducted, and whilst the affective conditions for learning are generating more attention [30], there is more research to be carried out that explores the potential utility of a range of emotional experience on learning and development. Finally, in an era where coaches of all ages and stages are wrestling with a variety of athlete welfare concerns, there is an essential need for careful, granular interrogation of concepts and the realities of the HP domain. As per the data in this sample, more consideration needs to be paid to the wants and needs of HP athletes and data from the participation setting should not be universally applied. PS is not a term that should be used loosely, it is a well-established and empirically supported concept in specific organisational settings [12]. This may prevent the confusion amongst practitioners in the HP sport community, which, anecdotally at least, seems to be causing such a range of issues between athletes and coaches.

Author Contributions: Conceptualization, J.T., M.A. and D.C.; methodology, J.T. and M.A.; formal analysis, J.T. and M.A.; investigation, J.T. and M.A.; writing—original draft preparation, J.T. and M.A.; writing—review and editing, J.T., M.A. and D.C. All authors have read and agreed to the published version of the manuscript.

Funding: This research received no external funding.

Institutional Review Board Statement: The study was conducted in accordance with the Declaration of Helsinki and approved by the Institutional Review Board (or Ethics Committee) of the University of Edinburgh (protocol code JT09022021 7 April 2021).

Informed Consent Statement: Informed consent was obtained from all subjects involved in the study.

Data Availability Statement: Not applicable.

Conflicts of Interest: The authors declare no conflict of interest.

References

1. Abbott, A.; Collins, D. Eliminating the dichotomy between theory and practice in talent identification and development: Considering the role of psychology. *J. Sports Sci.* **2004**, *22*, 395–408. [CrossRef] [PubMed]
2. Baker, J.; Wilson, S.; Johnston, K.; Dehghansai, N.; Koenigsberg, A.; De Vegt, S.; Wattie, N. Talent Research in Sport 1990–2018: A Scoping Review. *Front. Psychol.* **2020**, *11*, 607710. [CrossRef]
3. Cushion, C.J.; Armour, K.M.; Jones, R.L. Locating the Coaching Process in Practice: Models 'for' and 'of' Coaching. *Phys. Educ. Sport Pedagog.* **2006**, *11*, 83–99. [CrossRef]
4. Collins, D.; MacNamara, Á.; Cruickshank, A. Research and Practice in Talent Identification and Development—Some Thoughts on the State of Play. *J. Appl. Sport Psychol.* **2019**, *31*, 340–351. [CrossRef]
5. Gulbin, J.; Weissensteiner, J.; Oldenziel, K.; Gagné, F. Patterns Of Performance Development in Elite Athletes. *Eur. J. Sport Sci.* **2013**, *13*, 605–614. [CrossRef] [PubMed]
6. Neelen, M.; Kirschner, P. *Evidence-Informed Learning Design: Use Evidence to Create Training which Improves Performance*; KoganPage: London, UK, 2020.
7. Johnston, K.; Baker, J. The Complex and (Sometimes) Conflicting Beliefs about Talent: A Case Study of Elite Distance Running Coaches. *J. Expertise* **2022**, *5*. Available online: https://journalofexpertise.org/articles/volume5_issue1/JoE_5_1_Johnston_Baker.pdf (accessed on 26 March 2022).
8. Swann, C.; Moran, A.; Piggott, D. Defining Elite Athletes: Issues in the Study of Expert Performance in Sport Psychology. *Psychol. Sport Exerc.* **2015**, *16*, 3–14. [CrossRef]
9. Baker, J.; Schorer, J.; Wattie, N. Defining expertise: A taxonomy for researchers in skill acquisition and expertise. In *The Routledge Handbook of Sport Expertise*; Baker, J., Farrow, D., Eds.; Routledge: London, UK, 2015; pp. 145–155.
10. Kahn, W.A. Psychological Conditions of Personal Engagement and Disengagement at Work. *Acad. Manag. J.* **1990**, *33*, 692–724.
11. Edmondson, A. Psychological Safety and Learning Behavior in Work Teams. *Adm. Sci. Q.* **1999**, *44*, 350–383. [CrossRef]
12. Edmondson, A.C. *The Fearless Organization: Creating Psychological Safety in the Workplace for Learning, Innovation, and Growth*; Wiley: Hoboken, NJ, USA, 2018.
13. Vella, S.A.; Mayland, E.; Schweickle, M.J.; Sutcliffe, J.T.; McEwan, D.; Swann, C. Psychological safety in sport: A systematic review and concept analysis. *Int. Rev. Sport Exerc. Psychol.* **2022**, 1–24. [CrossRef]
14. Taylor, J.; Collins, D.; Ashford, M. Psychological Safety in High Performance Sport: Contextually applicable? *Front. Sports Active Living* **2022**, 169. [CrossRef]
15. Deng, H.; Leung, K.; Lam, C.K.; Huang, X. Slacking Off in Comfort: A Dual-Pathway Model for Psychological Safety Climate. *J. Manag.* **2017**, *45*, 1114–1144. [CrossRef]
16. Pearsall, M.J.; Ellis, A.P.J. Thick as thieves: The effects of ethical orientation and psychological safety on unethical team behavior. *J. Appl. Psychol.* **2011**, *96*, 401–411. [CrossRef] [PubMed]
17. Taylor, J.; Collins, D. Getting in the Way: Investigating Barriers to Optimising Talent Development Experience. *J. Expertise* **2021**, *4*, 315–332.
18. Eubank, M.; Nesti, M.; Cruickshank, A. Understanding high performance sport environments: Impact for the professional training and supervision of sport psychologists. *Sport Exerc. Psychol. Rev.* **2014**, *10*, 30–37.
19. Gosai, J.; Jowett, S.; Nascimento-Júnior, J.R.A.D. When leadership, relationships and psychological safety promote flourishing in sport and life. *Sports Coach. Rev.* **2021**, 1–21. [CrossRef]
20. Morgan, P.B.C.; Fletcher, D.; Sarkar, M. Developing team resilience: A season-long study of psychosocial enablers and strategies in a high-level sports team. *Psychol. Sport Exerc.* **2019**, *45*, 101543. [CrossRef]
21. Smittick, A.L.; Miner, K.N.; Cunningham, G.B. The "I" in team: Coach incivility, coach gender, and team performance in women's basketball teams. *Sport Manag. Rev.* **2019**, *22*, 419–433. [CrossRef]

22. Rees, T.; Hardy, L.; Güllich, A.; Abernethy, B.; Côté, J.; Woodman, T.; Montgomery, H.; Laing, S.; Warr, C. The Great British Medalists Project: A Review of Current Knowledge on the Development of the World's Best Sporting Talent. *Sports Med.* **2016**, *46*, 1041–1058. [CrossRef]
23. Collins, D.; MacNamara, Á.; McCarthy, N. Super Champions, Champions, and Almosts: Important Differences and Commonalities on the Rocky Road. *Front. Psychol.* **2016**, *6*, 2009. [CrossRef]
24. Collins, D.; MacNamara, Á. The Rocky Road to the Top: Why Talent Needs Trauma. *Sports Med.* **2012**, *42*, 907–914. [CrossRef] [PubMed]
25. Williams, G.G.; MacNamara, Á. Challenge is in the eye of the beholder: Exploring young athlete's experience of challenges on the talent pathway. *J. Sports Sci.* **2022**, *40*, 1078–1087. [CrossRef] [PubMed]
26. Sarkar, M.; Fletcher, D. Adversity-related experiences are essential for Olympic success: Additional evidence and considerations. In *Progress in Brain Research*; Walsh, V., Wilson, M., Parkin, B., Eds.; Elsevier: Amsterdam, The Netherlands, 2017; Volume 232, pp. 159–165.
27. Till, K.; Cobley, S.; Morley, D.; O'hara, J.; Chapman, C.; Cooke, C. The influence of age, playing position, anthropometry and fitness on career attainment outcomes in rugby league. *J. Sports Sci.* **2016**, *34*, 1240–1245. [CrossRef] [PubMed]
28. McCarthy, N.; Collins, D.; Court, D. Start Hard, Finish Better: Further Evidence for the Reversal of the RAE Advantage. *J. Sports Sci.* **2016**, *34*, 1461–1465. [CrossRef]
29. Taylor, J.; Collins, D. The Highs and the Lows—Exploring the Nature of Optimally Impactful Development Experiences on the Talent Pathway. *Sport Psychol.* **2020**, *34*, 319–328. [CrossRef]
30. Taylor, J.; Collins, D. Navigating the Winds of Change on the Smooth Sea—The Interaction of feedback and emotional disruption on the talent pathway. *J. Appl. Sport Psychol.* **2021**, 1–27. [CrossRef]
31. Martindale, R.J.; Collins, D.; Daubney, J. Talent Development: A Guide for Practice and Research Within Sport. *Quest* **2005**, *57*, 353–375. [CrossRef]
32. Webb, V.; Collins, D.; Cruickshank, A. Aligning The Talent Pathway: Exploring The Role And Mechanisms Of Coherence In Development. *J. Sports Sci.* **2016**, *34*, 1799–1807. [CrossRef]
33. Collins, D.; MacNamara, Á.; McCarthy, N. Putting the Bumps in the Rocky Road: Optimizing the Pathway to Excellence. *Front. Psychol.* **2016**, *7*, 1482. [CrossRef]
34. Taylor, J.; Collins, D.; Cruickshank, A. Too Many Cooks, Not Enough Gourmets: Examining Provision and Use of Feedback for the Developing Athlete. *Sport Psychol.* **2021**, *36*, 89–100. [CrossRef]
35. Lara-Bercial, S.; Mallett, C.J. The Practices and Developmental Pathways of Professional and Olympic Serial Winning Coaches. *Int. Sport Coach. J.* **2016**, *3*, 221–239. [CrossRef]
36. Bray, S.R.; Beauchamp, M.R.; Eys, M.A.; Carron, A.V. Does the Need for Role Clarity Moderate the Relationship between Role Ambiguity and Athlete Satisfaction? *J. Appl. Sport Psychol.* **2005**, *17*, 306–318. [CrossRef]
37. Din, C.; Paskevich, D.; Gabriele, T.; Werthner, P. Olympic Medal-Winning Leadership. *Int. J. Sports Sci. Coach.* **2015**, *10*, 589–604. [CrossRef]
38. Noddings, N. An Ethic of Caring and Its Implications for Instructional Arrangements. *Am. J. Educ.* **1988**, *96*, 215–230. [CrossRef]
39. Noddings, N. Identifying and responding to needs in education. *Camb. J. Educ.* **2005**, *35*, 147–159. [CrossRef]
40. Cronin, C.; Knowles, Z.R.; Enright, K. The challenge to care in a Premier League Football Club. *Sports Coach. Rev.* **2020**, *9*, 123–146. [CrossRef]
41. Cronin, C.; Armour, K. Care theory and sport coaching. In *Care in Sport Coaching: Pedagogical Cases*, 1st ed.; Routledge: London, UK, 2018.
42. Knust, S.K.; Fisher, L.A. NCAA Division I Female Head Coaches Experiences of Exemplary Care Within Coaching. *Int. Sport Coach. J.* **2015**, *2*, 94–107. [CrossRef]
43. Potrac, P.; Jones, R.L.; Gilbourne, D.; Nelson, L. 'Handshakes, BBQs, and bullets': Self-interest, shame and regret in football coaching. *Sports Coach. Rev.* **2012**, *1*, 79–92. [CrossRef]
44. Roderick, M. Adding insult to injury: Workplace injury in English professional football. *Sociol. Health Illn.* **2006**, *28*, 76–97. [CrossRef]
45. Collins, D.; Bailey, R.; Ford, P.A.; MacNamara, Á.; Toms, M.; Pearce, G. Three Worlds: New directions in participant development in sport and physical activity. *Sport Educ. Soc.* **2012**, *17*, 225–243. [CrossRef]
46. Cronin, C. Caring Coaching Examining the Notion of 'cruel to be kind' and Other Caring Myths. In *Myths of Sport Coaching*; Whitehead, A., Coe, J., Eds.; Sequoia Books: Anchorage, AK, USA, 2022.
47. Taylor, J.; Collins, D. Shoulda, Coulda, Didnae—Why Don't High-Potential Players Make it? *Sport Psychol.* **2019**, *33*, 85–96. [CrossRef]
48. Giacobbi, P.R.; Poczwardowski, A.; Hager, P. A Pragmatic Research Philosophy for Sport and Exercise Psychology. *Sport Psychol.* **2005**, *19*, 18–31. [CrossRef]
49. Jenkins, S.P.R. Beyond 'crude pragmatism' in sports coaching: Insights from C.S. Peirce, William James and John Dewey. *Int. J. Sports Sci. Coach.* **2017**, *12*, 8–19. [CrossRef]
50. Silverman, D. *Interpreting Qualitative Data: Methods for Analyzing Talk, Text, and Interaction*, 3rd ed.; Sage: London, UK, 2006.
51. Strean, W.B. Possibilities for Qualitative Research in Sport Psychology. *Sport Psychol.* **1998**, *12*, 333–345. [CrossRef]

52. Berger, R. Now I see it, now I don't: Researcher's position and reflexivity in qualitative research. *Qual. Res.* **2013**, *15*, 219–234. [CrossRef]
53. Cruickshank, A.; Collins, D. Beyond 'crude pragmatism' in sports coaching: Insights from C.S. Peirce, William James, and John Dewey: A commentary. *Int. J. Sports Sci. Coach.* **2017**, *12*, 70–72. [CrossRef]
54. Jiménez, T.R.; Orozco, M. Prompts, Not Questions. Four Techniques for Crafting Better Interview Protocols. *Qual. Sociol.* **2021**, *44*, 507–528. [CrossRef]
55. Savage, J.; Collins, D.; Cruickshank, A. Exploring Traumas in the Development of Talent: What Are They, What Do They Do, and What Do They Require? *J. Appl. Sport Psychol.* **2017**, *29*, 101–117. [CrossRef]
56. Braun, V.; Clarke, V.; Terry, G.; Hayfield, N. Thematic analysis. In *Handbook of Research Methods in Health and Social Sciences*; Liamputtong, P., Ed.; Springer: Berlin/Heidelberg, Germany, 2018; pp. 843–860.
57. Braun, V.; Clarke, V. One size fits all? What counts as quality practice in (reflexive) thematic analysis? *Qual. Res. Psychol.* **2020**, *18*, 328–352. [CrossRef]
58. Braun, V.; Clarke, V. Using Thematic Analysis in Psychology. *Qual. Res. Psychol.* **2006**, *3*, 77–101. [CrossRef]
59. Braun, V.; Clark, V. *Thematic Analysis: A Practical Guide*; Sage: London, UK, 2022.
60. Smith, B.; McGannon, K.R. Developing Rigor In Qualitative Research: Problems And Opportunities Within Sport And Exercise Psychology. *Int. Rev. Sport Exerc. Psychol.* **2018**, *11*, 101–121. [CrossRef]
61. Nowell, L.S.; Norris, J.M.; White, D.E.; Moules, N.J. Thematic Analysis: Striving to Meet the Trustworthiness Criteria. *Int. J. Qual. Methods* **2017**, *16*, 1609406917733847. [CrossRef]
62. Henriksen, K.; Larsen, C.H.; Christensen, M.K. Looking at Success from its Opposite Pole: The Case of a Talent Development Golf Environment in Denmark. *Int. J. Sport Exerc. Psychol.* **2014**, *12*, 134–149. [CrossRef]
63. Eccles, D.W.; Balk, Y.; Gretton, T.W.; Harris, N. "The forgotten session": Advancing research and practice concerning the psychology of rest in athletes. *J. Appl. Sport Psychol.* **2022**, *34*, 3–24. [CrossRef]
64. Collins, D.; Willmott, T.; Collins, L. Periodization and Self-Regulation in Action Sports: Coping With the Emotional Load. *Front. Psychol.* **2018**, *9*, 1652. [CrossRef]
65. Roussin, C.J.; MacLean, T.L.; Rudolph, J.W. The Safety in Unsafe Teams:A Multilevel Approach to Team Psychological Safety. *J. Manag.* **2016**, *42*, 1409–1433.
66. Abraham, A.; Collins, D. Taking the Next Step: Ways Forward for Coaching Science. *Quest* **2011**, *63*, 366–384. [CrossRef]
67. Jowett, S. Coaching Effectiveness: The Coach–Athlete Relationship At Its Heart. *Curr. Opin. Psychol.* **2017**, *16*, 154–158. [CrossRef]
68. Bjørndal, C.T.; Ronglan, L.T. Orchestrating Talent Development: Youth Players' Developmental Experiences in Scandinavian Team Sports. *Sports Coach. Rev.* **2018**, *7*, 1–22. [CrossRef]
69. Nash, C.; Collins, D. Tacit Knowledge in Expert Coaching: Science or Art? *Quest* **2006**, *58*, 465–477. [CrossRef]
70. Collins, L.; Collins, D. Decision Making and Risk Management in Adventure Sports Coaching. *Quest* **2013**, *65*, 72–82. [CrossRef]
71. Côté, J.; Gilbert, W. An Integrative Definition of Coaching Effectiveness and Expertise. *Int. J. Sports Sci. Coach.* **2009**, *4*, 307–323. [CrossRef]
72. Martin, S.; Marks, J. *Messengers: Who We Listen To, Who We Don't, and Why*; Random House: New York, NY, USA, 2019.
73. Jones, R.L.; Corsby, C. A Case for Coach Garfinkel: Decision Making and What We Already Know. *Quest* **2015**, *67*, 439–449. [CrossRef]
74. Cronin, C.; Armour, K.M. 'Being' in the coaching world: New insights on youth performance coaching from an interpretative phenomenological approach. *Sport Educ. Soc.* **2017**, *22*, 919–931. [CrossRef]
75. Cronin, C.; Walsh, B.; Quayle, L.; Whittaker, E.; Whitehead, A. Carefully supporting autonomy—Learning coaching lessons and advancing theory from women's netball in England. *Sports Coach. Rev.* **2019**, *8*, 149–171. [CrossRef]
76. Carless, D.; Boud, D. The Development Of Student Feedback Literacy: Enabling Uptake Of Feedback. *Assess. Eval. High. Educ.* **2018**, *43*, 1315–1325. [CrossRef]
77. Jowett, S.; Cockerill, I.M. Olympic Medallists' Perspective of the Athlete-Coach Relationship. *Psychol. Sport Exerc.* **2003**, *4*, 313–331. [CrossRef]
78. Felton, L.; Jowett, S.; Begg, C.; Zhong, X. A multistudy examination of the complementarity dimension of the coach–athlete relationship. *Sport Exerc. Perform. Psychol.* **2021**, *10*, 27–42. [CrossRef]
79. Gilbert, W.; Côté, J. *Defining Coaching Effectiveness. Routledge Handbook of Sports Coaching*; Routledge: London, UK, 2012.
80. Abraham, A.; Collins, D.; Martindale, R. The coaching schematic: Validation through expert coach consensus. *J. Sports Sci.* **2006**, *24*, 549–564. [CrossRef]
81. Carless, D. Feedback Loops And The Longer-Term: Towards Feedback Spirals. *Assess. Eval. High. Educ.* **2019**, *44*, 705–714. [CrossRef]
82. Coutinho, P.; Mesquita, I.; Fonseca, A.M. Talent development in sport: A critical review of pathways to expert performance. *Int. J. Sports Sci. Coach.* **2016**, *11*, 279–293. [CrossRef]
83. Drasch, K.; Matthes, B. Improving retrospective life course data by combining modularized self-reports and event history calendars: Experiences from a large scale survey. *Qual. Quant.* **2013**, *47*, 817–838. [CrossRef]
84. Cronin, C.; Armour, K.; Enright, K. Caring through science and autonomy: Terry's story. In *Care in Sport Coaching: Pedagogical Cases*, 1st ed.; Cronin, C., Armour, K., Eds.; Routledge: London, UK, 2018.
85. Bowman, C. *Strategy in Practice*; Prentice Hall Europe: London, UK, 1998.

86. Collins, D.; MacNamara, Á. *Talent Development: A Practitioner Guide*; Routledge: London, UK, 2017.
87. Alder, J.P. Team Culture and Athlete Centred Coaching. In *Perspectives on Athlete-Centred Coaching*; Pill, S., Ed.; Routledge: Abingdon, UK, 2018; pp. 57–69.
88. Nembhard, I.M.; Edmondson, A.C. Psychological safety: A foundation for speaking up, collaboration, and experimentation. In *The Oxford Handbook of Positive Organizational Scholarship*; Spreitzer, G.M., Cameron, K.S., Eds.; Oxford University Press: Oxford, UK, 2012; pp. 490–503.
89. Rudolph, J.W.; Raemer, D.B.; Simon, R. Establishing a Safe Container for Learning in Simulation: The Role of the Presimulation Briefing. *Simul. Healthc.* **2014**, *9*, 339–349. [CrossRef] [PubMed]
90. Bjork, R.A.; Bjork, E.L. Desirable Difficulties in Theory and Practice. *J. Appl. Res. Mem. Cogn.* **2020**, *9*, 475–479. [CrossRef]

sports MDPI

Review

Methodological Approaches to Talent Identification in Team Sports: A Narrative Review

Sam Barraclough [1,2,*], Kevin Till [1,3], Adam Kerr [2] and Stacey Emmonds [1]

1 Carnegie School of Sport, Leeds Beckett University, Leeds LS6 3QS, UK; k.till@leedsbeckett.ac.uk (K.T.); s.emmonds@leedsbeckett.ac.uk (S.E.)
2 Leeds United Football Club, Walton Road, Thorp Arch, Leeds LS23 7BA, UK; adam.kerr@leedsunited.com
3 Leeds Rhinos Rugby League Football Club, Bridge Road, Kirkstall, Leeds LS5 3BW, UK
* Correspondence: a.s.barraclough@leedsbeckett.ac.uk

Abstract: Talent identification (TID) and talent development (TD) continue to receive significant investment from team sports organisations, highlighting their importance in attempting to identify potential elite athletes. Accompanying this continual pursuit to unearth future talent is an ever-increasing body of research aiming to provide solutions and strategies to optimise TID and TD processes. Therefore, the aim of this review is to provide a summary and critical synthesis of the methodological approaches applied to TID in team sports and present considerations for future TID research. Specifically, this review highlights three key areas for consideration: (1) the timespan of the research design; (2) the use of monodisciplinary or multidisciplinary variables; and (3) the fidelity of the methodological approaches to the assessment of talent. The review highlights the benefits of longitudinal, multidisciplinary, and ecologically valid research designs for TID within team sports.

Keywords: talent identification; talent development; youth; sport

Citation: Barraclough, S.; Till, K.; Kerr, A.; Emmonds, S. Methodological Approaches to Talent Identification in Team Sports: A Narrative Review. *Sports* **2022**, *10*, 81. https://doi.org/10.3390/sports10060081

Academic Editor: Aaron Scanlan

Received: 27 April 2022
Accepted: 20 May 2022
Published: 24 May 2022

1. Introduction

The lure of success and financial reward is big business for sporting organisations. At the highest levels of team sports, exorbitant investments are made for the recruitment of the most talented athletes who can help organisations accomplish their goals. With this in mind, sporting organisations continue to invest in the identification and development of young talented athletes within their academy systems, with the hope of unearthing or developing potential world class elite athletes of their own [1]. As a product of this investment, youth sport is becoming increasingly professionalised, with organisations now supplying considerable resources for talent identification (TID). TID is defined as "recognising players participating in the sport who have the potential to excel" [2] (p. 1). Once identified, organisations aim to provide appropriate learning environments so that such athletes have the opportunity to realise their potential by maximising the training and development opportunities of prospective talents with the greatest potential for success [3,4]. This occurs through talent development (TD), defined as a "relatively systematic combination of coaching, support, training, and match play designed to progress players" [2] (p. 1).

Due to the popularity and growth of elite athlete development programmes, there is a vast and diverse quantity of TID research available across multiple sports. The variety and depth of such research has been important in establishing an evidence base, providing valuable reference data across sports in multiple disciplines (e.g., technical, tactical, physical, psychosocial), that may be used to distinguish between performance levels. Yet, this volume of research has potentially led to contrasting opinions and widespread misconceptions of talent in high performance team sport settings [5]. For example, it is acknowledged that TID is a complicated process, with the question of "what is talent?" alone proving to be a highly divisive and contradictory topic [6,7]. Due to a lack of consensus on a definition and objective measure of talent, TID (for the purpose of this review) refers to recognising current participants with the potential to progress or to become an elite athlete [2,8]. TID has typically inferred potential based on current performance level [9], yet Bergkamp et al. [10] argued that using performance level as an outcome for TID (i.e., elite vs. non-elite) may be misleading. Performance level is a consequence of one or more (de)selection decisions, and therefore, may only reflect a perception of talented and less talented individuals, rather than an objective measure of talent or potential. Without a clear measure for identifying future elite players, TID has become a significant contributor to research on youth team sport athletes; however, with such a substantial amount of literature, issues emerge relating to the diversity of research methods.

As a key area for research within team sports, several recent review articles [5,10–15] raised issues regarding current approaches to TID. These include a lack of longitudinal research designs, the use of monodisciplinary research designs (i.e., physical, psychological, technical, tactical, etc.), and low-fidelity performance characteristics (e.g., isolated sprint testing). However, these methodological issues have yet to be examined collectively, with specific application to team sports as a whole. To date, there are no real consensuses on the underpinning methodology for talent identification or which characteristics of talent may best distinguish athletes which are most likely to progress to the elite level. Researchers and practitioners continue the search for the distinctive characteristics responsible for achieving sporting excellence, but is it fair to say that no consensus may ever be reached, considering the ever-evolving complex and dynamic nature of team sports and the subjective opinions of what constitutes successful performance [16]. Given the ever-expanding volume of interest, research and applied practice surrounding TID, this narrative review aims to provide a summary and critical synthesis of the methodological approaches to talent identification in team sports and to present considerations for future TID research.

2. Talent Identification Research Designs

2.1. Cross-Sectional Research

Cross-sectional research designs are the most common methodological approach in TID research (i.e., 68% of studies according to a recent scoping review by Baker et al. [11]). Cross-sectional studies often measure specific characteristics within different disciplines (e.g., speed, endurance (physical), passing, dribbling (technical), motivation, confidence (psychological), game intelligence, and general tactics (tactical)) at a one-off timepoint and make comparisons across two or more distinct groups. Previous research has included comparisons of elite vs. non-elite athletes [17], selected vs. non-selected regional athletes [18], academy vs. school athletes [19] or regional vs. national athletes [20]. This type of research is often used to measure the characteristics believed to be linked to successful performance in a cross-section of the sample of interest [21]. Such cross-sectional research designs provide a "snapshot" of performance at a moment in time, which is perhaps indicative of an individual's expertise or talent.

Cross-sectional study designs have been used in TID across multiple team sports, including soccer [22], rugby union [19], Australian football [23], netball [24], rugby league [25], basketball [26], and field hockey [27]. Whilst this research is of value, the efficacy of cross-sectional designs in identifying talented youth athletes remains in question. For example, research by Gil et al. [28] examined the selection process of a professional soccer club in Spain to identify the physical characteristics of players who were selected into the club's academy. Players who were selected between the ages of 9 and 10 years were leaner (48.9 mm vs. 66.2 mm sum of skinfolds, $p < 0.01$), quicker (4.96 s vs. 5.53 s in a 30-m sprint test, $p < 0.001$), more agile (5.81 s vs. 6.38 s in a 30-m agility test, $p < 0.001$), jumped higher (29.1 cm vs. 26.9 cm in a countermovement jump test, $p < 0.01$) and possessed greater aerobic endurance (618 m vs. 464 m in the yoyo intermittent recovery level 1 test, $p < 0.01$) than a control group from an open soccer camp who were not selected to train in the club's academy. If physical advantages at a young age, as observed by Gil et al. [28], are used in TID and selection processes, this seems heavily reliant on the assumption that any physical advantages would remain consistent within individuals across childhood and adolescence, and transfer to adult performance. This fails to account for the influences of individual growth and maturation [29–32] and the effects of development (i.e., practice, coaching and training) [2]. Similarly, research by Zuber and Conzelmann [33] demonstrated elite youth ice hockey players with higher intrinsic motivation (assessed via 5 motivational constructs–win orientation, goal orientation, hope for success, fear or failure and self-determination), were rated as better players by their coaches (using a 1–100 scale) when judging game performance, in comparison to their less motivated counterparts. Therefore, a key limitation of a cross-sectional research design as a methodological approach is that assessing performance, at a singular time-point, as an indicator of talent, provides limited information on future potential. This is partly due to the non-linear and dynamic nature of development in talented elite youth athletes [34,35], where variables that correlate with a performance advantage at young ages (e.g., an early developing basketball athlete with greater height) may not necessarily be the same factors explaining adult performance or that the individual's height may be an advantage in adulthood [6]. Research evidence shows the disparate development among youth athletes. For example, a longitudinal case study by Moran et al. [36] displayed substantial fluctuations in academy soccer player's sprint and jump performances over a 6-year period. Such research confirms that one off performance measures are likely temporary representations of athletic capabilities, where current performance is interpreted as a proxy for potential [9].

In summary, whilst cross-sectional data used in TID is advantageous for comparisons between groups or athletes at a singular timepoint, the inclusion of cross-sectional data in identification or de(selection) decisions within long-term TID/TD programmes can be considered imprudent, as it may prematurely exclude late-developing athletes, given the non-linear development of certain characteristics that may affect performance (e.g., speed, [36]). A more suitable approach is likely to be based on serial measurements of these characteristics over time, to better understand the trajectory of an elite youth team sport athlete's development [37].

2.2. Longitudinal Research

Longitudinal research has been used to follow a cohort of athletes and assess changes in characteristics at two or more time-points [38]. Through taking repeated measurements of an athlete or group of athletes, a longitudinal research design can assess the characteristics that may be linked to performance whilst also assessing changes and development over time [39]. In practice, longitudinal research has greater affinity than cross-sectional research to TD, where regular assessments can serve as a monitoring tool for a group of athletes. Longitudinal research surrounding TID is less common, research that does exist has demonstrated variations in the long-term development of certain characteristics between differing groups, in several sports including rugby league [40], field hockey [41], handball [42], soccer [43], and Australian rules football [44]. Key findings of such studies are summarised in Table 1. Studies were selected as being representative of a variety of team sports, having a minimum of three measurement occasions and a study period of at least 12 months in order to represent longitudinal change between groups that was not attributable to short-term intervention.

Table 1. Examples of Longitudinal Research for TID in Team Sports.

Authors/Sport	Sample/Timeframe	Objectives	Key Findings
Till et al., 2013 [40]/rugby league	81 male junior rugby league players from under 13-under 15/3 consecutive years.	Compare longitudinal development of physical and anthropometric characteristics considering position and selection level in junior rugby league players.	1. Selection level (national vs. regional) had a significant overall main effect on physical and anthropometric characteristics. 2. Players who moved up in selection level significantly improved sprint speed and were the quickest at under 15 age category. 3. There was a significant interaction between maturation and time for sprint speed, vertical jump, and medicine ball throw.
Matthys et al., 2013 [42]/handball	94 youth handball players from under 14-under 18/3 consecutive seasons.	Assess longitudinal changes in anthropometry and physical performance between elite and non-elite handball players.	1. Elite players did not improve their physical performance more rapidly than non-elites and had similar anthropometric profiles. 2. Elite players performed significantly better on the intermittent endurance, speed, and coordination items. It was revealed Yo-Yo distance and coordination with and without ball discriminated most between the two playing levels.
Roescher et al., 2010 [43]/soccer	130 male youth soccer players aged under 14-under 18/5 consecutive years with the exception of 1 year.	Investigate the development of intermittent endurance capacity, the underlying mechanisms affecting this development and attained adult playing level in talented youth soccer players.	1. From 15 years of age players who reach professional status show a faster development pattern than non-professionals. 2. Both hours spent in soccer-specific training and hours spent in additional training were positively related to the development of intermittent endurance capacity.
Elferink-Gemser et al., 2007 [41]/field hockey	30 elite and 35 sub-elite male and female youth field hockey players from under 14-under 16/3 consecutive years.	Identify the performance characteristics that may help identify future elite hockey players.	1. Both male and female elite players scored better than sub-elite on technical and tactical variables. 2. Female elite players also scored better on interval endurance capacity, motivation, and confidence. 3. Male and female elite players improved more than their sub-elite counterparts on interval endurance capacity and slalom dribble across the study period.
Pyne et al., 2005 [44] /Australian rules football	283 Australian rules football players/ 3 consecutive years.	Determine the relationships between anthropometrics and physical fitness tests and subsequent career progression.	1. Drafted players were faster (5, 10 and 20-m), had higher estimated VO2 max and a faster agility run performance than non-drafted players. 2. No substantial differences in anthropometric or jump tests were found between drafted and non-drafted players.

Whilst cross-sectional data can provide differences in characteristics between two distinct groups at singular timepoints, longitudinal research [45,46] provides practitioners with a measure of athlete progression to assess the effectiveness of TID/TD processes [31]. However, one major methodological challenge to longitudinal research is participant dropout, where repeated measures cannot be taken of athletes who are not afforded the opportunity to progress. This is highlighted in the work of Moran et al. [36] who's final sample of 6 athletes (from an initial 140) were the only individuals to achieve the longevity required for the 6-year period of study on longitudinal monitoring of physical characteristics within a single professional soccer academy. In such cases, a more thorough estimation of sample size requirements that accounts for participant attrition and expected drop out rates may help overcome such methodological challenges.

Most longitudinal research measures change on a group level, possibly sacrificing insight into changes on an individual level, which may provide a more in-depth understanding of development. Through monitoring longitudinal changes in the characteristics that underpin successful performance, researchers and practitioners are likely to be provided with a more valid, continuous indicator of an athlete's potential to progress based on that athlete frequently achieving the necessary characteristics to be retained within a TD programme. For example, an athlete who progresses through an academy and avoids deselection is likely to possess superior characteristics in one or more disciplines (physical, technical, tactical, psychological) at multiple timepoints, from both an objective (standardised assessments) and/or subjective (coach's perceptions) perspective, in comparison to their deselected peers. This allows them to continue in the pathway and have an opportunity to reach the professional level in their sport [6], rejecting the notion of TID as a transient process.

2.3. *Prospective/Retrospective Research Designs*

When discussing methodological issues surrounding TID in soccer, Bergkamp et al. [10] stated a key focus of TID research is to evaluate the predictive value of performance characteristics, not just to identify such characteristics. Research has attempted to both prospectively track an athlete's development into professional status [47], as well as retrospectively examine their development once professional status has been attained [48]. Approaching TID through prospective and retrospective research designs, often leads to TID being conceptualised as a direct relationship between a factor (e.g., height) and adult performance in a particular team sport (e.g., volleyball). For example, research in soccer players who went on to play at international or professional levels as adults, displayed superior performance in several anthropometrical and fitness measures at under 14 to under 16 age groups (i.e., height, body mass, maximal anaerobic power, countermovement jump, 40-m sprint time) [49]. More recent research supports such findings showing that future professional soccer players outperformed their non-professional counterparts in measures of speed (5/10/20-m sprint times), power (countermovement jump height), and endurance (distance covered in yoyo intermittent recovery test level 1) from age ~13/14 years onward [39]. Similar findings have also been shown when investigating psychological [50], tactical [51] and technical [52] characteristics, as well as multidimensional research designs [53]. For instance, Forsman et al. [53], found future elite players outscored non-elite players, at 15 years of age, in tests of dribbling and passing, passing and centering (technical), speed, agility, endurance (physical), motivation (psychological), and "acting in changing situations" (tactical). Whilst these examples of research may aid in establishing characteristics associated with future success (i.e., having better characteristics), research still fails to provide insight into the individual, non-linear developmental patterns of such characteristics [48].

A methodological approach that considers the dynamic nature of TID/TD as a long-term process, whilst also considering future career outcome, allows practitioners and researchers to further understand and examine the relationships and individual developmental trajectories that may influence the future career attainment of the most talented team sport athletes [48]. Studies using such an approach (i.e., longitudinal retrospective) are uncommon in the literature, with some exceptions [38,47,48]. For example, Till et al. [38] retrospectively examined the development of physical characteristics between 13–15 years of age for those players who attained professional, academy and amateur status in rugby league. It was found that the enhanced development of sitting height, speed, change of direction speed and estimated maximal oxygen consumption (VO_2 max) between 13–15 years of age could differentiate between career attainment outcome of professional and amateur players. Similar findings in soccer [48] showed different patterns of development in tests of vertical jumping and slalom agility when prospectively tracking future professionals and non-professionals, with professionals improving at a faster rate between 12–18 years of age. In contrast, Leyhr et al. [47] found no significant interactions between speed and technical skill development and future adult performance level (i.e., professional vs. non-professional). It should be noted however, inconsistencies in definitions of professional status were observed between the studies, with Leyhr et al. [47] limiting their scope to professional players only within Germany. These contrasting findings potentially suggest a lack of generalisability outside of their respective environments (e.g., sport, country), but also to the wider population due to the restriction in the range present in the respective samples typified by the homogeneity of groups (i.e., selection bias of team sport athletes selected to some form of TID programme [10]). Additionally, the selected studies tended to assess longitudinal development and career attainment interactions at a group level, where a case-by-case individual analysis of players may provide more insight [47].

As such, research designs may aim to identify characteristics important for successful performance, track the fluctuating development of these characteristics through periods of adolescence/maturity, and evaluate their relevance in future career outcomes assessed on an individual level. It should also be noted that due to the complex, myriad of factors responsible for team sport performance, research that is mono-disciplinary in nature (i.e., only examining one component of performance, such as physical characteristics) cannot provide a complete picture of TID. As an extension, research that incorporates an array of potential future successful performance characteristics, and their interactions, into a longitudinal evaluation of the player, appears to be the optimal approach for TID/TD purposes [37].

3. A Multidisciplinary Approach

One proposed component of talent is its multi-dimensional nature [6]. Whilst the call for research to adopt a multidisciplinary approach is a recurring message [4,8,13,32,35,54], both current and previous research surrounding TID in team sports has often adopted monodisciplinary designs [19,39,50–52]. This was highlighted by an underrepresentation of multidimensional designs in a recent review [5]. The lack of multidimensional designs is perhaps due to the fact that, in reality, the identification of talented individuals is difficult to objectively explain [6]. This is accompanied with the associated methodological challenges of needing to measure variables from each discipline in their entirety [46], combine these into a tool for TID purposes, and implement this across large samples (e.g., nationwide TID processes [55,56]).

Given the challenges of a multidisciplinary research design, a mono-disciplinary design is often utilised. Despite some of the limitations highlighted above, this approach can still provide rich insights for both researchers and practitioners. For example, research from several sports has solely examined physical qualities in relation to TID [57–61]. Additional mono-disciplinary research has shown the value of assessing tactical [51,62], technical [52,63], psychological [33,64], and even genetic traits [65] within TID. In such cases, it may be interpreted that through mono-disciplinary evaluations, an individual's superiority in one characteristic (e.g., speed) can potentially compensate for weakness in others (e.g., technical/tactical) [35]. Whilst such examples provide a valuable source of information for TID/TD, a mono-disciplinary approach to research, where the outcome variable is related to only one discipline of performance (e.g., physical characteristics), may not fully explain the intricacies of individual talent and development, as it fails to consider "*the interaction of many different elements spinning in the contextual web of final performance*" [66] (p. 2).

The interactions of such elements can also be problematic during the decision-making process for coaches [9]. Namely, the use of multiple sources of information across disciplines in TID decision-making can lead to athlete's having similar summative scores (across all characteristics) but very different individual performance scores. Figure 1 provides three hypothetical examples of different athlete types where such challenges may occur. In such cases, the decision to de(select) athletes becomes more complex. Here, each athlete has a very similar summative score, creating a choice between those with "the overall package" (even scores across all characteristics—Athlete 1) or those with "something special" (greater scores in specific characteristics—Athlete's 2 and 3), who's weaknesses could potentially be masked or substituted by other players within a team sport [9]. As each athlete's individual profile is unique to them, a multidisciplinary approach allows the identification of an athlete's ability in various disciplines and characteristics relevant for performance in their sport, whilst also allowing support staff within the environment to evaluate such strengths and weaknesses in order to facilitate a more individualised plan of development [67].

Figure 1. Hypothetical performance comparison for 3 athletes.

In this regard, a multidisciplinary approach in research to TID may allow for a more holistic profile of youth team sport athletes and increase the utility of TID [23,41,68,69]. Some examples of multidisciplinary research from various team sports are presented in Table 2.

Table 2. Examples of Multidisciplinary TID Research.

Authors/Sport	Sample	Variables	Disciplines	Key Findings
Dimundo et al., 2021 [70]/Rugby Union	74 elite under 15 male youth rugby union players.	Height, body mass, 10-m and 20-m sprint time, counter-movement jump, isometric hip extension, dominant handgrip strength, date of birth, perceptual-cognitive video simulation.	Physical, tactical	1. Selected players to an academy outperformed those not selected in body mass, handgrip strength, isometric hip extension and 20-m sprint ($p < 0.05$). 2. No significant differences were apparent for the perceptual-cognitive test.
(Elferink-Gemser et al., 2004) [27]/Field Hockey	126 elite male and female youth field hockey players aged 11–16.	Height, body mass, percentage body fat, peak shuttle sprint, repeated shuttle sprint, slalom sprint performance, interval endurance capacity, peak shuttle dribble, repeated shuttle dribble, slalom dribble, general tactics, tactics for possession of the ball, tactics for non-possession of the ball, motivation, confidence, anxiety control, mental preparation, team emphasis and concentration.	Physical, technical, tactical, psychological	1. Stepwise discriminant analysis predicted better tactics for possession of the ball, being younger, having a higher motivation, and a quicker slalom dribble could best discriminate between elite and sub-elite players. 2. Elite youth players scored better than sub-elite youth players on technical (peak dribble and repeated dribble shuttle run), tactical (general tactics, tactics for possession and non-possession of the ball) and psychological variables (motivation) ($p < 0.05$).
Falk et al., 2004 [45]/Water Polo	24 male youth water polo players aged 12–14.	50, 100, 200 and 400-m freestyle swim, 100-m breaststroke, 100-m butterfly, 50-m dribbling, throwing at goal, throwing for distance, vertical jump from water, game intelligence.	Physical, technical, tactical	1. Two years before selection to the junior national team, players who were selected outperformed those non-selected on game-intelligence, 50-m dribbling and all swim tasks except 50-m freestyle and 100-m breaststroke ($p < 0.05$). 2. Using an average rank score, predictions for 67% of players were in agreement with final selections.
Sieghartsleitner et al., 2019/Soccer	117 elite under 14 youth soccer players.	Age, relative age, age at peak height velocity, height, body mass, in-game performance, YoYo intermittent recovery test level 1, 40-m sprint, agility, dribbling, passing, juggling, achievement motive, achievement goal orientation, self-determination, importance of football within family, parent's priority of sport vs. school, financial investment, time investment, practice and play up to age 12.	Physical, technical, psychological, sociological	1. A holistic model combining all predictor variables had the greatest accuracy (88%) in correctly predicting who would achieve professional vs. non-professional status 5 years later.
Woods et al., 2015 [71]/Australian Rules Football	84 elite under 18 Australian rules football (AF) athletes.	Standing height, dynamic vertical jump height on non-dominant leg, 20-m multistage fitness test, kicking, handballing, video decision-making.	Physical, technical, tactical	1. Those selected for state representation ("talent identified") outperformed non-talent identified on each test ($p < 0.05$). 2. Using a summative score receiver operating characteristics were able to correctly classify 95% of talent identified and 86% non-talent identified participants (AUC = 95.4%).

As highlighted by some of the selected research in Table 2, whilst a multidisciplinary approach is becoming more prominent in TID research within team sports, research within specific disciplines appears more common (e.g., physical). For example, Dimundo et al. [70] utilised seven physical characteristics in comparison to one tactical, when investigating differences in selected and non-selected academy rugby union players, a finding that appears common across selected TID research with physical characteristics more routinely measured [27,45,71]. This is perhaps partly due to the difficulty in assessing some characteristics (e.g., assessing an athlete's tactical knowledge through retrospective video analysis [70,71]), compared to the ease of assessing others where the application of physical testing batteries and anthropometric measurements are commonplace within TID/TD environments. Where including characteristics from all disciplines in order to provide a balanced, comprehensive approach is not viable, research might aim to evaluate the relative importance of each characteristic relative to their sport. A case study by Jones et al. [72] utilised such an approach, i.e., the perceived importance of various fitness tests from a coach and player's perspective as a weighting factor for ranking the importance of certain physical qualities for individual players. Again, however, such research is limited to physical discipline, and further research across other disciplines is required.

Despite the multidimensional nature of the studies listed in Table 2, each used a cross-sectional research design [45,70,71,73] or only observed mean performance across two time-points [74], thus failing to understand if the longitudinal development of any of the investigated characteristics influenced TID decisions. Nevertheless, adopting a multidisciplinary approach to TID research appears more valid and applicable to team sports, as team sports require the interaction of multiple characteristics across disciplines [4]. From this perspective it becomes clear that performance in team sports is not synonymous with one set of characteristics from a single discipline, and yet the dominant approach within research is to assess perceived characteristics of relevance within disciplines in isolation [75].

4. Signs and Samples

4.1. Signs

A large volume of research across various team sports has recognised the multidisciplinary nature of sports performance, but often in TID research the isolated circumstances in which an athlete's characteristics are assessed bears little resemblance to performance itself. For example, some predictors of performance in numerous team sports include physical (i.e., speed, strength, and endurance characteristics [19,60,61]), psychological (i.e., achievement motive, motivation, self-confidence and concentration [27,33,73]), technical (dribbling, kicking and shooting [41,71,76]), and tactical (positioning and deciding, pattern recognition [53,64]). Such characteristics are commonly measured in discrete, controlled circumstances such as laboratory or field based-tests in order to obtain reliable and standardised results—a far cry from the open and often chaotic environment in which these characteristics are utilised during team sport performance.

Using a term borrowed from psychology literature, characteristics measured in this way can be termed as "signs" and are said to be conceptually related predictors of the future behaviour or performance of interest [77]. Sign-based tests are said to lack "fidelity" [10], in that they are distinct characteristics measured in a dissimilar task and context to that of the criterion behaviour (team sport performance). For example, assessing speed as a physical characteristic deemed important for differentiating talented and less-talented individuals in terms of their future sport performance using a signs approach may take the form of a 20-metre sprint test (see [22]). Here athletes would be expected to complete multiple trials of a linear sprint, commonly from a stationary start, over a pre-defined distance and with adequate rest-periods to reduce any potential elements of fatigue. In comparison, during actual performance, an athlete would most likely be already moving or adopting a different body position, may need to sprint in a curvilinear fashion and/or include changes of direction and is likely fatigued from prior actions performed. This is then further compounded by the interactions with moving opponents and team-mates, and the perceptual-cognitive and decision-making requirements of such a task. Therefore, a key methodological concern of a signs-based approach is that whilst providing a reliable and valid measure of a specific characteristic for each athlete in that setting, it is clear such an approach lacks resemblance in terms of task and context to how such characteristics would be utilised during on-field team sport performance. In contrast, given the complex, multi-faceted nature of team sport and the inherent difficulty of measuring individual team sport performance, breaking down performance into predictors from various disciplines and investigating their impact on predicting success and future performance makes sense from a practical perspective [10]. Particularly when many of these predictors have been shown to discriminate between performance levels [17,19,70,78,79].

4.2. Samples

If performance, skill, or expertise is viewed as the end-goal or outcome (Baker et al. [6]), then it would seem logical for TID research measuring the precursors to these outcomes, to attempt to mimic these criterion behaviours as closely as possible [75]. Such an approach can be termed as "sample" based, in that researchers sample a behaviour in a highly representative context, providing a higher fidelity measure. This sample is more analogous to the criterion (performance) and therefore likely has greater utility in TID for assessing those with greater potential for future performance, particularly in homogenous groups such as team sports [75]. As talent can be viewed as a complex and dynamic construct where future behaviours stem from the combination of psychological, technical, tactical, and physical characteristics [35], a samples approach does appear more valid within TID research in order to investigate how such multidisciplinary characteristics interact and combine to predict or measure actual team sport performance.

Examples of establishing a samples-based approach can be seen from recent research in soccer, where small-sided games (SSGs) have been investigated as potential tools for TID, as they obtain performance under similar task, environmental and behavioural conditions [80–82] and have been validated showing moderate-to-large relationships to actual 11 v 11 performance [82]. Fenner et al. [80] investigated player performance in SSGs (subjective scoring of technical aspects rated by the coaches) and match result. There was a significant and large relationship between players judged to have higher technical scores within the SSGs and those found to have more success in SSGs based on an accumulation of points for goals scored and match outcome ($r = 0.76$, $p < 0.001$). In addition, Bennett et al. [81] showed that higher skilled players (trained within a professional academy) had a significantly greater number of attempted and completed skill involvements in SSGs compared with low-level players (trained within a local academy) ($p < 0.01$). Further research within American Football demonstrated that samples of previous performance, measured via position specific in-game statistics (e.g., percentage pass completion for a quarterback), across a 1-year period at college level, was a statistically significant predictor ($p < 0.05$) of subsequent performance in the National Football League (NFL), whereas signs of performance (i.e., physical tests in the NFL Combine) failed to demonstrate predictive power of future NFL performance [83]. Equally, in Australian Football, O'Connor et al. [64] demonstrated a significant difference in recent match-play performance (sample) between selected and non-selected athletes into a national programme ($p < 0.001$). Recent match performance was also identified as a predictor variable that could discriminate between selected and non-selected, with a large standardised coefficient (0.851), indicating its importance. It should be noted however, that recent match performance in this study was based upon a coded variable indicating selection for participation in regional camps and tournaments and thus this sample of behaviour may reflect perceived match performance as opposed to actual performance.

4.3. Subjective Expert Opinion

Given the complexity of sampling performance in its entirety, one method utilised in order to provide a samples-based assessment is the inclusion of a subjective expert opinion (SEO), where a coach or practitioner can provide a holistic rating of player performance (e.g., a score from 1 to 4, [84]. Research has shown that inclusion of subjective ratings from coaches improves predictive models within TID in comparison to objective data alone [55,74]. However, the basis of and validity of such ratings is yet to be established with research showing a lack of agreement between coaches [85], an inability for coaches to accurately rate performance within specific disciplines (e.g., physical, [86,87]) and suggestions that ratings are potentially biased [82] and could be based on a coaches' perceived ability to influence and develop a player rather than solely on athlete ability alone [88,89]. Evidence of such biases has shown subjective ratings may vary based on an individual's stage of maturation and rate of growth, with a trend for ratings to decline for players around the time of their growth spurt, before increasing again post growth spurt [84]. Equally, it may be expected that maturity timing (e.g., late vs. early) may influence coach ratings, as early maturing players typically have physical advantages in size, strength, and speed versus their less mature counterparts [90]. In such scenarios, a samples approach where individuals are grouped relative to their biological age (i.e., "bio-banding") may remove such physical biases, allowing later maturing players more opportunity to exhibit their tactical and technical proficiency [91], potentially facilitating a more valid sample of performance through SEO. Due to the lack of evidence on the validity and reliability of SEO's, there are concerns regarding the use of coach ratings alone, as they may lack a shared and explicit criterion upon which ratings are based. Given such information, TID should attempt to utilise both objective and subjective profiling information to help inform their decision-making processes rather than solely rely on clinical judgement [75].

5. Conclusions

The current review highlights three key methodological approaches relevant to TID research, namely, the time-course of the research design (i.e., cross-sectional or longitudinal; prospective or retrospective), the disciplines of interest (i.e., mono or multi-disciplinary designs), and the assessment method applied (i.e., signs or samples). These methodological approaches have a range of strengths and limitations regarding TID research, and remain pertinent within research related to any team sport due to the dynamic, multidimensional, and complex demands of such sports.

To summarise, cross-sectional designs fail to account for the non-linear development of youth athletes and the emergenic, dynamic and symbiotic conceptualisation of talent [6]. This may potentially lead to misrepresentations of an individual's potential when undertaking (de)selection decisions, as different characteristics will evolve and develop at different rates for each individual athlete, in conjunction with the potentially confounding effects of growth, maturation, and development [31,61,92]. With this in mind, it is proposed that a longitudinal approach to TID research may be more beneficial, as it may provide insights into the individual developmental changes of indicators of talent and their effect on (de)selection decisions.

Equally, although team sports are complex, dynamic, and multi-dimensional in nature [74], TID research is often monodisciplinary. This is perhaps due, in part, to the relative ease of examining certain characteristics (i.e., anthropometric and physical characteristics) which are often routinely measured within embedded TID programmes (i.e., pre-season testing). In this regard, a multidisciplinary approach to TID is recommended to provide a more holistic evaluation of an athlete, accounting for their strengths and weaknesses in multiple aspects of performance, which can further facilitate TD and (de)selection processes.

Finally, the context in which indicators of TID are measured must be questioned. Discrete and controlled tests ("signs"), whether conducted in the laboratory or field, lack ecological validity and transference to actual performance within team sports. Accordingly, a samples-based approach may be more appropriate in TID programmes, where judgements are made based on assessments that more closely mimic the context, environment, and task of team sport performance [23,64,81,83], including the subjective expert opinions of relevant staff [84,86,93].

6. Directions for Future Research

Regardless of the sport, TID is and will remain a key area of interest within both research and practice. Despite the plethora of methodological approaches, the current review highlights and reinforces some key considerations for future research:

- Future TID research should strive to adopt a longitudinal research design in order to provide regular and comprehensive evaluations of athlete's performance in relevant characteristics and their individual rates of change as possible indicators of potential.
- A multidisciplinary approach to research would allow for more comprehensive athlete profiling and serve not only as a potential tool for TID but to also augment TD processes within team sport environments.
- Investigating both objective and subjective data through a combined approach of signs, samples and subjective expert opinions would allow researchers to bridge the gap between relevant characteristics and their transfer to performance, with an added perspective from "the coach's eye".

7. Practical Applications

As well as providing recommendations for future research, these methodological considerations should also serve as a comprehensive framework to athlete profiling, thus informing TID, TD and talent selection processes.

A comprehensive approach to athlete profiling should:

- Identify key actions for successful match play and the underpinning multidisciplinary characteristics required to perform such actions.
- Profile the actions and relevant characteristics through multiple methods—signs, samples, and subjective expert opinions.
- Repeat the profiling longitudinally to account for non-linear development whilst also examining the trend of development as an indicator of potential i.e., showing the capacity to successfully perform such actions in the future.

Author Contributions: Conceptualisation, S.B., S.E., K.T. and A.K.; writing—original draft preparation, S.B.; writing—review and editing, S.E., K.T. and A.K.; supervision, S.E., K.T. and A.K. All authors have read and agreed to the published version of the manuscript.

Funding: This research received no external funding.

Institutional Review Board Statement: Not applicable.

Informed Consent Statement: Not applicable.

Data Availability Statement: Not applicable.

Conflicts of Interest: The authors declare no conflict of interest.

References

1. Larkin, P.; Reeves, M.J. Junior-Elite Football: Time to Re-Position Talent Identification? *Soccer Soc.* **2018**, *19*, 1183–1192. [CrossRef]
2. Williams, A.M.; Ford, P.R.; Drust, B. Talent identification and development in soccer since the millennium. *J. Sports Sci.* **2020**, *38*, 1199–1210. [CrossRef] [PubMed]
3. Woods, C.T.; Joyce, C.; Robertson, S. What are talent scouts actually identifying? Investigating the physical and technical skill match activity profiles of drafted and non-drafted U18 Australian footballers. *J. Sci. Med. Sport* **2015**, *19*, 419–423. [CrossRef] [PubMed]
4. Vaeyens, R.; Lenoir, M.; Williams, A.M.; Philippaerts, R.M. Talent Identification and Development Programmes in Sport Current Models and Future Directions. *Sports Med.* **2008**, *38*, 703–714. [CrossRef]
5. Johnston, K.; Wattie, N.; Schorer, J.; Baker, J. Talent Identification in Sport: A Systematic Review. *Sports Med.* **2017**, *48*, 97–109. [CrossRef]

6. Baker, J.; Wattie, N.; Schorer, J. A proposed conceptualization of talent in sport: The first step in a long and winding road. *Psychol. Sport Exerc.* **2018**, *43*, 27–33. [CrossRef]
7. Issurin, V.B. Evidence-Based Prerequisites and Precursors of Athletic Talent: A Review. *Sports Med.* **2017**, *47*, 1993–2010. [CrossRef]
8. Williams, A.M.; Reilly, T. Talent identification and development in soccer. *J. Sports Sci.* **2000**, *18*, 657–667. [CrossRef]
9. Baker, J.; Schorer, J.; Wattie, N. Compromising Talent: Issues in Identifying and Selecting Talent in Sport. *Quest* **2017**, *70*, 48–63. [CrossRef]
10. Bergkamp, T.L.G.; Niessen, A.S.M.; Hartigh, R.J.R.D.; Frencken, W.G.P.; Meijer, R.R. Methodological Issues in Soccer Talent Identification Research. *Sports Med.* **2019**, *49*, 1317–1335. [CrossRef]
11. Baker, J.; Wilson, S.; Johnston, K.; Dehghansai, N.; Koenigsberg, A.; De Vegt, S.; Wattie, N. Talent Research in Sport 1990–2018: A Scoping Review. *Front. Psychol.* **2020**, *11*, 607710. [CrossRef] [PubMed]
12. Dimundo, F.; Cole, M.; Blagrove, R.; Till, K.; McAuley, A.; Hall, M.; Gale, C.; Kelly, A. Talent Identification and Devel-opment in Male Rugby Union: A Systematic Review. *J. Expertise* **2021**, *4*, 33–55.
13. Wrang, C.M.; Rossing, N.N.; Agergaard, S.; Martin, L.J. The missing children: A systematic scoping review on talent identification and selection in football (soccer). *Eur. J. Sport Soc.* **2021**, 1–17. [CrossRef]
14. Kelly, A.L.; Williams, C.A. Physical Characteristics and the Talent Identification and Development Processes in Male Youth Soccer: A Narrative Review. *Strength Cond. J.* **2020**, *42*, 15–34. [CrossRef]
15. Sarmento, H.; Anguera, M.T.; Pereira, A.; Araújo, D. Talent Identification and Development in Male Football: A Systematic Review. *Sports Med.* **2018**, *48*, 907–931. [CrossRef]
16. Fortin-Guichard, D.; Huberts, I.; Sanders, J.; van Elk, R.; Mann, D.L.; Savelsbergh, G.J. Predictors of selection into an elite level youth football academy: A longitudinal study. *J. Sports Sci.* **2022**, *40*, 984–999. [CrossRef]
17. Matthys, S.P.J.; Vaeyens, R.; VandenDriessche, J.; Vandorpe, B.; Pion, J.; Coutts, A.J.; Lenoir, M.; Philippaerts, R.M. A multidisciplinary identification model for youth handball. *Eur. J. Sport Sci.* **2011**, *11*, 355–363. [CrossRef]
18. Silva, M.J.C.; Figueiredo, A.J.; Simões, F.; Seabra, A.; Natal, A.; Vaeyens, R.; Philippaerts, R.; Cumming, S.P.; Malina, R.M. Discrimination of U-14 Soccer Players by Level and Position. *J. Sports Med.* **2010**, *31*, 790–796. [CrossRef]
19. Jones, B.; Weaving, D.; Tee, J.; Darrall-Jones, J.; Weakley, J.; Phibbs, P.; Read, D.; Roe, G.; Hendricks, S.; Till, K. Bigger, stronger, faster, fitter: The differences in physical qualities of school and academy rugby union players. *J. Sports Sci.* **2018**, *36*, 2399–2404. [CrossRef]
20. Till, K.; Cobley, S.; O'Hara, J.; Brightmore, A.; Cooke, C.; Chapman, C. Using anthropometric and performance characteristics to predict selection in junior UK Rugby League players. *J. Sci. Med. Sport* **2011**, *14*, 264–269. [CrossRef]
21. Gratton, C.; Jones, I. *Research Methods for Sports Studies*, 2nd ed.; Routledge: London, UK, 2010.
22. Murtagh, C.F.; Brownlee, T.E.; O'Boyle, A.; Morgans, R.; Drust, B.; Erskine, R.M. Importance of Speed and Power in Elite Youth Soccer Depends on Maturation Status. *J. Strength Cond. Res.* **2018**, *32*, 297–303. [CrossRef] [PubMed]
23. Tribolet, R.; Bennett, K.; Watsford, M.L.; Fransen, J. A multidimensional approach to talent identification and selection in high-level youth Australian Football players. *J. Sports Sci.* **2018**, *36*, 2537–2543. [CrossRef] [PubMed]
24. McKenzie, C.R.; Whatman, C.; Brughelli, M. Performance Profiling of Female Youth Netball Players. *J. Strength Cond. Res.* **2020**, *34*, 3275–3283. [CrossRef]
25. A Pearce, L.; Sinclair, W.; Leicht, A.S.; Woods, C.T. Passing and tackling qualities discriminate developmental level in a rugby league talent pathway. *Int. J. Perform. Anal. Sport* **2019**, *19*, 985–998. [CrossRef]
26. Arede, J.; Esteves, P.; Ferreira, A.P.; Sampaio, J.; Leite, N. Jump higher, run faster: Effects of diversified sport participation on talent identification and selection in youth basketball. *J. Sports Sci.* **2019**, *37*, 2220–2227. [CrossRef] [PubMed]
27. Elferink-Gemser, M.T.; Visscher, C.; Lemmink, K.A.P.M.; Mulder, T.W. Relation between multidimensional performance characteristics and level of performance in talented youth field hockey players. *J. Sports Sci.* **2004**, *22*, 1053–1063. [CrossRef]
28. Gil, S.M.; Zabala-Lili, J.; Bidaurrazaga-Letona, I.; Aduna, B.; Lekue, A.B.; Santos-Concejero, J.; Granados, C. Talent identification and selection process of outfield players and goalkeepers in a professional soccer club. *J. Sports Sci.* **2014**, *32*, 1931–1939. [CrossRef]
29. Peña-González, I.; Javaloyes, A.; Cervelló, E.; Moya-Ramón, M. The maturity status but not the relative age influences elite young football players' physical performance. *Sci. Med. Footb.* **2022**, 1–8. [CrossRef]
30. Towlson, C.; Salter, J.; Ade, J.D.; Enright, K.; Harper, L.D.; Page, R.M.; Malone, J.J. Maturity-associated considerations for training load, injury risk, and physical performance in youth soccer: One size does not fit all. *J. Sport Health Sci.* **2020**, *10*, 403–412. [CrossRef]
31. Till, K.; Cobley, S.; Hara, J.O.; Cooke, C.; Chapman, C. Considering maturation status and relative age in the longitudinal evaluation of junior rugby league players. *Scand. J. Med. Sci. Sports* **2013**, *24*, 569–576. [CrossRef]
32. Meylan, C.; Cronin, J.; Oliver, J.; Hughes, M. Talent Identification in Soccer: The Role of Maturity Status on Physical, Physiological and Technical Characteristics. *Int. J. Sports Sci. Coach.* **2010**, *5*, 571–592. [CrossRef]
33. Zuber, C.; Conzelmann, A. Motivational Patterns as an Instrument for Predicting Performance Not Only in Football? A Replication Study with Young Talented Ice Hockey Players. *Front. Psychol.* **2019**, *10*, 2357. [CrossRef] [PubMed]
34. Cobley, S.P.; Till, K.; O'Hara, J.; Cooke, C.; Chapman, C. Variable and Changing Trajectories in Youth Athlete Develop-ment: Further Verification in Advocating a Long-Term Inclusive Tracking Approach. *J. Strength Cond. Res.* **2014**, *28*, 1959–1970. [CrossRef] [PubMed]

35. Abbott, A.; Button, C.; Pepping, G.-J.; Collins, D. Unnatural selection: Talent identification and development in sport. *Nonlinear Dyn. Psychol. Life Sci.* **2005**, *9*, 61–88.
36. Moran, J.; Paxton, K.; Jones, B.; Granacher, U.; Sandercock, G.R.; Hope, E.; Ramirez-Campillo, R. Variable long-term developmental trajectories of short sprint speed and jumping height in English Premier League academy soccer players: An applied case study. *J. Sports Sci.* **2020**, *38*, 2525–2531. [CrossRef]
37. Randell, R.K.; Clifford, T.; Drust, B.; Moss, S.L.; Unnithan, V.B.; Croix, M.B.A.D.S.; Datson, N.; Martin, D.; Mayho, H.; Carter, J.M.; et al. Physiological Characteristics of Female Soccer Players and Health and Performance Considerations: A Narrative Review. *Sports Med.* **2021**, *51*, 1377–1399. [CrossRef]
38. Till, K.; Morley, D.; O'Hara, J.; Jones, B.L.; Chapman, C.; Beggs, C.B.; Cooke, C.; Cobley, S. A retrospective longitudinal analysis of anthropometric and physical qualities that associate with adult career attainment in junior rugby league players. *J. Sci. Med. Sport* **2017**, *20*, 1029–1033. [CrossRef]
39. Dugdale, J.H.; Sanders, D.; Myers, T.; Williams, A.M.; Hunter, A.M. Progression from youth to professional soccer: A longitudinal study of successful and unsuccessful academy graduates. *Scand. J. Med. Sci. Sports* **2021**, *31*, 73–84. [CrossRef]
40. Till, K.; Cobley, S.; O'Hara, J.; Chapman, C.; Cooke, C. A longitudinal evaluation of anthropometric and fitness characteristics in junior rugby league players considering playing position and selection level. *J. Sci. Med. Sport* **2013**, *16*, 438–443. [CrossRef]
41. Elferink-Gemser, M.T.; Visscher, C.; Lemmink, K.A.P.M.; Mulder, T. Multidimensional performance characteristics and standard of performance in talented youth field hockey players: A longitudinal study. *J. Sports Sci.* **2007**, *25*, 481–489. [CrossRef]
42. Matthys, S.P.; Vaeyens, R.; Fransen, J.; Deprez, D.; Pion, J.; VandenDriessche, J.; Vandorpe, B.; Lenoir, M.; Philippaerts, R. A longitudinal study of multidimensional performance characteristics related to physical capacities in youth handball. *J. Sports Sci.* **2013**, *31*, 325–334. [CrossRef] [PubMed]
43. Roescher, C.R.; Elferink-Gemser, M.T.; Huijgen, B.C.H.; Visscher, C. Soccer Endurance Development in Professionals. *Endoscopy* **2010**, *31*, 174–179. [CrossRef] [PubMed]
44. Pyne, D.; Gardner, A.; Sheehan, K.; Hopkins, W. Fitness testing and career progression in AFL football. *J. Sci. Med. Sport* **2005**, *8*, 321–332. [CrossRef]
45. Falk, B.; Lidor, R.; Lander, Y.; Lang, B. Talent identification and early development of elite water-polo players: A 2-year follow-up study. *J. Sports Sci.* **2004**, *22*, 347–355. [CrossRef] [PubMed]
46. Zuber, C.; Zibung, M.; Conzelmann, A. Holistic Patterns as an Instrument for Predicting the Performance of Promising Young Soccer Players—A 3-Years Longitudinal Study. *Front. Psychol.* **2016**, *7*, 1088. [CrossRef]
47. Leyhr, D.; Kelava, A.; Raabe, J.; Höner, O. Longitudinal motor performance development in early adolescence and its relationship to adult success: An 8-year prospective study of highly talented soccer players. *PLoS ONE* **2018**, *13*, e0196324. [CrossRef]
48. Saward, C.; Hulse, M.; Morris, J.G.; Goto, H.; Sunderland, C.; Nevill, M.E. Longitudinal Physical Development of Future Professional Male Soccer Players: Implications for Talent Identification and Development? *Front. Sports Act. Living* **2020**, *2*, 578203. [CrossRef]
49. Le Gall, F.; Carling, C.; Williams, M.; Reilly, T. Anthropometric and fitness characteristics of international, professional and amateur male graduate soccer players from an elite youth academy. *J. Sci. Med. Sport* **2010**, *13*, 90–95. [CrossRef]
50. Saward, C.; Morris, J.G.; Nevill, M.E.; Minniti, A.M.; Sunderland, C. Psychological characteristics of developing excellence in elite youth football players in English professional academies. *J. Sports Sci.* **2019**, *38*, 1380–1386. [CrossRef]
51. Kannekens, R.; Elferink-Gemser, M.T.; Visscher, C. Positioning and deciding: Key factors for talent development in soccer. *Scand. J. Med. Sci. Sports* **2010**, *21*, 846–852. [CrossRef]
52. Huijgen, B.C.H.; Elferink-Gemser, M.T.; Post, W.J.; Visscher, C. Soccer Skill Development in Professionals. *Laryngo-Rhino-Otologie* **2009**, *30*, 585–591. [CrossRef] [PubMed]
53. Forsman, H.; Blomqvist, M.; Davids, K.; Liukkonen, J.; Konttinen, N. Identifying technical, physiological, tactical and psychological characteristics that contribute to career progression in soccer. *Int. J. Sports Sci. Coach.* **2016**, *11*, 505–513. [CrossRef]
54. Rees, T.; Hardy, L.; Güllich, A.; Abernethy, B.; Côté, J.; Woodman, T.; Montgomery, H.; Laing, S.; Warr, C. The Great British Medalists Project: A Review of Current Knowledge on the Development of the World's Best Sporting Talent. *Sports Med.* **2016**, *46*, 1041–1058. [CrossRef] [PubMed]
55. Höner, O.; Murr, D.; Larkin, P.; Schreiner, R.; Leyhr, D. Nationwide Subjective and Objective Assessments of Potential Talent Predictors in Elite Youth Soccer: An Investigation of Prognostic Validity in a Prospective Study. *Front. Sports Act. Living* **2021**, *3*, 638227. [CrossRef]
56. Höner, O.; Votteler, A. Prognostic relevance of motor talent predictors in early adolescence: A group- and individual-based evaluation considering different levels of achievement in youth football. *J. Sports Sci.* **2016**, *34*, 2269–2278. [CrossRef]
57. Emmonds, S.; Till, K.; Jones, B.; Mellis, M.; Pears, M. Anthropometric, Speed and Endurance Characteristics of English Academy Soccer Players: Do They Influence Obtaining a Professional Contract at 18 Years of Age? *Int. J. Sports Sci. Coach.* **2016**, *11*, 2212–2218. [CrossRef]
58. Till, K.; Jones, B.; Geeson-Brown, T. Do Physical Qualities Influence the Attainment of Professional Status within Elite 16–19 Year Old Rugby League Players? *J. Sci. Med. Sport* **2015**, *19*, 585–589. [CrossRef]
59. Gonaus, C.; Müller, E. Using physiological data to predict future career progression in 14- to 17-year-old Austrian soccer academy players. *J. Sports Sci.* **2012**, *30*, 1673–1682. [CrossRef]

60. Jennings, J.; Wundersitz, D.W.; Sullivan, C.J.; Cousins, S.D.; Tehan, G.; Kingsley, M.I. Physical testing characteristics better explain draft outcome than in-game movement profile in junior elite Australian rules football players. *J. Sci. Med. Sport* **2021**, *24*, 1284–1289. [CrossRef]
61. Torres-Unda, J.; Zarrazquin, I.; Gil, J.; Ruiz, F.; Irazusta, A.; Kortajarena, M.; Seco, J.; Irazusta, J. Anthropometric, physiological and maturational characteristics in selected elite and non-elite male adolescent basketball players. *J. Sports Sci.* **2012**, *31*, 196–203. [CrossRef]
62. Elferink-Gemser, M.T.; Visscher, C.; Richart, H.; Lemmink, K.A.P.M. Development of the Tactical Skills Inventory for Sports. *Percept. Mot. Ski.* **2004**, *99*, 883–895. [CrossRef] [PubMed]
63. Gabbett, T.J.; Georgieff, B. The Development of a Standardized Skill Assessment for Junior Volleyball Players. *Int. J. Sports Physiol. Perform.* **2006**, *1*, 95–107. [CrossRef] [PubMed]
64. O'Connor, D.; Larkin, P.; Williams, A.M. Talent identification and selection in elite youth football: An Australian context. *Eur. J. Sport Sci.* **2016**, *16*, 837–844. [CrossRef] [PubMed]
65. Santiago, C.; Gonzalez-Freire, M.; Serratosa, L.; Morate, F.J.; Meyer, T.; Gomez-Gallego, F.; Lucia, A. ACTN3 genotype in professional soccer players. *Br. J. Sports Med.* **2007**, *42*, 71–73. [CrossRef]
66. Buekers, M.; Borry, P.; Rowe, P. Talent in sports. Some reflections about the search for future champions. *Mov. Sports Sci. Sci. Et Mot.* **2014**, *88*, 3–12. [CrossRef]
67. Kelly, A.; Williams, C.; Wilson, M. Developing a Football-Specific Talent Identification and Development Profiling Concept—The Locking Wheel Nut Model. *Appl. Coach. Res. J.* **2018**, *2*, 32–41.
68. Vaeyens, R.; Malina, R.M.; Janssens, M.; Van Renterghem, B.; Bourgois, J.; Vrijens, J.; Philippaerts, R.M.; e Silva, M.J.C. A multidisciplinary selection model for youth soccer: The Ghent Youth Soccer Project * Commentary. *Br. J. Sports Med.* **2006**, *40*, 928–934. [CrossRef]
69. Junior, D.B.R.; Werneck, F.Z.; Oliveira, H.Z.; Panza, P.S.; Ibáñez, S.J.; Vianna, J.M. From Talent Identification to Novo Basquete Brasil (NBB): Multifactorial Analysis of the Career Progression in Youth Brazilian Elite Basketball. *Front. Psychol.* **2021**, *12*, 617563. [CrossRef]
70. Dimundo, F.; Cole, M.; Blagrove, R.C.; McAuley, A.B.T.; Till, K.; Kelly, A.L. Talent Identification in an English Premiership Rugby Union Academy: Multidisciplinary Characteristics of Selected and Non-selected Male Under-15 Players. *Front. Sports Act. Living* **2021**, *3*, 688143. [CrossRef]
71. Woods, C.T.; Raynor, A.J.; Bruce, L.; McDonald, Z.; Robertson, S. The application of a multi-dimensional assessment approach to talent identification in Australian football. *J. Sports Sci.* **2015**, *34*, 1340–1345. [CrossRef]
72. Jones, B.; Till, K.; Manley, A.J.; McGuigan, M.R. A Multidisciplinary Approach to the Profiling and Interpretation of Fitness Testing Data: A Case Study. In Proceedings of the British Association of Sport and Exercise Sciences (BASES) Conference, St George's Park, Burton upon Trent, UK, 1–2 December 2015.
73. Huijgen, B.C.H.; Elferink-Gemser, M.T.; Lemmink, K.A.P.M.; Visscher, C. Multidimensional performance characteristics in selected and deselected talented soccer players. *Eur. J. Sport Sci.* **2012**, *14*, 2–10. [CrossRef] [PubMed]
74. Sieghartsleitner, R.; Zuber, C.; Zibung, M.; Conzelmann, A. Or Coaches' Eye?—Both! Beneficial Collaboration of Multidimensional Measurements and Coach Assessments for Efficient Talent Selection in Elite Youth Football. *J. Sports Sci. Med.* **2019**, *18*, 32–43. [PubMed]
75. Hartigh, R.J.R.D.; Niessen, A.S.M.; Frencken, W.G.P.; Meijer, R.R. Selection procedures in sports: Improving predictions of athletes' future performance. *Eur. J. Sport Sci.* **2018**, *18*, 1191–1198. [CrossRef] [PubMed]
76. Figueiredo, A.J.; Gonçalves, C.E.; Coelho-E-Silva, M.; Malina, R.M. Youth soccer players, 11–14 years: Maturity, size, function, skill and goal orientation. *Ann. Hum. Biol.* **2008**, *36*, 60–73. [CrossRef]
77. Niessen, A.S.M.; Meijer, R.R.; Tendeiro, J. Predicting Performance in Higher Education Using Proximal Predictors. *PLoS ONE* **2016**, *11*, e0153663. [CrossRef]
78. Deprez, D.N.; Fransen, J.; Lenoir, M.; Philippaerts, R.M.; Vaeyens, R. A Retrospective Study on Anthropometrical, Physical Fitness, and Motor Coordination Characteristics That Influence Dropout, Contract Status, and First-Team Playing Time in High-Level Soccer Players Aged Eight to Eighteen Years. *J. Strength Cond. Res.* **2015**, *29*, 1692–1704. [CrossRef]
79. Fenner, J.S.J.; Iga, J.; Unnithan, V. The evaluation of small-sided games as a talent identification tool in highly trained prepubertal soccer players. *J. Sports Sci.* **2016**, *34*, 1983–1990. [CrossRef]
80. Bennett, K.J.M.; Novak, A.R.; Pluss, M.A.; Stevens, C.J.; Coutts, A.J.; Fransen, J. The use of small-sided games to assess skill proficiency in youth soccer players: A talent identification tool. *Sci. Med. Footb.* **2017**, *2*, 231–236. [CrossRef]
81. Bergkamp, T.L.G.; den Hartigh, R.J.R.; Frencken, W.G.P.; Niessen, A.S.M.; Meijer, R.R. The validity of small-sided games in predicting 11-vs-11 soccer game performance. *PLoS ONE* **2020**, *15*, e0239448. [CrossRef]
82. Lyons, B.D.; Hoffman, B.J.; Michel, J.W.; Williams, K.J. On the Predictive Efficiency of Past Performance and Physical Ability: The Case of the National Football League. *Hum. Perform.* **2011**, *24*, 158–172. [CrossRef]
83. Hill, M.; Scott, S.; McGee, D.; Cumming, S. Coaches' Evaluations of Match Performance in Academy Soccer Players in Relation to the Adolescent Growth Spurt. *J. Sci. Sport Exerc.* **2020**, *2*, 359–366. [CrossRef]
84. Wiseman, A.C.; Bracken, N.; Horton, S.; Weir, P.L. The Difficulty of Talent Identification: Inconsistency among Coaches through Skill-Based Assessment of Youth Hockey Players. *Int. J. Sports Sci. Coach.* **2014**, *9*, 447–455. [CrossRef]

85. Dugdale, J.H.; Sanders, D.; Myers, T.; Williams, A.M.; Hunter, A.M. A case study comparison of objective and subjective evaluation methods of physical qualities in youth soccer players. *J. Sports Sci.* **2020**, *38*, 1304–1312. [CrossRef] [PubMed]
86. McCormack, S.; Jones, B.; Elliott, D.; Rotheram, D.; Till, K. Coaches' Assessment of Players Physical Performance: Subjective and Objective Measures are needed when Profiling Players. *Eur. J. Sport Sci.* **2021**, 1–11. [CrossRef] [PubMed]
87. Christensen, M.K. "An Eye for Talent": Talent Identification and the "Practical Sense" of Top-Level Soccer Coaches. *Sociol. Sport J.* **2009**, *26*, 365–382. [CrossRef]
88. Roberts, A.H. The Coaches' Eye: Exploring Coach Decision-Making during Talent Identification. Ph.D. Thesis, Edith Cowan University, Perth, Australia, 2021.
89. Cumming, S.P.; Lloyd, R.S.; Oliver, J.L.; Eisenmann, J.C.; Malina, R.M. Biobanding Applications. *Strength Cond. J.* **2017**, *39*, 34–47. [CrossRef]
90. Lüdin, D.; Donath, L.; Cobley, S.; Romann, M. Effect of bio-banding on physiological and technical-tactical key performance indicators in youth elite soccer. *Eur. J. Sport Sci.* **2021**, 1–9. [CrossRef]
91. Toum, M.; Tribolet, R.; Watsford, M.L.; Fransen, J. The confounding effect of biological maturity on talent identification and selection within youth Australian football. *Sci. Med. Footb.* **2020**, *5*, 263–271. [CrossRef]
92. Hill, M.; Scott, S.; McGee, D.; Cumming, S.P. Are relative age and biological ages associated with coaches' evaluations of match performance in male academy soccer players? *Int. J. Sports Sci. Coach.* **2020**, *16*, 227–235. [CrossRef]
93. Hendry, D.T.; Williams, A.M.; Hodges, N.J. Coach ratings of skills and their relations to practice, play and successful transitions from youth-elite to adult-professional status in soccer. *J. Sports Sci.* **2018**, *36*, 2009–2017. [CrossRef]

sports

MDPI

Article

Happy Birthday? Relative Age Benefits and Decrements on the Rocky Road

Neil McCarthy [1,*], Jamie Taylor [2,3,4], Andrew Cruickshank [2] and Dave Collins [2,4]

1 Premiership Rugby Ltd., 7th Floor, Regal House, Twickenham TW1 3QS, UK
2 Grey Matters Performance Ltd., Stratford upon Avon CV37 9TQ, UK; jamie@greymattersuk.com (J.T.); andrew@greymattersuk.com (A.C.); dave@greymattersuk.com (D.C.)
3 School of Health and Human Performance, Faculty of Science and Health, Dublin City University, D09 Dublin, Ireland
4 Holyrood Campus, Moray House School of Education and Sport, The University of Edinburgh, Edinburgh EH8 8AQ, UK
* Correspondence: nmccarthy@premiershiprugby.com

Abstract: (1) Background: There is abundant literature in talent development investigating the relative age effect in talent systems. There is also growing recognition of the reversal of relative age advantage, a phenomenon that sees significantly higher numbers of earlier born players leaving talent systems before the elite level. However, there has been little investigation of the mechanisms that underpin relative age, or advantage reversal. This paper aimed to investigate (a) the lived experience of relative age in talent development (TD) systems, (b) compare the experience of early and late born players, and (c) explore mechanisms influencing individual experiences. (2) Methods: interviews were conducted with a cohort of near elite and elite rugby union players. Data were subsequently analysed using reflexive thematic analysis and findings considered in light of eventual career status. (3) Results: challenge was an ever-present feature of all players journeys, especially at the point of transition to senior rugby. Psycho-behavioural factors seemed to be a primary mediator of the response to challenge. (4) Conclusions: a rethink of approach to the relative age effect is warranted, whilst further investigations of mechanisms are necessary. Relative age appears to be a population-level effect, driven by challenge dynamics.

Keywords: talent identification; talent development; challenge

Citation: McCarthy, N.; Taylor, J.; Cruickshank, A.; Collins, D. Happy Birthday? Relative Age Benefits and Decrements on the Rocky Road. *Sports* **2022**, *10*, 82. https://doi.org/10.3390/sports10060082

Academic Editors: Adam Leigh Kelly, Sergio L. Jiménez Sáiz, Sara Diana Leal dos Santos and Alberto Lorenzo Calvo

Received: 19 April 2022
Accepted: 18 May 2022
Published: 24 May 2022

1. Introduction

Effective and efficient talent identification and talent development (TD) processes are a significant part of the strategic management of TD systems. Increasing curiosity and investigation of such elements is a significant challenge for many national governing bodies (NGBs). TD systems are under increasing scrutiny, with data challenging established paradigms in relation to many TD dynamics [1]. Of significant debate are the dynamics pertaining to selection and development of athletes as they journey into, through and out of talent systems [2,3]. Whilst the accurate prediction of future performance has been a topic of significant research, practical application of this is significantly challenged by the biopsychosocial complexities of development [4,5]. This is especially so in the earlier years of talent development, with a variety of dynamics apparent, especially at selection gateways [6].

One such factor suggested as underpinning these selection biases is the relative age effect (RAE) [7,8]. The inevitable chronological grouping of children as they enter the education system has been shown to promote early advantage for those born just before or after the academic cut-off date (11). This mechanism for selecting children continues as they enter organised sport and talent systems. An abundance of literature highlighting asymmetric birthdates during selection processes has linked RAEs to maturation and the comparative advantages and/or disadvantages of being born one side or the other of the selection cutoff date within sport (typically Sept 1st in the United Kingdom, for a review see [9]). To date, much of the literature has focused on the disproportionate volume of players born in the first two quartiles (Q1 and Q2) of the selection year in comparison to those born towards the end of the selection year (Q3 and Q4).

Explanations for these effects have tended to focus on advanced physical maturation offering relatively older individuals up to 12 months advantage over their relatively younger peers [8,10]. Less reported, however, are the advantages/disadvantages that have been identified in other domains such as the cognitive and emotional disruptions observed during formative developmental periods [11–15]. Perhaps most importantly, we know little about how these biopsychosocial dimensions manifest in TD systems, especially given that RAE and maturation are acknowledged as separate constructs [16].

The orientation for the majority of RAE literature in sport has led to a focus on the potential negative effects age groupings have on the identification of individuals and their experiences [17]. These studies have generally focused on a specific moment in time for data collection (e.g., selection into talent systems) and thus offer limited perspective on long term effects. The general consensus is an assumption that the RAE is something to eradicate, to prevent large numbers of performers being excluded [18].

1.1. RAE Advantage Reversal

More recently, the literature has challenged these assumptions and has begun to report a potential positive that emerges from the attritional and/or challenging experiences of the relatively young. This has separately been identified as the 'underdog hypothesis' [19] and 'advantage reversals' [1,20], identifying that whilst a disproportionately high number of early birthday athletes are initially selected, the relatively young are proportionately more likely to reach senior elite status. This finding appears robust across a wide range of sporting contexts: in handball [21], cricket [22], ice hockey [23] and across male elite sport [24]. Indeed, highlighting the robustness of the finding, replications have consistently shown the same finding. For example, recent findings [8] show evidence of the same RAE advantage reversal previously found in a single academy [20] and across international pathways in rugby union and cricket [1]. Therefore, it appears that whilst those with early advantages are being selected into the initial stages of talent systems in greater proportions, earlier born athletes are leaving in far higher numbers than later born. Importantly, this 'advantage reversal' does not suggest a reversal of the RAE. Instead, it shows that those born later in the selection year are less likely to be deselected than their earlier born counterparts.

1.2. Mechanisms

This would suggest that the current literature base is limited by some key assumptions and by a lack of mechanistic focus. Those who are born earlier in the selection year are more likely to be selected for a TD system. It also appears that, at the population level, their relatively younger peers are more likely to continue through the TD system to elite performance. Yet, to this point, much of the extant literature has focused on 'solving' a variety of early advantage effects by focusing on levelling the playing field, for example: bio-banding [25]; age order shirt banding [26]; birthday banding [27], performance banding [28] and corrective adjustment procedures [29]. Yet, very little attention has been paid to the dropout rates of those with earlier advantage [30] and investigation of underpinning mechanisms is disappointingly sparse. As a result, we know that the RAE exists and that there is likely to be a reversal of advantage, but we do not know why this happens. This is a key barrier for the practitioner seeking to optimise TD processes. McCarthy and Collins [1] suggested a potential mechanism could be the initial impact of negative selection experiences, with these early disadvantages being facilitative of greater psychological 'growth' and/or acting as a mechanism for a more intrinsically focused and longer-term motivational orientation This hypothesis suggested that RAE advantage reversals may be driven by the motivational orientation of individuals and how that is anchored through formative experiences.

Motivation is a significant factor in sports participation, progression and drop out [31–33]. One underpinning feature of an individual's motivation is perceived competence. Perceived competence acts as a domain-specific indicator of self-esteem that contributes to and is affected by the individual's motivational orientation [34,35]. For example, the relatively old may progress rapidly as a result of early challenge-free experiences, arriving at early selection gateways with a high degree of perceived competence. This may inadvertently develop an individual with extrinsically anchored motivational orientation. Conversely, those athletes not afforded this advantage may develop a more intrinsic motivational orientation, becoming more likely to remain and persevere within TD systems. As such, later drop out from TD systems is proportionately higher from relatively older cohorts [8], suggesting these individuals may not be sufficiently orientated and/or equipped to cope with and prosper motivationally through transitions when early advantages begin to disappear [20,36]. Notably, this hypothesis seems to marry with other research suggesting a complex interaction between challenge and psycho-behavioural skills [2] and the risk posed to progression when there is a mismatch between the two [33].

Accordingly, there appears to be increasing evidence pointing to the interaction of challenge and psycho-behavioural skills [37–39]. This need for athletes to be challenged in their development has been accepted from a variety of research perspectives [30,40,41]. Key differences between these positions notwithstanding, it appears that performers who bring a variety of psycho-behavioral resources to challenging periods will be more likely to cope with and learn from their experience [42]. In this regard, recent investigations have suggested that challenge-filled sporting pathways are an essential feature of developmental journeys [43]. Further, it appears that sporting 'traumas' and/or challenging experiences, rather than being directly causative of 'psychological growth', instead act to test, prove and encourage previously developed psychological skills [42]. Indeed, perceptions of control, confidence and perspective, underpinned by psycho-behavioural skills [38], along with appropriate reflection and social support, appear particularly important in this regard [37,44]. This also appears to be the case amongst the limited populations where the reversal of relative advantage has been tested [45].

Reflecting these complexities, and beyond establishing RAEs and consequently advantage reversals in different contexts, there is a need to understand the mechanisms at play [23]. This is especially the case for the applied practitioner or policy maker who needs to make decisions regarding sporting systems, placing the individual performer's experience as a primary concern [43]. Accordingly, this paper aimed to (a) generate a deeper understanding of the lived experience of the RAE in TD systems across groups of more and less successful athletes, (b) compare the experience of early and late born players, and (c) explore the mechanisms influencing individual experiences.

2. Materials and Methods

2.1. Research Philosophy

Grounded in the real-world factors covered in our introduction, alongside our desire to deepen knowledge in RAE for practical purposes, a pragmatic philosophy was adopted for the present study [46]. The primary objective of pragmatic research is to generate knowledge that is practically useful for the individuals and groups that it studies, plus the practitioners who support them [47]. Ontologically, pragmatism therefore requires researchers to avoid seeking universal truths or entirely subjective constructions and to instead identify processes and mechanisms that shape common experiences in specific settings at specific times [46]. Epistemologically, pragmatism is also based on the idea that a continuum exists between more objective and more subjective perspectives. Rather than posing questions against a pre-set epistemology, pragmatists therefore place their questions at the heart of a study and select an epistemological position and methods that are appropriate to answering it [48].

Based on the aims established in our introduction, including the need to move understanding of RAE beyond statistical phenomena, an interpretivist epistemology and qualitative strategy were selected for the present study [49]. More specifically, these approaches reflected our intention to understand experiences of the RAE, from the views of a relevant—and internally diverse—group of individuals [50,51]. Importantly, pragmatism also views researchers as part of the world they explore and encourages them to actively interact with the experiences of their participants in the knowledge generation process [46,52,53]. In this respect, all parts of this study were aided by the research team's record of performing and working in elite sport and—with direct relevance to the participants in this study—elite rugby union specifically. Most notably, the first author was involved in elite rugby union as a player, then coach and TD practitioner; the second author as a coach, TD practitioner and coach developer; and the third and fourth authors as psychologists and coach developers.

2.2. Participants

To explore experiences of the RAE, eight male players who had entered the academy system in English Premiership rugby union and reached the transition point to the professional game (i.e., the final academy phase) were purposefully sampled via the contacts of the first author. At the time, he was an Academy Director at a professional club. To avoid the risks of collecting data overly influenced by situational factors, participants were also selected on the basis that they had gone through the academy system across various periods of time (rather than all coming from one cohort, or close cohorts). As such, data were collected from participants who had transitioned out of the academy programme at different times over the course of eight seasons. For sufficient comparison of experience (as per our third aim) individuals were also identified on the basis that their birthday was at either end of the sport's selection year (i.e., Q1: between September and November; Q4: between June and August).

At the time of interview, all players were aged between 21–32 years of age ($M = 26.5$, $N = 8$) and actively involved as professional players either at an English Premiership (if retained) or Championship team (if released after reaching the end of the academy phase). Details relating to each specific participant's birth quartile, initial, mid and overall career status (the latter using criteria from two) are provided in Table 1. Initial career status was determined by contract status when leaving the academy programme. Mid-career status was determined by analyzing each participant 5 years post transition from the academy. In all instances: 'Championship' refers to the second division of English rugby, 'Premiership' is the highest level of the domestic game in England and 'senior test' is a player who has played at international level. Please note additional information is limited to protect anonymity.

Table 1. Participant information.

	Birth Quartile	Initial Career Status	Mid Career Status	Overall Career Status	Eventual Career Status
Player 1	Q1	Retained	Senior Test	Senior Test	Super Champion
Player 2	Q1	Retained	Senior Test	Senior Test	Super Champion
Player 3	Q1	Released	Championship	Championship	Almost
Player 4	Q1	Released	Championship	Championship	Almost
Player 5	Q4	Released	Championship	Premiership	Champion
Player 6	Q4	Retained	Premiership	Senior Test	Super Champion
Player 7	Q4	Retained	Premiership	Senior Test	Super Champion
Player 8	Q4	Released	Championship	Championship	Almost

2.3. Data Collection

Prior to data collection, ethical approval was obtained from the first author's institutional ethics committee and informed consent gained from each participant. All interviews were conducted by the first author who began by asking participants to plot their career trajectory on a gridded timeline. More specifically, the X-axis spanned the participant's first involvement in sport all the way to the date of interview; and the Y-axis represented the participant's perceived level of development and performance throughout this time [54,55]. Participants were then asked to highlight particularly critical periods and events along this timeline [37]. Using these timelines to minimize the limitations of retrospective recall [56], particularly for those who had moved from an academy to full professional contract a number of years previously, interviews were constructed against a semi-structured guide focused on key transitional periods and individual experiences for each participant. Consisting of open-ended questions and follow-up probes and prompts that were informed by RAE and TD literature, the guide was designed to offer flexibility for participants to describe their experiences in bespoke ways, while ultimately remaining focused on the study's aims and principles reported in prior research [57]. Examples of main interview questions were: 'Looking at your timeline can we discuss the points in your journey that were challenging and the points where you were finding it easy?' and 'What are your reflections on your progression now you have transitioned through the academy?'. All interviews took place face to face, lasting between 45 and 90 min (M = 66.4), and were audio recorded.

2.4. Data Analysis

Following data collection, all interviews were transcribed verbatim and analyzed using QSR NVivo software. Coherent with our desire to understand lived RAE experiences, plus the meanings attributed to this by participants, a reflexive thematic analysis (TA) was chosen as the specific analytic strategy [50]. Similarly, TA was also coherent with our pragmatic philosophy in that this form of analysis recognizes that researchers are a resource to support the interpretive process [58]. For the purpose of comparison between birth quartiles, players were grouped based on selection and school year. For participants from the first birth quartile ($N = 4$) and the fourth quartile ($N = 4$), players were further grouped according to initial career status ($N = 2$, retained and $N = 2$, released) from each birth quartile.

Based on the established TA process [59], analysis was undertaken in a recursive and blended fashion (based on the experience of the research team: [58]) and began with the first author reading through each transcript to optimize familiarity with the data and note early points of interest. This was followed by the application of codes to meaningful sections of raw data. More specifically, these codes were either semantic (to capture surface meaning) or latent (utilizing pre-exiting theories to interpret meaning: [50]). The third step saw the generation of initial themes, with significant codes being promoted to a theme, or similar codes being clustered together as patterns of shared meaning [50]. The fourth step involved a review of initial themes, after which the fifth and final step was taken to generate overarching themes and the final thematic map [60]. Importantly, whilst the first five phases of analysis were completed in the period following the collection of data, the sixth and final stage of TA, the write up, was delayed until each participant was 30 years old to take account of eventual career status of the participants.

2.5. Trustworthiness

As well as the approaches detailed in Sections 2.3 and 2.4, numerous others were applied to enhance the trustworthiness of the research process and ultimate findings. Regarding data collection—and given the importance of rapport between interviewer and interviewee [61]—the quality of data was supported by the pre-existing relationships between the first author and all participants [62]. In addition, all interviews were undertaken in a private, quiet location at the training ground of the player's club to aid comfort and openness. Of course, these advantages had to be balanced with measures to protect against any imbalances of power and the limitations of familiarity (e.g., the provision of socially desirable responses). Specifically, such issues were mitigated through the retrospective nature of the interview (i.e., the first author had no live management or selection influence as all participants were no longer academy players) as well as adherence to the components of ethical research by Hewitt [63]. The data collection process was further supported by a pilot study with two athletes who met the same inclusion criteria as detailed in Section 2.2 (M = 23.4). This work led to adjustments to the interview guide, with specific prompts altered and some jargon removed from questions.

Regarding data analysis, trustworthiness was enhanced by the first author's use of a reflexive journal to document methodological and analytical considerations, the rationale behind decisions, and the interaction of the research team's assumptions and biases [64]. In addition, the second, third, and fourth authors acted as critical friends across the full analysis. In particular, the fourth author provided critical feedback on selected procedures, while the second and third authors focused primarily on the use and outcomes of these procedures. As an accepted approach at the time of data collection, member checking was also used by returning transcripts to participants for them to assess the extent to which these accurately, fairly, and respectfully reflected their experiences [61]; a process which resulted in no significant changes. While a request for further member reflections could have added an extra dimension to our analysis [65], the checking process provided a degree of assurance on data fidelity.

3. Results

Addressing the first aim of the study and considering the different experiences of challenge through the pathway for each participant, Figure 1 shows the graphic timelines that were drawn by participants prior to the collection of interview data. They show the overall trajectory of athletes, representing their lived experiences of development and performance.

Across the sample it appeared that, regardless of birthdate in the selection year, there did not appear to be a pattern in the volume or intensity of the challenges faced on the journey to the professional game. Importantly however, this did not appear to be the case for players that were subsequently released. It appears that players born in Q1 and subsequently released experienced a challenge-free journey prior to academy entry at 16 years of age. Players born in Q4, and subsequently released, plotted their experiences in a similar manner. This contrasted significantly with the experiences of retained athletes with first and fourth quartile birthdates, who plotted a consistent series of bumpy challenging experiences prior to and through the academy system.

3.1. Player Perceptions of Challenge

As athletes progressing through the talent system, all identified a variety of challenges such as selection dynamics, peer to peer competition and increased stress from managing competing demands. Many of these challenges were associated with maturation dynamics. For example, consider the experience of these retained players:

> I was tiny between 14 and 16. When I turned up at the academy at 16, I was 70 kg and still very small . . . I was always very small all the way until I was 16 or 17, that was when I actually really grew . . . I would never be able to physically dominate anyone at all. The only hope I had was to use my feet and pace which I think really, really helped and it's probably why I ended up at 9 I think (Player 1: Q1—Retained).

> It was only the fact that at 18 or 19 I found a bit of pace that kind of gave me that X factor to try to compensate a little bit for not being the strongest or the most physical. Physicality is one thing that has always been brought out with me in any review (Player 2: Q1—Retained).

> I wasn't physically muscular I don't think I was strong I think compared to the others but I was quite tall and slim but I wasn't massive, I don't think I stood out

from the crowd in any manner I was just a bit taller or you know probably in the top third of height—things like that at that age (Player 7: Q4—Retained).

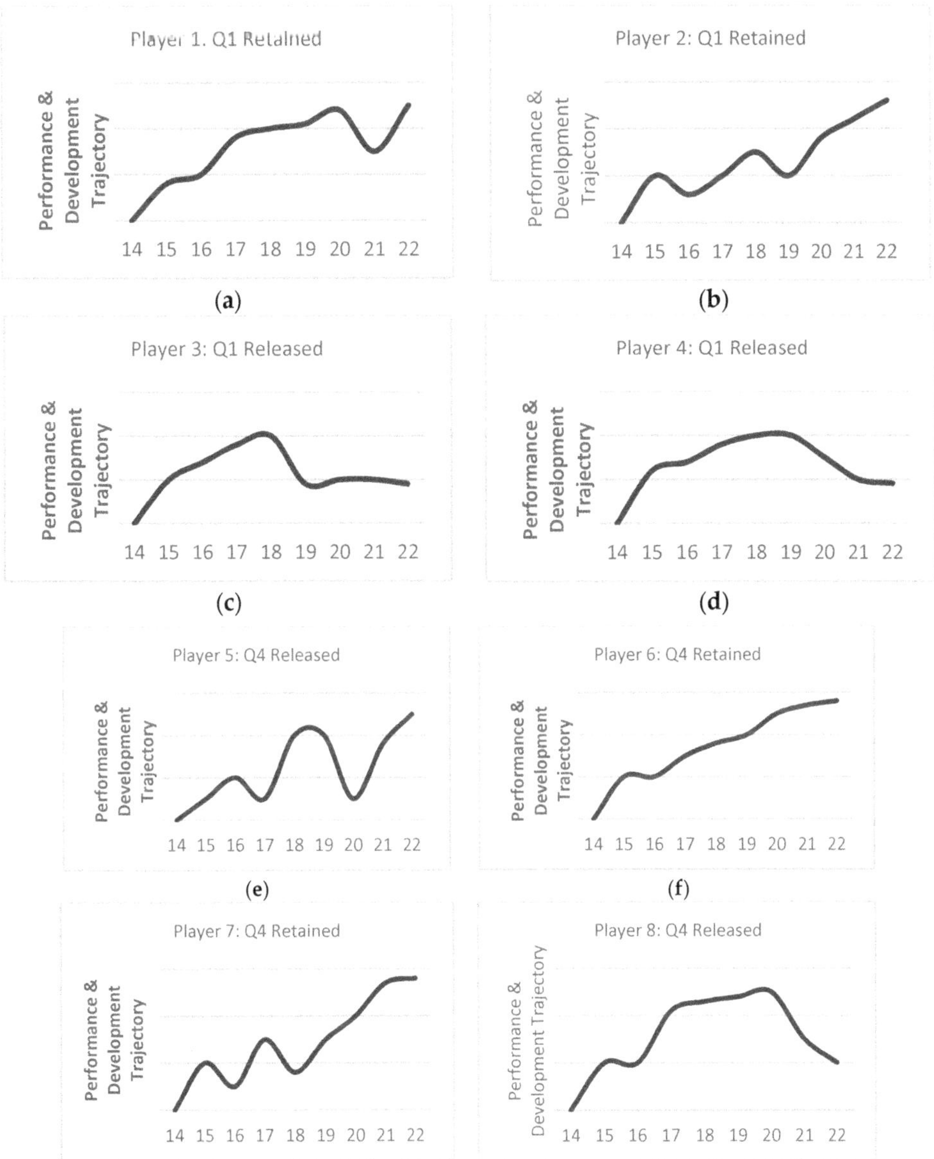

Figure 1. Graphic timelines of participants ((**a**–**h**), retained and released).

For all eight participants, awareness of maturational status was a significant feature of their journey. These perceptions were emotionally laden and appeared to influence a wide range of behaviours. Despite self-identified, later-maturing players being present across quartiles, those who were retained did not express a perception of disadvantage, irrespective of birth quartile. Indeed, whilst reflecting on the consequences of later maturation, players appeared to perceive this as an enabling factor and something to work with over the long term. This contrasted with many released athletes who experienced early advantages through maturation:

> I was bigger so I could run through people and get around people. It was easier to play because I was a little bit bigger. Skillset-wise I seemed to be a little bit behind (Player 8: Q4—Released).

> It seemed seamless to me [transition into senior environment] to be honest I think it was because I was brought in to play in the second team games and then you come to some of the first team training sessions as well and then eventually, they brought you in full time ... it was a good transition, it was easy (Player 4: Q1—Released).

> I was bigger and taller than a lot of them, that's the main bit. I had always been taller than everyone my own age (Player 5: Q4—Released).

As players progressed through the academy, significant differences emerged in their ability to deal with challenges. Across the retained group, players appeared to have the ability to utilise and reflect on past experiences of being disadvantaged. This perspective-taking appeared to influence perceptions of competence and control:

> I knew I was better than players I was playing with at school level, but then you come somewhere like (club) ... you know you are far from where you think you are. I kept my head down and worked hard, but I never felt like I didn't deserve to be in the academy. I sort of felt that I deserved a chance to be in it and give it a shot, but when you get here you sort of realise there are 18-year olds who are way more physical ... but that is good. At 16 you strive because you think I have got to catch him up, you know, it gives you goals (Player 1: Q1—Retained).

> There was like older guys there as well ... so we had 18/19-year-olds who were a lot more physically developed and experienced and better players than us so we were exposed to that and trained with that day in day out, at times like it was difficult ... I had to deal with some right XXXX and eventually you start to find your way (Player 6: Q4—Retained).

> I was still very small, I was still told probably too small to be a rugby player ... it was never a thought of mine to be a professional rugby player but I was always going to be a small skinny player as far as I was concerned (Player 7: Q4—Retained).

In contrast, the released group struggled to cope with the increased range and intensity of challenges they faced. The response to these challenges was often perceived outside the player's locus of control, with problems attributed to external factors. This contrasted with the retained group who saw challenges as obstacles to overcome by deploying a range of psycho-behavioural resources. As a consequence, released players appeared less equipped to cope with and learn from challenging periods:

> It was frustrating that I could not do things that I used to do at 14/15 (years old), running and scoring plenty of tries, but my game changed a lot and I turned into a very different player due to that and it was frustrating (Player 8: Q4—Released).

> It was quite scary because I had not played much in senior rugby, I just did not really know ... so it was just quite scary not knowing where I was going to be and not knowing what I was going to do ... I just lost direction (Player 3: Q1—Released).

> Obviously, it was a lot more physically demanding and nothing you were sort of used to before. It was really tough ... Just a lot more intense, a lot more volume with the actual rugby skill development and the strength and conditioning development. I'd never had it before, wasn't really expecting it either (Player 4: Q1—Released).

> The first couple of months it really p*ssed me off. You feel like you are standing still and you are desperate to play at that age, and then I remember just speaking to [brother] and my old man, and he just said 'work hard and make sure that when you do get a go, you are ready to go' ... it was then a case of looking at it from a different angle and saying I need to keep working at my passing and my kicking, the gym, the speed (Player 1: Q1—Retained).

As players continued their journey, these features seemed to become even more prominent, with overcoming a range of challenges seemingly a key differentiator.

3.2. *Mechanisms Impacting Player Experience*

Finally, in addition to exploring the individual players' responses to various challenges, we also sought to understand the mechanisms that seemed to influence their overall experience.

3.2.1. Nature of Commitment to the Sport

There appeared to be significant differences between the nature of the commitment to the sport between those who were retained and those released. Retained athletes seemed to engage and play with a focus on progression and enjoyment of the developmental process, rather than winning or domination of the game at earlier stages. In contrast, the released group seemed to heavily invest, from an early stage, with a focus on playing and winning matches, rather than engagement in other sports or training:

> I moved there because it was the best team, the team I was with wasn't that great and the mini set up was like fading out rather than like getting stronger and at the time (community club) had a strong mini section, so I joined that (Player 5: Q4—Released).

Released players also tended to focus solely on the outcome of selection in the short term, either for international rugby, or inclusion in an academy programme. This contrasted with retained players who seemed to focus on improving themselves rather than on an end goal of selection:

> I was very rugby-focused not thinking too much about school. The next level for me was to get into the [club] academy and play for England at under 16s level, that was my driving goal (Player 8: Q4—Released).

> The only thing I fixed on was decision making, two on ones, three on ones and obviously the backs were doing something different, all those skills I think that really benefited me and I remember thinking just focus on getting this stuff done... you'll be better (Player 1: Q1—Retained).

3.2.2. Nature and Influence of Support

Many of the retained group reflected on the use of experiences as a platform for reflection. Significantly, this seemed to be promoted by various supportive influences. This guiding of reflection seemed to be a key factor by which players were able to maintain perceptions of control during challenging experiences. In essence, this support seemed to be more facilitative, when compared to direct and 'driving' input of those who did not progress:

> Mum and Dad used to make me clean my boots and that, I had one pair of boots and they had to last me for a season, so I was always told to polish them and look after them and make sure they did not split (Player 1: Q1—Retained).

> I think by the time it came around to me playing they were supportive but not until I was about 16 ... I pretty much had to make sure that I got myself sorted for everything (Player 6: Q4—Retained).

Some subtle differences were observed between the retained and released groups in terms of the nature of the support from parents as they began to progress towards the transition to the senior game. In contrast to the retained players, the released group experienced far higher levels of support than those who were retained, indeed, something that did not appear to change, even as athletes progressed.

> My Mum and Dad were so supportive that I didn't need anything. They sort of volunteered and bought me wherever I needed to go—it was literally all for me (Player 8: Q4—Released).

3.2.3. How Players Learned from Challenge

In addition, there appeared significant differences in the response of players to challenge and also how they learned from their experiences. Amongst the retained group, there appeared to be a greater perception of control during periods of challenge. As a result, players seemed to have the confidence to deploy previously developed skills and capitalize on the emotional experience of challenge. When this was not the case, especially amongst the released group, it appeared to be a barrier to long term progression. For player 8 reflecting on his early transition to the senior team and the changing perceptions of his earlier size advantage, led to the regret that he was unable to deploy the necessary skills to navigate the challenge:

> You have got guys who were probably 20 kg heavier than me ... I think a lot of it may have come down to confidence and I didn't integrate well going into a first team environment ... holding back a little bit more than I should have (Player 8: Q4—Released).

> The differential response appeared to be a result of a lack of previous experience, reflection on, or development of the skills to cope with or learn from challenge. In contrast, amongst the retained group, players seemed to actively seek out challenging experiences. For example, player 1 deliberately chose to play in an age group beyond his chronological age as a means of increasing his challenge: "I was too young for that age group so at Sunday rugby I always played a year above" (Player 1: Q1—Retained).

We can also consider player 1′s perceptions of challenge as he progressed into the senior squad:

> There was the likes of XXX and, a lot of the senior players who either were playing or had just retired and were coaching, really kind of nurtured me along the way ... it was pretty tough period and I just kept focusing on getter better ... yeah tough (Player 1: Q1—Retained).

4. Discussion

The specific aims of this study were firstly to generate a deeper understanding of the lived experience of the RAE, secondly to compare the lived experiences of early and late birth players, and finally to understand the mechanisms that influenced individual experience. We responded to criticisms of the existing body of research in RAE which has focused at the population level with limited use of qualitative methodologies to understand underpinning mechanisms.

4.1. Challenge

Whilst our exploratory approach set out to understand the impact of challenge in relation to RAE, what emerged was the impact of challenge irrespective of RAE. That is, later born players did not necessarily experience higher levels of challenge, nor did increased challenge necessarily lead to greater psychological growth [16,45]. As such, at the individual level, whilst players in this sample were drawn from the full spectrum of an age band, their relative advantage or disadvantage prior to the senior level seemed to have long lasting and significant effects. It appears that RAE is not in itself a mechanism. Instead, perceptions of and response to challenge seem more impactful than when an individual is born. Further, it suggests that, at the population level, whilst an early birthdate in the selection year is associated with early advantage, the degree to which this advantage persists is dependent on the ability of the individual to navigate/exploit future challenges. Indeed, these data suggest that the experience of significant challenge was an omnipresent feature of these athletes' pathways, in the latter stages of an academy journey and whilst transitioning to the senior team, irrespective of birthdate [37].

4.2. Push and Pull Factors

Consequently, whilst relative age did not appear to be a mechanism in itself, there appeared to be three core factors that influenced player's perceptions of control, confidence and overall perspective [38]. Participants who were better able to cope with and learn from the inevitable highs and lows of development seemed better able to orient their focus in a manner that would help them continue to progress. The ability to do this seemed to depend on skills that were developed prior to significant challenges and previous navigation of challenge often highlighted through maturational differences [66]. Moreover, the early development of skills impacted the player's ability to cope with and learn from highly challenging experiences later in the pathway [67]. Data further highlighted the impact of these experiences and the skills deployed in the retained group's reaction to challenge in comparison to the released group. This was consistently highlighted by the way each player was able to make sense of and process challenges as they occurred. This appeared to have significant impact on each athlete as they faced a series of emotionally laden challenges. Furthermore, this manifested in differences in the nature of the player's commitment which continued as each one of them progressed [33]. Early advantages (often as a result of advanced maturation) seemed to drive an external focus (selection and winning). In contrast, early disadvantage seemed to promote a more internal focus on personal development. In turn, this suggests a reframing of RAE as a population-level effect, one that indicates a deeper phenomenon rather than having a direct effect.

By exploring the relative advantages and disadvantages of players at stages of their TD journey, against their later career success, we show that birth quartile number means very little without a deeper understanding of individual biopsychosocial context. What did appear critical for players to make the most of high challenge and subsequent emotional disturbance was the use of an appropriate range of psycho-behavioural skills [43,67]. Relative early advantage (experienced proportionately more frequently in earlier born groups) generates push-like effects. Push factors (pushing the player forwards), whilst allowing for early high performance relative to peers (and perhaps encouraging selection), seemed to retard later progress when missing skills were exposed [33]. Importantly, these push factors, whether they were high levels of parental input, or low levels of early challenge, were experienced across birth quartiles and seemed to have long lasting effects. In all but one case, the inability to overcome early push factors acted to prevent initial entry to the professional game and, even in light of eventual career status, only one player was able recover from deselection to play in the first tier of senior rugby. In contrast, those players who experienced more pull factors earlier in their TD journey (e.g., size disadvantages pulling them back), seemed to have more developmentally appropriate experiences that helped to prepare them for later challenges [42,44].

Of course, no research is without limitations. In this particular case, a common criticism of a pragmatic approach is that it risks provincialism, that is knowledge that is simply located in a particular context [68]. Indeed, whilst it is clearly not the prerogative of the epistemological approach used, or that of qualitative research in general, the same could be said for this study in adopting a relatively small sample of participants in a particular context. As a result, we ask the reader to focus on the principles underpinning our data and the possible transferability of findings to their own unique context [69]. Additionally, there is a risk that the relationship between participants and the first author as a leading professional in a rugby union academy programmes may have been a factor. This was mitigated by the individual participant timeline and interviews being conducted when participants had gone through the process of transition and retention and release status was established. There does however remain the risk of a player offering answers perceived as socially desirable. These risks were mitigated by adhering to the components of ethical research suggested by Hewitt [63] in acknowledging bias, developing rigor, a genuine level of rapport with participants, respect for their autonomy and the complete avoidance of exploitation.

5. Applied Implications

The evidence presented here raises the intriguing question of the extent to which various push and pull factors can be deliberately implemented in the experience of the athlete and at what point might they be appropriate. We would suggest that a sustained consideration (and balance) of push and pull factors should be a key feature of TD and, perhaps, participation environments [70]. This is especially the case for those aiming to support the development of athletes over the long term, not only developing junior career success (e.g., [71]). To be clear, this is not to suggest that an abundance of pull factors will always be a positive for overall development. Indeed, the large number of studies that explore the RAE show that greater numbers of pull factors may prevent athletes across sports getting selected in the first place [72]. In practice, these perspectives begin to challenge the hypothesis that relatively younger athletes will always benefit from playing against relatively older counterparts throughout development [8,20]. Consequently, for both research and practice we suggest that a rethink of our perceptions of RAE and its use as a metric in understanding TD is warranted. Quantitative methodologies have both offered insight into the impact of early advantages in terms of selection and the statistical consequences of challenge dynamics [1,8]. Across the domain, however, there is a need to complete more mechanistically focused research [73,74]. These insights are notable, not only considering the player's initial retained or released status, but also the extent of their overall achievement. In taking a novel approach, we were able to consider cross-sectional data with the additional benefit of understanding a player's long term career status. In the present sample, this allowed for demonstration that a number of these players progressed their careers to becoming the most elite players in the world. We would suggest that future research may benefit from adopting a similar approach, where cross-sectional data might be analysed considering long-term career status. In addition, we would suggest that research in RAE begins to move beyond further identification of RAE and advantage reversal in even more populations. Instead, a more granular consideration of the mechanisms at play is essential to truly understand the effect, an approach that has been taken in other areas of TD research [75]. From an applied perspective, this is essential if research is to make a difference in the real world and help the field think beyond simplistic solutions.

For many TD systems, it has been assumed that an appropriate target for selection has been a balance across quartiles, ensuring an appropriate number of Q4s are given opportunities [18]. We, nor any other researcher, can suggest what an 'optimal' balance of selection would look like, however, especially if a talent system was looking for an outcome marker of effective processes [73]. The evidence presented here should challenge simplistic narratives and the drive to 'do something about RAE'. Instead, we suggest the need to focus on the individual in TD practice [76]. In addition, our data highlight the need for TD practitioners to have a central focus on the perceptions and needs of the individual athlete's curriculum [77]. Our data also suggest that, whilst attempts to dampen, control and do-away-with the RAE are well intentioned, the unintended consequences of not exploring the complexity of this phenomenon may divert us from optimising TD practice. This is especially the case with top down systemic interventions that, by nature, require the simplification of complex processes [78]. Previous recommendations seeking to implement blanket strategies to mediate against disproportionately high pull factors seem overly simplistic. Strategies such as bio-banding or birthday banding may be easily implemented at the policy level but lack a holistic consideration of the biopsychosocial factors that influence relative advantage or disadvantage. As an example, there is growing recognition of social circumstances that may act as pull factors [79]. In essence, we are suggesting that if we are to offer truly practical implications to support the growth of a research-informed profession [80], the field should begin considering relative advantage or disadvantage on a holistic biopsychosocial basis, rather than using discreet indicators (e.g., maturation and/or relative age) alone. Notably, recent evidence has taken steps towards alternative approaches to levelling of the playing field with players being banded by technical competence [28]. In addition, there have been suggestions that coaches use a variety of methods for the grouping of players to provide a broad range of experiences for the player [28]. For practical purposes, we would suggest that coaches are better off implementing an approach built on individual periodising of challenge [36,81]. This could be achieved through greater flexibility of age bandings, allowing athletes to be offered appropriate competitive and training opportunities based on individual needs. As an example, it appears that a common practice for selection has become selecting players purely based on their birth quartile, with the assumption that Q3 and Q4 athletes will automatically possess a better psychological skillset. At a minimum, this paper should serve to challenge such simplistic narratives. Indeed, we would suggest that there is a core need for practitioners to begin focusing at a more individual level. The dynamics presented in this small sample, when compared to previous data [8,20], suggest the need for a far more individual approach in practice. This means that rather than resorting to blanket strategies, we need a more fine-grained approach to the grouping of individuals and management of challenge than previously advocated [25,27].

6. Conclusions

This study has considered the lived experience of relative age amongst a cohort of elite and near-elite rugby union players, analysed in light of eventual career status. Data presented clearly challenge beliefs held by the field of both researchers and practitioners. We suggest the need for a rethink of assumptions in the field, including the idea that RAE should be tackled with blanket policies. It is likely that RAE is a statistical outcome of challenge dynamics at the population level. We would therefore suggest a broader consideration of the dynamics of challenge, with a focus on the various push and pull factors that an athlete may be exposed to.

Author Contributions: Conceptualization, N.M. and D.C.; methodology, N.M.; formal analysis, N.M.; data curation, N.M.; writing—original draft preparation, N.M. and J.T.; writing—review and editing, N.M., J.T., A.C. and D.C.; supervision, D.C. All authors have read and agreed to the published version of the manuscript.

Funding: This research received no external funding.

Institutional Review Board Statement: The study was conducted in accordance with the Declaration of Helsinki, and approved by the Institutional Review Board (or Ethics Committee) of University of Central Lancashire (BAHSS 200—5 SEP 2014).

Informed Consent Statement: Informed consent was obtained from all subjects involved in the study.

Data Availability Statement: The data are not publicly available owing for the need to maintain the anonymity of participants.

Conflicts of Interest: The authors declare no conflict of interest.

References

1. McCarthy, N.; Collins, D.; Court, D. Start Hard, Finish Better: Further Evidence for the Reversal of the RAE Advantage. *J. Sports Sci.* **2016**, *34*, 1461–1465. [CrossRef] [PubMed]
2. Collins, D.; MacNamara, Á.; McCarthy, N. Super Champions, Champions, and Almosts: Important Differences and Commonalities on the Rocky Road. *Front. Psychol.* **2016**, *6*, 2009. [CrossRef] [PubMed]
3. Güllich, A. Selection, de-selection and progression in German football talent promotion. *Eur. J. Sport Sci.* **2014**, *14*, 530–537. [CrossRef] [PubMed]
4. Bailey, R.; Collins, D. The Standard Model of Talent Development and Its Discontents. *Kinesiol. Rev.* **2013**, *2*, 248–259. [CrossRef]
5. Bailey, R.P.; Collins, D.; Ford, P.A.; MacNamara, Á.; Pearce, G.; Toms, M. Participant development in sport: An academic literature review. *Comm. Rep. Sports Coach. UK. Leeds Sports Coach. UK* **2010**, *4*, 1–134.

6. Johnston, K.; Baker, J. Waste Reduction Strategies: Factors Affecting Talent Wastage and the Efficacy of Talent Selection in Sport. *Front. Psychol.* **2020**, *10*, 2925. [CrossRef]
7. Barnsley, R.H.; Thompson, A.H.; Barnsley, P.E. Hockey success and birthdate: The relative age effect. *Can. Assoc. Health Phys. Educ. Recreat. J. (CAHPER)* **1985**, *51*, 23–28.
8. Kelly, A.L.; Till, K.; Jackson, D.; Barrell, D.; Burke, K.; Turnnidge, J. Talent Identification and Relative Age Effects in English Male Rugby Union Pathways: From Entry to Expertise. *Front. Sports Act. Living* **2021**, *3*, 12. [CrossRef]
9. Musch, J.; Grondin, S. Unequal Competition as an Impediment to Personal Development: A Review of the Relative Age Effect in Sport. *Dev. Rev.* **2001**, *21*, 147–167. [CrossRef]
10. Dimundo, F.; Cole, M.; Blagrove, R.C.; Till, K.; McAuley, A.B.; Hall, M.; Gale, C.; Kelly, A.L. Talent Identification and Development in Male Rugby Union: A Systematic Review. *J. Expert.* **2021**, *4*, 33–55.
11. Armstrong, H.G. A comparison of the performance of summer and autumn-born children at eleven and sixteen. *Br. J. Educ. Psychol.* **1966**, *36*, 72–76. [CrossRef] [PubMed]
12. Bell, J.F.; Daniels, S. Are Summer-born Children Disadvantaged? The Birthdate Effect in Education. *Oxf. Rev. Educ.* **1990**, *16*, 67–80. [CrossRef]
13. Hauck, A.L.; Finch, A.J., Jr. The effect of relative age on achievement in middle school. *Psychol. Sch.* **1993**, *30*, 74–79. [CrossRef]
14. Martin, R.P.; Foels, P.; Clanton, G.; Moon, K. Season of Birth Is Related to Child Retention Rates, Achievement, and Rate of Diagnosis of Specific LD. *J. Learn. Disabil.* **2004**, *37*, 307–317. [CrossRef]
15. Doncaster, G.; Medina, D.; Drobnic, F.; Gómez-Díaz, A.J.; Unnithan, V. Appreciating Factors Beyond the Physical in Talent Identification and Development: Insights From the FC Barcelona Sporting Model. *Front. Sports Act. Living* **2020**, *2*, 91. [CrossRef] [PubMed]
16. McCarthy, N. (Ed.) Non -linear development of talent: The relative age effect and new perspectives. In *Rugby Football Union (RFU) Player Development Conference: The future Player*; Twickenham Stadium: Twickenham, UK, 2015.
17. Kelly, A.L.; Williams, C.A. Physical Characteristics and the Talent Identification and Development Processes in Male Youth Soccer: A Narrative Review. *Strength Cond. J.* **2020**, *42*, 15–34. [CrossRef]
18. Webdale, K.; Baker, J.; Schorer, J.; Wattie, N. Solving sport's 'relative age' problem: A systematic review of proposed solutions. *Int. Rev. Sport Exerc. Psychol.* **2020**, *13*, 187–204. [CrossRef]
19. Gibbs, B.G.; Jarvis, J.A.; Dufur, M.J. The Rise of the Underdog? The Relative Age Effect Reversal Among Canadian-born NHL Hockey Players: A Reply to Nolan and Howell. *Int. Rev. Sociol. Sport* **2011**, *47*, 644–649. [CrossRef]
20. McCarthy, N.; Collins, D. Initial identification & selection bias versus the eventual confirmation of talent: Evidence for the benefits of a rocky road? *J. Sports Sci.* **2014**, *32*, 1604–1610. [CrossRef]
21. Bjørndal, C.T.; Luteberget, L.S.; Till, K.; Holm, S. The relative age effect in selection to international team matches in Norwegian handball. *PLoS ONE* **2018**, *13*, e0209288. [CrossRef]
22. Jones, B.D.; Lawrence, G.P.; Hardy, L. New evidence of relative age effects in "super-elite" sportsmen: A case for the survival and evolution of the fittest. *J. Sports Sci.* **2018**, *36*, 697–703. [CrossRef] [PubMed]
23. Fumarco, L.; Gibbs, B.G.; Jarvis, J.A.; Rossi, G. The relative age effect reversal among the National Hockey League elite. *PLoS ONE* **2017**, *12*, e0182827. [CrossRef] [PubMed]
24. Ford, P.R.; Williams, A.M. No relative age effect in the birth dates of award-winning athletes in male professional team sports. *Res. Q Exerc. Sport* **2011**, *82*, 570–573. [CrossRef] [PubMed]
25. Malina, R.M.; Cumming, S.P.; Rogol, A.D.; Coelho-e-Silva, M.J.; Figueiredo, A.J.; Konarski, J.M.; Kozieł, S.M. Bio-Banding in Youth Sports: Background, Concept, and Application. *Sports Med.* **2019**, *49*, 1671–1685. [CrossRef]
26. Mann, D.L.; van Ginneken, P.J.M.A. Age-ordered shirt numbering reduces the selection bias associated with the relative age effect. *J. Sports Sci.* **2017**, *35*, 784–790. [CrossRef]
27. Kelly, A.L.; Jackson, D.T.; Taylor, J.J.; Jeffreys, M.A.; Turnnidge, J. "Birthday-Banding" as a Strategy to Moderate the Relative Age Effect: A Case Study Into the England Squash Talent Pathway. *Front. Sports Act. Living* **2020**, *2*, 145. [CrossRef]
28. Moran, J.; Cervera, V.; Jones, B.; Hope, E.; Drury, B.; Sandercock, G. Can discreet performance banding, as compared to bio-banding, discriminate technical skills in male adolescent soccer players? A preliminary investigation. *Int. J. Sports Sci. Coach.* **2022**, *17*, 325–333. [CrossRef]
29. Cobley, S.; Abbott, S.; Eisenhuth, J.; Salter, J.; McGregor, D.; Romann, M. Removing relative age effects from youth swimming: The development and testing of corrective adjustment procedures. *J. Sci. Med. Sport* **2019**, *22*, 735–740. [CrossRef]
30. Collins, D.; MacNamara, Á. The Rocky Road to the Top: Why Talent Needs Trauma. *Sports Med.* **2012**, *42*, 907–914. [CrossRef]
31. Boiché, J.C.S.; Sarrazin, P.G. Proximal and distal factors associated with dropout versus maintained participation in organized sport. *J. Sports Sci. Med.* **2009**, *8*, 9–16.
32. Baron-Thiene, A.; Alfermann, D. Personal characteristics as predictors for dual career dropout versus continuation—A prospective study of adolescent athletes from German elite sport schools. *Psychol. Sport Exerc.* **2015**, *21*, 42–49. [CrossRef]
33. Taylor, J.; Collins, D. Shoulda, Coulda, Didnae—Why Don't High-Potential Players Make it? *Sport Psychol.* **2019**, *33*, 85–96. [CrossRef]
34. Harter, S. Effectance Motivation Reconsidered. Toward a Developmental Model. *Hum. Dev.* **1978**, *21*, 34–64. [CrossRef]
35. Losier, G.F.; Vallerand, R.J. The Temporal Relationship between Perceived Competence and Self-Determined Motivation. *J. Soc. Psychol.* **1994**, *134*, 793–801. [CrossRef] [PubMed]

36. Collins, D.; MacNamara, Á.; McCarthy, N. Putting the Bumps in the Rocky Road: Optimizing the Pathway to Excellence. *Front. Psychol.* **2016**, *7*, 1482. [CrossRef] [PubMed]
37. Taylor, J.; Collins, D. Navigating the Winds of Change on the Smooth Sea—The Interaction of feedback and emotional disruption on the talent pathway. *J. Appl. Sport Psychol.* **2021**, 1–27. [CrossRef]
38. Savage, J.; Cruickshank, A.; Collins, D. Perspective, Control, and Confidence: Perceived Outcomes of Using Psycho-Behavioral Skills in the Developmental Trauma Experience. *Int. J. Sport Exerc. Psychol.* **2021**, *20*, 377–396. [CrossRef]
39. Barnes, F.B.; Fletcher, D.; Neely, K.C. Stress-Related Growth in Elite Sport Performers: Qualitative Differentiators in Psychosocial Mechanisms. *Sport Psychol.* **2021**, *1*, 1–2. [CrossRef]
40. Rees, T.; Hardy, L.; Güllich, A.; Abernethy, B.; Côté, J.; Woodman, T.; Montgomery, H.; Laing, S.; Warr, C. The Great British Medalists Project: A Review of Current Knowledge on the Development of the World's Best Sporting Talent. *Sports Med.* **2016**, *46*, 1041–1058. [CrossRef]
41. Sarkar, M.; Fletcher, D.; Brown, D.J. What Doesn't Kill Me . . . : Adversity-Related Experiences Are Vital In The Development Of Superior Olympic Performance. *J. Sci. Med. Sport* **2015**, *18*, 475–479. [CrossRef]
42. Savage, J.; Collins, D.; Cruickshank, A. Exploring Traumas in the Development of Talent: What Are They, What Do They Do, and What Do They Require? *J. Appl. Sport Psychol.* **2017**, *29*, 101–117. [CrossRef]
43. Taylor, J.; Collins, D. The Highs and the Lows—Exploring the Nature of Optimally Impactful Development Experiences on the Talent Pathway. *Sport Psychol.* **2020**, *34*, 319–328. [CrossRef]
44. Williams, G.G.; MacNamara, Á. Challenge is in the eye of the beholder: Exploring young athlete's experience of challenges on the talent pathway. *J. Sports Sci.* **2022**, *40*, 1078–1087. [CrossRef]
45. Cumming, S.P.; Searle, C.; Hemsley, J.K.; Haswell, F.; Edwards, H.; Scott, S.; Gross, A.; Ryan, D.; Lewis, J.; White, P.; et al. Biological maturation, relative age and self-regulation in male professional academy soccer players: A test of the underdog hypothesis. *Psychol. Sport Exerc.* **2018**, *39*, 147–153. [CrossRef]
46. Giacobbi, P.R.; Poczwardowski, A.; Hager, P. A Pragmatic Research Philosophy for Sport and Exercise Psychology. *Sport Psychol.* **2005**, *19*, 18–31. [CrossRef]
47. Badley, G. The Crisis in Educational Research: A Pragmatic Approach. *Eur. Educ. Res. J.* **2003**, *2*, 296–308. [CrossRef]
48. McCaslin, M. Pragmatism. In *The SAGE Encyclopedia of Qualitative Research Methods*; Given, L.M., Ed.; SAGE Publications: Thousand Oaks, CA, USA, 2012; pp. 672–675.
49. Denzin, N.K.; Lincoln, Y.S. *Strategies of Qualitative Inquiry*, 3rd ed.; Sage: Los Angeles, CA, USA, 2008.
50. Braun, V.; Clarke, V.; Terry, G.; Hayfield, N. Thematic analysis. In *Handbook of Research Methods in Health and Social Sciences*; Liamputtong, P., Ed.; Springer: Berlin/Heidelberg, Germany, 2018; pp. 843–860.
51. Strean, W.B. Possibilities for Qualitative Research in Sport Psychology. *Sport Psychol.* **1998**, *12*, 333–345. [CrossRef]
52. Bryant, A. Grounded Theory and Pragmatism: The Curious Case of Anselm Strauss. *Forum Qual. Soc. Res.* **2009**, *10*. [CrossRef]
53. Morgan, D.L. Paradigms Lost and Pragmatism Regained: Methodological Implications of Combining Qualitative and Quantitative Methods. *J. Mix. Methods Res.* **2007**, *1*, 48–76. [CrossRef]
54. Cruickshank, A.; Collins, D.; Minten, S. Culture Change in a Professional Sports Team: Shaping Environmental Contexts and Regulating Power. *Int. J. Sports Sci. Coach.* **2013**, *8*, 271–290. [CrossRef]
55. Ollis, S.; Macpherson, A.; Collins, D. Expertise and talent development in rugby refereeing: An ethnographic enquiry. *J. Sports Sci.* **2006**, *24*, 309–322. [CrossRef] [PubMed]
56. Drasch, K.; Matthes, B. Improving retrospective life course data by combining modularized self-reports and event history calendars: Experiences from a large scale survey. *Qual. Quant.* **2013**, *47*, 817–838. [CrossRef]
57. Smith, J.A. Semi-structured Interviewing and Qualitative Analysis. In *Rethinking Methods in Psychology*; Smith, J.A., Harré, R., Langenhove, L.V., Eds.; Sage: London, UK, 1995; pp. 9–26.
58. Braun, V.; Clarke, V. One size fits all? What counts as quality practice in (reflexive) thematic analysis? *Qual. Res. Psychol.* **2020**, *18*, 328–352. [CrossRef]
59. Braun, V.; Clarke, V. Using Thematic Analysis in Psychology. *Qual. Res. Psychol.* **2006**, *3*, 77–101. [CrossRef]
60. Braun, V.; Clarke, V. (Mis)conceptualising Themes, Thematic Analysis, and other Problems with Fugard and Potts' (2015) Sample-Size Tool for Thematic Analysis. *Int. J. Soc. Res. Methodol.* **2016**, *19*, 739–743. [CrossRef]
61. Sparkes, A.C.; Smith, B. Judging the Quality of Qualitative inquiry: Criteriology and Relativism in Action. *Psychol. Sport Exerc.* **2009**, *10*, 491–497. [CrossRef]
62. Braun, V.; Clarke, V. *Successful Qualitative Research: A Practical Guide for Beginners*; Sage: London, UK, 2013.
63. Hewitt, J. Ethical Components of Researcher—Researched Relationships in Qualitative Interviewing. *Qual. Health Res.* **2007**, *17*, 1149–1159. [CrossRef]
64. Nowell, L.S.; Norris, J.M.; White, D.E.; Moules, N.J. Thematic Analysis: Striving to Meet the Trustworthiness Criteria. *Int. J. Qual. Methods* **2017**, *16*, 1609406917733847. [CrossRef]
65. Smith, B.; McGannon, K.R. Developing Rigor in Qualitative Research: Problems and Opportunities within Sport and Exercise Psychology. *Int. Rev. Sport Exerc. Psychol.* **2018**, *11*, 101–121. [CrossRef]
66. MacNamara, Á.; Button, A.; Collins, D. The Role of Psychological Characteristics in Facilitating the Pathway to Elite Performance Part 1: Identifying Mental Skills and Behaviors. *Sport Psychol.* **2010**, *24*, 52–73. [CrossRef]

67. Hill, A.; MacNamara, Á.; Collins, D. Psychobehaviorally Based Features of Effective Talent Development in Rugby Union: A Coach's Perspective. *Sport Psychol.* **2015**, *29*, 201–212. [CrossRef]
68. Rosiek, J.L. Pragmatism and post-qualitative futures. *Int. J. Qual. Stud. Educ.* **2013**, *26*, 692–705. [CrossRef]
69. Tracy, S.J. Qualitative Quality: Eight "Big-Tent" Criteria for Excellent Qualitative Research. *Qual. Inq.* **2010**, *16*, 837–851. [CrossRef]
70. Nash, C.; Taylor, J. 'Just let them play': Complex dynamics in youth sport, why it isn't so simple. *Front. Psychol.* **2021**, *12*, 700750. [CrossRef]
71. Li, P.; De Bosscher, V.; Pion, J.; Weissensteiner, J.R.; Vertonghen, J. Is international junior success a reliable predictor for international senior success in elite combat sports? *Eur. J. Sport Sci.* **2018**, *18*, 550–559. [CrossRef]
72. Wattie, N.; Schorer, J.; Baker, J. The Relative Age Effect in Sport: A Developmental Systems Model. *Sports Med.* **2015**, *45*, 83–94. [CrossRef]
73. Collins, D.; MacNamara, Á.; Cruickshank, A. Research and Practice in Talent Identification and Development—Some Thoughts on the State of Play. *J. Appl. Sport Psychol.* **2019**, *31*, 340–351. [CrossRef]
74. Baker, J.; Mosher, A.; Fraser-Thomas, J. Is it too early to condemn early sport specialisation? *Br. J. Sports Med.* **2020**, *55*, 179–180. [CrossRef]
75. Van Nieuwstadt, M.H.H.; Das, M.; Elferink-Gemser, M.T. Mechanisms explaining the birthplace effect for male elite football players. *J. Sports Sci.* **2021**, *39*, 576–582. [CrossRef]
76. Martindale, R.J.; Collins, D.; Daubney, J. Talent Development: A Guide for Practice and Research Within Sport. *Quest* **2005**, *57*, 353–375. [CrossRef]
77. Taylor, J.; Collins, D. The Talent Development Curriculum. In *Practical Sport Coaching*, 2nd ed.; Nash, C., Ed.; Routledge: London, UK, 2022.
78. Bjørndal, C.T.; Ronglan, L.T. Athlete Development in Norwegian Handball. In *Talent Identification and Development in Sport: International Perspectives*, 2nd ed.; Baker, J., Cobley, S., Schörer, J., Eds.; Routledge: New York, NY, USA, 2020. [CrossRef]
79. Book, R.T.; Henriksen, K.; Stambulova, N.; Storm, L.K. "All they have seen is a model for failure": Stakeholder's perspectives on athletic talent development in American underserved communities. *J. Appl. Sport Psychol.* **2021**, 1–21. [CrossRef]
80. Neelen, M.; Kirschner, P. *Evidence-Informed Learning Design: Use Evidence to Create Training Which Improves Performance*; KoganPage: London, UK, 2020.
81. Moodie, G.; Taylor, J.; Collins, D. Teach, Test, Tweak, Repeat: Reflecting on the development of psycho-behavioural skills. *J. Appl. Sport Psychol. In Review*.

sports

Article

An Examination of Relative Age and Athlete Dropout in Female Developmental Soccer

Kristy L. Smith * and Patricia L. Weir

Department of Kinesiology, University of Windsor, Windsor, ON N9B 3P4, Canada; weirl@uwindsor.ca
* Correspondence: smith43@uwindsor.ca

Abstract: Sport dropout rates among children and youth are a concern for researchers and policy makers. The impact of relative age effects (RAEs) on dropout trends has not been adequately examined in female samples. The purpose of this study was to longitudinally examine dropout in a female soccer cohort in Ontario, Canada. Registration entries for a one-year cohort were examined across a seven-year period (n = 9908; age 10–16 years). A chi-square analysis established the presence of RAEs in the initial year of registration. Survival analyses assessed the impact of relative age, competition level, and community size on athlete dropout. A median survival rate of four years was observed for players born in the first quartile, while all remaining quartiles had a median survival of three years. Community size did not predict dropout in this analysis; however, competition level was a significant predictor, with competitive players being more likely to remain engaged vs. recreational players (55.9% vs. 20.7%). The observed trends are likely to have a significant impact from both a healthy development and systems perspective (e.g., economic/market loss). Intervention is needed to mitigate current dropout trends in female athletes. Practical applications are discussed.

Keywords: relative age effects; athlete dropout; sport dropout; female; soccer; competition level; sport development

Citation: Smith, K.L.; Weir, P.L. An Examination of Relative Age and Athlete Dropout in Female Developmental Soccer. *Sports* **2022**, *10*, 79. https://doi.org/10.3390/sports10050079

Academic Editor: Adam Baxter-Jones

Received: 23 March 2022
Accepted: 10 May 2022
Published: 20 May 2022

1. Introduction

Sport dropout rates among children and youth present a growing concern for researchers and policy makers alike. From a healthy development perspective, organized sport participation is associated with a variety of physical, psychological, and social benefits [1–3]. For example, youth who engage in organized sport may experience greater social competence [4] and fewer depressive symptoms [5,6], and they may be more likely to develop fundamental movement skills that promote physical engagement in alternative sport contexts and healthy leisure pursuits across their lifespan [7,8]. From a talent development perspective, sport dropout causes a reduction in potential talent for future advancement in sport, as the development of expertise is theoretically predicated by ongoing participation. High rates of organized sport participation have been reported. For instance, seventy-seven percent of Canadian children and youth aged 5–19 years old participate in organized physical activity or sport, as reported by their parents [9]. However, high dropout levels have also been observed and are estimated to be between 30 and 35% per year [10,11], although current estimates are unavailable and likely vary by sex, sport context, and chronological age [12].

A systematic review examining organized sport dropout identified that intrapersonal (e.g., lack of enjoyment) and interpersonal (e.g., parental pressure) constraints are commonly associated with disengagement among children and youth [13]. This review also highlighted a potential connection between frequently cited dropout factors and relative age, that being physical factors (e.g., maturation) and perceptions of competence. Commonly known as Relative Age Effects (RAEs), this term refers to the (dis)advantages resulting from subtle variations in chronological age and thus lived experience and physical/psychological development in age-grouped peers [14]. Within sport, RAEs are believed to advantage those who are relatively older (i.e., born earlier and closer to an organization-imposed cut-off date for grouping similar-age athletes) by providing increased access to higher levels of competition, training, and coaching [15,16].

The underlying mechanisms contributing to RAEs are likely multi-factorial in nature and include a variety of individual, task, and environmental contributors [16,17]; the "maturation-selection" hypothesis is most commonly cited by researchers [18–20]. Briefly, this hypothesis suggests that advanced chronological age is accompanied by greater anthropometric (e.g., stature) and physical attributes (e.g., muscular strength and endurance), which provide performance advantages in many sport contexts. These differences are further exacerbated during adolescence. Consequently, relatively older children who are likely to be further along in terms of maturational development receive more attention from coaches and may experience a higher likelihood of selection for elite levels of sport competition, which ultimately furthers their athletic development. Conversely, relatively younger participants may not have the same opportunity to develop and are more likely to struggle with perceptions of competence and self-worth. In Crane and Temple's review [13], five of the six studies identifying maturation as a contributing factor to dropout suggested that RAEs were a factor, although the reviewers also noted that more research was needed to understand the connection between competing with chronologically older peers and experiences leading to dropout.

Sport popularity has also been associated with RAEs [16,21,22], and this consideration may have a connection to another variable of interest in the sport development literature, that being community size or the "birthplace effect" [23]. Previous research reports have documented increased rates of participation in small to medium-sized communities that are large enough to support youth sport leagues but not so densely populated that the competition for sport facilities, team membership, etc., is detrimental to participation [24–27]). However, the metrics used to assign community size categories have been somewhat inconsistent, and the role of this variable in sport dropout is unknown. Furthermore, the exact nature of the interaction between community size and RAEs as well as the role these variables play in athlete development outcomes have been somewhat elusive.

Initial observations of an association between relative age and sport dropout were made several decades ago by Barnsley and colleagues [28,29], who suggested that relatively older Canadian ice hockey players were more likely to remain engaged in the sport when compared to the relatively younger participants. Similarly, an examination of male youth soccer in Belgium indicated that a higher dropout rate was present among later-born players at 12 years of age [15]. Large-scale, cross-sectional studies of French soccer and basketball provided further evidence of increased rates of dropout amongst the youngest players over a one-year period [30–32]. These trends were consistent across a variety of pre-up to post-adolescent age groups in both male and female samples, leading the researchers to suggest that the over-representation of relatively older participants often observed in sport samples may be in part due to a greater number of relatively younger athletes among the "dropouts" [32].

Two longitudinal examinations of relative age and dropout rate have also been conducted. Figueiredo and colleagues [33] reported the inconsistent tracking of participation by birth quartile for male soccer players at two- and ten-year timepoints after baseline analyses (i.e., at 11 and 13 years of age); playing status could not be predicted by birth quartile. However, this study was limited by a small sample size (n = 112). Lemez et al. [34] provided a more substantial analysis of male athletes by examining 14,325 registrants in Canadian ice hockey over a five-year period (age 10–15 years). Relatively younger participants born in the fourth quartile were found to be 17% more likely to drop out than their first-quartile counterparts (OR 1.175, 95% CI 1.054, 1.309). Subsequent analyses attempted to unravel the impact of player movement between competition levels on the observed patterns of dropout. The observations suggested that dropout players were more likely to remain at the same level of competition prior to disengagement from the sport.

While the weight of the evidence in the published literature points to a higher risk of sport dropout for the relatively younger players, one exception to this pattern has been noted. Wattie and colleagues [35] observed increased odds of reported dropout among relatively older female participants at the recreational level in German youth sport clubs, with no comparable effect in the male sample. This finding may have been driven by a high proportion of athletes participating in artistic or individual sport contexts (e.g., gymnastics) within the sample, with smaller physical size providing a competitive advantage. However, these findings also raise questions about the possibility of sex differences in dropout trends. Vincent and Glamser [36] suggested that the "maturation-selection" hypothesis may exemplify the male sporting experience to a greater extent than that of females due to the associated disadvantages that maturation brings to female athletes (e.g., shorter legs and wider hips [37]) as compared to the physical advantages afforded to early maturing males (e.g., increased speed, power, and endurance in motor skills [38]). The findings of Wattie and colleagues may also implicate the role played by talent identification and development processes, as the athletes examined participated in recreational contexts [35]. Indeed, entry into competitive contexts at young ages—known as *early specialization*—has been associated with negative sport experiences (e.g., sport withdrawal, burnout [39–41]).

Given the consistent presence of RAEs at the introductory levels and the related evidence with respect to dropout, it is necessary to continue to evaluate participation trends across various age, sport, and competitive levels in a longitudinal manner. Sport participation likely varies across the lifespan, and many factors may contribute to an athlete's decision to participate in a certain sport context. Consequently, the primary objective of this study was to retrospectively examine dropout in a female cohort across a seven-year period (i.e., covering the pre-adolescent to post-adolescent transition years) with respect to relative age. Given the trends observed in past work that examined sport dropout [30,34] and the consistent reporting of RAEs in soccer (see Smith et al. for a review [42]), it was hypothesized that the relatively older athletes would be more likely to remain engaged in sport across the pre- to post-adolescent years; however, the magnitude could potentially vary based on relevant contextual factors. Thus, additional variables found to influence participation were also evaluated, including community size [24,43] and competition level [34,44].

2. Materials and Methods

Following institutional ethics approval, an anonymized dataset of all female members of a one-year cohort registered with *Ontario Soccer* from the age of ten years was obtained from the provincial organization. This dataset included all subsequent registrations across a six-year period for the initial cohort of members (i.e., up to and including existing registration entries at 16 years of age). A total of 38,248 registration entries for 9915 participants were available. Prior to analysis, the participant data were screened for inconsistent and/or missing information with respect to birth month. Twenty-three registration entries were corrected upon confirmation of birth month with a minimum of two other entries for the participant (0.0006% of original sample). One participant was removed because the month of birth could not be confirmed (a total of seven registration entries); one participant was removed because the entries were believed to be a duplicate set (a total of five registration entries); five additional participants were removed because they had an "inactive" status at the age of 10 years and no subsequent registrations beyond that year. Therefore, 99.9% of registration entries were retained (n = 9908 participants).

The remaining participants' birthdates were coded according to birth quartile (i.e., Quartile One-Q1: January—March; Quartile Two-Q2: April—June; Quartile Three-Q3: July—September; Quartile Four-Q4: October—December) in consideration of the December 31st cut-off date employed in Ontario youth soccer. The data were also coded for two other potential determinants of participation. Community size was coded according to *census subdivision*. Census subdivision corresponds to the municipality structure that would determine funding for local sport facilities in Canada [45]. It is a well-established metric used by Statistics Canada and refers to a municipality (as determined by provincial/territorial legislation) or areas treated as municipal equivalents for statistical purposes. Categories that have been employed in previous research were utilized (1: >1,000,000 people; 2: 500,000–999,999; 3: 100,000–499,999; 4: 30,000–99,999; 5: 10,000–29,999; 6: 5000–9999; 7: 2500–4999; 8: 1000–2499; 9: <1000; e.g., [23,25]).

The level of play at the time of the athlete's last registration (i.e., competition level prior to disengaging from the sport or at age 16 years) was coded according to the Ontario Soccer organization structure (1: Mini outdoor; 2: Recreational; 3: Competitive). Mini Outdoor is a small-sided game, typically for players 12 years and under. Beyond age 12, players are typically categorized as being at the recreational level (e.g., house league, where selection processes are absent and any child or youth can theoretically participate) or the competitive level (e.g., representative or more elite-level players who gain membership through selection processes or "tryouts"). All registered participants engage in some form of match/game play, although the amount may vary. This structure is recommended by Ontario Soccer and may or may not be followed at the local level (e.g., players may be classified as recreational or competitive prior to age 12 years). These classifications were provided by representatives from Ontario Soccer [46].

A preliminary chi-square analysis and a visual inspection of the birth distribution were conducted to ascertain whether RAEs might be present during the initial year of the registration entries, at the age of ten years. The observed number of participants born in each quartile was compared to the number expected based on the number of days in each quartile. Traditionally, an equal distribution of 25% has been utilized as the expected proportion of participants for each birth quartile in RAE research. Delorme and Champely [47] argue that this method inflates the risk of Type I error. Thus, the actual distribution of the population from which the sample was taken should be utilized, and in the absence of this information, the expected distribution should be adjusted to the number of days present in each birth quartile. For this study, the birth distribution for the overall population of Ontario female soccer players was not available; therefore, the expected distribution was calculated by dividing the number of days in each quartile by 365. A statistically significant chi-square value ($p < 0.05$) was used to calculate the w effect size statistic to determine the strength of the relationship. The w effect size statistic is calculated by taking the value of chi-square divided by the number of subjects and taking the square root ($w = \sqrt{(\chi 2 / n)}$) [48]. Cohen [48] proposed that w values of 0.1, 0.3, and 0.5 represent small, medium, and large effect sizes, respectively. The calculation of standardized residuals was planned for a chi-square analysis producing w values ≥ 0.1, with a value of ≥ 1.96 indicating an over-representation and a value of ≤ -1.96 indicating an under-representation in terms of relative age distribution.

Survival analyses were then carried out to assess the impact of relative age on dropout from developmental soccer between the ages of 10 and 16 years. Dropout was identified using the last registration entry that was present in the longitudinal dataset provided by Ontario Soccer. Thus, a participant who last registered at the age of 10 through 15 years would be coded as a "dropout", and a participant who had a registration entry at the age of 16 years would be coded as "engaged". A Kaplan–Meier analysis was used to investigate the dropout rate with respect to relative age by birth quartile. The log-rank test assessed the null hypothesis of a common survival curve. This was followed by a Cox Regression to further evaluate the impact of birth quartile, with a consideration of community size and competition level. The proportional hazards assumption was tested using the goodness-of-fit approach [49]. This assumption states that the hazard (i.e., risk of dropping out) for one individual must be proportional to the hazard for any other individual, and that the hazard ratio must be constant over time [49].

3. Results

3.1. General Findings–Relative Age

Results from the preliminary chi-square analysis are presented in Table 1. An over-representation of relatively older participants was observed in the initial sample (χ^2 (3) = 182.972, $p < 0.001$) with a small effect size ($w = 0.14$). Quartile 2 had the highest number of participants at ten years of age, followed by Q1, Q3, and Q4. The Kaplan–Meier analysis revealed that 23.3% of the initial cohort remained until the end of the seven-year period. The survival curve for each birth quartile is available in Figure 1. The log-rank test indicated that the null hypothesis should be rejected (χ^2 (3) = 26.321, $p < 0.001$). A median survival rate of four years was observed for players born in the first quartile over the subsequent six years of registration; this differed from a median survival of three years for players born in the remaining quartiles (outlined further in Table 2).

Table 1. Results from the preliminary chi-square analysis.

Birth Quartile	Observed (*n*)	Expected	Standardized Residual	Odds Ratio	95% CI Lower	95% CI Higher
Quartile 1 (Q1)	2674	2443.1	**4.56**	**1.13**	1.07	1.19
Quartile 2 (Q2)	2803	2470.2	**6.74**	**1.19**	1.13	1.25
Quartile 3 (Q3)	2472	2497.4	−0.476	0.99	0.92	1.05
Quartile 4 (Q4)	1959	2497.4	**−10.745**	**0.73**	0.67	0.80

Note: Values in bold indicate an over-representation (i.e., ≥ 1.96) or under-representation (i.e., ≤ −1.96) with respect to relative age distribution by quartile.

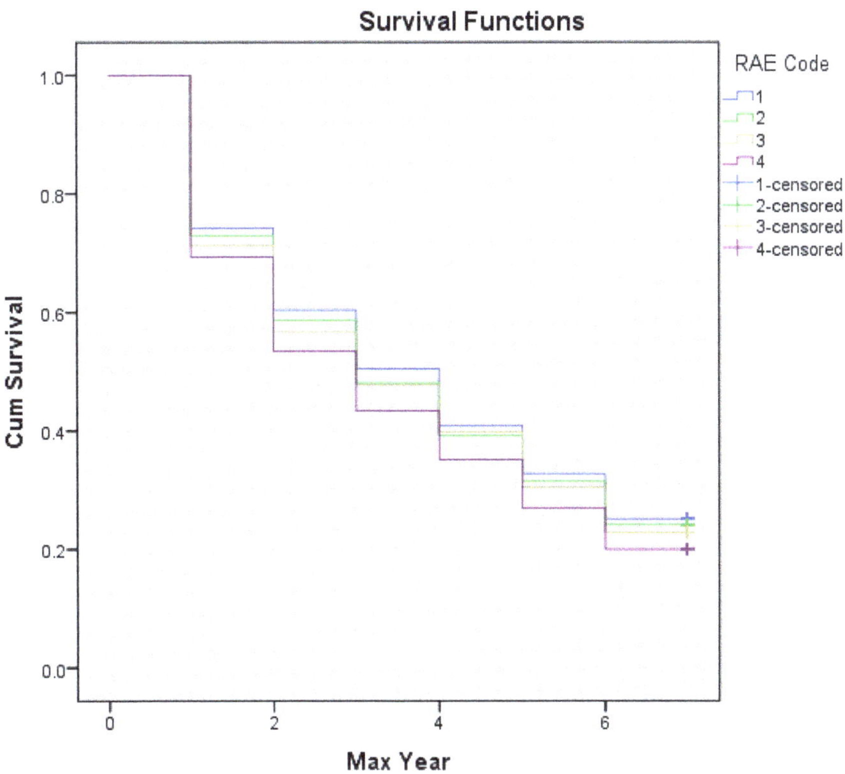

Figure 1. Survival curve for each birth quartile, indicating the highest cumulative survival over the seven-year period.

Table 2. Results from the Kaplan–Meier survival analysis: mean and median values for survival time.

	Mean				Median			
Birth Quartile	Est.	Std. Error	95% CI Lower	95% CI Upper	Est.	Std. Error	95% CI Lower	95% CI Upper
Q1	3.840	0.046	3.750	3.929	4.000	0.097	3.809	4.191
Q2	3.748	0.045	3.660	3.835	3.000	0.097	2.811	3.189
Q3	3.688	0.048	3.595	3.782	3.000	0.119	2.767	3.233
Q4	3.483	0.052	3.381	3.586	3.000	0.086	2.831	3.169
Overall	3.705	0.024	3.659	3.752	3.000	0.054	2.894	3.106

Note: Estimation is limited to the largest survival time if it is censored.

3.2. Additional Factors—Competition Level and Community Size

Prior to conducting the Cox Regression, it was recognized that players who dropped out during "mini outdoor" would bias the survival analysis as any player who was classified in this category (i.e., coded according to last registration entry) would theoretically drop out by the age of 12 years, according to Ontario Soccer's organizational structure. Thus, only players coded as "competitive" (n = 2327) and "recreational" (n = 4836) at the time of their last registration were included in the Cox Regression (overall n = 7163). The findings are presented in Table 3. The analysis indicated that birth quartile was not statistically significant ($p > 0.05$) when the impact of community size and competition level were considered. Community size did not predict dropout in this analysis; however, competition level was observed to be a significant predictor of continued sport involvement ($p < 0.001$).

Table 3. Results from the Cox Regression survival analysis (overall).

	Regression Coefficient	Std. Error	$p > \lvert z \rvert$	Hazard Ratio	95% CI Lower	95% CI Upper
Q1	0.015	0.043	0.717	1.016	0.934	1.104
Q2	0.005	0.042	0.901	1.005	0.926	1.092
Q3	0.025	0.043	0.565	1.025	0.942	1.116
CS	0.003	0.002	0.080	1.003	1.000	1.007
Comp. Level	0.953	0.035	0.000	2.593	2.419	2.779

Notes: Quartile 4 used as reference category. Community size (CS) divided by 100,000 for analysis purposes. Confidence intervals that include a value of 1.0 indicate equivalence in the hazard rate (i.e., not statistically significant).

The survival and hazard functions using the mean for competition level can be found in Figure 2a,b, respectively. By percentage, 55.9% of competitive players were still registered with Ontario Soccer at the age of 16 years, while only 20.7% of recreational-level players remained (see Table 4 and Figure 2a). Descriptively, this corresponds to a yearly dropout rate of more than 30% of recreational players each year. Competitive players were more than twice as likely to remain engaged in soccer until the age of 16 years when compared to recreational-level participants (Hazard ratio 2.593, 95% CI 2.419, 2.779; see Figure 2b). In consideration of the significance of competition level, a graphical representation of the quartile distributions for each year was generated for both the competitive and the recreational streams to inspect the transient relative age distribution. The competitive trajectory (see Figure 3a) showed a classic RAE, with Quartile 1 consistently over-represented and Quartile 4 consistently under-represented across the seven-year period; on the other hand, the recreational stream (see Figure 3b) showed an over-representation in Quartile 2 and an under-representation in Quartile 4.

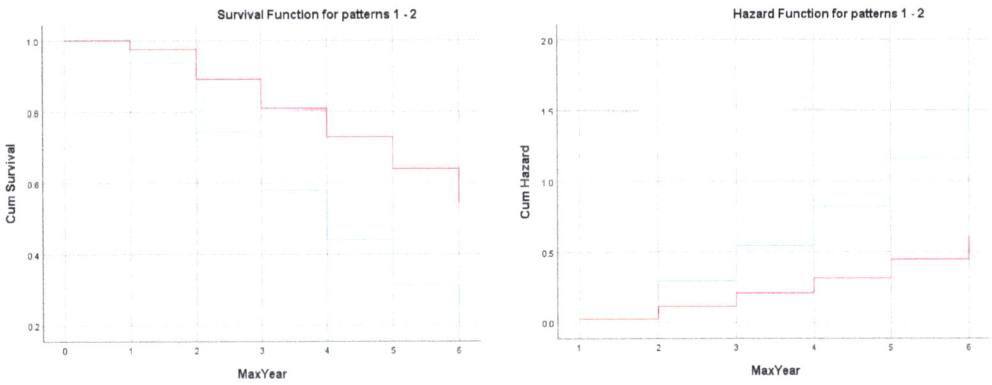

(**a**) (**b**)

Figure 2. Competitive (red) and recreational (blue): (**a**) Survival function at the mean for the competition level. The vertical axis shows the probability of survival. The horizontal axis represents time-to-event data. (**b**) Hazard function at the mean for the competition level. The vertical axis shows the cumulative hazard, equal to the negative log of the survival probability. The horizontal axis represents time-to-event data.

Table 4. Results from the Cox Regression survival analysis (competition level).

Competitive Level	Dropout before Age 16 Years (*n*)	Engaged at Age 16 Years (*n*)	Engaged at Age 16 Years (%)	Overall *n*
Competitive	1027	1300	55.9%	2327
Recreational	3835	1001	20.7%	4836

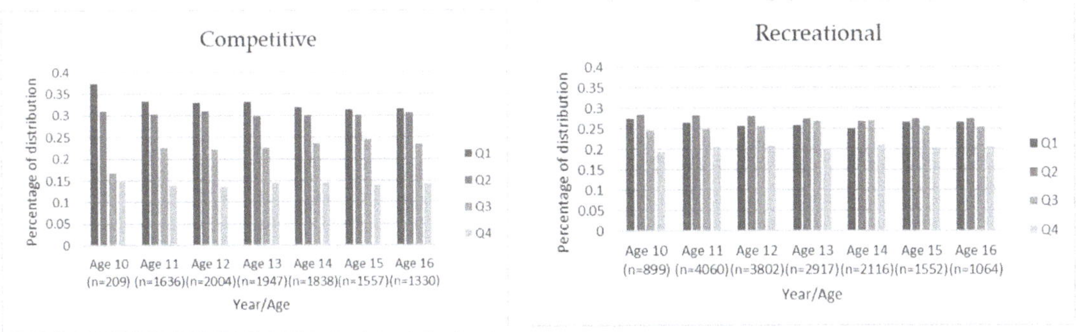

(**a**) (**b**)

Figure 3. Birth distribution by quartile and chronological age (10–16 years). (**a**) Competitive players; (**b**) Recreational players. Note: The majority of participants ages 10–12 years would be classified as "mini outdoor" according to Ontario Soccer's organizational structure and are therefore not represented.

4. Discussion

The primary objective of this study was to retrospectively examine athlete dropout with respect to birth quartile in a female cohort for a total of seven years: beginning at the age of ten years, and subsequently followed across a six-year period. Thus, this study provides a longitudinal snapshot of the pre-adolescent to post-adolescent transition years within female soccer in Ontario. A significant RAE was observed in the initial cohort with the relatively oldest participants (i.e., those born earlier in the same-age cohort) having the highest rates of participation at age ten years. The participants born in the first quartile were found to have a greater likelihood of continued engagement in youth soccer during the examination period, as inferred by a median survival rate of one additional year when compared to their peers. However, birth quartile was not found to be a significant factor when competition level and community size were considered as part of the analysis. Thus, the preliminary hypotheses were generally supported by the results of the analyses.

The outcome of this study suggests that female dropout patterns in Ontario Soccer are comparable to previous findings in team sport contexts, with the relatively youngest exhibiting higher rates of disengagement. The one noted exception in the literature [35] may be differentiated by the artistic/individual sport contexts in which the participants engaged. Physical contact is inherent in the sport of soccer, providing an advantage to those with advanced growth and/or maturational status. Additionally, the team context might also emphasize physical differences as comparisons between players occur on the field and are generally based on more subjective evaluations of participants by coaches as opposed to objective measures that are more commonly associated with individual sports (e.g., a 100-meter swim time [50]). The aforementioned sample [35] was also considered to be "recreational" in nature. Interestingly, competitive level was observed to be an important variable in the current analysis, negating the impact of birth quartile when included in the analysis.

If considered to be an accurate estimate, the findings of this study suggest that approximately 7200 participants (or 73%) of this one-year, provincial cohort (n = 9908) are at risk of dropping out one year earlier because of their birthdate position with respect to an arbitrary, age-group cut-off. This statistic is alarming from both a healthy development perspective (i.e., continued participation is associated with positive outcomes; see examples discussed in the Introduction) and a systems perspective (i.e., continued growth of the sport). For example, a significant reduction in participation contributes to an economic/market loss [51]; that is, a high rate of dropout contributes to a reduction in game interest, loss of membership fees, and a reduced talent pool for future advancement in sport. Furthermore, youth sport is predominantly run by volunteers. Individuals who disengage from a sport during childhood or adolescence may be less likely to transition to a contributive role in their adult years.

These findings also highlight the potential impact of competitive streaming on sport dropout. While a greater proportion of competitive-level players were engaged at the age of 16 years (55.9% vs. 20.7% for recreational-level players), a more biased birthdate distribution favoring the relatively older players was also evident in the competitive context when evaluated by year of registration (see Figure 3a). This may suggest that RAEs resulting from initial growth differences are being perpetuated by talent selection processes [42], and is consistent with the available research examining female youth soccer players [52,53]. At no point during the seven-year period were the relatively youngest athletes observed to "catch-up" despite the culmination of maturational processes within the examined timeframe. While the recreational stream had a more evenly distributed birth representation (see Figure 3b), the high disengagement of athletes over the seven-year period may highlight a concerning trend for recreation-level athletes. This is somewhat surprising given the reduced demands of playing at the recreational level as compared to higher levels of competition, where the increased demands of additional training and performance might conflict with other priorities for this age demographic (e.g., schoolwork, part-time employment, social activities). However, it may also be indicative of athletes choosing to prioritize alternative forms of sport participation.

Community size did not appear to be a significant factor with respect to sport dropout in this sample. This finding differs from previous research studies (e.g., [24–27]) that have found increased rates of participation in small to medium-sized communities that are large enough to support youth sport leagues but not so densely populated that the competition for sport facilities, team membership, etc., is detrimental to participation. The survival analyses utilized in this study may not have detected subtle trends related to sport dropout in this sample due to the large range of community sizes in Ontario (i.e., census subdivisions range from 5 to 2,615,060 inhabitants). The impact of community size in this sample is evaluated further in a separate study that used geospatial mapping and odds ratio analyses [54].

Although not a primary goal of this work, this study documented the over-representation of the second quartile in the initial cohort at ten years of age (followed by Q1, Q3, and Q4); this provided the first RAE observed in a Canadian soccer sample. This pattern differs from the classic, linear RAE pattern (Q1 > Q2 > Q3 > Q4) that would be expected, based purely on chronological age differences. Female samples have been associated with a Q2 over-representation in previous studies, particularly in Canadian ice hockey at developmental and national levels [44,55]; but also observed in post-adolescent [56] and adult [22,30] female soccer samples.

The cause of this Q2-trend has largely been undetermined to date. Previous hypotheses have suggested that the "best" Q1-born, female athletes may be playing in male sport to gain a competitive advantage or are perhaps engaged in a more popular sport, leaving those born in the second quartile to experience success in the context under examination. This study adds evidence against the latter hypothesis in consideration of the Canadian Heritage Sport Participation 2010 report [57], which identified soccer as the most highly played sport by Canadian children. However, it was noted that the Q2 over-representation in this study was primarily driven by registration numbers in the recreational context when the sample was evaluated according to the competitive stream (cf. Figure 3a vs. Figure 3b), suggesting that the relatively oldest were experiencing greater success within the context of soccer at both the competitive and recreational levels.

Underlying patterns observed in a sample compiled for a recent meta-analysis of female athletes provide evidence that the effect might possibly be associated with early specialization opportunities for Q1-born athletes and consequent burnout, injury, and/or sport withdrawal (see Smith et al. [42] for further discussion). This hypothesis might partially explain the observed trends in this sample. However, the birth quartile distribution showed essentially the same pattern of representation across all years examined at both the competitive (i.e., Q1 over-representation) and the recreational (i.e., Q2 over-representation) levels; no transitional RAEs were observed. Thus, the underlying mechanisms of these trends requires further examination, and the exact contributor in this sample and others remains unknown.

The dropout rates observed in this longitudinal analysis are reflective of the high rates of dropout that have been observed in other samples (e.g., [10,11,58,59]). Sport administrators should seek to organize sport in a way that promotes the personal development of all its members, with varying levels of ability and motivation [3,60]. Strategies that support recreational-level athletes appear to be particularly needed. Future applied research should evaluate whether the provision of opportunities for skill development and other experiences that competitive players have (e.g., tournaments, inter-city play, skill development initiatives, team building events) would encourage engagement in recreational streams with increasing chronological age while still maintaining the reduced time demands (vs. competitive levels) that are likely to be desirable for high-school-age athletes. The recent trend towards sport-specific academies (i.e., academic institutions offering combined athletic and academic curricula) may be a promising avenue for continued sport engagement into the adolescent/post-adolescent years as they offer access to facilities/coaching and a flexible academic schedule. However, continued alignment between these academies and existing sport governing bodies is needed [61,62].

This study adds to the limited pool of research on female soccer athletes. A review of female RAEs found small but consistent RAEs in this sport context [42]; however, the existing work has primarily focused on elite competitors in post-adolescent and adult age groups (see Smith et al. [42] for a review) as opposed to the more developmental levels of the sport. Thus, the documentation of RAEs in pre-adolescent/adolescent and recreational-level athletes is important. The majority of work in soccer has focused on RAEs for male athletes [22,63], and additional examinations of RAEs for females at all levels of the sport are needed in order to inform meaningful interventions that reduce the inequities in athlete development. Furthermore, this study adds to the limited literature available that examines relative age and dropout in a longitudinal manner within a youth sport sample. To date, dropout from organized sport with respect to relative age has not been adequately studied, and a continued evaluation of the patterns that exist in different sport contexts (i.e., team vs. individual, competitive vs. recreational), across age groups, and between the sexes is required. Following a one-year cohort through the pre-adolescent to post-adolescent transition was an important element in this analysis, as adolescence has been identified as a critical timepoint for overall declines in physical activity levels [64]. However, information is still lacking with respect to participants who declined participation prior to the age of ten years and beyond 16 years of age. An evaluation of a broader age range and a comparative male sample from Ontario youth soccer would be beneficial.

Future studies also need to consider the longitudinal nature of sport participation along with the dynamic nature of athletic development. For instance, Cobley et al. [65] identified transient relative age advantages among national-level Australian swimmers, with the relatively oldest and youngest being over-represented at different time points (i.e., age 12 and 18 years, respectively); this suggests that detailed examinations to increase knowledge and understanding of relative age mechanisms are justified. A multi-level systems perspective should be maintained [66–68] in these future investigations, as athlete development does not occur within a *vacuum* [17].

The use of survival analysis provided an alternative way for assessing dropout, that being the use of time-to-event data. Traditional statistical methods of assessing the birth date distributions of athlete samples, such as chi-square analysis and linear regression, cannot handle the censoring of events (i.e., when survival time is unknown). However, as discussed above, a survival analysis may not be sensitive enough to pick up community size-related variations, and this variable will require a deeper level of examination in future studies. A consistent approach was taken to the coding of each participant's registration entry by census subdivision due to the correlation of this variable with municipal funding for sport facilities; this consistency was lacking in previous research on community size. However, this approach still has limitations as the census subdivision may not be the true size of the community and does not account for the proximity of neighboring communities, which might provide additional options for sport club membership, opportunities for training, an enhanced pool of competition, etc. Finally, this analysis is limited in the same manner as many studies using relative age; the evaluation of quantitative trends cannot answer the questions of "why" and "how" relative age influences dropout. Mixed-method approaches and person-centered analyses are needed in future research to learn more about the athletes on an individual level [17]. Researchers should seek to gain a better understanding of the developmental experiences of individuals who succeed despite a disadvantageous relative-age position within a cohort in order to inform sport engagement strategies and promote positive sport experiences and the equitable distribution of opportunities for all athletes.

5. Conclusions

Relative age effects are present in developmental-level, female soccer in Ontario. A higher risk of dropout is incurred by the relatively youngest and recreational-level players. Future research is needed to confirm the exact mechanism(s) contributing to these trends and to determine effective methods of supporting at-risk athletes.

Author Contributions: K.L.S. and P.L.W. conceptualized the study. K.L.S. prepared the data, conducted the analyses, and drafted the manuscript in consultation with P.L.W. All authors have read and agreed to the published version of the manuscript.

Funding: Funding for this study was received through a Social Sciences and Humanities Research Council Doctoral Fellowship (K. L. Smith).

Institutional Review Board Statement: This study was reviewed and approved by the Office of the Research Ethics Board, University of Windsor (REB #16-179) on 12 September 2016.

Informed Consent Statement: Written, informed consent from the participants' legal guardian was not required to participate in this study, in accordance with the national legislation and institutional requirements.

Data Availability Statement: The data analyzed in this study were obtained from Ontario Soccer. Access to the data is not possible due to ethical considerations as they contain personal information.

Acknowledgments: The data used in this analysis were provided by Ontario Soccer. The contents of this manuscript first appeared in the author's dissertation thesis [69].

Conflicts of Interest: The authors declare no conflict of interest.

References

1. Eime, R.M.; Young, J.A.; Harvey, J.T.; Charity, M.J.; Payne, W.R. A Systematic Review of the Psychological and Social Benefits of Participation in Sport for Children and Adolescents: Informing Development of a Conceptual Model of Health through Sport. *Int. J. Behav. Nutr. Phys. Act.* **2013**, *10*, 98. [CrossRef] [PubMed]
2. Janssen, I.; LeBlanc, A.G. Systematic Review of the Health Benefits of Physical Activity and Fitness in School-Aged Children and Youth. *Int. J. Behav. Nutr. Phys. Act.* **2010**, *7*, 40. [CrossRef] [PubMed]
3. Lagestad, P. Differences between Adolescents Staying in and Dropping out of Organized Sport: A Longitudinal Study. *J. Phys. Educ. Sport* **2019**, *19*, 444–452. [CrossRef]
4. Howie, L.D.; Lukacs, S.L.; Pastor, P.N.; Reuben, C.A.; Mendola, P. Participation in Activities Outside of School Hours in Relation to Problem Behavior and Social Skills in Middle Childhood. *J. Sch. Health* **2010**, *80*, 119–125. [CrossRef]
5. Boone, E.M.; Leadbeater, B.J. Game On: Diminishing Risks for Depressive Symptoms in Early Adolescence Through Positive Involvement in Team Sports. *J. Res. Adolesc.* **2006**, *16*, 79–90. [CrossRef]
6. Sanders, C.E.; Field, T.M.; Diego, M.; Kaplan, M. Moderate Involvement in Sports Is Related to Lower Depression Levels among Adolescents. *Adolescence* **2000**, *35*, 793–797. [PubMed]
7. Lubans, D.R.; Morgan, P.J.; Cliff, D.P.; Barnett, L.M.; Okely, A.D. Fundamental Movement Skills in Children and Adolescents: Review of Associated Health Benefits. *Sports Med. Auckl.* **2010**, *40*, 1019–1035. [CrossRef]
8. Stodden, D.F.; Goodway, J.D.; Langendorfer, S.J.; Roberton, M.A.; Rudisill, M.E.; Garcia, C.; Garcia, L.E. A Developmental Perspective on the Role of Motor Skill Competence in Physical Activity: An Emergent Relationship. *Quest Natl. Assoc. Kinesiol. High. Educ.* **2008**, *60*, 290–306. [CrossRef]
9. Canadian Fitness and Lifestyle Research Institute (CFLRI). *Bulletin 02: Participation in Organized Physical Activity and Sport*; Canadian Fitness and Lifestyle Research Institute (CFLRI): Ottawa, ON, Canada, 2016.
10. Balish, S.M.; McLaren, C.; Rainham, D.; Blanchard, C. Correlates of Youth Sport Attrition: A Review and Future Directions. *Psychol. Sport Exerc.* **2014**, *15*, 429–439. [CrossRef]
11. Petlichkoff, L.M. The Drop-out Dilemma in Youth Sports. In *The Child and Adolescent Athlete*; Bar-Or, O., Ed.; Blackwell Science: Oxford, UK, 1996; Volume 6, pp. 418–430.
12. Butcher, J.; Lindner, K.J.; Johns, D.P. Withdrawal from Competitive Youth Sport: A Retrospective Ten-Year Study. *J. Sport Behav.* **2002**, *25*, 145–163.
13. Crane, J.; Temple, V. A Systematic Review of Dropout from Organized Sport among Children and Youth. *Eur. Phys. Educ. Rev.* **2015**, *21*, 114–131. [CrossRef]
14. Wattie, N.; Cobley, S.; Baker, J. Towards a Unified Understanding of Relative Age Effects. *J. Sports Sci.* **2008**, *26*, 1403–1409. [CrossRef]
15. Helsen, W.F.; Starkes, J.L.; Van Winckel, J. The Influence of Relative Age on Success and Dropout in Male Soccer Players. *Am. J. Hum. Biol.* **1998**, *10*, 791–798. [PubMed]
16. Musch, J.; Grondin, S. Unequal Competition as an Impediment to Personal Development: A Review of the Relative Age Effect in Sport. *Dev. Rev.* **2001**, *21*, 147–167. [CrossRef]
17. Wattie, N.; Schorer, J.; Baker, J. The Relative Age Effect in Sport: A Developmental Systems Model. *Sports Med.* **2015**, *45*, 83–94. [CrossRef] [PubMed]
18. Baker, J.; Cobley, S.; Montelpare, W.J.; Wattie, N.; Faught, B.E. Exploring Proposed Mechanisms of the Relative Age Effect in Canadian Minor Hockey. *Int. J. Sport Psychol.* **2010**, *41*, 148–159.
19. Cobley, S.; Baker, J.; Wattie, N.; McKenna, J. Annual Age-Grouping and Athlete Development: A Meta-Analytical Review of Relative Age Effects in Sport. *Sports Med.* **2009**, *39*, 235–256. [CrossRef]

20. Lovell, R.; Towlson, C.; Parkin, G.; Portas, M.; Vaeyens, R.; Cobley, S. Soccer Player Characteristics in English Lower-League Development Programmes: The Relationships between Relative Age, Maturation, Anthropometry and Physical Fitness. *PLoS ONE* **2015**, *10*, e0137238. [CrossRef]
21. Wattie, N.; Baker, J.; Cobley, S.; Montelpare, W.J. A Historical Examination of Relative Age Effects in Canadian Hockey Players. *Int. J. Sport Psychol.* **2007**, *38*, 178–186.
22. Baker, J.; Schorer, J.; Cobley, S.; Bräutigam, H.; Büsch, D. Gender, Depth of Competition and Relative Age Effects in Team Sports. *Asian J. Exerc. Sport Sci.* **2009**, *6*, 1–7.
23. Côté, J.; Macdonald, D.J.; Baker, J.; Abernethy, B. When "Where" Is More Important than "When": Birthplace and Birthdate Effects on the Achievement of Sporting Expertise. *J. Sports Sci.* **2006**, *24*, 1065–1073. [CrossRef]
24. Fraser-Thomas, J.; Côté, J.; MacDonald, D.J. Community Size in Youth Sport Settings: Examining Developmental Assets and Sport Withdrawal. *Phys. Health Educ. Acad. J.* **2010**, *2*, 1–9.
25. Curtis, J.E.; Birch, J.S. Size of Community of Origin and Recruitment to Professional and Olympic Hockey in North America. *Sociol. Sport J.* **1987**, *4*, 229–244. [CrossRef]
26. Balish, S.M.; Rainham, D.; Blanchard, C. Community Size and Sport Participation across 22 Countries. *Scand. J. Med. Sci. Sports* **2015**, *25*, e576–e581. [CrossRef] [PubMed]
27. Imtiaz, F.; Hancock, D.J.; Vierimaa, M.; Côté, J. Place of Development and Dropout in Youth Ice Hockey. *Int. J. Sport Exerc. Psychol.* **2014**, *12*, 234–244. [CrossRef]
28. Barnsley, R.H.; Thompson, A.H.; Barnsley, P.E. Hockey Success and Birth Date: The Relative Age Effect. *Can. Assoc. Health Phys. Educ. Recreat.* **1985**, *51*, 23–28.
29. Barnsley, R.H.; Thompson, A.H. Birthdate and Success in Minor Hockey: The Key to the NHL. *Can. J. Behav. Sci. Can. Sci. Comport.* **1988**, *20*, 167–176. [CrossRef]
30. Delorme, N.; Boiché, J.; Raspaud, M. Relative Age Effect in Female Sport: A Diachronic Examination of Soccer Players. *Scand. J. Med. Sci. Sports* **2010**, *20*, 509–515. [CrossRef]
31. Delorme, N.; Boiché, J.; Raspaud, M. Relative Age and Dropout in French Male Soccer. *J. Sports Sci.* **2010**, *28*, 717–722. [CrossRef]
32. Delorme, N.; Chalabaev, A.; Raspaud, M. Relative Age Is Associated with Sport Dropout: Evidence from Youth Categories of French Basketball. *Scand. J. Med. Sci. Sports* **2011**, *21*, 120–128. [CrossRef]
33. Figueiredo, A.J.; Coelho-e-Silva, M.J.; Cumming, S.P.; Malina, R.M. Relative Age Effect: Characteristics of Youth Soccer Players by Birth Quarter and Subsequent Playing Status. *J. Sports Sci.* **2019**, *37*, 677–684. [CrossRef] [PubMed]
34. Lemez, S.; Baker, J.; Horton, S.; Wattie, N.; Weir, P. Examining the Relationship between Relative Age, Competition Level, and Dropout Rates in Male Youth Ice-Hockey Players. *Scand. J. Med. Sci. Sports* **2014**, *24*, 935–942. [CrossRef]
35. Wattie, N.; Tietjens, M.; Cobley, S.; Schorer, J.; Baker, J.; Kurz, D. Relative Age-Related Participation and Dropout Trends in German Youth Sports Clubs. *Eur. J. Sport Sci.* **2014**, *14*, S213–S220. [CrossRef] [PubMed]
36. Vincent, J.; Glamser, F.D. Gender Differences in the Relative Age Effect among US Olympic Development Program Youth Soccer Players. *J. Sports Sci.* **2006**, *24*, 405–413. [CrossRef] [PubMed]
37. Malina, R.M. The Young Athlete: Biological Growth and Maturation in a Biocultural Context. In *Children and Youth in Sport: A Biopsychosocial Perspective*; Smoll, F.L., Smith, R.E., Eds.; Brown & Benchmark: Dubuque, IA, USA, 1996; pp. 161–186.
38. Malina, R.M.; Bouchard, C.; Bar-Or, O. Maturity-Associated Variation in Growth and Performance. In *Growth, Maturation, and Physical Activity*; Malina, R.M., Bouchard, C., Bar-Or, O., Eds.; Human Kinetics: Champaign, IL, USA, 2004; pp. 337–365.
39. Côté, J.; Abernethy, B. A Developmental Approach to Sport Expertise. In *The Oxford Handbook of Sport and Performance Psychology*; Murphy, S.M., Ed.; Oxford University Press: New York, NY, USA, 2012; pp. 435–447.
40. Fraser-Thomas, J.; Côté, J.; Deakin, J. Examining Adolescent Sport Dropout and Prolonged Engagement from a Developmental Perspective. *J. Appl. Sport Psychol.* **2008**, *20*, 318–333. [CrossRef]
41. Wall, M.; Côté, J. Developmental Activities That Lead to Dropout and Investment in Sport. *Phys. Educ. Sport Pedagog.* **2007**, *12*, 77–87. [CrossRef]
42. Smith, K.L.; Weir, P.L.; Till, K.; Romann, M.; Cobley, S. Relative Age Effects Across and Within Female Sport Contexts: A Systematic Review and Meta-Analysis. *Sports Med.* **2018**, *48*, 1451–1478. [CrossRef]
43. Turnnidge, J.; Hancock, D.J.; Côté, J. The Influence of Birth Date and Place of Development on Youth Sport Participation: Contextual Factors and Sport Participation. *Scand. J. Med. Sci. Sports* **2014**, *24*, 461–468. [CrossRef]
44. Smith, K.L.; Weir, P.L. An Examination of the Relative Age Effect in Developmental Girls' Hockey in Ontario. *High Abil. Stud.* **2013**, *24*, 171–184. [CrossRef]
45. Smith, K.; (University of Windsor, Windsor, ON, Canada); Morin, G.; (Statistics Canada, Ottawa, ON, Canada). Personal communication, 2017.
46. Smith, K.; (University of Windsor, Windsor, ON, Canada); Wilkinson, T.; Miller, G.; (Ontario Soccer, Vaughan, ON, Canada). Personal communication, 2016.
47. Delorme, N.; Champely, S. Relative Age Effect and Chi-Squared Statistics. *Int. Rev. Sociol. Sport* **2015**, *50*, 740–746. [CrossRef]
48. Cohen, J. Statistical Power Analysis. *Curr. Dir. Psychol. Sci.* **1992**, *1*, 98–101. [CrossRef]
49. Kleinbaum, D.G.; Klein, M. *Survival Analysis: A Self-Learning Text*, 3rd ed.; Springer: New York, NY, USA, 2012.
50. Baker, J.; Janning, C.; Wong, H.; Cobley, S.; Schorer, J. Variations in Relative Age Effects in Individual Sports: Skiing, Figure Skating and Gymnastics. *Eur. J. Sport Sci.* **2014**, *14*, S183–S190. [CrossRef] [PubMed]

51. Cobley, S.P.; Till, K. Participation Trends According to Relative Age across Youth UK Rugby League. *Int. J. Sports Sci. Coach.* **2017**, *12*, 339–343. [CrossRef]
52. Sedano, S.; Vaeyens, R.; Redondo, J.C. The Relative Age Effect in Spanish Female Soccer Players. Influence of the Competitive Level and a Playing Position. *J. Hum. Kinet.* **2015**, *46*, 129–137. [CrossRef] [PubMed]
53. Lagestad, P.; Steen, I.; Dalen, T. Inevitable Relative Age Effects in Different Stages of the Selection Process among Male and Female Youth Soccer Players. *Sports* **2018**, *6*, 29. [CrossRef]
54. Smith, K.L.; Weir, P.L. Female Youth Soccer Participation and Continued Engagement: Associations With Community Size, Community Density, and Relative Age. *Front. Sports Act. Living* **2020**, *2*, 552597. [CrossRef]
55. Weir, P.L.; Smith, K.L.; Paterson, C.; Horton, S. Canadian Women's Ice Hockey—Evidence of a Relative Age Effect. *Talent Dev. Excell.* **2010**, *2*, 209–217.
56. Korgaokar, A.D.; Farley, R.S.; Fuller, D.K.; Caputo, J.L. Relative Age Effect Among Elite Youth Female Soccer Players across the United States. *Sport Mont* **2018**, *16*, 37–41. [CrossRef]
57. Canadian Heritage. Sport Participation 2010. 2013. Available online: https://publications.gc.ca/site/archivee-archived.html?url=https://publications.gc.ca/collections/collection_2013/pc-ch/CH24-1-2012-eng.pdf (accessed on 2 June 2021).
58. Bélanger, M.; Gray-Donald, K.; O'Loughlin, J.; Paradis, G.; Hanley, J. When Adolescents Drop the Ball: Sustainability of Physical Activity in Youth. *Am. J. Prev. Med.* **2009**, *37*, 41–49. [CrossRef]
59. Telama, R.; Yang, X. Decline of Physical Activity from Youth to Young Adulthood in Finland. *Med. Sci. Sports Exerc.* **2000**, *32*, 1617–1622. [CrossRef]
60. Thedin Jakobsson, B. What Makes Teenagers Continue? A Salutogenic Approach to Understanding Youth Participation in Swedish Club Sports. *Phys. Educ. Sport Pedagog.* **2014**, *19*, 239–252. [CrossRef]
61. Balderson, D. Sport Academies: A Growing Phenomenon in Canadian Schools. *Phys. Health Educ. J.* **2015**, *80*, 27–29.
62. Way, R.; Repp, C.; Brennan, T. Sport Schools in Canada: The Future Is Here. *Can. Sport Cent. Pac.* **2010**, *83*.
63. Sierra-Díaz, M.; González-Víllora, S.; Pastor-Vicedo, J.; Serra-Olivares, J. Soccer and Relative Age Effect: A Walk among Elite Players and Young Players. *Sports* **2017**, *5*, 5. [CrossRef] [PubMed]
64. Nader, P.R.; Bradley, R.H.; Houts, R.M.; McRitchie, S.L.; O'Brien, M. Moderate-to-Vigorous Physical Activity From Ages 9 to 15 Years. *JAMA J. Am. Med. Assoc.* **2008**, *300*, 295–305. [CrossRef] [PubMed]
65. Cobley, S.; Abbott, S.; Dogramaci, S.; Kable, A.; Salter, J.; Hintermann, M.; Romann, M. Transient Relative Age Effects across Annual Age Groups in National Level Australian Swimming. *J. Sci. Med. Sport* **2018**, *21*, 839–845. [CrossRef]
66. Bronfenbrenner, U. Towards an Experimental Ecology of Human Development. *Am. Psychol.* **1977**, *32*, 513–531. [CrossRef]
67. Bronfenbrenner, U. Developmental Ecology through Space and Time: A Future Perspective. In *Examining Lives in Context: Perspectives on the Ecology of Human Development*; Moen, P., Elder, G.H., Jr., Lüscher, K., Eds.; American Psychological Association: Washington, DC, USA, 1995; pp. 619–647.
68. Bronfenbrenner, U. Environments in Developmental Perspective: Theoretical and Operational Models. In *Measuring Environment across the Life Span: Emerging Methods and Concepts*; Friedman, S.L., Wachs, T.D., Eds.; American Psychological Association: Washington, DC, USA, 1999; pp. 3–28.
69. Smith, K. *The Role of Relative Age, Community Size, and Positive Youth Development on Female Youth Soccer Participation*; ProQuest Dissertations Publishing: Ann Arbor, MI, USA, 2019.

sports MDPI

Article

Changes in Athletic Performance in Children Attending a Secondary School with a Physical Activity Profile

Tomas Peterson [1,*], Staffan Ek [2,3], Ola Thorsson [2,3], Magnus K. Karlsson [4], Magnus Dencker [2,3] and Per Wollmer [2,3]

1 Department of Sport Sciences, Malmö University, 205 02 Malmö, Sweden
2 Department of Translational Medicine, Lund University, 221 00 Lund, Sweden; staffanek@gmail.com (S.E.); ola.thorsson@med.lu.se (O.T.); magnus.dencker@skane.se (M.D.); per.wollmer@med.lu.se (P.W.)
3 Department of Medical Imaging and Physiology, Skåne University Hospital, 205 02 Malmö, Sweden
4 Clinical and Molecular Osteoporosis Research Unit, Department of Clinical Sciences and Orthopaedics, Lund University, Skåne University Hospital, 205 02 Malmö, Sweden; magnus.karlsson@med.lu.se
* Correspondence: tomas.peterson@mau.se; Tel.: +46-709-655370

Abstract: The longitudinal and multidisciplinary research project Malmö Youth Sport Study measured the sports results achieved by two cohorts of pupils using a variable named ACHIEVE, dividing the pupils into three categories (an elite group competing at the national or international level, a group competing at the district level, and a third group either not competing at all or below district level). This was assessed three and six years after baseline at age 13. An additional hypothetical measure, based on information from the athletes' trainers, predicted the category the pupils were expected to belong to after twelve years (age 25). Social variables related to the ACHIEVE variable are sex, socio-economic position of the parents, ethnicity, completed secondary sports school, sports capital, and quartile of birth. After three years, 28% of the pupils belonged to the elite group and after six years, 26%. Thirty-two and 48%, respectively, had abandoned their elite efforts. The elite group remained fairly stable over time but fewer girls than boys advanced to the elite group. The pupils at the school have a homogenous middle-class background. We found little evidence that socio-economic factors affected ACHIEVE. Nearly all parents had been engaged in sports, either competing or as coaches. On admission to the school, there was a pronounced relative age effect (RAE). This remained after three years as the age was significantly different between the three groups but was reduced after six years. According to the prognosis made by the coaches, the elite group would be considerably smaller when the subjects reached the age of 25. The RAE was again significant in the prognosis. A further follow-up when the subjects are 25 years old will reveal not only what proportion of subjects are actively competing, but also if they are engaged in recreational sports, to what extent the RAE is present, and how accurately coaches can predict success.

Keywords: sports school; talent identification; longitudinal; elite sports; multidisciplinary

Citation: Peterson, T.; Ek, S.; Thorsson, O.; Karlsson, M.K.; Dencker, M.; Wollmer, P. Changes in Athletic Performance in Children Attending a Secondary School with a Physical Activity Profile. *Sports* **2022**, *10*, 71. https://doi.org/10.3390/sports10050071

Academic Editor: Kevin Till

Received: 12 April 2022
Accepted: 27 April 2022
Published: 2 May 2022

1. Introduction

Lower secondary schools with a physical activity profile have become popular in Sweden, both among youngsters and local politicians. This trend has been both acclaimed and criticized [1–9]. Although the purpose of the school is not clearly defined, the focus seems to be on "good grades, good health and good sport results".

Malmö Youth Sport Study (MYSS) involves two cohorts of girls and boys (N = 156) that attended Malmö Idrottsgrundskola (lower secondary school) between the ages of 13 and 16. After Idrottsgrundskolan, 60 percent of the pupils were admitted to a sports gymnasium (upper secondary school) between the ages of 17 and 19. The aim of this longitudinal and multidisciplinary study is to increase knowledge of how to create an activity that is inclusive and enables a commitment to sport and a lifelong interest in physical activity and sport (a health perspective) and at the same time a successful talent identification and talent development (an elite perspective). The project was formed in 2012 by researchers from the Universities of Halmstad, Lund, and Malmö. The researchers represent different disciplines and faculties, and hence different angles related to the research questions. MYSS is based on baseline data and a three-year, a six-year, and (eventually) a twelve-year follow-up. The project includes social, psychological, and physiological studies, as well as gender aspects [10–21]. In this article, talent development after secondary school (the three-year follow-up) and upper secondary school (the six-year follow-up) are analyzed, as well as an additional, *hypothetical* measure, based on information from the athletes' trainers, predicting the category the pupils can be expected to belong to at the age of 25. The analysis is based on the variable ACHIEVE (Achieved set goals for elite effort). Additional variables related to ACHIEVE that have been analyzed are sex/gender, socio-economic position of the parents, ethnicity, sports capital, completed secondary sports school, and quartile of birth [22].

2. Material and Methods

2.1. Participants

Every year 48 boys and 30 girls are admitted to Idrottsgrundskolan. The reason for the difference in numbers is a historical one. The school started out of the Youth Academy of Malmö Football Club (MFF), so every male player from this club was, and is, attending the school. Eventually, more individuals representing different sports were admitted. In the years of the two cohorts of the Malmö Youth Sport Study, twelve sports were represented: football, swimming, diving, tennis, squash, figure skating, badminton, track and field, ice hockey, basketball, floor boll, and gymnastics. However, every year at least 20 football-playing boys representing Malmö FF are admitted. The MYSS group of two cohorts thus contains 44 football-playing boys and 18 football-playing girls (in all 62 out of 156, i.e., 40%).

To become an elite athlete, you must be talented. To develop your talent, first, it has to be identified. Over two consecutive years, 593 girls and boys applied for the sports school at the age of 13. One out of four—156—were admitted. Their marks from the sixth grade were not used in the selection process, only their achieved sports results and a forecast of future sports development. The selection was made by local or regional representatives of the different sports federations together with the school officials. The pupils were accepted on the basis of a favorable prognosis of becoming elite athletes. In the fierce competition, everyone admitted was thus forecasted to achieve results on a national and/or international level.

2.2. Measures

The MYSS project includes social, psychological, and physiological studies, as well as gender aspects. The sociological part of the study consists of two questionnaires, one for pupils and one for parents. Only the parent questionnaire was used in this study. It contains questions about with whom the child lives and the family form of housing; in which country the parents were born; the number of siblings; if the family had access to a car, boat (with possibilities to sleep in), summer house, or camper; the education and occupation of the parents; as well as the family's sports experience.

2.3. Procedure

To measure the sports results objectively, the variable of ACHIEVE was used in the study. To "Achieve set goals for elite effort" means that you accomplish what you set out to do in terms of sports results when you started your education. The trainers and/or representatives of the district organizations were interviewed five times during 2015–2020 (in all 62 interviews). The pupils were divided on the basis of ACHIEVE, which grouped the pupils into a dropout group, a group not achieving set goals, a group performing at a district level, and one group performing at a national/international level. The individuals in the dropout group abandoned their sports efforts but they continued to attend the

Malmö Idrottsgrundskola. Three out of four groups are empirically visible and normatively neutral. The second group (which did not achieve set goals) is, however, built by subjective assessments. However, this group can be objectively defined in another sense: it contains those who do not belong to one of the other three groups. For practical as well as statistical reasons, in a second step, the dropout group and the group not achieving set goals were merged into one group. The three groups are hereafter called "under district", "district", and "elite". In the statistical analysis, the groups were assigned 1, 2, and 3, respectively. In the analysis, we used the ACHIEVE category obtained three years after admission to the school and six years after admission when the subjects had finished upper secondary school. In addition, we asked the trainers if they could give a hypothetical "informed guess" as to which ACHIEVE group every individual will likely belong to at the age of 25 (prognosis).

2.4. Data Analysis Plan

We analyzed the ACHIEVE variable in relation to the socio-economic position of the parents, ethnicity, sports capital, progression to an upper secondary sports school, sex, and the relative age effect. Information about the parents' level of education, occupation, and country of birth was obtained from the questionnaire and, when possible, compared to statistics for the population of Malmö. We scored education from 0 to 3 (0: not completed basic education, 1: completed basic education, 2: completed upper secondary school, 3: college or university exam). The scores for the parents were added. Occupation was categorized according to the socio-economic divisions (SEI) from Statistics Sweden (Mis 1982:4) and scored from 1 to 3, with 1 being the least-qualified occupation. The SEI is based mainly on occupations and is divided into workers, civil servants, and business owners. The scores for the two parents were added.

Three questions in the parental questionnaire covered the families' experience with sports. First, whether the parents themselves had been actively engaged in sports, secondly, whether they had served as coaches/leaders in a sports club, and thirdly, whether any sister or brother of the pupil had been active in sports. A score was constructed in which one point was awarded for each of the items, giving a maximum of 5 points (both parents previously active and also serving as coaches/leaders and at least one active sibling).

Information about progression to upper secondary sports school was obtained from the trainers and school officials.

Finally, we analyzed ACHIEVE in relation to age on admission to the sports school. It is well known that recruitment by elite sports groups of youngsters is weighted to the older subjects within a given age range, a phenomenon known as the relative age effect (RAE). We have previously shown that the RAE is a determining factor for admission to Idrottsgrundskolan and that this is related to physical maturity [10]. RAE was analyzed relative to birth quarter as well as to age as a continuous variable.

Statistical analysis was mainly performed by non-parametric methods as specified in each case.

3. Results

As the pupils were all accepted on the basis of a favorable prognosis of becoming elite athletes on a national and/or international level, there was only one baseline position for the pupils. After three years this was only confirmed for 28% of the pupils (Table 1); after six years for 26%. The grouping according to ACHIEVE thus included:

Table 1. ACHIEVE divided into elite group, district group, and under district group ($N = 156$).

Group	3 Years	6 Years
Results at a national and/or International level (3)	44	41
Results at a district level (2)	62	40
Not achieved set goals (below district) (1)	50	75

When groups were assigned 1–3, the mean value (±SD) for the ACHIEVE variable was 1.96 ± 0.78 after three years and 1.78 ± 0.84 after six years. The decline was statistically significant by Wilcoxon's signed-rank test ($p = 0.04$).

Transitions between the ACHIEVE groups are shown in Figure 1. The members of the elite group who lived up to expectations on admission to the school, by and large, kept their position over time. Out of 44 pupils, after three years 31 were elite athletes, and after six years a further 10 advanced from the district group. The middle group lost 40% of its members at three years to the below-district group, whereas ten pupils advanced to the elite group. Twelve percent of the youngsters in the below-district group advanced to the middle group; none advanced to the elite group.

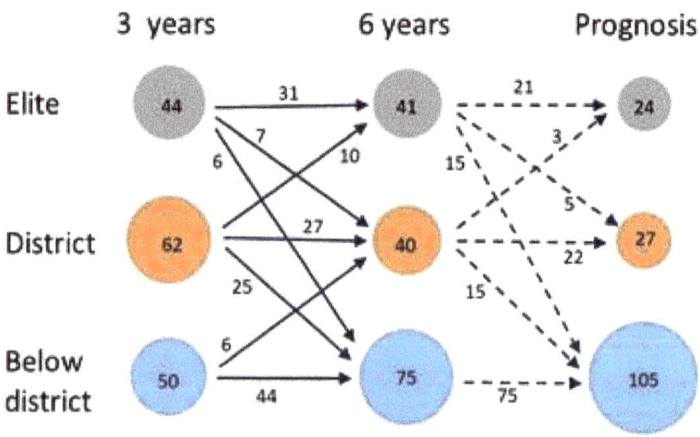

Figure 1. Changes from the three-year follow-ups, the six-year follow-ups, and from the last follow-up to adulthood, as predicted by the coaches (prognosis), based on the ACHIEVE variable ($N = 156$).

The informed guesses made by the coaches indicate that a substantial number of elite athletes will abandon their efforts before the age of 25. An equal number in the district group is predicted to be lost to the below-district group, whereas only three athletes are expected to advance from the district to the elite level.

The flow between the groups is clearly seen in Figure 2, showing changes between the three-year follow-ups and the prognosis for the age of 25.

The results of the follow-ups can be summarized as follows:

- Two-thirds (65.3%) of all pupils belonged to the same group at the ages of 16 and 19.
- The elite group lost and gained roughly the same number of individuals from and to the district group.
- The district group lost 40% of its members to the under-district level group.
- There were no "late bloomers". Among those belonging to the under-district group after three years, no one made national/international results up to the age of 19.

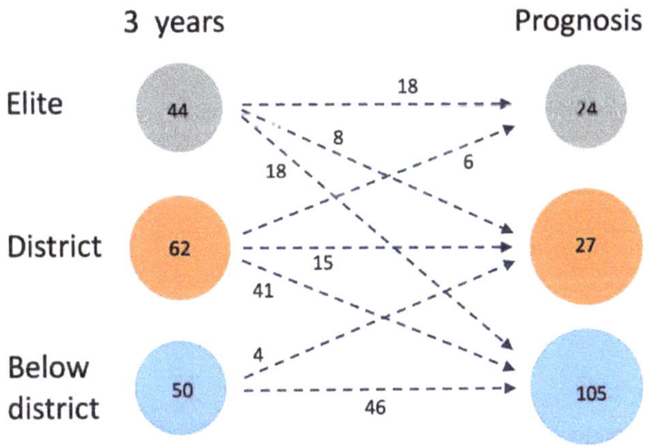

Figure 2. Changes from the three-year follow-ups to the prognosis for the twelve-year follow-ups, based on the ACHIEVE variable ($N = 156$).

3.1. Correlation between ACHIEVE and Other Variables

According to their trainers, 24 out of the 44 pupils will remain elite after twelve years of elite effort. Given the grouping of the pupils at the three-year follow-ups and the six-year follow-ups, are there any correlations, positive or negative, between ACHIEVE and the other variables?

3.2. The Socio-Economic Positions of the Parents

We have previously shown that the socio-economic position dominating the parental group is that of the upper-middle class, characterized by parents who were largely born in Sweden, have a high educational and occupational level, and are occupationally dominated by senior officials and employers who live in well-defined families in detached or terraced houses. These characteristics strongly influenced who was admitted into the Idrottsgrundskolan [20]. The only sociological variable found to be related to ACHIEVE in this analysis was the occupation of the parents. The parents of children in the under-district group consistently had the most advanced jobs and the parents of the district group the least advanced. The differences were, however, small, the median score for occupation being 4 for the elite district as well as the below-district groups at both 3 and 6 years with interquartile ranges (IQR) of 4–5, 4–4, and 4–5, respectively, both at three and at six years ($p = 0.02$ by Kruskal–Wallis test). We found no relationship between the parents' level of education and ACHIEVE.

3.3. Ethnicity

Of the MYSS mothers, 82 percent were born in Sweden [20]. Among the others, 22 different countries were stated, albeit no country more than three times. The fathers born in Sweden also constituted 82 percent. Among the rest, 23 different countries were stated, none, however, more than three times. The proportion of children (aged 0–17) in Malmo with a foreign background in 2008 was 56% [23]. Our results thus show that neither the foreign-born mothers nor fathers represent the composition of all foreign-born Malmö female citizens. Malmö's population represents 183 different foreign countries. In all, the foreign-born MYSS parents were not representative of the foreign-born population of Malmö—neither proportionally, nor regarding place of birth. Thus, ethnicity is one of the variables that formed the composition of the group admitted to the Idrottsgrundskolan. It did not, however, affect the grouping of ACHIEVE between baseline and the three-year or six-year follow-ups.

3.4. *Sports Capital*

The MYSS parents and siblings possess an extensive sports capital, the median score being 3 with an IQR of 3–4. Only eight pupils out of the 153 we have information on have parents that lack experience with sports. Seven out of these have one or more siblings that have been active in a sports club, and some of them still are. Thus, only one child had a family that lacked sports experience. The amount of sports capital possessed by the family was also one of the variables that formed the composition of the group admitted to the Idrottsgrundskolan. It did not, however, affect the grouping of ACHIEVE between baseline and the three-year or six-year follow-ups.

3.5. *Completed Upper Secondary Sport School*

A large proportion of the subjects in this study might be expected to apply for Idrottsgymnasiet (the upper secondary sports school) after having completed the Idrottsgrundskolan. In fact, 60% applied—31 out of 60 girls (52%) and 63 out of 96 boys (66%).

To qualify that assumption, this should be true for those belonging to the ACHIEVE elite group and district level group. The better sports results during Idrottsgrundskolan, the more likely they applied for and were admitted to the upper secondary sports school. On the other hand, after Idrottsgrundskolan, close to four out of ten pupils had ended their elite efforts. This seems to have affected their way of choosing the next school form, as a majority from the below-district group did not attend an upper secondary sports school, whereas essentially the same proportion of subjects in the district and the elite groups did attend an upper secondary sports school (Figure 3).

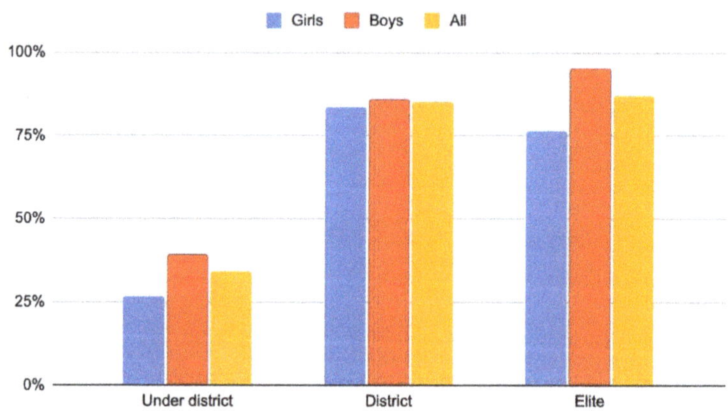

Figure 3. The MYSS girls and boys attending Sport Upper Secondary School (idrottsgymnasium), based on three-year follow-ups of ACHIEVE (%).

3.6. *Sex*

There are differences between girls and boys regarding ACHIEVE. The transitions between the groups for boys are shown in Figure 4. Two-thirds of the elite group continue to compete at the same level after six years. The elite group loses one-third of its members to the district group and an approximately equal number are recruited from the district group. About 40% of the district group continue at the same level after six years and the same proportion drops to the below district group. The latter group appears fairly stable, retaining more than 80% of its members while gaining a substantial number from the district group.

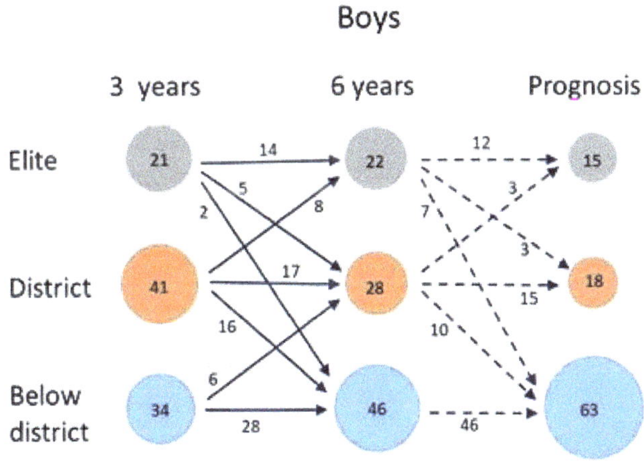

Figure 4. Changes from the three-year to the six-year follow-ups, based on the ACHIEVE variable for the boys (n = 96).

Among the girls, the loss from the elite group from three to six years is somewhat smaller than for the boys (Figure 5) but only two girls make the transition from the district to the elite group. The elite group of girls has, in contrast to the corresponding group of boys, thus shrunk by 17%. The district group retains about half of its members, whereas an almost equal number are lost to the below-district group. The latter group retains all its original members and, therefore, grows considerably.

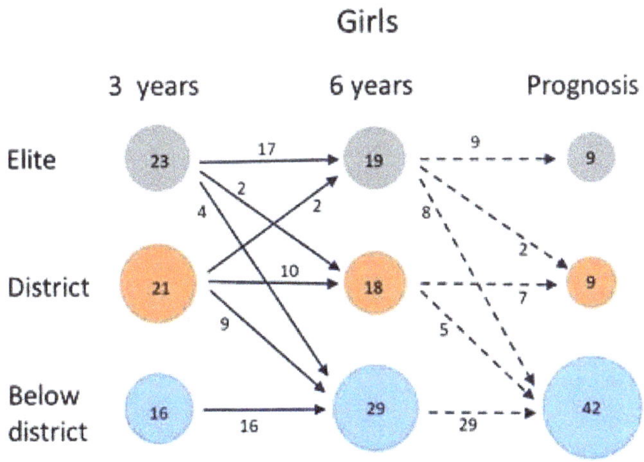

Figure 5. Changes from the three-year to the six-year follow-ups, based on the ACHIEVE variable for the GIRLS (n = 60).

The changes in ACHIEVE are summarized in Table 2. Very few girls travel upward. However, a larger proportion of girls than boys are in the elite group after three years and retain that position after six years. Boys travel upward to a larger extent. An approximately equal proportion of girls and boys (25.0 and 24.0%, respectively), made a downward journey. More girls traveled downward and more boys traveled upward.

Table 2. Changes in ACHIEVE during follow-up.

3 → 6 Years	Increasing	No Change	Decreasing
All	16 (10%)	102 (65%)	38 (24%)
Boys	14 (15%)	59 (61%)	23 (24%)
Girls	2 (3%)	43 (72%)	15 (25%)
6 Years → Prognosis			
All	3 (2%)	118 (76%)	35 (22%)
Boys	3 (3%)	73 (76%)	20 (21%)
Girls	0 (0%)	45 (75%)	15 (25%)

The sex differences can be summarized as follows:

- After three years, the proportion of girls in the elite group is larger than for the boys (38 vs. 22%).
- After six years, the number of girls in the elite group has declined, but proportionally the elite group of girls is still larger than the boys (32% vs. 23%).
- All girls in the under-district group after three years are still in the under-district group after six years (6 out of 34 boys moved into the district group).
- Almost half of the pupils belong to the under-district group after six years. This is the case both for girls and boys.
- There were no "late bloomers", neither among the boys nor the girls. All girls in the under-district group after three years were still in the under-district group after six years. Six boys were upgraded from the under-district level to the district level.
- Thus, attending the sports gymnasium (upper secondary school) does not improve the development of the girls' sports results.

3.7. Relative Age Effect

The distribution of the birth quarter of the pupils admitted to Idrottsgrundskolan is shown in Figure 6. The distribution is skewed ($p < 0.05$ by X^2 test) with very few subjects born in the fourth quarter.

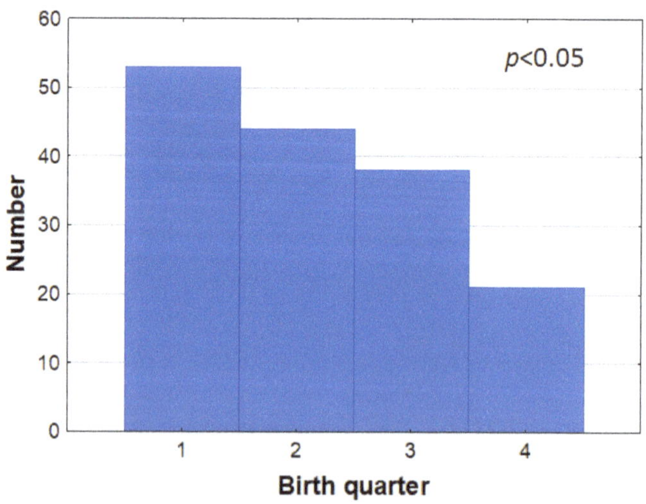

Figure 6. Frequency distribution of birth quarter at admission.

The relative age effect was significantly influencing not only the admission to the Idrottsgrundskolan but also the forming of the ACHIEVE groups as well as the reshaping of them during the three years spent in the school (Table 3). The same was not the case during the period at the upper secondary sports school, but in the prognosis for their next six years a significant pattern once again appears. Even if the difference is small, the effect is considered moderate as assessed by Hedge's g (0.60 at 3 years and 0.73 for the prognosis).

Table 3. ACHIEVE—RAE.

Follow-Up	Mean Age at Admission (years)	SD	*p* (ANOVA)
3 years			
Elite	13.6 *	0.36	
District	13.5	0.26	0.008
Below district	13.4 *	0.31	
6 years			
Elite	13.6	0.32	
District	13.4	0.33	0.096
Below district	13.5	0.29	
Prognosis			
Elite	13.6 *	0.36	
District	13.4 *	0.36	0.010
Below district	13.5	0.28	

* Significant differences between groups by Tukey's test.

4. Discussion

In this study, we analyze sports development in a lower secondary sports school. The relationship between school and sport is seldom problematized. Many claims have been made about the mutual benefits of school results and sports results [24]. Most people think of it in a win-win manner: school is good for sport and sport is good for the school. In society as well as in research materials, there is a general understanding that PA is good [25–27]. Research conclusions have been drawn that PA, beyond the physiological effects, a supportive environment and context can lead to persistence, discipline, commitment, wellbeing, self-esteem, anti-depression, stress reduction, the feeling of belonging, networks, status, and an increased ability to concentrate [25]. Increasing weekly PA over nine years has been associated with improved academic achievement in boys [28,29]. All these effects are in the favor of good school results. At the same time, sports schools are believed to help the pupil´s sports careers by strengthening their international competition capacity, facilitating dual careers, strengthening the professionalization of sports, and supporting public health [2,4,30].

A large amount of knowledge combined with good subject-specific knowledge, as well as characteristics such as disciplined handling of studying, concentration capacity, and a talent for long-term thinking, are all assets that are regarded as being in favor of sports development. At the same time, scientific results claim that intensive physical activity works in the favor of good school results.

Based on a statistical analysis, we have discussed the variables of ACHIEVE to understand the differences in talent development for two cohorts of athletes attending Idrottsgrundskolan in Malmö. Three sets of conclusions are drawn. The first one relates to changes in the groupings between baseline, at three years, at six years, and at hypothetically twelve years of development. The second is related to the effects of societal variables on the outcomes. The third one is related to the biological variables affecting the outcomes.

The overall changes in the ACHIEVE groups could be described as a gradual polarization: a group where every pupil was defined as elite at baseline, became a large middle group after three years, and then a large under-district group after six years, at the same time as the elite group became notably stable but somewhat smaller. This process continues—hypothetically—for up to twelve years. Either there was an incorrect prognosis at baseline in seven out of ten cases, or things happened in seven out of ten cases related to, for instance, injuries, effects on the sports development on behalf of the athlete or his/her trainer, or social factors.

The polarization can be understood in relation to a number of factors that influence talent development during lower and upper secondary school, including school-related factors, sports-related factors, and social factors. Some factors are logically linked to the talent-development process, for example, physical and psychological factors [11,15,17]. Another such factor, detectable through the relative age effect is physical maturity [10]. As very few individuals became internationally successful, one factor influencing the outcome is an unrealistic optimism in the younger age groups, which is often eventually replaced by a more sober judgement about the meaning as well as the effects, of the commitment. The competition logic pushes the athletes in different directions; both wins and losses have to be handled. In addition, the impact of injuries is obvious, as is the influence—good or bad—of trainers and coaches. The role of coaches has been demonstrated in a study by Witkowski et al. [31].

If the outcome in terms of ACHIEVE, from the point of view of a lower secondary sports school, is to be regarded as either a successful or unsuccessful development, it can be discussed in the form of "whether the glass is half full or half empty". On the one hand, it is difficult to claim that a special sports school form is needed to produce elite athletes on a national and/or international level if only 28 percent had succeeded when they entered the Idrottsgrundskolan at the age of 16. On the other hand, 45 pupils from one lower secondary school in the third-largest city in a small country, having achieved sports results at a national and/or international level at the age of 16, can be argued as being a very successful development. Out of these—hypothetically—24 could remain elite athletes up to senior age. However, the counterargument needs to be problematized by the fact that half of these 24 elite athletes were admitted to Idrottsgrundskolan because they belonged to the most advanced and selective Football Youth Academy in Sweden, Malmö FF. They did not need Idrottsgrundskolan to acquire the amount of support—quantitatively or qualitatively—needed to become successful football players. Similarly for a number of the other half of the elite athletes. All in all, the results do not present a strong case for the conclusion that this lower secondary sports school is needed to identify and develop elite athletes on a national and/or international level. However, to draw that conclusion we still have to wait for the official twelve-year follow-ups.

After three years of talent development and then three more years of upper secondary school, we cannot rule out that the pupils could still become elite athletes as adults. Only the dropouts seem like "lost cases". On the other hand, behind the label "dropout" we have found that many of the boys and girls who have stopped exercising for elite sports themselves are coaching and training others, or they dropped out because of sports injuries. Nevertheless, after three years of development, one out of three has stopped doing sports or has not achieved set goals for elite efforts. After six years, half of the pupils belonged to this group.

The background factors regarded within social sciences as being of profound importance when dividing citizens into different groups, which are later affected in different ways and with different results, are age, sex/gender, social class/group (educational level and occupation), ethnicity, and geographical location/population density [32,33]. The MYSS parent questionnaire includes questions about most of these factors as well as questions about sports habits [34,35]. Sport and physical exercise habits are clearly marked socially, socioeconomically, socio-geographically, and culturally [20,34–37]. The socio-economic position dominating the parental group in this study is that of the upper-middle class, characterized by parents who were largely born in Sweden, have a high educational and occupational level, and are occupationally dominated by senior officials and employers who live in well-defined families in detached or terraced houses. Obviously, these characteristics strongly influenced who was admitted into the Idrottsgrundskolan [20]. The sports capital for the children at Idrottsgrundskolan was very large, the IQR being 3–4. This means that for 50 % of the group, both parents had a sports background (active and/or leader).

However, only the occupation showed significance in relation to the dividing of the pupils into the ACHIEVE groups. One reason for not finding more significant associations is that the group was very homogeneous. It seems like the effects of socio-economic variables diminished once the school group was formed. After that, other variables must have decided who changed ACHIEVE groups. The main variables found in this study were both biologically related—sex and differences in physical maturity, known as the relative age effect (RAE).

Differences in sports achievements between boys and girls can be related both to biological factors (sex) and to socially constructed factors (gender) [15,38]. When it comes to gender, this is discussed in length by Larneby [15]. Here we have focused on questions such as: What are the group compositions for the girls and the boys? How do they change between three and six years? How many girls and boys keep their original groups? How many girls and boys make a journey upward? How many girls and boys make a journey downward? There are differences between girls and boys regarding ACHIEVE. All girls in the under-district group after three years were still in the under-district group after six years, whereas 6 out of 34 boys moved into the district group. Thus, attending upper secondary sports school did not improve the girls' sports results development. Very few girls traveled upward. However, a larger proportion of girls than boys were in the elite group after three years and retain that position after six years. More girls traveled downward, and more boys traveled upward, particularly if they (the boys) were born in quartiles 3 and 4. The effects of RAE are larger for the boys between three and six years.

The relative age effect is a widely recognized effect of selection systems within competitive children and youth sports across a variety of sports and countries all over the world. RAE significantly influenced not only admission to the Idrottsgrundskolan but also the forming of the ACHIEVE groups as well as the reshaping of them during the three years spent at the school. The same was not the case during the upper secondary sports school, but in accordance with the prognosis for their next six years a significant pattern will once again appear.

The major strengths of this study include its longitudinal design and that the information has been obtained for all subjects at all time points. The limitations include the moderate number of subjects studied and the lack of diversity of the sports represented. Individual as well as team sports are included, which may entail differences.

5. Conclusions

The children attending the Idrottsgrundskolan are characterized as having an upper-middle-class background and extensive sports capital. After completion of the lower secondary sports school, three groups of youngsters can be identified: those competing at a national/international level, those competing at a district level, and those having more or less abandoned their elite efforts. The elite group, thereafter, remains fairly stable, whereas the district group is progressively reduced. Few children make an upward journey, i.e., there are no late bloomers. Few social variables affect sports achievements probably owing to the selection mechanisms for admission to the school. Sex affected sports development—a larger proportion of girls than boys reached the elite group after lower and upper secondary school, but fewer girls than boys made an upward journey during the period. The relative age effect profoundly affected admission to the school as well as achievements after lower secondary school but less so after upper secondary school.

Author Contributions: Conceptualization, T.P., O.T., M.K.K., M.D. and P.W.; methodology, T.P. and P.W.; validation, T.P., S.E. and P.W.; formal analysis, T.P. and P.W.; investigation, T.P. and S.E.; resources, T.P., M.K.K., M.D., O.T. and P.W.; data curation, T.P., S.E. and P.W.; writing—original draft preparation, T.P.; writing—review and editing, all authors; visualization, T.P. and P.W.; supervision, T.P., O.T., M.K.K. and M.D.; project administration, T.P., M.D. and P.W.; funding acquisition, T.P. All authors have read and agreed to the published version of the manuscript.

Funding: The authors express their thanks to The Swedish Research Council for Sport Science for financial support to the Malmö Youth Sport Study (MYSS) project.

Institutional Review Board Statement: The study was approved by the Regional Ethical Review Board at Lund (2020-04290).

Informed Consent Statement: All pupils and their parents gave written, informed consent to the study.

Data Availability Statement: The data presented in this study are available on request from the corresponding author.

Conflicts of Interest: The authors declare no conflict of interest. The funders had no role in the design of the study; in the collection, analyses, or interpretation of data; in the writing of the manuscript, or in the decision to publish the results.

References

1. Eliasson, I.; Ferry, M.; Olofsson, E. *Dörrarna Öppnade för Mera*; Riksidrottsförbundet: Stockholm, Sweden, 2012.
2. Fahlström, P.G.; Glemne, M.; Linnér, S. *Goda Idrottsliga Utvecklingsmiljöer*; Riksidrottsförbundet: Stockholm, Sweden, 2016.
3. Ferry, M.; Meckbach, J.; Larsson, H. School sport in Sweden: What it is, and how did it come to be? *Sport Soc.* **2013**, *16*, 805–818. [CrossRef]
4. Ferry, M. *Idrottsprofilerad Utbildning–I Spåren av en Avreglerad Skola*; Göteborgs Universitet: Göteborg, Sweden, 2014.
5. Lund, S. Regulation and deregulation in education policy. *Sport Educ. Soc.* **2014**, *19*, 241–257. [CrossRef]
6. Peterson, T. When the field of sport crosses the field of physical education. In *Educare - Vetenskapliga Skrifter*; pp. 83–97. ISSN 1653-1868. Available online: https://mau.diva-portal.org/smash/record.jsf?dswid=2660&pid=diva2%3A1400535&c=1&searchType=SIMPLE&language=en&query=&af=%5B%5D&aq=%5B%5B%7B%22localid%22%3A%227225%22%7D%5D%5D&aq2=%5B%5B%5D%5D&aqe=%5B%5D&noOfRows=50&sortOrder=author_sort_asc&sortOrder2=title_sort_asc&onlyFullText=false&sf=all (accessed on 10 April 2022).
7. Peterson, T. *Talangutveckling Eller Talangavveckling?* SISU Idrottsböcker: Malmö, Sweden, 2011.
8. Schenker, K. För en Kritisk Didaktik Inom Idrottslärarutbildningen. 2009. Available online: https://www.google.com/url?sa=t&rct=j&q=&esrc=s&source=web&cd=&ved=2ahUKEwj7g43Qsbv3AhVKSvEDHfzxAqMQFnoECAQQAQ&url=http%3A%2F%2Fidrottsforum.org%2Farticles%2Fschenker%2Fschenker090429.pdf&usg=AOvVaw14LVXkBzd3v5x4laFfaL3B (accessed on 10 April 2022).
9. Stambulova, N. *På Väg Mot Att Bli "Vinnare i Långa Loppet"*. Swedish Sports Confederation Report 2013:3. 2013. Available online: https://docplayer.se/123907-Pa-vag-mot-att-bli-vinnare-i-langa-loppet.html (accessed on 10 April 2022).
10. Ek, S.; Wollmer, P.; Karlsson, M.K.; Peterson, T.; Thorsson, O.; Olsson, M.C.; Malmborg, J.S.; Dencker, M. Relative Age Effect of Sport Academy Adolescents, a Physiological Evaluation. *Sports* **2020**, *8*, 5. [CrossRef] [PubMed]
11. Ingrell, J.; Larneby, M.; Johnson, U.; Hedenborg, S. Student athletes' beliefs about athletic ability: A longitudinal and mixed method gender study. *Scand. Sport Stud. Forum* **2019**, *10*, 117–138.
12. Ingrell, J. What does it Take to be Successful Here? Ph.D. Thesis, Malmö Studies in Sport Sciences, Malmo, Sweden, 2019. Volume 32.
13. Lahti, A.; Rosengren, B.E.; Nilsson, J.Å.; Peterson, T.; Karlsson, M.K. Age, gender and family-related factors were the most important socio-ecological associations with physical activity in children with a mean age of eight. *Brief Rep. Acta Paediatr.* **2019**, *109*, 853–854. [CrossRef] [PubMed]
14. Larneby, M Transcending gender hierarchies? Young people and floorball in Swedish school sport. *Sport Soc. Cult. Commer. Media Politics* **2016**, *19*, 1–12.
15. Larneby, M. Vi Och de Andra. Ph.D. Thesis, Malmö Studies in Sport Sciences, Malmö, Sweden, 2020. Volume 36.
16. Malmborg, J.S.; Olsson, M.C.; Bergman, S.; Bremander, A. Musculoskeletal pain and its association with maturity and sports performance in 14-year-old sport school students. *BMJ Open Sport Exerc Med.* **2018**, *4*, e000395. [CrossRef] [PubMed]
17. Malmborg, J.S. Pain and Health in Adolescents and Young Adults–Pieces of a Puzzle. Ph.D. Thesis, Högskolan i Halmstad, Halmstad, Sweden, 2020.
18. Malmborg, J.S.; Bremander, A.; Bergman, S.; Haglund, E.; Olsson, M.C. Musculoskeletal pain and its association with health status, maturity, and sports performance adolescent sport school students: A 2-year follow up. *BMC Sports Sci. Med. Rehabil.* **2022**, *14*, 43. [CrossRef] [PubMed]
19. Peterson, T. Malmö Youth Sport Study. 2017. Available online: https://www.google.com/url?sa=t&rct=j&q=&esrc=s&source=web&cd=&ved=2ahUKEwiGupSPqrv3AhWaSfEDHRnuDHcQFnoECAIQAQ&url=https%3A%2F%2Fwww.diva-portal.org%2Fsmash%2Fget%2Fdiva2%3A1399838%2FFULLTEXT01.pdf&usg=AOvVaw0mclHx4UBNwdaXoG2zfH1z (accessed on 10 April 2022).
20. Peterson, T.; Wollmer, P. Who Are Admitted to a Sport School? 2020. Available online: https://idrottsforum.org/wp-content/uploads/2020/01/peterson-wollmer200127.pdf (accessed on 10 April 2022).
21. Sæther, S.A.; Peterson, T.; Matin, W. The relative age effect, height and weight characteristics among lower and upper secondary school athletes in Norway and Sweden. *Sports* **2017**, *5*, 92. [CrossRef] [PubMed]

22. Cobley, S.; Baker, J.; Wattie, N.; McKenna, J. Annual Age-Grouping and Athlete Development. A Meta-Analytical Review of Relative Age Effects in Sport. *Sports Med.* **2009**, *39*, 235–256. [CrossRef] [PubMed]
23. Malmö kommun: Välfärdsredovisningen 2013, p. 30. Available online: http://docplayer.se/1479773-Redovisning-stadskontoret-malmo-stads-valfardsredovisning-2013.html (accessed on 10 April 2022).
24. Hedberg, M. *Idrotten Sätter Agendan*; Acta Universitatis Gothoburgensis: Götoborg, Sweden, 2014.
25. Bailey, R.; Armour, K.; Kirk, D.; Jess, M.; Pickup, I.; Sandford, R. The educational benefits claimed for physical education and school sport: An academic review. *Res. Pap. Educ.* **2009**, *24*, 1–27. [CrossRef]
26. Bailey, R.; Hillman, C.; Arent, S.; Petitpas, A. Physical activity: An underestimated investment in human capital. *J. Phys. Act. Health* **2013**, *10*, 289–308. [CrossRef] [PubMed]
27. Ericsson, I.; Karlsson, M. Motor Skills and School Performance in Children with Daily Physical Education in School–A Nine-Year Intervention Study. *Scand. J. Med. Sci. Sports* **2012**, *24*, 273–278. [CrossRef] [PubMed]
28. Cöster, M.E.; Fritz, J.; Karlsson, C.; Rosengren, B.E.; Karlsson, M.K. Extended physical education in children aged 6–15 years was associated with improved academic achievement in boys. *Acta Pædiatrica* **2018**. [CrossRef] [PubMed]
29. Fritz, J.; Cöster, M.E.; Rosengren, B.E.; Karlsson, C.; Karlsson, M.K. Daily School Physical Activity Improves Academic Performance. *Sports* **2020**, *8*, 83. [CrossRef] [PubMed]
30. Department of Education. *Idrottsutbildningar i Gymnasie-Och Gymnasiesärskolan: U2020/04134/GV*; Regeringskansliet: Stockholm, Sweden, 2020.
31. Witkowski, K.; Proskura, P.; Piepiora, P. The role of a combat sport coach in the education of youth–a reference to the traditional standards and perception of understanding the role of sport in life of an individual and society. *Arch. Budo Sci. Martial Art Extrem. Sport* **2016**, *12*, 123–130.
32. Engström, L.-M. *Idrott som Social Markor*; HLS Forlag: Goteborg, Sweden, 1999.
33. Schelin, B. Den Ojämlika Idrotten. In *Om Idrottsstratifiering, Idrottspreferens och val av idrott*; Department of Sociology, Lund University: Lund, Sweden, 1985.
34. Bourdieu, P. Program for a Sociology of Sport. *Sociol. Sport J.* **1987**, *5*, 153–161. [CrossRef]
35. Engström, L.-M. *Smak for Motion: Fysisk Aktivitet som Livsstil och Social Markor*; Stockholms Universitets Forlag: Stockholm, Sweden, 2010.
36. Ferry, M.; Lund, S. Pupils in upper secondary school sports: Choices based on what? *Sport Educ. Soc.* **2018**, *23*, 270–282. [CrossRef]
37. Seippel, Ø.; Strandbu, Å.; Aaboen Sletten, M. *Ungdom og Trening. Endringover Tid og Sosiale Skillelinjer, Norsk Institutt for Forskning om Oppvekst, Velferd og Aldring. NOVA Rapport*. 2011. Available online: https://docplayer.me/111905-Ungdom-og-trening-endring-over-tid-og-sosiale-skillelinjer-nr-3-11-ornulf-seippel-ase-strandbu-mira-aaboen-sletten.html (accessed on 10 April 2022).
38. Grahn, K. *Flickor och Pojkar i Idrottens Laromedel*; Acta Universitatis Gothoburgensis: Göteborg, Sweden, 2008.

sports MDPI

Article

Early Sport Specialization and Relative Age Effect: Prevalence and Influence on Perceived Competence in Ice Hockey Players

Vincent Huard Pelletier [1,2,*] and **Jean Lemoyne** [1,2]

[1] Department of Human Kinetics, Université du Québec à Trois-Rivières, 3351 Boulevard des Forges, Trois-Rivières, QC G9A 5H7, Canada; jean.lemoyne@uqtr.ca
[2] Laboratoire de Recherche sur le Hockey UQTR, 3351 Boulevard des Forges, Trois-Rivières, QC G9A 5H7, Canada
* Correspondence: vincent.huard.pelletier@uqtr.ca

Abstract: The relative age effect (RAE) and early sport specialization (ESS) have been of growing interest in the sports world, especially in ice hockey, because of their potential adverse effects. However, little is known about their distribution within each level of play in Canadian minor ice hockey, or whether they influence young people's perceived competence, a variable of interest in long-term sports development. A sample of elite adolescent players (N = 204) and a sample of recreational and competitive players (N = 404) were used to measure these constructs, and chi-square tabulations were conducted to compare their distribution. Our results reveal that RAE (χ^2 = 20.03, $p < 0.01$, Cramer's V = 0.13) and ESS (χ^2 = 66.14, $p < 0.001$, Cramer's V = 0.24) are present, but there are apparently no gender differences in their distributions. Neither the level of RAE nor ESS seems to affect the perceived competence of the players, regardless of gender. The results of this study highlight the presence of RAE and ESS in Canadian minor ice hockey, especially at the elite level, but indicate that they do not affect the self-perception of ice hockey players. Additional research on these concepts is needed to obtain a complete picture of their potential impact on sports development.

Keywords: relative age effect; early sports specialization; perceived competence; ice hockey; youth sport

Citation: Huard Pelletier, V.; Lemoyne, J. Early Sport Specialization and Relative Age Effect: Prevalence and Influence on Perceived Competence in Ice Hockey Players. *Sports* **2022**, *10*, 62. https://doi.org/10.3390/sports10040062

Academic Editors: Adam Leigh Kelly, Sergio L. Jiménez Sáiz, Sara Diana Leal dos Santos and Alberto Lorenzo Calvo

Received: 29 March 2022
Accepted: 13 April 2022
Published: 18 April 2022

1. Introduction

The benefits of sustained physical activity on the health of individuals, regardless of age, have long been known and promoted [1]. In Canada, a very popular form of physical activity, especially among the youth, is participation in organized sports. Indeed, 77% of parents reported that their child aged 5 to 19 participated in at least one organized sport, while 66% of young people aged 10 to 15 reported being involved in at least one organized sports activity [2]. Organized sport also has its own set of health benefits. At the physical level, it protects against the chances of suffering from a long-term health problem [3]. We note that young athletes have an aerobic capacity above the 90th percentile, have an advantageous body composition, demonstrate greater strength and enjoy a better general physical condition than their non-sporting peers [4]. The benefits continue at the psychosocial level. Organized sports participation prevents school dropout [5], depressive symptoms and suicidal thoughts [6], while promoting positive self-perceptions and the acquisition of life skills that can be used later in adult life [7]. In Canada, ice hockey offers an interesting example, in that it is the national sport, has a strong governance structure, occupies an important place in popular culture and currently includes over 600,000 players of all ages and profiles [8]. Young people involved in ice hockey, especially at the competitive level, appear to be more physically active than their peers [9] and display signs of good health into adulthood [10]. Unfortunately, organizational factors such as early sports specialization or relative age effect can negatively impact the benefits associated with organized sports participation.

Youth sport has greatly changed over the past 20 years to become a hyper-organized and competitive environment primarily driven by adults [11]. An increasingly prevalent issue in both the research community [12] and the mainstream media is early sports specialization (ESS), a developmental pathway choice generally recognized to have several negative consequences [13]. ESS can be defined using three criteria [14]: (1) intensive participation in organized training or competitive activities over a period of at least 8 months per year; (2) participation in a single sport that interferes with the pursuit of other physical activities (active leisure, other sports); and (3) involvement of prepubescent children or those under the age of 12. Although the disadvantages of this pathway choice are now well known, the fact remains that a large proportion of young ice hockey players continue to engage in it [15]. The reasons for engaging in ESS can be historical, economic or simply due to mistaken beliefs. Papers such as Chase and Simon's 1973 famous study [16], Bloom's Developing Talent in Young People [17], or Ericson's Deliberate Practice Theory [18] have all been used as a justification to encourage the early and repeated practice of an activity to achieve a high level of expertise instead of sport sampling, a much more recommended pathway [14]. Today, the main reason for early specialization is players' and parents' belief that it gives them an advantage when seeking to be selected by junior or varsity sports teams or considered for a sports scholarship [19,20]. By definition, ESS significantly increases the chances of falling into overtraining [14]. In addition to promoting premature withdrawal from sports in adolescence, this doubles the risk of suffering an overuse injury [21]. Overuse injuries cause at least 20% of all sport dropouts in elite sport; they are the most important reason for dropout, the first being pressure to perform, another consequence of ESS [22,23]. These are not the only harmful outcomes: young athletes who specialize prematurely risk developing an unhealthy dependency on their coach [12], which predisposes them to abuse. In addition, they experience more stress, are prone to psychological problems [12] and gradually lose their love for physical activity, placing them at greater risk of becoming sedentary adults [24]. The prevalence of specialization has recently been measured in Quebec minor ice hockey [15], but it is not yet known to what extent this phenomenon is present for minors at the highest levels of performance, including national team selections.

Another organizational factor observed in multiple sports for decades is the relative age effect (RAE), which can be defined as the age difference between individuals in the same group [25]. To bring together young people at the same stage of development, ensure that differences in performance are limited and establish a certain level of safety, 1- or 2-year chronological age groups are common in youth sport [26]. Thus, it is possible to end up with two athletes who are 23 months apart playing against each other. This may not seem very much to an adult, but it represents a substantial difference for teenagers. Relative age effect therefore refers to the disadvantages and implications of the interaction between the athlete's date of birth and the cut-off date used by the sports federation to define age categories [26]. The real problem occurs when this advantage persists over time, offering relatively older players a better chance to participate and develop in the sport, a phenomenon known as the Mathew Effect [27]. For example, a parent may decide not to enroll their child in an organized sport for fear the child would be disadvantaged compared with another who was born earlier and is therefore more physically imposing. With respect to the prevalence of relative age across several game calibers, recent ice hockey data suggest that there is no significant difference between recreational and competitive calibers; however, an RAE is reported in elite selections [28,29]. Hancock states that one reason for the presence of a strong RAE in elite males is that, besides being potentially more imposing physically because of their older age, they may have been exposed to a higher level of competition when young and may therefore possess more quality experience when entering adolescence [30].

As important as RAE or ESS may be in terms of their direct impacts on the development of young athletes, they may also affect another important construct in sports practice: perceived competence. Perceived competence refers to an individual's perception of their capacities in a certain performance area [31]. Developed by Harter [32], the theory of competence motivation holds that young individuals who consider themselves adept in a specific domain of competence tend to devote more energy to further developing their skills in that area. As a result, skills improvement leads to an increase in perceived competence, resulting in higher autonomous motivation towards an activity [32,33]. RAE has previously been shown to have a slightly negative effect on the perceived competence of young ice hockey players, possibly because relatively younger players are also at a physical disadvantage compared to their peers [28]. ESS also has a negative impact on perceived competence insofar as highly specialized young people tend to compare themselves more to other elite players and receive more frequent corrective and negative feedback from coaches assuming a greater role than they should at this stage of young players' sports development [34]. Both RAE and ESS, and their influence on perceived sports competence, could theoretically affect the long-term athletic commitment of ice hockey players, whether or not they are involved at the elite level.

This study has two objectives. The first is to measure the prevalence of RAE and ESS within a selection of elite adolescent teams compared with its prevalence in non-elite teams. The second is to establish the effects of RAE and ESS on players' perceived competence within an elite environment. These objectives are the basis for five hypotheses.

Hypothesis 1 (H1). *RAE will be present in higher proportions in the elite selection compared with the non-elite teams.*

Hypothesis 2 (H2). *ESS will be present in the same proportions in the elite selection and the non-elite teams.*

Hypothesis 3 (H3). *ESS and RAE will be present in the same proportions in both genders.*

Hypothesis 4 (H4). *Positive perceived competence will not be present in the same proportion based on level of ESS.*

Hypothesis 5 (H5). *Positive perceived competence will not be present in the same proportion based on birth quartiles.*

2. Materials and Methods

2.1. Sample and Procedures

For this study, 1 primary sample was used to measure the perceived competence, ESS, and RAE of elite ice hockey players, and 2 secondary samples were used to compare these with non-elite ice hockey players. Data for the main sample were collected at a Team Québec selection camp for men (U15) and women (U17) in summer 2021. During this camp, players came to the arena, participated in physical tests and were then evaluated on the ice 2 times a day for 4 days for a chance to make the provincial team. They were asked to complete a questionnaire measuring their perceived competence and ESS during a free period of about 1 h on the second day to ensure they were in a good mental condition. During this collection, 89 males and 115 females were surveyed in compliance with the ethics committee of the researchers' institution. Players came from all corners of Quebec, all spoke French, and a binary definition of gender was used, as it currently is in the federation. This was the last step in a year-long selection process during which the number of candidates fell from about 204 in the provisional teams to about 45–50 players per gender at the end of this camp. Thus, these players were the best of their cohort, at least according to those involved in the selection process. The second sample had already been used in a research project by the authors of this article. Data for the sample used as a comparison for the prevalence of ESS and RAE in lower-level ice hockey (N = 202 competitive, N = 202 recreative, 15.4 ± 1.9 years old, no female players) were collected in a previous project, and the details surrounding the collection are discussed in this article [9]. In brief, we used two data collection methods during the 2017–2018 hockey season. For the first approach, we identified teams via the website of Quebec's ice hockey federation, and coaches were contacted to ask for their permission to talk to their players. After explaining the project, the research team issued online or paper questionnaires to those who agreed to participate. For the second approach, we first asked some tournament directors for permission to meet with the participating teams at the registration desk to meet and inform the coaches, who then discussed the project with their players. Research staff members then distributed questionnaires to players who agreed to participate, and they had to be completed during the weekend in their free time, not immediately before or after a game.

2.2. Measurement

2.2.1. Early Sports Specialization

The same measurement method for ESS was employed for both data collections [9]. To calculate an ESS score, 3 questions were formulated based on Laprade's definition, which is the one most frequently used [14]. The first question asked if athletes had participated intensively in ice hockey (development camps, spring league, summer training program) for (1) more than or (2) less than eight months per year. The coded score was 0 for 8 months or less, and 1 for more than 8 months. The second question asked if their ice hockey practice prevented them from participating in other sports activities. The coded score was 0 for no, and 1 for yes. The third question asked for how many years they had been involved intensively in ice hockey. The answer was then deducted from their age to discover if players had this level of commitment before the age of 12. The coded score was 1 for yes, and 0 for no. A 3-point composite score was then created by adding the results of the 3 questions. Three categories of specialization were obtained: <2 = little or no ESS, ≥ 2 but <3 = moderate ESS and 3 = high ESS.

2.2.2. Relative Age Effect

We measured the prevalence of RAE using the same method as in a previous article [28]. Each birth date was coded into birth quartiles according to the federation cutoff date: Q1 = January–February–March, Q2 = April–May–June, Q3 = July–August–September, and Q4 = October–November–December. We then compared it with the numbers obtained from a national survey for the 2007 year of birth [35].

2.2.3. Ice Hockey Perceived Competence

We measured perceptual competence in ice hockey specifically using a recently validated questionnaire that showed robust psychometrics values ($\chi^2_{(154)} = 204.160$, $p = 0.004$, RMSEA = 0.021, CFI = 0.997, TLI = 0.993) [36]. The questionnaire consisted of 26 items measured by a 5-point Likert scale and assessed 6 dimensions of ice hockey competence: physical involvement (4 items, $\omega = 0.95$, e.g., *"I win my "one-on-one" battles"*), ice hockey IQ (7 items, $\omega = 0.97$, e.g., *"I take good decision with the puck"*), offensive skills (4 items, $\omega = 0.95$, e.g., *"I have good offensive creativity"*), skating abilities (4 items, $\omega = 0.95$, e.g., *"I am a fast skater"*), resilience (3 items, $\omega = 0.92$, *"I see coaches' comments as an opportunity to improve"*), and leadership (4 items, $\omega = 0.94$, e.g., *"I stay confident even if playing time is diminished"*). A composite score of all dimensions was then calculated in addition to an overall competence score.

2.2.4. Statistical Analyses

Chi-square cross tabulations were complketed, and Cramer's V was calculated to assess effect size and interpreted using Volker's recommendations [37] and in accordance with Delorme [38]. A significant result ($p < 0.05$) indicated a different distribution between the different variables studied, and the interpretation of Cramer's V was as follows: >0 = very weak, >0.05 = weak, >0.10 = moderate, >0.15 = strong and >0.20 = very strong [39]. We also tested each category with its previous group to confirm if distributions differ from one category to another. Percentage deviation was measured to show the degree to which an observed chi-square cell frequency differs from the value expected on the basis of the null hypothesis.

3. Results

Table 1 shows the distribution of RAE by gender and playing level. As can be seen, there is a significant difference in the distribution of RAE by gender ($p < 0.05$) and playing level ($p < 0.001$), with a moderate size effect (V > 0.10) in both cases. These results confirm Hypothesis 1 and partially confirm Hypothesis 3. We also note that the elite players show different distributions regarding RAE; more players are born in Q1 and fewer in Q4 at this playing level compared with other playing levels and the Canadian population of this age.

Table 1. RAE prevalence across playing level and gender.

Birth Quartile	Canadian Birth Percentage (2007)	Gender N (Percentage Variation)		χ^2	Playing Level N (Percentage Variation)			χ^2
	Both Genders *	Male	Female		Recreative ***	Competitive	Elite **	
Q1	23	38 (−6.7)	56 (+5.2)	2.46	64 (−9.3)	58 (−19.8)	94 (+28.6)	20.03
Q2	25	23 (+17.9)	22 (−13.7)	($p = 0.48$)	49 (−3.2)	61 (+17.5)	45 (−14.2)	($p < 0.01$)
Q3	27	20 (+7.3)	23 (−5.6)	Cramer's	44 (+5.2)	41 (−4.4)	43 (−0.7)	Cramer's
Q4	25	7 (−23.1)	14 (+17.7)	V = 0.11	39 (+18.2)	41 (+21.2)	21 (−38.5)	V = 0.13

* There is a significant difference ($p < 0.01$) between Canadian birth and male ($p < 0.01$, Cramer's V = 0.28) and female ($p < 0.01$, Cramer's V = 0.28) elite players. ** Elite is significantly different from recreative ($p = 0.01$, Cramer's V = 0.16) and competitive ($p < 0.001$, Cramer's V = 0.21). *** The competitive and recreative samples contain only male participants.

Table 2 shows the distribution of RAE by gender and playing level. We note there is no significant difference in the distribution of ESS level according to gender ($p > 0.05$). However, a substantial difference in magnitude ($p < 0.001$) exists in ESS levels based on the playing level of the ice hockey players. Specifically, there are fewer competitive players in the high ESS group and more in the low ESS group than at the other two playing levels. Elites also have a significantly larger proportion of highly specialized athletes and a lower proportion of athletes with low ESS. These results refute Hypothesis 2, but partially confirm Hypothesis 3.

Table 2. ESS prevalence across playing level and gender.

ESS Score	Gender N (% Variation)		χ^2	Playing Level N (% Variation)			χ^2
	Male	Female		Recreative ***	Competitive *	Elite **	
Low	3 (−23.3)	8 (+16.5)	$\chi^2 = 0.98$	40 (−5.5)	73 (+72.5)	11 (−72.1)	$\chi^2 = 66.14$ *
Moderate	27 (+1.4)	36 (−1.0)	Cramer's	61 (−8.3)	71 (+6.7)	63 (+1.7)	Cramer's
High	48 (+1.5)	66 (−1.1)	V = 0.07	101 (+8.4)	58 (−37.7)	114 (+31.5)	V = 0.24

* Competitive differs significantly from recreative ($p < 0.001$, Cramer's V = 0.23) and elite ($p < 0.001$, Cramer's V = 0.41). ** Elite differs significantly from recreative ($p < 0.001$, Cramer's V = 0.21). *** The competitive and recreative samples contain only male participants.

Table 3 compares the composite scores for the dimensions of ice hockey competence among elite players. It shows that there is only one significant difference: female players feel less competent in terms of their offensive abilities ($p < 0.001$).

Table 4 represents the distribution of scores regarding the 6 dimensions of perceived ice hockey competence based on SSE and RAE levels. It shows there is no significant difference in perceived competence for these two variables. These results therefore refute Hypotheses 4 and 5.

Table 3. RAE association with ice hockey perceived competence.

Construct	Mean Scores	
	Male	Female
Perceived offensive abilities	4.30	4.15 **
Perceived skating abilities	4.27	4.39
Perceived physical imnvolvement	4.25	4.16
Perceived hockey sense	4.29	4.21
Perceived ability to lead	4.47	4.50
Perceived ability to face adversity	4.22	4.23

** Significant difference at 0.001.

Table 4. ESS and RAE association with ice hockey perceived competence.

Perceived Competence	ESS				χ^2	RAE				χ^2
	Scores	Low	Moderate	High		Q1	Q2	Q3	Q4	
Offensive abilities	3.0	0	0	0		0	0	0	0	
	3.5	0	5	7	χ^2 = 6.546	7	1	3	1	χ^2 = 6.189
	4.0	1	13	32	Cramer's	23	12	7	6	Cramer's
	4.5	7	23	32	V = 0.140	31	17	8	6	V = 0.110
	5.0	3	17	28		21	10	13	5	
Defensive abilities	3.0	0	0	4		1	1	0	4	
	3.5	0	2	8	χ^2 = 9.170	6	0	2	10	χ^2 = 11.605
	4.0	1	13	26	Cramer's	22	7	3	42	Cramer's
	4.5	3	19	21	V = 0.164	19	12	7	44	V = 0.150
	5.0	7	24	42		34	12	7	73	
Physical involvement	3.0	0	0	1		0	1	0	0	
	3.5	0	2	11	χ^2 = 8.487	7	3	2	1	χ^2 = 13.168
	4.0	5	16	34	Cramer's	23	13	16	4	Cramer's
	4.5	4	17	30	V = 0.59	30	9	8	6	V = 0.160
	5.0	2	22	24		22	14	5	7	
Hockey sense	3.0	0	4	5		5	2	2	0	
	3.5	0	5	9	χ^2 = 4.777	9	3	3	0	χ^2 = 6.762
	4.0	3	16	24	Cramer's	19	11	10	4	Cramer's
	4.5	2	15	32	V = 0.119	25	10	9	6	V = 0.114
	5.0	6	18	31		24	14	8	9	
Ability to lead	3.0	0	0	0		0	0	0	0	
	3.5	0	2	8	χ^2 = 4.135	4	1	4	1	χ^2 = 8.818
	4.0	2	8	18	Cramer's	10	8	7	3	Cramer's
	4.5	2	20	27	V = 0.111	24	13	9	3	V = 0.131
	50	7	27	48		44	17	12	12	
Ability to face adversity	3.0	0	2	8		3	2	5	0	
	3.5	0	4	8	χ^2 = 4.846	7	1	5	0	χ^2 = 15.557
	4.0	3	18	31	Cramer's	24	14	9	6	Cramer's
	4.5	2	10	21	V = 0.119	17	9	4	4	V = 0.173
	5.0	6	24	33		31	14	9	9	

4. Discussion

This study, which was conducted using two samples of adolescent ice hockey players, aimed to describe the prevalence of ESS and RAE at several levels of play and relate them to perceived ice hockey skill, a variable that impacts sports development. We can now review the five hypotheses put forward earlier. First, Hypothesis 1 was confirmed because the distribution of the RAE differed between elite male and female players. It also differed between those playing at other playing levels and in the Canadian population of the same age. The presence of RAE is hardly surprising, as the concept has been part of Canadian ice hockey for over 40 years [25]. Our data are also consistent with the situation in minor

Russian ice hockey, where 65% of players were born in the first 6 months of the year [40]. What sets our study apart, however, is that both genders were measured in the elite level and that we also assessed recreative and competitive players. Contrary to the findings of a previous study [28], the proportion of ice hockey players born in the first 2 quartiles of the year was greater in elite players than in the other playing levels, where the RAE was not significantly present. These results make sense to say the very least, because we know that the presence of RAE is strongly influenced by the level of competition. So, it is logical that they are found at a greater proportion in the elite leagues, where players compete fiercely to gain small advantages over their peers and where a difference of 6 or 9 months can have a significant impact. Hypothesis 3 must be partially confirmed regarding the presence of an RAE in both genders, as no significant differences were found within the sample of elite players participating in the national team camps. That ice hockey is Canada's national sport and is popular with both genders most likely raises the level of competition, even in the female sample, which may well have had less RAE based on previous work [25]. This finding is positive in that it confirms that women's field hockey has become very competitive in Canada, but is negative in that we are starting to see the same problems as in men's ice hockey.

Next, we disproved Hypothesis 2, which predicted that the proportion of specialized players would be the same, regardless of playing level. However, the opposite was observed: each playing level had a different ESS distribution than the others. Compared with competitive and recreational players, elite players seem to have a significantly smaller proportion of low or no specialized players and a larger proportion of highly specialized players. Competitive players have many more non-specialized or less specialized players and fewer highly specialized players than other playing levels. At the recreational level, patterns of specialization fall between those observed at the elite and competitive levels. As Quebec ice hockey already demonstrated [15], competitive players show healthier specialization tendencies than recreational players, which can be explained in two ways as follows: (1) these players are usually coached by qualified coaches as opposed to volunteers who are generally involved in recreational play, and (2) they may have better athletic abilities than those playing at a lower level, which predisposes them to a wider range of activities. Elite players, on the other hand, have by far the largest proportion of highly specialized players. This may appear surprising, as they are coached at least as well as athletes at the competitive level and are expected to be even more athletically gifted and well rounded. The reason for their high level of early specialization may be that young players and their parents believe in the importance of being labeled "talented" at a young age in order to gain access to better teams, receive extensive training and compete at a higher level in preparation for an eventual athletic career [12]. Another argument maintains that more specialized young people tend to receive athletic scholarships more often at the college level, as Hockett indicates [19]. While the physical and psychological benefits of sports sampling are well-known, the short-term benefit of ESS can be compelling to many, especially in a major camp such as the national team. The fact that the camp includes 14-year-old ice hockey players may also accentuate this phenomenon, as the selection occurs early in the sport's development process compared with the junior (15–16) and professional (17–18) drafts.

Regarding SSE, this study's results partly confirm Hypothesis 3 as well, because no significant difference was found between male and female players. Previous work suggests that girls are more prone to experience SSE, but this may be largely because most studies deal with early developing individual sports such as synchronized swimming or gymnastics [41]. The percentage of highly specialized girls is lower in team sports such as ice hockey; the results of this study therefore support this trend.

The last two hypotheses addressed perceived competence in ice hockey and its relationship with the two constructs under study, RAE and ESS. Contrary to what one might have expected from the literature, the proportion of field hockey players with a positive perception of their skills did not change according to ESS or RAE levels. Indeed, previous work established a link, albeit a relatively weak one, between perception of endurance and

ESS [15], while another paper linked perception of physical strength and RAE [28]. In both cases, a broader physical self-perception tool was used, highlighting the importance of using an accurate, sport-specific measure. Alternatively, the literature also suggests that ESS may interfere with the normal development of young athletes by isolating them from their peers, causing overly close bonds with their coaches, and placing them at risk for overtraining [14]. Thanks to the present study, we believe that in the context of Quebec elite ice hockey, even highly specialized players evaluate themselves positively in terms of their skills. This may be because ice hockey is a sport with approximately 20 players per team, where the cost of ice rental makes it more difficult to practice in isolation. Less isolation and the esteem-enhancing qualities of a healthy team environment are conducive to very positive self-perception, which is to be expected from a group of elite players [42,43]. These findings moderate the associations between RAE, SSE and perceived competence, which may be somewhat reassuring for the psychological health of elites who choose to invest in a single sport at a young age or for those disadvantaged by their birthdate.

A strength of this work is the number of concerns it addresses relative to ESS and RAE research. Although these issues have attracted increasing attention in the last 20 years, there is still an absence of data regarding their representation in a particular team sport such as ice hockey at all youth playing levels [13,15]. Furthermore, this study is in line with a recent consensus statement on ESS that highlights the need for conducting research to differentiate gender-based patterns, limit the use of adult samples to reduce the possible drawbacks of a multiple-year recall, and further explore the impacts of ESS [44]. A second strength of our study is that it shows that RAE and ESS have apparently no impact on the perceived competence of ice hockey players. This improves our understanding of these issues, but also raises specific questions concerning ice hockey. First, might the relationship with perceived competence be different if it were measured at ages 18, 20, and 25? Do players with low ESS playing at competitive levels tend to become more successful later in their development, or does the advantage that those with high ESS gain in early adolescence carry over into adulthood? Although both of these questions could be answered using a longitudinal research design, such was unfortunately not possible in this project. Additionally, a more precise operational definition of ESS could better categorize young athletes' level of specialization. LaPrade's current three-indicator definition is slightly too broad and does not necessarily point to potentially negative behaviors and variables about training load [14]. For example, is it safe to say that a 10-year-old playing ice hockey for 10 months and enrolling in soccer for 2 months while continuing to train for ice hockey (non-specialized by the current definition) is at less risk of adverse psychological or physical consequences than a player who does not enroll in soccer, but has the same practice in ice hockey (specialized by the current description)? Additional data regarding active behaviors and the physical and psychological impacts of SSE could be added to better identify youth with a genuinely problematic sports development [44].

RAE has been present for at least a hundred years in practically all sports, and it affects both men and women at an equivalent level of competition [45]. This is a difficult phenomenon to work around logistically, since separating players by birthdates and having them play together in age-based groups of 24 months as in ice hockey is a simple, safe and inexpensive solution that ensures a relatively homogeneous performance level [25]. However, a very high RAE between the Canadian population and elite level ice hockey players may be concerning, especially because there is a RAE between the elites and other levels of play as well. The visibility and access to qualified coaches these players enjoy in comparison to their peers confer a distinct advantage. Thus, initiatives such as tournaments reserved for the best players born in Q3 and Q4 could partially reduce the RAE by offering these neglected players additional opportunities to showcase their skills. At the very least, we may discover that players disadvantaged by their later birthdate do not have a lower perceived competence than others. This opens the door to research on other psychological variables of interest, including perceived enjoyment, stress, burnout, or social support, to reach solid conclusions about the real impact of RAE and ESS on youth sports development [46].

5. Conclusions

In conclusion, this study explores the distribution of the relative age effect as well as early sports specialization within Canadian minor ice hockey in function of the different playing levels and the two genders. It confirms that these two issues are present in Canadian elite ice hockey, especially at the elite level, but do not affect players' perception of competence. Although this research adds to the present ESS and RAE literature, it is purely descriptive and somewhat limited by its cross-sectional design. Further research is needed to investigate the additional impacts they may have on the athletic development and long-term physical activity of these young athletes. Awareness of the harmful effects of ESS and RAE must continue in youth sport, especially at the elite level.

Author Contributions: Conceptualization: V.H.P.; methodology: V.H.P.; software: V.H.P.; validation: J.L. and V.H.P.; formal analysis: V.H.P.; investigation: V.H.P.; resources: J.L.; data curation: J.L.; writing—original draft preparation: J.L.; writing—review and editing: V.H.P.; visualization, J.L.; supervision: J.L.; project administration: J.L. All authors have read and agreed to the published version of the manuscript.

Funding: This research was funded by the Ministère de l'Éducation et de l'Enseignement Supérieur du Québec Projet Synergique-Hokey Québec, l'Institut National du Sport-Québec (National Sport Institute-Quebec) and Université du Québec à Trois-Rivières. This project was also funded under the MITACS Elevation program (15-04-21), with the financial collaboration of Optimize® as financial partner.

Institutional Review Board Statement: The study was conducted according to the guidelines of the Declaration of Helsinki and approved by the Institutional Review Board (or Ethics Committee) of Université du Québec à Trois-Rivières (CER-21-275-07.04, 15 April 2021).

Informed Consent Statement: Informed consent was obtained from all participants involved in the study. Participants gave their written consent to publish this paper.

Data Availability Statement: The authors are committed to providing all data used to conduct this study upon request.

Acknowledgments: The authors thank to Hockey Québec direction for supporting the project. Authors are also thankful to all hockey players who took part in the data collection for their collaboration.

Conflicts of Interest: The authors declare no conflict of interest.

References

1. Malm, C.; Jakobsson, J.; Isaksson, A. Physical activity and sports—Real health benefits: A review with insight into the public health of Sweden. *Sports* **2019**, *7*, 127. [CrossRef] [PubMed]
2. ParticipACTION. *Le Rôle de la Famille dans L'Activité Physique, les Comportements sÉdentaires et le Sommeil des Enfants et des Jeunes. L'édition 2020 du Bulletin de L'Activité Physique Chez les Enfants et les Jeunes de ParticipACTION*; ParticipACTION: Toronto, ON, Canada, 2020.
3. Janssen, I.; LeBlanc, A.G. Systematic review of the health benefits of physical activity and fitness in school-aged children and youth. *Int. J. Behav. Nutr. Phys. Act.* **2010**, *7*, 40. [CrossRef] [PubMed]
4. Cordingley, D.M.; Sirant, L.; MacDonald, P.B.; Leiter, J.R. Three-year longitudinal fitness tracking in top-level competitive youth ice hockey players. *J. Strength Cond. Res.* **2019**, *33*, 2909–2912. [CrossRef] [PubMed]
5. Ruiz-Ariza, A.; Grao-Cruces, A.; de Loureiro, N.E.M.; Martinez-Lopez, E.J. Influence of physical fitness on cognitive and academic performance in adolescents: A systematic review from 2005–2015. *Int. Rev. Sport Exerc. Psychol.* **2017**, *10*, 108–133. [CrossRef]
6. Taliaferro, L.A.; Rienzo, B.A.; Miller, D.M.; Pigg, M.R.; Dodd, V.J. Potential mediating pathways through which sports participation relates to reduced risk of suicidal ideation. *Res. Q. Exerc. Sport* **2010**, *81*, 328–339. [CrossRef]
7. Chinkov, A.E.; Holt, N.L. Implicit transfer of life skills through participation in Brazilian jiu-jitsu. *J. Appl. Sport Psychol.* **2016**, *28*, 139–153. [CrossRef]
8. International Ice Hockey Federation. Survey of Players. 2022. Available online: https://www.iihf.com/en/static/5324/survey-of-players (accessed on 1 November 2020).

9. Pelletier, V.H.; Girard, S.; LeMoyne, J. Adolescent hockey players' predispositions to adopt sport and exercise behaviours: An ecological perspective. *PLoS ONE* **2020**, *15*, e0228352. [CrossRef]
10. Kitchen, P.; Chowhan, J. Forecheck, backcheck, health check: The benefits of playing recreational ice hockey for adults in Canada. *J. Sports Sci.* **2016**, *34*, 2121–2129. [CrossRef]
11. Feeley, B.T.; Agel, J.; Laprade, R.F. When is it too early for single sport specialization? *Am. J. Sports Med.* **2016**, *44*, 234–241. [CrossRef]
12. Malina, R.M. Early sport specialization: Roots, effectiveness, risks. *Curr. Sports Med. Rep.* **2010**, *9*, 364–371. [CrossRef]
13. Popkin, C.A.; Bayomy, A.F.; Ahmad, C.S. Early sport specialization. *J. Am. Acad. Orthop. Surg.* **2019**, *27*, e995–e1000. [CrossRef] [PubMed]
14. LaPrade, R.F.; Agel, J.; Baker, J.; Brenner, J.S.; Cordasco, F.A.; Côté, J.; Hewett, T.E. AOSSM early sport specialization consensus statement. *Orthop. J. Sports Med.* **2016**, *4*, 2325967116644241. [CrossRef] [PubMed]
15. Pelletier, V.H.; Lemoyne, J. Exploring the impacts of early sport specialization among Québec's adolescent hockey players. *J. Expertise* **2020**, *3*, 41–54.
16. Chase, W.G.; Simon, H.A. Perception in chess. *Cogn. Psychol.* **1973**, *4*, 55–81. [CrossRef]
17. Bloom, B. *Developing Talent in Young People*; Ballantine Books: New York, NY, USA, 1985.
18. Ericsson, K.A.; Krampe, R.T.; Tesch-Römer, C. The role of deliberate practice in the acquisition of expert performance. *Psychol. Rev.* **1993**, *100*, 363. [CrossRef]
19. Hockett, M.P. *Impact of Early Sport Specialization on College Student Athletes*; University of Minnesota Digital Conservancy: Duluth, MI, USA, 2018. Available online: https://hdl.handle.net/11299/200481 (accessed on 1 November 2020).
20. Ahlquist, S.; Cash, B.M.; Hame, S.L. The Associations of Early Sport Specialization and Training Volume with Injury Rates in NCAA Division I Athletes. *Orthop. J. Sports Med.* **2019**, *7*, 2325967119S2325900195. [CrossRef]
21. Bell, D.R.; Post, E.G.; Biese, K.; Bay, C.; McLeod, T.V. Sport specialization and risk of overuse injuries: A systematic review with meta-analysis. *Pediatrics* **2018**, *142*, e20180657. [CrossRef]
22. Butcher, J.; Lindner, K.J.; Johns, D.P. Withdrawal from competitive youth sport: A retrospective ten-year study. *J. Sport Behav.* **2002**, *25*, 145–163.
23. Gould, D. Early sport specialization: A psychological perspective. *J. Phys. Educ. Recreat. Danc.* **2010**, *81*, 33–37. [CrossRef]
24. Russell, W.D.; Limle, A.N. The relationship between youth sport specialization and involvement in sport and physical activity in young adulthood. *J. Sport Behav.* **2013**, *36*, 82–98.
25. Musch, J.; Grondin, S. Unequal competition as an impediment to personal development: A review of the relative age effect in sport. *Dev. Rev.* **2001**, *21*, 147–167. [CrossRef]
26. Romann, M.; Cobley, S. Relative age effects in athletic sprinting and corrective adjustments as a solution for their removal. *PLoS ONE* **2015**, *10*, e0122988. [CrossRef] [PubMed]
27. Merton, R.K. The Matthew effect in science: The reward and communication systems of science are considered. *Science* **1968**, *159*, 56–63. [CrossRef]
28. Lemoyne, J.; Huard Pelletier, V.; Trudeau, F.; Grondin, S. Relative Age Effect in Canadian Hockey: Prevalence, Perceived Competence and Performance. *Front. Sports Act. Living* **2021**, *3*, 14. [CrossRef]
29. de Subijana, C.L.; Lorenzo, J. Relative age effect and long-term success in the Spanish soccer and basketball national teams. *J. Hum. Kinet.* **2018**, *65*, 197. [CrossRef] [PubMed]
30. Hancock, D.J.; Ste-Marie, D.M.; Young, B.W. Coach selections and the relative age effect in male youth ice hockey. *Res. Q. Exerc. Sport* **2013**, *84*, 126–130. [CrossRef] [PubMed]
31. Horn, T.S. Developmental Perspectives on Self-Perceptions in Children and Adolescents. In *Developmental Sport and Exercise Psychology: A Lifespan Perspective*; Weiss, M.R., Ed.; Fitness Information Technology: Morgantown, WV, USA, 2004; pp. 101–143.
32. Harter, S. Pleasure derived from challenge and the effects of receiving grades on children's difficulty level choices. *Child Dev.* **1978**, *49*, 788–799. [CrossRef]
33. Babic, M.J.; Morgan, P.J.; Plotnikoff, R.C.; Lonsdale, C.; White, R.L.; Lubans, D.R. Physical activity and physical self-concept in youth: Systematic review and meta-analysis. *Sports Med.* **2014**, *44*, 1589–1601. [CrossRef]
34. Horn, T.S. Social Psychological and Developmental Perspectives on Early Sport Specialization. *Kinesiol. Rev.* **2015**, *4*, 248–266. Available online: https://journals.humankinetics.com/view/journals/krj/4/3/article-p248.xml (accessed on 18 February 2022). [CrossRef]
35. Statistique Canada. Tableau 13-10-0415-01. Naissances Vivantes, Selon le Mois. Available online: https://doi.org/10.25318/1310041501-fra (accessed on 1 November 2021).
36. Huard Pelletier, V. Ice Hockey Perceived Competence and Athletic Development. Unpublished Ph.D. Thesis, Université du Québec à Trois-Rivières, Trois-Rivières, QC, Canada, 2022.
37. Volker, M.A. Reporting effect size estimates in school psychology research. *Psychol. Sch.* **2006**, *43*, 653–672. [CrossRef]
38. Delorme, N.; Champely, S. Relative age effect and chi-squared statistics. *Int. Rev. Soc. Sport* **2015**, *50*, 740–746. [CrossRef]
39. Akoglu, H. User's guide to correlation coefficients. *Turk. J. Emerg. Med.* **2018**, *18*, 91–93. [CrossRef] [PubMed]
40. Bezuglov, E.; Shvets, E.; Lyubushkina, A.; Lazarev, A.; Valova, Y.; Zholinsky, A. Relative Age Effect in Russian Elite Hockey. *J. Strength Cond. Res.* **2020**, *34*, 2522–2527. [CrossRef] [PubMed]

41. Jayanthi, N.; Dugas, L. The Risks of Sports Specialization in the Adolescent Female Athlete. *Strength Cond. J.* **2017**, *2*, 20–26. [CrossRef]
42. Pedersen, S.; Seidman, E. Team sports achievement and self-esteem development among urban adolescent girls. *Psychol. Women Q.* **2004**, *28*, 412–422. [CrossRef]
43. Edwards, D.J.; Edwards, S.D.; Basson, C.J. Psychological well-being and physical self-esteem in sport and exercise. *Int. J. Ment. Health Promot.* **2004**, *6*, 25–32. [CrossRef]
44. Kliethermes, S.A.; Marshall, S.W.; LaBella, C.R.; Watson, A.M.; Brenner, J.S.; Nagle, K.B.; Beutler, A.I. Defining a research agenda for youth sport specialisation in the USA: The AMSSM Youth Early Sport Specialization Summit. *Br. J. Sports Med.* **2021**, *55*, 135–143. [CrossRef]
45. Jakobsson, J.; Julin, A.L.; Persson, G.; Malm, C. Darwinian Selection Discriminates Young Athletes: The Relative Age Effect in Relation to Sporting Performance. *Sports Med. Open* **2021**, *7*, 1–18. [CrossRef]
46. Gould, D. The Current Youth Sport Landscape: Identifying Critical Research Issues. *Kinesiol. Rev.* **2019**, *8*, 150–161. Available online: https://journals.humankinetics.com/view/journals/krj/8/3/article-p150.xml (accessed on 25 February 2022). [CrossRef]

sports

MDPI

Article

An Examination of the Experiences of Practitioners Delivering Sport Psychology Services within English Premier League Soccer Academies

Francesca Dean [1,*], Emma Kavanagh [2], Amanda Wilding [1] and Tim Rees [1]

[1] Department of Rehabilitation and Sport Sciences, Bournemouth University, Bournemouth BH12 5BB, UK; awilding@bournemouth.ac.uk (A.W.); trees@bournemouth.ac.uk (T.R.)

[2] Department of Sport and Event Management, Bournemouth University, Bournemouth BH12 5BB, UK; ekavanagh@bournemouth.ac.uk

* Correspondence: fdean@bournemouth.ac.uk

Abstract: Sport psychology has become increasingly recognized and accepted within professional sports, including soccer. To date, there is a lack of research that examines the provision of sport psychology within elite soccer, particularly from the experience of applied practitioners working within the field. The current study adopted a qualitative, inductive approach, to examine the experiences of practitioners responsible for sport psychology delivery within elite soccer academies in England. Seven participants (four females; three males), working within academies in the English Premier League, took part in semi-structured interviews about their experience of delivering sport psychology services within elite soccer academies. Results demonstrated that the provision of sport psychology is continually evolving, yet there are a number of factors that appear to inhibit the full integration of the discipline into academy soccer. Six key themes were identified: The breadth of sport psychology provision; what is sport psychology; the stigma surrounding sport psychology services; psychological literacy; the elite youth soccer environment; and the delivery of sport psychology under the Elite Player Performance Plan. Participants identified a lack of psychological literacy among coaches and academy staff, as well as a low level of guidance regarding the provision of psychology within the England Football Association's guiding document—the *Elite Player Performance Plan*—leading to considerable variation in the nature of the sport psychology provision. Future research would do well to also sample from a range of staff working within English soccer academies, in order to assess their perception of the level of provision and understanding of psychology.

Keywords: soccer; sport psychology; academy soccer; EPPP; psychological literacy; experiences of a sport psychologist; sport psychology integration

Citation: Dean, F.; Kavanagh, E.; Wilding, A.; Rees, T. An Examination of the Experiences of Practitioners Delivering Sport Psychology Services within English Premier League Soccer Academies. *Sports* **2022**, *10*, 60. https://doi.org/10.3390/sports10040060

Academic Editor: Iain Greenlees

Received: 28 January 2022
Accepted: 8 April 2022
Published: 13 April 2022

1. Introduction

Although integration of psychological services within the preparation of soccer players has tended historically to lag behind a focus on physical, technical, and tactical aspects, sport psychology has become increasingly recognized and accepted within soccer [1,2] as it has in sport in general [3–5]. Indeed, although there is still some resistance towards employing sport psychology services [6,7], psychological skills are increasingly understood as fundamental for success and career progression in sports such as soccer [2,8,9]. As such, there is now an appreciation that successful performance is as much about a player's overall psychological state and the development and execution of advanced psychological skills as it is about physical prowess.

Since 2012, the English Premier League (EPL) has, via its Elite Player Performance Plan (EPPP), mandated the inclusion of psychology within the development program for English soccer players. Despite this initiative, the integration of sport psychology remains a challenge [2], and the discipline is not always positively received by all [10].

Additionally, there is limited research describing the experience of applied sport psychology practitioners working in the sport. The present study sought to gain insight into the delivery of sport psychology within the framework of the EPPP through capturing the experiences of sport psychology practitioners currently working within EPL soccer academies in the United Kingdom.

The EPPP was instigated in 2012, supported by a budget of GBP 320 million [11]. The main aim underpinning the development of the EPPP was to "increase the number and quality of [English] players gaining professional contracts and playing first team soccer at the highest level" [12]. This ambition would be realized through the development of athletes within the Premier League and Soccer League academies (the key development environment for young players in England), which are designed to provide young players with expert coaching, support, and education. Academies are audited every three years and must provide evidence of individualized coaching interventions that address each player's physical, technical, tactical, and social and psychological development. Academies are subsequently awarded a categorization of 1 to 4 (1 being the highest), which then dictates the level of provision that is expected of them. As part of the EPPP guidelines, academies must make a sport science and medicine program available to all players. Such programs include physical testing and measurement, physiotherapy, medical services, nutrition, performance analysis, and psychology.

Although guidelines regarding the delivery of sport psychology within the EPPP were initially quite limited, the EPPP guidelines have more recently progressed, such that there are now more clearly defined minimum requirements. These include psychological testing, lifestyle management, and the delivery of mental skills education, as well as appropriate qualifications for the person delivering the psychological support. The latter means that those delivering sport psychology in English soccer academies must be qualified to a master's degree-level in psychology or sport psychology, and registered on either the British Psychology Society or British Association of Sport and Exercise Sciences accreditation pathways.

Beyond the above minimum requirements, however, soccer club academies are afforded flexibility in delivery of psychological provision, such that there is no uniform content or standard of delivery. Some choose to draw on work focusing on the 5Cs of confidence, control, concentration, commitment, and communication [13,14], but otherwise evidence for the specific breakdown of psychology provision is scant [15]. The purpose of the 5Cs psychological intervention was to ensure that youth soccer players were provided with a greater amount of psychological education and development within training sessions [15]. Harwood et al. [15] suggested that their model was derived from Lerner et al.'s 6Cs of youth development—character, confidence, connection, competence, caring, and contribution [16]—developed with reference to the skills necessary to develop a more civilized society [15–17]. It is important to note, however, that it is difficult to assess the overall effectiveness of this model because there is no record of the number of clubs using it. Furthermore, the research examining the model is only being conducted by a small group of researchers [2,15,17,18].

More widely, coaches have long been considered a key source of support [19], who are often the "first line of defense" in dealing with psychological concerns. In this respect, there has been recent interest in the need for coaches and support staff in sporting organizations to develop their psychological literacy. The importance of psychological literacy, defined as "the ethical application of psychological skills and knowledge" [20] should not be underestimated. Although increasing the reach and impact of psychology through non-professionals (not just trained practitioners) would be a goal of developing psychological literacy, Murdoch [20] also noted the potential risks of non-professionals sharing psychological knowledge that may not be accurate. Indeed, Pain and Harwood [18] previously observed a lack of understanding and knowledge of sport psychology by coaches and support staff in English Soccer Academies, meaning they were potentially ill-equipped (even if they sensed the need) to draw on psychology within their coaching sessions. Thus, although the integration of psychological literacy across clubs is a positive steppingstone, it should not overtake the importance of having certified practitioners delivering psychological services to young players. The current research draws upon psychological literacy, that is, an individual's ability to apply psychological understanding to their everyday (personal, social, and work) lives. The purpose of drawing upon psychological literacy is to provide relevant context to the study in relation to the understanding of psychology and how non-psychology practitioners then apply that knowledge.

The purpose of the present study was therefore twofold: first, to explore the role and experiences of practitioners working in sport psychology within elite soccer academies in England; and second, to understand the nature and implementation of sport psychology within the auspices of the EPPP. Furthermore, this study sought to explore these practitioners' perceptions of what is successful and what needs further development in relation to the EPPP and how the academies are integrating sport psychology into their youth programs.

2. Method

An interpretivist paradigm [21] was adopted to explore the experiences of practitioners responsible for sport psychology delivery within soccer academies in England.

2.1. Participants

Participants (four females, three males; aged 31 to 41 years, mean = 34.8) were British (all white), representing six elite English soccer academies at four Premier League (the highest-level league in England) and two Championship (the second tier of English football) clubs. Of the six academies, five had (at the time the research was conducted) Category 1 status (the highest-level academy in England, running on a scale from 1 to 3, as stipulated by the EPPP academy audit system); one had Category 2 status. All participants held a master's degree or higher, three were British Association of Sport and Exercise Sciences (BASES) accredited, two participants were BASES Sport and Exercise Psychology Accreditation Route (SEPAR) accredited, and the same two were also Health and Care Professions Council (HCPC) registered. Participants' experience of working as practitioners within a psychology unit ranged from 4 to 15 years.

2.2. Sampling

For inclusion in the study, participants were required to meet one key judgment criterion: participants needed to be responsible for delivering the sport psychology program to youth performers at an elite soccer academy with EPPP category status in England. In order to ensure that all those working within psychology departments at soccer academies had an equal chance to take part in the study, purposive chain sampling [22] was initiated during the data collection process. Participants recommended other practitioners who met the criterion of the study and who might be willing to take part.

2.3. Procedure

Following ethical approval from an institutional research committee, an internet search using terms including "EPPP", "soccer academies", and "English soccer" was undertaken to identify all elite English soccer clubs with an academy holding EPPP status. Clubs which held the contact details for those working in the psychology department at each of the academies were listed, and clubs that did not display these contact details were removed. Potential participants were then contacted and invited to be interviewed for the study through an initial email. If a response was not received, follow-up contact was attempted, through email, two weeks later. Those who failed to respond were excluded from the study. Those who did respond and agree to be interviewed were sent a participant information pack, which contained details regarding the study aims, as well as information regarding anonymity and confidentiality. Participants were informed that they could withdraw from the study at any time, until the point of data analysis and write up. Prior to each interview, participants were provided with the opportunity to review the participant information sheet and give further verbal consent for participation.

2.4. Data Collection

Semi-structured interviews were undertaken to allow participants freedom to clearly explain their perspective and experiences, but the direction of the interview was still guided by pre-determined questions [21,23]. The interview guide was flexible, in order to allow each participant to share their own experience of providing psychological support, yet it centered around a number of key areas of interest. For example, the initial questions ascertained demographic information such as the participants' role, their qualifications and years of experience within the field, and their experience of delivering sport psychology services to youth performers. Questions such as "Can you tell me about your career in sport psychology?" and "Can you tell me about the academy structure at your club and where you sit within it?" were used. The guide probed further in order to explore participants' view of sport psychology and its current delivery in English Academy settings. It further examined the role of the EPPP in guiding such psychological provision.

As each interview concluded, each participant was asked if there was anything that they wanted to add that had not previously been discussed; once answered, the interviewee thanked the participant and closed the interview. A number of probes, where appropriate, were used in order to elicit more depth from the participants, in order to gain a more detailed understanding of their perceptions and experiences. Interviews lasted between 60 and 90 min and were audio recorded. The interviews took place virtually, via video calls (rather than in person), due to restrictions on in person meetings due to COVID-19.

In order to ensure the trustworthiness of the data, the judgement criteria proposed by Smith and McGannon [24] were followed throughout the design and implementation of the study. It is anticipated that the sample will provide naturalistic generalizability across any other Category 1, 2, or 3 academy club that has EPPP status. However, due to the elite nature of the environment that those interviewed worked in, the study would not be transferable to grass roots or recreational soccer because the same dimensions and processes may not exist in those environments. Although qualitative research does not infer transferability per se, naturalistic generalizability of the results may be inferred through offering an insight into a variety of psychologists working across English elite soccer academies.

2.5. Data Analysis

In order to identify common themes within the data collected, in line with the interpretivist approach, thematic analysis was used as the framework of analysis [25]. Each interview was recorded and transcribed verbatim into a word document. Any irrelevant data (welcomes, inaudible audio, pleasantries) were removed in order to ensure the data analysis was meaningful and relevant to the study. As suggested by Braun and Clarke [26], a six-phase thematic analysis process was employed: (a) data familiarization and writing familiarization notes was achieved by listening to the interviews and transcribing them; (b) systematic data coding involved grouping common themes emerging from the raw data; (c) generating initial themes from coded and collated data was achieved through the codes being re-examined and patterns identified among the data; (d) developing and reviewing themes was completed when the identified themes were compared to the raw data, and looked over by the supervisory team, to ensure that they were appropriate; (e) refining, defining, and naming themes occurred once the themes were agreed upon; and, finally, (f) writing the report meant that all themes were presented and then discussed in detail. Although the process consists of six phases, each phase does not have to be isolated. According to Braun and Clarke [26], as the researcher becomes more familiar with thematic analysis, the six phases can become blended and carried out simultaneously. Following the interviews and transcription, in order to maintain anonymity, each participant was given a pseudonym. Using inductive thematic analysis, six key themes were identified. Within these key themes, raw data quotations were coded and grouped to form sub-themes.

3. Results and Discussion

From the interviews, 50 raw data codes were extracted and organized into six key themes: The breadth of sport psychology provision, what is sport psychology, the stigma surrounding sport psychology services, psychological literacy, the elite youth soccer environment, and the delivery of sport psychology under the Elite Player Performance Plan. Across these six themes, a further six sub-themes were generated which highlight additional contributing factors to two of the key themes (see Table 1).

Table 1. Themes and sub-themes described by the sport psychology practitioners.

Theme	Sub-Theme	Description
The breadth of sport psychology provision	The provision of sport psychology in academies	Psychological practice is diverse, encompassing team-based and individually-tailored support, the delivery of psychological education, psychological skills training, and counselling services, alongside clinical referral or mental health support for players.
	The application of psychological skills	There is value in both classroom and practical sessions, especially when used concurrently. Classroom sessions provide theory, while practical sessions allow for the development of skills in a controlled setting.
	Barriers to the implementation of sport psychology	Clubs need to provide sufficient resources to deliver a successful program. At present, there is an apparent lack of staffing and financial resources dedicated to sport psychology.
Understanding of sport psychology		Although sport psychology may be progressing in terms of awareness and acceptance, there is still a level of resistance to the implementation of sport psychology and the role of practitioners, linked to a narrow understanding of the breadth and scope of psychological support.
The stigma surrounding sport psychology services		There still appears to be a stigma attached to the role of psychology and sport psychologists. The role of the psychologist is still unclear to players and is often seen as a way to solve problems, rather than as a holistic aspect of player development and performance programs.
Psychological literacy		There is an awareness of sport psychology, but this remains narrow. Psychological literacy is currently lacking in academy settings, especially within core support staff, such as coaches.
The elite youth soccer environment		Academy cultures have an impact upon the delivery and reception of psychological support. It is evident that the academy soccer environment is challenging, highly competitive, and places great demands on both athletes and support staff.
Delivery of sport psychology under the EPPP	Lack of guidance	It was commonly reported that the EPPP fails to include sufficient detail to scaffold delivery of standardized provision of sport psychology within soccer academies.
	Freedom in delivery	In some cases, the lack of detail and specificity of the EPPP was viewed as beneficial, because it gives psychology practitioners greater flexibility and ownership over their delivery of sport psychology, allowing them to better cater to the needs of the academy and players.
	Room for development	Further development of the EPPP guidance is required, with greater input from experts in the field, in order to ensure the guidance is relevant and detailed.

3.1. *The Breadth of Sport Psychology Provision*

The initial theme which arose from participants was the breadth of sport psychology provision. This included three sub-themes: (a) the provision of sport psychology in academies, (b) the application of psychological skills, and (c) the barriers to the implementation of sport psychology. The current theme relates to the content that is being delivered to academy players and the way in which that is happening, as well as discussing the barriers that have the biggest effect on this.

With regard to the provision of sport psychology across the six academies, it was apparent that the levels of provision varied greatly. When discussing differences between academy soccer and other youth sports, participants often discussed individual sports rather than other team sports. This is because there appeared to be more obvious differences in the provision of sport psychology in individual sports, and it was a common assumption that individual sports were significantly ahead of academy soccer in the provision of sport psychology.

3.1.1. The Provision of Sport Psychology in Academies

When discussing the provision of sport psychology, it is important to articulate the meaning of provision. For the purpose of the study, provision is referred to as the amount of sport psychology services that is provided to the players. This includes the form of delivery that is used, how much time is dedicated to delivering sport psychology, and the content that is covered within the sport psychology program.

Samantha, who at the time of the study, had worked in the setting for eight years, explained:

> 9–11 [the age group], we do half and half, so half out on the pitch and half in the classroom. With the 12 s it's again a mixture, so it's half out on the pitch and half in the classroom. With the 13 s and 14 s at the moment it's all classroom-based, just because of resources. 15 s, 16 s, again it's a mixture, because we've got one member of staff who mainly focuses on those two, so it's mainly in the classroom and out on the pitch. 18 s, their two and a half sessions a week is, err, classroom based.

Similarly, Tony, who has been at his current club for three seasons, described the way the sport psychology program is delivered to different age groups within an academy:

> For the younger 9 s and 10 s, I'd run a monthly workshop, where I get them in weekly or bi-weekly. For the other age groups going up, where something I like to do is, and they have field focus as well. So, I don't want psych to be in the class at all times, so I end up on the field. ... So, I have, uh, about four periods where I'll run through a couple of like challenges or a couple of skills and get all the players involved, with the main objective being a psychological one. Not looking at a textbook, although that's working them as well, but they're still working on their soccer. But the main objective would be that, working on communication, for example, or confidence when they want all the players to be the most confident player in the world, kind of thing. It seemed to buy into that a lot more than when it's in a classroom. And then I say one-to-ones would be more for 18–23 s.

The way sport psychology is delivered appears to vary from academy to academy, and even from age group to age group. Dave explained that their sport psychology program had been designed with an over-arching framework in mind, thus evidencing more of a structural approach to delivery:

> So, we have our kind of, I suppose, bespoke profile, if you want to call it that, that we use to identify key characteristics we believe are important to make it as a first team player within our club. So, we use that as our over-arching framework, and then from that we kind of filter it in, into learning plans, into targets, into goals, based on strengths and development areas as well.

Harry, who has been working as a practitioner in soccer academies since 2012, went further to explain that their sport psychology program had been designed to be a key part of the academy's environment, taking a more holistic approach:

> When it's performance-related, and essentially, we do a series of things, we do workshops with players, with staff, we do individual support, we try and influence kind of the environment in some ways, so some of the projects we work on are more ... kind of, systemic, so we try and influence the ... I don't want to

> go as far as to say the culture, that's a bit extreme, but to kind of influence the staff to then influence the players, so we have an indirect impact, so we approach it in different ways, there's lots of different strategies that, but yeah, some of the main things we use would be workshops, interactive workshops, player support, and yeah, kind of just creating, hopefully an environment whereby psychology is talked about and utilized as much as possible.

It was apparent from the interviews that the practice of psychology is diverse, encompassing team-based and individually tailored support, the delivery of psychological education, psychological skills training, and counselling services, alongside clinical referral or mental health support for players.

3.1.2. The Application of Psychological Skills

Although sport psychology is often delivered through classroom-based workshops, an increasing number of academies are now focusing on integrating sport psychology during on-pitch sessions. Therefore, although the amount of official time allocated to sport psychology may lag behind other sport science disciplines, there are novel developments with regard to how sport psychology is being delivered. As Karen noted:

> When we are out there, and we are talking, and we are doing things and then we might reflect afterwards and just try to attach anything to the actual game and making it as engaging and as real as physically possible.

Although there is an increasing focus on delivering sport psychology during on-pitch sessions, Dave explained the benefit of retaining classroom-based sessions. Dave stated that the academy players are educated regarding psychological skills and strategies, and then given the opportunity to practice them within a classroom setting before transferring them into training and game situations:

> What we try to do is practice the techniques in the classroom. We might do different games, might do, err, just a drive through of the technique, then we'll actually say ok when are these situations or what situations will arise where we might be able to use this technique? And then we go practice it for two or three weeks in training, and they can start to manipulate and practice it in games as well.

From these findings, it is evident that there is value in both classroom and practical sessions, especially when used concurrently; the classroom sessions provide a chance for information and techniques to be explained in detail, while the practical sessions allow for these techniques to be practiced within a controlled environment that can then be manipulated to reflect real match scenarios.

3.1.3. Barriers to the Implementation of Sport Psychology

Barriers towards the use of sport psychology were reported by all participants, thus evidencing that these barriers are not just isolated to one club or category status, but instead are relevant across English soccer academies. Multiple barriers were reported including financial resources, staffing resources, and the relatively slow uptake of sport psychology.

There was evidence in support of Pain and Harwood's [18] assertion that a lack of financial resources was the most commonly perceived barrier to the implementation of sport psychology. As Dave explained:

> So, I think probably limited resources would be the one area ... I think a lot of it, in my, probably being controversial here, it's down to funding a little bit, because if you do have the money you can buy better facilities, you can go abroad and potentially invite players—some clubs don't have that luxury.

This was similarly noted by Harry:

> I have experienced resources as a barrier. You know, you don't need loads of money to implement psychology, I'm not saying that. But to do things in a certain

> way, having kind of funding backing, even if it's to get things designed nicely or in a professional way.

As well as financial resources proving to be a barrier, it was reported that time and staff resources also had an impact on the delivery of sport psychology. Samantha stated, "With the 13 s and 14 s at the moment it's all classroom-based, just because of resources. We haven't got the time or the resources to be able to implement it out on the pitch all the time and with them." In the same vein, Harry explained:

> I think that the resources you have can sometimes dictate the delivery you can give. So, my previous experience was of me being alone as kind of being the single practitioner within that environment, so my time was spread a lot thinner, and so my strategy and approach had to be a lot different. I think with more staff and resources, you can do more with the players, because that's when you can have a direct impact.

A number of participants reported that sport psychology still lagged behind other sport science disciplines in its level of integration into academy soccer development programs. As Lily, who at the time of the study had worked in the environment for seven years, stated, "Sport psychology is still late to the table, and we are still not there yet." However, with the introduction of the EPPP, academies are now required to provide psychological support. Nonetheless, Lily did not credit the EPPP initiative for increasing such support, leaving questions regarding its impact upon the integration of sport psychology services into academy development programs. Samantha did, however, note a greater integration of sport psychology within soccer academies:

> Probably over the last 18 months or so, maybe a little bit longer, psychology has been very much more integrated. So we are in the audit meetings, we are in the MDT [Multi-Disciplinary Team] meetings, we have our say, we are now working with the analysis team looking at the psychological behavior analysis and linking that up with the analysis team and clipping that footage so it can go into the prebrief and debrief. A lot of the things that coaches and the MDT are saying is psychological language ... a lot of their CPD [Continuing Professional Development] incorporates psychology.

Samantha then added:

> So, we are kind of riding the wave with that. I think we're probably where sports science was five years ago, where S&C [Strength and Conditioning] was a bit of a taboo subject, but it's on the horizon and it's kind of on the up-and-coming, and I think psychology is the same at the minute. It's still a bit of a taboo subject, but give it two or three years we'll be up and coming, we'll be forefront just like S&C.

Despite the latter quote concerning one specific club, it offers some evidence that the integration of sport psychology within academy soccer is progressing.

These results suggest that, in order for sport psychology to thrive within academy soccer, it still needs to take greater precedence, and to do so a lack of staffing and financial resources must be addressed. That is, clubs need to provide sufficient resources to deliver a successful program. That said, improving resources alone will not solve the issue of sport psychology integration—there also needs to be a greater level of acceptance from coaches and players in order to minimize the apparent resistance against sport psychology.

3.2. Understanding of Sport Psychology

According to Konter et al. [2], soccer psychology is multi-faceted, comprising social psychology, developmental psychology, clinical psychology, health psychology, and the psychology of coaching. Exposure to such a variety of psychological sub-disciplines may influence the way in which practitioners interpret the purpose and meaning of sport psychology. This theme's key finding mirrors historical debates in the literature [27], in that, while being concerned with the psychological strategies and techniques to improve performance, sport psychology is also concerned with players' overall wellbeing. As Lily stated, "For me, sport psychology, performance psychology, is like the interconnection of wellbeing and performance". Similarly, Harry stated:

> So, for me, sport psychology is about educating and supporting athletes and staff within a sporting environment. So, it's about informing and upskilling them in psychology principles. It's about, it's about educating them in an appropriate way around what will impact their performance and what will impact the performance of the athletes they work with ... and it's about supporting appropriately so whether that's supporting performance enhancement, and I also think it includes supporting wellbeing as well. So, I think both are really important, and I know different practitioners have different philosophies on that. For me they're both completely linked, they're both intertwined, performance and wellbeing. You can't have one without the other. So, for me, both of those are really relevant to the role of a sport psych.

This quote thus underpins that, although sport psychology is of course concerned with athletes' performance, its focus may also be taking a more holistic wellbeing approach [28,29]. Indeed, Sara noted that although sport psychology is concerned with enhancing performance, it is also about understanding athlete mental health:

> Sport psychology is just looking at the mental aspect of their performance. ... Sport psychology is not purely about how their mental side affects their sort of training and their performance on the pitch but also the wider aspects around the mental health, just like their daily wellbeing as well. It kinda (sic) just incorporates everything that they [the athletes] are. ... The sport psychology will look more at the performance aspects, so looking at kind of like peak performance, optimal performance and those kinds of things.

The findings in this theme support the work of Moore and Bonagura [30], who stated that sport psychology is not solely for the purpose of performance enhancement, but instead seeks to support performers, their families, and their related organizations to optimize their level of functioning in a multitude of areas. Thus, while practitioners might think deeply about what psychology means to them [31], discussion then moved on to the level of psychological literacy among players and staff and the stigma that is still associated with the discipline. This demonstrates that, while sport psychology may be progressing in terms of awareness and acceptance, there is still a level of resistance from non-practitioners in relation to the implementation of sport psychology and the role of practitioners.

3.3. *The Stigma Surrounding Sport Psychology Services*

Participants referred to a stigma associated with the use of sport psychology and the role of the practitioner that still persists in academy soccer. This supports the findings of Champ et al. [10], who stated that sport psychology is not always received positively by everyone. Lily noted experiences of such stigma: "They [Premier League personnel] said, well a sport psych is somebody that a player goes to for twenty minutes to be forced to see and then lies to them for twenty minutes and then leaves." This negative attitude that players are "forced" to see a psychology practitioner and then "lie" suggests that players may still feel the need to avoid psychological support and would be reluctant to be honest when with the psychology practitioner. This showcases that players either believe that there is little value in speaking to a psychology expert, or they do not want the psychology practitioner to know the truth about how they are really feeling. Regardless, this demonstrates that a level of stigma persists around seeking psychological support. Such stigma was deemed to not just be present with academy players but also higher up in the professional soccer environment. Lily then explained:

> I still think that the reason there aren't as many sport psychologists is because it is still the unknown ... so [coaches and players] will say we really value it, we really think it's important, and we really buy into it, and we talk about it all the

> time and, you know. . . . However, we still are quite scared about what we do with it.

The above quote demonstrates that although coaches and players recognize the value of sport psychology, there remains a level of resistance to its adoption in soccer academies [6,7]. Harry reported that he still experiences a level of reluctance towards sport psychology: "I think psych is always that one that's got a bit of, I guess stigma's a word you could use for it, so there have [sic] been in my experience some reluctance." Additionally, Samantha highlighted that players are still exposed to the stigma that seeking psychological support means that there is a problem, rather than the notion that the use of psychology can lead to performance gains:

> So even though all the players know psychology is part of it [performance development] and all of that, you've still got that stigma behind that, go and see the psychologist because you've got an issue, and I think that's still potentially what it is, rather than know that we are performance psychs (sic).

Participants noted that because players spend most of their time with their coach(es), these coaches have the potential to have more influence on the players than a sport psychologist. Sara stated, "It [the integration of sport psychology into training sessions] comes down to the coaches' own philosophy." Thus, if coaches do not consider sport psychology to be valuable, they may not include it. Karen provided insight as to why some coaches may be less likely to incorporate sport psychology into their training sessions: "Maybe sometimes you get the sort of coaches, who are typically in the professional development phase, so 18 s, the 23 s, seem to be ex-professionals, they are the ones that tend to be less open to change." Sara noted: "Although you know what you should be doing, if the coach doesn't then allow you to do it, it becomes like a little bit of conflict between what you know you should be doing versus what you're actually doing".

Our findings show how there still appears to be a stigma attached to the role of psychology and working with a sport psychologist. As Sara stated, "They still see it as they go to the shrink. . . . You go and see them when you've got a problem". Additionally, it seems that the role of sport psychology is generally unclear to players. As Konter et al. [2] noted, sport psychology is not perceived as a regular performance aspect of development, but rather more often as a strategy for overcoming problems and helping the "problem athlete".

3.4. *Psychological Literacy*

Psychological literacy was noted by the participants as the level of understanding a person has of sport psychology and their ability to apply that understanding. In this case, discussing the current level of psychological literacy participants believed coaches and players had, Lily stated, "Actually, the awareness and psychological awareness and literacy, I don't think it's, I don't think it's where it could be. I think it could be improved, hundred percent, on what it is". Although Sara suggested that coaches' psychological literacy is poor, she had a different stance on where the problem lies: "So, their [coaches'] awareness of it [sport psychology] is high, their understanding of it is low . . . but if you ask them, they would say they have a very high understanding of it. So, I don't think they realize quite how complex it is". As this quote demonstrates, it would appear that coaches have a high level of awareness of sport psychology, in that they are exposed to sport psychology within academy soccer. However, as Sara stated, these same coaches may also mistakenly believe they have a high understanding of sport psychology. This highlights the value of coach education and workshops and the need to educate coaches about the fundamentals of sport psychology, how to use it, and the benefits that surround it [32]. This in turn could provide coaches with a greater understanding of sport psychology and the tools to successfully incorporate it into their coaching sessions.

In a similar vein, Karen explained a lack of psychological literacy in simpler terms: "I think there will be some staff where it's quite new and the topics are not gonna (sic) be particularly familiar . . . they might have heard of the terms but not know what they mean and what it looks like in practice". This is an important consideration that is supported by Pain and Harwood [18], who stated that those coaches with a low knowledge base and understanding of sport psychology are ill-equipped to deliver such skills within their training sessions. As recognized by Murdoch [20], this has the potential to be problematic, because inappropriate (and incorrect) information and skills can be passed onto the athletes, which can even be detrimental for their performance and wellbeing. The present study's findings demonstrate an awareness of sport psychology but also a lack of understanding among coaches in relation to the application of sport psychology within their coaching practice. As will be discussed in the later sub-theme "Freedom in delivery", coach and player understanding of sport psychology may be affected by the varied psychological programs employed across club academies, meaning there is no standardization in exposure of coaches and players to sport psychology.

3.5. The Elite Youth Soccer Environment

As it is in sport in general [33], elite youth soccer is characterized by a highly pressurized climate for success [34]. This characteristic remains true of the English soccer academies highlighted in the current research, with a primary focus on results, potentially at the expense of player development.

Participants noted that although the development of players is important, the focus of academy soccer remains on results and winning. This is evidenced by Sara:

> I think the reality of it is that it's still very competitively orientated. It's still very ego oriented, it's still very much a win culture. Although they are looking to develop the players, um, it's not purely developmental orientated. Whilst they are developing the players, I think the winning aspect is still the primary focus.

Although Sara stated that there is consideration of player development within the academies, the fact that the primary focus is on results may be potentially problematic, because this may mean that other factors, such as player wellbeing, may be neglected. In a similar vein to Sara, Lily stated that the level of competitiveness is related to the idea of academies operating as businesses.

> It's much more of a business than I gave it credit for when I started, there's a massive element of, you know, is it gonna (sic) make us money, is it gonna (sic), is he gonna (sic) go to our first team, are we gonna (sic) be able to sell him for a profit, much more so than I probably wanted to believe when I started.

This statement echoes previous work noting that soccer clubs maintain a primary focus on achieving the best possible results and winning each game [35].

In a similar vein, Harry stated the importance of understanding that some staff and players may find the academy soccer environment significantly challenging:

> It can be a really difficult environment as well. I think there's, it's, it would be remiss to ignore that some people really struggle with this environment. . . . There's a lot of pressure, whether it's from themselves [academy players], whether it's from their parents, whether it's from other people involved in their journey, to kind of be a certain way, to perform a certain way, and for some players, that's not for everyone. . . . It's a high expectation environment that's not for everyone, and again some staff thrive on it, some staff don't. So I think for both it's one of those that can go either way.

This notion is further supported by Tony, who explained the difficulty that staff working within the environment encounter:

> The environment, soccer generally, as in the organization, is a very cut-throat environment. It's yeah, it's not too forgiving. But at the same time, within your organization everyone's quite close. Like when you have a lot of, yeah, relationships between people can be really good. But then at the same time, if

> those people don't do well, then potentially you could lose your job. You know. Which might have no relation to you. So that's part of the environment. I think it can be quite hard I think for the coaches as well, all staff, not just coaches.

It is evident that the academy soccer environment is challenging, highly competitive, and places considerable demands on both athletes and support staff [36].

3.6. Delivery of Sport Psychology under the EPPP

When conducting research around sport psychology within English soccer academies, it is important to discuss the presence of the EPPP, because, as the overarching framework for player development in English academies, it mandates the minimum requirements for psychology provision. When discussing the EPPP and the delivery of sport psychology within soccer academies, three sub-themes emerged: (a) lack of guidance; (b) freedom in delivery; and (c) room for development.

3.6.1. Lack of Guidance

Participants expressed their sense of the inadequacy of the EPPP as a guidance framework for standardized provision of sport psychology within soccer academies. Many of the participants shared concerns surrounding the EPPP as a guiding document, given its lack of (age appropriate) guidelines. As Champ et al. [36] purported, to date the EPPP has not yet achieved what it originally set out to in terms of guiding psychological practice in academies. This is reflected in the following extract from Sara:

> I guess, it's [the psychology section of the EPPP] vastly inadequate, in terms of when we get audited and I look at the one-page Premier League document, whereby it says, have you seen it, you've seen it right? 'phychology' (sic) ... do I need to say any more?

This was supported by Samantha, "Within the audit process, they spelt psychology wrong, so I mean they obviously don't have a great understanding. They want to include it, but I don't think they have a full understanding of it". Lily then continued to explain why she believed the psychology section of the EPPP to be inadequate:

> It was written by a sport scientist, so the language and everything like that, in terms of testing is very physically driven, and all they've done is they've picked up the sport science kind of language, if you like, and they've dumped it onto a page that's 'psychology', and they can't even spell psychology right, so let's not go there.

The above quote suggests a lack of expertise in the creation of the psychology section of the EPPP. All of our participants suggested that the EPPP currently fails to include sufficient detail to scaffold the delivery of standardized provision of sport psychology across academy settings.

3.6.2. Freedom in Delivery

Alongside criticisms of the EPPP as a guiding document, some participants reframed this more positively, because it gave them freedom in their delivery of sport psychology. As Karen noted:

> I think that the psychology section of it [EPPP] is one of the most vague ... and I think that might frustrate people, but I think it might also be a good thing. ... I think the pure performance psychology section is sort of one page, so it gives us a little bit of a framework and the minimal thing we've got to do, but how we do it is entirely up to us.

Lily similarly stated, "There was a lot of freedom given to clubs on how they wanted to do it, so that is based on the philosophy of the club and resources, of course, and funding and where they want to do things". The freedom noted by Karen and Lily might be problematic, if all clubs deliver psychology differently, without any form of standardization.

However, Karen further stated that such an approach may be preferable than a stricter set of guidelines: "I think if it [EPPP] had a lot more detail and was a lot more prescript (sic), I think you'd lose that autonomy as a practitioner, and I think the role would be a lot less enjoyable because you'd feel like you were ticking boxes all the time".

Although it was noted that the EPPP lacked detail and specificity, in some cases, this was viewed as beneficial, because this gave psychology practitioners greater flexibility and ownership over their delivery of sport psychology, allowing them to better cater to the needs of the academy and players. In some ways this mirrors the concept noted by others (e.g., see Daley et al. [37]), that although psychological models and frameworks can be useful as a starting point for applied practice, practitioners should indeed tailor such content to the context in which they are working. That said, this lack of detail and specificity may also lead to a lack of standardization across academies in terms of the psychological content and support that players receive, and thus our results suggested there was still room for development regarding the EPPP and its sport psychology guidance.

3.6.3. Room for Development

Participants noted that implementing the EPPP guidelines could become a "tick box exercise". As Lily stated, "The first thing that always springs to my mind [in relation to the EPPP], rightly or wrongly, is tick box". Samantha noted that this can lead to academies not going further in their provision of psychology in their programs: "Clubs are then using it [EPPP] as a tick box exercise, so they never fully integrate it themselves". Dave did not feel constrained by the guidelines, however, stating:

> Well, it is embedded, but I wouldn't go in to a one-to-one with a player or a workshop thinking, 'oh is this going to be ok for EPPP?' Uh, I think I kind of have faith in the program that we have meets EPPP requirements, rather than trying to make it fit in to the EPPP.

Conversely, Sara stated, "So, I would say it's given us a little bit more structure, but in terms of what we deliver and how we deliver it, it has no impact whatsoever".

Overall then, these findings suggest that the guidance for sport psychology within the EPPP is considered somewhat sub-optimal, but that the lack of detail and specificity may not necessarily be a negative. Indeed, with this lack of detail and specificity, sport psychology practitioners may have more freedom to design their psychology program to best meet the needs of their academy. However, even with the above said, according to this study's participants, at least, further development of the EPPP guidance appears warranted, in particular with greater input from experts in psychology.

4. Conclusions

In addressing the purpose of our study, we sought to examine the experiences of sport psychology practitioners working within elite soccer academies in England. Across the six themes presented, our findings highlight the rich and diverse nature of psychological service delivery within these settings. Psychological practice encompassed team-based and individually tailored support, the delivery of psychological education, psychological skills training, and counselling services, alongside clinical referral or mental health support for players. Practitioners highlighted their responsibility for both performance enhancement and player welfare—a critical shift in service delivery. Ecological frames of reference were adopted, whereby participants acknowledged that while they worked with athletes this happened within academy environments with cultures and norms that have an impact on their practice and the behavior(s) of the athletes with whom they work. As such, participants recognized how, in modern settings, psychology practitioners are responsible for supporting wider cultural development in clubs and academies, alongside education of support staff and coaches. It is evident that practitioners working in soccer academy settings do so in dynamic environments—psychological practice is varied and ever evolving, and there remains no single way of being a psychologist in this setting [38].

The results demonstrated that, although the provision of sport psychology is improving and continuing to develop, there remain a number of factors that inhibit a full integration of the discipline into the academy development program. The most prominent factors identified in this study were the level of psychological literacy among coaches, and the attention and resources dedicated to the discipline. That is, in order for the integration of sport psychology into the wider development program to be successful, those who are responsible for aiding the promotion and delivery of psychological content require a greater level of understanding than is currently present. In line with the findings of Murdoch [20], our study found that the level of psychological understanding among coaches appeared to be lower than the coaches perceived it to be. This is potentially problematic due to the level of impact and influence that coaches have on players [39], and the fact that they are such a key source of support [19]. If, due to a lack of psychological literacy, coaches are promoting sub-optimal psychological advice, then players may be at risk of developing poor psychological habits. Thus, in order for sport psychology to be successfully and safely provided to academy soccer players, practitioners should be employed to deliver such content, but further should have a significant role in the education of coaches and other support staff about sport psychology and the importance of delivering sound psychological advice [40]. Additionally, from the findings, it is evident that there is still a stigma attached to the discipline of sport psychology. This stigmatized thinking and the fear of the unknown is still causing a level of resistance from some coaches and players when seeking psychological support [41]. Therefore, a recommendation of this study would be to provide education to players and coaches around sport psychology and the way in which it can benefit performance, in order to help combat such thinking.

As well as examining the general experiences of sport psychology practitioners, we also sought to explore participant experiences of working under the auspices of the EPPP—the English Premier League's guiding document concerning player development—of which psychology is a core component of sport science support. Theme six specifically reflected this experience. Although participants acknowledged that, since its inception, the guidance for sport psychology delivery had been improved, they felt that it still lagged behind other sport science disciplines in depth and/or quality of the syllabus, thereby limiting its potential to guide the delivery of psychological services. A key suggestion echoed by all participants was that in order to improve the integration of sport psychology services within academy soccer, the psychological guidance within the EPPP would benefit from being significantly revised. The evidence from the present study suggests that the EPPP documentation, in its current format, provides inadequate guidance in relation to the content that should be delivered and how it should be delivered. Such revision of the EPPP guidelines would help to extend the current minimum standards of psychological provision and would also create a level of uniformity across academies within each category, thereby increasing the focus on developing psychological literacy in players. Along these lines, participants noted their preference for the EPPP guidance for sport psychology to be revised and re-written by leading practitioners who are familiar with the discipline and who have experience of working within academy settings. This would enable the adoption of correct terminology as well as providing clear guidance with a contextual understanding of the demands of such settings. By way of example, further guidance with regard to employing a holistic, player-centered approach, which focuses on well-being as well as performance, would be important. In addition, the provision of one-to-one support alongside more general educational sessions, and on pitch-based sessions, with monitoring of player psychological development, would be important [36,42]. Increasing guidance with regard to the emotional and cognitive maturation of players, alongside an understanding of adolescent psychological development, would provide the opportunity for psychological service delivery that meets the changing demands of youth performers across critical stages of their (player) lifespan. The EPPP, with revision, thus has potential to encourage service provision that genuinely supports youth players within and beyond soccer environments.

The results of this study could aid in the future refinement of the psychological elements of the EPPP. Indeed, a strength of this study is that data were collected from seven practitioners, working across a number of different academies, with representation of both male and female participants. Thus, the results from this study should transfer to any academy with EPPP Category 1 or 2 status, or indeed to Category 3 academies attempting to reach Category 2 status. With the above said, a limitation is that a number of practitioners whom we approached either (a) did not respond to our invitation or (b) did respond but did not participate. Thus, our sample was to an extent self-selected, such that our results may simply reflect those more willing to speak out about their experiences [43]. Overall, this study has helped to deepen our understanding of the role of practitioners responsible for the delivery of sport psychology within English academy soccer. Future research would do well to also examine perceptions from a range of staff working within English soccer academies, in order to assess their perception of the level of provision and understanding of psychology.

Author Contributions: Conceptualization, F.D., T.R., A.W. and E.K.; methodology, F.D., T.R., A.W. and E.K.; formal analysis, F.D.; investigation, F.D.; data curation, F.D.; writing—original draft preparation, F.D. and T.R.; writing—review and editing, F.D., T.R., E.K. and A.W.; supervision, T.R., A.W. and E.K.; project administration, T.R. and A.W.; funding acquisition, T.R. and A.W. All authors have read and agreed to the published version of the manuscript.

Funding: This work was supported by funding awarded to the first author by AFC Bournemouth and Bournemouth University.

Institutional Review Board Statement: The study was conducted in accordance with the Declaration of Helsinki, and approved by the Institutional Review Board (or Ethics Committee) of Bournemouth University (protocol code 27313, 30/07/2019).

Informed Consent Statement: Informed consent was obtained from all participants involved in the study.

Data Availability Statement: The authors confirm that the data supporting the findings of this study are available within the article.

Conflicts of Interest: The authors declare no conflict of interest.

References

1. Heidari, J.; Pelka, M.; Beckmann, J.; Kellmann, M. A practitioner's perspective on psychological issues in soccer. *Sci. Med. Soccer* **2018**, *3*, 169–175. [CrossRef]
2. Konter, E.; Beckmann, J.; Loughead, T. *Football Psychology: From Theory to Practice*; Routledge: Oxon, UK, 2019.
3. Barker, S.; Winter, S. The practice of sport psychology: A youth coaches' perspective. *Int. J. Sports Sci. Coach.* **2014**, *9*, 379–392. [CrossRef]
4. Hings, R.; Wagstaff, C.; Anderson, V.; Gilmore, S.; Thelwell, R. Better preparing sports psychologists for the demands of applied practice: The emotional labor training gap. *J. Appl. Sport Psychol.* **2019**, *32*, 335–356. [CrossRef]
5. Sly, D.; Mellalieu, S.; Wagstaff, C. "It's psychology Jim, but not as we know it!": The changing face of applied sport psychology. *Sport Exerc. Perform. Psychol.* **2020**, *9*, 87–101. [CrossRef]
6. Kremer, J.; Moran, A. *Pure Sport: Practical Sport Psychology*, 2nd ed.; Taylor and Francis: Hove, UK, 2013.
7. Ong, N.; Harwood, C. Attitudes toward sport psychology consulting in athletes: Understanding the role of culture and personality. *Sport Exerc. Perform. Psychol.* **2018**, *7*, 46–59. [CrossRef]
8. Gledhill, A.; Harwood, C.; Forsdyke, D. Psychosocial factors associated with talent development in football: A systematic review. *Psychol. Sport Exerc.* **2017**, *31*, 93–112. [CrossRef]
9. Sarmento, H.; Anguera, M.; Pereira, A.; Araújo, D. Talent identification and development in male football: A systematic review. *Sports Med.* **2018**, *48*, 907–931. [CrossRef]
10. Champ, F.; Nesti, M.; Ronkainen, N.; Tod, D.; Littlewood, M. An exploration of the experiences of elite youth soccer: The impact of organizational culture. *J. Appl. Sport Psychol.* **2018**, *32*, 146–167. [CrossRef]
11. Horrocks, D.; McKenna, J.; Whitehead, A.; Taylor, P.; Morley, A.; Lawrence, I. Preparation, structured deliberate practice and decision making in elite level football: The case study of Gary Neville (Manchester United FC and England). *Int. J. Sports Sci. Coach.* **2016**, *11*, 673–682. [CrossRef]
12. Premier League. *Elite Player Performance Plan—Goal Reports*. goalreports.com. 2011. Available online: https://www.goalreports.com/EPLPlan.pdf (accessed on 2 September 2021).

13. Pain, M. Mental interventions. In *Soccer Science*, 1st ed.; Strudwick, T., Ed.; Human Kinetics: Champaign, IL, USA, 2016; pp. 389–413.
14. Godfrey, R.; Winter, S. Winning Mentality: A Reflective Account of Delivery to a Professional Soccer Academy. *Case Stud. Sport Exerc. Psychol.* **2017**, *1*, 63–75. [CrossRef]
15. Harwood, C.G.; Barker, J.B.; Anderson, R. Psychosocial development in youth soccer players: Assessing the effectiveness of the 5Cs intervention program. *Sport Psychol.* **2015**, *29*, 319–334. [CrossRef]
16. Lerner, R.; Fisher, C.; Weinberg, R. Toward a science for and of the people: Promoting civil society through the application of developmental science. *Child Dev.* **2000**, *71*, 11–20. [CrossRef]
17. Steptoe, K.; King, T.; Harwood, C. The consistent psycho-social development of young footballers. In *Football Psychology: From Theory to Practice*; Beckmann, J., Loughead, T., Eds.; Routledge: Oxon, UK, 2019; pp. 61–73.
18. Pain, M.; Harwood, C. Knowledge and perceptions of sport psychology within English soccer. *J. Sports Sci.* **2004**, *22*, 813–826. [CrossRef]
19. Rees, T.; Freeman, P.; Bell, S.; Bunney, R. Three generalizability studies of the components of perceived coach support. *J. Sport Exerc. Psychol.* **2012**, *34*, 238–251. [CrossRef]
20. Murdoch, D. Psychological literacy: Proceed with caution, construction ahead. *Psychol. Res. Behav. Manag.* **2016**, *9*, 189–199. [CrossRef]
21. Hennink, M.; Hutter, I.; Bailey, A. *Qualitative Research Methods*; Sage: Riverside County, CA, USA, 2020.
22. Patton, M. *Qualitative Research and Evaluation Methods*, 2nd ed.; Sage: Riverside County, CA, USA, 1990.
23. Jones, I.; Gratton, C. *Research Methods for Sports Studies*, 3rd ed.; Routledge: Oxon, UK, 2015.
24. Smith, B.; McGannon, K.R. Developing rigor in qualitative research: Problems and opportunities within Sport and Exercise Psychology. *Int. Rev. Sport Exerc. Psychol.* **2017**, *11*, 101–121. [CrossRef]
25. Herzog, C.; Handke, C.; Hitters, E. Analyzing talk and text II: Thematic analysis. In *The Palgrave Handbook of Methods for Media Policy Research*, 1st ed.; Van den Bulck, H., Puppis, M., Donders, K., Van Audenhove, L., Eds.; Palgrave Macmillan: Cham, Switzerland, 2019; pp. 385–401.
26. Braun, V.; Clarke, V. One size fits all? What counts as quality practice in (reflexive) thematic analysis? *Qual. Res. Psychol.* **2020**, *18*, 328–352. [CrossRef]
27. Brady, A.; Maynard, I. Debate: At an elite level the role of a sport psychologist is entirely about performance enhancement. *Sport Exerc. Psychol. Rev.* **2010**, *6*, 59–66.
28. Bissett, J.; Kroshus, E.; Hebard, S. Determining the role of sport coaches in promoting athlete mental health: A narrative review and Delphi approach. *BMJ Open Sport Exerc. Med.* **2020**, *6*, e000676. [CrossRef]
29. Yousuf, S. The Evolving Field of Sport Psychology in Human Performance. 2019. Available online: https://leadersinsport.com/performance/shameema-yousuf-1/ (accessed on 2 September 2021).
30. Moore, Z.; Bonagura, K. Current opinion in clinical sport psychology: From athletic performance to psychological well-being. *Curr. Opin. Psychol.* **2017**, *16*, 176–179. [CrossRef]
31. Poczwardowski, A.; Sherman, C.P.; Ravizza, K. Professional philosophy in the sport psychology service delivery: Building on theory and practice. *Sport Psychol.* **2004**, *18*, 445–463. [CrossRef]
32. Villalon, C.A.; Martin, S.B. High school coaches' coaching efficacy: Relationship with sport psychology exposure and gender factors. *J. Appl. Sport Psychol.* **2019**, *32*, 64–80. [CrossRef]
33. McDougall, M.; Nesti, M.; Richardson, D. The challenges of sport psychology delivery in elite and professional sport: Reflections from experienced sport psychologists. *Sport Psychol.* **2015**, *29*, 265–277. [CrossRef]
34. Sagar, S.S.; Busch, B.K.; Jowett, S. Success and failure, fear of failure, and coping responses of adolescent academy football players. *J. Appl. Sport Psychol.* **2010**, *22*, 213–230. [CrossRef]
35. Larsen, C. Bringing a knife to a gunfight: A coherent consulting philosophy might not be enough to be effective in professional soccer. *J. Sport Psychol. Action* **2017**, *8*, 121–130. [CrossRef]
36. Champ, F.; Ronkainen, N.; Tod, D.; Eubank, A.; Littlewood, M. A tale of three seasons: A cultural sport psychology and gender performativity approach to practitioner identity and development in professional football. *Qual. Res. Sport Exerc. Health* **2020**, *13*, 847–863. [CrossRef]
37. Daley, C.; Ong, C.W.; McGregor, P. Applied psychology in academy soccer settings: A systems-led approach. In *The Psychology of Soccer*; Dixon, J.G., Barker, J., Thelwell, R.C., Mitchell, I., Eds.; Routledge: New York, NY, USA, 2020. [CrossRef]
38. Diment, G.; Henriksen, K.; Larsen, C.H. Team Denmark's sport psychology professional philosophy 2.0. *Scand. J. Sport Exerc. Psychol.* **2020**, *2*, 26–32. [CrossRef]
39. Mossman, G.J.; Robertson, C.; Williamson, B.; Cronin, L. Coaches, parents, or peers: Who has the greatest influence on sports participants' Life Skills Development? *J. Sports Sci.* **2021**, *39*, 2475–2484. [CrossRef]
40. Henriksen, K.; Larsen, C.H.; Storm, L.K.; Ryom, K. Sport psychology interventions with young athletes: The perspective of the sport psychology practitioner. *J. Clin. Sport Psychol.* **2014**, *8*, 245–260. [CrossRef]
41. Souter, G.; Lewis, R.; Serrant, L. Men, mental health and elite sport: A narrative review. *Sports Med.-Open* **2018**, *4*, 57. [CrossRef]
42. Keegan, R.; Stoljarova, S.; Kessler, L.; Jack, S. Psychological support for the talent pathway: Qualitative process evaluation of a State Sport Academy's Psychology Service. *J. Appl. Sport Psychol.* **2020**, 34. [CrossRef]
43. Sharma, G. Pros and cons of different sampling techniques. *Int. J. Appl. Res.* **2017**, *3*, 749–752. [CrossRef]

sports

MDPI

Article

The Birthplace Effect in 14–18-Year-Old Athletes Participating in Competitive Individual and Team Sports

Zohar Maayan, **Ronnie Lidor** * and **Michal Arnon**

Wingate Institute, The Academic College at Wingate, Netanya 4290200, Israel; zoharm@wincol.ac.il (Z.M.); michalar2@gmail.com (M.A.)
* Correspondence: lidor@wincol.ac.il

Abstract: The birthplace (the place where an athlete was born) effect (BPE) has been found to be one of the environmental variables associated with early talent development and the achievement of a high level of proficiency in sport. The purpose of the current study is twofold: (1) to calculate the BPE in 14–18-year-old athletes who participated in individual and team sports and (2) examine how coaches perceived this effect. The participants were 1397 athletes (390 females and 1007 males) who competed in 5 individual (gymnastics, judo, swimming, tennis, and track and field) and 5 team (basketball, soccer, team handball, volleyball, and water polo) sports, as well as 147 coaches who provided their preliminary thoughts about the BPE. Data analyses revealed that although the BPE was not found to be associated with cities of a similar size, it was observed that growing up in cities of small and medium sizes was more beneficial than growing up in towns or cities of other sizes. Most of the coaches believed that certain characteristics of the place or city where the athlete grew up (e.g., proximity to sport facilities) could contribute positively to the athlete's development. We discuss how the BPE data can aid policymakers in developing a sport policy associated with early phases of talent development.

Keywords: birthplace; individual sports; team sports; youth sport programs; sport policy

Citation: Maayan, Z.; Lidor, R.; Arnon, M. The Birthplace Effect in 14–18-Year-Old Athletes Participating in Competitive Individual and Team Sports. *Sports* **2022**, *10*, 59. https://doi.org/10.3390/sports10040059

Academic Editors: Adam Leigh Kelly, Sergio L. Jiménez Sáiz, Sara Diana Leal dos Santos and Alberto Lorenzo Calvo

Received: 22 January 2022
Accepted: 7 April 2022
Published: 11 April 2022

1. Introduction

One of the environmental factors linked to the development of expertise in sport is the place where an athlete was born (i.e., the birthplace effect (BPE)) (see, for example, [1,2]). A typical quantitative study on the BPE examines the association between the distribution of athletes' city of development (determined by birthplace or city of registration) within the sport program or system and the distribution within the general population using census data (e.g., [1–4]). The BPE was examined in a series of studies on both male and female athletes in various individual and team sports (see, for example, [3,5]).

BPE data were collected on male athletes in major sports, among them being baseball, basketball, golf, hockey [2], football [6], and junior ice hockey [7]. In addition, BPE data were obtained from female elite performers in soccer and golf [8]. The findings from these studies indicated that male and female athletes who were born in cities of small-to-medium sizes were more likely to play for professional leagues and attain a higher level of sport participation than athletes who were born in larger cities. In other words, athletes who were born in cities of a small or medium size had a greater chance of reaching the highest level in their given sports.

Support for the above-mentioned BPE findings was also found in quantitative studies using an alternative approach to examining this effect: population density. In one study conducted on male Danish team handball and soccer players, it was reported that the overall participation rates were higher among players who were born in low-density communities rather than high-density communities [9]. In another study on Portuguese volleyball players, it was observed that the birthplace population density was significantly lower for players who played for First League teams (the highest level of soccer competition in the country) than players who played for Third League (a lower level of competition) teams [10].

A number of arguments has been made to explain why it was more beneficial for young athletes to grow up in cities of a small-to-medium size than in cities of a different size. Although large cities can provide children and youths with enhanced conditions such as well-designed and equipped sporting facilities, as well as better coaching guidance, it appears that the large cities' sport programs are heavily structured around and hindered by the lack of space and time in which young athletes can participate [11,12]. The physical environments of smaller cities or towns allow for instructional and social benefits that cannot be gained in larger cities, among them being (1) a greater amount of independent mobility and physical safety, (2) an integrative approach to sport participation involving schools, families, and the community at large, and (3) a more personal relationship between athletes and coaches. As noted by MacDonald, King, and their colleagues [8], " ... the developmental opportunities for nurturing sporting talent offered by small towns and cities may be somehow superior to the development opportunities of larger cities" (p. 234).

Similar explanations for the benefits athletes can gain in cities of small-to-medium sizes were also provided in a number of qualitative studies [13,14]. In a case study, Balish and Côté [13] performed interviews and analyzed documents in order to explore how a small and successful sport community (646 residents) contributed to the development of the local athletes. Three themes emerged from this study: (1) developmental experiences concerning youths engaging in organized and unorganized sport activities where teammates remained stable throughout development, (2) community influences concerning the interdependence between the local schools and the community, and (3) sociocultural influences concerning youths who possess a collective identity coupled with an intense inter-community rivalry.

Of particular interest to the aims of the current BPE study are two studies in which BPE data were collected on elite male and female Israeli athletes. In one study [15], data were obtained from 521 male ball game players who played in Division 1 (the highest division for competitive ball games in Israel). Mixed findings for the BPE across sports were found. For the soccer players, it was indicated that the likelihood of players who were born in very small communities or a small city to play for teams in Division 1 was lower than for players who were born in a city of a different size. In addition, the likelihood of soccer players who were born in a medium-sized city to play in Division 1 was higher than that for players who were born in a town or city of another size.

Growing up in a city of a medium size was also found to be advantageous for team handball players. The likelihood of players who were born in a medium city to play in Division 1 was higher than that for players who were born in a city of a different size. For the volleyball players, it was indicated that the likelihood of players who were born in a small place (<2000 people) to play in Division 1 was higher than that for players who were born in a city of a different size. The main finding of Lidor et al.'s [15] study revealed that the likelihood of reaching the highest level of competition in team sports, such as soccer and team handball, was higher among those who were born in cities of a medium size than among those who were born in cities or towns of a different size.

In another study on female elite athletes [16], BPE data were collected on 389 ball game individuals. The main findings of this study were that for the team handball players, the likelihood for players who were born in a medium-sized city to play in Division 1 was higher than that for players who were born in a city of a different size. In addition, for the volleyball players, it was found that the likelihood of players who were born in a small place (<2000 people) to play in Division 1 was higher than that for players who were born in a place of a different size. These volleyball data were similar to the volleyball data obtained in Lidor et al.'s [15] study. Finally, for the basketball players, team handball players, and volleyball players, the likelihood of players who were born in a small city to play for teams in Division 1 was lower than that for players who were born in a city of a larger size.

Our main aim in the current study was to assess the BPE in young Israeli athletes (ages: 14–18 years) who participated in competitive sport programs. We selected 14–18-year-old athletes because this age category represents two important periods in the early careers of young competitive athletes. According to Côté's *Developmental Model of Sport Participation* (DMSP) (see [2,17]), the ages of 13–15 are considered to be the turning point from the developmental years to the specialization years. The age of 18 years is considered to represent the beginning of a transitional phase from being part of a youth sport program to becoming a member of an elite adult sport program in most programs in Israel [15,18].

It has already been argued that evidence-based quantitative and qualitative BPE data can assist those professionals who work with young athletes, such as policymakers, program directors, coaches, instructors, and sport psychology consultants, in developing sport policies that focus on positive and enjoyable experiences for children in the early phases of talent development (see [5]). In the current BPE study, we collected data on children and youths who had been part of competitive sport programs for a number of years but had not yet reached the highest level of competition in the local sport structure. Since Israel is considered to be a small nation (see [19]). where only a small portion of the children and youths are involved in competitive individual and team sport programs [18,20], evidence-based data on the size of the place of birth of the young athlete can help those professionals who work regularly with them to increase their understanding of how to recruit children to sports, as well as how to motivate them to maintain their participation in the selected sport program(s) for longer periods of time. In line with previous findings on the BPE (e.g., [7,8,15,16]), we assumed in the current study that the benefits associated with growing up in cities of a medium size would result in a high distribution of 14–18-year-old athletes who participated in sport programs compared with their distribution within the general population.

In order to enable the analysis of long-term developmental trends that are related to competitive sport programs in a given city or country, data on core environmental factors (e.g., the BPE) should be collected regularly and analyzed across different phases of development [18,21]. In the current study, we compared the BPE of young athletes in competitive sports programs with BPE data collected in two previous studies [15,16]. The aim was to conduct a further examination of the BPE in 14–18-year-old individuals over a 10-year period.

To complement the BPE data collected on the young athletes, we also attempted to explore how coaches who worked with children and youths perceived the effect. More specifically, we aimed at examining how coaches valued the contribution of the city or place where the athletes grew up to their development. In order to conceptually understand the contribution of BPE to the development of young athletes, it might be beneficial to collect data on how professionals who work regularly with these athletes perceive the existence or nonexistence of the effect, such as coaches who play a major role in the long-term developmental pathways of the athletes (see [22,23]). Such qualitative data could provide additional insights into the BPE.

As such, the purpose of the current study was twofold: (1) to examine the contribution of Israeli athletes' (ages 14–18 years) city sizes (i.e., population) to sports participation and performance and (2) collect preliminary data on how the coaches perceived this effect. The data on the BPE of the young athletes obtained in the current study were compared with the BPE data collected in two previous studies on elite male [15] and female [16] Israeli ballplayers and to examine specific BPE trends in one local sport structure from a perspective of approximately one decade.

2. Methods

2.1. Participants

The BPE was assessed in 1397 14–18-year-old athletes (mean age = 16.37 years; *SD* = 1.54), among them being 390 females (mean age = 16.28 years; *SD* = 1.57) and 1007 males (mean age = 16.40 years; *SD* = 1.53). The athletes were part of 10 sports: 5 individual (gymnastics, judo, swimming, tennis, and track and field) and 5 team (basketball, soccer, team handball, volleyball, and water polo) sports. The number of athletes in each sport is described in Table 1. The low number of females who took part in our study demonstrates the low number of young active female athletes in the country. For example, according to the Central Bureau of Statistics, only 19% of the active athletes 13–18 years of age in Israel are females (see [24]). We were aware not only of the low number of female athletes who participated in the current study but also the low number of male athletes in some of the sports (e.g., only 18 male gymnasts took part in the study). Our findings relating to the BPE are presented while taking into account the restrictions of the relatively small sample size. In a previous study, we used this sample of young athletes to calculate their relative age effect [18].

Table 1. The number of male and female athletes across sports.

Sport	Male Athletes	Female Athletes
Individual Sports		
Gymnastics	18	46
Judo	175	51
Swimming	104	50
Tennis	127	52
Track and field	58	27
Team Sports		
Basketball	158	76
Soccer	215	21
Team handball	51	35
Volleyball	66	29
Water polo	35	3
Total	1007	390

The participants in the current study were recruited from leading individual and team sport programs. We did not recruit athletes from all existing sport programs in the country to ensure that the young participants in the current study were all members of well-organized sport programs; that is to say, the female and male athletes in this study participated in between four and six sessions of practice on a weekly basis. The majority of the sport programs in Israel in both individual and team sports permit registration at the age of eight, and therefore, the participants in our study who were aged 14–18 years had at least 6 years of experience in training and competition. In a number of sports, such as gymnastics, swimming, and soccer, children can enter the sport program even earlier than the age of eight, and thus, some of them had undergone at least 9 years of training and competition.

Coaches. In addition to the players, 147 coaches (82 of individual sports and 65 of team sports; mean coaching experience = 14.20 years; median = 12 years) participated in the study. Among these coaches, 68 (14 females and 54 males) worked directly with the 14–18-year-old athletes who participated in our study, and 79 worked in the same sports programs as the other coaches but did not work with the athletes who participated in the current study. All the coaches were certified by their sports federations.

2.2. Procedure

This study was approved by the ethics committee of the Academic College at Wingate (Ethic Code # 112).

Information about the athletes' birthplaces, genders, and types of sport was obtained via questionnaires, which were given to the athletes by their coaches. Each director of the sport program where the athletes who participated in the study practiced received a letter providing the background and objectives of this study. In addition, informed consent was obtained from the parents of the participants. After approval was obtained, the first author approached the coaches of the athletes (n = 68) and sent them the questionnaires via electronic mail.

To examine the BPE in the 14–18-year-old athletes participating in our study, similar procedures to the ones performed in previous studies (e.g., [2,16]) were implemented. The city size of the BPE of each athlete from each of the 10 sports was based on Israel Census data obtained from a demographics website (see www.cbs.gov.il/shnaton60/st02_11x.pdf (accessed on 10 February 2022)). Four categories of city size were used in the study, as performed in previous BPE studies on Israeli athletes (for more details, see [15,16]). The designation of these four categories allowed us to examine the BPE in four sizes of residential areas: small towns and small, medium, and large cities.

The selection procedure we used in the current study was also applied in two previous studies on Israeli elite male [15] and female [16] ball game players, which therefore allowed us to compare the data obtained from the same categories of city size across a period of about a decade. It should be noted that information was unavailable on the migration between large cities and small cities or between small cities and large cities. It was assumed that the net movements between large cities and small cities or between small cities and large cities were likely to be equal.

Since the mean age of the participants (females and males) was 16.32 years and the majority of the data in our study were collected in 2016, we used the census data presented in 2014. That year was the most appropriate one to reflect the ages of 14 of the athletes (i.e., a turning point from developmental years to specialization years) who took part in our study.

In order to collect preliminary information on how the BPE is perceived by coaches, 150 coaches were approached by the first author. The coaches were asked to answer in writing a closed question related to the BPE: "Do you think that the athletes' places of birth affect their development?" There were two options for the answer: "yes" or "no". Only if the answer was "yes" was the coach asked another (open) question: "How does the place of birth contribute to the athletes' development?" The coaches were asked to outline how the BPE contributed to the athletes' development. Out of the 150 coaches who were approached, 147 responded (rate of responsiveness = 98%).

2.3. Data Analysis

To test the BPE, odds ratios (ORs) were calculated to determine the likelihood of participating in each of the sport programs (compared to the distribution of the population at age 14; year 2014) for each city size, and 95% confidence intervals (CIs) were calculated around each OR. An OR greater than 1 (with upper and lower limits higher than 1) implied that an athlete born in the given city size was more likely to become a participant in the sport program in the given sport than if they had been born in any other city size. An OR of less than 1 (with upper and lower limits less than 1) implied that an athlete born in the given city size was less likely to become a participant in the sport program than if they had been born in a city of a different size. ORs including the value of 1 within their CI range were not considered to be statistically significant.

As indicated previously, we were aware of the fact that the sample size of the female athletes was relatively small (390 participants). However, it was our aim to analyze their data across the different sports in order to strengthen our understanding of the existence of the BPE in the various sport programs available to female children and youths in the country. The responses of the 147 coaches who provided exploratory information about the BPE were descriptively analyzed. The coaches' responses were also thematically analyzed.

3. Results

The results are presented separately for the BPE and the responses of the coaches.

3.1. BPE Data

The representations of the Israeli population aged 14–18 years, the male athletes, and ORs and CIs across cities of different sizes are presented in Table 2 (individual sports) and Table 3 (team sports). The information on the female athletes who participated in our study is presented in Table 4 (individual sports) and Table 5 (team sports). Mixed results were found for the BPE in the male and female athletes and across the 10 sports.

3.1.1. Male Athletes

Data are presented separately for individual sports and team sports.

Individual Sports

In only two sports was the BPE found to be significant: gymnastics and swimming. In gymnastics, it was found that small towns of up to 2000 people yielded an OR significantly higher than 1. This means that the likelihood of participating in a competitive sport program for those gymnasts who were born in a small place was higher than that for those who were born in a city of a different size. In addition, cities with a population of 50,000–200,000 yielded an OR significantly lower than 1. The likelihood for gymnasts who were born in a medium-sized city to be part of a competitive sport program was lower than that for those who were born in a city with a smaller or larger population.

Similar results were obtained in swimming; small towns of up to 2000 people yielded an OR significantly higher than 1. This means that the likelihood of participating in a competitive swimming program for swimmers who were born in a small place was higher than that for those who were born in a town or city of a different size. In addition, cities with a population of 50,000–200,000 yielded an OR significantly lower than 1. The likelihood for swimmers who were born in a medium city to be part of a swimming program was lower than that for those who were born in a city with a smaller or larger population.

Team Sports

The BPE was found to be significant in each sport; however, mixed results were observed. In basketball, cities with a population of 50,000–200,000 yielded an OR significantly higher than 1. This means that the likelihood for basketball players who were born in a medium-sized city to be part of a basketball program was higher than that for those who were born in a city of a larger or smaller size. In soccer, the findings for cities of the same size (50,000–200,000) were found to be the opposite; cities of this population size yielded an OR significantly lower than 1, and the likelihood for the soccer player who was born in a medium-sized city to participate in a competitive soccer program was lower than that for those who were born in a city with a smaller or larger population.

For the team handball players, cities with a population of 50,000–200,000 yielded an OR significantly higher than 1. This means that the likelihood for team handball players who were born in a medium-sized city to be part of a team handball program was higher than that for those who were born in a city of a larger or smaller size. It was also found that small towns of up to 2000 people yielded an OR significantly lower than 1. The likelihood of participating in a competitive sport program for those team handball players who were born in a small place was lower than that for those who were born in a place with a higher population.

For the volleyball players, small towns of up to 2000 people yielded an OR significantly lower than 1. This means that the likelihood of participating in a competitive sport program for those volleyball players who were born in a small place was lower than that for those who were born in a city of a different size. Finally, for the water polo players, cities with a population of more than 200,000 people yielded an OR significantly higher than 1. This means that the likelihood for water polo players who were born in a big city to be part of a water polo program was higher than that for those who were born in a city of a different size.

Table 2. Representation of the Israeli population, male individual sport athletes [a], and ORs and CIs across cities of different sizes.

City Size	Israel Pop [b]	Gymnastics			Judo			Swimming			Tennis			Track and Field		
		%	OR	CI	%	OR	CI	%	OR	CI	%	OR	CI	%	OR	CI
>2000	8.57	27.78	4.10	(1.80, 9.37) *	7.39	0.85	(0.31, 2.37)	25.96	3.74	(1.63, 8.59) *	14.17	1.76	(0.72, 4.32)	13.79	1.71	(0.69, 4.21)
2000–50,000	34.9	33.33	0.93	(0.52, 1.67)	38.07	1.15	(0.64, 2.04)	31.73	0.87	(0.48, 1.56)	33.07	0.92	(0.51, 1.66)	31.03	0.84	(0.47, 1.51)
50,000–200,000	29.48	5.56	0.14	(0.05, 0.37) *	24.43	0.77	(0.41, 1.45)	17.31	0.50	(0.26, 0.98) *	24.41	0.77	(0.41, 1.45)	20.69	0.62	(0.33, 1.19)
<200,000	27.06	33.33	1.35	(0.74, 2.47)	30.11	1.16	(0.63, 2.15)	25.00	0.90	(0.48, 1.69)	28.35	1.07	(0.57, 1.98)	34.48	1.42	(0.78, 2.60)

Note: OR = odds ratio; CI = confidence interval. * Significant difference. [a] Percentage of male athletes who participated in individual sport programs in 2016 and who grew up in each of the subdivisions of the 2014 census. [b] Israel population = percentage of males under the age of 14 in each of the subdivisions of the 2014 Israel Census.

Table 3. Representation of the Israeli population, male team sport players [a], and ORs and CIs across cities of different sizes.

City Size	Israel Pop [b]	Basketball			Soccer			Team Handball			Volleyball			Water Polo		
		%	OR	CI	%	OR	CI	%	OR	CI	%	OR	CI	%	OR	CI
>2000	8.57	7.55	0.87	(0.31, 2.42)	15.74	1.99	(0.82, 4.82)	0.00	0.00	-	1.52	0.16	(0.03, 0.95) *	14.29	1.78	(0.73, 4.36)
2000–50,000	34.9	25.16	0.63	(0.34, 1.15)	46.76	1.64	(0.93, 2.89)	37.25	1.11	(0.62, 1.97)	27.27	0.70	(0.38, 1.28)	0.00	0.00	–
50,000–200,000	29.48	44.65	1.93	(1.08, 3.46) *	13.89	0.39	(0.19, 0.79) *	50.98	2.49	(1.39, 4.45) *	42.42	1.76	(0.98, 3.16)	22.36	0.71	(0.38, 1.34)
<200,000	27.06	22.64	0.79	(0.41, 1.5)	23.61	0.83	(0.44, 1.58)	11.76	0.36	(0.17, 0.76) *	28.79	1.09	(0.59, 2.02)	62.36	4.56	(2.51, 8.31) *

Note: OR = odds ratio; CI = confidence interval. * Significant difference. [a] Percentage of male athletes who participated in team sport programs in 2016 and who grew up in each of the subdivisions of the 2014 census. [b] Israel population = percentage of males under the age of 14 in each of the subdivisions of the 2014 Israel Census.

Table 4. Representation of the Israeli population, female individual sport athletes [a], and ORs and CIs across cities of different sizes.

City Size	Israel Pop [b]	Gymnastics			Judo			Swimming			Tennis			Track and Field		
		%	OR	CI	%	OR	CI	%	OR	CI	%	OR	CI	%	OR	CI
>2000	8.57	60.87	16.59	(0.81, 37.2)	28.00	4.14	(1.81, 9.46) *	10.00	1.18	(0.45, 3.0)	5.77	0.65	(0.21, 1.95)	18.52	2.42	(1.02, 5.74) *
2000–50,000	34.9	19.57	0.45	(0.23, 0.86) *	34.00	0.96	(0.53, 1.72)	28.00	0.72	(0.39, 1.32)	32.69	0.90	(0.50, 1.62)	48.15	1.73	(0.98, 3.05)
50,000–200,000	29.48	6.52	0.16	(0.06, 0.41) *	20.00	0.59	(0.31, 1.14)	22.00	0.67	(0.35, 1.27)	21.15	0.64	(0.33, 1.22)	11.11	0.29	(0.14, 0.63) *
<200,000	27.06	13.04	0.40	(0.19, 0.83)	18.00	0.59	(0.30, 1.16)	40.00	1.79	(0.99, 3.2)	40.38	1.82	(1.0, 3.31)	22.22	0.77	(0.40, 1.46)

Note: OR = odds ratio; CI = confidence interval. * Significant difference. [a] Percentage of female athletes who participated in individual sport programs in 2016 and who grew up in each of the subdivisions of the 2014 census. [b] Israel population = percentage of females under the age of 14 in each of the subdivisions of the 2014 Israel Census.

Table 5. Representation of the Israeli population, female team sport players [a], and ORs and CIs across cities of different sizes.

City Size	Israel Pop [b]	Basketball			Soccer			Team Handball			Volleyball		
		%	OR	CI	%	OR	CI	%	OR	CI	%	OR	CI
>2000	8.57	9.21	1.08	(0.40, 2.86)	4.76	0.53	(0.16, 1.69)	0.00	0.00	-	20.69	2.78	(1.18, 6.51) *
2000–50,000	34.9	21.05	0.49	(0.26, 0.93) *	47.62	1.69	(0.96, 2.99)	5.71	0.11	(0.04, 0.28) *	34.48	0.98	(0.54, 1.75)
50,000–200,000	29.48	27.63	0.91	(0.49, 1.68)	14.29	0.39	(0.19, 0.80) *	48.57	2.25	(1.26, 4.04) *	27.59	0.91	(0.49, 1.68)
<200,000	27.06	17.24	1.96	(1.08, 3.54) *	33.33	1.34	(0.73, 2.47)	45.71	2.26	(1.25, 4.09) *	17.24	0.56	(0.28, 1.10)

Note: OR = odds ratio; CI = confidence interval. * Significant difference. [a] Percentage of female athletes who participated in team sport programs in 2016 and who grew up in each of the subdivisions of the 2014 census. [b] Israel population = percentage of females under the age of 14 in each of the subdivisions of the 2014 Israel Census.

3.1.2. Female Athletes

Data are presented separately for individual sports and team sports.

Individual Sports

The BPE was found to be significant in three sports: gymnastics, judo, and track and field. For the gymnasts, cities with a population of 2000–50,000 or 50,000–200,000 people yielded an OR significantly lower than 1. This means that the likelihood for gymnasts who were born in small-to-medium-sized cities to be part of a competitive sport program was lower than that for those who were born in a city of a different size. For the judokas, it was found that small towns of up to 2000 people yielded an OR significantly higher than 1. This means that the likelihood of participating in a competitive judo program for those individuals who were born in a small place was higher than that for those who were born in a city with a larger population.

For the track and field athletes, being born in a small town was also found to be beneficial; small towns of up to 2000 people yielded an OR significantly higher than 1. This means that the likelihood of participating in a competitive sport program for those track and field athletes who were born in a small place was higher than that for those who were born in a city with a larger population. In addition, it was found that cities with a population of 50,000–200,000 people yielded an OR significantly lower than 1. The likelihood for those who were born in medium-sized cities to be part of a competitive sport program was lower than that for those who were born in a city of a different size.

Team Sports

The BPE was found to be significant in all team sports but water polo. (The number of the active players was very small, and therefore, BPE analyses were not performed). Mixed results were indicated. For the basketball players, cities with a population of more than 200,000 people yielded an OR significantly higher than 1. This means that the likelihood for basketball players who were born in a big city to be part of a basketball program was higher than that for those who were born in a city with a smaller population. In addition, cities with a population of 2000–50,000 people yielded an OR significantly lower than 1. This means that the likelihood for those who were born in small cities to be part of a competitive sport program in basketball was lower than that for those who were born in a city of a different size.

For the soccer players, cities with a population of 50,000–200,000 yielded an OR significantly lower than 1. The likelihood for soccer players who were born in a medium-sized city to be part of a competitive sport program was lower than that for those who were born in a city of a different size.

For the team handball players, cities with a population of 50,000–200,000 people or more than 200,000 people yielded an OR significantly higher than 1. The likelihood for team handball players who were born in medium- and large-sized cities to be part of a competitive sport program was higher than that for those who were born in a city of a smaller size. In addition, cities with a population of 2000–50,000 people yielded an OR significantly lower than 1. This means that the likelihood for those who were born in small cities to be part of a competitive team-handball program was lower than that for those who were born in a city of a different size.

For the volleyball players, small towns of up to 2000 people yielded an OR significantly higher than 1. This means that the likelihood to participate in a competitive volleyball program for those who were born in a small place was higher than that for those who were born in a place with a higher population.

3.2. *Coaches' Preliminary Thoughts on the BPE*

The coaches' responses associated with the BPE are presented in Table 6. More than 80% of the coaches (about 85% of the coaches who worked with individual sport athletes and 81% of those who coached team sport players) answered "yes" when they were asked if the athletes' places of birth contributed to their development. Three main observations associated with the BPE emerged: (1) proximity to sport facilities, where the coaches argued that living near sports facilities is beneficial for the young athlete. Among the benefits they outlined were that the athletes spent less time on the roads, there was no need for adults to transport them to practices, and the athletes had a greater feeling of security; (2) the socioeconomic statues of the living place, where the coaches claimed that wealthy places or cities can provide better instructional support and more formal and informal learning opportunities for athletes to develop their athletic abilities and skills than what less wealthy towns or cities can offer; and (3) sport popularity in a given place, where the coaches argued that it is easier for children to select a given sport if they grow up in a town or city where the sport is popular and supported by the community.

Table 6. Coaches' BPE preliminary reflections.

Question: Do You Think That the Athletes' Places of Birth Affect Their Development?		
	Individual Sports	**Team Sports**
Yes	85.5% (71)	81.3% (52)
No	14.5% (12)	18.8% (12)
If yes, how does the place of birth contribute to the athletes' development?		
Proximity to sport facilities		
Socioeconomic status of the living place		
Sport popularity		

4. Discussion

The discussion is composed of three parts. In the first part, we discuss the BPE data from the 14–18-year-old athletes, and in the second part, we examine the reflections of the coaches on the contribution of the place of residence to the young athletes' development. Based on the BPE data obtained in our study, in the third part, we elaborate upon a number of aspects associated with the establishment of a national sport policy for talent detection and early phases of talent development.

4.1. BPE Data

Mixed BPE results were found for the male and female athletes across the individual and team sports. The main BPE findings in the male athletes were that (1) for those who participated in individual sport programs—gymnastics and swimming—growing up in a small place (less than 2000 people) had greater benefits than any other sized place of residence; (2) for the athletes who were part of competitive basketball and team handball programs, it was more beneficial to grow up in cities of a medium size (50,000–200,000 people) than in cities of other sizes; and (3) only for the water polo players was it more beneficial to grow up in a large city (more than 200,000 people).

The main BPE findings in the female athletes were that (1) for those who participated in individual sport programs—gymnastics, judo, and track and field—it was more beneficial to grow up in a small place or in a city of a medium size (gymnastics) than in a place of a different size, and (2) for the athletes who played team sports—basketball and team handball—it was more beneficial to grow up in a large city than in a city of any other size. For the team handball players, it was also beneficial to grow up in a medium-sized city. Living in a small place was found to beneficial for only the volleyball players.

The mixed BPE results of our study on the 14–18-year-old athletes were similar to the ones reported in Lidor et al.'s studies [15,16] on elite athletes; that is, no clear-cut observations could be made for the contribution of one city of one size to the development of the athletes across sports. Different cities of different sizes were found to be associated with achieving a high level of proficiency in the 20 (10 for the males and 10 for the females) analyzed sport programs. However, in 12 out of the 20 programs, it was found that growing up in a very small place or a city of a medium size had greater benefits. Living in a city of a large size was indicated to be beneficial to athletes in only three sport programs. In only 7 out of the 20 sport programs was it found to be detrimental to grow up in cities of small or medium sizes.

Therefore, the contribution of small places and cities of a medium size to the development of the young athletes in our study can presumably be explained by the instructional and social characteristics of these places, as discussed in previous BPE studies (e.g., [6,13,16]). Towns and cities of these particular sizes may be more likely to encourage athletes and parents toward certain sports and may have unique resources, such as the availability of sport facilities, that can enhance athletic development (see, for example, [3,5,6,25]). Therefore, the contribution of growing up in small- or medium-sized towns up until the age of 14 to the development of young athletes can be observed not only in large-population countries (e.g., Canada, Germany, or the USA) but also in small countries such as Israel.

From a 10-year perspective, it appears that in Israel, it has been more beneficial for athletes to grow up in towns and cities of a small-to-medium size [15,16]. Even in soccer, which is considered the most popular sport in Israel [20], growing up in large cities was not found to be beneficial for young soccer players. Although the best professional soccer clubs in Israel are situated in large cities (i.e., Haifa, Jerusalem, and Tel-Aviv), our data show that joining a soccer program in a large city may not provide the optimal instructional-psychological conditions needed for attaining proficiency in the game of soccer.

However, the explanations favoring growing up in places or cities of small or medium sizes may not tell the whole story of the findings obtained in our study, since we also found that athletes who participated in 7 sport programs (out of the 20 programs) did not benefit from living in places or cities of small or medium sizes. It appears that additional characteristics of the place or city, rather than only its size, may be associated with the athletes' development. The analysis of the coaches' BPE reflections can add another dimension to the quantitatively analyzed BPE data in our study.

4.2. Coaches' Reflections

Most of the coaches pointed out that the place of birth did in fact contribute to the athletes' development; however, they did not focus on the size of the city but rather on other characteristics. A number of their reflections indeed strengthened the finding that it is more beneficial to grow up in small- and medium-sized cities than in cities of a greater size. For example, the coaches emphasized ways that athletes can benefit from living close to the sports facilities: they can save time on the roads, arrive independently to the facility, and feel secure on their way to and from practice. These observations are typically associated with small towns and cities of a medium size and therefore provide support for the benefits of such places to a young athlete's development (see, e.g., [5,25]).

However, the coaches also mentioned that the socioeconomic status of the place or city where the athlete grew up had the potential to influence his or her development. More specifically, they claimed that children and youths are provided with both more formal and informal opportunities to develop their sport skills in wealthy places or cities rather than those available to them in poorer places or cities. The coaches did not relate to cities of a given size. In addition, they did not mention a specific sport program that might be negatively or positively influenced by the socioeconomic status of the place or city where the athlete lived. These thoughts of the coaches reflect the idea that certain conditions, among them the availability of sport facilities, instructional support, and social support (typically associated with wealthy places or cities), resulted in a higher rate of sport participation among children [26]. In addition, prior research (e.g., [27,28]) highlighted the consistency of the socioeconomic status limitations to participation in high-performance sports.

Another perspective that was shared by the coaches was the one related to how popular a sport program was in a given city. The coaches felt that children are initially attracted to popular sport activities; the more popular the sport, the more chances there are that the children will select it for themselves. For example, it was found in our study that the likelihood of participating in competitive sport programs for the female volleyball players who were born in a small place (up to 2000 people) where the game was popular was higher than that for those who were born in a city of a different size. These data can be explained by the fact that in Israel, the game of volleyball was originally played in places like the Kibbutz (cooperative farming settlements) and small towns [18], and therefore, the game of volleyball has been popular in such places.

The combined qualitative reports given by the coaches and the quantitative BPE analysis provide further support for the notion that small-to-medium-sized towns and cities, rather than large cities, are considered to be more suitable sports environments for children and youths (aged 14–18 years old, females and males) to develop their sports skills. The coaches felt that urban conditions, such as living in close proximity to the sports facilities, and the socioeconomic status of the city are detrimental contributors to a young athlete's early development phases. The coaches argued that such conditions can help young athletes focus on the training program.

Finally, the observations made by the coaches in our study can add dimensions to the concept of BPE, other than the size of the city or the population density. As noted by Wattie et al. [25], " ... further research is needed in this area to understand contextual differences related to birthplace" (p. 378). Obtaining data on specific cultural and social characteristics of places or cities of different sizes can increase the understanding of possible mechanisms involved in the association between the BPE and achieving success in sports.

4.3. BPE and the Establishment of a (National) Sports Policy in a Small Nation

Up to now, a national sports policy associated with early phases of talent identification and development has not been established in Israel. In one study evaluating a number of aspects of the elite sports policy in Israel [29], only one third of the directors of the elite sports federations who participated in the study claimed that their federation had a well-developed policy for early development in sports. In addition, according to the directors of these federations, there is a lack of support and involvement by sports scientists, multidimensional support services appropriate to the age and level of young athletes, and nationally coordinated support for the combination of sports development and scientific study. De Bosscher and her colleagues summarized that " ... priority sports need to include a long-term talent development training plan" (p. 154).

Based on the data of the current study, as well as on similar data collected in previous BPE studies on adult Israeli athletes [15,16] and on athletes who grew up in small communities in other countries [13,14], we discussed two issues associated with the BPE and the development of sports policy for young athletes: how to increase the number of children in individual and team sport programs and the prioritization of sport programs.

In order to increase the number of children who choose to participate in a sport program, it is proposed to sport policymakers that they further develop sport programs in individual and team sports in small- and medium-sized cities by increasing (1) the number of sports facilities available to the public at large and (2) the subsidization of children's participation in the sport program(s). The public investment in sport program(s) in small- and medium-sized cities may not only benefit the young members of the local community but also those children who live in a large city which is located near the small community. In many areas in Israel, the geographical distance between small communities and large cities is small (e.g., 2–3 km), and therefore, those who live in a large city (e.g., more than 200,000 people) may also benefit from the sport programs taking place in a nearby small- or medium-sized city.

BPE data can also be used by policymakers when decisions about prioritization of sport programs are made (see, for example, [30]). For example, Israeli judokas (females and males) have won medals in several Olympic Games (e.g., in the 2016 and 2020 Games) and World Championships (e.g., in the year 2019). Up to now, Israeli athletes have won 13 Olympic medals, 6 of them by judokas (5 individual medals and 1 team medal). In fact, judo has become the individual sport in Israel with the most achievements at the Olympic level. Looking at the data of our BPE study, it was indicated that only for the female judokas is the likelihood of participating in a competitive judo program for those who were born in a small place higher than that for those who were born in a city of a larger size. If the goal of the Ministry of Culture and Sports and the Judo Federation is to attract more children and youths—not only females but also males—to join judo programs in light of the international success achieved by Israeli elite judokas, then it might be of relevance to develop more judo programs in small- and medium-sized cities. It is assumed that the greater investment in judo programs in cities of these sizes, the higher the potential to increase the popularity of the sport in these communities, and subsequently the higher the chances of increasing the number of active young judokas.

In summary, mixed BPE findings were obtained from our study. No clear-cut conclusions can be made on the contribution of one city of one size to the development of athletes across the observed individual and team sports; that is to say, different sizes of cities were indicated to be linked to attaining success in individual and team sports.

4.4. Practical Implications for Policymakers

It might be of interest to policymakers in Israel to develop a national sports policy in order to maximize the potential to recruit children to sport programs and provide them with the appropriate conditions to achieve success. If this is the case, knowing about the contribution of the BPE to an athlete's success can help policymakers implement these evidence-based data in the decision-making processes. However, if such a national sport policy is not the main interest of the ministry due to its motivation to enable the different sport federations to establish their own sport policies, then the data of the BPE can be used by the federations as well.

Author Contributions: Conceptualization, R.L. and Z.M.; Methodology, R.L., Z.M. and M.A.; Software, M.A.; Formal Analysis, M.A. and Z.M. Investigation, Z.M.; Writing and editing, R.L. and Z.M. Visualization, R.L. All authors have read and agreed to the published version of the manuscript.

Funding: This research received no external funding.

Institutional Review Board Statement: The study was conducted according to the guidelines of the Declaration of Helsinki and approved by the ethics committee of the Academic College at Wingate (Ethic Code # 112; approved date: 19 July 2016).

Informed Consent Statement: Informed consent was obtained from all participants involved in the study.

Conflicts of Interest: The authors declare no conflict of interest.

References

1. Baker, J.; Schorer, J.; Cobley, S.; Schimmer, G.; Wattie, N. Circumstantial development and athletic excellence: The role of birth date and birth place. *Eur. J. Sport Sci.* **2009**, *9*, 329–339. [CrossRef]
2. Côté, J.; MacDonald, D.J.; Baker, J.; Abernethy, B. When "where" is more important than "when": Birthplace and birthdate effects on the achievement of sporting expertise. *J. Sports Sci.* **2006**, *24*, 1065–1073.
3. MacDonald, D.J.; Baker, J. Circumstantial development: Birthdate and birthplace effects on athlete development. In *Conditions of Children's Talent Development in Sport*; Côté, I.J., Lidor, R., Eds.; Fitness Information Technology: Morgantown, WV, USA, 2013; pp. 197–208.
4. Turnnidge, J.; Wright, E.; Matthews, A. Organizational structures in sport. In *Birth Advantages and Relative Age Effects in Sport*; Kelly, A.L., Côté, J., Jeffreys, M., Turnnidge, J., Eds.; Routledge: London, UK, 2021; pp. 30–56.
5. Hancock, D.J.; Murata, A.; Côté, J. Parents' roles in creating socio-environmental birth advantages for their children. In *Birth Advantages and Relative Age Effects in Sport*; Kelly, A.L., Côté, J., Jeffreys, M., Turnnidge, J., Eds.; Routledge: London, UK, 2021; pp. 207–221.
6. Macdonald, D.J.; Cheung, M.; Côté, J.; Abernethy, B. Place but not Date of Birth Influences the Development and Emergence of Athletic Talent in American Football. *J. Appl. Sport Psychol.* **2009**, *21*, 80–90. [CrossRef]
7. Baker, J.; Logan, A.J. Developmental contexts and sporting success: Birth date and birthplace effects in national hockey league draftees 2000–2005. *Br. J. Sports Med.* **2007**, *41*, 515–517. [CrossRef] [PubMed]
8. Macdonald, D.J.; King, J.; Côté, J.; Abernethy, B. Birthplace effects on the development of female athletic talent. *J. Sci. Med. Sport* **2009**, *12*, 234–237. [CrossRef] [PubMed]

9. Rossing, N.N.; Nielsen, A.B.; Elbe, A.M.; Karbing, D.S. The role of community in the development of elite handball and football players in Denmark. *Eur. J. Sport Sci.* **2016**, *16*, 237–245. [CrossRef] [PubMed]
10. Hancock, D.J.; Coutinho, P.; Côté, J.; Mesquita, I. Influences of population size and density on birthplace effects. *J. Sports Sci.* **2018**, *36*, 33–38. [CrossRef] [PubMed]
11. Brown, T.; Kelly, A.L. Relative access to wealth and ethnicity in professional cricket. In *Birth Advantages and Relative Age Effects in Sport*; Kelly, A.L., Côté, J., Jeffreys, M., Turnnidge, J., Eds.; Routledge: London, UK, 2021; pp. 184–206.
12. Fraser-Thomas, J.; Côté, J.; MacDonald, D.J. Community size in youth sport settings: Examining developmental assets and sport withdrawal. *Rev. Phéneps/PHEnex J.* **2010**, *2*, 1–9.
13. Balish, S.; Côté, J. The influence of community on athletic development: An integrated case study. *Qual. Res. Sport Exerc. Health* **2013**, *6*, 98–120. [CrossRef]
14. McCalpin, M.; Evans, B.; Côté, J. Young Female Soccer Players' Perceptions of Their Modified Sport Environment. *Sport Psychol.* **2017**, *31*, 65–77. [CrossRef]
15. Lidor, R.; Côté, J.; Arnon, M.; Zeev, A.; Cohen-Maoz, S. Relative age and birthplace effects in Division 1 players—Do they exist in a small country? *Talent. Dev. Excell.* **2010**, *2*, 181–192.
16. Lidor, R.; Arnon, M.; Maayan, Z.; Gershon, T.; Côté, J. Relative age and birthplace effects in Division 1 female players—The relevance of sport-specific factors. *Int. J. Sport Exerc. Psychol.* **2014**, *12*, 19–33. [CrossRef]
17. Côté, J. The Influence of the Family in the Development of Talent in Sport. *Sport Psychol.* **1999**, *13*, 395–417. [CrossRef]
18. Lidor, R.; Maayan, Z.; Arnon, M. Relative age effect in 14- to 18-year-old athletes and their initial approach to this effect—Has anything changed over the past 10 years? *Front. Sports Act. Living* **2021**, *3*, 39. [CrossRef] [PubMed]
19. Sam, M. Youth sport policy in small nations. In *Routledge Handbook of Youth Sport*; Green, K., Smith, A., Eds.; Routledge: London, UK, 2016; pp. 535–542.
20. Lidor, R.; Blumenstein, B. Soccer as a mediator for fostering relationships and building peace among Jewish and Arab players. In *Sport for Development, Peace, and Social Justice*; Schinke, R.J., Hanrahan, S.J., Eds.; Fitness Information Technology: Morgantown, WV, USA, 2012; pp. 39–55.
21. Baker, J.; Schorer, J.; Wattie, N. Compromising Talent: Issues in Identifying and Selecting Talent in Sport. *Quest* **2018**, *70*, 48–63. [CrossRef]
22. Collins, D.; MacNamara, A. *Talent Development—A Practitioner Guide*; Routledge: London, UK, 2018.
23. Ericksen, K.; Gilbert, W. Coach-athlete interactions in children's sport. In *Conditions of Children's Talent Development in Sport*; Côté, J., Lidor, R., Eds.; Fitness Information Technology: Morgantown, WV, USA, 2013; pp. 139–156.
24. Central Bureau of Statistics. Population of Israel on the Eve of 2021. 2021. Available online: https://www.cbs.gov.il/en/mediarelease/Pages/2020/Population-of-Israel-on-the-Eve-of-2021.aspx (accessed on 10 December 2021).
25. Wattie, N.; MacDonald, D.J.; Cobley, S. Birthdate and birthplace effects on expertise attainment. In *Routledge Handbook of Sport Expertise*; Baker, J., Farrow, D., Eds.; Routledge: London, UK, 2015; pp. 373–382.
26. Fraser-Thomas, J.; Falcão, W.; Wolman, L. Understanding take-up, drop-out and drop-off in youth sport. In *Routledge Handbook of Youth Sport*; Green, K., Smith, A., Eds.; Routledge: London, UK, 2016; pp. 227–242.
27. Beamish, R. The persistence of inequality: An analysis of participation patterns among Canada's high performance athletes. *Int. Rev. Sociol. Sport* **1990**, *25*, 143–152. [CrossRef]
28. Beamish, R. Towards a socio-cultural profile of Canada's high performance athletes. *Int. Rev. Sociol. Sport* **1992**, *27*, 279–288. [CrossRef]
29. De Bosscher, V.; de Rycke, J.; Truyens, J.; Lidor, R.; Helberstein, D.; Yanovich, E. *Elite Sport Policy Evaluation in Israel*; Vrige Universiteit: Brussel, Belgium, 2017.
30. Ministry of Culture and Sports. Summary of the 2014's Activity (in Hebrew). 2014. Available online: https://foi.gov.il/sites/default/files (accessed on 10 December 2021).

sports

MDPI

Article

Talent Identification in Elite Adolescent Ice Hockey Players: The Discriminant Capacity of Fitness Tests, Skating Performance and Psychological Characteristics

Jean Lemoyne [1,2,*], Jean-François Brunelle [2,3], Vincent Huard Pelletier [1,2], Julien Glaude-Roy [2,3] and Gaëtan Martini [1,2]

1 Department of Human Kinetics, Université du Québec à Trois-Rivières, 3351 Boulevard des Forges, Trois-Rivières, QC G9A 5H7, Canada; vincent.huard.pelletier@uqtr.ca (V.H.P.); gaetan.martini@uqtr.ca (G.M.)

2 Laboratoire de Recherche sur le Hockey de l'UQTR, 3351 Boulevard des Forges, Trois-Rivières, QC G9A 5H7, Canada; jean-francois.brunelle@uqtr.ca (J.-F.B.); julien.glaude-roy@uqtr.ca (J.G.-R.)

3 Service de l'Activité Physique et Sportive, Université du Québec à Trois-Rivières [UQTR], 3351 Boulevard des Forges, Trois-Rivières, QC G9A 5H7, Canada

* Correspondence: jean.lemoyne@uqtr.ca

Abstract: Background: The process of talent identification in ice hockey occurs during middle adolescence when players are selected to participate in "off-season" evaluation camps, where coaches observe their fitness levels and status of development. Recently, the Quebec ice hockey federation opted for a holistic approach by evaluating players based on three criteria: (1) fitness, (2) skating abilities and (3) personality traits and psychological assets. This study aimed to analyze the discriminant validity of a multi-dimensional talent identification testing protocol in competitive ice hockey. Method: Data were collected from 160 adolescent hockey players who took part in Team Quebec summer evaluation camps. Off-ice fitness, skating abilities and psychological variables were measured on two consecutive days. Descriptive statistics, group comparisons (gender, positions) and discriminant analyses (selected versus non-selected) were performed. Results: No differences were observed among males in which selected players were similar to non-selected. Results from discriminant analyses also showed no discriminant function for male players. For females, selected players displayed higher fitness, on-ice agility and psychological characteristics. Nine performance markers were significantly discriminant. Conclusions: A holistic evaluation protocol allows for the discrimination of selected and non-selected players in elite ice hockey. Developing more discriminant tests is a promising avenue of research in male ice hockey. Knowing the factors that are associated with team selection in competitive ice hockey allow to focus on the specific attributes to work with young promising players.

Keywords: performance; athletic development; ice hockey expertise; fitness assessment

Citation: Lemoyne, J.; Brunelle, J.-F.; Huard Pelletier, V.; Glaude-Roy, J.; Martini, G. Talent Identification in Elite Adolescent Ice Hockey Players: The Discriminant Capacity of Fitness Tests, Skating Performance and Psychological Characteristics. *Sports* **2022**, *10*, 58. https://doi.org/10.3390/sports10040058

Academic Editors: Adam Leigh Kelly, Sergio L. Jiménez Sáiz, Sara Diana Leal dos Santos and Alberto Lorenzo Calvo

Received: 15 March 2022
Accepted: 4 April 2022
Published: 8 April 2022

1. Introduction

Ice hockey involves high-intensity, intermittent actions that require players to perform at superior technical and tactical levels. Accordingly, excelling in ice hockey necessitates a vast repertoire of physical attributes, technical–tactical skills [1] and psychological assets [2]. Regarding the physical aspect, attributes such as aerobic capacity, anaerobic systems, strength, power, speed and agility are required [3]. More specific to ice hockey, Mascaro [4] stipulates that skating speed and agility figure among the most important skills to possess, even if they are used for very short periods [5]. Psychological assets, such as grit [6] and type of personality [7], are also considered when it comes time to adapt to the demands of ice hockey and distinguish the best talents. According to Tarter and colleagues [8], the best way to predict the transition to professional hockey (e.g., National Hockey League) is to consider indicators from different categories such as game performance, skill observation and evaluation, fitness, personality traits and perceptivo-cognitive assets. Indeed, these assumptions suggest that to be efficient, the talent identification process should be designed holistically.

In the last two decades, ice hockey has become increasingly popular internationally and is now played by some 1.5 million registered players from more than 75 different countries. With over 900,000 players under the age of 20 (https://www.iihf.com/en/static/5324/survey-of-players; accessed on 17 December 2021) and hundreds of professional leagues across the world, access to professional hockey is among young hockey players' long-term objectives. As for most sports, national team selections, world championships and Olympic Games performance and the number of drafted players in the National Hockey League (NHL and others, such as the KHL) were the criteria used by hockey federations to evaluate the quality of their development model. There is a wealth of past research on the fitness status of professional ice hockey players [9,10], which has led to an improved understanding of the game, helped to establish standards and played a role in the evolution of strength and conditioning approaches [11]. However, less is known about the contribution of each component of performance to the definition of sports talent. Can we assume that players who perform better in fitness tests and other assessments, such as on-ice sprint tests and psychometric tests, are prioritized in team selection or professional drafts? Vescovi [12] offers an excellent example of the shortcomings of the NHL combine regarding the associations between players' fitness and draft status over 3 years. Because certain associations between off-ice and on-ice fitness are plausible [13], further research is needed to verify whether these components can help to determine the best athletes.

In line with these standards and the complex nature of the sport, the pathway to elite ice hockey is a long-term process that needs to start at a young age [14]. As with most sports development models, the expertise acquisition phase is planned for late puberty, between ages 14 and 16, with the result that many evaluation and development camps become an important step in the talent identification process. The International Ice Hockey Federation (IIHF) development camp (IIHF Development Camp: https://www.iihf.com/en/statichub/4625/development; accessed on 10 December 2021), where the most talented players display their skills before hockey coaches and stakeholders, is a good example. The world's top hockey countries (e.g., Canada, United States, Finland) also have development camps where players are evaluated at the early stages of sports expertise. Indeed, many countries have specific protocols to conduct periodic evaluation or development camps that serve to guide hockey federations in terms of future national team selections (e.g., under-18, under-20, Olympic Games). Camps such as these are first-rate opportunities to develop performance standards for both off-ice fitness and on-ice skills [15]. In addition, evaluating players at this stage of development is useful to establish the age-group and position-specific standards needed to monitor athletes during their development. Valuable work in this regard has been done in the field of ice hockey. Roczniok and colleagues described and identified variables, such as relative VO_{2max}, relative peak power and height, that determined success in the Polish men's national team selection camp [15]. Recently, Vigh-Larsen et al. [16] conducted a cross-sectional study comparing the best U20 (junior elite) and professional Danish players with a similar sample of U20 Finnish players in terms of anthropometric measures and off-ice and on-ice fitness tests. Interestingly, the most important differences were those observed for anthropometric measures, indicating that professional athletes were taller and heavier, whereas no differences (in fitness and skating) were observed relative to players' position. This demonstrates that the talent identification process is a complex one [17] and that analyzing the profile of U20 players may have limitations since it very likely starts at earlier stages of development [18,19]. As a result, we believe that older players had already experienced the selection process, thus hindering the real value of testing players based on the multiple factors that determine talent for ice hockey. A recent review by Huard Pelletier and colleagues showed that the transfer from fitness testing to on-ice skating performance is relatively well established [20]. However, few authors have verified the value of fitness testing results in the settings of the talent identification process. Despite the popularity of fitness testing in the athlete development processes, little is known about the relative contribution of each test for athlete selection and/or talent identification. In other words, what factors help to differentiate the best players (or prospects) from the average ones? Bracko [5] offered some insights on the issue by showing that there were significant differences between the skating abilities of elite and non-elite players. However, as Johnston reported [18], the concept of sports talent is multifaceted and should be examined further to better define it from a more global perspective. In this regard, viewing sports talent holistically enables us to identify and combine multiple methods and approaches that are conducive to assessing (and evaluating) its multiple components. Therefore, we define sports talent based on hypotheses from past research to the effect that sports success results from a combination of physical, technical, tactical and psychological assets [8,21].

Despite the multiple assessment methods and their usefulness in the talent identification process, less is known about the discriminant capacity of psychological variables. Mustafovic and colleagues [22] suggested that talent identification and selection processes could benefit from the incorporation of more psychological variables to get a better sense of players' true worth. A good example is rugby, where several psychological variables, including perfectionism, learning strategies, motivation and mental health, are part of the talent identification process [23]. With regard to ice hockey, Gábor [24] offered some interesting insights by testing the discriminant ability of on-/off-ice motor skills and psychological assets to explain success and performance among the best Hungarian players under 18 years old. The measurement of psychological characteristics using validated tools is far superior to the qualitative observation of players that is often used by recruiters or coaches and lacks consistency [25]. Consequently, we believe that a more holistic approach that includes testing protocols that consider the multidimensionality of sports talent is a promising way to improve our understanding of talent identification in ice hockey. The addition of psychological variables measured with validated tools would complete the profile of these athletes, making sure that the best are ultimately retained while allowing stakeholders to better supervise the players and ensure they are mentally healthy.

This study aimed to investigate the athlete evaluation process in Canadian hockey (e.g., Quebec) in the early phase of the talent identification process. Because ice hockey is Canada's national winter sport, excellence is achieved through the rigorous observation and monitoring of player development. Every four years, each province develops a team composed of players under 15 years old for males and under 16 years old for females to compete at the Canada Games. In other years, Hockey Canada asks the provincial federations to organize development camps regarding the evaluation process for all international events, including the Hlinka Gretzky Cup for men, the Summer Series against USA for women and, of course, the World Junior Championships (https://www.iihf.com/en/tournaments; accessed on 4 January 2022). The purpose of this study is twofold. First, it aimed to describe the fitness, on-ice skating abilities and psychological characteristics of the male and female adolescent players that were pre-selected to take part in the 2021 Équipe Québec development evaluation camp: https://www.hockey.qc.ca/fr/page/excellence/equipe_quebec.html; accessed on 21 July 2021).

The camp serves as the first phase of team selection for national competitions involving provincial teams. Second, it aimed to verify the discriminant capacity of each attribute tested in the evaluation camps. In summary, this investigation can potentially strengthen our understanding of the talent identification process in the early stages of competitive hockey. Since sports talent is multidimensional [8,23], we hypothesized that some key variables would be discriminant and help to differentiate selected and non-selected players at the end of the camp.

2. Materials and Methods

2.1. Sample and Procedures

This study was developed in collaboration with researchers and the governing bodies of Hockey Quebec, the province's ice hockey federation. A total of 199 players between the ages of 14 and 16 years (86 boys: 43%, 14 years old; 113 girls: 57%, 16 years old) were invited to the Team Quebec evaluation camp. The criteria for invitation refer to players' regular season performances. Indeed, both prospect (e.g., male and female) camps serve as an important talent identification stage to determine those who will represent Quebec in national competitions. A week before each camp, players were informed about the research project during an online information meeting. Those who agreed to participate were asked to sign a consent form (if <16 years old, the parents signed). The project was approved by the ethics board of the researchers' institution (CER-21-278-07.09). The full protocol included three categories of measures (as described in the following section): (1) off-ice fitness tests (8 tests), (2) on-ice skating abilities (2 tests) and (3) psychological attributes (6 measures). The testing procedure was completed on three separate days (day 1—pre-camp questionnaire, day 2—off-ice fitness tests, day 3—on-ice tests) during one weekend camp.

2.2. Measures

Measures were selected based on two criteria. The first was related to Hockey Canada's standards to ensure that some tests were aligned with those of other provinces. The second was related to the scientific literature on testing and refers to specific components identified as potential determinants of performance (or talent) in ice hockey. For off-ice fitness, ten variables were measured and divided into four categories: (1) anthropometric measures, (2) lower limb power, (3) running and VO_{2max} and (4) upper limb power. For on-ice fitness, two skating tests were used to measure skating speed and agility. Finally, two categories of psychological measures were assessed: personality traits and grit.

2.2.1. Off-Ice Fitness Tests

Description of Testing Session

Anthropometric measures are commonly used in player evaluation to assess the body composition of a cohort of ice hockey players [15,16,26–28]. To measure height, the athlete stood on the stadiometer platform with their shoes off. Their feet were together and heels were supported on the base of the device, with their arms alongside the body. The participant was instructed to gaze outward, and the measurement was taken following a maximum respiration rate to the nearest 0.5 cm. We assessed body weight simultaneously by instructing the athlete to step on the scale (with shoes off). Results were collected in kilograms (to the nearest 0.1 kg).

Off-ice fitness testing protocol (see Table 1): Before the testing session, a warm-up was administered and supervised by strength conditioning coaches and trainers who were certified kinesiologists. The warm-up consisted of a 15 min session that was a combination of short runs (8–10 min) and plyometric exercises (5–7 min). Athletes were then invited to perform dynamic stretches for both the upper and lower limbs, as well as a few core activation workouts. The warm-up was performed by both cohorts.

Muscular power (upper and lower body): Since skating is a core element of performance, several studies have examined lower body power as a predictor of performance [29–32]. The vertical jump and horizontal jump tests are often cited in studies and appear frequently in the NHL's combine tests list (NHL Central Scouting Combine fitness results: https://link.nhl.com/centralscouting/public/; accessed on 10 December 2021). Table 1 describes the protocols that were implemented to collect these data. For the vertical jump test, a Vertec (Power Systems, Knoxville, TN, USA) (Power Systems: https://www.power-systems.com/shop/product/vertec, accessed on 10 December 2021) was used to assess this variable. Upper body strength, power and endurance also play a natural role in ice hockey [33]. Indeed, we chose grip strength and pull-ups considering that they are omnipresent in the NHL combine research (NHL Central Scouting Combine fitness results: https://link.nhl.com/centralscouting/public/; accessed on 10 December 2021) [12]. The seated medicine ball throw is now also part of the NHL's combine and is known as a reliable way to assess upper body muscular power [34]. A Ballistic Ball™ (Assess2Perform, Petoskey, MI, USA) (Move Factor X Ballistic Ball A2P: https://movefactorx.com/, accessed on 10 December 2021) was used in this regard.

Table 1. Off-ice tests explained.

Variable Test	Measures (Units) [Instrumentation]	Protocol
Lower body power Vertical jump	Height of jump (cm) [Vertec]	Athlete stands under the testing device to measure their maximum height with their arm at maximum flexion. Once the height is recorded, athlete stands at an elbow's distance from the testing device and performs one pre-trial jump. Athlete bends their legs and pauses for a second before jumping while reaching out as high as possible. The best of 3 attempts is retained.
Broad jump	Length of jump (cm) [Tape on floor]	Athlete positions both feet behind the line. Legs are bent quickly and arms are swung back and forth to initiate the jump. Athlete must land in control, and once stable on their feet, the foot that travels the least distance is measured. The best of 3 attempts is retained.
Aerobic capacity Léger 20 m Shuttle	Maximum oxygen consumption (mL/kg/min) [Tool kit]	The protocol involves the repetition of 20 m shuttle runs to maximal fatigue. Athlete must stand 1 m behind the line before the next beep before changing direction and continue to run They must wait for the signal before starting. The speed increases with each level. Athlete must follow the rhythm of the soundtrack.
Running speed/agility 30 m sprint	Time (s) Acceleration curve (m/s) [Swift timing gates] [Stalker Radar]	Athlete stands with their foot behind the starting line. When the Swift Gate turns green and beeps, athlete can initiate the sprint at any time. They sprint as quickly as possible to the finish line. Athlete makes two attempts separated by a 3 min break to achieve their best time. Athlete must be informed that they should decelerate only after crossing the sprint distance to obtain the best possible time.
5-10-5 agility drill	Time (s) [Swift timing gates]	Athlete places their foot on starting line. They then move as quickly as possible to a line from the end past the photocells and touche the line with their hand. Must move as quickly as possible to the line at the other end. Athlete touches the line with their hand before returning to the center as quickly as possible (best of 2 trials).
Upper body strength Grip strength	Sum of grip test (kg) [Dynamometer]	Dynamometer is set to 0 before each test. Athlete grabs dynamometer in a neutral grip and keeps it close to hips. Exhales heavily and compresses it as hard as possible. Athlete then returns the dynamometer to the evaluator who records the result and resets it at 0 for the next test. Two attempts per hand are allowed and the best result per side is retained.
Vertical pull up	Maximum reps (*n*) [Bar for hanging]	Athlete hangs from the bar with arms straight and performs an arm/back pull to raise the body and lift the chin over the bar. They then descend, fully unlocking the elbows before the next repetition. Count to 2 consecuitve repetitions before stopping the test. Swings and help from the legs are not allowed.
Seated medicine ball-throw	Power (W) [Move Factor Ballistic Ball A2P]	Athlete sits on the ground with back against wall and feet separated by a distance of 60 cm. They hold the ball against their chest at the sides, a little behind the center, with forearms parallel to the ground. They then throw the ball as hard as possible while keeping their back against the wall. The distance of the throw is recorded with the accelerometer in the ball. Three attempts are allowed and the best result is retained.

Aerobic capacity, speed and agility: In addition to lower body power, aerobic capacity [35] and running agility [36] are also often assessed in both the scientific literature and the field of testing in ice hockey. We measured running speed by administering the 30 m sprint [37] with a Stalker Pro Radar II (Stalker Sport, Richardson, TX, USA) (Stalker Sport: https://stalker.sport/pro-ii/, accessed on 10 December 2021) combined with

Swift Speedlight photocells (Swift Performance, Northbrook, IL, USA) (Swift Performance: https://swiftperformance.com/, accessed on 10 December 2021). For the off-ice agility, we used the Pro-Agility Drill 5-10-5 [38] with a second set of Swift Speedlight photocells.

2.2.2. On-Ice Skating Tests

Only two on-ice tests were chosen for the study. In the interest of efficient time management, this part of the testing protocol was limited to 20 to 24 players on the ice for 50 min of ice time. The 44 m sprint and the Finnish Vierumaki's ice hockey centre of excellence skating agility test were selected. Four sessions of ice time were needed to screen all the participants, both males and females.

Skating speed and acceleration: The 44 m sprint [39] was measured with a Stalker Pro Radar II (Stalker Sport, Richardson, TX, USA) (Stalker Sport: https://stalker.sport/pro-ii/ (accessed on 10 December 2021)) combined with three pairs of Swift Speedlight photocells (Swift Performance, Northbrook, IL, USA) (Swift Performance: https://swiftperformance.com/ (accessed on 10 December 2021)), one at the start, a second at 6 m to assess the 0–6 m acceleration and a third at 44 m. Figure 1 illustrates the setup of the 44 m sprint test. In the sprint test, the athlete is in the standing position with one foot behind the starting line. After the Swift Gate turns green and beeps, the athlete can initiate a sprint at any time. They skate as quickly as possible to the 6 m gates, then to the 44 m gates, in a straight line. Two attempts are separated by a 3 min break to achieve the best time. The athlete must be informed that they decelerate only after crossing the sprint distance to obtain the lowest possible time.

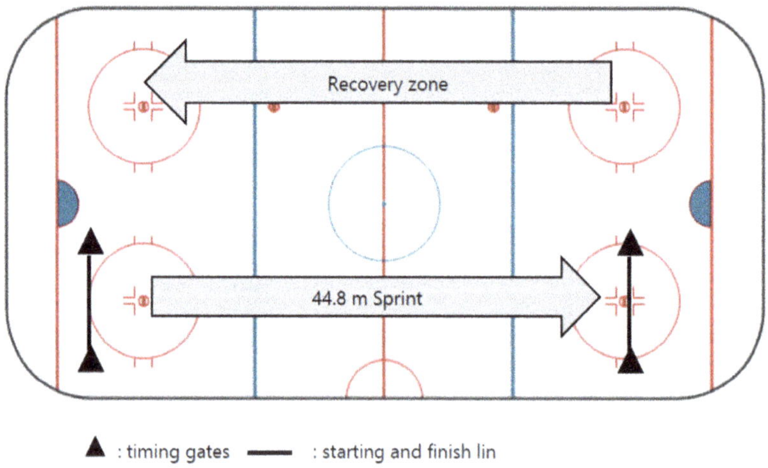

Figure 1. The 44.8 m skating sprint test.

Skating agility: We used one of Vierumaki's Ice Hockey Centre of Excellence's skating tests (https://iihce.fi/suomeksi/Testaaminen/Pohjola-leiritestit/tabid/1150/Default.aspx#/material/872/2401; accessed on 4 January 2022) (illustrated in Figure 2), which measures agility for a wide range of movements in the hockey player's repertoire [40]: explosive start, braking, short sprints, sharp/tight turns, open pivots and backward skating (Figure 2). Illustrates the four-step design of the test. For the skating agility test, the six cones are placed in a rectangle shape, with three pairs of two cones and each pair separated horizontally by 9 m and vertically by 7 m. Step 1: Athlete stands with their foot placed behind the starting line. After the Swift Gate turns green and beeps, they can initiate the test at any time. The athlete first skates all the way to the other end of the circuit. Athlete brakes and sprints back to the line of the second pair of cones. Step 2: After this second step, the athlete aims to their right, outside the cone to perform two consecutive short turns around the two cones of the third pair. Step 3: After the short turns, the athlete aims for the second pair of cones, where they perform two open pivots (facing the starting line side) around the cones. Step 4: After the open pivots, the athlete aims for the first pair of cones. Turning around the first one, they pivot completely and skate backward to the third pair, performing a slalom with the second pair. After completing one side of backward skating, they move forward to the other side and repeat the pivot for the backward slalom. After the second side is done, the athlete sprints back to the starting line. The best of two trials is recorded.

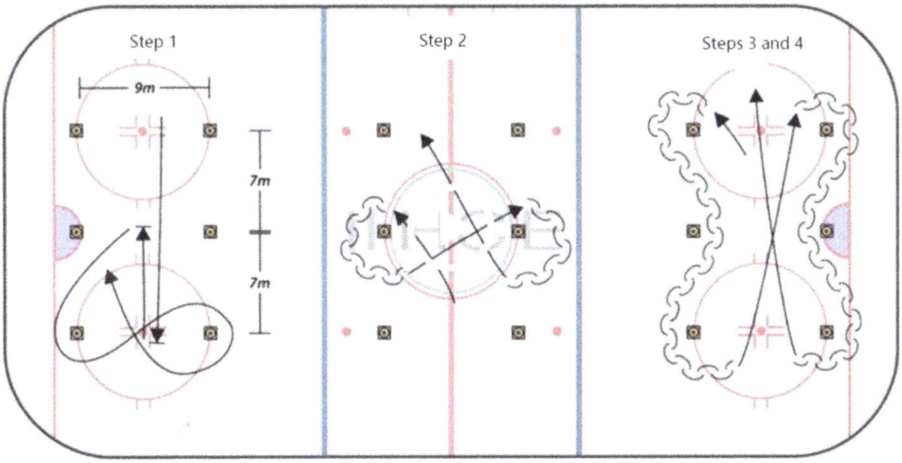

Figure 2. The skating agility test (adapted from the International Ice Hockey Centre of Excellence, Vierumaki, Finland). Illustration shows the three-step procedures, performed on a single circuit. (Video demonstration: https://iihce.fi/suomeksi/Testaaminen/Pohjola-leiritestit/tabid/1150/Default.aspx#/material/872/2401, accessed on 4 January 2022).

2.2.3. Psychological Measures: Personality Traits and Grit

Regarding psychological measures, data were collected by means of online questionnaires (Qualtrics software) and participants were given one week before camp to ensure they were in the right frame of mind. Two psychological constructs were measured: personality traits [41,42] and grit [43,44]. Personality traits were assessed with the French version of the Big Five Inventory (BFI-Fr), which is a 44 item Likert-type scale (1 to 5) that has been used for several years in research and clinical settings and is validated in several different languages and populations. It includes conscientiousness, agreeableness, extraversion, neuroticism and openness to experience, which are traits that are significantly associated with sports performance. For example, a high level of extroversion and conscientiousness and a low level of neuroticism would be expected of an elite athlete in a team sport like ice hockey [43]. Grit is a psychological asset that refers to an athlete's ability to face competition and not give up [44]. We assessed grit with the Short Grit Scale, which is an 8-item Likert-type questionnaire that has shown solid psychometric properties in past research [44].

2.3. *Statistical Analyses*

We calculated descriptive statistics for all data and verified the distributions to identify potential outliers. We also verified the distribution and checked whether assumptions of normality had been violated. The preliminary results indicated no major deviation from normality regarding all off-ice and on-ice measures. Few data were missing because all players took part in the evaluation camp. To verify the contribution of each part of the protocol related to talent identification, as well as the discriminant capacity of each group of components, discriminant analyses were conducted in three consecutive phases. All analyses were performed using SPSS software and based on the recommendations of

Tabachnik and Fidell [44]. Discriminant analysis [45] makes it possible to verify which variables are related to group membership (e.g., selected versus non-selected) by assigning a discriminant function to each variable under study. In this regard, group membership was accorded following the evaluation camp, where 44 males (out of 86) and 54 females (out of 107) were retained for the selection process. Those retained were categorized as "selected" (value = 1) and those not retained as "not selected" (value = 0). After the selection process was completed, analyses were conducted on five sets of variables: (1) anthropometric measures (height and weight), (2) off-ice fitness (lower and upper body strength-power), (3) running ability (aerobic capacity, speed and agility), (4) on-ice skating (skating speed and agility) and (5) psychological factors (personality traits and grit). We conducted the analyses separately to prevent the overlapping of factors and verify the discriminant capacity of each category of measures. For each model, we conducted direct discriminant analyses based on our interpretations based on Box's M and Wilks' lambda statistics. Finally, we analyzed the discriminant function for each set of predictors and verified the proportion (%) of correct categories regarding the selected players.

3. Results

3.1. Sample Characteristics

Table 2 presents the sample's characteristics. The players invited to each selection camp came from two age-group categories. Males were taller than females ($F_{height} = 68.03$, $p < 0.001$), but no differences were observed regarding body weight. Proportions for player positions were similar across groups owing to the specificity of ice hockey in which a usual roster is mainly composed of 47 as forwards, 31 as defense and 8 as goaltenders. For the male cohort, goaltenders tended to be taller and heavier than defense and forwards ($F_{height} = 6.75$, $p = 0.002$; $F_{weight} = 6.38$, $p = 0.003$). No player-position-related differences were observed regarding the anthropometric profile of the female cohort.

Table 2. Sample characteristics (mean (M) and standard deviation (SD)).

		n	Age (M ± SD)	Height (m)	Weight (kg)
Males $^{+}$	Goaltenders	8	13.81 ± 0.40	1.79 ± 5.41 *	71.68 ± 7.54 *
	Defense	31		1.73 ± 7.13	67.35 ± 9.81
	Forwards	47		1.70 ± 6.98	62.09 ± 7.69
	Total	86		1.72 ± 7.37 $^{++}$	64.88 ± 9.03
Females	Goaltenders	11	14.97 ± 0.92	1.65 ± 3.72	63.72 ± 8.88
	Defense	41		1.65 ± 5.71	64.18 ± 8.73
	Forwards	61		1.64 ± 5.97	61.41 ± 8.48
	Total	113		1.64 ± 5.67	62.64 ± 8.63

* $p < 0.01$. $^{+}$ Males were taller than females ($p < 0.001$); $^{++}$ Goaltenders were significantly taller.

3.2. Objective 1: Fitness, On-Ice Abilities and Personality Traits of Elite Adolescent Hockey Players

Table 3 shows the descriptive statistics of each of the testing protocol's variables. Males displayed higher scores than females on all the off-ice fitness tests (all group comparisons were significant at $p < 0.01$ or 0.001). The same pattern was observed regarding the on-ice skating tests, which showed that boys were faster, had better acceleration and were more agile. Male goaltenders also performed better than female goaltenders on the shuffle tests. However, different patterns were observed when comparing participants' personality traits. The female cohort displayed higher scores for three traits: extraversion ($t_{(df)} = 2.41_{(187)}$, $p = 0.017$), conscientiousness ($t_{(df)} = 2.16_{(187)}$, $p = 0.032$) and openness ($t_{(df)} = 3.55_{(187)}$, $p < 0.001$). Comparisons between the selected and non-selected players (males and females) were presented in the sections on discriminant analyses. Finally, we compared all measures according to the players' positions (goaltenders, defense and forwards) and found no significant differences (all p-values > 0.05), suggesting that players' fitness and abilities were similar for males and females.

Table 3. Sample characteristics regarding off-ice tests for players who took part in Team Quebec's summer selection camp.

	Males				Females			
	G (n = 8)	D (n = 31)	F (n = 47)	T (n = 86) **	G (n = 11)	D (n = 41)	F (n = 61)	T (n = 113)
Off-ice (fitness)								
VO_2 (mL/kg/min)	46.69 ± 5.05	49.63 ± 4.87	50.35 ± 4.76	49.75 ± 4.88	44.59 ± 4.08	43.86 ± 5.54	44.49 ± 5.74	44.28 ± 5.49
Broad jump (cm)	226.33 ± 14.97	226.33 ± 14.87	227.62 ± 15.25	226.72 ± 14.44	194.10 ± 10.12	193.07 ± 15.00	196.27 ± 15.79	194.88 ± 15.04
Vertical jump (cm)	47.70 ± 5.19	49.04 ± 6.35	48.28 ± 7.11	48.46 ± 6.62	43.64 ± 5.66	42.87 ± 5.82	44.58 ± 6.27	43.87 ± 6.05
Grip strength (kg)	102.50 ± 19.53	99.71 ± 16.46	92.79 ± 16.23	95.83 ± 16.79	71.09 ± 11.88	74.10 ± 10.39	71.78 ± 15.39	72.56 ± 13.40
Chin-ups (n)	6.63 ± 4.14	4.19 ± 9.13	9.19 ± 3.22	8.93 ± 3.71	1.36 ± 2.01	2.68 ± 3.11	3.07 ± 2.95	2.76 ± 2.95
Ball throw (W)	165.50 ± 18.88	167.35 ± 19.24	171.85 ± 30.82	169.64 ± 26.08	129.64 ± 17.93	141.27 ± 18.83	134.80 ± 18.39	136.66 ± 18.74
30 m sprint (s)	4.79 ± 0.22	4.75 ± 0.19	4.79 ± 0.21	4.78 ± 0.20	5.19 ± 0.23	5.13 ± 0.23	5.05 ± 0.23	5.09 ± 0.23
Agility 5-10-5 (s)	5.29 ± 0.17	5.36 ± 0.18	5.29 ± 0.19	5.31 ± 0.19	5.71 ± 0.30	5.73 ± 0.27	5.65 ± 0.26	5.69 ± 0.27
On-ice (skating)								
0–6 m accel (s)	n/a	1.31 ± 0.08	1.32 ± 0.07	1.32 ± 0.07	n/a	1.42 ± 0.78	1.43 ± 0.09	1.42 ± 0.09
44 m sprint (s)	n/a	6.25 ± 0.17	6.28 ± 0.25	6.27 ± 0.22	n/a	6.71 ± 0.29	6.77 ± 0.31	6.75 ± 0.30
Agility (s)	n/a	35.54 ± 1.12	35.34 ± 1.40	35.42 ± 1.29	n/a	37.73 ± 1.53	38.27 ± 2.30	38.05 ± 2.04
Shuffles $^{\delta}$ (s)	18.31 ± 1.58 *	n/a	n/a	n/a	19.80 ± 1.19	n/a	n/a	n/a
Recoveries $^{\delta}$ (s)	28.25 ± 2.87 *	n/a	n/a	n/a	31.51 ± 3.24	n/a	n/a	n/a
Psychological variables (5-point scales)								
Extraversion	3.57 ± 0.37	3.33 ± 0.28	3.31 ± 0.60	3.34 ± 0.34	3.49 ± 0.58	3.52 ± 0.34	3.44 ± 0.39	3.47 ± 0.39 ***
Conscientiousness	3.70 ± 0.21	3.66 ± 0.30	3.46 ± 0.28	3.56 ± 0.30	3.69 ± 0.43	3.67 ± 0.34	3.65 ± 0.31	3.66 ± 0.33
Neuroticism	2.62 ± 0.37	2.69 ± 0.41	2.64 ± 0.52	2.66 ± 0.47	2.75 ± 0.60	2.79 ± 0.49	2.69 ± 0.46	2.73 ± 0.48
Agreeableness	2.86 ± 0.24	3.19 ± 0.54	3.10 ± 0.38	3.11 ± 0.44	3.13 ± 0.51	3.15 ± 0.51	3.10 ± 0.50	3.12 ± 0.50 ***
Openness	3.30 ± 0.35	3.26 ± 0.59	3.28 ± 0.45	3.28 ± 0.50	3.55 ± 0.43	3.63 ± 0.46	3.45 ± 0.48	3.53 ± 0.47 ***
Grit	4.13 ± 0.61	3.98 ± 0.45	3.93 ± 0.61	3.97 ± 0.56	4.08 ± 0.61	3.94 ± 0.54	3.76 ± 0.61	3.85 ± 0.60

G: goaltenders; D: defense; F: forwards; T: total. * Boys displayed better scores on goaltenders' tests: $p < 0.05$. ** Boys displayed better scores on fitness and on-ice tests: $p < 0.001$. *** Girls displayed higher scores regarding psychological measures: $p < 0.05$. $^{\delta}$ Goaltenders' on-ice tests: in the female group, 9 goaltenders completed the test.

3.3. *Objective 2: Discriminant Capacity for Each Facet of the Talent Identification Protocol*

3.3.1. U15 Male Selection

Table 4 illustrates the results of the discriminant analyses for the male cohort. As mentioned, no significant results were found (see Appendix A, Table A1). Indeed, no significant differences were observed, even when the mean scores of the "selected" versus the "non-selected" players were compared, with the exception of the med-ball throw, where differences favored the 44 selected players ($F_{(df)} = 2.80_{(2)}$, $p < 0.10$). No significant discriminant functions were noted in each of the five proposed models. In model 4, none of the variables showed a significant discriminant function, but the selected players had higher scores for grit ($F_{(df)} = 3.05_{(1,75)}$, $p < 0.10$). We retained loadings higher than 0.3 to identify the most discriminating variables, which are identified in Table 5. Each of the five models classified 48 to 60% of selected players.

3.3.2. U16 Female Selection

As shown in Table 5, we found significant results for the female cohort, where some variables had a discriminant effect on the selection process. Group comparisons (see Appendix B, Table A2) revealed that selected players displayed higher scores in 10 indicators, in which 9 of these 10 variables had a significant discriminant function. Except for anthropometric measures (where we found no significant results), other measures such as off-ice fitness, skating abilities and personality traits revealed a significant discriminant function. With regard to off-ice fitness, aerobic capacity ($VO^2{}_{max}$), leg power, speed and agility were identified as discriminant factors. On the ice, skating speed and agility had a significant discriminant function. For psychological measures, agreeability and grit have a significant discriminant function. Each of the five models correctly classified 52–74% of selected players. In summary, models 2,3,4 and 5 displayed significant discriminant functions.

Table 4. Results from the discriminant analyses for the 44 selected male players (standardized discriminant function coefficients).

Variables	Box's M	Wilk's Lambda ($\chi^2{}_{(df)}$)	Discriminant Function		Classification
Model 1: Anthropometry	1.341 ^{ns}	0.963 (13.780$_{(2)}$) ^{ns}	Weight	1.116	49%
			Height	−0.157	
Model 2: Off-ice fitness	32.620 **	0.963 (2.926$_{(5)}$) ^{ns}	Broad jump	0.463	56%
			Vertical jump	−0.015	
			Med-ball throw	0.954	
			Grip strength	0.404	
			Chin-ups	−0.105	
Model 3: Running ability	3.727 ^{ns}	0.994 (0.490$_{(5)}$) ^{ns}	VO_{2max}	0.950	48%
			30 m sprint	0.080	
			50-10-5 agility	0.124	
Model 4: Skating performance	8.098 ^{ns}	0.995 (0.346$_{(3)}$) ^{ns}	44 m sprint	0.154	51%
			6 m accel	0.361	
			Agility	0.952	
Model 5: Psychological measures	55.073 **	0.889 (8.457$_{(6)}$) ^{ns}	Extraversion	−0.184	60%
			Agreeableness	0.318	
			Conscientiousness	−0.287	
			Neuroticism	−0.036	
			Openness	0.427	
			Grit	0.571 *	

** $p < 0.01$; * $p < 0.10$: Selected players displayed higher grit scores. Bold text: loadings > 0.30 represent the most discriminating variables. ^{ns} non significant.

Table 5. Results of the discriminant analyses for the 54 selected female players (standardized discriminant function coefficients).

Variables	Box's M	Wilk's Lambda ($\chi^2_{(df)}$)	Discriminant Function		Classification
Model 1: Anthropometry	1.743 ns	0.996 (0.464 $_{(2)}$) ns	Weight	**0.892**	52%
			Height	0.203	
Model 2: Off-ice fitness	20.661 ns	0.876 (13.889$_{(5)}$) **	Broad jump	**0.879** **	62%
			Vertical jump	**0.401**	
			Med-ball throw	0.283	
			Grip strength	0.281	
			Chin ups	**0.753** **	
Model 3: Running ability	7.797 ns	0.711 (35.946$_{(5)}$) **	VO_{2max}	**0.829** **	74%
			30 m sprint	**0.665** **	
			50-10-5 agility	**0.859** **	
Model 4: Skating performance	29.497 **	0.651 (40.161$_{(3)}$) **	44 m sprint	**0.651** **	73%
			6 m accel	**0.494** **	
			Agility	**0.975** **	
Model 5: Psychological measures	60.351 **	0.930 (7.921$_{(6)}$) **	Extraversion	−0.110	61%
			Agreeableness	**0.844** **	
			Conscientiousness	**0.323**	
			Neuroticism	**0.351**	
			Openness	−0.213	
			Grit	−0.043	

** $p < 0.001$: Selected players displayed higher mean scores for the selected variables. Bold text: loadings > 0.30 represent the most discriminant variables. ns non significant.

4. Discussion

In ice hockey, team selection is a complex process that involves the observation and assessment of multiple variables. According to Tarter [8], the combination of game performance, fitness, on-ice attributes and psycho-cognitive factors is probably the best way to determine a player's potential. This study sheds some light on talent identification for elite adolescent Canadian hockey players. Given the importance of ice hockey in Canada and this country's culture of evaluation camps, we believe the present study offers an excellent opportunity to refine our understanding of the factors that distinguish players who are prioritized or categorized as "top prospects" in their field. Additionally, this study offers substantial knowledge about the level of Canadian adolescent elite hockey players, especially because of the multiple determinants that were observed (19 variables) and the population studied, which consisted of male and female hockey players.

The first part of our study shows that fitness measures have similar patterns when male and female participants are compared. As expected, boys displayed higher scores than girls, which can be explained by the physical maturation and physiological aspects that favor male athletes at this stage [46]. With the exception of male goaltenders' height, we found no fitness differences related to the players' positions. This result appears to contradict the findings of previous studies, such as that of Daigle et al. [13], who showed potential position-related differences for certain fitness components. In our opinion, young athletes' stage of development may explain the lack of position-specific differences regarding players' fitness. Measurements took global fitness measures into account, which was probably insufficient to detect position-specific skills. These results suggest that introducing "position-specific" skills, such as backward skating (for defense) and skating skills with puck control (for forwards) would be relevant for future research. In terms of differences in personality traits, female players scored significantly higher than male players on extraversion, agreeableness and openness, which is consistent with the scientific literature. However, the score for neuroticism was not significantly higher in the female sample, contrary to what was expected [47,48].

The second part of the study was designed to verify the discriminant capacity of each variable tested in the selection process. Results from our analyses offer interesting insights, particularly because the substantial differences they revealed were based on gender. In the male cohort, we found that not all five models were discriminatory. At first glance, this suggested that the current protocol with the current group of players did not contribute definitively to the selection process in the male sample. However, certain measures, including muscular strength, aerobic capacity, skating agility/acceleration and grit, tended to be more strongly associated with the players selected during the process. The picture was different in the female sample, as the fitness tests, running skills and skating performance successfully classified between 62 and 74% of the selected athletes. On a different note, the higher grit score of the selected male players was to be expected since this was noted in several studies. However, a higher conscientiousness score would also have been expected regarding these players, as it is the strongest predictor of success in sports and is generally associated with grit. Interestingly, the protocol's discriminant capacity was significant in the female cohort. The three personality traits that appeared to distinguish the selected female players were agreeableness, conscientiousness and neuroticism, which is plausible in a team sport context. These three traits are generally associated with success in team sports, as conscientiousness predisposes players to invest more effort, a low level of neuroticism provides greater emotional stability under pressure and agreeableness promotes healthy relationships with teammates [49]. The fact that the training camp contained high-level athletes only very likely reduced the discriminatory power of personality traits; a comparison with recreational athletes would possibly have demonstrated much more obvious results. In general, both male and female participants demonstrated the expected profile of low neuroticism and high conscientiousness and agreeableness [48–50].

In our view, the divergent results (related to the discriminant capacity of each protocol) can be explained from a gender perspective using socio-demographics. Even if the initial samples (selected and non-selected) were similar in size, they were not similar regarding the representation of the province's pool of elite players since Quebec had over 450 male players classified as U15–U18 elite players [50]. The numbers were significantly lower for the female group, where 275 (U15–U18) players were categorized as elite. The development structure suggested that 80 boys represented slightly less than 20% of the highly competitive players as compared with 30% of competitive female players. These numbers indicate that the testing protocol was probably less effective for determining the best prospects in more homogeneous groups, as was the case for the male cohort. Our study has certain practical implications. We believe that two approaches are conducive to the development of improved talent identification protocols for differentiating selected and non-selected athletes. The first is to increase the number of tested players (regarding the male cohort) while considering factors such as birthplace and birth month, which may prove that current tests can effectively identify the best U15 prospects. Teoldo and Cardos discuss these issues with regard to Brazilian soccer [51], which is definitely comparable to Canadian ice hockey in terms of the sport's popularity. The second approach is to integrate tests that could help to differentiate the most gifted players. In this regard, we think that attributes such as the ability to repeat sprints [35] and skating agility in multiple contexts [52] should be considered in future protocols. We also believe the said protocols should include multiple game performance indicators [13] in view of the association between fitness, on-ice skills and game performance.

Despite its contribution, this study had limitations. The first was the potential selection bias. To participate in the study, players had to be registered in a league governed by Hockey Quebec. This means that the participants were engaged in a very similar pattern of sports development, possibly leading to an almost identical athletic profile, reducing the discriminant capacity of the protocols, especially among males. The Hockey Quebec development model for males was implemented in 2011, whereas its implementation for females is far more recent (2015). Another limitation of the sample concerns the players' age groups and physical maturation. Even if further analyses from the current database confirm the physical maturation of most participants, we believe the maturation process is an important factor to consider, especially in sports like ice hockey, where physicality prevails. Additionally, we recognize that certain key talent indicators are missing in the present investigation.

As Tarter stipulates [8], game observation is crucial to the talent identification process. Despite the challenges involved, establishing associations between fitness and on-ice skills with game performance is a promising avenue for future research. In practical terms, knowing which factors are related with team selection is useful for coaches and stakeholders (e.g., scouts, program directors, strength conditioning coaches) in a way that it allows to identify talent at early stages of hockey expertise. In addition, it seems that these factors might vary according players' gender, which means that same testing protocols are not assessing talent in the same way. Focusing on more specific ice hockey attributes (e.g., repeated sprints, skills with the puck, anaerobic capacity) for male selections might be an interesting alternative. For female talent selection, our results suggest that the suggested protocol is sufficient to select the best players. Even if game observation was outside the scope of this study, we think that systematic observation and the use of advanced technologies for assessing performance in real settings will be prioritized in subsequent stages of this research. We also believe that the inclusion of perceptual-cognitive markers such as reaction time and decision making could contribute to talent identification [53]. Although some promising work has been done in the field, further developments are needed to establish associations with game performance in real settings.

5. Conclusions

In Canadian hockey, talent identification starts in late adolescence and is a crucial stage in the development of sport expertise. Player development in Quebec is monitored in off-season (summer) evaluation camps designed to identify the province's best prospects for future competitions. Knowledge about the discriminant capacity of each component of talent is crucial for stakeholders (e.g., coaches, scouts, program directors) in talent detection camps. This study refined our understanding of talent identification and revealed that the discriminant capacity of protocols differs based on gender. Interestingly, it showed that multiple components of sport talent help to classify the female players who emerged in the first phase of the selection process. Lower body power, upper body strength, running abilities and on-ice skating efficiency are useful components of the selection process. Inversely, such protocols are less conclusive for male cohorts, suggesting that further research is needed in this area. Future investigations can build on this study by focusing on variables assessed through game observation and other key factors for distinguishing the most promising talents.

Author Contributions: Conceptualization, J.L. and J.-F.B.; literature review, G.M.; methodology, J.L., J.-F.B., J.G.-R. and V.H.P.; software, J.L. and V.H.P.; validation, V.H.P.; formal analysis, J.L.; investigation, all; data curation, J.L. and J.G.-R.; writing—original draft preparation, J.L., J.-F.B. and V.H.P.; writing—review and editing, J.L.; visualization, J.L. and J.-F.B.; supervision, J.L.; project administration, J.L.; funding acquisition, J.L. All authors have read and agreed to the published version of the manuscript.

Funding: This research was funded by the Ministère de l'Éducation et de l'Enseignement Supérieur du Québec Projet Synergique-Hokey Québec, l'Institut National du Sport-Québec (National Sport Institute-Quebec) and Université du Québec à Trois-Rivières.

Institutional Review Board Statement: The study was conducted according to the guidelines of the Declaration of Helsinki and approved by the Institutional Review Board (or Ethics Committee) of the Université du Québec à Trois-Rivières (CER-21-278-07.09; delivered 8 July 2021).

Informed Consent Statement: Informed consent was obtained from all subjects involved in the study. All players signed a consent form for participation in this research. All parents were informed about the project and accepted their child's participation.

Data Availability Statement: Data in the actual form was used by researchers and collaborators in the actual project. Researchers agreed to give access to data upon request. Those interested should inform the corresponding author at: jean.lemoyne@uqtr.ca.

Acknowledgments: The authors wish to thank the management of Hockey Quebec: Marcel Patenaude (hockey development manager) for his collaboration on the research design, along with Katerine Aubry-Hébert (female hockey development) and Jonathan Marcotte-Levasseur (male hockey development). We are also grateful to all players who completed the protocols and the coaches who collaborated in the study. Furthermore, the authors acknowledge the contribution of the Service de l'Activité Physique et Sportive de l'UQTR regarding materials and staff.

Conflicts of Interest: The authors declare no conflict of interest.

Appendix A

Table A1. Comparisons between selected and non selected male players.

Indicators	Group	Mean ± SD	Confidence Intervals (CI 95%)	*F*	Eta^2 (CI 95%)
VO_2 (mL/kg/min)	NS (*n*) S (*n*)	49.51 ± 4.91 49.99 ± 4.88	48.00–51.03 48.45–51.53	0.19	0.02 (0.00–0.06)
Broad jump (cm)	NS S	225.45 ± 14.22 228.05 ± 14.72	221.01–229.88 223.34–232.75	0.66	0.01 (0.00–0.08)
Vertical jump (cm)	NS S	48.49 ± 5.27 48.45 ± 7.92	46.87–50.11 45.91–50.98	0.01	0.00 (0.00–0.00)
Ball throw (Watts)	NS S	169.79 ± 21.43 169.47 ± 30.46	163.27–176.31 159.98–178.97	0.01	0.00 (0.00–0.01)
Grip strength (kg)	NS S	94.65 ± 18.54 97.04 ± 14.85	89.01–100.29 92.49–101.67	0.43	0.01 (0.00–0.07)
Chin-ups (*n*)	NS S	8.86 ± 3.29 9.00 ± 4.13	7.86–9.86 7.71–10.28	0.03	0.00 (0.00–0.04)
30-m sprint (s)	NS S	4.77 ± 0.20 4.77 ± 0.20	4.71–4.83 4.71–4.84	0.01	0.00 (0.00–0.01)
Agility 5-10-5 (s)	NS S	5.31 ± 0.19 5.31 ± 0.18	5.25–5.37 5.25–5.37	0.008	0.00 (0.00–0.02)
0-6 m accel (s)	NS S	1.31 ± 0.07 1.31 ± 0.06	1.29–1.34 1.29–1.33	0.07	0.00 (0.00–0.06)
44m sprint (s)	NS S	6.27 ± 0.20 6.27 ± 0.23	6.20–6.33 6.18–6.35	0.00	0.00 (0.00–0.00)
Agility (s)	NS S	35.52 ± 1.26 35.31 ± 1.31	35.11–35.94 34.87–35.76	0.48	0.01 (0.00–0.09)
Shuffles (s) **	NS (*n* = 7) S (*n* = 7)	18.80 ± 1.79 17.82 ± 1.41	15.95–21.65 15.57–20.07	0.73	0.11 (0.00–0.51)
Recoveries (s) **	NS S	29.90 ± 1.67 26.60 ± 3.02	27.23–32.57 21.79–31.41	3.65 *	0.38 (0.00–0.68)
Extraversion	NS S	3.36 ± 0.32 3.32 ± 0.34	3.25–3.46 3.21–3.43	0.26	0.01 (0.00–0.07)
Agreeableness	NS S	3.06 ± 0.23 3.16 ± 0.57	2.98–3.14 2.97–3.34	0.89	0.01 (0.00–0.10)
Conscientiousness	NS S	3.59 ± 0.27 3.52 ± 0.31	3.50–3.68 3.42–3.62	1.03	0.01 (0.00–0.10)
Neuroticism	NS S	2.66 ± 0.42 2.65 ± 0.50	2.53–2.80 2.49–2.81	0.03	0.00 (0.00–0.04)
Openness	NS S	3.19 ± 0.46 3.35 ± 0.52	3.04–3.34 3.18–3.51	2.06	0.02 (0.00–0.13)
Grit	NS S	3.04 ± 0.33 3.00 ± 0.41	2.93–3.15 2.87–3.13	0.25	0.01 (0.00–0.07)

* $p = 0.10$; ** Goaltenders only; NS: not selected players; S: Selected players.

Appendix B

Table A2. Comparisons between selected and non selected female players.

Indicators	Group	Mean ± SD	Confidence Intervals (CI 95%)	F	Eta² (CI 95%)
VO_2 (mL/kg/min)	NS (*n*)	41.93 ± 5.34	40.51–43.35	26.71 *	0.19 (0.79–0.32)
	S (*n*)	46.80 ± 4.45	45.57–48.03		
Broad jump (cm)	NS	190.27 ± 14.45	186.48–194.06	12.75 *	0.11 (0.21–0.22)
	S	200.01 ± 14.12	196.08–203.95		
Vertical jump (cm)	NS	42.89 ± 5.22	41.52–44.27	3.12	0.03 (0.00–0.11)
	S	44.89 ± 6.71	43.06–46.72		
Ball throw (Watts)	NS	133.81 ± 17.96	129.13–138.49	2.92	0.03 (0.00–0.11)
	S	139.83 ± 19.24	134.52–145.13		
Grip strength (kg)	NS	72.40 ± 12.05	69.26–75.54	0.02	0.00 (0.00–0.02)
	S	72.72 ± 14.84	68.67–76.77		
Chin-ups (*n*)	NS	2.01 ± 2.56	1.34–2.68	8.39 *	0.07 (0.01–0.18)
	S	3.58 ± 3.15	2.71–4.45		
30-m sprint (s)	NS	5.17 ± 0.19	5.12–5.23	19.21 *	0.15 (0.05–0.27)
	S	4.99 ± 0.23	4.93–5.06		
Agility 5-10-5 (s)	NS	5.80 ± 0.24	5.74–5.87	31.05 *	0.22 (−0.09–0.34)
	S	5.55 ± 0.22	5.49–5.61		
0-6 m accel (s)	NS	1.44 ± 0.08	1.42–1.47	12.42 *	0.12 (0.02–0.24)
	S	1.39 ± 0.08	1.36–1.41		
44m sprint (s)	NS	6.86 ± 0.20	6.80–6.92	21.57 *	0.18 (0.06–0.31)
	S	6.60 ± 0.33	6.50–6.70		
Agility (s)	NS	39.17 ± 1.73	38.69–39.65	50.19 *	0.34 (0.19–0.47)
	S	36.79 ± 1.57	36.32–37.25		
Shuffles (s) **	NS (n = 8)	20.12 ± 1.27	15.95–21.65	1.32	0.16 (0.00–0.53)
	S (n = 8)	19.17 ± 1.27	15.58–20.07		
Recoveries (s) **	NS	32.28 ± 3.78	28.31–36.25	1.02	0.13 (0.00–0.50)
	S	29.97 ± 0.92	27.67–32.26		
Extraversion	NS	3.48 ± 0.33	3.39–3.58	0.18	0.01 (0.00–0.05)
	S	3.45 ± 0.43	3.33–3.57		
Agréabilité	NS	2.98 ± 0.27	2.91–3.06	8.60 *	0.07 (0.01–0.18)
	S	3.25 ± 0.62	3.08–3.42		
Conscientiousness	NS	3.62 ± 0.24	3.56–3.69	1.16	0.01 (0.00–0.08)
	S	3.69 ± 0.40	3.58–3.80		
Neuroticism	NS	2.68 ± 0.42	2.56–2.79	1.24	0.01 (0.00–0.08)
	S	2.78 ± 0.53	2.64–2.93		
Openness	NS	3.55 ± 0.48	3.41–3.68	0.33	0.01 (0.00–0.06)
	S	3.49 ± 0.45	3.37–3.62		
Grit	NS	2.90 ± 0.35	2.80–2.99	4.86 *	0.04 (0.00–0.014)
	S	3.08 ± 0.50	2.95–3.22		

* $p < 0.01$; ** Goaltenders only; NS: not selected players; S: Selected players.

References

1. Burr, J.F.; Jamnik, R.K.; Baker, J.; Macpherson, A.; Gledhill, N.; McGuire, E.J. Relationship of physical fitness test results and hockey playing potential in elite-level ice hockey players. *J. Strength Cond. Res.* **2008**, *22*, 1535–1543. [CrossRef]
2. Botteril, C. The psychology of professional hockey. *Athl. Insight* **2004**, *6*, 17–24.
3. Montgomery, D.L. Physiological profile of professional hockey players-a longitudinal comparison. *Appl. Physiol. Nutr. Metab.* **2006**, *31*, 181–185. [CrossRef] [PubMed]
4. Mascaro, T.; Seaver, B.L.; Swanson, L. Prediction of skating speed with off-ice testing in professional hockey players. *J. Orthop. Sports Phys. Ther.* **1992**, *15*, 92–98. [CrossRef]
5. Bracko, M.R. On-ice performance characteristics of elite and non-elite women's ice hockey players. *J. Strength Cond. Res.* **2001**, *15*, 42–47. [CrossRef]
6. Willis, K.L. *Hockey Grit, Grind & Mind: Your Playbook for Increasing Toughness, Focus, Drive, Resilience, Confidence, and Consistency in Today's Game*; Morgan James Publishing: New York, NY, USA, 2018.
7. Miller, S. *Hockey Tough: A Winning Mental Game*; Human Kinetics Publishers: Champaign, IL, USA, 2003.
8. Tarter, B.C.; Kirisci, L.; Tarter, R.E.; Weatherbee, S.; Jamnik, V.; McGuire, E.J.; Gledhill, N. Use of aggregate fitness indicators to predict transition into the National Hockey League. *J. Strength Cond. Res.* **2008**, *23*, 1828–1832. [CrossRef]

9. Quinney, H.A.; Dewart, R.; Game, A.; Snydmiller, G.; Warburton, D.; Bell, G. A 26-year physiological description of a National Hockey League team. *Appl. Physiol. Nutr. Metab.* **2008**, *33*, 753–760. [CrossRef]
10. Stanula, A.J.; Roczniok, R.K.; Gabrys, T.T.; Szmatlan-Gabrys, U.B.; Ozimek, M.J. Relationships of physical fitness test results and player performance indicators in national-level ice hockey players. *Soc. Integration. Educ. Proc. Int. Sci. Conf.* **2018**, *4*, 264–273. [CrossRef]
11. Ransdell, L.B.; Murray, T.M.; Gao, Y. Off-ice fitness of elite female ice hockey players by team success, age, and player position. *J. Strength Cond. Res.* **2013**, *27*, 875–884. [CrossRef]
12. Vescovi, J.D.; Murray, T.M.; Fiala, K.A.; VanHeest, J.L. Off-ice performance and draft status of elite ice hockey players. *Int. J. Sports Physiol. Perform.* **2006**, *1*, 207–221. [CrossRef]
13. Daigle, A.P.; Brunelle, J.F.; Belanger, S.; Lemoyne, J. Functional performance tests, on-ice testing and game performance in elite junior ice hockey players. *J. Hum. Kinet.* **2022**, in press.
14. Balyi, I.; Way, R.; Higgs, C. *Long-Term Athlete Development*; Human Kinetics: Champaign, IL, USA, 2013.
15. Roczniok, R.; Maszczyk, A.; Stanula, A.; Czuba, M.; Pietraszewski, P.; Kantyka, J.; Starzyński, M. Physiological and physical profiles and on-ice performance approach to predict talent in male youth ice hockey players during draft to hockey team. *Isokinet. Exerc. Sci.* **2013**, *21*, 121–127. [CrossRef]
16. Vigh-Larsen, J.F.; Beck, J.H.; Daasbjerg, A.; Knudsen, C.B.; Kvorning, T.; Overgaard, K.; Mohr, M. Fitness characteristics of elite and subelite male ice hockey players: A cross-sectional study. *J. Strength Cond. Res.* **2019**, *33*, 2352–2360. [CrossRef] [PubMed]
17. Wiseman, A.C.; Bracken, N.; Horton, S.; Weir, P.L. The difficulty of talent identification: Inconsistency among coaches through skill-based assessment of youth hockey players. *Int. J. Sports Sci. Coach.* **2014**, *9*, 447–455. [CrossRef]
18. Johnston, K.; Wattie, N.; Schorer, J.; Baker, J. Talent identification in sport: A systematic review. *Sports Med.* **2018**, *48*, 97–109. [CrossRef] [PubMed]
19. Baker, J.; Cobley, S.; Schorer, J. Talent identification and development in sport: International perspectives. *Int. J. Sports Sci. Coach.* **2012**, *7*, 177–180. [CrossRef]
20. Huard Pelletier, V.; Glaude-Roy, J.; Daigle, A.P.; Brunelle, J.F.; Bissonnette, A.; Lemoyne, J. Associations between Testing and Game Performance in Ice Hockey: A Scoping Review. *Sports* **2021**, *9*, 117. [CrossRef] [PubMed]
21. Baker, J.; Wattie, N.; Schorer, J. A proposed conceptualization of talent in sport: The first step in a long and winding road. *Psychol. Sport Exerc.* **2019**, *43*, 27–33. [CrossRef]
22. Mustafovic, E.; Causevic, D.; Covic, N.; Ibrahimovic, M.; Alic, H.; Abazovic, E.; Masic, S. Talent Identification in Youth Football: A Systematic Review. *J. Anthropol. Sport Phys. Educ.* **2020**, *4*, 37–43. [CrossRef]
23. Dimundo, F.; Cole, M.; Blagrove, R.; Till, K.; McAuley, A.; Hall, M.; Kelly, A. Talent identification and development in male rugby union: A systematic review. *J. Expertise* **2021**, *4*, 33–55.
24. Gábor, G. *Success and Talent Development as Indicated by Motor Tests and Psychometric Variables of U18 Ice Hockey Players*; Semmelweis University Sport Sciences Doctoral School: Budapest, Hungary, 2009.
25. Guenter, R.W.; Dunn, J.G.; Holt, N.L. Talent identification in youth ice hockey: Exploring "intangible" player characteristics. *Sport Psychol.* **2019**, *33*, 323–333. [CrossRef]
26. Boland, M.; Delude, K.; Miele, E.M. Relationship between physiological off-ice testing, on-ice skating, and game performance in Division 1 female hockey players. *J. Strength Cond. Res.* **2019**, *33*, 1619–1628. [CrossRef] [PubMed]
27. Chiarlitti, N.; Delisle-Houde, P.; Reid, R.; Kennedy, C.; Andersen, R. The Importance of Body Composition in the National Hockey League Combine Physiologic Assessments. *J. Strength Cond. Res.* **2017**, *32*, 3135–3142. [CrossRef] [PubMed]
28. Peterson, B.J.; Fitzgerald, J.S.; Dietz, C.C.; Ziegler, K.S.; Ingraham, S.J.; Baker, S.E.; Snyder, E.M. Division I hockey players generate more power than Division III players during on- and off-ice prformance tests. *J. Strength Cond. Res.* **2015**, *29*, 1191–1196. [CrossRef]
29. Delisle-Houde, P.; Chiarlitti, N.A.; Reid, R.E.R.; Andersen, R.E. Predicting On-Ice Skating Using Laboratory- and Field-Based Assessments in College Ice Hockey Players. *Int. J. Sports Physiol. Perform.* **2019**, *14*, 1184–1189. [CrossRef]
30. Runner, A.R.; Lehnhard, R.A.; Butterfield, S.A.; Shihfen, T.U.; O'Neill, T. Predictors of speed using off-ice measures of college hockey players. *J. Strength Cond. Res.* **2016**, *30*, 1626–1632. [CrossRef]
31. Janot, J.M.; Beltz, N.M.; Dalleck, L.D. Multiple Off-Ice Performance Variables Predict On-Ice Skating Performance in Male and Female Division III Ice Hockey Players. *J. Sports Sci. Med.* **2015**, *14*, 522–529.
32. Boucher, V.G.; Parent, A.-A.; Miron, F.S.-J.; Leone, M.; Comtoisi, A.S. Comparison between power off-ice test and performance on-ice anaerobic testing. *J. Strength Cond. Res.* **2020**, *34*, 3498–3505. [CrossRef]
33. Bežák, J.; Přidal, V. Upper body strength and power are associated with shot speed in men's ice hockey. *Acta Gymnica* **2017**, *47*, 78–83. [CrossRef]
34. Beckham, G.; Lish, S.; Disney, C.; Keebler, L.; DeBeliso, M.; Adams, K.J. The Reliability of the Seated Medicine Ball Throw as Assessed with Accelerometer Instrumentation. *J. Phys. Act. Res.* **2019**, *4*, 108–113. [CrossRef]
35. Stanula, A.; Roczniok, R.; Maszczyk, A.; Pietraszewski, P.; Zając, A. The role of aerobic capacity in high intensity intermittent efforts in ice hockey. *Biol. Sport* **2014**, *31*, 193–199. [CrossRef] [PubMed]
36. Cordingley, D.M.; Sirant, L.; MacDonald, P.B.; Leiter, J.R. Three-Year Longitudinal Fitness Tracking in Top-Level Competitive Youth Ice Hockey Players. *J. Strength Cond. Res.* **2019**, *33*, 2909–2912. [CrossRef] [PubMed]

37. Haukali, E.; Tjelta, L.I. Correlation between "off-ice" variables and skating performance among young male ice hockey players. *Int. J. Appl. Sports Sci.* **2015**, *27*, 26–32. [CrossRef]
38. Stewart, P.F.; Turner, A.N.; Miller, S.C. Reliability, factorial validity, and interrelationships of five commonly used change of direction speed tests. *Scand. J. Med. Sci. Sports* **2014**, *24*, 500–506. [CrossRef]
39. Perez, J.; Guilhem, G.; Brocherie, F. Reliability of the force-velocity-power variables during ice hockey sprint acceleration. *Sports Biomech.* **2019**, *22*, 56–70. [CrossRef]
40. International Ice Hockey Centre of Excellence Skating Agility Test (From Finnish Ice Hockey Federation). Available online: https://iihce.fi/suomeksi/Testaaminen/Pohjola-leiritestit/tabid/1150/Default.aspx#/material/872/2401 (accessed on 10 February 2022).
41. Plaisant, O.; Courtois, R.; Réveillère, C.; Mendelsohn, G.A.; John, O.P. Validation par analyse factorielle du Big Five Inventory français (BFI-Fr). Analyse convergente avec le NEO-PI-R. *Ann. Médico-Psychol.* **2010**, *168*, 97–106. [CrossRef]
42. Steca, P.; Baretta, D.; Greco, A.; D'Addario, M.; Monzani, D. Associations between personality, sports participation and athletic success. A comparison of Big Five in sporting and non-sporting adults. *Personal. Individ. Differ.* **2018**, *121*, 176–183. [CrossRef]
43. Duckworth, A.L.; Quinn, P.D. Development and validation of the Short Grit Scale (GRIT–S). *J. Personal. Assess.* **2009**, *91*, 166–174. [CrossRef]
44. Tabachnick, B.G.; Fidell, L.S. *Using Multivariate Statistics: International Edition*; Pearson: London, UK, 2013.
45. Lachenbruch, P.A.; Goldstein, M. Discriminant analysis. *Biometrics* **1979**, *35*, 69–85. [CrossRef]
46. Malina, R.M. Physical growth and biological maturation of young athletes. *Exerc. Sports Sci. Rev.* **1994**, *22*, 280–284. [CrossRef]
47. Weisberg, Y.J.; DeYoung, C.G.; Hirsh, J.B. Gender differences in personality across the ten aspects of the Big Five. *Front. Psychol.* **2011**, *2*, 178. [CrossRef] [PubMed]
48. Cormier, D.L.; Ferguson, L.J.; Gyurcsik, N.C.; Briere, J.L.; Dunn, J.G.; Kowalski, K.C. Grit in sport: A scoping review. *Int. Rev. Sport Exerc. Psychol.* **2021**, 1–38. [CrossRef]
49. Allen, M.S.; Greenlees, I.; Jones, M. Personality in sport: A comprehensive review. *Int. Rev. Sport Exerc. Psychol.* **2013**, *6*, 184–208. [CrossRef]
50. Ligue de Hockey d'Excellence du Québec (LHEQ). Available online: http://www.lheq.ca (accessed on 4 January 2022).
51. Teoldo, I.; Cardoso, F. Talent map: How demographic rate, human development index and birthdate can be decisive for the identification and development of soccer players in Brazil. *Sci. Med. Footb.* **2021**, *5*, 293–300. [CrossRef]
52. Martini, G.; Brunelle, J.F.; Trudeau, F.; Lemoyne, J. Measuring ice hockey skills in a repeated measures testing context: The effects of fatigue on skating efficiency, passing, agility, and shooting. *Sport J.* **2018**, *21*, 1–16.
53. Satrapová, P.; Rulík, T.P.A. Virtual Reality as A Training Tool for the Development of Ice Hockey Thinking. In Proceedings of the Fourth International Conference on Advanced Research in Teaching and Education, Budapest, Hungary, 6–8 August 2021; Available online: https://www.dpublication.com/wp-content/uploads/2021/08/10-4010.pdf (accessed on 9 January 2022).

sports

MDPI

Article

Is There a Sex Difference in Technical Skills among Youth Soccer Players in Norway?

Arne Sørensen *, Emma C. Haugen and Roland van den Tillaar

Department of Sport Sciences and Physical Education, Nord University, 7600 Levanger, Norway; emma.c.haugen@student.nord.no (E.C.H.); roland.v.tillaar@nord.no (R.v.d.T.)

* Correspondence: arne.sorensen@nord.no

Abstract: Female soccer has recently experienced an impressive increase in the number of players, and an impressive improvement in the quality of elite matches. Still, studies show sex differences in match statistics on passing accuracy and the ability to control the ball in international matches, which is explained by a lower skill of level in female soccer players as compared to male players. Therefore, the aim of this study was to evaluate if female youth soccer players had bridged the gap in technical skills to reach the level that boys have traditionally attained. Sixteen male and 17 female youth soccer players of the same age and experience level took part in technical skill tests of reception of the ball on the ground and long passes. The results show a significant difference between the sexes in reception performance in favour of the male players ($p < 0.05$, ES = 1.09), but no significant difference in the long pass test ($p = 0.11$, ES = 0.43). This leads to the conclusion that the lower score on ball reception is probably the result of experience in small-sided self-organised soccer games during childhood among the male players, which influences reception skills but not the ability to make accurate long passes.

Keywords: soccer; technical skills; technical test; long passes; reception

Citation: Sørensen, A.; Haugen, E.C.; van den Tillaar, R. Is There a Sex Difference in Technical Skills among Youth Soccer Players in Norway? *Sports* **2022**, *10*, 50. https://doi.org/10.3390/sports10040050

Academic Editor: Silvio Lorenzetti

Received: 16 February 2022
Accepted: 24 March 2022
Published: 29 March 2022

1. Introduction

Organised female soccer started in several European countries in the mid-1960s, and it has continued to experience an impressive increase in participants. The latest information from Federation Internationale de Football Association (FIFA) shows that worldwide around 30 million female senior and youth players take part in a football club [1]. Even though the quality of female soccer is improving, media interest and the economy is not in the same league as male soccer [2]. The critique of female soccer has often been about low pace and poor skills in comparison to male soccer [3]. Research has proven that there is a difference in performance between sexes in all sports with high demands for physical capacity [4,5].

High performance in soccer is determined by skill levels of players, with skills defined as a combination of technical and tactical ability in the rapidly changing environment of soccer games [6]. Analyses of the top matches of clubs and national teams for both sexes show a similarity with regard to the number of ball touches and time in possession of the ball [7]. Differences have also been detected, such as female players lost the ball more often [8], had a higher percentage of mistakes in reception and passes, and made more long passes from the defensive area [9]. Suggested reasons for sex differences in match performance are that female players experience more fatigue than male players during matches [7], that game tactics in female soccer do not focus as much on possession of the ball as do male game tactics and that female players make more long passes from the defensive area [9]. This has led to a lower number of passes and more passes with a higher possibility of interference. Finally, these differences could also mean that female players have lower technical and tactical soccer skills than male players.

Studies that have evaluated the sex differences in general motor control show that young men perform better in gross motor skills and young women are better at mastering activities which involve fine motor skills [10,11]. Furthermore, young men show significantly better performance than young women on tests of object control, such as throwing and kicking [12].

Differences in dribble speed have been found in youth soccer players [13,14], and it has been argued that much of the difference can be explained by sex difference in sprinting speed [13]. Analyses of the ability to rapidly change direction (COD), which is a vital part of dribbling speed, reveal relative strength to be decisive for high performance [15]. One study reported no significant sex differences among youth players in important technical soccer skills, such as long passes, heading and shooting [14], while differences in favour of male players in technical passing skills were found in another [16].

Little research on senior soccer has been published related to sex differences in soccer skills. Two separate studies have shown large differences in the Loughborough Soccer Passing Test [17,18] and dribble speed in favour of male players [19]. According to Thomas and French [20], the reason for these sex differences in motor control is more likely to be environmental than biological, as the superiority of boys in gross motor skills and differences in technical skills can probably be explained by the fact that males have done more of this type of motor skill training than have females.

The observed improvement in the quality of female international soccer and the increasing popularity for girls to start soccer training at an early age [1] leads to an interesting question: Have youth female soccer players bridged the gap in soccer skills to reach the level that boys have traditionally attained? Findings from previous studies on this question are unclear because most have used dribble speed as a sign of skill level. Since dribble speed is correlated with sprint speed [13] and COD which depend on muscular strength [15], these tests will favour male players. Therefore, it is important to bring new attention to the subject and evaluate other soccer skills in which the sex difference in physical strength is reduced. The ability to control the ball and make accurate passes has been shown to be different in elite matches for males and females [7,8]. Our purpose is, therefore, to examine if there is a sex difference in such skills as long passes and ball reception among youth soccer players. The hypothesis of this study was that youth female players are at the same technical level as youth male players.

2. Materials and Methods

2.1. Method

It is a challenge to design a relevant test to evaluate a possible sex difference in skill level in soccer because the quality of the opponent plays a decisive role in the performance in matches [21]. It is difficult to test soccer skills with high reliability and construct validity [22], and the measurement of decision making by means of an iPad [23] provides uncertain information on players' tactical abilities. Therefore, we chose to test player technical levels to evaluate if there is a sex difference in the technical part of soccer skills. This design will not directly measure the complete skill level, but since level of technical ability in soccer is related to level of competition [24] and technical tests could provide high reliability [6], we decided to use those tests in this study.

2.2. Participants

Seventeen female outfield soccer players (age 17.8 $\pm$ 0.6 years; height 1.69 $\pm$ 0.05 m; body mass 62.9 $\pm$ 6.04 kg) and 16 male outfield soccer players (age 17.8 $\pm$ 0.7 years; height 1.78 $\pm$ 0.07 m; body mass 71.1 $\pm$ 6.7 kg) participated in the study. The inclusion criteria were that the participants were active players in soccer clubs, and the exclusion criteria were that they did not have any injuries or illness on the test days. All male players competed in teams in the first division of the regional U18 league, and female players competed in senior teams at the third and fourth levels. The tests were executed in November and December, where the soccer season was finished in the Norwegian soccer leagues. In this period, the players have a relatively low training load, which will not influence the skill results as much. The players signed a written consent form to take part in this study, in accordance with regulations of the Norwegian Centre for Research Data. Approval to use the data and to conduct the study was given by the Norwegian Centre for Research Data (reference code nr. 835109), on 13 August 2020.

2.3. *Procedures*

Two technique tests were conducted to evaluate if there was a sex difference in technical skills in passing and reception of the ball in youth soccer players. During the test of reception of the ball on the ground, player performances were recorded using an AV video camera (Sony PXW-Z90, Sony, Tokyo, Japan).

All tests were conducted in an indoor soccer hall with artificial turf, and similar balls (size 5) with identical air pressure were used (0.6 bar). The players took part in a general standardised warm-up (15 min) of running and dribbling with each ball (5 min) and passing and reception from different distances (10 min) before the test. Information about their soccer histories, daily training, match frequency, and level of match play was gathered on test day from the players.

After the warm-up, the participants stood 8 m away from the ball projection machine (Sport Tutor, Burbank, CA, USA) and were instructed to receive the ball in the way they wanted for the purpose of controlling the ball inside a marked area in front of them (1 × 1.5 m, see Figure 1). The participants were instructed to make one touch in the reception and then one touch to pass the ball in the direction of the ball projection machine. The participants completed 10 attempts to receive the ball on the ground with a ball speed at 19 km/h (measured with a laser gun (Stalker Pro II+, Richardson, TX, USA)) and were instructed to make the receive every second time to the right and left to make a pass with the right or left foot. The test was adapted from [25] (Table 1). Two educated soccer coaches, one with a UEFA b-licence and with more than 10 years of experience as a soccer coach and one with a UEFA c-licence with more than 5 years of experience, scored every reception of the ball based on set criteria [25].

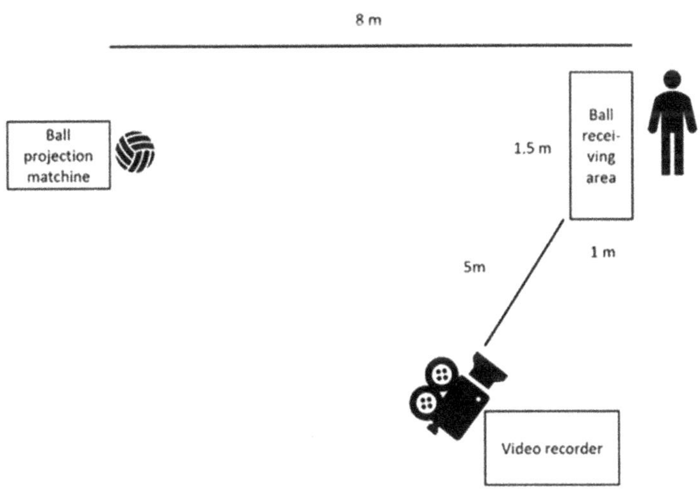

Figure 1. The set up for the ball reception test.

Table 1. The criteria for scoring the reception of the ball from the ball projection machine at a speed of 19 km/h from, adapted from [25].

Score	Criterium, Ball Reception, on the Ground, 19 km/t
1	The participant does not manage to stop the ball.
2	The participant loses control over the ball, and the ball goes outside the area in front of or beside him (1 × 1.5 m) before the player manages to pass the ball.
3	The participant manages to control the ball in the area, but the ball is not controlled in the correct direction (left or right).
4	The participant manages to control the ball in the area, in the correct direction (left or right), but is too close or too far from the player, resulting in difficulty in making the pass.
5	The participant manages to control the ball within the area, in the correct direction (left or right), and the receiving is perfect, so the pass is easy to perform.

After the ball reception test, the participants started with the ball 30 m away from a target (a large cone) and were instructed to take one touch of the ball inside a 5 m area followed by a long pass in the air with their best foot (see Figure 2). The participants made the pass when the ball was rolling. The aim was to pass the ball through the air and hit the ground as close to the marker as possible. There were four circles with 2 m increase in diameter around the cone with the highest score closest to the cone and the lowest outside the largest circle (5-1 points; Figure 2), modified after [26]. Two judges were standing 5 m from the marker and scored each trial; if the ball bounced at the line, the highest score was given. The participants made two familiarisation trials, and then 15 trials in the test. Players' average scores were used for further analysis. A relatively short distance was chosen (25 m) for the long pass test to minimise the effect of sex difference in strength on the results.

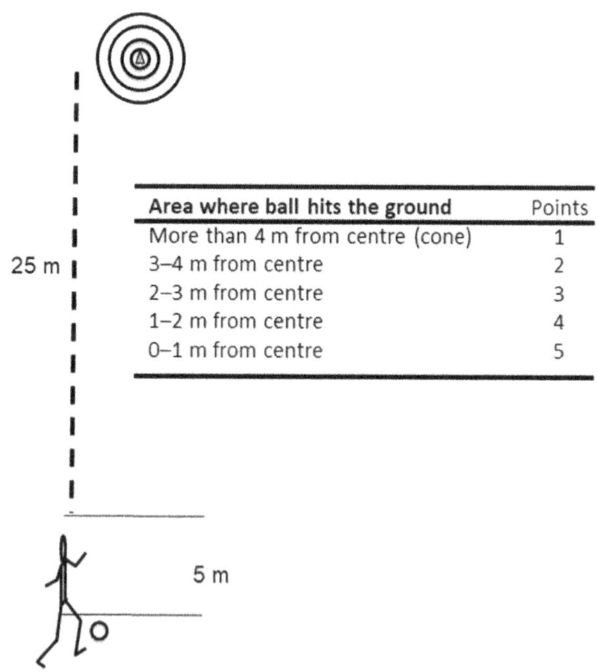

Figure 2. Test on long passes with criteria for scoring the test.

2.4. Statistical Analysis

Players' average scores for each test were used for further analysis. The data were expressed as mean ± standard derivation (SD), and we also reported Cohen's d effect size for sex differences in both technical tests. An effect size of 0.2 was considered small, 0.5 medium, and 0.8 large [27]. The difference between sexes was analysed by independent sample T-test with a significance level of $p \leq 0.05$. A Pearson's correlation was used for all participants and per sex to investigate correlations between the scores of the ball reception and long passes. Threshold values for the correlation coefficients' interpretation as an effect size were 0.1–0.3 (trivial), 0.3–0.5 (moderate), 0.5–0.7 (large), and 0.7–0.9 (very large; Hopkins et al., 2009).

A repeatability analysis was performed on a subset of 50 randomly chosen samples and assessed at two different points in time by two coaches to evaluate the accuracy of the judgements of the test ratings. Mean estimates with 95% confidence intervals (CI) were reported for an intraclass correlation coefficient (ICC). Interpretation was as follows: <0.50 poor; from 0.50 to 0.75 fair; from 0.75 to 0.90 good; and above 0.90 excellent. The ICC for inter-rater reliability between coaches was good to excellent at 0.90. All statistical analyses were performed using SPSS Statistical Analysis Software for Windows® (SPSS, version 25, Chicago, IL, USA).

3. Results

No significant differences were found between sexes for soccer experience and training experience per week ($t \leq 1.46$, $p \geq 0.155$, ES ≤ 0.5, Table 2).

Table 2. Participation in a soccer club, matches per week, and soccer training per week.

	Female Players	Male Players
Years in a soccer club	11 ± 1.3	11.2 ± 1.8
Matches per week	1	1
Soccer training in school and club (h/week)	8.8 ± 2.3	9.2 ± 0.9
Self-organised soccer training (h/week)	0.9 ± 0.9	1.5 ± 1.5

A significant difference between the sexes was found in the ball reception test ($t = 3.1$, $p = 0.004$, ES = 1.09) but not in the long passes test ($t = 1.2$, $p = 0.22$, ES = 0.43, Figure 3). The individual scores for ball reception varied from 2.4 to 4.4 points, while the long passes' scores varied from 1.53 to 3.8 (Figure 3).

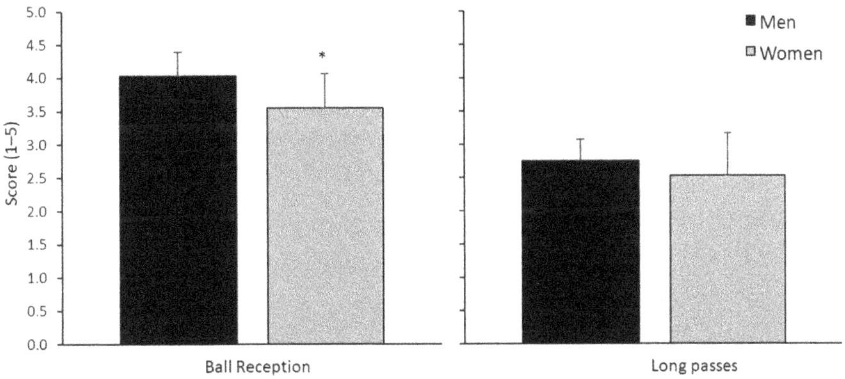

Figure 3. Mean (±SD) score for the ball reception and long passes tests for male and female youth soccer players; * indicates a significant difference between the sexes on a $p \leq 0.05$ level.

A significant moderate positive correlation (r = 0.38, p = 0.028) between the scores for the ball reception and long passes was found for the whole group, but when specified per men (r = 0.14, p = 0.61) and women (r = 0.38, p = 0.125), only trivial and moderate nonsignificant correlations were found (Figure 4).

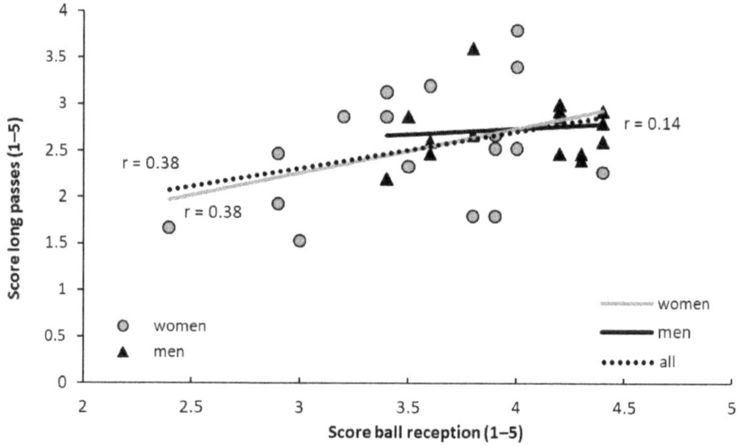

Figure 4. Correlation between ball reception and long passes tests for the whole group and specified per sex.

4. Discussion

The aim of the study was to examine if there was a sex difference in the skills in long passes and reception of the ball in youth soccer players. The main findings were a significantly higher accuracy in reception of the ball on the ground for male youth players compared to female youth players, but not in long passes. Thereby, the hypothesis was partly correct. The participants in this study were equal in terms of age, years in a soccer club, and hours of soccer training in school and club, and both male and female players took part in one match per week during the season (Table 2). Theoretically, only environmental differences should explain the sex differences in the test [20]. The explanation of the findings could be that—as the authors of this study have experienced—far more boys than girls participate at a young age (6–12 years) in soccer play in school and in their spare time. This is a crucial period to improve skills in soccer, and research has shown that players who take a greater part in these activities increase their chances of becoming a professional soccer player (Ford et al., 2009).

Our finding of no sex difference in long passes was not in line with a study of Finish youth soccer players who showed significant differences in passing performance [16]. It is worth noting that these results are from surveys conducted between 2002 and 2010. The development of female soccer players' skill levels has increased after that. Our findings of sex difference in accuracy in reception are in contrast to Perroni [14], who found no significant differences between sexes on tests of other technical skills in soccer such as passing, heading, and shooting for youth players. These authors also found that female players performed significantly better than males in one juggling test. This is rather surprising given that the male players in Perroni et al.'s study had been part of a football club for a longer period of time than the females and took part in more football practice each week.

We found no significant difference between sexes in the test of accuracy in long passes. This indicated that the distance of the long pass of 25 m did not discriminate against the female players with regard to their lower muscular strength. The reason for no significant sex difference could be the equality in their soccer history. Both sexes started with soccer when they were approximately 6 years old and have played on a soccer team for about 10 years.

The reason for the significant difference in performance of receptions but not for long passes could possibly be explained by the 'early engagement hypothesis' [28,29] and the specificity principle of training [30]. If both sexes practised long passes for an equal time, the improvement should be the same [20]. Better performance in reception by the male players indicates a large amount of training of the actual skill, probably through more soccer play in childhood. Both social and cultural reasons could explain the observed issue that boys take more part in unorganised soccer play in the period from 6 to 12 years [31]. There is reason to believe that soccer play in small groups (SSG) will have more influence on the ability of reception than on long passes [32] since such activities are often performed in a limited area (school yards) where the need to make long passes is low.

The individual results of reception (Figure 4) showed that one female player scored points equal to that of the best performing male players. Furthermore, a female player had the best performance in the test of long passes. This indicates that with enough training, female players could score just as well as male players in the technical tests used in the present study.

A moderate significant correlation between the two technical tests was found when the whole group performance was used while only trivial and moderate nonsignificant correlations for each sex were found. This shows that high quality performance in one technical skill in soccer does not automatically indicate high performance on another. These findings are in accordance with the theory about the transfer of motoric learning [33]. If players, for example, practice long passes, they will probably improve their performance, but this will not improve their ability to make accurate receptions. The reason is a low degree of similarity in the two skills [33].

The study has some limitations as the scoring system on both technical tests was from 1 to 5 points. A wider scoring system could have increased the sensitivity of the test to provide a better separation of player quality. A relatively limited number of attempts (10 and 15 in ball reception and long pass test, respectively) could have led to coincidences affecting the results. To ensure a higher reliability in tests, the quantity of attempts could have been higher [34]. Furthermore, it is possible that by adding more participants to the study, the results could have been different. In addition, the soccer level of participants could have affected the results, even though both groups were categorised as medium level players in the area. Care should be taken in generalisation of the results, since experience, level of competitiveness, and age can affect the technical level of players.

Future studies on this subject should evaluate technical skills in top soccer players, both youth players and adult players, to detect a possible sex difference. If there is a sex difference in technical skills so that male players perform better on technical tests, it would be reasonable to believe that the sex difference in match statistics would continue.

5. Conclusions

Youth female soccer players scored significantly lower on technical performance in reception but not in long passes. As reception is a vital skill in soccer, and poor quality will influence the opportunity to make a precise pass or a good shot. Thereby, the reason for sex differences in match statistics could be explained by differences in ability to make accurate receptions of the ball, which could influence the percentage of successful passing. The two groups of youth players had similar football history in a soccer club, but a hypothetical explanation of the finding could be more self-organised soccer play in childhood among the male players, as small-sided games influence reception skill rather than the ability to make accurate long passes. Thereby, sex differences still occur in ball reception.

Author Contributions: Conceptualisation, E.C.H. and A.S.; methodology, E.C.H. and A.S.; data curation A.S. and R.v.d.T.; writing—original draft preparation, E.C.H. and A.S.; writing—review and editing, A.S. and R.v.d.T., supervision, R.v.d.T. All authors have read and agreed to the published version of the manuscript.

Funding: This research received no external funding.

Institutional Review Board Statement: The ethics of the study was completed according to the institutional requirements and approval for data security and handling was obtained from the Norwegian Centre for Research Data project nr: 835109. The study was conducted in line with the current ethical regulations for research and with the latest Declaration of Helsinki.

Informed Consent Statement: Patient consent was waived as the study was answered anonymously.

Data Availability Statement: The data presented in this study are available on request from the corresponding author. The data are not publicly available due to rules of Norwegian Center for Research Data.

Conflicts of Interest: The authors declare no conflict of interest.

References

1. (FIFA), F.I.d.F.A. Women's Football Survey. Available online: https://www.icsspe.org/system/files/FIFA%20-%20Womens%20Football%20Survey.pdf (accessed on 25 August 2021).
2. Hovden, J.; von der Lippe, G. The gendering of media sport in the Nordic countries. *Sport Soc.* **2019**, *22*, 625–638. [CrossRef]
3. Hjelseth, A.; Hovden, J. Negotiating the status of women's football in Norway. An analysis of online supporter discourses. *Eur. J. Sport Soc.* **2014**, *11*, 253–277. [CrossRef]
4. Sandbakk, Ø.; Solli, G.S.; Holmberg, H.-C. Sex Differences in World-Record Performance: The Influence of Sport Discipline and Competition Duration. *Int. J. Sports Physiol. Perf.* **2018**, *13*, 2. [CrossRef] [PubMed]
5. Thibault, V.; Guillaume, M.; Berthelot, G.; Helou, N.E.; Schaal, K.; Quinquis, L.; Nassif, H.; Tafflet, M.; Escolano, S.; Hermine, O.; et al. Women and Men in Sport Performance: The Gender Gap has not Evolved since 1983. *J. Sports Sci. Med.* **2010**, *9*, 214–223. [PubMed]
6. Ali, A. Measuring soccer skill performance: A review. *Scand. J. Med. Sci. Sports* **2011**, *21*, 170–183. [CrossRef] [PubMed]
7. Bradley, P.; Dellal, A.; Mohr, M.; Castellano, J.; Wilkie, A. Gender differences in match performance characteristics of soccer players competing in the UEFA Champions League. *Hum. Movem. Sci.* **2013**, *33*, 159–171. [CrossRef]
8. Casal, C.; Losada, J.; Dios, R.; Arda, A. Gender differences in technical-tactical behaviour of La Liga Spanish football teams. *J. Hum. Sport Exerc.* **2020**, *16*, 37–52. [CrossRef]
9. Hjelm, J. The bad female football player: Women's football in Sweden. *Soccer Soc.* **2011**, *12*, 143–158. [CrossRef]
10. Rodrigues, P.; Ribeiro, M.; Barros, R.; Lopes, S.; Sousa, A. Performance on the movement assessment battery for children: A systematic review about gender differences. [Desempeño en la batería de evaluación del movimiento para niños: Una revisión sistemática sobre las diferencias de género]. *Rev. Int. Cienc. Desporte* **2019**, *55*, 72–87. [CrossRef]
11. Dorfberger, S.; Adi-Japha, E.; Karni, A. Sex differences in motor performance and motor learning in children and adolescents: An increasing male advantage in motor learning and consolidation phase gains. *Behav. Brain Res.* **2009**, *198*, 165–171. [CrossRef] [PubMed]
12. Barnett, L.M.; van Beurden, E.; Morgan, P.J.; Brooks, L.O.; Beard, J.R. Gender Differences in Motor Skill Proficiency From Childhood to Adolescence. *Res. Quart. Exerc. Sport* **2010**, *81*, 162–170. [CrossRef] [PubMed]
13. O'Brien-Smith, J.; Bennett, K.J.M.; Fransen, J.; Smith, M.R. Same or different? A comparison of anthropometry, physical fitness and perceptual motor characteristics in male and female youth soccer players. *Sci. Med. Football* **2020**, *4*, 37–44. [CrossRef]
14. Perroni, F.; Gallotta, M.C.; Pisano, S.; Reis, V.M.; Emerenziani, G.P.; Guidetti, L.; Baldari, C. Gender differences in anthropometric parameters and technical performance of youth soccer players. *Sport Sci. Health* **2018**, *14*, 399–405. [CrossRef]
15. Peterson, M.; Alvar, B.; Rhea, M. The Contribution of Maximal Force Production to Explosive Movement Among Young Collegiate Athletes. *J. Strength Cond. Res.* **2006**, *20*, 867–873. [CrossRef] [PubMed]
16. Nunome, H.; Drust, B.; Dawson, B. Analysis of Finnish young soccer players' passing and dribbling skills. In *Science and Football VII*; Routledge: London, UK, 2013; pp. 239–244.
17. Ali, A.; Foskett, A.; Gant, N. Validation of a Soccer Skill Test for Use with Females. *Int. J. Sports Med.* **2008**, *29*, 917–921. [CrossRef]
18. Ali, A.; Williams, C.; Hulse, M.; Strudwick, A.; Reddin, J.; Howarth, L.; Eldred, J.; Hirst, M.; McGregor, S. Reliability and validity of two tests of soccer skill. *J. Sports Sci.* **2007**, *25*, 1461–1470. [CrossRef]
19. Mujika, I.; Santisteban, J.; Impellizzeri, F.M.; Castagna, C. Fitness determinants of success in men's and women's football. *J. Sports Sci.* **2009**, *27*, 107–114. [CrossRef]
20. Thomas, J.; French, K. Gender Differences Across Age in Motor Performance. A Meta-Analysis. *Psychol. Bullet* **1985**, *98*, 260–282. [CrossRef]
21. Liu, H.; Gómez, M.-A.; Gonçalves, B.; Sampaio, J. Technical performance and match-to-match variation in elite football teams. *J. Sports Sci.* **2016**, *34*, 509–518. [CrossRef]
22. Russell, M.; Benton, D.; Kingsley, M. Reliability and construct validity of soccer skills tests that measure passing, shooting, and dribbling. *J. Sports Sci.* **2010**, *28*, 1399–1408. [CrossRef]
23. Bennett, K.; Novak, A.; Pluss, M.; Coutts, A.; Fransen, J. Assessing the validity of a video-based decision-making assessment for talent identification in youth soccer. *J. Sci. Med. Sport* **2018**, *22*, 729–734. [CrossRef] [PubMed]

24. Pedersen, A.V.; Lorås, H.; Norvang, O.P.; Asplund, J. Measuring soccer technique with easy-to-administer field task in female soccer players from four different comeptitive levels. *Perc. Mot. Skills* **2014**, *119*, 961–970. [CrossRef] [PubMed]
25. Sørensen, A.; Sørensen, V.; Dalen, T. A Novel Approach for Comparison of Reception Performance in a Technique Test and Small-Sided Games. *Sports* **2021**, *9*, 66. [CrossRef] [PubMed]
26. Kelly, A.; Wilson, M.R.; Jackson, D.T.; Williams, C.A. Technical testing and match analysis statistics as part of the talent development process in an English football academy. *Int. J. Perf. Anal. Sport* **2020**, *20*, 1035–1051. [CrossRef]
27. Cohen, J. *Statistical Power Analysis for the Behavioral Sciences*; Lawrence Erlbaum Associates: Hillsdale, NJ, USA, 1988.
28. Ford, P.R.; Ward, P.; Hodges, N.J.; Williams, A.M. The role of deliberate practice and play in career progression in sport: The early engagement hypothesis. *High Abil. Stud.* **2009**, *20*, 65–75. [CrossRef]
29. Erikstad, M.K.; Høigaard, R.; Johansen, B.T.; Kandala, N.-B.; Haugen, T. Childhood football play and practice in relation to self-regulation and national team selection; a study of Norwegian elite youth players. *J. Sports Sci.* **2018**, *36*, 2304–2310. [CrossRef]
30. Sharkey, B.J.; Gaskill, S.J. *Sport Physiology for Coaches*; Human Kinetics: Champaign, IL, USA, 2006.
31. Trost, S.; Pate, R.; Dowda, M.; Saunders, R.; Felton, G. Gender Differences in Physical Activity and Determinants of Physical Activity in Rural Fifth Grade Children. *J. School Health* **1996**, *66*, 145–150. [CrossRef]
32. Owen, A.; Wong, D.P.; Paul, D.; Dellal, A. Physical and Technical Comparisons between Various-Sided Games within Professional Soccer. *Int. J. Sports Med.* **2014**, *35*, 286–292. [CrossRef]
33. Magill, R.A.; Anderson, D.I. *Motor Learning and Control: Concepts and Applications*, 11th ed.; McGraw-Hill: New York, NY, USA, 2017.
34. Pedersen, A.V.; Lorås, H. When Is a Test Score Fair for the Individual Who Is Being Tested? Effects of Different Scoring Procedures across Multiple Attempts When Testing a Motor Skill Task. *Front. Psychol.* **2017**, *8*, 619. [CrossRef]

sports

MDPI

Article

Looking Back and Looking Around: How Athletes, Parents and Coaches See Psychosocial Development in Adolescent Performance Sport

Sergio Lara-Bercial [1,2,*] and Jim McKenna [1]

[1] Research Centre for Sport Coaching, Carnegie School of Sport, Leeds Beckett University, Leeds LS6 3QT, UK; j.mckenna@leedsbeckett.ac.uk
[2] Global Coaching Office, International Council for Coaching Excellence, Leeds Beckett University, Leeds LS6 3QT, UK
* Correspondence: s.lara-bercial@leedsbeckett.ac.uk

Abstract: Sport has the potential to support psychosocial development in young people. However, extant studies have tended to evaluate purpose-built interventions, leaving regular organised sport relatively overlooked. Moreover, previous work has tended to concentrated on a narrow range of outcomes. To address these gaps, we conducted a season-long ethnography of a youth performance sport club based on a novel Realist Evaluation approach. We construed the club as a social intervention within a complex system of agents and structures. The results are published in this special issue as a two-part series. In this first paper, we detail the perceptions of former and current club parents, players and coaches, using them to build a set of programme theories. The resulting network of outcomes (i.e., self, emotional, social, moral and cognitive) and generative mechanisms (i.e., the attention factory, the greenhouse for growth, the personal boost and the real-life simulator), spanning across multiple contextual layers, provides a nuanced understanding of stakeholders' views and experiences. This textured perspective of the multi-faceted process of development provides new insights for administrators, coaches and parents to maximise the developmental properties of youth sport, and signposts new avenues for research in this area.

Keywords: positive youth development; youth sport; realist evaluation; life skills; personal development; psychosocial development

Citation: Lara-Bercial, S.; McKenna, J. Looking Back and Looking Around: How Athletes, Parents and Coaches See Psychosocial Development in Adolescent Performance Sport. *Sports* **2022**, *10*, 47. https://doi.org/10.3390/sports10040047

Academic Editor: Michael Chia

Received: 29 December 2021
Accepted: 17 March 2022
Published: 22 March 2022

1. Introduction

Over the last 40 years, sport has become integral to the lives of many children and young people (CYP). Beyond the typical array of physical and mental health and wellbeing benefits ascribed to participation (i.e., enhanced cardiovascular and musculoskeletal function and lower risk of obesity, diabetes and depression [1–3]), sport is also regularly presented as a tool to foster psychosocial development [4,5]. Given these potential benefits, governments and communities across the world continue to invest large amounts of financial and human capital in the promotion and provision of sport [6,7]. Whilst the physical and mental wellbeing benefits of participation have been studied extensively, a number of authors have contested the assumption that sport participation inevitably and automatically translates into psychosocial development [8,9]. Concerns have been raised that the extant literature fails to clearly define and fully understand the nature and extent of this phenomenon [8,10,11].

The study of sport as a tool for psychosocial development has been approached from three related yet distinctive perspectives, namely sport for development (SFD), life-skills development (LSD) and positive youth development (PYD). SFD is broadly defined as encompassing programmes that use 'sport as a tool for development and peace', aiming to maximise 'the potential of sport as a tool to reach personal, community, national and international development objectives' [12]. LSD typically focuses on the capacity for sport to teach youth the essential skills needed to successfully negotiate the demands of everyday life [13,14]. Finally, PYD has emerged as a broader field than LSD, focusing on the contribution that sport can make to foster positive and developmental behaviours and attitudes in young people, allowing them to thrive and transition successfully into young adulthood [15,16]. An in-depth review of this broad and extensive literature is beyond the scope of this paper (for this, see [17,18]). Nonetheless, a brief critique is presented to set the context for the current paper.

Research emerging from these three perspectives has enhanced our understanding of the properties of sport as a development agent. However, a number of concerns have been raised. First, there is no consensus regarding what constitutes personal development through sport [8]. Second, there is a tendency to study discrete outcomes using experimental designs to the detriment of establishing an integrated view of the process in natural settings [19]. Therefore, some authors have called for more process-based research focused on understanding the underpinning conditions and mechanisms that lead to positive psychosocial outcomes [13,20]. Finally, the field has been criticised for generally placing the focus on evaluating purpose-built interventions, thus overlooking regular organised sport in which most children take part [21]. As a result, some researchers suggest that many of the claims made for the psychosocial benefits of sport participation are only partially substantiated and lack ecological validity [8,22,23].

Youth sport researchers are thus being challenged to develop a deeper, systemic and integrated understanding of the factors and processes that lead to positive development [8,17,20,22]. To the best of our knowledge, a comprehensive exploration of sport-based psychosocial development meeting these criteria has not been conducted to date. The current study aims to address this gap in the literature through the ethnographic investigation of a youth performance sport club. In doing so, the new evidence base it produces will enhance the capacity of sport psychologists, coaches, parents, clubs and schools to foster positive growth. This evidence base will also be timely; it will contribute to providing essential political and financial support for youth sport to address the many harms accompanying the response to, and aftermath of, the COVID-19 pandemic.

2. Study Design and Methods

2.1. Theoretical Framework

To address the gaps identified above, this study adopted a novel realist evaluation (RE) approach [24,25]. RE typically aims to establish how well a given purpose-built social intervention achieves its expected outcomes. RE's unique features have particular utility in the study of psychosocial development in organised youth sport when sport is construed and positioned as an organic social intervention embedded within a complex social system. In other words, environments where, with or without the awareness, intention and volition of those within it, a series of mechanisms interact, leading to explicit, implicit, desired and undesired developmental outcomes for all involved.

At an ontological and epistemological level, RE espouses a realist ontology and epistemology that accepts the existence of the social and material reality we interact with, yet states that the claims we can make about knowledge are always partial and never final. From this perspective, the role of the researcher is to constantly improve the existing available knowledge [24]. RE emphasises the fundamental role of theory in driving social research, and the importance of looking beyond quantification and correlation to shed light on the generative mechanisms affecting choices, behaviours and, ultimately, the outcomes of a given intervention. RE is thus concerned with the notion and nature of causality, focusing on "what works for whom, under what circumstances and why" [25] (p. 29).

Despite this focus on explanation and causality, RE accepts that social programmes take place within fluid and changing systems, meaning they feature flux, self-transformation and repatterning. Explained as 'morphogenesis' [26], RE implies that social interventions are never implemented in the same way twice, nor under similar conditions. Central to this variability is the intervention context. Given that social programmes are embedded into specific social systems, context is pivotal for how any programme works (or does not). From this perspective, programme outcomes emerge from a combination of agency and structure, the interface between the choices and capacities of contributing individuals and the collective resources at their disposal [24]. The fluidity and dynamism of the social world makes it challenging for the social evaluator to speak in absolutes or to make generalisations. Instead, RE recognises that certain elements and processes are relatively stable, and that these demi-regularities are the "subject matter of realist evaluation" [25] (p. 6). RE, therefore, subtly balances generalisability with specificity.

At a practical level, social programming and evaluation must start by making explicit the theories informing programme development and implementation, also known as programme theories (PTs) [24]. In RE, PTs comprise context–mechanism–outcome (CMO) configurations that allow researchers to make explicit not only the objectives of a programme (i.e., the expected outcomes of the intervention) within a specific context(s), but also the underlying assumptions about how a programme works (i.e., the generative mechanisms). As blueprints, PTs facilitate looking into the black box of programmes to establish not only if they work, but for whom, how and why. Having clearly articulated the PTs, researchers can explore how well these are realised in the "live" environment of the intervention; if they are not, or they are reached in unexpected ways, researchers can offer alternative explanations to create an "adaptive theory" [27], which is ever-evolving in light of new data [28].

Although RE has become a popular philosophical and methodological framework in social science in the last 30 years, this trend did not reach the sport literature until recently. RE-based approaches have started to proliferate in sport psychology [29], sport coaching [28,30], coach development [31], and sport policy [32]. Yet, it has not been used to explore young people's personal development in and through sport. In this respect, RE not only allows, but also requires, researchers to integrate the wide number of 'everyday' factors that constitute ordinary sport club functioning into nuanced and fine-grained accounts of how psychosocial development may materialise. The resulting accounts will help sport psychologists, coaches and club administrators to become more effective, proactive and deliberate in their efforts to support it.

2.2. Study Design

To achieve the above, we conducted a case study investigation of psychosocial development in a youth performance basketball club in north-west England. Case studies entail a detailed examination of an individual unit, case or system, allowing researchers to achieve a deep understanding of their internal structure and operation [33]. By focusing on a single unit, case studies make it possible to address the complexity and individuality of social phenomena [34] in a live context [35]. In the context of previous research in this area, the case study approach allows researchers to address the need for reality, complexity and depth in the investigation of psychosocial development because the case being studied already exists—it is not manufactured for the purpose of the research [33].

Criterion-based sampling was used to select this setting [36]. The criteria included an environment featuring: (i) a high intensity and frequency of participant engagement combined with high future stakes (i.e., a professional career); (ii) a hybrid club ethos combining community- and performance-based values and provision, leading to a wide range of experiences and interactions; and (iii) an explicit humanistic philosophy combined with a prolonged high level of success at the national and international level. It is important to note the lead researcher's familiarity with the setting, where he has coached for the last 20 years. The pros and cons of this element were carefully weighed regarding the balance of insider versus outsider positionality; the research team decided that the benefits outweighed the potential negative effects. Note also that none of the research subjects were or had ever been coaches coached by the lead author. On these bases, permission to progress with the study was provided by the club Chairman. Approval was subsequently granted by the Ethics Committee of Leeds Beckett University.

2.3. Participants

Purposeful sampling led to the identification of participants from three distinct samples: (i) former parents and players who had left the club at least 10 years ago; (ii) current parents and players; and (iii) current coaches. This aimed to generate two views of the same phenomenon. The former parents and players offered 'updated yet retrospective views'. By contrast, the current parents and players provided 'contemporary prospective views' of their perception of day-to-day development, as well as anticipation of future benefits. The coaches, having worked at the club for multiple seasons, straddled a full range of these viewpoints. From an RE perspective, the depth and width of the accounts generated by such a group allows for a close understanding of the structure and impact of the whole programme by drawing on personal insights, reflecting different time frames of engagement and reflecting distinctive experience and expertise [1].

Five club coaches were selected (all male; mean age = 47; SD = 20.61). In addition, eight former parents (5 male and 3 female; mean age = 58.25; SD = 1.56), and six former players (all male; mean age = 30.83; SD = 1.77) were purposefully sampled based on their diverse trajectories after leaving the club (gaining a college scholarship in the US, turning pro, playing recreationally, etc.). Finally, 10 current parents (5 male and 5 female; mean age = 46,5; SD = 2.24) and 10 current athletes (all male; mean age = 13.9; SD = 1.57) were purposefully selected from the under 13 (U13) and under 16 (U16) teams (for full details of participants, see Table S1a–e in Supplementary Materials).

2.4. Data Collection Methods

A variety of data collection methods were used to elicit the PTs from the three stakeholder groups. Coaches and former parents and players were asked to take part in in-depth semi-structured interviews lasting an average of 70 minutes (range of 35 to 121). All interviews were tape-recorded and transcribed verbatim, producing over 350 pages of double-spaced text. The interviews commenced with a broad question: 'In your view, beyond gaining physical and basketball-related skills, what is the impact of participation in the club programme at a personal level for the players?'. The interviewees were encouraged to elaborate on their responses, with examples, to explore different areas. Given the RE approach, the researcher explicitly probed interviewees to go beyond the identification of developmental outcomes and to articulate the how and why of these processes, asking questions like 'Would you be able to explain to me how this happened?' or 'What factors do you think are responsible for this?'. The interviewees were also explicitly asked to consider the same issues from a negative standpoint (i.e., 'From a personal development perspective, do you see any negative outcomes of being involved in the club?').

By contrast, current parents' and players' PTs were identified through a series of focus groups (FGs). FG participants were selected working with the team coaches who were asked to identify a group of individuals with contrasting backgrounds and experiences to provide a greater variety of accounts (socioeconomic status, ethnicity, squad status, etc.). All in all, four separate FGs took place: two with U13 (n = 5) and U16 parents (n = 5), and two with U13 (n = 5) and U16 (n = 5) players. These lasted an average of 52.25 minutes (range 40–72). The FGs included similar questions to the in-depth interview, but in order to elicit as high a number of responses as possible, the participants were asked to initially write each developmental outcome on a sticky note and to submit all of their answers into an opaque container. Subsequently, the researcher took one item at a time out of the box and asked the participants to come forward to explain what they meant, or if no one came forward, asked the others if they had a similar experience. Throughout the FGs, the researcher encouraged elaboration to explore the features of the context and mechanisms that brought about specific outcomes. The FGs were tape-recorded and transcribed verbatim, producing over 60 pages of double-spaced text.

2.5. Data Analysis

The interviews and FGs were analysed using a deductive–inductive iterative process facilitated by NVivo 10 software [37]. A three-stage process was implemented. First, the interview transcripts were read once to become familiarised with the content. Second, a deductive phase ensued in which the transcripts were read a second time, focusing on the identification of developmental outcomes. In line with RE tenets, this step was theory-driven and guided by a psychosocial outcomes framework (POF) created by the researchers through a substantial review of the developmental psychology literature (Figure 1; the full framework and review is available in the Supplementary Materials). Outcomes that did not fit into any of the categories of the framework were grouped as 'Miscellaneous'. In the final, third stage, the analysis transitioned to an inductive approach focused on identifying generative mechanisms and their links to specific outcomes and to key contextual features. This process aimed to build a deep understanding of how coaches, parents and players construed the impact of participation in sport on psychosocial development.

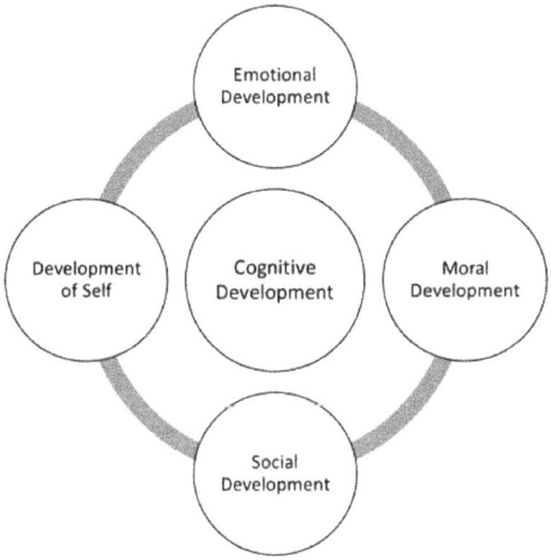

Figure 1. Psychosocial outcomes framework.

3. Results

The study results are presented in four sections. The first two introduce the outcomes and generative mechanisms identified by all stakeholders using the POF as an organising tool. In the third section, the proposed mechanisms are explored in detail and a new practitioner-oriented categorisation, based on four broad families, is proposed. Finally, the fourth section details how the context modulated the impact of participation. For economy, given the broad range of findings, a combination of summative tables and text is used to introduce the findings, and interview and FG quotes are used sparingly. Full quotes can be found in Table S2a,b in the Supplementary Materials.

3.1. Participation-Based Outcomes

The analysis of the interviews with coaches, parents and players elicited a wide array of both positive and negative outcomes (Table 1).

Table 1. Developmental outcomes of participation in youth performance sport.

Category	Positive Outcomes	Definition	Negative Outcomes	Definition
Development of the Self	Positive identity Sense of hope and life purpose Self-confidence	A healthy view of oneself in the world A positive outlook of the future and that one's life has meaning A belief in one's capability to deal with day-to-day issues	Quashed individuality Low self-confidence	Submission to the group's needs above one's own A reduced belief in one's capability to deal with day-to-day issues
Social Development	Interpersonal skills Sense of belonging Social capital Cooperation skills Leadership attributes A broader worldview	Competence to interact positively with others A feeling of being part of something bigger than oneself A network of people around oneself which contributes to positive outcomes Ability to work with others for a common purpose Capacity to take initiative and to influence others An understanding of different perspectives and experiences	Social isolation Selfishness	A sense of being cut off from friends and relatives Putting one's personal needs and desires above those of the group
Emotional Development	Emotional wellbeing Emotional literacy Emotional control	Overall high mental health Understanding of a range of emotions Being able to emotionally self-regulate	Low emotional wellbeing General demotivation	Reduced mental health Lack of motivation to do things beyond sport
Cognitive Development	Higher learning ability Enhanced decision-making Improved communication	Capacity to engage in learning opportunities Ability and proactivity to make decisions Capacity to express oneself clearly and publicly	None recorded	
Moral Development	Respect for others Moral decision-making	Respecting others' rights and feelings Being able to tell right from wrong and decide accordingly	Uncontrolled aggression Disrespect for others Bullying Occasional drinking	Bouts of aggression and hostility towards others Lack of respect for others' rights and feelings Purposefully trying to hurt others physically or emotionally Episodes of drinking with teammates
Miscellaneous	Work ethic Competitiveness Self-reliance The springboard effect	Ability to work hard for long periods Desire to do well and succeed Capacity and proactivity to resolve problems independently Advantages later in life facilitated by their sport experiences		

A total of 21 positive and 11 negative outcomes were identified across the five development areas of the POF and an additional miscellaneous category. The social development category contained the highest number of positive outcomes (n = 6), whilst moral development contained the most negative outcomes (n = 4). With regard to single positive outcomes, a positive identity, a sense of hope and life purpose, a sense of belonging, a broader worldview, higher learning ability and work ethic were the most widely reported by interviewees. The below quotes illustrate these outcomes:

- Positive identity: "'I'm pretty popular in my school. I even get away sometimes with stuff others don't because I play in the basketball team." (Kyle, current player)
- Sense of hope and life purpose: "This is what he wants to do, and this is the best possible place for him to do it in. He is determined to do it and he really feels like this place helps him get closer to his dream." (Jerome, current parent)
- Sense of belonging: "Look, at the time, we lived in [name of town], it was a shithole, we couldn't even go play outside and everyone was into football or rugby. Driving into [name of bigger city] to come to the centre was massive for me, having friends outside [name of town] was a salvation for me." (Darren, former player)
- A broader worldview: "It was just great to see Sid interacting with all these different people. Where we live and where he goes to school most people are white middle class and he built some great relationships with kids that he would have never met otherwise, and I think that has stood him in great stead going forward to uni and now work." (Mark, former parent)
- Higher learning ability: "It improves their concentration span. For him, it has gone through the roof, he has to pay attention to what coach is saying, to what's going on around him, it really has helped him by being constantly exposed to information and coaching." (James, current parent)
- Work ethic: "[. . .] sheer hard work, the boys learn how much harder they can work than they thought before. Coaches expect players to be on time, to do what they have to do, and to respect them and all around them. It builds a great work ethic." (Coach George)

Conversely, low self-confidence, selfishness and uncontrolled aggression were the most discussed negative outcomes. The quotes below display examples of these components:

- Low self-confidence: "When they lose an important game or when they don't play that much it's hard, they are proper down and lose confidence. All you can do is comfort them and wait for them to bounce back." (Sophie, current parent)
- Selfishness: "You always get two or three in a team that think they are the bee's knees, and they become a problem." (Coach Carl)
- Uncontrolled aggression: "I have said it before, they are a pack of alpha dogs trying to establish who the uber-alpha is going to be. They all come here having been the best in their local clubs, and now they have to work out who is top dog. And that can get hairy sometimes." (Chloe, current parent)

3.2. Generative Mechanisms

Guided by the RE orientation of the study, identifying the generative mechanisms proposed by stakeholders was prioritised. Interviews elicited 68 mechanisms involved in the generation of the 21 positive outcomes (Table 2). Some of these were found to be more salient due to their involvement in the creation of more than one outcome: love for the game ($n = 6$), the inspirational coach ($n = 6$), success/winning ($n = 5$), playing other club roles ($n = 5$) and diversity ($n = 4$) were the most salient. Similarly, 16 mechanisms leading to negative outcomes were reported (Table 3). Internal competition ($n = 5$) and negative coaching behaviours ($n = 3$) were the most prevalent in the generation of multiple outcomes. Please note the addition of a miscellaneous category to account for outcomes that did not fit the POF.

Table 2. Generative mechanisms of positive outcomes of participation.

Development of Self	Social	Emotional	Cognitive	Moral	Miscellaneous
Positive Identity Success/winning Height as positive Being cool Love for the game **Sense of Hope and Life Purpose** Success/winning Inspirational coach Perception of progress Love for the game **Self-Confidence** Success and failure Competence Being cool Belonging to wider family	**Interpersonal Skills** Constant interaction Banter Social diversity Including new players Co-education Playing up an age group Positive use of social media Playing other club roles Coach as facilitator Interface with adults **Sense of belonging** Community spirit Belonging to wider family Soft hierarchy Common goals Looking out for others Love for the game Second home Sense of hope and life purpose Success/winning Inspirational coach All family involved High family time Family sacrifices Parent as volunteer One special friend **Social Capital** Belonging to wider family Diversity Playing up an age group Contact beyond basketball Positive use of social media **Cooperation Skills** Common goals Need to cooperate Understanding role and hierarchy From 'I' to 'Us' **Leadership Attributes** Being the captain Helping young players Doing the right thing **A Broader Worldview** Social diversity Travel opportunities	**Emotional Wellbeing** Caring environment Structure and routine The social network All family involved Letting off steam Competence Love for the game **Emotional Literacy** High exposure to range of emotions Inspirational coach Diverse responses to events **Emotional Control** Standards and expectations High pressure Coping with setbacks Putting team first Inspirational coach Parental support Parental presence Strong social network Steam release Being ready to learn and perform	**Higher Learning Ability** Constantly taught Coaching behaviours Constant feedback Setting personal goals **Enhanced Decision-Making Capacity** High-paced decision-making demands Love for the game **Improved Communication** Interface with adults Coach as role model Constant interaction	**Respect** Club values Putting team first Competition Being ready to learn and perform Playing other club roles Social diversity Skill diversity **Moral Decision Making** Club values Love for the game Competition Keeping busy	**Work Ethic** Club values Inspirational coach Internal competition Success/winning Family sacrifices Career ambitions Playing other club roles **Competitiveness** Inspirational coach High pressure **Self-Reliance** Standards and expectations Lack of parental support Career ambitions Playing other club roles **The Springboard Effect** Playing other club roles Strong social network Perceived added value of sport

Table 3. Generative mechanisms of negative outcomes of participation.

Self	Social	Emotional	Cognitive	Moral	Miscellaneous
Quashed Individuality Putting team first **Low Self-Confidence** Internal competition External competition Negative coach behaviours The parent coach	**Social Isolation** Lack of social time Being different in school **Selfishness** Burning desire to win Internal competition Parental behaviours	**Lowered Emotional Wellbeing** Pressure Lack of social time Lack of study time Negative coach behaviours Intra-team bullying **General Demotivation** Love for the game	None recorded	**Occasional Drinking** Steam release **Uncontrolled aggression** Internal competition External competition Parental behaviours Negative coach behaviours **Disrespect and Bullying** Daily internal competition Being different in school	None recorded

3.3. Classifying Generative Mechanisms

Further to identifying the generative mechanisms and their associated outcomes, the mechanisms were grouped into related families. This exercise aimed to simplify the emerging complex picture and to start building a useable framework for practitioners. Four major groups of mechanisms were identified and named: (i) The Greenhouse for Growth; (ii) The Personal Boost; (iii) The Attention Factory; and iv) The Real Life Simulator. These families are defined below and presented in Tables 4 and 5.

Table 4. Summary of generative mechanisms leading to positive outcomes by family.

The Greenhouse for Growth	The Personal Boost	The Attention Factory	The Real Life Simulator
Club Ethos Club humanistic philosophy and values High standards and expectations **Coaches' Behaviours** Coach as inspiration Coach as facilitator Coach as role model **Parental Support/Influence** All family involved Family sacrifices Parental presence (or lack of) and contribution Proactive parental management by the club **Social Support/Influence** Soft hierarchy Community spirit and belonging Common goals Looking out for others One special friend The social network Second home Interface with adults	**Experience of Success** Exposure to success and failure Perception of progress and competence Sense of hope and purpose **Athletic Kudos** Being cool Height as positive **Steam release** Letting off steam	**Love for the Game** Personal infatuation with the game Collective infatuation with the game **A Purposeful Life** Career ambitions Structure and routine Sense of hope and purpose Short- and mid-term personal goals Keeping busy	**Competition** High pressure Coping with setbacks Internal and external competition Playing up **The Team** Putting team first Understanding roles and hierarchy Constant interaction Healthy banter Including players Looking out for others Do the right thing Chance to be a captain **Learning** Being constantly taught Constant feedback Setting personal goals High-pace decision-making **Diversity** Social diversity Geographical diversity **Mini-Workplace** Playing other club roles

Table 5. Summary of generative mechanisms leading to negative outcomes by family.

The Greenhouse for Growth	The Personal Boost	The Attention Factory	The Real Life Simulator
Club Ethos Putting team first (quashed individuality) **Coaches' Behaviours** Negative teaching Negative emotions **Parental Support/Influence** Parental negative influence The overinvolved parent **Social Support/Influence** **Being different** Lack of social time Bullying	**Experience of Success** Exposure to success (feeling superior to others leading to disrespectful behaviours and a general sense of entitlement) **Steam Release** Letting off steam (leading to negative behaviours such as uncontrolled aggression or occasional drinking)	**Love for the Game** Personal infatuation with the game (leading to general demotivation to do anything other than sport)	**Competition** High pressure Pathological desire to win (leading to immoral decisions) Internal competition External competition

3.3.1. Greenhouse for Growth

'The Greenhouse for Growth' relates to the built-in features of the setting that lay the foundation for the club to become a source of personal growth rather than just a sporting venue. Our 'greenhouse' allusion is intended to reflect generalised notions of a nurturing environment and climate. Four sub-categories were identified (i.e., club ethos, coaches' behaviours, parental support/influence and social support influence), incorporating a further 17 mechanisms.

3.3.2. Personal Boost

'The Personal Boost' focuses on the capacity of participation in performance sport to generate elevated states of mind (i.e., happiness, joy, satisfaction, elation, pride, etc.) leading to increased player wellbeing and a strengthened drive to stay involved in the sport and the club. Three main sub-categories were associated to this family (i.e., experience of success, athletic kudos, and steam release) containing six mechanisms.

3.3.3. Attention Factory

'The Attention Factory' revolves around the notion of sport participation providing athletes with a clear focus in life that (i) leads to positive behaviours and outcomes and (ii) individuals use to confirm their personal agency. This attentional focus also acts as a protective shield and deterrent against negative sport-based attitudes and behaviours and insulates against engaging in rivalrous conducts seen (by most adults) as undesirable. Two sub-categories of mechanisms were created (i.e., love for the game and a purposeful life) encapsulating a further seven mechanisms.

3.3.4. Real Life Simulator

Finally, 'The Real Life Simulator' relates to the idea that participation in youth performance development sport micro-replicates elements of the 'adult world'. From this perspective, participation offers inexperienced young athletes multiple opportunities to practise facing and resolving challenges and situations that may recur in their future lives at university and/or in work. This family contains five sub-categories (i.e., competition, the team, learning, diversity and mini-workplace), encompassing 19 single mechanisms.

Tables 4 and 5 summarise the mechanisms and sub-mechanisms in each major theme. This secondary analysis, aimed at increasing the practical application of these findings for coaches and club administrators, led to the further distillation of the mechanisms presented in Tables 2 and 3, with some of them merging into a single theme for a new total of 49 mechanisms. A full depiction of the four families of mechanisms with sample quotes can be found in the Supplementary Materials.

3.4. Importance of Context

In keeping with the RE approach, the study also reflected on the impact of the specific context in which the club operated on the range of outcomes and mechanisms discussed by the contributors. Following the work of Pawson [25] (p. 37), we addressed four contextual layers: (i) individuals (i.e., the characteristics and capacities of the various stakeholders in the programme); (ii) interpersonal relationships (i.e., the stakeholder relationships that carry the programme); (iii) institutional settings (i.e., the rules, norms and customs local to the programme); and (iv) infrastructure (i.e., the wider social, economic and cultural setting of the programme). Individuals and their relationships will be treated jointly, as the interviewees reported them as inseparable.

3.4.1. Individuals and Interpersonal Relationships

Four main groups of stakeholders were identified: club officials, coaches, parents and players. The club officials and coaches were primarily ex-high school teachers or past club players with a stated 'genuine' disposition to care for others, supplemented by years of successfully working with young people in school and sport settings. Their unity drove 'what the club was about': prioritising positive relationships and personal development over winning and external success. The parents typically felt this grouping of officials and coaches was key to the club operating as it did: this group provided the smooth alignment that integrated everyone at the club, while activating every player and every team to develop individually in positive ways.

Parents also played a significant part in the workings of the club. As a diverse cohort with a broad range of backgrounds (i.e., ethnic, educational and socioeconomic), parental involvement was typically and mainly seen as positive. Three profiles were described by the parents and coaches during the interviews and focus groups: (i) 'peripheral parents', who tended to stay outside of the activities and maintained a predominantly transactional relationship with the club; (ii) 'rarely-seen parents', who were typically working shifts, having to look after younger siblings or simply living too far away and having no transport; and (iii) 'core parents', who became club volunteers, playing roles such as team manager, mini-bus driver, table official, fund raiser, or match steward.

Parental dispositions reflected the levels and styles of engagement. Playing-based and player-based outcome expectations moderated this experience. Most parents recognised the slim chance of their son making a living out of basketball; they instead emphasised the 'added value' of participation as their main concern. For fewer parents, winning national titles, playing for Great Britain and getting a college scholarship were the main drivers for involvement. The 'added value' group deployed more adaptive and supportive attitudes; they focused more on the positive developmental outcomes of participation and less on their son's playing time or performance. They were relatively content to 'belong' and to feel 'included' in the club environment. This attachment style may be moderated by the characteristics of the players, as described below.

The players, like their families, were also highly heterogeneous. Two main elements differentiated player experience and outcomes: (i) the level of parental support; and (ii) the pecking order within the club in terms of their expected progress. High parental support was linked to greater confidence, emotional wellbeing and emotional control. However, it was also connected to a sense of entitlement, or, in the case of overinvolved parents, to lowered confidence and outbursts of bad temper. Consistently 'low' parental support often created space for players to develop their resilience and self-responsibility, leading them to take charge of their own management and progress.

In relation to the playing pecking order, having high prospects and extended playing time was related to positive identity, a strong sense of hope and life purpose and higher confidence. Yet, a high ranking was also linked to taking a different interpersonal stance;

selfishness and disrespect for lower-ranked players were noted. By contrast, players ranked lower risked developing low self-esteem, decreased emotional wellbeing and were prone to selfish acts to compensate. Despite this risk, powerful integrative and developmental daily routines (i.e., individualised training plans) coupled with a strong sense of identity meant that only few lower-ranked players exhibited these problematic responses.

3.4.2. Institutional Settings

The institutional settings incorporate the rules, norms and customs local to the programme. Four main features were reported: (i) the explicit humanistic ethos of the club—rooted in the personal beliefs of the club founders; (ii) inclusivity—the club hosted both a national league programme and community leagues, and operated a low-cost policy to facilitate access; (iii) member expectations—all club members were expected to contribute to its smooth running by embodying its positive principles and carrying out the required chores; and (iv) the role of the basketball centre—the building itself created a socially rich environment that contributed to the development of a strong sense of belonging and enhanced wellbeing. As a result, the perception amongst all interviewees was that club members tended to spend much more time at the facility than was required by their training schedule. In effect, the basketball club seemed to operate as a *de facto* community centre where players and parents gathered to share in their love of the sport and each other's company.

3.4.3. Infrastructure

With regard to the wider social, economic and cultural setting in which the programme is embedded, a number of elements warrant special attention: (i) set in an inner-city neighbourhood of a large city in the north of England, the centre attracts a broad range of families from multiple ethnicities and socioeconomic backgrounds. Inner-city young people from lower income families shared the space with high income youths living in suburbs as far as 40 miles away. This diversity was seen by the interviewees as a catalyst for positive development, mirroring the real world and providing the children with a wider perspective and broader horizons; (ii) the northern location strongly translated into a sense of 'northern identity' characterised by grit, toughness and a need to prove oneself to 'The South'; and (iii) the minority-sport nature of basketball in the UK underpinned the development of a marked 'cult-like identity'. Belonging to a very unique 'baller' community, with a distinct lifestyle, leisure profile and sense of fashion, created a common identity. The low national profile of the sport with few professional opportunities also meant that the players and their families typically aspired to gain a scholarship to a USA college on completion of their secondary education. This facilitated a greater focus by the club on the long-term development of the players and on the prioritisation of academic achievement by players and their families.

In sum, the above description of the multiple layers of the context and their impact on the generative mechanisms and developmental outcomes serves to re-emphasise the integrated, multi-dimensional and highly individualised nature of the sport experience for young players. Table 6 provides a summary of the most salient contextual features.

Table 6. Most salient contextual features.

Most Salient Contextual Features
Individual and Interpersonal Level
Personal values and beliefs
Socioeconomic status
Coaches' professional background
Parental involvement and attitudes
Pecking order
Institutional Level
Club ethos
Club status
Club owned facility
High contact time
Infrastructure
Minority sport
Lack of professional pathways

4. Discussion

This study used an RE approach to study psychosocial development in youth sport in a novel way. This part one account has detailed the perceptions of former and current club parents, players and coaches, using them to build a set of programme theories: configurations of outcomes, context and mechanisms. The resulting network of outcomes and generative mechanisms spanning multiple contextual layers provides a nuanced understanding of stakeholders' views and experiences. To respond to the many and dynamic needs of young people and the people who support them, the club appears to respond with a similarly complex offer, embodied by expectations and accountabilities built into its routines, practices and rituals. In line with the previous literature, no silver bullets for positive development were reported in this club [8,9]; instead, the high expectations placed on club members, and a group of stakeholders who were energised and committed to sacrifice and to hard work, done well and persistently, appeared to drive the outcomes [10,17,38]. Innovatively, this paper offers a textured perspective of the multi-faceted process of development. In doing so, it moves the study of psychosocial development in sport beyond the limitations of the existing research and identifies the depth and integration of what goes into creating developmental experiences in sport [19,20].

The results highlight the complex and systemic nature of the development process wherein a variety of factors and elements combine, interact and catalyse in multiple, non-linear ways to produce highly individualised outcomes. Within this context, and in line with previous research, the figure of the coach was revealed as a key modulator and catalyst [19,28,39]. Finally, in line with the previous literature, the study found that, despite a broad range of possible outcomes resulting from sport participation, the players' internal and external assets and the personal narrative they attached to the experience all played a powerful role in determining the scale of their development [20,40].

Central to the novelty of this study, the use of an RE methodology addresses previous calls for using a more systemic and process-based research approach [13,19,20]. Identifying an emerging network of outcomes and generative mechanisms, filtered by individual, relational, institutional, and infrastructure contextual features, as hinted at by Kochanek and Erickson [41], this study moves beyond the typically reductionist approach taken in existing work. In doing so, it adds depth and practicality to current theoretical conceptualisations [13,20,42–44]. Likewise, this investigation answers the need to conduct ecologically valid research [21,23] by studying psychosocial development in a live organised youth sport setting, where the majority of the participants and practitioners operate. Moreover, this paper problematised development in youth sport by highlighting the potential for both positive and negative outcomes [8,9,22].

Most importantly, however, the strength of this study lies in highlighting the intricacy of youth sport as a developmental system. This is linked to the complexity of human development [45] and the myriad actors and potential generative mechanisms that combine—in multiple ways—to interact with the young person's existing assets and personal narratives [38]. Unsurprisingly, this leads to highly individualised outcomes. Despite this inherent complexity, as called for in previous literature [21,23], using the RE framework has helped to make these intricate processes more accessible to practitioners [25]. By identifying and then classifying a clear set of developmental outcomes and mechanisms, sport psychologists, coaches and programme leaders are provided with a concrete menu of options to facilitate programme design and implementation. Similarly, the study highlights the return on investment from deploying a tailored, deliberate and integrative approach to personal development in organised sport [5,10]. This calls for all stakeholders to reconsider how their existing programmes address this wide range of features in their day-to-day processes.

Notwithstanding these findings, this paper represents only one half of the RE process. Once the stakeholder's PTs are established 'on paper', it is necessary to test how these elements and processes feature in the 'live' environment. This 'coalface' analysis may reveal previously unseen components and relationships that challenge existing participants' views; it may also unearth important new areas of inquiry and implementation. Layder refers to this process as the development of "adaptive theory" [27], the never-ending cycle of refining existing theory based on new emerging data to arrive at new insights. To address this, part two of the study, also published in this special issue, details the full-season ethnographic immersion of the lead author into the club's environment. Through this multi-stage process, part two will also present an evidence-based, theory-driven and practitioner-oriented model of psychosocial development in youth sport, together with recommendations for future research.

Supplementary Materials: The following are available online at https://www.mdpi.com/article/10.3390/sports10040047/s1, Table S1a–e: Participant demographic data. Table S2a,b: Generative mechanisms quotes examples.

Author Contributions: Conceptualization, S.L.-B. and J.M.; methodology, S.L.-B. and J.M.; software, S.L.-B.; formal analysis, S.L.-B. and J.M.; investigation, S.L.-B.; data curation, S.L.-B. and J.M.; writing—original draft preparation, S.L.-B.; writing—review and editing, S.L.-B. and J.M.. All authors have read and agreed to the published version of the manuscript.

Funding: This research received no external funding.

Institutional Review Board Statement: The study was conducted according to the guidelines of the Declaration of Helsinki and approved by the Ethics Committee of Leeds Beckett University (6 February 2014).

Informed Consent Statement: Informed consent was obtained from all subjects involved in the study.

Acknowledgments: The authors would like to thank the sport club in which this research took place for their generosity, and all those who took part in the research for their willingness to contribute and for the candidness of their responses.

Conflicts of Interest: The lead author declares a previous relationship with the studied sport club where he was a coach in previous years. However, the lead author had no current role at the club at the time of the research taking place.

References

1. Alberga, A.S.; Sigal, R.J.; Goldfield, G.; Homme, D.P.; Kenny, G.P. Overweight and obese teenagers: Why is adolescence a critical period? *Pediatr. Obes.* **2012**, *7*, 261–273. [CrossRef] [PubMed]
2. Biddle, S.J.H.; Asare, M. Physical activity and mental health in children and adolescents: A review of reviews. *Br. J. Sports Med.* **2011**, *45*, 886–895. [CrossRef]
3. Smith, J.J.; Eather, N.; Morgan, P.J.; Plotnikoff, R.C.; Faigenbaum, A.D.; Lubans, D.R. The Health Benefits of Muscular Fitness for Children and Adolescents: A Systematic Review and Meta-Analysis. *Sports Med.* **2014**, *44*, 1209–1223. [CrossRef] [PubMed]

4. Holt, N.L.; Camiré, M.; Tamminen, K.A.; Pankow, K.; Pynn, S.R.; Strachan, L.; Macdonald, D.J.; Fraser-Thomas, J. PYDSportNET: A knowledge translation project bridging gaps between research and practice in youth sport. *J. Sport Psychol. Action* **2017**, *9*, 132–146. [CrossRef]
5. Pierce, S.; Kendellen, K.; Camiré, M.; Gould, D. Strategies for coaching for life skills transfer. *J. Sport Psychol. Action* **2016**, *9*, 11–20. [CrossRef]
6. European Commission. *Expert Group on Skills and Human Resources Development in Sport Guidelines Regarding the Minimum Requirements in Skills and Competences for Coaches*; European Commission: Brussels, Belgium, 2020.
7. Sport England. *Uniting the Movement*; Sport England: London, UK, 2021.
8. Coakley, J. Youth Sports: What Counts as "Positive Development? *J. Sport Soc. Issues* **2011**, *35*, 306–324. [CrossRef]
9. Coalter, F. *Sport for Development: What Game Are We Playing?* Routledge: Abingdon, UK, 2013.
10. Bean, C.; Kramers, S.; Forneris, T.; Camiré, M. The Implicit/Explicit Continuum of Life Skills Development and Transfer. *Quest* **2018**, *70*, 456–470. [CrossRef]
11. Turnnidge, J.; Côté, J.; Hancock, D.J. Positive Youth Development From Sport to Life: Explicit or Implicit Transfer? *Quest* **2014**, *66*, 203–217. [CrossRef]
12. Sport and Development. What is Sport and Development? *2017*. Available online: https://www.sportanddev.org/en/learn-more/what-sport-and-development (accessed on 20 December 2021).
13. Gould, D.; Carson, S. Life skills development through sport: Current status and future directions. *Int. Rev. Sport Exerc. Psychol.* **2008**, *1*, 58–78. [CrossRef]
14. Hodge, K.; Danish, S. Promoting Life Skills for Adolescent Males through Sport. In *Handbook of Counseling Boys and Adolescent Males: A Practitioner's Guide*; Horne, A.M., Kiselica, M.S., Eds.; SAGE Publications Inc.: Thousand Oaks, CA, USA, 1999; pp. 55–71.
15. Benson, P.L.; Leffert, N.; Scales, P.C.; Blyth, D.A. Beyond the 'Village' Rhetoric: Creating Healthy Communities for Children and Adolescents. *Appl. Dev. Sci.* **2012**, *16*, 3–23. [CrossRef]
16. Lerner, R.M.; Almerigi, J.B.; Theokas, C.; Lerner, J.V. Positive Youth Development A View of the Issues. *J. Early Adolesc.* **2005**, *25*, 10–16. [CrossRef]
17. Holt, N.L.; Neely, K.C.; Slater, L.G.; Camiré, M.; Côté, J.; Fraser-Thomas, J.; Macdonald, D.; Strachan, L.; Tamminen, K. A grounded theory of positive youth development through sport based on results from a qualitative meta-study. *Int. Rev. Sport Exerc. Psychol.* **2017**, *10*, 1–49. [CrossRef] [PubMed]
18. Whitley, M.A.; Massey, W.V.; Camiré, M.; Blom, L.C.; Chawansky, M.; Forde, S.; Boutet, M.; Borbee, A.; Darnell, S.C. A systematic review of sport for development interventions across six global cities. *Sport Manag. Rev.* **2019**, *22*, 181–193. [CrossRef]
19. Holt, N.L.; Neely, K.C. Positive Youth Development through Sport: A Review. *Revista De Iberoamericana De Psicología Del Ejercicio Y El Deporte* **2011**, *6*, 299–316.
20. Pierce, S., Gould, D.; Camiré, M. Definition and model of life skills transfer. *Int. Rev. Sport Exerc. Psychol.* **2017**, *10*, 186–211. [CrossRef]
21. Holt, N.L.; Gareth, W.J. Future Directions for Positive Youth Development and Sport Research. In *Positive Youth Development Through Sport*; Holt, N.L., Ed.; Routledge: Abingdon, UK, 2008; pp. 122–131.
22. Hartmann, D.; Kwauk, C. Sport and Development. *J. Sport Soc. Issues* **2011**, *35*, 284–305. [CrossRef]
23. Kay, T. Developing through sport: Evidencing sport impacts on young people. *Sport Soc.* **2009**, *12*, 1177–1191. [CrossRef]
24. Pawson, R.; Tilley, N. *Realistic Evaluation*; SAGE Publications Inc.: London, UK, 1997.
25. Pawson, R. *The Science of Evaluation: A Realist Manifesto*; Sage: London, UK, 2013.
26. Archer, M.S. *Realist Social Theory*; Cambridge University Press (CUP): Cambridge, UK, 1995.
27. Layder, D. *Sociological Practice*; SAGE Publications Inc.: Thousand Oaks, CA, USA, 1998.
28. North, J. *Sport Coaching Research and Practice: Ontology, Interdisciplinarity and Critical Realism*; Routledge: Abingdon, UK, 2017.
29. Lara-Bercial, S.; McKenna, J. PsyTool Supplement Suppl 3. *J. Sport Psychol.* **2018**, *27*, 53–58.
30. North, J. A Critical Realist Approach to Theorising Coaching Practice. In *Sports Coaching*; Routledge: London, UK, 2015; pp. 133–144.
31. Dohme, L.-C.; Rankin-Wright, A.J.; Lara-Bercial, S. Beyond Knowledge Transfer: The Role of Coach Developers as Motivators for Lifelong Learning. *Int. Sport Coach. J.* **2019**, *6*, 317–328. [CrossRef]
32. Lusted, J. A critical realist morphogenetic approach to researching sport policy: Reflections on a large-scale study of policy implementation in grassroots English football. *Int. J. Sport Policy Politics* **2018**, *10*, 705–719. [CrossRef]
33. Creswell, J.W. *Research Design: Qualitative, Quantitative and Mixed Methods Approaches*, 4th ed.; SAGE Publications Inc.: Thousand Oaks, CA, USA, 2013.
34. Schwandt, T. *Qualitative Enquiry: A Dictionary of Terms*, 4th ed.; SAGE Publications Inc.: Thousand Oaks, CA, USA, 2015.
35. Simons, H. *Case Study Research in Practice*; SAGE Publications: London, UK, 2009.
36. Sparkes, A.C.; Smith, B. *Qualitative Research Methods in Sport, Exercise and Health: From Process to Product*; Routledge: London, UK, 2013.
37. QRS International. *N-VIVO Qualitative Analysis Software*; QRS International: Burlington, NJ, USA, 2015.
38. Benson, P.L. *All Kids Are Our Kids: What Communities Must Do to Raise Caring and Responsible Children and Adolescents*, 2nd ed.; Jossey-Bass: San Francisco, CA, USA, 2006.
39. Newland, A.; Newton, M.; Moore, E.W.G.; Legg, W.E. Transformational Leadership and Positive Youth Development in Basketball. *Int. Sport Coach. J.* **2019**, *6*, 30–41. [CrossRef]

40. Pierce, S.; Erickson, K.; Dinu, R. Teacher-Coaches' Perceptions of Life Skills Transfer from High School Sport to the Classroom. *J. Appl. Sport Psychol.* **2019**, *31*, 451–473. [CrossRef]
41. Kochanek, J.; Erickson, K. Interrogating Positive Youth Development Through Sport Using Critical Race Theory. *Quest* **2020**, *72*, 224–240. [CrossRef]
42. Martinek, T. Enhancing youth development programs through logic model assessment. [Aumentar los programas de desarrollo juvenil a través de la evaluación del modelo lógico]. *Revista Internacional De Ciencias Del Deporte* **2017**, *13*, 302–316. [CrossRef]
43. Bronfenbrenner, U. *Making Human Beings Human: Bioecological Perspectives on Human Development*; SAGE Publications Inc.: Thousand Oaks, CA, USA, 2005.
44. Wells, M.S.; Arthur-Banning, S.G. The Logic of Youth Development: Constructing a Logic Model of Youth Development through Sport. *J. Park Recreat.* **2008**, *26*, 189–202.
45. Petitpas, A.J.; Cornelius, A.E.; Van Raalte, J.L.; Jones, T. A Framework for Planning Youth Sport Programs That Foster Psychosocial Development. *Sport Psychol.* **2005**, *19*, 63–80. [CrossRef]

sports

MDPI

Article

Roots to Grow and Wings to Fly: An Ethnography of Psychosocial Development in Adolescent Performance Sport

Sergio Lara-Bercial [1,2,*] and Jim McKenna [1]

1 Research Centre for Sport Coaching, Carnegie School of Sport, Leeds Beckett University, Leeds LS6 3QT, UK; j.mckenna@leedsbeckett.ac.uk
2 Global Coaching Office, International Council for Coaching Excellence, Leeds Beckett University, Leeds LS6 3QT, UK
* Correspondence: s.lara-bercial@leedsbeckett.ac.uk

Abstract: This study aimed to explore the potential for sport to support psychosocial development in young people in a youth performance setting using a novel realistic evaluation approach. Part 1 of this two-paper series published in this Special Issue identified the programme theories—how the programme is supposed to work. A wide and deep network of context, generative mechanisms and outcomes responsible for psychosocial development in this youth performance basketball club emerged. The first paper also concluded that the outcomes and the experience are highly contextual and individualised. In this second part, the stakeholder's programme theories were tested during a full-season ethnography of the same club. Immersion in the day-to-day environment generated a fine-grain analysis of the processes involved, including: (i) sustained attentional focus; (ii) structured and unstructured skill-building activities; (iii) deliberate and incidental support; and (iv) feelings indicating personal growth. Personal development in and through sport is thus shown to be conditional, multi-faceted, time-sensitive and idiosyncratic. The findings of this two-part study are considered to propose a model of psychosocial development in and through sport. This heuristic tool is presented to support sport psychologists, coaches, club administrators and parents to deliberately create and optimise developmental environments.

Keywords: positive youth development; youth sport; realist evaluation; life skills; personal development

Citation: Lara-Bercial, S.; McKenna, J. Roots to Grow and Wings to Fly: An Ethnography of Psychosocial Development in Adolescent Performance Sport. *Sports* **2022**, *10*, 48. https://doi.org/10.3390/sports10040048

Academic Editor: Michael Chia

Received: 29 December 2021
Accepted: 21 March 2022
Published: 22 March 2022

1. Introduction

Sport has the potential to support positive outcomes in young people. Part 1 of this two-paper series provided a detailed review of existing literature in this area and of its shortcomings. By way of a summary, extant search has positively linked sport participation to physical and mental health and wellbeing. Likewise, prior studies have also highlighted the capacity for sport to foster psychosocial development in young people. However, our review also showed that previous studies have tended to evaluate purpose-built interventions, leaving regular organised sport relatively overlooked. Moreover, previous work has tended to concentrate on a narrow range of outcomes and methodologies. To address these gaps, we conducted a season-long ethnography of a youth performance sport club using a novel realist evaluation (RE) approach [1,2]. This methodology has never been used before in the study of psychosocial development in youth sport.

RE typically aims to establish how well a given purpose-built social intervention achieves its expected outcomes. RE's unique features have particular utility in the study of psychosocial development in organized youth sport when sport is construed and positioned as an organic social intervention embedded within a complex social system. In other words, these are environments where, with or without the awareness, intention and volition of those within it, a series of mechanisms interact, leading to explicit, implicit, desired and undesired developmental outcomes for all involved. RE emphasises the fundamental role of theory in driving social research and the importance of looking beyond quantification and correlation to shed light on the generative mechanisms affecting choices, behaviours and, ultimately, the outcomes of a given intervention. RE is thus concerned with the notion and nature of causality.

Part 1 of this two-paper series focused on identifying key stakeholder programme theories (PTs) related to psychosocial development in a youth performance basketball club in the North of England. Former and current players (n = 16), their parents (n = 18), and current coaches (n = 5) took part in in-depth semi-structured interviews and focus groups. Current players and parents belonged to the Under 13 (U13) and Under 16 (U16) squads. The analysis elicited a deep network of interrelated positive and negative developmental outcomes, generative mechanisms and salient features of the context. The findings also highlighted the complex and systemic nature of the PYD process, where factors and elements combine, interact and catalyse in multiple, non-linear ways to produce highly individualised outcomes.

In line with the adopted RE approach, this second part of the study tested the previously proposed stakeholder PTs in the day-to-day workings of the club. This 'coalface' analysis was deployed to establish, or refute, the validity of the respective PTs and reveal previously unseen components and relationships that may challenge the espoused accounts. Layder [3] describes this "adaptive theory" as the never-ending cycle of refinement of existing theory based on new emerging data to arrive at new insights [4]. This second phase also aimed to develop and present an integrative model of psychosocial development to support sport psychologists, coaches, club administrators and parents to create and optimise other developmental environments.

2. Study Design and Methods

2.1. Study Design

As part of the case study methodology described in part 1, the second part of the study was comprised of the lead author conducting a full-season ethnography of a youth performance basketball club. To the best of our knowledge, no ethnography accounts of psychosocial development in youth sport exist in the literature. Wolcott [5] describes ethnography as the process of learning about culture as manifested through distinct and observable patterns of socially shared behaviours. Ethnographers, therefore, aim to gain a deep understanding of the functioning of social groups and of the implications for the group and its members. However, this understanding of culture must come from the viewpoint of the group members [6]. Therefore, central to all different interpretations of ethnography is that it includes "a commitment to first-hand experience" [7] (p. 34) in the shape of "direct and sustained social contact with agents" [8] (p. 5).

The study used Wolcott's experiencing-enquiring-examining framework [5]. Experiencing revolves around participant observation and, in some cases, the researchers' own participation. Enquiring requires that we ask participants about "what's going on"(p. 49). Examining involves the study of documents and artefacts produced in the environment. As a result, ethnographic researchers need to remain flexible in using a wide variety of research methods depending on how social life unfolds [9].

Notwithstanding this need for methodological elasticity, Wolcott's framework indicates that ethnographers must start with intent: the researcher creates in advance a clear understanding of what is to be studied. For Wolcott, "fieldwork must be preceded by mindwork" (p. 70). From this perspective, ethnography acts more as a philosophy of how to conduct research rather than a method or technique to be employed [9]. Ethnography is thus theory-driven yet dynamic and adaptable to deal with the varying demands of the 'live' environment. An ethnographic approach thus fits well within the broader RE theoretical framework of the study and its focus on explanation and causality in the complex and multi-layered context of a social intervention.

2.2. Study Setting

As described in part 1, a youth performance development basketball club in North West England was identified following a criterion-based sampling process. Criteria included an environment featuring: (i) high intensity and frequency of participant engagement combined with high future stakes (i.e., a professional career); (ii) a hybrid club ethos combining community- and performance-based values and provision, leading to a wide range of experiences and interactions; and (iii) an explicit humanistic philosophy combined with a prolonged high level of success at the national and international level. It is important to note the lead researcher's familiarity with the setting, where he has coached for the last 20 years. The pros and cons of this element were carefully weighed regarding the balance of insider versus outsider positionality; the research team decided that the benefits outweighed the potential negative effects. Note also that none of the research subjects were or had ever been coaches coached by the lead author. On these bases, permission to progress with the study was provided by the club Chairman. Approval was subsequently granted by the Ethics Committee of Leeds Beckett University.

2.3. Participants

The study focused on the U13 and U16 boys squads. A single-gender approach was chosen due to the female section of the club being nearly completely run separately from the male side. The squads were comprised of 15 players each, with fluctuations over the course of the season due to injuries and late additions. Due to the location of the club in an inner-city and its performance-based nature, the club attracts a broad range of young people and families from varying socioeconomic statuses and ethnicities. Whereas part 1 focused on a small number of participants from each squad, this second part took into consideration the experiences of all squad members and their families and coaches (for full details of participants, see Paper 1 of the series and its Supplementary Materials from Paper 1). Moreover, given the full immersion of the lead author within the club, other inhabitants of the setting also contributed to the study by interacting with the researcher purposefully and explicitly, or in passing, during his visits to the club.

2.4. Data Collection

The lead author spent a full season (August to May) in the field. The emerging collection of context-mechanism-outcome networks (CMONs) described in part 1 was used to guide the researcher during the immersive period. The researcher purposefully sought to ascertain, refute and/or refine the presence of these networks in the club environment. During the immersion period, a variety of data collection methods were used. These are described using Wolcott's experiencing-enquiring-examining framework.

Experiencing. The researcher attended 32 practices and 16 matches involving the U13 and U16 squad over the course of the season (approximately half of the total number of training sessions and matches during this period). A mix of targeted observations—guided by the CMONs—were alternated with more casual engagements, where the researcher adopted a broader outlook. This mix of approaches helped the researcher to "make the familiar strange, and the strange familiar" [5] (p. 89).

Enquiring. Over the season, the researcher regularly spoke to coaches, parents and players, with conversations ranging from one-liners said in passing and casual 5–10 min courtside conversations to formal in-depth interviews. Some specific interactions were initiated by the researcher to address a specific agenda and shed light on a particular CMON. All observations and interactions were recorded as fieldwork notes and subsequently transcribed using a word processor to generate over 250 pages of double-spaced data.

Examining. The researcher reviewed all available documentation of the club's history, goals, philosophy and values from the club website (i.e., club statement, club philosophy, coach, parent and player codes of conduct, club safeguarding policy, etc.). Additional club promotional materials were also collected (i.e., club leaflets, holiday camp brochures, and information circulars). All these supplementary sources of information were analysed in relation to the identified CMONs looking for corroboration or discrepancies.

2.5. Data Analysis

Once data were collected, the analysis took place. It progressed from the generic to the specific and back to the generic. First, the researchers interrogated the extent to which the proposed categories of mechanisms presented in part 1 were seen in the day-to-day activities of the club. A deductive-inductive analysis was conducted using NVIVO-10 [10]. Once specific mechanisms were established, the researcher attempted to link them to the outcomes—positive and negative—and contextual features described in stage one. The final analytical step integrated the findings from both stages of the project, aiming to develop a holistic, systemic understanding that supported the presentation of a theory-led, practitioner-oriented model. Within this was a commitment to identifying generic good practice, higher-level principles that might inform specific developmental practices in other sport-based environments.

2.6. Reflexivity

In conducting the ethnography, the issue of reflexivity was also explored and considered. Reflexivity is the process and practice of becoming aware of how researchers' personal histories, expectations, resource opportunities and constraints impact the way they conduct and interpret research [9]. The issue of the researcher as insider/actor/agent, far from denuding the value of ethnography-based research, offers the opportunity to explore and attempt to grasp situations and occurrences which other approaches cannot begin to apprehend [5]. It is important, however, that researchers using this approach place themselves within the context of the investigation and openly acknowledge their relationship with others within it [9,11]. Although ethnography will always feature some idiosyncrasies [5], the ethnographer must present as much evidence as justifies making specific claims.

3. Results

Findings are presented in three sections. First, we introduce the confirmatory analysis of the generative mechanisms, contextual features and outcomes following the immersive period. Second, we describe the idiosyncratic nature of the experience and the contrasting ways in which each young person engages with the club's developmental systems. Finally, by bringing together the findings across both phases of the study, we aim to present an evidence-based model of psychosocial development in this setting. For economy, observational notes and participant quotes have been minimised. Pseudonyms and unique identifying events are used to prevent the identification of individuals. In the case of club co-founders Coach Jack and Coach George (pseudonyms), identifying them and their real name is easy to do, given that it is feasible to determine the lead author's club affiliation. Thus, explicit permission has been sought and granted from them to publish this paper.

3.1. Confirmation of Generative Mechanisms, Contextual Features and Outcomes

Part 1 identified four major families of mechanisms responsible for driving or precluding psychosocial development in sport: (i) The Greenhouse for Growth; (ii) The Personal Boost; (iii) The Attention Factory; and (iv) The Real-Life Simulator (Figure 1). 'The Greenhouse for Growth' revolves around the setting features that create a context in which young people secure personal growth. 'The Personal Boost' focuses on the distinctive capacity of participation in performance sport to generate elevated states of mind (i.e., happiness, joy, satisfaction, elation, pride, etc.). 'The Attention Factory' relates to the notion of the club's 'way' of doing sport participation, providing athletes with a clear focus in life that activates personal agency. Finally, 'The Real-Life Simulator' is linked to the idea that participation in youth performance development sport intentionally and unintentionally simulates elements of adult life.

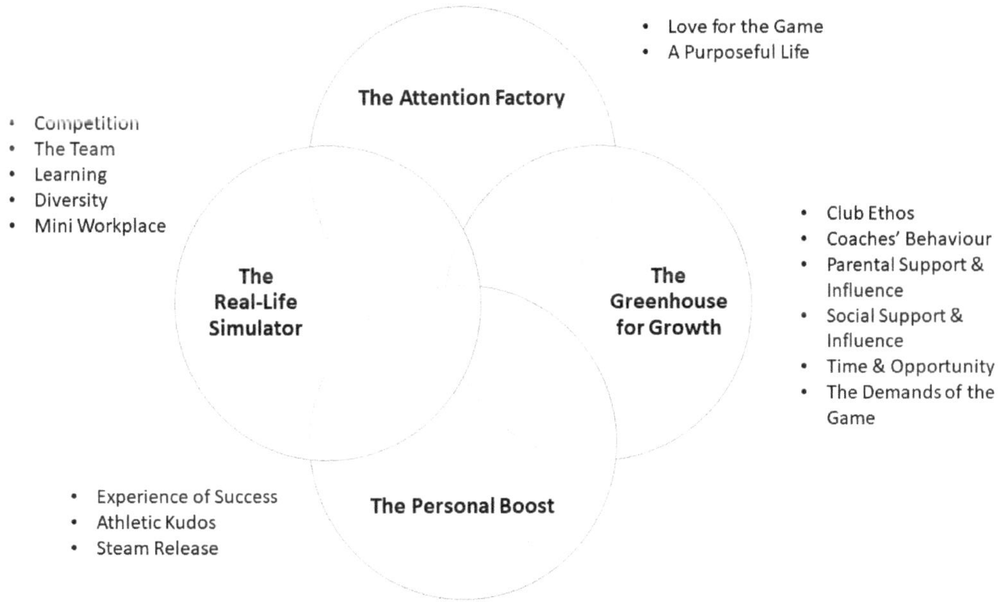

Figure 1. The four families of generative mechanisms of psychosocial development in sport (post Phase 1).

3.1.1. The Greenhouse for Growth

The notion of the club as a space endowed with fertile developmental 'soil' and optimal 'weather' for growth was confirmed during the club visits. A further two mechanisms were identified, namely, time and opportunity and the demands of the game (Table 1). In this section, we review each mechanism as they were encountered in the field.

Table 1. The Greenhouse for Growth—complete mechanisms post-immersion.

The Greenhouse for Growth
Club Ethos
Coaches' Behaviours
Parental Support/Influence
Social Support/Influence
Time and Opportunity (new)
The Demands of the Game (new)

The club ethos became an inescapable theme. Underpinned by the club's mission 'to provide opportunities for all sections of the community regardless of ability, background, gender or age', it could be said that the club's modus operandi revolved around two key elements: (i) prioritising the development of the human over the player; and (ii) high, non-negotiable expectations that all stakeholders contribute to this overall outcome. Observations revealed the key role played by the club founders in maintaining a focus on this big purpose. Former high school teachers, Coach Jack and his wife Mary, and Coach George (all of them in their 70s) set the tone for the rest of the club. Jack's attitude, especially, was repeatedly highlighted as the club's cornerstone. As a score-keeper and guardian of the club standards, Jack was known for upholding strong values and willingly enforcing them with no concern for his public image:

> "Look, he can be a pain in the backside, but someone has to be prepared to do that or the club will go down the drain. He was out for a few months because

> of [health issue] and you could see the club starting to slip up because no-one was prepared to do what he does. Not even Coach George (co-founder) would be able to do that." (Coach James)

In addition, coaches' behaviours were consistently seen as central to the greenhouse effect. Notwithstanding this, parents expressed a clear preference for the coaches they favoured for their children. Parents saw older coaches like Coach Jack, Coach George and Coach Dean as more encouraging and personable. In contrast, some of the younger coaches were deemed "too in your face and aggressive" (Sophie, parent), which risked damaging individual players' self-esteem, which was typically regarded as fragile. Observations during matches and training sessions confirmed the existence of these two relatively different approaches to coaching. As they got older, however, players favoured the more aggressive and "army-sergeant like" (Julius, parent) approach. Michael, one of the U16 players, said:

> "I love Coach Marvin's style. He is pushing us very hard because that's what we need now. We need to be in shape and be strong physically and mentally. If we can't cope with training, how are we going to cope with the games?"

Notwithstanding these preferences, coaches whose style met the needs of specific players were generally described as inspirational, facilitative and as great role models. Jennifer, one of the mothers, explained:

> "Coach George and Coach Dean were tremendous for my son. They turned a shy and timid boy into a happy and confident lad just by guiding him carefully and showing him a lot of love. They are wonderful role models."

Parental support and influence were also corroborated during the immersion period. Parents were constantly seen contributing to the functioning of the environment or, as Coach Jack put it, "mucking in". This included undertaking the duties of team manager, bus driver, fund-raiser, and many other chores. For some parents, this became almost a part-time occupation. U13 Team Manager Natalie is a prime example, as can be seen in the following:

Observer notes, 10 March—Training session 6–8 p.m. at the club:

Natalie, the team manager for the U13s, makes me giggle. Perhaps I'm being judgemental but it's amazing what people will do for their kids. Natalie is in her mid-40 s, quite overweight and sporting a bright spray tan, full make-up, extremely long nails and massive earrings. Surely, she doesn't belong on a basketball court, but here she is, running around the place in her heels collecting subs, sorting out transport for the weekend's game and organising kit and refreshments. And she is loving it and thriving in this role. When I ask her about it, all she says is: "I love it. I'd do anything to help my kid, but I actually enjoy dealing with all this, I'm quite bossy you know". Then she runs away chasing some parent who hasn't paid yet.

The notion of social support and influence also featured strongly during live observations. The community hub nature of the club manifested itself at different levels. Club members used the facility more as a traditional youth club than as a performance sport club. Young people were seen constantly hanging around the centre, spending inordinate amounts of time "just chilling and meeting people" (Kyle, player). Thanks to its open-door policy, players came and went constantly from the moment school finished to the moment the building closed. Often, players had to be chivvied off the premises by the coaches. Likewise, the co-educational nature of the club, which did not feature in Phase 1, became prominent in Phase 2. Parents saw this positively: "look, they are 15, they are going to be into girls, I'd much rather they do it here in this safe space where we know the girl and we know their family too and what they are into" (Megan, parent).

Two new processes were, however, elicited in Phase 2. 'Time and Opportunity' relates to the large amounts of time players spend at the club and the opportunities this afforded for positive change. During Phase 1, time was construed as a constraint with implications for school, social and family life. During the immersion period, interpretations of time shifted to one of being much more of an enabler of opportunity; it was because of time, especially 'better-than' time, that participation at the club produced positive personal development. Jerome, a parent, put it like this: "any time spent at the basketball centre is better than what they could be getting up to in the street corners or on the PlayStation".

The second novel mechanism elicited during Phase 2 related to the impact of the traditions and rules of the sport. As defaults, these acted as silent, organic mechanisms to support personal development. For instance, basketball rules place special importance on the behaviour of players and coaches, strictly enforced by referees in all games; this is accepted by all as part of the game. As one of the parents put it: "basketball people are nice; football is horrible, our son used to play football too and they get away with blue murder." (Nadine, parent). Moreover, the game mandates other positive traditions such as the respectful ritual handshakes pre- and post-game with opposition and referees and the relationship-building hosting of a home-made finger buffet for the visiting team after each game.

3.1.2. The Attention Factory

Participation in this basketball club was seen by stakeholders as providing young athletes with a clear and sustained focus. This sustained engagement was viewed as a protective shield, deterring against negative attitudes and lifestyles. In Phase 1, two major sets of sub-mechanisms were proposed: (i) love for the game and (ii) a purposeful life (Table 2). While these elements were corroborated across the immersive season, on-site observations and discussions with the stakeholders provided further insight and texture.

Table 2. The Attention Factory—complete mechanisms post-immersion.

The Attention Factory
Love for the Game
A Purposeful Life

Player's 'Love for the game' was the most discussed and observed mechanism throughout the ethnographic period. Parents regularly reported how their kids were enthralled by the sport. Jennifer, one of the mothers, shook her head while she said: "all my son wants to do in his spare time is basketball, come to the centre, in the back garden, in videogames, watching it on TV or YouTube, that's all he has time for". In this narrative, passion for the sport was seen as the first step in achieving a host of positive outcomes, including protection from typical dangers and distractions of the teenage years or the development of a healthy lifestyle. Nevertheless, for some parents, this passion became a liability in its own right. Sophie explained:

> "His newly found ability in basketball has given him a lot of confidence and self-belief, but at the same time, that's all he wants to do, he is not motivated to study more or put the same level of effort into anything else."

Likewise, the notion of basketball providing 'A purposeful life' was salient during the immersion period. Most players expressed the ambition of securing a US college scholarship to represent Great Britain and play basketball professionally. This desire provided clear life goals, focus, purpose and direction. With such ambitions, first steps were acknowledged; do well in their GCSE exams (i.e., end of compulsory education in the UK at 16 years of age) and do extra work to refine basketball and physical skills. With this 'purpose', both parents and players recognised the value of the club for providing and requiring engagement in demanding, positive structures and routines. Kenny, a player for the U16 squad, clearly describe this impact:

> "I know that the days I have training, I have to get well organised, come from school, get my homework done straight away, have some tea, rest, and go to

> training. I know that by the time I come back, I'll be knackered and good for nothing but food and bed."

3.1.3. The Personal Boost

The capacity of sport participation to generate elevated states of mind and increased player wellbeing and intrinsic motivation was regularly confirmed during the ethnography. Phase 1 interviews established three sub-themes: (i) experience of success; (ii) athletic kudos; and (iii) steam release. The immersion period elicited a further central component: fun and enjoyment (Table 3).

Table 3. The Personal Boost—complete mechanisms post-immersion.

The Personal Boost
Experience of Success
Athletic Kudos
Steam Release
Fun and Enjoyment (new)

The experience of success was consistently reported as imperative for securing many of the positive outcomes achieved by these young athletes; for instance, it was reported to be necessary to the development of the self (i.e., self-esteem, positive identity) and for enhanced social development (i.e., belonging). At this club, success was experienced at various levels. Of course, winning matters at this club. The club's record of over 60 national titles in the last 20 years is unmatched nationwide. Success was also enjoyed vicariously by parents. Julius, a U16 parent, said in this respect: "every day (they) walk into the centre and look up and see the banners on the wall with all the titles. They want to have their team's name on that banner next year, every year". The presence of these banners and the attention paid to them provided further reinforcement about the importance of committing to working hard to play and do well.

Success was also experienced at the individual level through the constant provision of opportunities for players to improve (and to prove) their competence. Club officials continually looked to create extra opportunities for high-quality competition to stretch the players. These include opportunities to 'play up', playing against older and more experienced teams, organising competitive tournaments, and creating their own internal leagues when external competitions were insufficiently challenging. Over the course of a season, coaches regularly and deliberately put players in situations well beyond the players' comfort zone; knowing that coaches only did this with adaptable players helped to create player confidence, which was then amplified by good experiences and positive reflections on the experience.

In relation to athletic kudos, players reported that playing basketball increased their 'street cred' (Israel, U16 player). Being recognised as a good basketball player afforded them a powerful social cache. For instance, Kenny, a bubbly and lanky 15-year old with a reputation for being difficult to manage, saw basketball as his 'saving grace' in school:

> "I got into a bit of trouble in school, being a bit stupid you know, trying to get attention, but because I am a good ball player they let me get away with some of it, they need me for the basketball team you know."

Players reported that being good at basketball made them unique amongst their peers. For example, when they met new people through non-basketball friends, they were introduced along the lines of "this is the basketball guy I told you about", and this, for them, was "cool" (Israel, U16 player). Often, their superior height, which may have been a negative factor in younger years, had become a real asset in their teens. Jayden (U16) jokingly explained it as becoming "a bit of a chick-magnet".

As for the potential for sport to act as a release valve, players were observed to arrive at the centre shortly after school finished to spend in excess of three to four hours a day on the premises. This was most common in the older groups, who were free to travel independently; they would use that time in team training, doing self-training, 'shooting around' or just chatting to others until 10 p.m. Training was characterised by high levels of exertion, regardless of age group. The fatigue this produced, and the need for quality sleep, helped to ensure that players would have little time and energy to dedicate to other activities (positive or negative).

Finally, the quality of time spent at the club highlighted the overall experience as 'better than' many alternatives, especially for experiencing 'Fun and Enjoyment'. Players were observed during self-training sessions, normally in pairs or small groups, genuinely enjoying practicing the sport and regularly smiling and 'laughing out loud'. This fun and enjoyment was also evident in organised sessions or competitive matches. Observations of coach-led training highlighted how coaches strived to balance the fine-tuning of technical and tactical work and drills with activities where the main goal was to offer free play so players could enjoy themselves. In this context, it was common to witness player's exhilaration and excitement.

3.1.4. The Real-Life Simulator

Many of the features of participation at the club resembled elements of the real world. 'Adult-life-like' situations were presented regularly and often to players at earlier ages than most of their non-club contemporaries. These experiences were laden with real meaning and consequences but were possible because of the psychological safety built through regular exposure to the club environment. The major themes proposed during phase one were confirmed in stage two and included: Competition; The Team; Learning; Diversity; and Mini-Workplace (Table 4).

Table 4. The Real-Life Simulator—complete mechanisms post-immersion.

The Real-Life Simulator
Competition
The Team
Learning
Diversity
Mini-Workplace

Parents and players saw the club environment as "loaded with competitive pressure from multiple angles" (Amy, parent). Pressurising elements included internal competition (i) for securing selection to the weekend squad, then (ii) for playing minutes during matches, and (iii) to perform well during training and matches. Wider forms of 'pressure' arose from the uncertainties of the annual review to retain squad membership in the following season, competitive selection for Great Britain squads, and the challenges around securing a US scholarship. While acknowledging the inherent difficulties of dealing with this volume of pressure at early ages, both parents and players felt it fostered positive development because it was framed by the support provided by the club's day-to-day ways of working. Many offered a relatively stoic view of this phenomenon:

> " ... pressure is part of life. They have to get used to it and learn to deal with it, with the failures and the disappointments. Better to do it in this safe environment than waiting to be out in the real world." (Eva, parent)

Players were equally pragmatic and stressed that "if you can't take the pressure you shouldn't play national league" (Jayden).

With regard to being part of a team, the opportunity to learn to work as part of a group was widely valued. Understanding different roles, hierarchy, and the importance of discipline and doing one's job well to maximise the chance of group success were all highlighted. Coaches consistently reinforced good team behaviours and readily reprimanded negative alternatives. For example, during one training session, a player threw a tantrum after a series of bad plays. The coach immediately stopped the session and shouted: "Who do you think you are? This is not about you, this is about the team, and what you are doing now is hurting the team!". Being part of a team also created opportunities to lead others. This experience was highly valued by parents and players, as explained by U16 player Mikael:

> "This year I have had to step up. I wasn't the new guy anymore and there was less slack. I feel I have become much more of a leader; I have learnt to talk to people and reason with them to get them to do what they need to do for the team."

The immersion period not only confirmed but also offered extra detail on how participation increased players' learning capacity. A number of recurrent practices, however, did not appear to align with current learning theory; coaches often used lengthy verbal explanations, and many chose to dictate every move, allowing little room for players to contribute. New technologies (e.g., iPads, video analysis, etc.) were rarely used in teaching/coaching; instead, coaches relied on formal instruction. During 'coaching episodes', players were generally in one of two groups: those who seemed fully engaged, and those who appeared disengaged and even bored. Often, coaches made no clear effort to engage the disengaged players. Likewise, some players displayed sub-optimal learning attitudes, including poor listening skills and/or reluctance to accept external feedback.

Paradoxically, however, players relished every chance to direct their own learning when they were away from coaches and regular team sessions. By observation and in conversations, players were deeply committed to improving their skills. Players were seen videoing each other so they could analyse their own performance to find areas for improvement. Players watched YouTube videos of NBA player workouts to pick up drills they could try on their own. They were also regularly seen peer coaching and were comfortable with peer learning. Community leagues and holiday camps were also fertile ground for peer coaching and for players to explore their skills in a less controlled and structured setting.

Regarding diversity, Phase 2 showed the broad demographic range of individuals in the club environment. At the time of the study, club coaches came from the UK, USA, Iran, Lithuania and Spain, and the senior team included players from the UK, USA, Spain and Greece. In the national league programme, many players belonged to migrant families who came to the UK as a result of the 2008 recession (particularly from Poland, Spain and Portugal, where basketball is a mass sport). In addition, many players had roots in former British colonies, especially from the Caribbean and Africa. Moreover, diversity was enhanced by competing against, and meeting with, players from different English cities during national league games and in international tournaments. For parents and players, the developmental benefits of this situation included: (i) experiencing different cultural approaches to life; (ii) broadened horizons; and (iii) learning to deal with a variety of people and situations.

> "It was just great to see him interacting with all these different people. Where we live and where he goes to school most people are white middle class and he built some great relationships with kids that he would have never met otherwise, and I think that has stood him in great stead going forward to uni and now work." (Mark, parent)

Finally, regarding the club environment as a mini-workplace, the ethnographic period revealed that few players were taking direct advantage of these opportunities. Opportunities for work came with refereeing and coaching in the community leagues and camps, contributing to cleaning the centre, and volunteering during club events. A small group of players actively pursued these opportunities; through them, they could practice basic workplace skills and attitudes, including time-keeping, planning, responsibility and accountability.

The ethnographic period thus served to confirm, refute and refine the findings of Phase 1 of the case study. Although the mechanisms are presented as four discrete categories to ease understanding, on the ground, they represent a complex network of interdependencies and causal relationships. Mechanisms interacted in multiple ways to foster visible developmental outcomes. Two major principles appear to govern this network: (i) specific developmental outcomes are affected by multiple mechanisms; and (ii) specific mechanisms influence many networks, meaning they may simultaneously contribute to generating a number of developmental outcomes.

3.2. *The Idiosyncratic Nature of Development*

The process of apportioning causality is complex, and participant and researcher interpretation is required to connect the developmental dots. Findings from this current study confirm that personal development in this youth performance setting was significantly more nuanced and textured than previously described. The fine-grained detail provided by this case study enhances our understanding of this phenomenon. In the following section, focus is placed on the different ways in which sport was experienced and used by different athletes.

Phase 1 concluded that the experience of this programme was unique to each player. Phase 2 confirmed that the internal and external assets of the young person strongly influenced the activation of causal mechanisms and thus the generation of outcomes. Considering the overall findings, four processes were identified to explain this differential: (i) few mechanisms were universally available in equal measure; (ii) no available mechanism was optimised by all participants; (iii) psychosocial outcomes of participation were mediated and moderated by individuals' social environments; and (iv) critical life incidents conditioned the personal narrative and the meaning of the experience. These principles will be elaborated upon separately.

3.2.1. Few Mechanisms Were Universally Available in Equal Measure

Examples of mechanisms affected by this included parental support/influence, coach behaviours, social support/influence, athletic kudos, mini-workplace and diversity. We concluded that the availability and causal strength of mechanisms tended to be modulated by a series of factors. For instance, some parents were unable to regularly support their child for multiple reasons: i.e., being a single-parent family, caring for other siblings, work commitments, or compromised finances. This created impacts at different levels and had both positive and negative effects. The story of Aki illustrates this point.

Aki lives quite a few miles away from the basketball centre in a different town. His parents are immigrants from an African country, work evening jobs and do not own a car. He takes 3 separate buses (a 75-min journey in total) to get to basketball. Aki says that this has made him resourceful and self-reliant, because he cannot just sit around and wait for his mother and father to take him to baskestball. He wishes his parents came to basketball more, however, because he says they struggle to understand why he loves it so much and have tried to talk him out of it a few times; they think it may harm his education.

3.2.2. No Available Mechanism Was Optimised by All Participants

Specific conditions either initiated, postponed or prevented the activation and optimisation of some generative mechanisms. These conditions typically related to the young person's internal characteristics or assets. Playing level and progression potential are prime examples. High-ability players secured and enjoyed more playing time and, resultingly, higher status. In a virtuous spiral, these players had more exposure to success, meaning their popularity and public and self-esteem rose again. Likewise, coaches saw players' emotional and cognitive maturity as central to activating some positive mechanisms. Players could be exposed to similar contexts, yet their individual developmental status could produce widely differing outcomes.

> "They are all so different and react to things in such a different way that this helps them understand emotions much better. The kids come from very different backgrounds and have very different coping mechanisms." (Coach Carl)

3.2.3. Psychosocial Outcomes of Participation Were Mediated and Moderated by Individuals' Social Environments

The quality and intensity of the engagement with these additional contexts plays a powerful role in determining, mitigating or enhancing the impact of sport participation. The club environment either reinforced outcomes that were primed by or already well developed by other environments. For instance, U16 parent Jamal said: "Yes, the club has taught some very good values to Simon, but they ain't no different to the values we have tried to teach him at home. We wouldn't have him here if that weren't the case." In contrast, for other players, participation in the basketball club compensated for deficiencies in other settings. Antonio, a U14 player who had recently arrived in the UK from a southern European country, was a perfect example. His mum and dad explained:

> "Thank goodness for basketball. The poor child arrived with very basic English and terrified of having no friends in this cold country. He was a real sod on the plane and for the first couple of weeks of being in England, until we found this place. The moment he stepped through the doors here his face lit up. He has made lots of friends, regained a lot of confidence and learnt lots of English."

3.2.4. Critical Life Incidents Condition the Personal Narrative and the Meaning of the Experience

The second phase of the case study offered a privileged window into the personal journey of a number of players and their families who had experienced 'critical life incidents' (CLIs). These CLIs included loss of a sibling or parent, parental divorce, family illness, moving to a new country, bullying in school, socioeconomic deprivation, or social isolation, and each played a central role in their personal narrative and how they interpreted their sport experience. CLIs created a strong personal narrative whereby sport was framed as bringing a sense of accomplishment and justice. The narratives found at the club tended to fall into one of two categories: (i) 'reconstruction', which was defined as shifting from pain and suffering into happiness and accomplishment; and (ii) 'against the odds', which was defined as moving from humble beginnings or disadvantaged situations to high achievement. The below examples portray both narratives.

Reconstruction: Two families in the U13 squad, Amy and Andy and Patricia and Matthew, told harrowing stories. Amy and Andy's family endured first the loss of a sibling. Then, Amy developed cancer, and recently, a young sibling was badly burnt in a home accident. Matthew and Patricia went through a very traumatic divorce from Matthew's dad, a violent drug addict. In both cases, basketball became the vehicle through which they tried to rebuild their life. The three key mechanisms linked to this reconstruction process included spending as much time as possible at basketball, finding solace in the social network provided by the centre, and, especially for Andy, experiencing success and enjoyment on a regular basis.

Against the odds: Mikael, a U16 player that had recently migrated to the UK from another European country, provides an example of this narrative. His family originated from a war-torn African country and had fled to Europe looking for asylum. When he arrived in the UK, he spoke no English. Mikael explained how he had always felt he had to work twice as much as everyone else to show them all that he was a worthy human being. In basketball, he had found a welcoming environment where he could excel and be supported by likeminded young people that took him under his wing. Mikael threw everything into basketball, and that 'under siege' mentality never left him. Now 18, he is at university and working at a sports goods superstore in town. Basketball was the catalyst for him.

3.3. *A Summary Model of Psychosocial Development in a Performance Development Club*

Having established a wide range of outcomes, mechanisms and context networks and determined the idiosyncratic nature of development, the next section will present an integrative summary model of psychosocial development as seen in this basketball club

(Figure 2). The model brings together the findings of both phases of the study in a coherent and practical way to aid sport stakeholders in making sense of the full range of factors and processes involved.

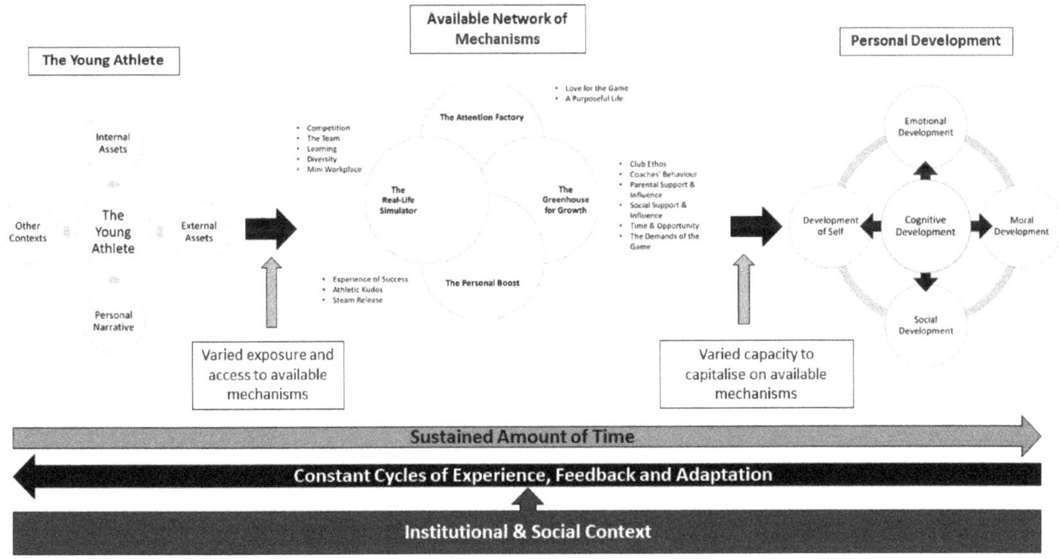

Figure 2. Psychosocial development model in a performance development basketball club.

3.3.1. The Young Athlete

The model starts by acknowledging that the outcome of the experience of sport is significantly influenced by the internal and external assets at the young person's disposal, the influence of other developmental contexts, and the personal narrative attached to the sport experience. Any of these factors may be activated to direct specific generative mechanisms to bring or inhibit valued developmental outcomes. Given the range of experiences and the ways in which club experiences may cause activation, every sport experience and trajectory is highly individualised.

3.3.2. A Network of Mechanisms and Outcomes

Four major families of mechanisms were elicited: (i) The Attention Factory; (ii) The Greenhouse for Growth; (iii) The Personal Boost; and (iv) The Real-Life Simulator. *The* Attention Factory comprises all mechanisms that direct and enhance the players' focus and thus provide clear direction and objectives in sport and beyond. The Greenhouse for Growth is linked to mechanisms that provide 'fertile ground conditions' for positive development, especially such as the club ethos, atmosphere and social support; all require player compliance, meaning that 'growth' is readily and easily tested and seen. The Personal Boost refers to features of the experience that increase self-worth, self-esteem and greater personal wellbeing. Lastly, The Real-Life Simulator relates to exposure to situations that replicate adult life. This experience, supported by club-based adults, prepares players to better resolve them in the future.

Although presented in four discrete categories, this case study shows that these mechanisms form a complex network. Mechanisms interact and catalyse in multiple ways to foster or preclude developmental outcomes. Two major principles govern this network: (i) single developmental outcomes are typically affected by a combination of multiple mechanisms; conversely, (ii) single mechanisms are characteristically involved in multiple networks responsible for different outcomes. Establishing causality is challenging and very likely futile if the aim is to find 'silver bullet' solutions. Psychosocial development in youth performance settings is a complex multi-level and multi-directional process. The nuanced analysis provided by this case study, however, contributes to making its components more accessible and intelligible.

3.3.3. Personal Development

This investigation confirmed the suitability of the psychosocial developmental outcomes framework created during the literature review stage (see part 1). The five categories of cognitive, emotional, social, moral and self-development appropriately encompass the majority of athletes' personal outcomes. The study also confirmed a significant level of interdependence between developmental outcomes whereby some outcomes are gatekeepers to the generation of others. Notwithstanding this, each developmental outcome was linked to a set of mechanisms and the preconditions required for their activation and utilisation. The traditional realist evaluation nomenclature of context-mechanism-outcome configurations and catalogues was substituted by the newly coined term 'context-mechanism-outcome networks'. This term brings to the fore the ambiguous reality of social life where multiple preconditions and generative mechanisms constantly interact in the production of outcomes over time.

3.3.4. Time and Recurrent Experiences

This research underlines the utility of additional time (frequency and duration) spent on engaging with the sport experience. For any type of change or adaptation to occur, sustained exposure to a stimulus is required. The club environment was stimulating in many ways, especially by being time-affluent and experience-rich. Moreover, this near-limitless supply of time facilitated players' engagement in endless cycles of experience, feedback and attitudinal and behavioural adaptation, while being diverted from other unhelpful behavioural choices. By spending more time at the club experiencing positive growth, they became progressively more skilled in capitalising on opportunities for growth. For players experiencing negative adaptations, opposite pathways prevailed.

3.3.5. The Context

A final piece in the development jigsaw is the multi-layered nature and textured influence of the context for creating and moderating conditions that lead to positive personal development. Pre-existing individual characteristics determine the starting point for growth at the point of joining this club. From there, players engage with the inevitabilities of the club's day-to-day routines, and this activates important developmental mechanisms. The number and nature of players' interpersonal relationships within the club and beyond the club combined to impact development. Unsurprisingly, coaches, parents and peers played an essential part in socially-based development. In addition, specific institutional characteristics, including the non-negotiables of a strongly humanistic club philosophy, its 'cross-roads' geographic location, and a longstanding commitment to competitiveness, all contributed systems that mandated responsiveness and, therefore, personal development. Likewise, the infrastructure, understood as the external conditions—including levels of funding—all impacted the way the sport was, and this coloured the way players developed. Specifically, the minority and non-professional character of basketball in the United Kingdom created additional drivers for self-sufficiency, resilience and proactivity, all important markers of personal development.

4. Discussion

4.1. The Nature of Development

The findings of this case study show psychosocial development as a conditional, multi-faceted, time-sensitive and highly individualised phenomenon. Given the relevance and the essential dynamism of context, offering a deterministic and infallible view of the development process is futile. Positive development can thus be considered a 'wicked problem' (i.e., a problem with multiple interdependencies, no single solution, and no clear stopping point) within a complex system [12]. Nevertheless, this study has shown the development process can still be reduced to a set of basic principles that can be transferred to diverse settings and contexts. A wide-lens approach has been adopted to distil these key principles and processes to generate a synoptic view of psychosocial development in sport. This evidence-based catalogue of options can inform sport psychologists, coaches and all other stakeholders in the creation of psychosocially 'competent environments' [13] to support development in young people.

The study shows the cyclical, repetitive and sustained nature of the development process—a system of repetition. It suggests a constant flow of interactions between the individual and the environment and acknowledges the mutable condition of the individual and the setting. Heraclitus' famous aphorism, "No man ever steps in the same river twice, for it is not the same river and he is not the same man" [14] (p. 29), could be adapted for youth sport to read "No young person goes to the same club twice, for it is not the same club, and they are not the same person." Psychosocial development is thus the result of ongoing and multi-level cycles of experience, feedback, reflection and adaptation players consciously and subconsciously engage with during their participation at the club which, over time, and in line with previous studies, lead to significant and long-lasting changes [15–19]. Notably, we show this affecting not only players but also parents and coaches.

4.2. *The Process of Development*

The present study contributes to advancing the field and supporting practitioners on the ground by identifying four processes (or families of mechanisms) that mediate development.

4.2.1. The Attention Factory (Attentional Focus)

The study suggests that the compelling effect of the sport—its power to grab the young players' attention and imagination—was central to its success. Without focused attention and commitment to the activity—deep attention—the impact is reduced due to lack of meaning and significance [20,21]. Innovatively, the study identified that attention is driven by two parallel processes: first, a basic stimulus-reward system, where certain 'reward-rich' activities and features 'hook' athletes to the setting; and second, the life-affirming nature of other mechanisms, which provided a clear and distinct life purpose. These were magnified by the socially satisfying elements of that experience (i.e., belonging to a cohesive group). Attention can thus be driven both autonomously, quasi-organically and deliberately through the systems put in place by key stakeholders.

4.2.2. The Real-Life Simulator (Structured and Unstructured Skill-Building Activities)

Once attentional focus is achieved, skill-building activities are essential to ensure development [22]. Some of these activities can be structured and planned to deliver clear personal development goals (i.e., classroom-based workshops, doing chores that contribute to the running of the club, etc.). Other activities may be more organic and unstructured in nature (i.e., inherent diversity at the club, exposure to teaching, the demands of the game, etc.). Significantly, the present study shows that repeated and consistent exposure to relatively low-key activities within this club was important and successful for building individuals' internal assets. This underlines the ongoing debate about the extent of development arising from serendipity versus 'by design' activities [23–27]. The evidence developed through this study suggests that both pathways coexist and should be addressed, if only because the pathways to impact are so complicated and hard to isolate [23].

4.2.3. The Greenhouse for Growth (Deliberate and Incidental Support)

Having a supply of personal and personalised support appears to modulate psychosocial development in sport [28]. An explicit humanistic philosophy embodied by all stakeholders, ensured by recruiting individuals with a genuine caring disposition into key positions, especially coaches and club administrators, guarantees that few players could consistently escape, or purposely avoid, this experience. Support may be provided in two ways. It can be deliberate and purposeful, built into the routines of the adults in the setting (i.e., welcome meetings, discussion sessions, general and casual check-ins, crisis management help, and regular communication with families). Other approaches may occur within the spontaneous functioning of the networks of club supporters, including parents (one's own and others') and all other social actors within the setting, including teammates and club members. Together, all these elements enhance personal wellbeing, encourage players to be ever-ready to capitalise on any development opportunity, and buffer against the negative effects of participation [23,29].

4.2.4. The Personal Boost (Feelings of Personal Growth)

The fourth developmental process involves regularly experiencing feelings of personal growth. A consistent sense of betterment and achievement has been shown to encourage recommitment to the activity, creating a virtuous cycle of engagement and re-engagement [30,31]. Over time, recurrent and multiple opportunities to interact with the various developmental mechanisms available in the club increase the chances of practicing different developmental behaviours [29]. These feelings of growth come from a variety of sources. Players were provided with regular opportunities to experience tangible success, personal kudos, and the increased sense of personal worth and identity gained from belonging to a large community of interest.

Taken together, the process of personal development can be summarised using the unattributed adage, "The best gift we can give our children is roots to grow and wings to fly". Basketball provided the attentional focus, and the deliberate and incidental support provided the roots. The ground conditions provide motivational nutrients and a stable climate to support growth. Complementing these conditions, both structured and unstructured skill-building opportunities and the regular provision of feelings of personal growth equipped players with the confidence to grow their wings and fly towards better versions of themselves.

4.3. *Differentiated Impact—The Young Person Who Was, Who Is, and Who Will Be*

Despite evidence of the co-existence of the above-described processes, the impact of the experience is widely divergent. In line with recent studies, this paper refutes the naïve belief of sport offering a 'magic bullet' with inherent properties that bring universal benefits regardless of how it is delivered and experienced [32–35]. Operating with distinctive thresholds and activation points, generative mechanisms were highly unique; often, it was not possible to predict their activation and, as a result, even less possible to know how players might react when they were activated [36]. In line with Bronfenbrenner's bioecological model of human development [37], the elements young people derived from participation reflected person-specific drives and experiences. In the current study, prominent person-specific factors were internal disposition, external assets and personal narratives. Further, these factors interacted to mould the young athletes' behaviours and interactions, depending on the settings and dynamic qualities of the context.

General and sport-specific psychosocial development literature has espoused Bronfenbrenner's views on human development [38–40]. However, research in sport has tended to emphasise environmental factors [41–43] and the actions of significant others like coaches, parents and peers [44,45]. Few studies have addressed how any players' initial personal characteristics influence subsequent developmental trajectories. Recent exceptions include Pierce et al.'s model of transfer in life skill development [18,36], and Holt et al.'s grounded theory model [39].

In these models, the starting point for understanding the transfer of life skills from sport to life is the internal and external assets of the young person, alongside their autobiographical experiences. The findings of the current case study align with these models. The current study also signals the need for coaches, administrators and parents to respond to individual characteristics and to create opportunities that: (i) increase the development of each player's internal disposition; (ii) buffer any potential negative effect of external assets; and (iii) allow players to become familiar with their own personal narratives and identities. Deliberately and momentarily shifting to consider the young person who 'was', how they appear 'now' and the young person who 'will be', is central to supporting the dynamic, idiosyncratic processes of development.

5. Conclusions

This case study shows that participation in performance sport affects young people's development. This occurs whether or not sport psychologists, coaches, programme designers, parents or players intentionally plan for it. The current findings, however, provide a comprehensive yet distinctively individualised and multi-layered picture of the conditions and processes that support positive development. They do not describe any young person's development. The anonymous adage, "The best gift we can give our children is roots to grow and wings to fly," summarises our findings. The generic principles distilled from this specific case provide a useful starting point for developing more programmes to support the positive development of more young players. In the wake of COVID-19, the findings may be especially relevant for supporting continued investment in and re-engagement with sport.

The study also showed the value of using a RE approach for the investigation of psychosocial development in sport. It helped to generate a detailed account of the process and also helped in understanding its idiosyncratic nature. This notwithstanding, in recognising its limitations, we can recommend future research directions. Longitudinal tracking and measuring specific developmental outcomes may yield new insights. Likewise, our limited focus on a single club, age group and gender calls for further studies. Variously, these may corroborate, refine or refute the programme theories presented in this paper. Similarly, studies focusing on the internal and external assets and personal narratives of individuals may provide additional awareness of how these elements condition the short- and long-term impact of the experience.

Author Contributions: Conceptualization, S.L.-B. and J.M.; methodology, S.L.-B. and J.M.; software, S.L.-B.; formal analysis, S.L.-B. and J.M.; investigation, S.L.-B.; data curation, S.L.-B. and J.M.; writing—original draft preparation, S.L.-B.; writing—review and editing, S.L.-B. and J.M. All authors have read and agreed to the published version of the manuscript.

Funding: This research received no external funding.

Institutional Review Board Statement: The study was conducted according to the guidelines of the Declaration of Helsinki and approved by the Ethics Committee of LEEDS BECKETT UNIVERSITY (6 February 2014).

Informed Consent Statement: Informed consent was obtained from all subjects involved in the study.

Acknowledgments: The authors would like to thank the sport club in which this research took place for opening their generosity and all those who took part in the research for their willingness to contribute and the candidness of their responses.

Conflicts of Interest: The lead author declares a previous relationship with the studied sport club where he was a coach in previous years. However, the lead author had no current role at the club at the time of the research taking place.

References

1. Pawson, R. *The Science of Evaluation: A Realist Manifesto*; SAGE Publications Inc.: London, UK, 2013.
2. Pawson, R.; Nick, T. *Realistic Evaluation*; SAGE Publications Inc.: London, UK, 1997.
3. Layder, D. *Sociological Practice: Linking Theory and Social Research*; SAGE Publications Inc.: Thousand Oaks, CA, USA, 1998. [CrossRef]
4. North, J. *Sport Coaching Research and Practice: Ontology, Interdisciplinarity and Critical Realism*; Routledge: Abingdon, UK, 2017.
5. Wolcott, H. *Ethnography: A Way of Seeing*; Rowman Altamira: Lanham, MD, USA, 2011.
6. Krane, V.; Baird, S.M. Using Ethnography in Applied Sport Psychology. *J. Appl. Sport Psychol.* **2005**, *17*, 87–107. [CrossRef]

7. Sparkes, A.; Brett, S. *Qualitative Research Methods in Port, Exercise and Health: From Process to Product*; Routledge: London, UK, 2013. [CrossRef]
8. Willis, P.; Trondman, M. Manifesto for Ethnography. *Ethnography* **2000**, *1*, 5–16. [CrossRef]
9. O'Reilly, K. *Ethnographic Methods*; Routledge: London, UK, 2012.
10. QRS International. *N-VIVO Qualitative Analysis Software*; QRS International: Burlington, VT, USA, 2015.
11. Madden, R. *Being Ethnographic: A Guide to the Theory and Practice of Ethnography*; SAGE Publications Ltd.: London, UK, 2010.
12. Cabrera, D.; Laura, C. *Systems Thinking Made Simple: New Hope for Solving Wicked Problems*; Odyssean Books: Toronto, ON, Canada, 2016.
13. Perpiñán, S. *Atención Temprana y Familia. Cómo Intervenir Creando Entornos Competente*; Narcea Ediciones: Madrid, Spain, 2009.
14. Heraclitus. *Fragments*; Penguin: London, UK, 2011.
15. Agnew, D. "I Think It's Going To Save Lives" Sport Administrator Perspectives on Youth Development Through Sport. *Sport J.* **2016**, *19*, 1–16.
16. Bean, C.; Forneris, T. Examining the Importance of Intentionally Structuring the Youth Sport Context to Facilitate Positive Youth Development. *J. Appl. Sport Psychol.* **2016**, *28*, 410–425. [CrossRef]
17. Hodge, C.J.; Kanters, M.A.; Forneris, T.; Bocarro, J.N.; Sayre-McCord, R. A Family Thing: Positive Youth Development Outcomes of a Sport-Based Life Skills Program. *J. Park Recreat. Adm.* **2017**, *35*, 34–50. [CrossRef]
18. Pierce, S.; Gould, D.; Camiré, M. Definition and Model of Life Skills Transfer. *Int. Rev. Sport Exerc. Psychol.* **2017**, *10*, 186–211. [CrossRef]
19. Storm, M.K.; Williams, K.N.; Shetter, E.M.; Kaminsky, J.; Lowery, C.M.; Gall, J.; Caldas, S.V.; Winch, P.J. Sink or Swim: Promoting Youth Development Through Aquatics Programs in Baltimore, Maryland. *J. Park Recreat. Adm.* **2017**, *35*, 66–79. [CrossRef]
20. Larson, R. Toward a Psychology of Positive Youth Development. *Am. Psychol.* **2000**, *55*, 170–183. [CrossRef]
21. Larson, R.W.; Hansen, D.M.; Moneta, G. Differing Profiles of Developmental Experiences across Types of Organized Youth Activities. *Dev. Psychol.* **2006**, *42*, 849–863. [CrossRef]
22. Eccles, J.S.; Barber, B.L. Student Council, Volunteering, Basketball, or Marching Band: What Kind of Extracurricular Involvement Matters? *J. Adolesc. Res.* **1999**, *14*, 10–43. [CrossRef]
23. Bean, C.; Kramers, S.; Forneris, T.; Camiré, M. The Implicit/Explicit Continuum of Life Skills Development and Transfer. *Quest* **2018**, *70*, 456–470. [CrossRef]
24. Côté, J.; Turnnidge, T.; Evans, B.M. The dynamic process of development through sport. *Kinesiol. Slov.* **2014**, *20*, 14–26.
25. Gould, D.; Carson, S. Life Skills Development through Sport: Current Status and Future Directions. *Int. Rev. Sport Exerc. Psychol.* **2008**, *1*, 58–78. [CrossRef]
26. Lara-Bercial, S.; Mckenna, J. PsyTool Supplement Suppl 3. *J. Sport Psychol.* **2018**, *27*, 53–58.
27. Turnnidge, J.; Côté, J.; Hancock, D.J. Positive Youth Development From Sport to Life: Explicit or Implicit Transfer? *Quest* **2014**, *66*, 203–217. [CrossRef]
28. Benson, P. *All Kids Are Our Kids: What Communities Must Do to Raise Caring and Responsible Children and Adolescents*, 2nd ed.; Jossey-Bass: San Francisco, CA, USA, 2006.
29. Pierce, S.; Kendellen, K.; Camiré, M.; Gould, D. Strategies for Coaching for Life Skills Transfer. *J. Sport Psychol. Action* **2018**, *9*, 11–20. [CrossRef]
30. Ryan, R.M.; Deci, E.L. Self-Determination Theory and the Facilitation of Intrinsic Motivation, Social Development, and Well-Being. *Am. Psychol.* **2000**, *55*, 68–78. [CrossRef]
31. Dweck, C. *Mindset: The New Psychology of Success*; Ballantine: New York, NY, USA, 2006.
32. Coakley, J. Youth Sports: What Counts as 'Positive Development?'. *J. Sport Soc. Issues* **2011**, *35*, 306–324. [CrossRef]
33. Coalter, F. *Sport for Development: What Game Are We Playing?* Routledge: Abingdon, UK, 2013.
34. Hartmann, D.; Kwauk, C. Sport and Development: An Overview, Critique, and Reconstruction. *J. Sport Soc. Issues* **2011**, *35*, 284–305. [CrossRef]
35. Theeboom, M. Sport and Education in Europe: Is Practice as Simple as It Is in Theory? In *European Sport Summit*; European University Sports Association: Arhus, Denmark, 2017.
36. Pierce, S.; Erickson, K.; Dinu, R. Teacher-Coaches' Perceptions of Life Skills Transfer from High School Sport to the Classroom. *J. Appl. Sport Psychol.* **2019**, *31*, 451–473. [CrossRef]
37. Bronfenbrenner, U. *Making Human Beings Human: Bioecological Perspectives on Human Development*; SAGE Publications Inc.: Thousand Oaks, CA, USA, 2005.
38. Holt, N.L.; Neely, N.C. Positive Youth Development through Sport: A Review. *Rev. Iberoam. Psicol. Ejerc. Deporte* **2011**, *6*, 299–316.
39. Holt, N.L.; Neely, K.C.; Slater, L.G.; Camiré, M.; Côté, J.; Fraser-Thomas, J.; Macdonald, D.; Strachan, L.; Tamminen, K.A. A Grounded Theory of Positive Youth Development through Sport Based on Results from a Qualitative Meta-Study. *Int. Rev. Sport Exerc. Psychol.* **2017**, *10*, 1–49. [CrossRef] [PubMed]
40. Lerner, R.M. Commentary: Studying and Testing the Positive Youth Development Model: A Tale of Two Approaches. *Child Dev.* **2017**, *88*, 1183–1185. [CrossRef] [PubMed]
41. National Research Council and Institute of Medicine; National Academy of Sciences. 2002. Available online: http://www.nasonline.org/ (accessed on 27 December 2021).
42. Fraser-Thomas, J.L.; Côté, J.; Deakin, J. Youth Sport Programs: An Avenue to Foster Positive Youth Development. *Phys. Educ. Sport Pedagog.* **2005**, *10*, 19–40. [CrossRef]

43. Strachan, L.; Côté, J.; Deakin, J. A New View: Exploring Positive Youth Development in Elite Sport Contexts. *Qual. Res. Sport Exerc. Health* **2011**, *3*, 9–32. [CrossRef]
44. Holt, N.L.; Knight, C.J. *Parenting in Youth Sport: From Research to Practice*; Routledge: Abingdon, UK, 2014.
45. Gould, D.; Collins, K.; Lauer, L.; Chung, Y. Coaching Life Skills through Football: A Study of Award Winning High School Coaches. *J. Appl. Sport Psychol.* **2007**, *19*, 16–37. [CrossRef]

sports MDPI

Article

Relative Age Effects in Male Cricket: A Personal Assets Approach to Explain Immediate, Short-Term, and Long-Term Developmental Outcomes

Adam L. Kelly [1,*], Thomas Brown [1,2], Rob Reed [3], Jean Côté [4] and Jennifer Turnnidge [4]

[1] Centre for Life and Sport Sciences (CLaSS), Faculty of Health, Education and Life Sciences, Birmingham City University, Birmingham B15 3TN, West Midlands, UK; thomas.brown7@mail.bcu.ac.uk
[2] Warwickshire County Cricket Club, Birmingham B15 3TN, West Midlands, UK
[3] Corsham Cricket Club, Corsham SN13 9EU, Wiltshire, UK; rob@f5design.co.uk
[4] PLAYS Research Group, School of Kinesiology and Health Studies, Queen's University, Kingston, ON K7L 3N6, Canada; jc46@queensu.ca (J.C.); jm123@queensu.ca (J.T.)
* Correspondence: adam.kelly@bcu.ac.uk

Abstract: The purpose of this study was to adopt the Personal Assets Framework (PAF) to examine the immediate, short-term, and long-term developmental outcomes associated with relative age effects (RAEs) in male cricket. As such, this study was comprised of three aims: (a) examine the birth quarter (BQ) distribution of players throughout the England and Wales Cricket Board (ECB) national talent pathway (i.e., Regional U15, Regional U17, England U19, England Lions, England T20, England ODI, and England Test; $n = 1800$; immediate timescale), (b) explore the youth-to-senior transitions based on BQ and skill-set (i.e., batters and bowlers; short-term timescale), and (c) analyse the average number of games played at senior levels based on BQ and skill-set (i.e., long-term timescale). A chi-square goodness of fit test, Cramer's V, odds ratios, and 95% confidence intervals were used to compare the BQ distributions of each cohort against the expected BQ distributions. In the immediate timescale, results showed that relatively older players were overrepresented throughout all the youth levels ($p < 0.05$, V = 0.16–0.30), whereas there were no differences at senior levels ($p > 0.05$, V = 0.05–0.15). In the short-term timescale, when the senior cohorts were compared to the expected BQ distributions based on the Regional U15 cohort, relatively younger players were more likely to transition from youth to senior levels ($p < 0.05$, V = 0.22–0.37). In the long-term timescale, relatively older batters were selected for more games ($p < 0.05$, V = 0.18–0.51), whereas relatively younger bowlers were selected for more games ($p < 0.05$, V = 0.17–0.39). Moving forward, it is important for researchers and practitioners to better understand how (bi)annual-age grouping shapes developmental outcomes in across different timescales (i.e., immediate, short-term, and long-term), as well as consider alternative grouping strategies and RAE solutions.

Keywords: talent identification; talent development; expertise; youth cricket; batting; bowling

Citation: Kelly, A.L.; Brown, T.; Reed, R.; Côté, J.; Turnnidge, J. Relative Age Effects in Male Cricket: A Personal Assets Approach to Explain Immediate, Short-Term, and Long-Term Developmental Outcomes. *Sports* **2022**, *10*, 39. https://doi.org/10.3390/sports10030039

Academic Editor: Adam Baxter-Jones

Received: 14 January 2022
Accepted: 22 February 2022
Published: 5 March 2022

1. Introduction

Identifying athletes with the potential to achieve expertise at adulthood is a contemporary challenge for many national governing bodies in sport [1]. In the context of cricket, the England and Wales Cricket Board (ECB) are tasked with the design, implementation, and evaluation of their respective national talent pathway in order to facilitate the next generation of senior international players. However, the difficulty of accurately predicting future performance abilities can result in biases during the selection process into talent pathways [2,3]. Particular selection and developmental biases that have been consistently highlighted in the literature are relative age effects (RAEs) [4]. Relative age effects illustrate that when athletes are banded according to (bi)annual-age groups, those who are born near the beginning of the cut-off date are often overrepresented in recreational and talent pathways compared to those who are born towards the end [5]. Indeed, it is important that stakeholders employed at the ECB and in youth cricket settings better understand how RAEs may affect their talent pathway, in order to ensure they are using resources most effectively, providing an equitable system, capture a wide pool of potential talent, and better understand how they may impact individual development based on skill-set (i.e., batters and bowlers). Possible explanations that have been offered for RAEs include the enhanced physiological and psychosocial skills of relatively older athletes, which allows them to outperform their relatively younger but age-matched peers [6–9]. More specifically, if relatively older athletes are selected because of their physiological and psychosocial qualities, they may gain access to more coaching and competition opportunities, which could allow them to become better athletes in the long-term [8–10]. In comparison, studies have shown detrimental effects for relatively younger athletes, such as limited selection opportunities, lower participation, and higher dropout rates [11,12].

Despite being vulnerable to RAEs, little research specifically in cricket exists. Indeed, where it does, early evidenced was offered through limited sources. For instance, RAEs in cricket were first studied in a Letter to the Editor [13], which analysed the birth distribution of British male county competition based on their skill-set during the 1990/91 season. They revealed RAEs in fast bowlers but not for spin bowlers, batters, or wicketkeepers. However, it should be noted that their cut-off dates were misinterpreted, since the April 1st to March 31st were used in-line with the cricket season, whereas the annual selection year representative of national normative data (i.e., September 1st to August 31st in the UK) is generally accepted as the appropriate analysis procedure [14]. Thereafter, using such approach, conference proceedings [15,16], as well as unpublished doctoral thesis data [17], found no evidence of RAEs in senior level cricket. Interestingly, however, RAEs across Under (U)12 to U17 English cohorts have been revealed, whereby those born earlier in the annual selection year were overrepresented compared to those who were born later (birth quarter [BQ]1 = 32–44% (i.e., born in September, October, or November) vs. BQ4 = 11–15% (i.e., born in June, July, or August)); although there were no statistically significant RAEs at U19 level (e.g., BQ1 = 29% vs. BQ4 = 20%) [17]. It is important to note that the number of years included in this study varied considerably and only included limited longitudinal data that may not fully capture the longevity of RAEs throughout the ECB talent pathway.

More recently, RAEs were examined in the male and female Australian youth national championships and senior state competition [18]. Results indicated that players born in the first quartile of the cricket season were significantly overrepresented in male U15, U17, and U19 levels, as well as female U15 and U18 levels (BQ1 = 34–38%), compared to the fourth quartile (BQ4 = 16–20%). In contrast, there were no significant differences at the senior state levels for either male or female cricketers. In comparison to senior levels, RAEs were investigated in 'super-elite' senior international test cricketers over a 20-year period according to eleven performance criteria [19]. Results revealed RAEs were prevalent when all skill-sets were combined, and was also observed for the batting and spin bowling skill-sets; although no RAEs were found for the pace bowling skill-set. Similarly, it was shown how England senior international spin bowlers were relatively older when compared to their England senior professional counterparts [20]. Moreover, career batting averages of England senior international players were used to create samples of high-performing and low-performing batters [21]. Results showed how 'high-performers' were 1.6 times more likely to be born in BQ1 compared to BQ4, whereas there was no significant difference in the BQ distribution for the low-performers. Contrastingly, however, other research has showed how RAEs did not discriminate between English senior international and professional batters [20]. Indeed, this supports other existing cricket research that illustrates how differential RAEs are contingent on batting or bowling skill-sets. Upon examining the youth-to-senior level transition, a 'reversal effect' of relative age was reported in the ECB national talent pathway [22]. They showed how relatively older players were significantly overrepresented at youth level (e.g., BQ1 = 36% vs. BQ4 = 16%), whereas a significantly greater proportion of relatively younger players successfully transitioned to senior international levels (e.g., BQ1 = 2.5% vs. BQ4 = 6.7%). Although these preliminary findings serve as a useful opening, they did not consider various competition levels, skill-sets, or performance outcomes at adulthood. Thus, it would be worthwhile further exploring the mechanisms of RAEs in the ECB national talent pathway in order to better understand who is at risk of RAEs during the immediate, short-term, and long-term timescales in cricket.

In an effort to better understand the far-reaching implications of RAEs, this phenomenon can be examined through the lens of youth development. Indeed, grounding relative age research in theory is imperative to advancing this field [23,24]. As an example, an initial model was offered by Hancock and colleagues [9], which focused on the role social agents (i.e., parents, coaches, and athletes) in creating and perpetuating RAEs. Shortly thereafter, Wattie and colleagues [25] suggested a constraints-based model (i.e., individual, environment, and task) to explain RAEs. Although there are several applicable youth development models, this current study puts forward the Personal Assets Framework (PAF [26–28]) as a representation of development in sport by drawing from the context of cricket. Based on work in developmental and sport psychology, the PAF suggests that there are three key 'dynamic elements' required for sport development to occur, including: (a) personal engagement in activities (i.e., the what), (b) appropriate settings and organisational structures (i.e., the where), and (c) quality social dynamics (i.e., the who). When these elements interact with each other, an *immediate* sporting experience is created that can influence *short-term* (e.g., competence, confidence, connection, and character; 'the 4Cs') and *long-term* (e.g., performance, participation, and personal development; 'the 3Ps') developmental outcomes [29]. By highlighting the key mechanisms (i.e., the dynamic elements) and desired outcomes (i.e., immediate, short-term, and long-term), the PAF provides a useful framework to summarize the potential implications of RAEs on developmental outcomes in sport.

The purpose of this study was to explore the immediate (e.g., selection at youth levels), short-term (e.g., youth-to-senior transitions), and long-term (e.g., games played at senior levels) developmental outcomes due to RAEs in male cricket based on skill-set (i.e., batters and bowlers) through the lens of the PAF. As such, this study was comprised of three aims, including: (a) examine the BQ distribution of players throughout the ECB national talent pathway (i.e., Regional U15, Regional U17, England U19, England Lions, England T20, England ODI, and England Test; immediate timescale), (b) explore the youth-to-senior transitions based on BQ and skill-set (i.e., batters and bowlers; short-term timescale), and (c) analyse the average number of games played at senior levels based on BQ and skill-set (i.e., long-term timescale).

2. Methods

2.1. Sample and Procedure

Participants included for this study were selected into the ECB national talent pathway between the years of 1998 to 2020 (n = 1800). The duration of years varied based on the data that was available for each cohort. The respective duration of years for each cohort are outlined in the tables in the results section. The full dataset that was available for each cohort was gathered to provide the most accurate possible representation. In-line with the ECB national talent pathway, participants were selected for either: (a) Regional U15 (n = 914; a total of 65% of birthdates were publicly available for the Regional U15 cohort), (b) Regional U17 (n = 296; a total of 94% of birthdates were publicly available for the Regional U17 cohort), (c) England U19 (n = 170), (d) England Lions (n = 131), (e) England T20 (n = 91), (f) England ODI (n = 103), and (g) England Test (n = 95). All data was extracted from the public website *Cricket Archive* [30]. Aligning with the English annual-age group cut-off dates, this methodology divided the year into four equal BQs, starting with September 1st as month one and ending with August 31st as month twelve [1]. Accordingly, each participant was allocated a BQ that aligned with their birthdate to create an observed BQ distribution within each of the ECB national talent pathway cohorts. The observed BQ distributions from each cohort were subsequently compared against National Norms (i.e., the expected BQ distribution calculated from average national live births; Office for National Statistics (ONS)) [31]. Birth quartiles were adjusted for those participants born outside of the UK and emigrated after the age of 16 years to align with their original schooling system (e.g., Australia, India, New Zealand, South Africa, and Zimbabwe apply January 1st as month one and December 31st as month twelve; $n = 17$).

To examine the likelihood of achieving senior international levels (i.e., England T20, England ODI, and England Test) following entry into the talent pathway at youth level [32], the senior BQ distributions were compared against the Regional U15 BQ distribution (i.e., entry vs. expertise [1]). Further, players within the senior cohorts were allocated into groups based on their skill-set (i.e., batters or bowlers). Batters were defined as players who had batted for the majority ($\geq$75%) of their innings in the top six of the batting order. Bowlers were defined as those who bowled at least one over in the majority ($\geq$75%) of games they played. The specific skill-set distributions included: (a) England T20 (batters $n = 35$; bowlers $n = 50$), (b) England ODI (batters $n = 33$; bowlers $n = 58$), and (c) England Test (batters $n = 46$; bowlers $n = 43$). The number of games played by each participant within the senior cohorts based on their BQ distribution was then collated to analyse career selection at adulthood.

2.2. Data Analysis

A chi-square (χ^2) goodness of fit test was used to compare the BQ distributions of each cohort against the expected BQ distributions, following procedures outlined by McHugh [33]. Cramer's V were used to highlight the magnitude of differences between BQ distributions. Conventional thresholds for the Cramer's V analysis were applied, whereby a value of 0.06 or more indicated a small effect size, 0.17 or more indicated a medium effect size, and 0.29 or more indicated a large effect size [34]. Odds ratios (ORs) and 95% confidence intervals (CIs) were calculated to compare the likelihood of each BQ being selected. Results were considered significant at $p < 0.05$.

3. Results

With regards to the immediate timescale, there was a significant difference between the Regional U15 BQ distribution when compared to the National Norms, with a large effect size (χ^2(df = 3) = 108.18, $p < 0.001$, V = 0.30; see Table 1). Significant ORs identified an increased likelihood of relatively older players being selected, with the highest being BQ1 vs. BQ4 (OR 3.78; CI 2.65–5.40). Similarly, there was a significant difference between the Regional U17 (χ^2(df = 3) = 29.81, $p < 0.001$, V = 0.23) and the England U19 (χ^2(df = 3) = 8.25, $p < 0.041$, V = 0.16) BQ distributions when compared to the National Norms, with medium and small effect sizes, respectively. The ORs identified an increased likelihood of relatively older players being selected, with the highest being BQ1 vs. BQ4 for both U17 (OR 2.50; CI 1.53–4.06) and U19 (OR 1.83; CI 0.99–3.39). In comparison, no senior cohort (i.e., England Lions, England T20, England ODI, and England Test) displayed any significant differences between BQ distributions when compared to the National Norms.

When comparing the senior cohorts BQ distributions with the expected BQ distributions based on the Regional U15 cohort to examine the short-term timescale, there was a significant difference, with medium to large effect sizes, in all cohorts: (a) England Lions (χ^2(df = 3) = 35.02, $p < 0.001$, V = 0.37), (b) England T20 (χ^2(df = 3) = 15.24, $p = 0.002$, V = 0.29), (c) England ODI (χ^2(df = 3) = 10.10, $p = 0.002$, V = 0.22), and (d) England Test (χ^2(df = 3) = 19.36, $p < 0.001$, V = 0.32). Significant ORs identified an increased likelihood of relatively younger players to transition to the senior cohorts, with the highest being BQ4 vs. BQ1 in England Lions (OR 3.9, CI 1.81–8.42), England T20 (OR 3.1, CI 1.21–7.41), England ODI (OR 2.5, CI 1.01–5.97), and England Test (OR 3.6, CI 1.43–9.15).

Table 1. The observed and expected BQ distributions of the ECB talent pathway and senior international cohorts.

Cohort (Percentage Representation)	BQ1 (25.16%)	BQ2 (24.17%)	BQ3 (24.65%)	BQ4 (25.42%)	Total (n)	χ^2 (df = 3)	p	Cramer's V	BQ1 vs. BQ4 OR (CI)	BQ4 vs. BQ1 OR (CI)
Regional U15 (2000–2018)	242 (40.5%)	162 (27.1%)	130 (21.7%)	64 (10.7%)	598	108.18	**<0.001**	0.30	3.78 (2.7–5.4)	-
Regional U17 (2013–2019)	105 (37.9%)	72 (25.9%)	59 (21.2%)	42 (15.1%)	278	29.81	**<0.001**	0.23	2.50 (1.5–4.1)	-
England U19 (1998–2020)	55 (32.4%)	47 (27.6%)	38 (22.4%)	30 (17.6%)	170	8.25	**0.041**	0.16	1.83 (1–3.4)	-
England Lions (2011–2020)	31 (23.8%)	31 (23.8%)	36 (27.7%)	32 (24.6%)	130	0.68	0.877	0.05	—	-
England T20 (2005–2020)	23 (25.3%)	23 (25.3%)	26 (28.6%)	19 (20.9%)	91	1.13	0.769	0.08	—	-
England ODI (2000–2020)	28 (27.2%)	32 (31.1%)	25 (24.3%)	18 (17.5%)	103	4.61	0.202	0.15	—	-
England Test (2000–2020)	21 (22.1%)	27 (28.4%)	27 (28.4%)	20 (21.1%)	95	2.11	0.549	0.11	—	-
England Lions (Expected distribution *)	31 (53)	31 (35)	36 (28)	32 (14)	130	35.02	**<0.001**	0.37	—	3.91 (1.9–8.4)
England T20 (Expected distribution *)	23 (37)	23 (25)	26 (19)	19 (10)	91	15.24	**0.002**	0.29	—	3.05 (1.2–7.9)
England ODI (Expected distribution *)	28 (42)	32 (28)	25 (22)	18 (11)	103	10.10	**0.002**	0.22	—	2.45 (1–5.9)
England Test (Expected distribution *)	21 (38)	27 (26)	27 (21)	20 (10)	95	19.36	**<0.001**	0.32	—	3.62 (1.4–9.1)

* Expected distribution calculated from Regional U15 BQ distribution. Bold font denotes statistically significant chi-square at $p < 0.05$.

When analysing the average number of games played at senior levels (i.e., England T20, England ODI, and England Test) to explore the long-term-timescale, each BQ distribution displayed a significant difference when compared to the National Norms, with small to medium effect sizes, favouring BQ1 in England T20 (χ^2(df = 3) = 17.37, $p < 0.001$, V = 0.08), and BQ2 in both England ODI (χ^2(df = 3) = 169.27, $p < 0.001$, V = 0.14) and England Test (χ^2(df = 3) = 289.43, $p < 0.001$, V = 0.25; see Table 2). With regards to skill-sets and games played, batters across the senior cohorts displayed a significant difference in their BQ distributions when compared to the National Norms, with medium to large effect sizes, favouring those who are relatively older: (a) England T20 (χ^2(df = 3) = 51.12, $p < 0.001$, V = 0.18), (b) England ODI (χ^2(df = 3) = 566.56, $p < 0.001$, V = 0.39), and (c) England Test (χ^2(df = 3) = 663.81, $p < 0.001$, V = 0.51). Significant ORs identified an increased likelihood of relatively older batters playing more games at senior levels, with the largest being BQ1 vs. BQ4 in England T20 (1.9, CI 1.40–2.46), and BQ2 vs. BQ4 in both England ODI (OR 6.9, CI 5.42–8.78) and England Test (OR 10.3, CI 7.62–13.82). In contrast, bowlers across the senior cohorts displayed a significant difference in their BQ distributions when compared to the National Norms, with large effect sizes, favouring those who are relatively younger: (a) England T20 (χ^2(df = 3) = 40.31, $p < 0.001$, V = 0.17), (b) England ODI (χ^2(df = 3) = 352.23, $p < 0.001$, V = 0.29), and (c) England Test (χ^2(df = 3) = 292.27, $p < 0.001$, V = 0.39). Significant ORs identified an increased likelihood of relatively younger bowlers playing more games at the senior levels, with the largest being BQ3 vs. BQ1 in both England T20 (1.9, CI 1.42–1.57) and England ODI (OR 3.7, CI 2.05–4.47), and BQ4 vs. BQ1 in England Test (OR 4.3, CI 3.29–5.73).

Table 2. The observed BQ distributions of ECB senior international cohorts based on number of games played and skill-set.

Cohort (Percentage Representation)	BQ1 (25.46%)	BQ2 (24.47%)	BQ3 (24.65%)	BQ4 (25.42%)	Total	χ^2 (df = 3)	p	Cramer's V	BQ1 vs. BQ4 OR (CI)	BQ4 vs. BQ1 OR (CI)
England T20	391 (28.25%)	349 (25.22%)	357 (25.79%)	287 (20.74%)	1384	17.37	**<0.001**	0.08	1.36 (1.1–1.7)	—
England ODI	926 (21.99%)	1283 (30.47%)	1193 (28.33%)	809 (19.21%)	4211	169.27	**<0.001**	0.14	1.14 (1–1.3)	—
England Test	278 (12.31%)	823 (36.43%)	602 (26.65%)	556 (24.61%)	2259	289.43	**<0.001**	0.25	—	2.00 (1.7–2.4)
England T20 (Batters)	277 (35.79%)	153 (19.77%)	195 (25.19%)	149 (19.25%)	774	51.12	**<0.001**	0.18	1.86 (1.4–2.5)	—
England ODI (Batters)	687 (37.46%)	711 (38.77%)	329 (17.93%)	107 (5.83%)	1834	566.56	**<0.001**	0.39	6.41 (5–8.2)	—
England Test (Batters)	171 (13.51%)	652 (51.54%)	376 (29.72%)	66 (5.22%)	1265	663.81	**<0.001**	0.51	2.59 (1.9–3.6)	—
England T20 (Bowlers)	128 (17.51%)	199 (27.22%)	237 (32.42%)	167 (22.85%)	731	40.31	**<0.001**	0.17	—	—
England ODI (Bowlers)	225 (10.76%)	446 (22.28%)	804 (38.43%)	597 (28.54%)	2092	352.23	**<0.001**	0.29	—	2.66 (2.2–3.2)
England Test (Bowlers)	105 (11.18%)	160 (17.04%)	219 (23.32%)	455 (48.46%)	939	292.27	**<0.001**	0.39	—	4.34 (3.3–5.7)

Bold font denotes statistically significant chi-square at $p < 0.05$.

4. Discussion

The purpose of this study was to explore the RAEs throughout the ECB national talent pathway through the lens of the PAF. In the immediate timescale, key findings revealed there was a systematic selection bias throughout all the youth levels (i.e., Regional U15, Regional U17, and England U19), whereby relatively older players were significantly over-represented when compared to their relatively younger peers. At senior levels (i.e., England Lions, England T20, England ODI, and England Test), however, there were no significant differences between BQ distributions. When the senior cohorts were compared to the expected BQ distributions based on the Regional U15 cohort to explore the short-term timescale, relatively younger players were significantly more likely to transition from youth to senior levels when compared to their relatively older counterparts. With regards to the long-term timescale, based on skill-sets, batters and bowlers displayed contrasting BQ distributions at the senior levels, whereby relatively older batters were selected for significantly more games, whereas relatively younger bowlers were selected for significantly more games. In an attempt to better understand the immediate, short-term, and long-term implications of RAEs within the ECB national talent pathway, this discussion contextualises these key results using the PAF.

4.1. Immediate Timescale: How Sport Experiences Are Shaped

When the three dynamic elements of the PAF (i.e., personal engagement in activities, appropriate settings and organizational structures, and quality social dynamics) interact with one another, an *immediate* sporting experience is created. This has a subsequent impact on developmental opportunities in youth cricket, and thus can help explain how RAEs occur during the immediate timescale.

4.2. Personal Engagement in Activities

Selection into a talent pathway in cricket can offer a range of immediate benefits, such as gaining access to quality coaching, facilities, higher competition levels, and holistic development opportunities [35]. If those young cricketers who are relatively older are provided with greater openings into talent pathways due to their age, they are inevitably going to be exposed to more fruitful training and developmental activities. In contrast, those who are not selected into talent pathways due to being relatively younger will lose out on engaging in such activities. Moreover, fun, enjoyment, and interest may be heightened for relatively older players due to outperforming and/or being perceived as superior compared to their relatively younger peers. Indeed, this could have a positive impact on immediate cricket experiences at young ages, which may explain the greater proportion of relatively older players at youth levels in this current study. It is also important to realise that RAEs are prevalent from early childhood in sport [5] and are amplified in environments that favour practice over play, specialisation over sampling, and early selection based on immediate performance over long-term potential [24]. Many developmental programmes in cricket have established systems that encourage earlier age-specialisation. As an example, selection into county cricket talent pathways in England often takes place as young as aged 9 years [36]. Interestingly, however, this approach has been widely associated with significant pitfalls. Specifically, questions remain over the lack of evidence to accurately predict future performance capabilities at adulthood based on early selection and performance [3]. For instance, following analysis of match performance data at every age group throughout a First-class County's Cricket pathway (U10 to U19) Brown and colleagues [37] showed how those bowlers who achieved professional status could not be differentiated from their released peers until U17. Moreover, possible drawbacks of engaging in specialised environments (e.g., burnout, injury, and overtraining) have also been associated with selection at young ages [38]. Thus, it is worthwhile exploring the existing organisational structures in order to create more appropriate settings and equitable opportunities for all BQs, as well as help moderate the RAEs shown in this study.

4.3. Appropriate Settings and Organisational Structures

The current results show how ORs and effect sizes of RAEs were dependent on competition level, which corresponds with the Australian context [18]. Specifically, RAEs were more pronounced at the youngest (bi)annual-age group (i.e., Regional U15) and slowly declined as age increased until it levelled out at adulthood (i.e., senior cohorts). As an example, BQ1s were 3.8, 2.5, and 1.8 times more likely to be selected when compared to BQ4s at U15 Regional (large effect size), U17 Regional (medium effect size), and England U19 (small effect size) youth levels, respectively. Evidently, the (bi)annual-age grouping policies may contribute to the presence of RAEs throughout the ECB national talent pathway. However, it is important to understand that RAEs are not naturally occurring phenomenon. Rather, they are created by social agents through their decisions, actions, and policies. Organisational structures in cricket can choose to adapt policies to meet the needs of those who participate to create more equitable competition and moderate RAEs. More specifically, they can change how young players are recruited, how competition is structured, and how they interact with players, coaches, parents, communities, and the environments where we engage in cricket [29].

Similar to other possible discriminatory factors such as ethnicity, gender, and religion, the ECB lists *age* as a protected characteristic in their *Anti-Discrimination Code* [39]. Therefore, it is important that organisational structures in cricket attempt to create the most appropriate settings for every young player in order for them to achieve their full potential [14]. Moreover, Jakobsson and colleagues [40] suggest another possible issue of selecting based on (bi)annual-age grouping is the violation of the guiding principles by the United Nations *Convention on the Rights of the Child* (CRC) [41], which was ratified by the UK in 1991 and is referred to in the UK courts in relation to the *Human Rights Act 1998* [42]. Here, *Article 3* in the CRC states how all decisions regarding a child should be made in the best interest of the child. Thus, not only are there possible discriminatory issues surrounding (bi)annual-age selection to consider, but there could also be potential lawful implications too. This leads us to perhaps the most important question: if we cannot make these changes to moderate RAEs in cricket (and sport in general) now, then when?

Since there appears to be pronounced RAEs throughout the ECB national talent pathway, it is important to consider possible relative age solutions and offer directions for future research. Indeed, a range of potential solutions have been proposed in previous studies, such as coach education [43], an age-ordered shirt numbering system [44], avoiding early deselection [45], a selection quota [46], and delaying the selection process [47]. Moreover, literature on alternative grouping strategies to moderate RAEs is limited when compared to the body of research demonstrating its prevalence. Where proposed grouping strategies have been suggested, little evidence has documented their effectiveness or directly implement those [48]. As an example, Kelly and colleagues [14] conceptualised a *flexible chronological approach*, whereby early birth quartiles (i.e., BQ1s) and late birth quartiles (i.e., BQ4s) should be offered the opportunity to 'play-up' [49,50] and 'play-down' annual-age groups, respectively. Moreover, Kelly and colleagues [51] introduced birthday-banding, where young athletes move up to their next birthdate group on their birthday with aim is to remove particular selection time-points and specific chronological age groups. Other useful strategies that may utiliased in cricket could be drawn from organisational policies incorporated in youth American football (e.g., *age and anthropometric banding* [52,53]) and youth soccer (e.g., *bio-banding* [54,55]). Despite these banding approaches yet to prove their value in reducing RAEs, both strategies appear to systematically address one of the key mechanisms of RAEs, whereby relatively older athletes may have an advanced physiological skill [6]. As such, future research is required to explore the practical implications of these relative age solutions within a youth cricket context.

4.4. *Quality Social Dynamics*

Parents, coaches, and athletes (i.e., social agents) can amplify or mitigate RAEs in youth sports [56]. The Social Agents Model highlights the processes by which social agents influence RAEs in youth sports through three theoretical principles. First, the *Matthew effect* [57], describes how individuals who are initially advantaged are provided with the means to continue their development and further their advantage, whereas those who are initially disadvantaged remain so. In the context of the current findings, batters who are introduced to the game earlier have time to develop complex technical skills that are associated with expertise in batting [58]. Subsequently, these cricketers are most likely to be selected for early talent development programmes. This could also explain the overrepresentation of batters from BQ1 and BQ2 within the senior cohorts. The second and third theoretical principles of the Social Agents Model are the *Pygmalion* [59] and *Galatea* [60] effects. Both these principles refer to the association between the initial expected outcomes and the observed results. However, these expectations differ, whereby the Pygmalion effect refers to external, whereas the Galatea effect refers to internal. With regards to the results of this study, the beliefs and actions of social agents towards successful cricketers are more likely to be positive when compared to those who are less successful [61]. This may further explain why there is an overrepresentation of BQ1 and BQ2s within the youth cohorts when compared to BQ3 and BQ4s.

In order to understand the impact of RAEs on social dynamics, it is important to recognise how these are directly influenced by the rules and regulations of organised youth cricket. For instance, literature examining mixed-age and play can be drawn upon, where evidence exists to suggest that older and younger players can draw unique benefits from playing with each other [51]. On one hand, relatively older players can experience opportunities for leadership and helping of younger peers [62,63]. On the other hand, relatively younger players may benefit from the opportunity to hone their skills and compete against older teammates [14,64]. However, since fixed (bi)annual-age grouping does not allow players to shift between BQs, they will not face a diverse range of experiences or gain the developmental opportunities offered through mixed-age play. Another potential drawback of fixed (bi)annual-age grouping is that it limits different types of social comparison environments. According to Wood and Wilson [65], social comparison theory suggests that athletes rely on peers as a frame of reference to compare themselves, which is used to build self-perceptions such as competence and identity. By limiting opportunities for social comparison due to remaining the same BQ throughout their respective youth trajectories, cricket players face the prospect of a linear pathway and a restricted resilient sense of self throughout development. Given the limited body of literature that has researched the impact of RAEs on mixed-age play, social comparisons, and leadership opportunities, further research is warranted to substantiate these suggestions in the context of youth cricket.

4.5. Short-Term Timescale: The 4Cs

In the short-term, relatively older players gradually engage an increased number of practice and competition opportunities organised youth cricket compared to relatively younger players, which may lead to more cricket-specific *competence*. Thus, in order to compete against relatively older batters and bowlers to gain entry into the ECB national talent pathway, relatively younger batters and bowlers may need to develop higher levels of cricket-specific competence. From a seasonal perspective, relatively older players may score more runs and take more wickets, as well as win more matches and league titles due to higher levels of competence (i.e., greater performance outcomes), which may lead to higher levels of *confidence* [66]. In the context of youth soccer, for instance, Augste and Lames [67] revealed that relatively older youth teams achieved higher league rankings in Germany, while Verbeek and colleagues [68] showed how relatively older youth teams accrued more points-per-game in Holland. In relation to youth cricket, if relatively older players are being selected due to their superior (bi)annual-age group performances in order to gain a competitive advantage (i.e., selection focused on winning rather than nurturing future senior players), it could result in limited selection opportunities, lower participation levels, and higher dropout for relatively younger players in the short-term [11,12]. Indeed, it has been suggested that cricketers who adapt quickest to the increased skill and psychological demands of transitioning to higher levels of performance, could be earmarked as being high potential by cricket officials, and transition across the talent pathway sooner as a result [69]. This could partly explain the RAEs that are present at youth levels in the ECB national talent pathway.

Selection and deselection are often conducted on a seasonal basis in cricket in which reveals consistent RAEs, with early success at each new level valued by coaches [69]. Since relatively older players may be perceived as more competent and are overrepresented in the ECB national talent pathway at youth levels, relatively older players could subsequently gain more time with coaches. Here, the coach-athlete *connection* may be influenced by current competence rather than potential [70]. Moreover, higher levels of athletic competence in relatively older players may lead to higher levels of peer acceptance [71]. In the context of the ECB national talent pathway, relatively older players may be unintentionally exposed to greater opportunities to foster positive connections with key stakeholders, allowing them to thrive in talent pathways compared to their relatively younger peers. In contrast, relatively younger players may be presented with greater challenge and are thus more likely to develop higher levels of resilience [22]. It is possible that this may help foster features of *character* that are required during the transition from youth player to established senior player. However, given the lack of evidence in this area, it offers a range of future research directions to examine the association between age and character-related constructs (e.g., social identity, moral engagement, and pro-social behaviours). Although character is often used as a youth selection criteria in many professional cricket clubs (e.g., [72,73]), this may be personified based on a players birth quartile. As a result, the potential pool of talent believed to have the 'right' character may be limited, while possible inaccurate decisions due to the subjective nature of what the 'right' character actually is could be of concern. It is also important to recognise that there are diverse characters that engage in various activities differently. For instance, selection criteria are often based on traditional factors (e.g., technical skills and physical performance), which could favour relatively older players [6]. In comparison, play-based metrics (e.g., creativity and interest) may help broaden coaches and practitioners' perceptions of an athlete's potential rather than focussing on performance-based characteristics [24]. Moving forward, further research is encouraged to help better understand how RAEs can influence character development and how this can differentiate between BQs.

When examining the youth-to-senior level transition, it was revealed that BQ4s were 3.9, 3.1, 2.5, and 3.6 times more likely to successfully transition to England Lions, England T20, England ODI, and England Test compared to BQ1s based on the Regional U15s BQ distribution, respectively. These findings resonate with [22] results, who found a greater proportion of relatively younger cricketers successfully transitioned from youth to senior international levels compared to their relatively older peers. These results may be due to *reversal effects* of relative age, which is a psychologically based explanation of greater 'growth' that relatively younger players experience, whereby they are initially disadvantaged during their development due to additional challenges they face [22]. Although RAEs may benefit a greater proportion of relatively younger players in the short-term during the youth-to-senior level transition, little is known about the long-term impact of reversal effects and how they may genuinely influence senior career performance and longevity.

4.6. Long-Term Timescale: The 3Ps

When considering skill-set and number of games played at senior level, relatively younger bowlers played more games compared to their relatively older teammates. These findings are contrast with Jones and colleagues [19] who found RAEs in senior international 'super-elite' spin bowlers favouring relatively older players, as well as no RAEs for pace bowlers. Gibbs and colleagues [64] put forward the *underdog hypothesis* to explain why more relatively younger senior players may be outperforming their relatively older equivalents. In the context of the current study, relatively younger bowlers may be benefitting from more competitive play against relatively older counterparts throughout their development. However, it is important to consider how to create a 'BQ4 effect' for *all* bowlers who may require such challenges during their development to facilitate long-term *performance* outcomes [14].

Interestingly, however, when analysing the number of games played for batters, relatively older players appear to have played more games compared to their relatively younger teammates. For instance, BQ2 batters were 6.9 and 10.3 times more likely to play an England ODI and England Test match when compared to BQ4 batters, respectively. *Long-lasting effects* of relative age have been found across numerous sports at senior international levels [74]. For instance, Lupo and colleagues [75] showed how athletes who were born at the beginning of the selection year were 1.57, 1.34, 2.69, 1.48, and 1.45 times more likely to reach the senior first and second divisions in Italian basketball, rugby union, soccer, volleyball, and water polo, respectively, when compared to those born at the end of year. In the context of cricket, these findings correspond with previous results [21], which also showed how 'high-performing' international batters born in BQ1 were significantly more likely to be selected at senior levels when compared to BQ4s. These long-lasting effects may be due to the early selection of relatively older players, who subsequently have greater access to facilities (e.g., access to bowling machines and specialist equipment) and coaches (e.g., creating connections with key stakeholders), to facilitate their long-term performance capabilities. Further, early selected batters are likely to accumulate more hours of random and varied batting-specific practice through adolescence, which [20] highlighted as a key discriminator in achieving 'super elite' status. On the other hand, since relatively younger players may not have been offered these same developmental opportunities, they may not have been able to achieve their full potential. Indeed, this also has a knock-on effect by creating a smaller pool of talent to select from at senior levels due to relatively younger players with potential to excel at adulthood being overlooked at younger ages. Possible explanations for the performance differences between skill-sets could lie in the greater physiological requirements for bowlers that may not be achieved until post-adolescence [76], thus making them less exposed to RAEs as relatively younger players have the opportunity to 'catch-up'. In contrast, batters may require a larger accumulation of practice to develop the necessary perceptual-cognitive skills required to excel at senior levels [77], and therefore making them more vulnerable to RAEs as relatively younger players are less likely to 'catch-up'. Overall, stakeholders employed in cricket settings should be cautious of the diverse trajectories as well as the long-term development and performance outcomes between batters and bowlers.

The necessity of examining RAEs at more than one point in time is something that was recently encouraged by Schorer and colleagues [78]. This suggestion is reinforced by the limited evidence that has explored the implications of RAEs on the long-term *participation* of relatively younger players. Moreover, despite relatively older players being more likely to be recruited into talent development pathways at youth levels, it also seems they comprise a greater quantity of players who are unsuccessful in achieving senior levels. Thus, although relatively younger players have been reported to drop out of youth sport at young ages due to RAEs [11], it may be suggested this is being replicated by relatively older players at the latter stages of development during the youth-to-senior level transition. Thus, it is important to create immediate sport experiences that foster rich developmental outcomes that help retain players in the long-term independent of their BQ. In addition to possible deselection and dropout, relatively older players are also exposed to possible injury and burnout due to early selection procedures [35]. For instance, McGrath and Finch [79] identified that fast bowlers are most likely to suffer from overuse injuries, such as stress fractures in the lower lumbar spine. Since injuries are more likely to occur during the adolescence growth spurt [80], it is plausible that those who are selected onto talent pathways from an early age are at an increased risk of developing such injuries due to high training loads and subsequently drop out at the latter stages of the talent pathways. Thus, it is important that organisational structures focus on long-term participation strategies when recruiting, developing, and deselecting young cricketers in their talent pathways, in order to avoid dropout, injury, burnout, and promote continued engagement in cricket. Further research is warranted to better understand the long-term implications of RAEs on participation in cricket and help substantiate these suggestions.

When compared to the possible performance and participation implications, little is known about the *personal development* outcomes due to RAEs. Obvious personal benefits for those who attain senior levels includes membership to the Professional Cricketers' Association (PCA), global tours, heightened media profile, and possibly higher monetary opportunities. Moreover, selected players may gain access to additional support throughout their development (e.g., psychology profiling, nutrition programming, and strength and conditioning), which may positively impact their abilities for adaptive and positive behaviour [81]. Indeed, it is surprising how RAEs can have such a direct impact on whether an individual gains these personal benefits and additional support that can positively shape their adult livelihood. Although the current evidence-base is scarce, these effects can also play an important role in developing psychosocial skills that characterise oneself [9]. Further quantitative (e.g., observational coding, questionnaires) and qualitative (e.g., content analysis, composite narratives) enquiry is needed to investigate RAEs on a broader range of psychosocial mechanisms, such as global wellbeing, leadership skills, moral disengagement, social behaviour, and social identity.

5. Limitations

There are important methodological and contextual limitations to consider when interpreting these current results. From a methodological perspective, the data available for the U15 Regional and U17 Regional groups did not provide an entire representation for these particular cohorts. Nevertheless, both these samples offer a large enough representation to draw valid conclusions from and should not be overlooked. Moreover, online data entry began at various different time points for each group, thus the number of years included in each cohort varied. However, we included all the data that was available and captured the longevity of RAEs throughout the ECB talent pathway by including a minimum of seven years within each cohort. In addition, the number of games played that was used as a proxy for performance may not provide an entire reflection of how well a player has consistently played, although it is plausible to suggest that continued selection is an important contributing factor towards success at adulthood.

From a contextual viewpoint, it was the initial aim of this study to capture both male and female pathways. However, in light of the lack of data available within the female pathway, we decided there was not sufficient evidence to include this population. This may be due to the fact that the first professional female cricket league has only recently started in England (i.e., 2020), thus the absence of data available is likely due to the female pathway being far less developed when compared to their male equivalents. As such, it is important to recognise the many relative age lessons learnt in the male pathway from this current study when designing the emerging female structures to ensure the same issues are not recreated, and instead use this as an opportunity to create contemporary organisational structures and more appropriate youth cricket settings [46]. Moreover, it is also important to consider other possible selection and development biases that are prevalent throughout the ECB talent pathways, such as ethnicity [82], relative access to wealth [83], and birthplace effects [84], that may create a recipe to exacerbate some of the existing RAEs [24,25]. Thus, further research adopting a multidimensional approach to explore these obstacles together is required.

6. Conclusions

There appears to be a complex relationship between the month a batter or bowler is born, the likelihood they are selected into a talent pathway, and their opportunities to successfully transition and compete at senior levels. Key results showed how RAEs were prevalent throughout all youth levels (i.e., Regional U15, Regional U17, and England U19), but not at senior levels (i.e., England Lions, England T20, England ODI, and England Test) during the immediate timescale. Moreover, when compared to the expected BQ distribution based on the Regional U15 cohort, BQ4s were between 2.5 to 3.9 times more likely to transition to senior levels when compared to BQ1s during the short-term timescale. In addition, when considering the number of games played at senior levels during the long-term timescale, relatively older batters were selected for more games, whereas relatively younger bowlers were selected for more games. We used the PAF to capture the possible immediate (i.e., dynamic elements), short-term (i.e., the 4Cs), and long-term (i.e., the 3Ps) timescales that are influenced by RAEs based on skill-set (see Figure 1).

Moving forward, it is important for researchers and practitioners to better understand how (bi)annual-age grouping shapes developmental outcomes in sport across different timescales (i.e., the PAF), as well as examine alternative group banding strategies (e.g., a flexible chronological approach, birthday-banding) and possible RAE solutions (e.g., age-ordered shirt numbering, monitoring growth and maturation status). By doing so, it may help create a more appropriate learning environment for every individual to achieve their potential.

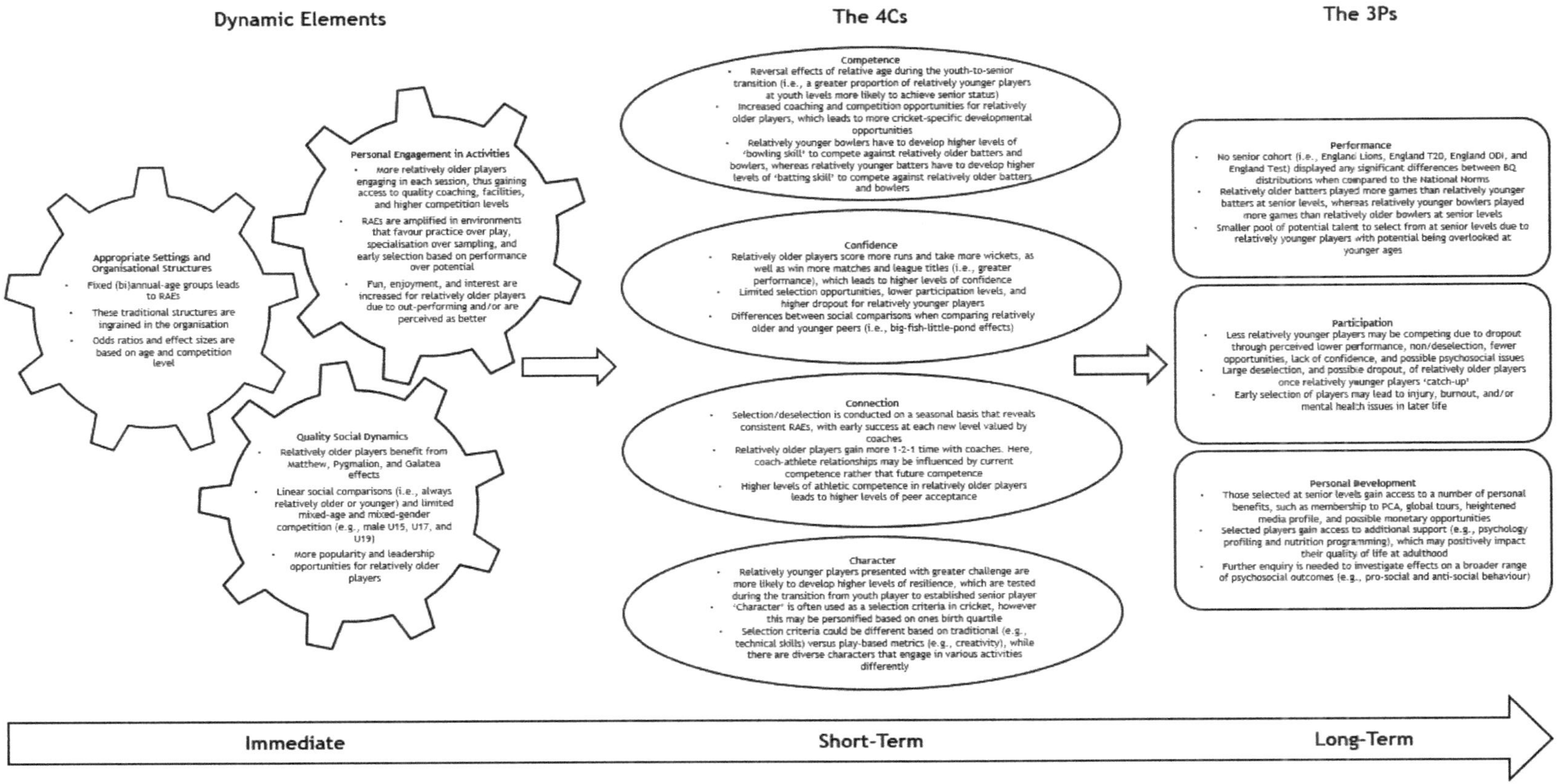

Figure 1. The Personal Assets Framework [26–28] as a representation of development in male cricket due to relative age effects.

Author Contributions: Conceptualization, A.L.K.; Data curation, R.R.; Formal analysis, T.B.; Writing—original draft, A.L.K.; Writing—review and editing, T.B., R.R., J.C. and J.T. All authors have read and agreed to the published version of the manuscript.

Funding: This research received no external funding.

Institutional Review Board Statement: Not applicable.

Informed Consent Statement: Not applicable.

Data Availability Statement: Not applicable.

Conflicts of Interest: The authors declare no conflict of interest.

References

1. Kelly, A.L.; Till, K.; Jackson, D.; Barrell, D.; Burke, K.; Turnnidge, J. Talent identification and relative age effects in English male rugby union pathways: From entry to expertise. *Front. Sports Act. Living* **2021**, *3*, 640607. [CrossRef] [PubMed]
2. Baker, J.; Schorer, J.; Wattie, N. Compromising talent: Issues in identifying and selecting talent in sport. *Quest* **2018**, *70*, 48–63. [CrossRef]
3. Till, K.; Baker, J. Challenges and [possible] solutions to optimizing talent identification and development in sport. *Front. Psychol.* **2020**, *11*, 664. [CrossRef]
4. Barnsley, R.H.; Thompson, A.H.; Barnsley, P.E. Hockey success and birthdate: The relative age effect. *CAHPER J.* **1985**, *51*, 23–28.
5. Smith, K.L.; Weir, P.L.; Till, K.; Romann, M.; Cobley, S. Relative age effects across and within female sport contexts: A systematic review and meta-analysis. *Sports Med.* **2018**, *48*, 1451–1478. [CrossRef] [PubMed]
6. Cobley, S.; Baker, J.; Wattie, N.; McKenna, J. Annual age-grouping and athlete development: A meta-analytical review of relative age effects in sport. *Sports Med.* **2009**, *39*, 235–256. [CrossRef]
7. Doncaster, G.; Medina, D.; Drobnic, F.; Gómez-Díaz, A.J.; Unnithan, V. Appreciating factors beyond the physical in talent identification and development: Insights from the FC Barcelona sporting model. *Front. Sports Act. Living* **2020**, *2*, 91. [CrossRef]
8. Hancock, D.J.; Adler, A.L.; Côté, J. A proposed theoretical model to explain relative age effects in sport. *Eur. J. Sport Sci.* **2013**, *13*, 630–637. [CrossRef]
9. Wattie, N.; Cobley, S.; Baker, J. Towards a unified understanding of relative age effects. *J. Sports Sci.* **2008**, *26*, 1403–1409. [CrossRef]
10. Furley, P.; Memmert, D. Coaches' implicit associations between size and giftedness: Implications for the relative age effect. *J. Sports Sci.* **2016**, *34*, 459–466. [CrossRef]
11. Delorme, N.; Chalabaev, A.; Raspaud, M. Relative age is associated with sport dropout: Evidence from youth categories of French basketball. *Scand. J. Med. Sci. Sports* **2011**, *21*, 120–128. [CrossRef] [PubMed]
12. Hancock, D.J.; Ste-Marie, D.M.; Young, B.W. Coach selections and the relative age effect in male youth ice hockey. *Res. Q. Exerc. Sport* **2013**, *84*, 126–130. [CrossRef] [PubMed]
13. Edwards, S. Born too late to win? *Nature* **1994**, *370*, 186. [CrossRef] [PubMed]
14. Kelly, A.L.; Wilson, M.R.; Gough, L.A.; Knapman, H.; Morgan, P.; Cole, M.; Jackson, D.T.; Williams, C.A. A longitudinal investigation into the relative age effect in an English professional football club: The 'underdog hypothesis'. *Sci. Med. Footb.* **2020**, *4*, 111–118. [CrossRef]
15. O'Donoghue, P.G.; Edgar, S.; McLaughlin, E. Season of birth bias in elite cricket and netball. *J. Sports Sci.* **2004**, *22*, 256–257.
16. Abernethy, B.; Farrow, D. Contextual factors influencing the development of expertise in Australian athletes. In Proceedings of the 11th World Congress of Sport Psychology, Sydney, Australia, 15–19 August 2005.
17. Barney, E.G. Preliminary Stages in the Validation of a Talent Identification Model in Cricket. Ph.D. Thesis, Bangor University, Bangor, UK, 2015.
18. Connor, J.D.; Renshaw, I.; Doma, K. Moderating factors influence the relative age effect in Australian cricket. *PeerJ* **2019**, *7*, e6867. [CrossRef]
19. Jones, B.D.; Lawrence, G.P.; Hardy, L. New evidence of relative age effects in "super-elite" sportsmen: A case for the survival and evolution of the fittest. *J. Sports Sci.* **2018**, *36*, 697–703. [CrossRef]
20. Jones, B.D.; Hardy, L.; Lawrence, G.; Kuncheva, L.I.; Brandon, R.; Bobat, M.; Thorpe, G. It ain't what you do—It's the way that you do it: Is optimizing challenge key in the development of super-elite batsmen? *J. Expertise* **2020**, *3*, 144–168.
21. Low, J.; Ford, P.R.; Williams, A.M. The effects of birth date and place on the development of expertise in professional cricket batters. *Int. J. Sport Psychol.* **2015**, *46*, 676–688.
22. McCarthy, N.; Collins, D.; Court, D. Start hard, finish better: Further evidence for the reversal of the RAE advantage. *J. Sports Sci.* **2016**, *34*, 1461–1465. [CrossRef]
23. Kelly, A.L.; Côté, J.; Turnnidge, J.; Hancock, D. Editorial: Birth advantages and relative age effects. *Front. Sports Act. Living* **2021**, *3*, 721704. [CrossRef] [PubMed]
24. Turnnidge, J.; Kelly, A.L. Organizational structures: Looking back and looking ahead. In *Birth Advantages and Relative Age Effects in Sport: Exploring Organizational Structures and Creating Appropriate Settings*; Kelly, A.L., Côté, J., Jeffreys, M., Turnnidge, J., Eds.; Routledge: London, UK, 2021; pp. 239–246.

25. Wattie, N.; Schorer, J.; Baker, J. The relative age effect in sport: A developmental systems model. *Sports Med.* **2015**, *45*, 83–94. [CrossRef] [PubMed]
26. Côté, J.; Turnnidge, J.; Evans, M.B. The dynamic process of development through sport. *Kinesiol. Slov. Sci. J. Sport* **2014**, *20*, 14–26.
27. Côté, J.; Turnnidge, J.; Vierimaa, M. A personal assets approach to youth sport. In *Handbook of Youth Sport*; Smith, A., Green, K., Eds.; Routledge: London, UK, 2016; pp. 243–256.
28. Côté, J.; Turnnidge, J.; Murata, A.; McGuire, C.; Martin, L. Youth sport research: Describing the integrated dynamics elements of the Personal Assets Framework. *Int. J. Sport Psychol.* **2020**, *51*, 562–578.
29. Kelly, A.L.; Erickson, K.; Turnnidge, J. Youth Sport in the Time of COVID-19: Considerations for Researchers and Practitioners. *Manag. Sport Leis.* **2020**. advanced online publication. [CrossRef]
30. Cricket Archive. Statistics [Online]. 2021. Available online: https://cricketarchive.com/ (accessed on 12 March 2021).
31. Office for National Statistics. Number of Live Births by Date, 1995 to 2014, in England and Wales [Online]. 2015. Available online: https://www.ons.gov.uk/peoplepopulationandcommunity/birthsdeathsandmarriages/livebirths/adhocs/005149numberoflivebirthsbydate1995to2014inenglandandwales (accessed on 12 May 2018).
32. Delorme, N.; Boiché, J.; Raspaud, M. Relative age effect in elite sports: Methodological bias or real discrimination? *Eur. J. Sport Sci.* **2010**, *10*, 91–96. [CrossRef]
33. McHugh, M.L. The chi-square test of independence. *Biochem. Med.* **2013**, *23*, 143–149. [CrossRef]
34. Cohen, J. *Statistical Power Analysis for the Behavioral Sciences*; L. Erlbaum Associates: Hillsdale, NJ, USA, 1988.
35. Lascu, A.; Spratford, W.; Pyne, D.B.; Etxebarria, N. Practical application of ecological dynamics for talent development in cricket. *Int. J. Sports Sci. Coach.* **2020**, *15*, 227–238. [CrossRef]
36. ECB. ECB Recommended Junior Formats [Online]. 2018. Available online: https://platform-static-files.s3.amazonaws.com/ecb/document/2019/03/20/45788f57-47ac-4812-b0de-1c4164bd6382/Junior-Formats-Infographic-Combined.pdf (accessed on 11 May 2021).
37. Brown, T.W.; Gough, L.; Khawaja, I.; Kelly, A.L. Performance Trajectories of Bowlers and Batters from Youth Level to Senior Professional Status in Cricket. *Int. J. Perform. Anal. Sport* **2021**, in press. [CrossRef]
38. Bergeron, M.F.; Mountjoy, M.; Armstrong, N.; Chia, M.; Côté, J.; Emery, C.A.; Faigenbaum, A.; Hall, G.; Kriemler, S.; Léglise, M.; et al. International Olympic Committee consensus statement on youth athletic development. *Br. J. Sports Med.* **2015**, *49*, 843–851. [CrossRef]
39. ECB. Anti-Discrimination Code [Online]. 2021. Available online: https://resources.ecb.co.uk/ecb/document/2021/03/16/f0036503-deaa-4b81-874d-7e027d7d4617/24.ECB-Anti-Discrimination-Code-2021-vF.pdf (accessed on 11 May 2021).
40. Jakobsson, J.; Julin, A.L.; Persson, G.; Malm, C. Darwinian selection discriminates young athletes: The relative age effect in relation to sporting performance. *Sports Med.-Open* **2021**, *7*, 16. [CrossRef] [PubMed]
41. United Nations. Convention on the rights of the child. *Treaty Ser.* **1989**, *1577*, 1–23.
42. Equality and Human Rights Commission. Understanding Human Rights [Online]. 2016. Available online: https://www.equalityhumanrights.com/en/secondary-education-resources/useful-information/understanding-human-rights (accessed on 11 May 2021).
43. Grossmann, B.; Lames, M. Relative age effect (RAE) in football talents: The role of youth academies in transition to professional status in Germany. *Int. J. Perform. Anal. Sport* **2013**, *13*, 120–134. [CrossRef]
44. Mann, D.L.; van Ginneken, P.J.M.A. Age-ordered shirt numbering reduces the selection bias associated with the relative age effect. *J. Sports Sci.* **2017**, *35*, 784–790. [CrossRef] [PubMed]
45. Tribolet, R.; Watsford, M.L.; Coutts, A.J.; Smith, C.; Fransen, J. From entry to elite: The relative age effect in the Australian football talent pathway. *J. Sci. Med. Sport* **2019**, *22*, 741–745. [CrossRef] [PubMed]
46. Bennett, K.J.M.; Vaeyens, R.; Fransen, J. Creating a framework for talent identification and development in emerging football nations. *Sci. Med. Footb.* **2018**, *3*, 36–42. [CrossRef]
47. Romann, M.; Rüeger, E.; Hintermann, M.; Kern, R.; Faude, O. Origins of relative age effects in youth football—A nationwide analysis. *Front. Sports Act. Living* **2020**, *2*, 591072. [CrossRef]
48. Webdale, K.; Baker, J.; Schorer, J.; Wattie, N. Solving sport's 'relative age' problem: A systematic review of proposed solutions. *Int. Rev. Sport Exerc. Psychol.* **2020**, *13*, 187–204. [CrossRef]
49. Goldman, D.E.; Turnnidge, J.; Kelly, A.L.; de Vos, J.; Côté, J. Athlete Perceptions of "Playing-Up" in Youth Soccer. *J. Appl. Sport Psychol.* **2021**. advanced online publication. [CrossRef]
50. Kelly, A.L.; Wilson, M.R.; Jackson, D.T.; Goldman, D.E.; Turnnidge, J.; Côté, J.; Williams, C.A. A multidisciplinary investigation into 'playing-up' in academy football according to age phase. *J. Sports Sci.* **2021**, *39*, 854–864. [CrossRef]
51. Kelly, A.L.; Jackson, D.T.; Taylor, J.J.; Jeffreys, M.A.; Turnnidge, J. "Birthday-banding" as a strategy to moderate the relative age effect: A case study into the England Squash Talent Pathway. *Front. Sports Act. Living* **2020**, *3*, 573890. [CrossRef] [PubMed]
52. Jones, C.; Visek, A.J.; Barron, M.J.; Hyman, M.; Chandran, A. Association between relative age effect and organisational practices of American youth football. *J. Sports Sci.* **2019**, *37*, 1146–1153. [CrossRef] [PubMed]
53. MacDonald, D.J.; Cheung, M.; Côté, J.; Abernethy, B. Place but not date of birth influences the development and emergence of athletic talent in American football. *J. Appl. Sport Psychol.* **2009**, *21*, 80–90. [CrossRef]

54. Thomas, C.; Oliver, J.; Kelly, A.L. Bio-banding in youth soccer: Considerations for researchers and practitioners. In *Birth Advantages and Relative Age Effects in Sport: Exploring Organizational Structures and Creating Appropriate Settings*; Kelly, A.L., Côté, J., Jeffreys, M., Turnnidge, J., Eds.; Routledge: London, UK, 2021; p. 31.
55. Bradley, B.; Johnson, D.; Hill, M.; McGee, D.; Kana-ah, A.; Sharpin, C.; Sharp, P.; Kelly, A.L.; Cumming, S.P.; Malina, R.M. Bio-banding in academy football: Player's perceptions of a maturity matched tournament. *Ann. Hum. Biol.* **2019**, *46*, 400–408. [CrossRef] [PubMed]
56. Hancock, D.J.; Côté, J. Birth advantages, social agents, and talent development in youth sport. In *Positive Human Functioning from a Multidimensional Perspective: Promoting High Performance*; Nova Science Publishers: Hauppauge, NY, USA, 2014; Volume 3, pp. 15–32.
57. Merton, R.K. The Matthew effect in science: The reward and communication systems of science are considered. *Science* **1968**, *159*, 56–63. [CrossRef] [PubMed]
58. Weissensteiner, J.; Abernethy, B.; Farrow, D. Towards the development of a conceptual model of expertise in cricket batting: A grounded theory approach. *J. Appl. Sport Psychol.* **2009**, *21*, 276–292. [CrossRef]
59. Rosenthal, R.; Jacobson, L. Pygmalion in the classroom. *Urban Rev.* **1968**, *3*, 16–20. [CrossRef]
60. Merton, R.K. Priorities in scientific discovery: A chapter in the sociology of science. *Am. Sociol. Rev.* **1957**, *22*, 635–659. [CrossRef]
61. Solomon, G.B. Coach expectations and differential feedback: Perceptual flexibility revisited. *J. Sport Behav.* **1998**, *21*, 298–310.
62. Côté, J.; Baker, J.; Abernethy, B. Practice and play in the development of sport expertise. In *Handbook of Sport Psychology*; Tenenbaum, G., Eklund, R.C., Eds.; Wiley: Hoboken, NJ, USA, 2007; pp. 184–202.
63. Jarvis, P. Dangerous activities within and invisible playground: A study of emergent male football play and teachers' perspectives of outdoor free play in the early years of primary school. *Int. J. Early Years Educ.* **2007**, *15*, 245–259. [CrossRef]
64. Gibbs, B.G.; Jarvis, J.A.; Dufur, M.J. The rise of the underdog? The relative age effect reversal among Canadian-born NHL hockey players: A reply to Nolan and Howell. *Int. Rev. Sociol. Sport* **2012**, *47*, 644–649. [CrossRef]
65. Wood, J.V.; Wilson, A.E. How important are social comparisons for self-evaluation? In *Handbook of Self and Identity*; Leary, M.R., Tangney, J.P., Eds.; Guilford Press: New York, NY, USA, 2003; pp. 344–366.
66. Hays, K.; Thomas, O.; Maynard, I.; Bawden, M. The role of confidence in world-class sport performance. *J. Sports Sci.* **2009**, *27*, 1185–1199. [CrossRef] [PubMed]
67. Augste, C.; Lames, M. The relative age effect and success in German elite U-17 soccer teams. *J. Sports Sci.* **2011**, *29*, 983–987. [CrossRef]
68. Verbeek, J.; Lawrence, S.; van der Breggan, J.; Kelly, A.L.; Jonker, J. The average team age method and its potential to reduce relative age effects. In *Birth Advantages and Relative Age Effects in Sport: Exploring Organizational Structures and Creating Appropriate Settings*; Kelly, A.L., Côté, J., Jeffreys, M., Turnnidge, J., Eds.; Routledge: London, UK, 2021; p. 18.
69. Jones, B.D.; Hardy, L.; Lawrence, G.; Kuncheva, L.I.; Du Preez, T.; Brandon, R.; Such, P.; Bobat, M. The identification of "game changers" in England cricket's developmental pathway for elite spin bowling: A machine learning approach. *J. Expert.* **2019**, *2*, 92–120.
70. Koz, D.; Fraser-Thomas, J.; Baker, J. Accuracy of professional sports drafts in predicting career potential. *Scand. J. Med. Sci. Sports* **2012**, *22*, 64–69. [CrossRef]
71. Weiss, M.R.; Duncan, S.C. The relationship between physical competence and peer acceptance in the context of children's sports participation. *J. Sport Exerc. Psychol.* **1992**, *14*, 177–191. [CrossRef]
72. Kent County Cricket Club. Kent Cricket Academy and Emerging Player Programme Annual Performance Plan [Online]. 2018. Available online: https://www.kentcricket.co.uk/wp-content/uploads/academy4.pdf (accessed on 9 May 2020).
73. Warwickshire County Cricket Club. Player Core Behaviours [Online]. 2020. Available online: https://edgbaston.com/warwickshire-ccc/pathway/player-core-behaviours/ (accessed on 30 June 2020).
74. Brustio, P.R.; Kearney, P.E.; Lupo, C.; Ungureanu, A.N.; Mulasso, A.; Rainoldi, A.; Boccia, G. Relative age influences performance of world-class track and field athletes even in the adulthood. *Front. Psychol.* **2019**, *10*, 1395. [CrossRef]
75. Lupo, C.; Boccia, G.; Ungureanu, A.N.; Frati, R.; Marocco, R.; Brustio, P.R. The beginning of senior career in team sport is affected by relative age effect. *Front. Psychol.* **2019**, *10*, 1465. [CrossRef] [PubMed]
76. Pyne, D.B.; Duthie, G.M.; Saunders, P.U.; Petersen, C.A.; Portus, M.R. Anthropometric and strength correlates of fast bowling speed in junior and senior cricketers. *J. Strength Cond. Res.* **2006**, *20*, 620–626.
77. Weissensteiner, J.; Abernethy, B.; Farrow, D.; Müller, S. The development of anticipation: A cross-sectional examination of the practice experiences contributing to skill in cricket batting. *J. Sport Exerc. Psychol.* **2008**, *30*, 663–684. [CrossRef] [PubMed]
78. Schorer, J.; Roden, I.; Büsch, D.; Faber, I. Relative age effects are developmental! The necessity of looking at more than one time point. In *Relative Age Effects in Sport: International Perspective*; Dixon, J.C., Horton, S., Chittle, L., Baker, J., Eds.; Routledge: London, UK, 2020; pp. 33–45.
79. McGrath, A.C.; Finch, C.F. *Bowling Cricket Injuries Over: A Review of the Literature*; Monash University Accident Research Centre: Clayton, Australia, 1996.
80. Niemeyer, P.; Weinberg, A.; Schmitt, H.; Kreuz, P.C.; Ewerbeck, V.; Kasten, P. Stress fractures in adolescent competitive athletes with open physis. *Knee Surg. Sports Traumatol. Arthrosc.* **2006**, *14*, 771–777. [CrossRef] [PubMed]
81. Larsen, C.H.; Alfermann, D.; Henriksen, K.; Christensen, M.K. Successful talent development in soccer: The characteristics of the environment. *Sport Exerc. Perform. Psychol.* **2013**, *2*, 190–206. [CrossRef]

82. Brown, T.W.; Greetham, P.; Powell, A.; Gough, L.; Khawaja, I.; Kelly, A.L. The sociodemographic profile of the England and Wales Cricket Board (ECB) talent pathways and first-class counties. *Manag. Sport Leis.* **2021**. advanced online publication. [CrossRef]
83. Brown, T.; Kelly, A.L. Relative access to wealth and ethnicity in professional cricket. In *Birth Advantages and Relative Age Effects in Sport: Exploring Organizational Structures and Creating Appropriate Settings*; Kelly, A.L., Côté, J., Jeffreys, M., Turnnidge, J., Eds.; Routledge: London, UK, 2021; pp. 184–206.
84. Turnnidge, J.; Hancock, D.J.; Cote, J. The influence of birth date and place of development on youth sport participation. *Scand. J. Med. Sci. Sports* **2014**, *24*, 461–468. [CrossRef]

sports

MDPI

Article

Psychosocial and Physiological Factors Affecting Selection to Regional Age-Grade Rugby Union Squads: A Machine Learning Approach

Julian Owen [1], Robin Owen [2,*], Jessica Hughes [1,3], Josh Leach [4], Dior Anderson [5] and Eleri Jones [3]

1 Institute of Applied Human Physiology, School of Human and Behavioural Science, Bangor University, Bangor LL572DG, UK; j.owen@bangor.ac.uk (J.O.); jlhughes02@outlook.com (J.H.)
2 School of Health and Sport Sciences, Liverpool Hope University, Liverpool L169JD, UK
3 Institute for the Psychology of Elite Performance, School of Human and Behavioural Science, Bangor University, Bangor LL572DG, UK; eleri.s.jones@bangor.ac.uk
4 Welsh Rugby Union, Colwyn Bay LL297SP, UK; jleach@wru.wales
5 Talent Pathway iD Ltd., Gaerwen LL606HE, UK; dior@talent-pathway.co.uk
* Correspondence: owenr@hope.ac.uk

Abstract: Talent selection programmes choose athletes for talent development pathways. Currently, the set of psychosocial variables that determine talent selection in youth Rugby Union are unknown, with the literature almost exclusively focusing on physiological variables. The purpose of this study was to use a novel machine learning approach to identify the physiological and psychosocial models that predict selection to a regional age-grade rugby union team. Age-grade club rugby players (n = 104; age, 15.47 ± 0.80; U16, n = 62; U18, n = 42) were assessed for physiological and psychosocial factors during regional talent selection days. Predictive models (selected vs. non-selected) were created for forwards, backs, and across all players using Bayesian machine learning. The generated physiological models correctly classified 67.55% of all players, 70.09% of forwards, and 62.50% of backs. Greater hand-grip strength, faster 10 m and 40 m sprint, and power were common features for selection. The generated psychosocial models correctly classified 62.26% of all players, 73.66% of forwards, and 60.42% of backs. Reduced burnout, reduced emotional exhaustion, and lower reduced sense of accomplishment, were common features for selection. Selection appears to be predominantly based on greater strength, speed, and power, as well as lower athlete burnout.

Keywords: talent identification; talent selection; psychological factors; physical performance; pattern recognition; Bayesian machine learning; youth rugby

Citation: Owen, J.; Owen, R.; Hughes, J.; Leach, J.; Anderson, D.; Jones, E. Psychosocial and Physiological Factors Affecting Selection to Regional Age-Grade Rugby Union Squads: A Machine Learning Approach. *Sports* **2022**, *10*, 35. https://doi.org/10.3390/sports10030035

Academic Editor: Kevin Till

Received: 17 December 2021
Accepted: 25 February 2022
Published: 28 February 2022

1. Introduction

Talent identification programmes assess the attributes of athletes, to guide talent selection programmes [1]. The aim of talent selection programmes is to select players with the potential to be the 'sporting superstars' of tomorrow and help clubs/governing-bodies achieve their long-term performance goals [2]. In furtherance of this long-term goal, selected players are usually integrated into talent development programmes which attempt to provide a learning environment that helps players achieve their potential [1]. However, talent selection programmes feature common problems. Firstly, youth performance is frequently used to predict success in adulthood when making selection decisions [3–5], despite youth performance offering low predictive accuracy [6–8]. For example, only 17% of male U18 sprinters who ranked among the top 50 highest performers internationally achieved the same ranking at senior level. Secondly, talent selection decisions are often made based on subjective criteria [1,3,4,9,10]. For example, interviews of national youth soccer coaches revealed that perceptions of talent and consequent selection decisions are primarily based on implicit coach preferences [9]. Consequently, current approaches to talent selection have been criticized [3,5] and more objective and evidence-based criteria are required to inform talent selection in sport [11–14].

Due to the physical nature of Rugby Union, players are generally required to have highly developed physical qualities [15]; players who are taller, heavier [16,17], faster [18], have greater strength [19], generate more power [20], and are relatively older compared to their peers [21], are more likely to be identified as having 'talent' and selected for development programmes. Within Rugby Union, there are also differences in the physical demands across positional units [22], especially within older age categories where position-specific fitness profiles are needed [23–25]. Notably, backs are involved in more high-intensity locomotor workload, whilst forwards perform more static high-intensity efforts than backs [26–29]. Accordingly, the physiological determinants of selection to talent development pathways appear to differ for forwards and backs; greater speed and agility has been shown to be an important talent selection criterion for backs, while greater upper-body strength, height, and mass are important talent selection criteria for forwards [16,30].

Although the physiological factors predicting selection to regional-age Rugby Union squads are relatively well understood, psychosocial factors (including personality) have received far less empirical attention [31]: with the exception being the work of Hill and colleagues [32,33]. It is generally accepted that selection/progression through elite performance pathways in other sports is facilitated by higher levels of emotional stability, coping strategies, perfectionism, optimism, extraversion, conscientiousness, emotional intelligence, agreeableness, discipline, self-confidence, resilience, and coachability [34–40]. Conversely, there is also evidence from other sports to suggest that certain psychosocial characteristics derail the development process of players [35]. For example, dysfunctional dispositions such as obsessive passion, maladaptive perfectionism, and dispositional optimism can negatively impact a player's progression [41–43]. In accordance with findings from other sports, Hill et al. [33] reported that coaches perceived youth Rugby Union players as more likely to succeed if they exhibited greater proactiveness, commitment, growth-mindset, realistic performance evaluations, and resilience. It was reasoned that these skills help negotiate key challenges and developmental opportunities. They may also be a factor in reducing the likelihood of burnout brought forward by excessive perfectionism [32]. For example, greater resilience may enable individuals to persevere and stay engaged despite initial failures [33].

A limitation of Hill and colleagues' [33] work is that conclusions were derived from retrospective coach opinions, rather than potentially more reliable player-based assessments. Additionally, their investigation primarily centred around progression through Rugby Union talent development programmes, forsaking psychosocial attributes' role in Rugby Union talent selection. To address these gaps in the literature and satisfy recent calls to further investigate the role of psychosocial factors in Rugby Union talent selection [21], we utilized extensive primary physiological and psychosocial test batteries to differentiate between selected and non-selected regional age-grade Rugby Union players in North Wales (i.e., under 16 and 18 age categories). A novel Bayesian pattern recognition technique was used to identify which attributes (termed features in the analysis) differentiate between selection and non-selection. Thus, the present investigation offers an arguably more comprehensive test of factors than previous studies into age-grade selection [21]; is the first attempt to objectively understand the currently subjective decision-making that determines selection to regional age-grade academies in Wales; and tests the role of physiological attributes via new and cutting-edge analytical methods. Specifically, the Bayesian pattern recognition analysis we utilized accounts for the complex interaction between multiple variables when constructing models [44], provides a rigorous/conservative method to test the feature models that predict group classification (i.e., selected vs. non-selected) [45,46], and provides a way to explore interactions without potentially misleading assumptions/hypotheses [47]. Given the large number of physiological and psychosocial variables collected and relative exploratory nature of machine learning techniques, precise a priori hypotheses were not formed. However, it was anticipated that feature selection stages would identify similar predictive physiological variables to previous investigations in rugby union [31] and similar psychosocial variables to previous investigations in other sports [34–40].

2. Materials and Methods

2.1. Participants

A total of 104 male U16 and U18 Rugby Union players (Mage = 15.47, SDage = 0.80; U16 $n = 62$; U18 $n = 42$) who attended one of two North Wales Rugby 'Talent Camps' in 2019 or 2020, volunteered to take part and gave informed consent in-line with institutional ethics guidelines. Of the 104 players who attended, 66 players were selected and 38 were not selected to the regional squads. Of the selected players, 37 were forwards (of which 16 = U16 and 21 = U18) and 29 were backs (of which 17 = U16 and 12 = U18). Of the non-selected players, 19 were forwards (of which 16 = U16 and 3 = U18) and 19 were backs (of which 13 = U16 and 6 = U18). These selections formed the six classification groups for analyses (i.e., selected players vs. non-selected players, selected forwards vs. non-selected forwards, and selected backs vs. non-selected backs).

2.2. Procedure

Players from regional squads and eligible age-grade clubs received an invitation to participate in a 1-day 'talent camp' in early spring 2019 or 2020, to assess their suitability for selection to a regional U16s or U18s rugby academy. Prior to these talent camps, players were advised to rest. During the talent camp, players completed a range of physiological and psychological assessments in a station-format which players rotated around until all tests were completed, followed by rugby matches. The selection decisions were made by regional coaches and based on subjective perceptions of performance during matches held on the talent days. For the purpose of this investigation, players were assessed on demographics, anthropometric, performance, and psychosocial measures (with the former 3 comprising 21 'physiological' variables and the latter 47 'psychosocial' variables) to identify differential features between those who were selected and not selected for the regional academy.

Physiological demographic measures included self-reported weekly physical activity hours (assessed in 5-h increments, starting at 1–5 and going up to 30 h+), self-reported weekly training frequency with the academy before the talent camp (to the nearest integer), self-reported incidence of a significant injury during their career (assessed as 'yes' or 'no'), and birth quarter (determined via birthday as: quartile 1 = September 1st to November 30th; quartile 2 = December 1st to February 28th/29th; quartile 3 = March 1st to May 31st; quartile 4 = June 1st to August 31st). For physiological anthropometric measures, players removed all heavy garments and footwear prior to recording measurements. Players' body mass (kg) was measured using electronic column scales (Seca 799, GmbH, Hamburg, Germany). Standing height and sitting height (cm) were measured using a portable stadiometer (HR001, Tanita Europe BV, Amsterdam, The Netherlands) and leg length was calculated as the difference between standing and sitting height (cm). Body Mass Index (BMI) was calculated as weight divided by height (in metres) squared. The Reciprocal Ponderal Index (RPI), also known as Sheldon's index [48], was calculated using the following equation: height (cm)/weight 0.333 (kg). Before measurement of physical performance measures, all participants completed a standardised (in terms of time and intensity) warm-up administered by regional strength and conditioning coaches and were briefed on how to execute each assessment. The counter movement jump was performed on a jump mat (JustJump, Probiotics Inc, Huntsville, AL, USA) indoors while wearing trainers, to assess jump height (cm) and peak anaerobic lower body power (W) using the Sayers Equation [49]; hands were positioned on the hips and the best jump height from three trials was recorded [50]. A hand grip strength test (Takei 5001 Grip-A Handgrip Dynamometer, Takei Scientific Instruments Co, Nigata, Japan) was used to infer strength (kg) within the dominant and non-dominant arm; participants stood with their back against a wall with their testing arm at 10°–15°

from the shoulder and elbow flexed at 90° with the highest score from two attempts (per arm) recorded [51]. Time (s) taken to sprint 10 and 40 m was recorded using timing gates (Brower Timing Systems, Draper, USA) on a 3G artificial grass pitch while wearing rugby shoes with studs; each sprint distance was completed twice with a 2-min rest between each repetition, with the fastest time recorded for each player. For the 40 m sprint: velocity was calculated as 40 m divided by the time taken to complete the 40 m; acceleration was calculated as velocity divided by the time taken to complete the 40 m sprint; force was calculated as acceleration multiplied by weight (kg); momentum was calculated as the velocity multiplied by the player's weight (kg); and average power was calculated using the Harman Formula [52].

The psychosocial questionnaires were administered in two questionnaires' packs to players during the 1-day 'talent camp'. Players were informed that their responses would not affect their selection. The first questionnaire pack gathered training behaviours (e.g., goal orientation, commitment, athlete identity). Players were also asked to report the number of hours of employed work they completed every week. The second questionnaire pack examined competitive experiences and personality traits (e.g., optimism, perfectionism, alexithymia). Questionnaires were chosen based on previous research which has identified these psychological constructs as important for athlete development [38]. In order to include several components and to circumnavigate issues with excessive questionnaire length, two items per construct were included. For complete information on the psychosocial variables collected, original sources, and items used, see Appendix A.

2.3. Data Analysis

To evaluate which features (i.e., predictor variables) best classified (i.e., determined) group membership (selected vs. non-selected), Bayesian pattern recognition was performed; a complete list of the features evaluated by the pattern recognition analysis (21 physiological and 47 psychosocial variables) can be found in Table 1. To explore the relative importance of factors within their topic-domains and player-positional categories and reduce the likelihood of machine learning overly reducing the features considered within datasets, we split analyses by positions (i.e., backs and forwards) and domains (i.e., physiological and psychosocial). Pattern recognition was performed using the open-source programming language R (R Core Development, 2021). Within this coding environment, the Tidyverse package [53] was used to perform advanced data manipulation, and the rWeka package [54]) was used to interface R with WEKA machine learning algorithms [55]. Analysis comprised three stages: first, features were standardised as part of data pre-processing; second, feature selection was performed to filter the dataset to a combination of its most predictive features, thus creating 'models' of features that best at differentiated group classification; and third, the classification accuracy of the created models was tested to evaluate how well the created models should predict group membership in future.

Table 1. Pattern recognition variable list.

Physiological Variables				
Weekly Physical Activity Hours	Weekly Academy Training Frequency	Injury Occurrence During Career	Birth Quarter	Height
Weight	Sitting Height	Leg Length	Reciprocal Ponderal Index	BMI
Counter Movement Jump	Dominant Hand Grip Strength	Non-Dominant Hand Grip Strength	10 m Sprint Time	40 m Sprint Time
40 m Sprint Momentum	40 m Sprint Velocity	40 m Sprint Acceleration	40 m Sprint Force	40 m Sprint Power
Peak Anaerobic Lower Body Power				

Table 1. *Cont.*

Psychosocial Variables				
Weekly Hours of Employed Work	Goal Orientation	Outcome Focus	Mastery Focus	Commitment to Training
Burnout	Exhaustion	Reduced Sense of Accomplishment	Sport Devaluation	Life Stress
Training Stress	Athlete Identity	Optimism	Difficulty Describing Feelings	Difficulty Identifying Feelings
Externally Orientated Feelings	Perfectionism	Perfectionistic Concerns	Perfectionistic Strivings	Self-Esteem
Extraversion	Agreeableness	Conscientiousness	Emotional Stability	Openness to New Experiences
Motivation	Amotivation	External Regulation	Introjected Regulation	Identified Regulation
Integrated Regulation	Intrinsic Motivation General	Resilience	Emotional Intelligence	Appraisal of own Emotions
Appraisal of Others' Emotions	Regulation of own Emotions	Regulation of Others' Emotions	Utilisation of Emotions	Coping Strategies
Coping with Adversity	Peaking under Pressure	Goal Setting and Mental Preparation	Concentration	Freedom from Worry
Confidence and Achievement Motivation	Coachability			

Note. A total of 21 physiological variables (comprising demographic, anthropometric, and performance measures) and 47 psychosocial variables were entered into the feature selection stage.

2.4. Pre-Processing

For all analyses, the data of U16 and U18 players were standardised and amalgamated. The raw U16 data were transformed into z-scores using the U16 means and standard deviations, and the raw U18 data was transformed into z-scores using the U18 means and standard deviations. Therefore, when the z-scored U16 and U18 data were amalgamated, z-scores indicated how much greater/less athletes scored on a feature (i.e., predictor variable) compared to their age-group peers. For data processing purposes, each z-scored feature was converted into a vector that went from 0 to 100 (with a player's score of 50 representing a score equivalent to the age-group mean and a score of 60 represented 1SD above the age-group mean, etc.). The purpose of amalgamating the data of U16 and U18 players was to: construct/evaluate classification models with greater accuracy via a larger dataset; identify features which determine overall 'age-grade' rugby union selection; and aid interpretation because similar features and model classification accuracies emerged for U16 and U18 players when analysed separately.

2.5. Feature Selection for Model Creation

Feature selection involved the use of correlation attribute evaluator [56], relief F attribute evaluator [57], gain ratio attribute evaluator [58], and info gain attribute evaluator [58], to identify (up to) 15 of the strongest features for determining group membership (i.e., selected vs. non-selected). Only features which were identified as being in the top 15 (this criterion was set arbitrarily based on the number of variables collected, prior to any data analysis) by at least two feature selection algorithms could become part of a 'model' and proceed in the analysis (some of the feature selection algorithms used can return less than 15 features if they are deemed as insufficiently predictive [56–58]). The resulting models were the combination of features within the dataset that best predicted group classification. Feature selection was performed a total of 6 times to create 6 models for 3 position conditions (all players, forwards, and backs) × 2 feature subsets (physiology features and psychosocial features).

2.6. Model Classification Accuracy

Each of the six models created by feature selection stage had its classification accuracy tested (i.e., how accurate, in percentage terms, a model is in predicting group membership) via the use of Naïve Bayes [59], J48 decision tree [60], Support Vector Machine [61], and K-nearest neighbour [62] classification algorithms. These algorithms assigned each player with an expected group membership (selected or non-selected) based on their score on features within the model. This process was iterated using a 'leave one out' cross-validation procedure wherein classification algorithms were performed repeatedly but with each of players' data left out once [45]. Thus, the final classification accuracy reported was the average percentage accuracy across each iteration. This 'leave one out' cross validation procedure was chosen over a training/validation sample-split to create the most accurate-possible models (i.e., via an as-large-as-possible dataset during feature selection), whilst still minimizing the overfitting of results to the specific dataset and preserving generalizability (i.e., via the conservative nature of the 'leave one out' method) [45].

3. Results

Table 2 contains the models created by feature selection and their overall classification accuracy. All models comprised between three and six features; naturally, less features were agreed on by more feature selection algorithms. Classification accuracy of the models ranged between 60 and 72% and was better than chance.

Table 2. Algorithm agreement among feature selection models and classification.

	Models					
	All Players		Forwards		Backs	
Number of feature selection algorithms in agreement	Physiological Features	Psychosocial Features	Physiological Features	Psychosocial Features	Physiological Features	Psychosocial Features
4		- Reduced sense of accomplishment.		- Life stress.		
3	- Power over 40 m.		- Force over 40 m; - Power over 40 m.		- Momentum;	
2	- 10 m sprint; - 40 m sprint; - Dominant hand grip strength; - Non-dominant hand grip strength.	- Burnout; - Exhaustion; - Introjected regulation.	- 40 m sprint; - Reciprocal Ponderal Index - Non-dominant hand grip strength; - 10 m sprint.	- Reduced sense of accomplishment; - Difficulty describing feelings;	- Birth quarter; - 40 m sprint; - Leg length; - Dominant hand grip strength.	- Reduced sense of accomplishment; - Introjected regulation; - Burnout
Classification accuracy	67.55%	62.26%	70.09%	73.66%	62.5%	60.42%

Note. The columns in this table contain the features within each group's model from feature selection. A greater number of algorithms agreeing that a feature is of great predictive validity, gives increased confidence in this feature belonging in the model. Classification accuracy is the mean accuracy (in percent) across all model levels (2, 3, and 4) and classification algorithms.

Table 3 displays the classification analysis of the physiological feature model created from all players. Overall, this model was able to correctly classify players 67.55% of the time. Specifically, selected players were faster in 10 m sprint (1.81 ± 0.13 vs. 1.83 ± 0.10 s), had greater power over 40 m (748.96 ± 131.91 vs. 600.72 ± 137.88 W), greater dominant (43.52 ± 6.54 vs. 39.78 ± 7.96 kg) and non-dominant hand grip strength (40.23 ± 7.07 vs. 36.97 ± 7.80 kg) than non-selected players.

Table 3. Classification breakdown of all players' physiological model.

Classifier	Classification Accuracy (%)	Sensitivity	Specificity	Area under ROC Curve
Naïve Bayes	67.31	0.45	0.80	0.63
Support Vector Machine	63.46	0	1	0.5
K Nearest Neighbour	69.23	0.45	0.83	0.68
J48 Decision Tree	70.19	0.34	0.91	0.43
Mean	67.55	0.31	0.89	0.56

Note. Sensitivity = 1—false positive. Specificity = 1—false negative. ROC = Receiver operating characteristic.

Table 4 displays the classification analysis of the psychosocial feature model created from all players. Overall, this model was able to correctly classify players 62.26% of the time. Specifically, selected players had lower reduced sense of accomplishment (10.20 ± 2.61 vs. 11.08 ± 2.19 questionnaire score), lower burnout (27.12 ± 5.79 vs. 29.37 ± 5.99 questionnaire score), lower exhaustion (9.15 ± 2.69 vs. 10.24 ± 3.55 questionnaire score), and lower introjected regulation' (4.12 ± 2.53 vs. 4.82 ± 3.14 questionnaire score) than non-selected players.

Table 4. Classification breakdown of all players' psychosocial model.

Classifier	Classification Accuracy (%)	Sensitivity	Specificity	Area under ROC Curve
Naïve Bayes	64.42	0.37	0.80	0.61
Support Vector Machine	63.46	0	1	0.5
K Nearest Neighbour	59.61	0.24	0.80	0.54
J48 Decision Tree	61.53	0.50	0.68	0.57
Mean	62.26	0.28	0.82	0.56

Note. Sensitivity = 1—false positive. Specificity = 1—false negative. ROC = Receiver operating characteristic.

Table 5 displays the classification analysis of the physiological feature model created from forwards. Overall, this model was able to correctly classify players 70.09% of the time. Specifically, compared to non-selected players, selected players had faster 10 m (1.85 ± 0.14 vs. 1.86 ± 0.10 s) and 40 m sprint times (5.74 ± 0.32 vs. 5.90 ± 0.36 s), expressed greater force (106.96 ± 14.65 vs. 83.60 ± 10.82 N) and power over 40 m sprints (747.92 ± 112.56 vs. 570.04 ± 89.89 W), had lower reciprocal ponderal index (40.63 ± 2.03 vs. 41.78 ± 1.79 cm kg 0.333) and greater non-dominant hand grip strength (40.18 ± 7.24 vs. 36.26 ± 7.57 kg).

Table 5. Classification breakdown of the forwards' physiological model.

Classifier	Classification Accuracy (%)	Sensitivity	Specificity	Area under ROC Curve
Naïve Bayes	69.64	0.53	0.78	0.74
Support Vector Machine	76.79	0.53	0.89	0.71
K Nearest Neighbour	73.21	0.53	0.84	0.73
J48 Decision Tree	60.71	0.58	0.62	0.52
Mean	70.09	0.54	0.78	0.68

Note. Sensitivity = 1—false positive. Specificity = 1—false negative. ROC = Receiver operating characteristic.

Table 6 displays the classification analysis of the psychosocial feature model created from forwards. Overall, this model was able to correctly classify players 73.66% of the time. Specifically, selected players had lower life stress (6.97 ± 2.22 vs. 8.89 ± 2.85 questionnaire score), lower reduced sense of accomplishment (10.05 ± 2.73 vs. 11.21 ± 2.74 questionnaire score) and lower difficulty describing feelings (3.84 ± 1.64 vs. 5.05 ± 1.93 questionnaire score) than non-selected players.

Table 6. Classification breakdown of the forwards' psychosocial model.

Classifier	Classification Accuracy (%)	Sensitivity	Specificity	Area under ROC Curve
Naïve Bayes	75.00	0.53	0.86	0.67
Support Vector Machine	62.29	0	0.97	0.49
K Nearest Neighbour	73.21	0.47	0.86	0.65
J48 Decision Tree	82.14	0.53	0.97	0.51
Mean	73.16	0.38	0.92	0.58

Note. Sensitivity = 1—false positive. Specificity = 1—false negative. ROC = Receiver operating characteristic.

Table 7 displays the classification analysis of the physiological feature model created from backs. Overall, this model was able to correctly classify players 62.5% of the time. Specifically, selected players had faster 40 m sprint times (5.40 ± 0.26 vs. 5.61 ± 0.41 s), greater momentum in the 40 m sprints (540.40 ± 76.31 vs. 481.84 ± 82.26 $kg \cdot m \cdot s^{-1}$), were born in an earlier birth quarter (2.10 ± 1.01 vs. 2.68 ± 1.20) and had greater dominant hand grip strength (43.05 ± 5.99 vs. 40.84 ± 9.13 kg) than non-selected players.

Table 7. Classification breakdown of the backs' physiological model.

Classifier	Classification Accuracy (%)	Sensitivity	Specificity	Area under ROC Curve
Naïve Bayes	68.75	0.47	0.83	0.61
Support Vector Machine	54.17	0	0.90	0.45
K Nearest Neighbour	58.33	0.26	0.79	0.59
J48 Decision Tree	68.75	0.26	0.97	0.26
Mean	62.50	0.24	0.87	0.48

Note. Sensitivity = 1—false positive. Specificity = 1—false negative. ROC = Receiver operating characteristic.

Table 8 displays the classification analysis of the psychosocial feature model created from backs. Overall, this model was able to correctly classify players 60.42% of the time. Specifically, selected players had lower introjected regulation (4.10 ± 2.60 vs. 5.26 ± 3.19 questionnaire score), lower reduced sense of accomplishment (10.38 ± 2.50 vs. 10.95 ± 1.51

questionnaire score) and lower burnout (27.93 ± 5.66 vs. 30.16 ± 6.69 questionnaire score) than non-selected players.

Table 8. Classification breakdown of the backs' psychosocial model.

Classifier	Classification Accuracy (%)	Sensitivity	Specificity	Area under ROC Curve
Naïve Bayes	62.50	0.26	0.86	0.63
Support Vector Machine	58.33	0	0.97	0.48
K Nearest Neighbour	66.67	0.37	0.86	0.69
J48 Decision Tree	54.17	0	0.90	0
Mean	60.42	0.16	0.90	0.45

Note. Sensitivity = 1—false positive. Specificity = 1—false negative. ROC = Receiver operating characteristic.

4. Discussion

The aim of this study was to measure primary physiological and psychosocial factors in age-grade Rugby Union players and to utilize a novel Bayesian pattern recognition technique to identify which attributes differentiate between selection and non-selection to regional U16 and U18 performance pathways. The main findings of this investigation suggested that the generated physiological models correctly classified 67.55% of all players, 70.09% of forwards, and 62.50% of backs. Greater hand-grip strength, faster 10 m and 40 m sprint, and power were common features for selection. The generated psychosocial models correctly classified 62.26% of all players, 73.66% of forwards, and 60.42% of backs. Reduced burnout and emotional exhaustion, and lower reduced sense of accomplishment, were common features for selection. Selection appears to be predominantly based on greater strength, speed, and power, as well as lower athlete burnout. Of note, the greater specificity and lower sensitivity across all analyses suggests that non-selected players were easier for the algorithms to identify. This finding is logical when one considers that players who should not be selected likely stand out more (e.g., particularly slow/weak) compared to players who should be selected (i.e., where the margins may be finer). The present investigation offers an arguably more comprehensive test of factors than previous studies into age-grade rugby selection (e.g., [21]); is the first attempt to objectively understand the currently subjective decision-making that determines selection to regional age-grade academies in Wales; and tests the role of psychosocial and physiological attributes via new and cutting-edge analytical methods.

The findings of this investigation provide a unique insight into differences in psychosocial components between selected and non-selected players. The results suggest that selected players (generally across positions) reported lower levels of overall burnout and specifically lower exhaustion and lower reduced sense of accomplishment compared to non-selected players. Consistent with previous research in Rugby Union [32,33], these results suggest that burnout is a prominent factor in the sport. Interestingly, the present pattern recognition analysis did not support previously proposed theoretical explanations of burnout, exhaustion, and reduced sense of accomplishment, such as perfectionism and coping [32,63]; it is possible that the mechanisms producing these outcomes were too individualized within the present sample to be identified at the feature selection stage. Regardless of the precise mechanisms leading to burnout, results highlight the need for coaches to consider how it could ultimately derail athlete progression within talent selection and development [10].

The psychosocial results also reveal differences across forwards and backs. Forwards report lower life stress, which is logical when viewed in line with results on burnout [64]. The selected forwards also reported lower scores in difficulties describing feelings, which is a component of the personality trait Alexithymia [65]. Those high in Alexithymia often are unable to express and recognise their emotions leading to difficulties in regulating emotions and difficulties with interpersonal relations [65]. Alexithymia is relatively under-researched in athletes, but some research has linked those high in Alexithymia with risk-taking [66], and endurance sports [67]. From a forward's perspective, it could be argued that the lower scores related to difficulty describing emotions are indicative of greater emotional regulation, and their ability to resolve negative emotions that arose from stressful aspects of life [67] and physical demands of playing forward. This is again logical when considered with the reports of lower life stress and burnout. Future research should attempt to tease the present findings apart by further investigating the coping strategies that might differ across positions.

Whilst research has shown positional differences from a relative age effect [68] and physiological perspective [69], the present investigation is, to the authors' knowledge, the first to reveal positional differences in psychosocial components of selected vs. non-selected rugby players. Research by Dimundo et al. [21] included one measure of cognitive skills in rugby union players but found no significant differences across players and went on to call for future research to include psychosocial characteristics in talent identification/selection methodologies. Adopting a battery of psychological tests comprising fewer items in the present investigation provided an opportunity to assess players on a wide variety of relevant psychosocial components. Recent applied research [70,71] has also adopted this method of utilizing fewer items per construct to facilitate both a broad assessment and to encourage athlete engagement. Whilst we would recommend that any psychosocial investigations such as this are followed up on a more detailed basis between a sport psychologist and coaches, the method adopted here does facilitate a broad understanding of psychosocial component relevant to talent identification, selection, and development.

Physiological models correctly classified selected players in the range between 62.50% and 70.09% and were stronger predictors of selection than psychosocial models, which has been alluded to previously [21]. In addition, the common features for selection within our models are generally in agreement with previous research examining differences in physical and performance measures between selected and non-selected players. For example, in the present investigation, greater hand grip strength was a performance feature important for selection across all players and within positional categories. Others have confirmed that greater strength in general [72,73] and handgrip strength specifically [19,21,74] is a characteristic of selection to rugby performance pathways and distinguishes between standard of play in age-grade players [75].

Sprinting speed is an important physical quality in Rugby Union and is associated with many performance parameters such as evading opponents, line and tackle breaks and has been shown to distinguish between selected and non-selected age-grade players [18,31,73,75]. In the present investigation, selected players recorded faster sprint times over 10 m (all players and forwards) and 40 m (forwards and backs). Indeed, 10 m sprinting speed was one of the features within the model that correctly classified 67.55% of all players, and coaches and sport scientists should ensure the inclusion of these assessments into talent selection programmes.

Previous research has consistently shown that selection for Rugby Union performance pathways across U15–U21 age grades is biased towards taller [73,75] and heavier players [21,75]. This may explain the well-established selection bias towards relatively older players [31,76–78] and to some degree early maturing players [79] (i.e., the relative age effect). Notably, in the present investigation, stature and body mass were not common features of selection for all players regardless of positional category. Although earlier birth quartile (Q1 and Q2) was part of the model for selected backs, this was not a feature for selection in the physiological models covering all players and forwards and may partly explain these findings. Despite its absence as a direct feature for selection, body weight did appear to be an important factor when expressed as momentum (backs), force (forwards) and power during 40 m sprinting (all players and forwards). Further suggestion of the importance of body shape and size was evidenced via a lower Reciprocal Ponderal Index (RPI) as an important feature for selection for forward players. The RPI is an index of adiposity calculated as the relationship between standing height divided by the cube root of body weight and based on allometric modelling has a stronger mathematical foundation than BMI, as weight is a variable of cubic dimensions [48]. RPI has been associated with performance in sports such as soccer and tennis [80,81]. The lower RPI found in selected forwards in the present investigation would infer that greater body mass rather than a more linear (ectomorphic) body shape is an important factor in terms of selection within this positional category.

The methods used to derive the aforementioned psychosocial and physiological findings feature several strengths. Firstly, the present investigation was the first to directly assess the role of primary player-derived psychosocial attributes on talent selection in Rugby Union. Secondly, the novel pattern recognition analysis performed on physiological features revealed similar predictive features for talent selection in Rugby Union to previous correlational studies [31] which, importantly, gives confidence to the psychosocial features identified as predictive for the first time. Lastly, a rigorous and conservative 'leave one out' cross-validation classification procedure was used. This classification procedure facilitates more accurate feature models (i.e., via an as-large-as-possible dataset during feature selection) whilst minimizing the overfitting of results to the specific dataset (i.e., by testing classification accuracy on the entire sample, instead of a small validation-specific sample) [45,46]. The newfound knowledge from the present investigation can be used by coaches, managers, parents, and guardians in making sure youth Rugby Union players are adequately developed and supported for future success. Coaches may wish to prioritize the physiological development of relatively stronger and faster players, while parents and guardians may wish to monitor for signs and causes related to burnout and exhaustion. Such provision should position Rugby Union players optimally for selection by regional age-grade academies.

It is important to note however, some of the present investigation's limitations. Classification accuracies (60–74%) were less than those of studies utilizing similar machine learning approaches in other sporting domains [82,83]. However, this result can be expected for two reasons. The regional academy's subjective/intuition-based selection criteria likely introduce inevitable statistical 'noise', and the present investigation's conservative 'leave one out' cross-validation classification procedure likely resulted in lower classification accuracies. One method to increase classification accuracy despite this, could be the use of even more comprehensive test batteries (e.g., via evaluations of practice histories, technical ability, tactical ability, and performance history). For example, evidence to suggest that the features collected in our study do not capture the role of tactical/technical attributes in determining selection, can be seen in the backs' generally lower classification accuracy (~60%) compared to forwards' (~70). For backs in particular, tactical and technical skill may be a particularly important trait when academies evaluate players. Future studies are encouraged to collect ratings of players' tactical and technical ability from independent coaches, alongside developmental variables such as practice histories, which have demonstrated themselves as important factors in determining future success [83]. Additionally, subsequent investigations may wish to also evaluate the interactive role of aerobic fitness, a variable that was not possible to assess in the present investigation due to time constraints on the talent camp day but has previously demonstrated an ability to differentiate between selected and non-selected rugby union players [31].

5. Conclusions

This is the first study that has utilized a machine learning approach to examine the factors that determine selection to a regional age-grade Rugby Union academy in Wales. The present investigation offers an arguably more comprehensive analysis of factors than previous studies in this population and informs an objective understanding of the current subjective decision-making that determines selection to regional age-grade academies in Wales. From these findings, it appears that physiological factors are more predictive of selection. Specifically, the findings of this present investigation suggest that greater strength, speed, and power during sprint running were important factors for selection and should be included as routine assessments in talent selection for regional academies. Nevertheless, psychosocial factors were also shown to be important with reduced burnout and emotional exhaustion, and lower reduced sense of accomplishment, common features for selection. Indeed, this is the first study to comprehensively examine psychosocial factors important for selection to rugby academies and the findings add weight to the argument that these factors should be considered as part of a holistic selection framework in Rugby Union. Furthermore, practitioners should also consider position-specific differences in factors important for selection when planning talent selection frameworks. Future studies are encouraged to adopt a holistic approach to talent selection through investigating a comprehensive combination of physiological and psychosocial factors alongside tactical and technical ratings and developmental variables such as practice histories.

Author Contributions: Conceptualization, J.O., E.J., J.L. and J.H.; methodology, J.O., E.J., J.L. and J.H.; investigation, J.H. and J.L.; software, D.A.; formal analysis, R.O. and D.A.; writing—original draft preparation, R.O.; writing—review and editing, R.O., J.O., E.J., D.A.; funding acquisition, J.O. and E.J. All authors have read and agreed to the published version of the manuscript.

Funding: J.H. was supported by a European Social Fund, Knowledge Exchange Scholarship Scheme (KESS II) Masters by Research grant in collaboration with the Welsh Rugby Union.

Institutional Review Board Statement: The study was conducted in accordance with the Declaration of Helsinki and approved by the Institutional Review Board of Bangor University's School of Human and Behavioural Sciences (Ethics Protocol Code: S02-18/19, ehical approval date being 24 April 2019).

Informed Consent Statement: Informed consent was obtained from all subjects involved in the study.

Data Availability Statement: Data can be obtained from the authors.

Acknowledgments: The authors would like to thank Kevin Williams and Jason Edwards for their technical assistance during this project and the staff and players of Rygbi Gogledd Cymru (RGC) for their time and commitment during data collection.

Conflicts of Interest: The authors declare no conflict of interest.

Appendix A

Table A1. Summary of measures used in psychological questionnaire packs 1 and 2.

Measure and Item Origin	Subscale	Items from Original Construct	Factor Loading	Author
		TRAINING BEHAVIOURS		
Perception of Success (Roberts, Treasure, and Balague, 1998)	Outcome Focus	1. When doing sport, I feel successful when I beat other people.	0.66	Items taken from the ADFS (Dunn et al., 2019)
		2. When doing sport, I feel successful when I outperform my opponents.	0.62	
	Mastery Focus	1. When doing sport, I feel successful when I perform to the best of my ability.	0.62	
		2. When doing sport, I feel successful when I show clear personal improvements.	0.72	

Table A1. *Cont.*

Measure and Item Origin	Subscale	Items from Original Construct	Factor Loading	Author
		TRAINING BEHAVIOURS		
Quality of Training Inventory (Woodman et al., 2010)	Commitment to Training	1. I always produce a high-quality training session. 2. No matter what is going on in my life, I still turn in a good training session.		Items taken from the ADFS (Dunn et al., 2019)
Inclusion of Others in the Self Scale (Aron, Aron, and Smollan, 1992)	Athlete Identity	1. My sport is the most important thing in my life. 2. My sport offers me more than anything else in life (e.g., friends, family, relationships, money).		Items taken from the ADFS (Dunn et al., 2019)
Behavioural Regulation in Sport (Lonsdale, Hodge, and Rose, 2008)	Amotivation	1. but I question why I continue. 2. but the reason why are not clear to me anymore	0.90 0.89	Items taken from the BRSQ-6 (Lonsdale, Hodge, and Rose, 2008)
	External Regulation	1. because people push me to play 2. because I feel pressure from other people to play	0.85 0.84	
	Introjected Regulation	1. because I would feel guilty of I quit 2. because I fee; obligated to continue	0.78 0.88	
	Identified Regulation	1. because the benefits of sport are important to me 2. because it teaches me self-discipline	0.80 0.57	
	Integrated Regulation	1. because it's an opportunity to just be who I am 2. because what I do in sport is an expression of who I am	0.70 0.77	
	IM-General	1 because I enjoy it 2. because I like it	0.82 0.81	
Performance-based Self-Esteem (Hallsten, Josephson, and Torgén, 2005)	Self-Esteem	1. I think that I can sometimes try to prove my worth by being competent. 2. My self-esteem, is far too dependent on my daily achievements. 3. At times, I have to be better than others to be good enough myself. 4. Occasionally I feel obsessed to accomplish something of value.	Range from 0.70 to 0.84	Items taken from the Pbse-scale (Hallsten, Josephson, and Torgén, 2005)
Athlete Coping Skills Inventory-28 (Smith, et al., 1995)	Coping with Adversity	1. I maintain emotional control no matter how things are going for me. 2. When things are going badly, I tell myself to keep calm, and this works for me.	0.60 0.58	Items taken from the ACSI-28 (Smith, et al., 1995)

Table A1. *Cont.*

Measure and Item Origin	Subscale	Items from Original Construct	Factor Loading	Author
		TRAINING BEHAVIOURS		
	Performing Under Pressure	1. To me, pressure situations are challenges that I welcome.	0.77	
		2. The more pressure there is during a game, the more I enjoy it	0.71	
	Goal Setting/Mental Preparation	1. On a daily or weekly basis, I set very specific goals for myself that guide what I do.	0.69	
		2. I tend to do lots of planning about how to reach my goals.	0.68	
	Concentration	1. I handle unexpected situations in my sport very well.	0.63	
		2. When I am playing sports, I can focus my attention and block out distractions	0.68	
	Free from Worry	1. While competing, I worry about making mistakes or failing to come through (**).	0.76	
		2. I put a lot of pressure on myself by worrying how I will perform (**).	0.66	
	Confidence and Achievement Motivation	1. I feel confident that I will play well.	0.65	
		2. I get the most out of my talent and skills.	0.62	
	Coachability	1. If a coach criticizes or yells at me, I correct the mistake without getting upset about it.	0.77	
		2. I improve my skills by listening carefully to advice and instruction from coaches and manager	0.57	
Measure and Item Origin	**Subscale**	**Items from Original Construct**	**Factor Loading**	**Author**
		PERSONALITY TRAITS		
The Multidimensional Inventory of Perfectionism in Sport (Stoeber et al., 2006)	Perfectionistic Concerns	1. During training, I get completely furious if I make mistakes. 2. During training, I get frustrated if I do not fulfil my high expectations. 3. During competition, I get completely furious if I make mistakes. 4. During competition, I get frustrated if I do not fulfil my high expectations.	Range from 0.86 to 0.91	Items taken from the ADFS (Dunn et al., 2019)

Table A1. *Cont.*

Measure and Item Origin	Subscale	Items from Original Construct	Factor Loading	Author
		PERSONALITY TRAITS		
The Sport Multidimensional Perfectionism Scale 2 (Gotwals and Dunn, 2009)	Perfectionistic Strivings	1. I feel that other athletes generally accept lower standards for themselves in sport than I do. 2. I have extremely high goals for myself in sport.	0.63 0.53	Items taken from the ADFS (Dunn et al., 2019)
Big Five-Inventory-10 (Gosling, Rentfrow, and Swann, 2003)	Extraversion	1. I see myself as: extraverted, enthusiastic. 2. I see myself as: reserved, quiet.	0.77	Items taken from the ADFS (Dunn et al., 2019)
	Agreeableness	1. I see myself as critical, quarrelsome. 2. I see myself as: sympathetic, warm	0.71	
	Conscientiousness	1. I see myself as: dependable, self-disciplined. 2. I see myself as: disorganised, careless	0.76	
	Emotional Stability	1. I see myself as: anxious, easily upset. 2. I see myself as: calm, emotionally stable.	0.70	
	Openness to Experiences	1. I see myself as: open to new experiences, complex. 2. I see myself as: conventional, uncreative.	0.62	
Life Orientation Test, (Scheier, and Carver, 1985)	Optimism	1. In uncertain times, I usually expect the best. 2. I always look on the bright side of things. 3. I'm always optimistic about my future. 4. I'm a believer in the idea that "every cloud has a silver lining".	0.56 0.72 0.61 0.66	Items taken from the LOT (Scheier, and Carver, 1985)
The Brief Emotional Intelligence Scale (Davies, et al., 2010)	Appraisal of own emotions	1. I know why my emotions change. 2. I easily recognise my emotions as I experience them.	0.77 0.62	Items taken from the BEIS-10 (Davies, et al., 2010)
	Appraisal of others; emotions	1. I can tell how people are feeling by listening to the tone of their voice. 2. By looking at their facial expressions, I recognise the emotions people are experiencing.	0.72 0.65	
	Regulation of own emotions	1. I seek out activities that make me happy 2. I have control over my emotions	0.71 0.83	
	Regulations of others' emotions	1. I arrange events others enjoy. 2. I help other people feel better when they are down	0.91 0.68	

Table A1. *Cont.*

Measure and Item Origin	Subscale	Items from Original Construct	Factor Loading	Author
		PERSONALITY TRAITS		
	Utilisation of emotions	1. When I am in a positive mood, I am able to come up with new ideas.	0.65	
		2. I use good moods to help myself keep trying in the face of obstacles	0.68	
Toronto Alexithymia Scale—20 (Bagby, Parker, and Taylor, 1994)	Difficulty Identifying Feelings	1. I have feelings that I cannot quite identify	0.77	Items taken from the TAS-20 (Bagby, Parker, and Taylor, 1994)
		2. I do not know what is going on inside me	0.66	
	Difficulty Describing Feelings	1. It is difficult for me to find the right words for my feelings.	0.70	
		2. I find it hard to describe how I feel about people.	0.54	
	Externally Orientated Feelings	1. Being in touch with emotions is essential (**).	0.47	
		2. I find examination of my feelings useful in solving personal problems (**).	0.62	
Measure and Item Origin	**Subscale**	**Items from Original Construct**	**Factor Loading**	**Author**
		PSYCHOLOGICAL FACTORS		
Athlete Burnout Measure (Raedeke, and Smith, 2001)	Emotional Exhaustion	1. I feel so tired from my training that I have trouble finding energy to do other things.	0.66	Items taken from the ABQ (Raedeke, and Smith, 2001)
		2. I feel overly tired from my [sport] participation.	0.69	
		3. I feel "wiped out" from [sport].	0.70	
		4. I feel physically worn out from [sport].	0.63	
		5. I am exhausted by the mental and physical demand of [sport].	0.70	
	Reduce Sense of Accomplishment	1. I'm accomplishing many worthwhile things in [sport].	0.67	
		2. I am not achieving much in [sport].	0.60	
		3. I am not performing up to my ability in [sport].	0.57	
		4. It seems that no matter what I do, I don't perform as well as I should.	0.78	
		5. I feel successful at [sport].	0.66	
	Sport Devaluation	1. The effort I spend in [sport] would be better spent doing other things.	0.63	
		2. I don't care as much about my [sport] performance as I used to.	0.50	
		3. I'm not into [sport] like I used to be	0.82	
		4. I feel less concerned about being successful in [sport] as I used to be.	0.66	
		5. I have negative feelings towards [sport].	0.65	

Table A1. *Cont.*

Measure and Item Origin	Subscale	Items from Original Construct	Factor Loading	Author
		PSYCHOLOGICAL FACTORS		
Perceived Stress Scale (Cohen, et al., 1983)	Global Stress and Training Stress	1. In the last week, how often have you felt that you were unable to control the important things in your life? 2. In the last week, how often have you felt confident about your ability to handle your personal problems? (**). 3. In the last week, how often have you felt that things were going your way? (**). 4. In the last week, how often have you felt difficulties were piling up so high that you could not overcome them?	Range from 0.82 to 0.86	Items taken from the PSS (Cohen, et al., 1983)

Key: ** = Reverse Score (i.e., 1 = 5, 2 = 4, 3 = 3, 4 = 2 and 5 = 1).

References

1. Williams, A.M.; Reilly, T. Talent identification and development in soccer. *J. Sports Sci.* **2000**, *18*, 657–667. [CrossRef]
2. Till, K.; Emmonds, S.; Jones, B. Talent identification. In *Strength & Conditioning for Youth Athletes. Science and Practical Applications*, 2nd ed.; Lloyd, R.S., Oliver, J., Eds.; Routledge: London, UK, 2019.
3. Baker, J.; Schorer, J.; Wattie, N. Compromising talent: Issues in identifying and selecting talent in sport. *Quest* **2018**, *70*, 48–63. [CrossRef]
4. Vaeyens, R.; Lenoir, M.; Williams, A.M.; Philippaerts, R.M. Talent identification and development programmes in sport. *Sports Med.* **2008**, *38*, 703–714. [CrossRef] [PubMed]
5. Durandt, J.; Parker, Z.; Masimla, H.; Lambert, M. Rugby-playing history at the national U13 level and subsequent participation at the national Under 16s and Under 18s rugby tournaments. *S. Afr. J. Sports Med.* **2011**, *23*, 103–105. [CrossRef]
6. Boccia, G.; Cardinale, M.; Brustio, P.R. Performance progression of elite jumpers: Early performances do not predict later success. *Scand. J. Med. Sci. Sports* **2021**, *31*, 132–139. [CrossRef] [PubMed]
7. Boccia, G.; Cardinale, M.; Brustio, P.R. World-class sprinters' careers: Early success does not guarantee success at adult age. *Int. J. Sports Physiol. Perform.* **2020**, *16*, 367–374. [CrossRef]
8. Brustio, P.R.; Cardinale, M.; Lupo, C.; Varalda, M.; De Pasquale, P.; Boccia, G. Being a top swimmer during the early career is not a prerequisite for success: A study on sprinter strokes. *J. Sci. Med. Sport* **2021**, *24*, 1272–1277. [CrossRef]
9. Christensen, M.K. "An eye for talent": Talent identification and the "practical sense" of top-level soccer coaches. *Sociol. Sport J.* **2009**, *26*, 365–382. [CrossRef]
10. Till, K.; Baker, J. Challenges and [Possible] Solutions to Optimizing Talent Identification and Development in Sport. *Front. Psychol.* **2020**, *11*, 664. [CrossRef]
11. Johansson, A.; Fahlén, J. Simply the best, better than all the rest? Validity issues in selections in elite sport. *Int. J. Sports Sci. Coach.* **2017**, *12*, 470–480. [CrossRef]
12. Reilly, T. The physiology of rugby union football. *Biol. Sport* **1997**, *14*, 83–101.
13. Nicholas, C.W. Anthropometric and physiological characteristics of rugby union football players. *Sports Med.* **1997**, *23*, 375–396. [CrossRef] [PubMed]
14. Cupples, B.B. Holistic Athlete Development in Australian Rugby League: From a Better Understanding of Developmental Environments to Addressing Individual Challenges. Ph.D. Thesis, University of Sydney, Sydney, Australia, 2020.
15. Darrall-Jones, J.D.; Jones, B.; Till, K. Anthropometric, sprint, and high-intensity running profiles of English academy rugby union players by position. *J. Strength Cond. Res.* **2016**, *30*, 1348–1358. [CrossRef] [PubMed]
16. Sedeaud, A.; Vidalin, H.; Tafflet, M.; Marc, A.; Toussaint, J.F. Rugby morphologies: "bigger and taller", reflects an early directional selection. *J. Sports Med. Phys. Fit.* **2013**, *53*, 185–191. Available online: https://hal-insep.archives-ouvertes.fr/hal-01778398 (accessed on 5 September 2021).
17. Patton, D.A.; McIntosh, A.S.; Denny, G. A review of the anthropometric characteristics, grading and dispensation of junior and youth rugby union players in Australia. *Sports Med.* **2016**, *46*, 1067–1081. [CrossRef]
18. Fontana, F.Y.; Colosio, A.L.; Da Lozzo, G.; Pogliaghi, S. Player's success prediction in rugby union: From youth performance to senior level placing. *J. Sci. Med. Sport* **2017**, *20*, 409–414. [CrossRef]

19. Vaz, L.; Batista, M.; Honório, S.; Fernandes, H. Physical performance tests and anthropometric data to predict selection in U19 rugby union players. *J. Hum. Sport Exerc.* **2019**, *14*, S1250–S1252. [CrossRef]
20. Chiwaridzo, M.; Ferguson, G.D.; Smits-Engelsman, B.C.M. Qualities or skills discriminating under 19 rugby players by playing standards: A comparative analysis of elite, sub-elite and non-rugby players using the SCRuM test battery. *BMC Res. Notes* **2019**, *12*, 536. [CrossRef]
21. Dimundo, F.; Cole, M.; Blagrove, R.C.; McAuley, A.B.T.; Till, K.; Kelly, A.L. Talent identification in an English premiership rugby union academy: Multidisciplinary characteristics of selected and non-selected male Under-15 players. *Front. Sports Act. Living* **2021**, *3*, 688143. [CrossRef]
22. Owen, C.; Till, K.; Weakley, J.; Jones, B. Testing methods and physical qualities of male age grade rugby union players: A systematic review. *PLoS ONE* **2020**, *15*, e0233796. [CrossRef]
23. Duthie, G.; Pyne, D.; Hooper, S. Applied physiology and game analysis of rugby union. *Sports Med.* **2003**, *33*, 973–991. [CrossRef] [PubMed]
24. Darrall-Jones, J.; Jones, B.L.; Till, K. Anthropometric and physical profiles of English academy rugby union players. *J. Strength Cond. Res.* **2015**, *29*, 2086–2096. [CrossRef] [PubMed]
25. James, N.; Mellalieu, S.D.; Jones, N.M.P. The development of position-specific performance indicators in professional rugby union. *J. Sports Sci.* **2005**, *23*, 63–72. [CrossRef] [PubMed]
26. Durandt, J.; Du Toit, S.; Borresen, J.; Hew-Butler, T.; Masimla, H.; Jokoet, I.; Lambert, M. Fitness and body composition profiling of elite junior South African rugby players. *S. Afr. J. Sports Med.* **2006**, *18*, 38–45. [CrossRef]
27. Roe, G.; Halkier, M.; Beggs, C.; Till, K.; Jones, B. The use of accelerometers to quantify collisions and running demands of rugby union match-play. *Int. J. Perform. Anal. Sport* **2016**, *16*, 590–601. [CrossRef]
28. Quarrie, K.L.; Hopkins, W.G.; Anthony, M.J.; Gill, N.D. Positional demands of international rugby union: Evaluation of player actions and movements. *J. Sci. Med. Sport* **2013**, *16*, 353–359. [CrossRef]
29. Ball, S.; Halaki, M.; Orr, R. Movement demands of rugby sevens in men and women: A systematic review and meta-analysis. *J. Strength Cond. Res.* **2019**, *33*, 3475–3490. [CrossRef]
30. Vaz, L.; Kraak, W.; Batista, M.; Honório, S.; Miguel Fernandes, H. Using anthropometric data and physical fitness scores to predict selection in a national U19 rugby union team. *Int. J. Environ. Res. Public Health* **2021**, *18*, 1499. [CrossRef]
31. Dimundo, F.; Cole, M.; Blagrove, R.; Till, K.; McAuley, A.; Hall, M.; Gale, C.; Kelly, A. Talent Identification and Development in Male Rugby Union: A Systematic Review. *J. Expert* **2021**, *4*, 33–55.
32. Hill, A.P.; Appleton, P.R. The predictive ability of the frequency of perfectionistic cognitions, self-oriented perfectionism, and socially prescribed perfectionism in relation to symptoms of burnout in youth rugby players. *J. Sports Sci.* **2011**, *29*, 695–703. [CrossRef]
33. Hill, A.; MacNamara, Á.; Collins, D. Psychobehaviorally based features of effective talent development in rugby union: A coach's perspective. *Sport Psychol.* **2015**, *29*, 201–212. [CrossRef]
34. Butt, D.S. (Ed.) *Personality of the Athlete*; The Psychology of Sport; VNR: New York, NY, USA, 1987; pp. 95–105.
35. MacNamara, A.; Collins, D. Profiling, exploiting, and countering psychological characteristics in talent identification and development. *Sport Psychol.* **2015**, *29*, 73–81. [CrossRef]
36. Allen, M.S.; Greenlees, I.; Jones, M. Personality in sport: A comprehensive review. *Int. Rev. Sport Exerc. Psychol.* **2013**, *6*, 184–208. [CrossRef]
37. Roberts, R.; Woodman, T. Contemporary personality perspectives in sport psychology. In *Contemporary Advances in Sport Psychology: A Review*; Mellalieu, S.D., Hanton, S., Eds.; Routledge/Taylor & Francis Group: Abingdon, UK, 2015; pp. 1–27.
38. Rees, T.; Hardy, L.; Güllich, A.; Abernethy, B.; Côté, J.; Woodman, T.; Warr, C. The great British medalists project: A review of current knowledge on the development of the world's best sporting talent. *Sports Med.* **2016**, *46*, 1041–1058. [CrossRef]
39. Steca, P.; Baretta, D.; Greco, A.; D'Addario, M.; Monzani, D. Associations between personality, sports participation, and athletic success. A comparison of Big Five in sporting and nonsporting adults. *Pers. Individ. Differ.* **2018**, *121*, 176–183. [CrossRef]
40. Kruger, A.; Plooy, K.D.; Kruger, P. Personality profiling of South African rugby union players. *J. Psychol. Afr.* **2019**, *29*, 383–387. [CrossRef]
41. Vallerand, R.J.; Mageau, G.A.; Elliot, A.J.; Dumais, A.; Demers, M.A.; Rousseau, F. Passion and performance attainment in sport. *Psychol. Sport Exerc.* **2008**, *9*, 373–392. [CrossRef]
42. Flett, G.L.; Hewitt, P.L. The perils of perfectionism in sports and exercise. *Curr. Dir. Psychol. Sci.* **2005**, *14*, 14–18. [CrossRef]
43. Bush, N.; Salmela, J.H. The development and maintenance of expert athletic performance: Perceptions of World and Olympic Champions. *J. Appl. Sport Psychol.* **2002**, *14*, 154–171. [CrossRef]
44. Güllich, A.; Hardy, L.; Kuncheva, L.; Woodman, T.; Laing, S.; Barlow, M.; Lynne, E.; Tim, R.; Abernethy, B.; Côté, J.; et al. Developmental biographies of Olympic super-elite and elite athletes: A multidisciplinary pattern recognition analysis. *J. Expert* **2019**, *2*, 23–46. Available online: https://www.journalofexpertise.org/articles/volume2_issue1/JoE_2019_2_1_Gullich.html (accessed on 5 September 2021).
45. Vehtari, A.; Gelman, A.; Gabry, J. Practical Bayesian model evaluation using leave-one-out cross-validation and WAIC. *Stat. Comput.* **2017**, *27*, 1413–1432. [CrossRef]

46. Kuncheva, L.I.; Whitaker, C.J. *Pattern Recognition and Classification*; Wiley: Hoboken, NJ, USA, 2014.
47. Stebbins, R.A. *Exploratory Research in the Social Sciences*; Sage: London, UK, 2008; Volume 48.
48. Smalley, K.J.; Knerr, A.N.; Kendrick, Z.V.; Colliver, J.A.; Owen, O.E. Reassessment of body mass indices. *Am. J. Clin. Nutr.* **1990**, *52*, 405–408. [CrossRef] [PubMed]
49. Sayers, S.; Harackiewicz, D.; Harman, E.; Frykman, P.; Rosenstein, M. Cross-validation of three jump power equations. *Med. Sci. Sports Exerc.* **1999**, *31*, 572–577. [CrossRef] [PubMed]
50. Markovic, G.; Dizdar, D.; Jukic, I.; Cardinale, M. Reliability and factorial validity of squat and countermovement jump tests. *J. Strength Cond. Res.* **2004**, *18*, 551–555.
51. Cronin, J.; Lawton, T.; Harris, N.; Kilding, A.; McMaster, D.T. A brief review of handgrip strength and sport performance. *J. Strength Cond. Res.* **2017**, *31*, 3187–3217. [CrossRef]
52. Harman, E.A.; Rosenstein, M.T.; Frykman, P.N.; Rosenstein, R.M.; Kraemer, W.J. Estimation of Human Power Output From Vertical Jump. *J. Appl. Sport Sci. Res.* **1991**, *5*, 116–120.
53. Wickham, H.; Averick, M.; Bryan, J.; Chang, W.; McGowan, L.; François, R.; Grolemund, G.; Hayes, A.; Henry, L.; Hester, J.; et al. Welcome to the Tidyverse. *J. Open Source Softw.* **2019**, *4*, 1–6. [CrossRef]
54. Hornik, K.; Buchta, C.; Zeileis, A. Open-source machine learning: R meets Weka. *Comput. Stat.* **2009**, *24*, 225–232. [CrossRef]
55. Witten, I.H.; Frank, E.; Hall, M.A. Data Mining: Practical Machine Learning Tools and Techniques. *Ann. Phys.* **2002**, *31*, 76–77. [CrossRef]
56. Hall, M.A. Correlation-Based Feature Selection for Machine Learning. Ph.D. Thesis, The University of Waikato, Hamilton, New Zealand, 1999.
57. Kira, K.; Rendell, L.A. A Practical Approach to Feature Selection. *Mach. Learn. Proc.* **1992**, *1992*, 249–256. [CrossRef]
58. Gnanambal, S.; Thangaraj, M.; Meenatchi, V.T.; Gayathri, V. Classification algorithms with attribute selection: An evaluation study using WEKA. *Int. J. Adv. Netw. Appl.* **2018**, *9*, 3640–3644.
59. John, G.H.; Langley, P. Estimating continuous distributions in Bayesian classifiers. *arXiv* **1995**, arXiv:1302.4964.
60. Quinlan, J.R. *Programs for Machine Learning*; Morgan Kaufmann: San Francisco, CA, USA, 1993.
61. Keerthi, S.S.; Shevade, S.K.; Bhattacharyya, C.; Murthy, K.R.K. Improvements to Platt's SMO algorithm for SVM classifier design. *Neural Comput.* **2001**, *13*, 637–649. [CrossRef]
62. Aha, D.W.; Kibler, D.; Albert, M.K. Instance-based learning algorithms. *Mach. Learn.* **1991**, *6*, 37–66. [CrossRef]
63. Raedeke, T.D.; Smith, A.L. Coping resources and athlete burnout: An examination of stress mediated and moderation hypotheses. *J. Sport Exerc. Psychol.* **2004**, *26*, 525–541. [CrossRef]
64. Gustafsson, H.; Sagar, S.S.; Stenling, A. Fear of failure, psychological stress, and burnout among adolescent athletes competing in high level sport. *Scand. J. Sci. Med. Sport* **2017**, *27*, 2091–2102. [CrossRef]
65. Bagby, R.M.; Parker, J.D.A.; Taylor, G.J. The twenty-item Toronto Alexithymia Scale-I. Item selection and cross-validation of the factor structure. *J. Psychosom. Res.* **1994**, *38*, 23–32. [CrossRef]
66. Barlow, M.; Woodman, T.; Chapman, C.; Milton, M.; Stone, D.; Dodds, T.; Allen, B. Who takes risks in high-risk sport?: The role of alexithymia. *J. Sport Exerc. Psychol.* **2015**, *37*, 83–96. [CrossRef]
67. Woodman, T.; Welch, C. Alexithymia and the Anxiolytic Effect of Endurance Running. *Sport Psychol.* **2021**, *1*, 1–7. [CrossRef]
68. Jones, B.D.; Lawrence, G.P.; Hardy, L. New evidence of relative age effects in "super-elite" sportsmen: A case for the survival and evolution of the fittest. *J. Sports Sci.* **2018**, *36*, 697–703. [CrossRef]
69. Till, K.; Cobley, S.; O'Hara, J.; Cooke, C.; Chapman, C. Longitudinal evaluations in junior rugby. *Scand. J. Med. Sci. Sports* **2014**, *24*, 569–576. [CrossRef] [PubMed]
70. Hill, A.; MacNamara, Á.; Collins, D. Development and initial validation of the Psychological Characteristics of Developing Excellence Questionnaire version 2 (PCDEQ2). *Eur. J. Sport Sci.* **2019**, *21*, 517–528. [CrossRef] [PubMed]
71. Langham-Walsh, E.; Anderson, D.; Dunn, E.; Gottwald, V.; Hardy, J.; Hardy, L.; Lawrence, G.; Lowery, M.; Oliver, S.; Roberts, R.; et al. Initial steps in the validation of the athlete development formulation survey. *J. Hum. Sport Exerc.* **2019**, *51*, 115.
72. Grobler, T.D.; Shaw, B.S.; Coopoo, Y. Influence of physical fitness parameters on relative age effect on amateur secondary school rugby union players. *S. Afr. J. Res. Sport Phys. Educ. Recreat.* **2017**, *39*, 29–39. Available online: https://hdl.handle.net/10520/EJC-c37afb24a (accessed on 6 September 2021).
73. Pienaar, A.E.; Spamer, M.J. A longitudinal study of talented young rugby players as regards their rugby skills, physical and motor abilities and anthropometric data. *J. Hum. Mov. Stud.* **1998**, *34*, 13–32.
74. Nutton, R.W.; Hamilton, D.F.; Hutchison, J.D.; Mitchell, M.J.; Simpson, A.H.R.; MacLean, J.G. Variation in physical development in school boy rugby players: Can maturity testing reduce mismatch? *BMJ Open* **2012**, *2*, e001149. [CrossRef]
75. Jones, B.; Weaving, D.; Tee, J.; Darrall-Jones, J.; Weakley, J.; Phibbs, P.; Read, D.; Roe, G.; Hendricks, S.; Till, K. Bigger, stronger, faster, fitter: The differences in physical qualities of school and academy rugby union players. *J. Sports Sci.* **2018**, *36*, 2399–2404. [CrossRef]
76. Kearney, P.E. The influence of nationality and playing position on relativeage effects in rugby union: A cross-cultural comparison. *S. Afr. J. Sports Med.* **2017**, *29*, 1–4. [CrossRef]
77. Lewis, J.; Morgan, K.; Cooper, S.-M. Relative age effects in Welsh age grade rugby union. *Int. J. Sports Sci. Coach.* **2015**, *10*, 797–813. [CrossRef]

78. McCarthy, N.; Collins, D. Initial identification and selection bias versus the eventual confirmation of talent: Evidence for the benefits of a rocky road? *J. Sports Sci.* **2014**, *32*, 1604–1610. [CrossRef]
79. Howard, S.M.; Cumming, S.P.; Atkinson, M.; Malina, R.M. Biological maturity-associated variance in peak power output and momentum in academy rugby union players. *Eur. J. Sport Sci.* **2016**, *16*, 972–980. [CrossRef] [PubMed]
80. Nevill, A.; Holder, R.; Watts, A. The changing shape of "successful" professional footballers. *J. Sports Sci.* **2009**, *27*, 419–426. [CrossRef] [PubMed]
81. Gale-Watts, A.S.; Nevill, A.M. From endurance to power athletes: The changing shape of successful male professional tennis players. *Eur. J. Sport Sci.* **2016**, *16*, 948–954. [CrossRef] [PubMed]
82. Jones, B.; Hardy, L.; Lawrence, G.; Kuncheva, L.; Brandon, R.; Such, P.; Bobat, M. The Identification of 'Game Changers' in England Cricket's Developmental Pathway for 3 Elite Spin Bowling: A Machine Learning Approach. *J. Expert* **2019**, *2*, 92–120. Available online: https://www.journalofexpertise.org/articles/volume2_issue2/JoE_2019_2_2_Jones.html (accessed on 16 February 2021).
83. Jones, B.; Hardy, L.; Lawrence, G.; Kuncheva, L.; Brandon, R.; Bobat, M.; Thorpe, G. It Ain't What You Do—It's the Way That You Do It: Is Optimizing Challenge Key in the Development of Super-Elite Batsmen? *J. Expert* **2020**, *3*, 144–168. Available online: https://www.journalofexpertise.org/articles/volume3_issue2/JoE_3_2_Jones_etal.html (accessed on 16 February 2021).

sports

MDPI

Article

Proportions of Early Specializers Varies According to Methods and Skill Level

Alexandra Mosher [1,*], Jessica Fraser-Thomas [1], Melissa J. Wilson [2] and Joseph Baker [1]

[1] School of Kinesiology and Health Science, Faculty of Health, York University, Toronto, ON M3J 1P3, Canada; jft@yorku.ca (J.F.-T.); bakerj@yorku.ca (J.B.)
[2] Swimming Australia, Brisbane 4109, Australia; melissa.wilson@swimming.org.au
* Correspondence: moshera@yorku.ca

Abstract: Sport researchers have warned about the lack of a clear and consistent definition of early specialization, while others have raised concerns around the validity of methods used to classify athletes as 'specializers'. The current investigation includes two studies examining the implications of varying classification methods for exploring both specialization and early specialization in sport. Study 1 examined whether different approaches to defining and measuring specialization affected the classification of athletes throughout development and provided a 'profile' of the sample in terms of developmental milestones related to specialization. Results indicated the proportion of athletes classified as specializers varied depending on the method used and athletes generally met specialization milestones after the age of 12. Study 2 examined the proportions of athletes who achieved 'elite', 'pre-elite', and 'non-elite' status in adulthood who were early specializers as determined by different methods. Results showed the method used changed the proportion of athletes classified as specializers at each level and there was no clear advantage or disadvantage to being a specializer. Combined, these studies provide intriguing data regarding the implications of different measures for assessing specialization in young athletes.

Keywords: early specialization; development; milestones; classification

Citation: Mosher, A.; Fraser-Thomas, J.; Wilson, M.J.; Baker, J. Proportions of Early Specializers Varies According to Methods and Skill Level. *Sports* **2022**, *10*, 34. https://doi.org/10.3390/sports10030034

Academic Editor: Adam Baxter-Jones

Received: 8 December 2021
Accepted: 24 February 2022
Published: 28 February 2022

1. Introduction

Early specialization has been defined many ways, with little consistency between studies [1]. Collectively, however, these studies generally indicate early specialization involves dedicating large amounts of time and effort to one sport from a young age in pursuit of becoming an elite athlete, e.g., [2–5]. Precise determination of 'young age' and 'early' have yet to be established across the field [6]; however, the Developmental Model of Sport Participation (DMSP) suggests 12 years of age or earlier as a critical cut-off point [2].

The number of athletes following the path of early specialization has appeared to have seen an increase in recent years, arguably due to the professionalization of youth sport [7]. While there is much debate about the potential negative outcomes associated with *early* specialization, the need to eventually specialize is strongly supported [8].

Much of the theoretical rationale for early specialization is found in the deliberate practice framework [9]. Ericsson and colleagues used data from musicians as the foundation for the framework, noting that expert musicians spend more time in highly focused, effortful practice aimed at improving performance compared to their lesser skilled peers. The authors suggested engagement in this specific type of training (i.e., activities that are not inherently enjoyable, designed for the purpose of improving performance and not instantly gratifying, which they labeled 'deliberate practice') was the key mechanism explaining differences between those who achieve expertise and those who do not. Essentially, this framework is grounded in the notion that in order to become an expert, one must engage in a large quantity of deliberate practice; the greater the time spent in deliberate practice, the higher the attained level of performance. Importantly for our context, Ericsson and colleagues [9] suggested those who started deliberate practice at a later age (i.e., specialized later) were at a disadvantage compared to their peers who began earlier.

Despite the suggestions that early specialization in sport is increasing, recent studies have found later specialization (described as specialization after the age of 12) to be more common among elite athletes. The average age of specialization was 14 years of age in a study of elite hockey players, with only 12% of athletes specializing before age 12 [10]. Similarly, in a study of Olympic track and field athletes, 17 was the average age of specialization [11]. Despite these and other studies providing evidence against the necessity of specializing early, many parents and athletes still believe early specialization is the optimal way to become a top performing athlete. In a large study of 3090 athletes playing at high school, collegiate, and professional levels, 79.7%, 80.6%, and 61.7%, respectively, agreed that specializing in one sport helps an athlete play at a higher level [12]. Additionally, a study of over 900 youth athletes found 91% believed specialization in one sport increased their chances of getting better at their sport and 66% felt it would increase their chances of making a college team [13]. Unfortunately, early specialization is linked to an increased risk of injury and potential for burnout [4,14], which is why many sport organizations have advised against the practice (e.g., American Academy of Pediatrics) [7]. However, findings regarding early specialization should be interpreted with caution, as multiple authors, e.g., [12,15–17] have highlighted several methodological shortcomings.

For many years, sport researchers have warned about the lack of a clear and consistent definition of early specialization [12,15], while others have raised concerns around the validity of current methods used to classify athletes as 'specializers' [16,17]. A recent systematic review by Mosher et al. [1] found inconsistent definitions of early specialization across 48 empirical studies, with 18 different methods used to classify athletes as early specializers. 'Single sport participation' was the most common indicator of early specialization with ~73% of the 129 papers included in the review using this marker. Comparing single sport versus multi-sport athletes is one of the more common ways researchers have used to classify athletes as early specializers, e.g., [18]. Some have used a more comprehensive approach, collecting a complete history of an athlete's sporting background, e.g., [19] while others have used a single yes/no item to determine specialization, e.g., [20]. A popular classification system in sports medicine is the "Sport Specialization Scale" developed by Jayanthi et al. [4], which classifies athletes on a spectrum from low to high specialization based on specialization as "year-round intensive training in a single sport at the exclusion of other sports" (p. 795). This scale, while used in youth populations (ages 7–18), does not include a measure of age and, therefore, does not distinguish between *early* specialization and specialization more generally. This variation in methods used to determine specialization can lead to inconsistent classification of who is a specializer. The scientific value of measuring specialization based on a dichotomized definition using arbitrary nominal variables needs to be evaluated and researchers are advocating for more adequate approaches (e.g., measuring a variety of continuous variables) [21]. However, as it is common practice to define and measure specialization in this way, there is value in examining the implications of different methods.

Importantly, the lack of a consensus definition can change the relationship and severity of outcomes associated with specialization. For example, in a study on the prevalence of specialization and injury history, Bell and colleagues [22] had high school students in the United States complete two different specialization classification tools, including a self-report as a 'single sport' or 'multi-sport' athlete, and the Sport Specialization Scale; both methods have been used in prior research to distinguish specializers from non-specializers. They found little agreement between the classification methods, with only 12% of students being classified as both single-sport and highly specialized and 26% being classified as multi-sport and low specialization. More troubling, the method used to classify athletes affected whether there was an association found between injury history and specialization. Athletes who self-classified as a single sport participant had no association to injury history, whereas those classified moderate or high specialization were more likely to report a history of injury [22].

The differing outcomes from different classification methods within the same study raise concerns about the reliability and validity of research examining specialization in general and early specialization in particular. While Bell and colleagues' [22] study indicates specialization rates appear to be dependent on the classification method used, their study is only one among a rapidly growing research base focused on both early specialization and specialization more generally, with little consistency in the definition of the construct. Further, Bell and colleagues' [22] study was cross-sectional in design, focused on athletes between the ages of 13 to 18. Given that much of the debate around specialization is concerned with *early* specialization and the dangers of specializing too soon, studies of this phenomenon in athletes before the age of 13 are needed. To this end, this investigation includes two studies, described below, examining the implications of varying classification methods for exploring both specialization and early specialization in sport.

2. Study 1

Given the recommendations from key organizations and athlete development models to avoid early specialization (e.g., American Academy of Pediatrics; Long Term Development Model) [7,23] coupled with issues related to approaches for classifying athletes as specializers, greater examination of early specialization measurement in youth (12 years of age and under) samples appears to be warranted. In this first study, we had two objectives. First, we examined how different approaches to defining and measuring specialization affected the classification of athletes throughout development. Based on prior work, our hypothesis was that the method used to determine specialization status would affect the proportion of the overall sample classified as 'specializers'. Our second objective was to provide a 'profile' of the sample in terms of developmental milestones related to specialization. Few, if any, studies have provided individual milestones related to specialization, instead providing only the age at which specialization occurred. Examining the age that each component of specialization was met provides a more comprehensive picture of specialization patterns. Collectively, this study extends our understanding of the age at which youth meet different indicators of specialization and determines if *early* specialization is occurring in this sample.

2.1. Study 1 Materials and Methods

Participants

Participants included 362 athletes from one of the largest samples of athletes' developmental histories [24,25]. In the original study, participants were recruited from all competition levels (e.g., local, regional, national, international) via advertisements on sport organization websites, social media and newsletters, or invitations from their coach. The sample was comprised of 203 females and 159 males with an age range of 14–42 ($M = 20.8$, $SD = 4.7$) from 10 different countries. The majority of participants were from Australia ($n = 255$) and Canada ($n = 97$). The athletes represented 36 different sports with the most popular being soccer ($n = 77$) and basketball ($n = 46$).

2.2. Measures

The indicators used in our studies were chosen mainly on their common use in the literature and not on their inherent value or evidence-based rationale. Data for this project came from a larger study of athlete development conducted in 2010–2011, where participants completed the Developmental History of Athletes Questionnaire (DHAQ) [26,27]. The DHAQ is a comprehensive instrument that gathers quantitative information on several different areas of an athlete's history including their main sport practice history, other sport participation, family sporting history, and attainment of sporting milestones. The DHAQ has been previously validated [26] and used in samples with both able-bodied and para-athletes [24,28] While a number of measures included in the DHAQ were demonstrated to have questionable reliability or validity, the items analyzed for this study were deemed to have sufficient validity and reliability to provide valuable insights to the academic discussion of specialization in sport. A key strength of the DHAQ is that it allows collection and examination of key markers of specialization (outlined below) throughout athletes' development.

2.2.1. Single Sport Participation

As mentioned, single sport participation is the most commonly used indicator of specialization [1]. Subsequently, many studies have used this single qualifier as a method of classifying athletes as specializers [18,29]. While the merit of this method is debatable [22], due to its frequency of use, single sport participation was the first method used for comparison in this study. To operationalize this indicator, drawing from the DHAQ variables, we used the number of sports participated in at each age between 5 and 18 years.

2.2.2. Year-Round Single Sport

Another common indicator of specialization in previous work consists of year-round participation in one sport [1,20], and reflects a two-indicator method focusing on the 'single sport participation' captured in method one, described above, but within a fixed time frame (i.e., over the full year). In the current study, we consider these two indicators together as a method of classifying athletes as specializers. The questions/variables drawn from the DHAQ to operationalize this indicator were: (a) age of first participation in year-round training in main sport and (b) number of other sports participated in at each age (5–18).

2.2.3. Sport Specialization Scale Items

While the Sport Specialization Scale [4] was not explicitly used in the DHAQ, the three indicators that comprise the scale can be inferred from the data collected [26]. Specifically, the Sport Specialization Scale consists of three self-reported questions pertaining to whether an athlete is engaged in: (a) year-round training, (b) exclusion of other sports, and (c) participation in one main sport. The aligning questions/variables drawn from the DHAQ were: (a) age of first participation in year-round training in main sport, (b) age of deliberate exclusion of other sports, and (c) number of sports participated in at each age (5–18). On the Sport Specialization Scale, an athlete responds 'yes' or 'no' to each of the three questions; specialization is then scored as low, moderate, or high (i.e., 'yes' to one, two, or three indicators). While the three similar items of the SSS were used as indicators of early specialization, our analysis was somewhat different from how these data have been typically considered, as described below.

2.3. *Coding*

Information collected from the DHAQ was used to classify athletes as specializers at each age of development based on the three methods described above (i.e., single sport participation, year-round single sport, and the Sport Specialization Scale items). The proportion of athletes meeting the criteria of specialization based on each method at each age from 5 to 18 years was calculated. Unlike Jayanthi and colleagues' [4] scale, which allowed for any one of the three indicators resulting in a score from one to three, this study used specific combinations of indicators (i.e., single sport alone; single sport and year-round training only; and single sport, year-round training, and exclusion of other sports). For a visual explanation of coding see Figure 1.

2.3.1. Single Sport Participation

The number of other sports in which an athlete participated was coded to indicate whether the athlete was a single sport participant and coded as 1 (i.e., response of zero additional sports) or participated in more than one sport (i.e., $\geq$1) coded as 2 for each age of development. In order to distinguish single-sport participation from data for years that participants had not yet started sport participation, athletes' age of first participation in their main sport was established. If, for example, an athlete did not start participating in any sports until age seven, the zero response for other sport participation at age 5 and age 6 was coded as a 2 rather than a 1. Additionally, in order to distinguish single sport participation from those who could not fill out information because they were younger than the age of study (i.e., any age under 18), the age of the participant during the study was taken into consideration. This meant if an athlete was 15 years old at the time of the study, for the ages after 15 (16–18) a zero response for other sport participation was marked as 'N/A' rather than single sport participation. At each age of development, an athlete was deemed a specializer if they received a score of 1 on this measure and a non-specializer if they received a score of 2.

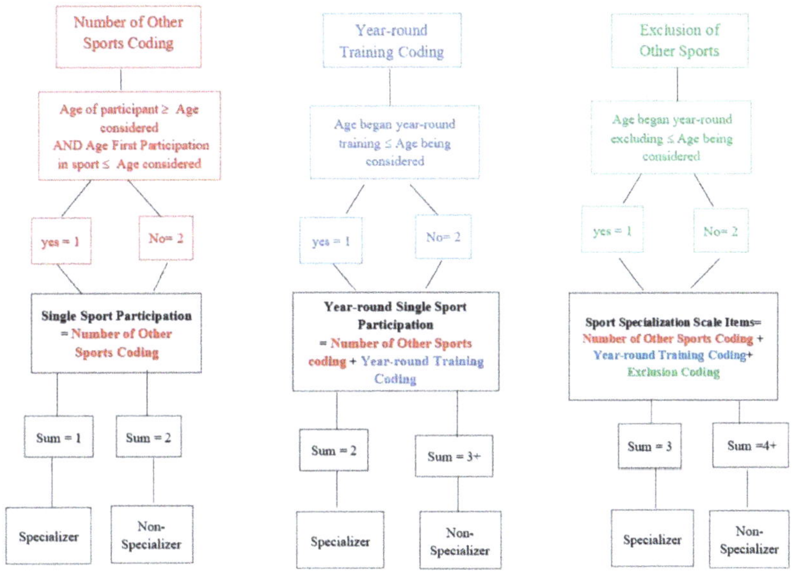

Figure 1. Breakdown of the coding for each variable and each method.

2.3.2. Year-Round Single Sport

Year-round participation was coded similarly for each age of development. If the age of first participation in year-round training was equal to or less than the age being considered in the analysis (e.g., if the age of first participation in year-round training was 6 and the age being considered was 6 or higher), the athlete received a coding of 1. If the age of first participation in year-round training was greater than the age being considered for analysis (e.g., if age of first participation in year-round training was 7 and the age being considered in the analysis was 6), the athlete was coded as a 2. The same coding described above was applied to determine single sport participation. The two measures were then summed, and an athlete was deemed an early specializer if they received a total score of 2 (e.g., as it would indicate a coding of one for both criteria); they were classified a non-specializer if they received a score of 3 or 4, indicating they did not meet both criteria.

2.3.3. Sport Specialization Scale items

The same coding of 1 or 2 was applied to age of deliberate exclusion of other sports at each age of development. Similar to year-round participation, if the age of deliberate exclusion of other sports was equal to or less than the age being considered (e.g., if the age being considered in the analysis was 8 and the age of deliberate exclusion of other sports was 8), the athlete received a coding of 1. If the age of deliberate exclusion of other sports was greater than the age being considered (e.g., if age being considered in the analysis was

8 and age of deliberate exclusion of other sports was 9) the athlete received a coding of 2. Year-round participation and single sport participation were coded as described above. After summing the three measures, an athlete was deemed a specializer if they received a total score of 3 (i.e., a coding of 1 for all three criteria).

2.4. Analyses

To address objective one, the proportions of athletes classified as specializers at each age were calculated using each of the methods described above. Objective two involved calculating the averages for the developmental milestones related to specialization. Averages were calculated for age of first participation in main sport, age of first participation in year-round training, and age of exclusion of other sports.

3. Study 1 Results

Analysis revealed athletes' average age of first participation in their main sport was 9.5 (*SD* = 5.1) years of age, the mean age of first participation in year-round training was 14.4 (*SD* = 3.9) years of age, and the mean age of exclusion of other sports was 15.1 (*SD* = 3.5) years of age. Only 16.3% of athletes began excluding participation in other sports by the age of 12, while 26.2% began year-round training by 12 years of age.

As expected, the number of athletes classified as specializers varied depending on the method used. Across every age, single sport participation resulted in the highest percentage of specializers. There was a large difference (20% or greater) between the proportion of specializers based on single sport participation compared to year-round single sport or the Sport Specialization Scale items between the ages of 5 to 12 (early specialization). By age thirteen (specialization), the differences between groups ranged from 20% to 6% with the greatest degree of convergence occurring at 18 years of age. Between the ages of 5 and 12 (early), the proportion of athletes classified as early specializers remained low for the Sport Specialization Scale items and year-round single sport ranging from 0% to 12%, while the proportion of specializers based on single sport participation started at a higher percentage (19%) at 5 years of age and continued to increase (to 32%) up to 12 years of age. After age 13, there was an increase in the number of specializers based on year-round single sport and the Sport Specialization Scale items from 30% up to 46%, while the proportion of specializers for single sport participation fluctuated between 46% up to 54%. For a full break down of percentages by classification method, see Table 1.

Table 1. Proportion of athletes classified as early specializers by method.

	Single Sport Participation	**Year-Round Single Sport**	**Sport Specialization Scale Items**
Age	% specializers (*n*)	% specializers (*n*)	% specializers (*n*)
Age 5	19.61 (71)	0.28 (1)	0.00 (0)
Age 6	23.48 (85)	0.83 (3)	0.55 (2)
Age 7	25.69 (93)	1.10 (4)	0.55 (2)
Age 8	29.01 (105)	2.21 (8)	0.83 (3)
Age 9	30.66 (111)	4.97 (18)	2.49 (9)
Age 10	30.66 (111)	7.73 (28)	3.59 (13)
Age 11	31.77 (115)	9.39 (34)	5.80 (21)
Age 12	32.87 (119)	12.43 (45)	7.73 (28)
Age 13	39.78 (144)	19.61 (71)	13.81 (50)
Age 14	45.86 (166)	30.11 (109)	25.14 (91)
Age 15	54.14 (196)	40.88 (148)	37.57 (136)
Age 16	**54.42 (197)**	45.58 (165)	42.54 (154)
Age 17	54.14 (196)	**46.13 (167)**	**45.30 (164)**
Age 18	50.83 (184)	44.48 (161)	44.20 (160)

Peak number of specializers indicated in bold.

4. Study 1 Discussion

Results from Study 1 highlight several implications for those studying early specialization. First, based on the average age of first participation in year-round training (~14 years), exclusion of other sports (~15 years) and the small percentage of athletes who met these milestones at 12 years of age or earlier, specialization (12 years of age or earlier) does not appear to be overly prevalent in this sample of athletes. This is in line with a study by Swindell et al. [30] that found approximately 17% of athletes had specialized at 12 years of age or earlier. While this is a meaningful minority and each athlete's experience and safety is important, these results suggest either the number of children meeting criteria for early specialization is not as large as the rhetoric would suggest, e.g., [7,31,32] or those who specialized early are no longer in the system at the age of this sample of athletes.

Second, our hypothesis suggesting differences in proportions of specializers based on method used was generally well supported, given large discrepancies found; the more indicators used, the lower the proportion of athletes classified as specializers.

At every age, using single sport participation as the sole indicator resulted in the highest number of athletes classified as specializers. Such variation in how early specialization is determined at younger ages [1] raises concerns regarding the methods used to generate the evidence for such strong condemnation of the dangers of early specialization, e.g., [33]. The choice of method has clear implications for how study results are positioned in the discourse around early specialization. At six years of age, for example, less than one percent of athletes were participating in *year-round-training* at the exclusion of other sports, yet approximately 23% were participating in only one sport. Our understanding of the appropriate level of sport engagement for youth participating in sport at this age is very limited. Many children at six years of age may simply be beginning sport participation, and it seems reasonable to attempt one sport at a time, yet these athletes would be classified as specializers according to the method used in many studies. The results from this study indicate that a measure with more indicators may be more suitable to classify athletes—particularly younger child-athletes—as specializers.

That the proportion of athletes categorized as specializers changes so dramatically based on the method used raises questions about the conclusions drawn from the evidence base on early specialization, particularly given the range of definitions reported in this literature [1] For example, how does the method used affect conclusions about the value of early specialization for becoming an elite athlete? We explore this issue in study 2.

5. Study 2

As a commonly agreed upon principle of early specialization is that it is a pathway often followed in pursuit of becoming an elite athlete [2], it is important to examine the relationship between athletes' sport development pathway (i.e., early specialization), and their attained skill level (i.e., elite). In this second study, we explored whether those who specialized in their youth became elite athletes in adulthood by examining the proportions of specializers (by each method used in Study 1) who achieved 'elite', 'pre-elite', and 'non-elite' status.

5.1. Study 2 Materials and Methods

5.1.1. Participants

A sub-sample of athletes from Study 1 was included in Study 2 (n = 237). Because this analysis focused on the highest level of skill attained, participants were limited to those *above* 18 years of age at the time of data collection. This age was chosen based on an assumption that generally, there is still room to improve and increase level of competition at age 16 or younger but by 18, if an athlete has not yet reached elite status, their chances of becoming elite decrease dramatically.

5.1.2. Measures

The same classification methods in Study 1 were used in Study 2 to determine specializers (i.e., single sport participation, year-round single sport, and the Sport Specialization Scale items). For single sport participation, athletes were either specializers meaning they met the single sport criterion (i.e., participated in one sport, sum of 1) or non-specializers (i.e., participated in more than one sport, sum of 2). For year-round single sport, athletes were considered specializers if they met both criteria (i.e., year-round and one sport, sum of 2) and non-specializers if they met one or none of the criteria (i.e., year-round or one sport or neither, sum of 3 or 4). Finally, in order to get a better understanding of the extremes of specialization compared to non-specialization, for the Sport Specialization Scale items, athletes had to meet all three criteria to be a specializer (i.e., year-round and exclusion of other sports and one main sport, sum of 3) but *none* of the criteria to be a non-specializer (as opposed to meeting some but not all criteria, sum of 6). This method provided a greater contrast than combining the remaining athletes (i.e., combining those who met one or two or none of the criteria) would have. Additionally, as there was overlap between all three methods (i.e., all use single sport, two use year-round), we felt much of the information would be provided by the other two methods and, therefore, the extreme comparisons for the Sport Specialization Scale items would be more valuable.

To determine highest level of competition and subsequently which skill group (elite, pre-elite, or non-elite) athletes belonged to, the milestones section of the DHAQ [24] was used. This section determines the age at which each athlete participated in different levels of competition (i.e., local, regional, national, international). Using the Athlete Development Triangle framework [24,34] athletes were categorized into three groups: (a) '*elite*' athletes had competed at a senior international level, (b) '*pre-elite*' athletes had competed at a junior international or senior national level, and (c) '*non-elite*' athletes were all those who competed in remaining lower levels of competition.

5.2. *Analyses*

We determined the percentage of specializers across ages by each method described above who achieved elite, pre-elite, and non-elite status. The number of athletes (n) in each group varied across each age as the number of specializers varied, as described in Study 1.

6. Study 2 Results

Results indicated differences between the proportions of elite, pre-elite, and non-elite athletes classified as specializers across ages based on the method used. For a complete profile of percentages across the method used, see Table 2 and Figure 2. Across all skill levels, single sport participation resulted in the highest percentage of athletes classified as specializers for each age. The largest differences between percentages classified as specializers based on method used across skill level occurred in the earlier ages (i.e., 5–12), with the difference generally decreasing with age.

Using the criterion of single sport participation, at the ages of 5–7, there was a higher percentage of specializers who became elite compared to pre-elite and non-elite, whereas when using both year-round single sport and the Sport Specialization Scale items, between the ages of 6 and 13 there was a higher percentage of specializers who became non-elite and pre-elite compared to elite. By all three methods, there was a higher percentage of athletes who were specializers at 18 who became elite compared to non-elite.

Table 2. Percentage of elite, pre-elite, and non-elite classified as specializers by method used.

	% Elite (*n*)	% Pre-Elite (*n*)	% Non-Elite (*n*)
Age 5	*n* = 26	*n* = 13	*n* = 7
Single sport participation	88% (23)	77% (10)	71% (5)
Year-round single sport	n/a	n/a	n/a
SSS items	n/a	n/a	n/a
Age 6	*n* = 33	*n* = 17	*n* = 10
Single sport participation	88% (29)	59% (10)	60% (6)
Year-round single sport	0% (0)	0% (0)	20% (2)
SSS	0% (0)	0% (0)	20% (2)
Age 7	*n* = 41	*n* = 23	*n* = 11
Single sport participation	73% (30)	65% (15)	55% (6)
Year-round single sport	0% (0)	4% (1)	18% (2)
SSS items	0% (0)	0% (0)	18% (2)
Age 8	*n* = 52	*n* = 29	*n* = 23
Single sport participation	63% (33)	66% (19)	54% (7)
Year-round single sport	2% (1)	7% (2)	15% (2)
SSS items	0% (0)	3% (1)	15% (2)
Age 9	*n* = 61	*n* = 33	*n* = 15
Single sport participation	57% (35)	64% (21)	53% (8)
Year-round single sport	5% (3)	9% (3)	20% (3)
SSS items	0% (0)	6% (2)	20% (3)
Age 10	*n* = 71	*n* = 42	*n* = 20
Single sport participation	49% (35)	48% (20)	50% (10)
Year-round one sport	6% (4)	14% (6)	15% (3)
SSS	0% (0)	7% (3)	15% (3)
Age 11	*n* = 77	*n* = 46	*n* = 23
Single sport participation	45% (35)	50% (23)	52% (12)
Year-round single sport	5% (4)	22% (10)	13% (3)
SSS items	1% (1)	10% (5)	13% (3)
Age 12	*n* = 85	*n* = 51	*n* = 27
Single sport participation	41% (35)	47% (24)	48% (13)
Year-round single sport	8% (7)	22% (11)	11% (3)
SSS items	1% (1)	14% (7)	11% (3)
Age 13	*n* = 92	*n* = 58	*n* = 29
Single sport participation	50% (46)	48% (28)	41% (12)
Year-round single sport	17% (16)	22% (13)	10% (3)
SSS items	10% (9)	15% (9)	10% (3)
Age 14	*n* = 102	*n* = 62	*n* = 33
Single sport participation	51% (53)	45% (28)	39% (13)
Year-round single sport	28% (29)	26% (16)	15% (5)
SSS items	19% (20)	20% (13)	15% (5)
Age 15	*n* = 110	*n* = 65	*n* = 33
Single sport participation	58% (64)	49% (32)	36% (12)
Year-round single sport	38% (42)	34% (22)	18% (6)
SSS items	32% (35)	28% (18)	18% (6)
Age 16	*n* = 114	*n* = 67	*n* = 33
Single sport participation	64% (73)	55% (37)	36% (12)
Year-round single sport	50% (58)	44% (30)	21% (7)
SSS items	45% (51)	39% (26)	21% (7)
Age 17	*n* = 116	*n* = 68	*n* = 34
Single sport participation	73% (85)	65% (44)	47% (16)
Year-round single sport	63% (73)	53% (36)	26% (9)
SSS items	60% (70)	53% (36)	26% (9)
Age 18	*n* = 119	*n* = 69	*n* = 34
Single sport participation	86% (102)	84% (58)	70% (24)
Year-round single sport	77% (92)	74% (51)	53% (18)
SSS items	76% (91)	74% (51)	53% (18)

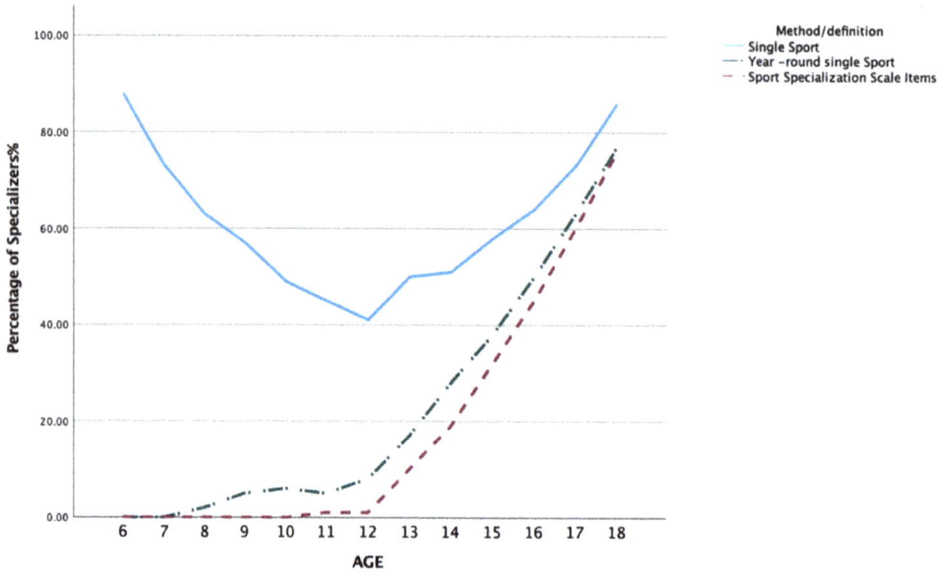

(A)

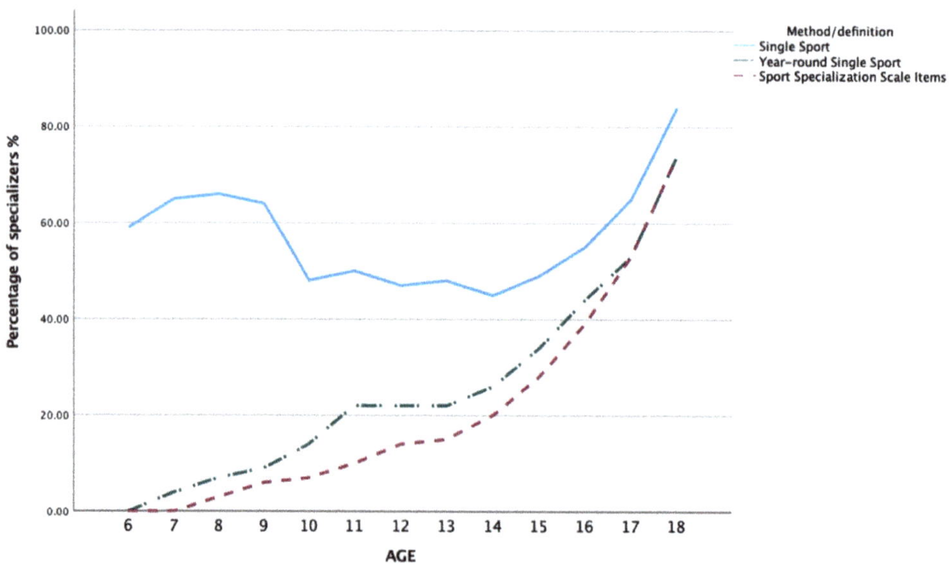

(B)

Figure 2. *Cont.*

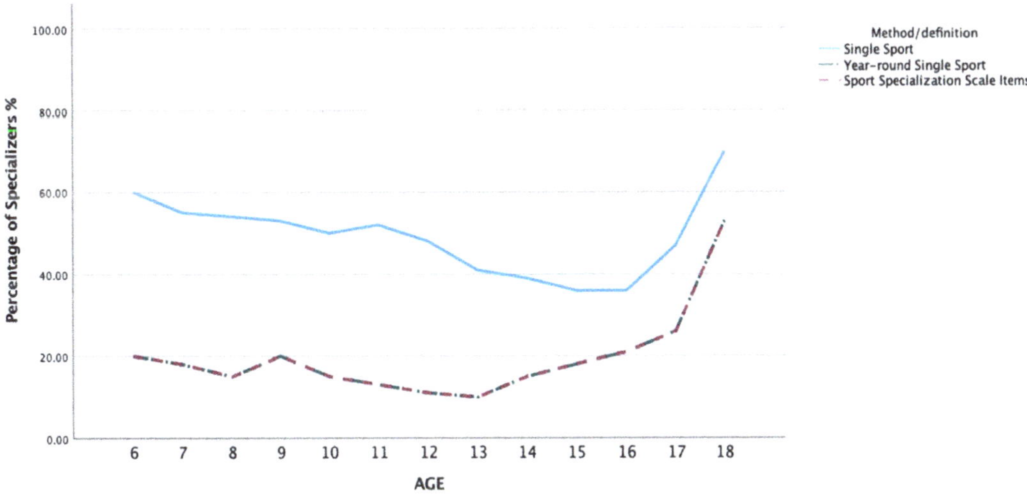

(C)

Figure 2. Percent of (**A**) Elite, (**B**) Pre-Elite, and (**C**) Non-Elite Defined as Specializers by Age and Method. Year-round single sport and the Sport Specialization Scale items have the same percentages and appear as one line.

7. Study 2 Discussion

The assumptions underpinning the need to specialize are that it improves an athlete's performance [13] and helps their chances of playing at a higher level [12]. The results of this study challenge aspects of these assumptions and highlight a range of implications. First, reinforcing the results from Study 1, there were large discrepancies between the proportion of athletes in each skill group based on the method used to classify athletes as specializers. As mentioned previously, single sport participation is not a nuanced measure of specialization status. However, even when using a more multi-dimensional measure (i.e., the Sport Specialization Scale items), the proportion of elite athletes versus pre-elite or non-elite was low until the age of fifteen. This suggests early specialization (i.e., specializing at 12 years of age or earlier) has limited benefit to performance and long-term elite attainment. This supports previous research by Wilhelm et al. [35] suggesting early sport specialization is *not a* requirement to compete at the most elite levels of sport. However, it is important to also recognize the large number of athletes who did go on to be elite, who had specialized before 18 (as measured by all three methods)—indicating that specialization in later adolescence is a common pathway to elite performance. More specifically, these results suggest specialization prior to 18 years of age may be required to become an elite athlete—but specialization does not need to or should not occur too early (i.e., not prior to 12–15 years). Further research is necessary to determine more precise optimal age(s) of specialization (and potential mitigating factors)—for athletes to reach top skill levels in adulthood.

8. General Discussion

Collectively, both studies demonstrate that single sport participation (i.e., the most commonly used indicator in prior research) resulted in the greatest proportion of the sample being classified as specializers. The large number of athletes classified as specializers based on the single sport participation method indicates there may be many athletes being classified as specializers who are not 'true' specializers (i.e., investing time and effort in one sport for the purpose of improving performance). There are many reasons a child may be participating in one sport, including parental time constraints, family financial constraints, and/or the child enjoying one sport over others. Simply asking the number of sports in which a child is participating fails to distinguish those who are deliberately choosing to play one sport to improve performance from those who participate in one sport for other reasons. The latter may be less of a concern as the child would likely be at a lower risk of injury associated with specializing for performance. Much of the increased injury risk associated with specialization is often attributed to the volume of training and overuse [36]. A child participating in one sport recreationally would likely not meet the volume of training that would warrant concern. In future work, if researchers choose to use single sport participation as a distinguishing factor it should be used in combination with other variables to ascertain why the athlete only participates in one sport in order to distinguish true specializers (i.e., the explicit decision to specialize to improve performance) from single sport athletes (e.g., those who may not be able to afford multiple sports or participate in low amounts in one sport at a recreational level).

When other variables were considered in addition to single sport participation (e.g., year-round participation), there were few to no athletes at the younger ages (i.e., 5–12 years) meeting the criteria for specialization. Much of the caution around specialization concerns the notion that specializing too soon is dangerous to the physical health and well-being of the athlete [37]. Collectively, both studies indicate that specialization is not prevalent at younger ages, but rather, is more common beginning in the early adolescent years (i.e., 13–16). Further research should focus on specialization in early adolescence, given similar concerns may arise (e.g., related to physical maturation, psycho-social outcomes) despite specialization being deemed more acceptable at this stage of development.

Our ultimate goal for this series of studies was to examine the varying approaches to measuring specialization in the same group of athletes. The inconsistency of classifications across different methods raises important issues about the validity and reliability of prior work on early specialization. This is particularly important given the value of research synthesis approaches (e.g., systematic reviews, meta-analyses) for generating patterns of evidence to reflect conclusions in a field. Many practitioners advise against specialization for youth athletes, but without a consistent classification method for specializers, the evidence behind these recommendations is unclear.

One important next step could be increasing the accuracy of the measure used to classify athletes by increasing the number of criteria an athlete has to meet. As we note above, there are potentially important elements of specialized engagement that are not captured in current approaches (e.g., reason for specialization). It should be noted, however, that increasing the number of criteria does not guarantee a more accurate measure. A recent study on the Sport Specialization Scale [16], for example, found that 30% of athletes were misclassified as moderately specialized, because they had only ever participated in one sport and, therefore, failed to meet the criteria of exclusion of other sports, when they should have been considered highly specialized. A strong rationale for item inclusion and consideration to the way new items are best measured are necessary when attempting to increase the accuracy of future measures.

A final consideration is the common dichotomizing of specialization. By dividing athletes into one or two groups, researchers may miss important information that could help elucidate the links between specialization and negative outcomes. For instance, some researchers have operationalized specialization as a simple yes or no question, such as "are you a specializer or not?" [20]. This method provides little information about the indicators chosen as key markers of specialization and, thus, makes understanding which aspect(s) of specialization is (are) most harmful or beneficial more difficult. The same issue applies when asking an athlete whether they are a single or multi-sport athlete. While this provides some information about the breadth of their participation, it does little to illuminate areas of potential concern. It would be more valuable to collect information on hours spent in each sport or months spent training to identify where overtraining may occur, rather than a dichotomized single versus multisport variable. While the Sport Specialization Scale [4] aimed to move away from this trend by creating degrees of specialization (low, medium, high), some have questioned the validity of the scale [16,17] as it does not account for volume of training and does not consider athletes who only ever played one sport. Overall, future research should move away from the simple single sport participation or dichotomous classification of early specialization and move towards a more comprehensive understanding of athletes' full participation history.

While both studies in this paper have several strengths, it is important to also discuss potential limitations. The data collected was retrospective not prospective in nature and, therefore, only captures the responses from 'survivors' in the sport system. This a common issue in athlete development research but it means it is possible those who dropped out from sport may have different participation trajectories that are not captured in this analysis. Moreover, the retrospective design also raises concerns regarding the accuracy of recall by participants, with participants attempting recall information from as far back as 30 years. Additionally, the specialization status of the sports was not considered; examining sports designed around early specialization (e.g., gymnastics, where peak performance can occur before 18 years for age) as well as later specializing sports (e.g., triathlon) is an important next step to gain more nuanced understanding of specialization. Moreover, our sample contained more elite athletes than pre-elite and non-elite athletes and, therefore, may not be generalizable to the average sporting population. Finally, development is best seen as a continuous process, which makes it hard to set 'cutoffs' for assessing development. We decided that having a cutoff would be more useful for analyses, and relevant for practitioners. Since the age of 18 is often used as an important marker of the transition from adolescent to adult, we felt this was an appropriate age to use as the cutoff in study two.

Despite these limitations, our results suggest important implications of the use of different definitions/measures for specialization in research studies. We are not alone in raising these concerns. Recently, there has been some progress towards a consensus definition of specialization [38] although whether this definition will be widely used remains to be seen [21]. While a consistent definition of specialization would be a positive step forward, this will need to be followed by addressing the methodological concerns highlighted in this study. Until there is a clear and consistent definition and aligning method used to classify specialization, position statements outlining risks should be interpreted with caution. In particular, future research should better measure and distinguish 'early' specialization from 'sport specialization'; this is essential when making specific and age-based recommendations. Despite the number of position statements and the passion of the rhetoric in this area, our understanding of the potential costs and benefits of early specialization is far from complete.

Author Contributions: Conceptualization, A.M., J.F.-T. and J.B.; Data curation, M.J.W.; Formal analysis, A.M.; Supervision, J.F.-T. and J.B.; Writing—original draft, A.M.; Writing—review & editing, A.M., J.F.-T., M.J.W. and J.B. All authors have read and agreed to the published version of the manuscript.

Funding: Support for initial data collection for this project was provided by a grant from the Ontario Ministry for Health Promotion's Research Program in Applied Sport Sciences.

Institutional Review Board Statement: The study was conducted in accordance with the Declaration of Helsinki, and approved by the Research ethics committee of Victoria University (HRETH 10/144. Date of approval 22 July 2010.

Informed Consent Statement: Informed consent was obtained from all subjects involved in the study.

Data Availability Statement: Data available upon request.

Conflicts of Interest: The authors declare no conflict of interest.

References

1. Mosher, A.; Fraser-Thomas, J.; Baker, J. What Defines Early Specialization: A Systematic Review of Literature. *Front. Sports Act. Living* **2020**, *2*, 596229. [CrossRef] [PubMed]
2. Côté, J.; Fraser-Thomas, J. Youth involvement and positive development in sport. In *Sport Psychology: A Canadian Perspective*, 3rd ed.; Crocker, P.R.E., Ed.; Pearson Prentice Hall: Toronto, ON, Canada, 2016; pp. 256–287.
3. Hill, G.M. Youth sport participation of professional baseball players. *Sociol. Sport J.* **1993**, *10*, 107–114. [CrossRef]
4. Jayanthi, N.A.; LaBella, C.R.; Fischer, D.; Pasulka, J.; Dugas, L.R. Sports-specialized intensive training and the risk of injury in young athletes: A clinical case-control study. *Am. J. Sports Med.* **2015**, *43*, 794–801. [CrossRef] [PubMed]
5. Wiersma, L.D. Risks and benefits of youth sport specialization: Perspectives and recommendations. *Pediatr. Exerc. Sci.* **2000**, *12*, 13–22. [CrossRef]
6. Baker, J.; Mosher, A.; Fraser-Thomas, J. Is it too early to condemn early sport specialization? *Br. J. Sports Med.* **2021**, *55*, 179–180. [CrossRef]
7. Brenner, J.S.; LaBella, C.R.; Brookes, M.A.; Diamond, A.; Hennrikus, W.; Kelly, A.K.W.; LaBotz, M.; Logan, K.; Loud, K.J.; Moffatt, K.A.; et al. Sports Specialization and Intensive Training in Young Athletes. *Pediatrics* **2016**, *138*, e20162148. [CrossRef]
8. Baker, J.; Young, B. 20 years later: Deliberate practice and the development of expertise in sport. *Int. Rev. Sport Exerc. Psychol.* **2014**, *7*, 135–157. [CrossRef]
9. Ericsson, K.A.; Krampe, R.T.; Tesch-Römer, C. The role of deliberate practice in the acquisition of expert performance. *Psychol. Rev.* **1993**, *100*, 363. [CrossRef]
10. Black, S.; Black, K.; Dhawan, A.; Onks, C.; Seidenberg, P.; Silvis, M. Pediatric Sports Specialization in Elite Ice Hockey Players. *Sports Health* **2019**, *11*, 64–68. [CrossRef]
11. Huxley, D.J.; O'Connor, D.; Larkin, P. The pathway to the top: Key factors and influences in the development of Australian Olympic and World Championship Track and Field athletes. *Int. J. Sports Sci. Coach.* **2017**, *12*, 264–275. [CrossRef]
12. Buckley, P.S.; Bishop, M.; Kane, P.; Ciccotti, M.C.; Selverian, S.; Exume, D.; Emper, W.; Freedman, K.B.; Hammoud, S.; Cohen, S.B. Early Single-Sport Specialization: A Survey of 3090 High School, Collegiate, and Professional Athletes. *Orthop. J. Sports Med.* **2017**, *5*, 2325967117703944. [CrossRef]
13. Brooks, M.A.; Post, E.G.; Trigsted, S.M.; Schaefer, D.A.; Wichman, D.M.; Watson, A.M.; McGuine, T.A.; Bell, D.R. Knowledge, Attitudes, and Beliefs of Youth Club Athletes Toward Sport Specialization and Sport Participation. *Orthop. J. Sports Med.* **2018**, *6*, 2325967118769836. [CrossRef]
14. Strachan, L.; Côté, J.; Deakin, J. "Specializers" versus "samplers" in youth sport: Comparing experiences and outcomes. *Sport Psychol.* **2009**, *23*, 77–92. [CrossRef]
15. Ferguson, B.; Stern, P.J. A case of early sports specialization in an adolescent athlete. *J. Can. Chir. Assoc.* **2014**, *58*, 377.
16. Miller, M.; Malekian, S.; Burgess, J.; LaBella, C. Evaluating a commonly used tool for measuring sport specialization in young athletes. *J. Athl. Train.* **2019**, *54*, 1083–1088. [CrossRef]
17. Smith, A.D.; Alleyne, J.M.; Pitsiladis, Y.; Schneider, C.; Kenihan, M.; Constantinou, D.; Webborn, N. Early sports specialization: An international perspective. *Curr. Sports Med. Rep.* **2017**, *16*, 439–442. [CrossRef]
18. DiCesare, C.A.; Montalvo, A.; Foss, K.D.B.; Thomas, S.M.; Hewett, T.E.; Jayanthi, N.A.; Myer, G.D. Sport specialization and coordination differences in multisport adolescent female basketball, soccer, and volleyball athletes. *J. Athl. Train.* **2019**, *54*, 1105–1114. [CrossRef]
19. DePhillipo, N.N.; Cinque, M.E.; Kennedy, N.I.; Chahla, J.; Moatshe, G.; LaPrade, R.F. Patellofemoral chondral defect in a preadolescent skier: A case report in early sport specialization. *Int. J. Sports Phys. Ther.* **2018**, *13*, 131. [CrossRef]
20. Russell, W.; Molina, S. A comparison of female youth sport specializers and non-specializers on sport motivation and athletic burnout. *J. Sport Behav.* **2018**, *41*, 330–350.
21. Güllich, A.; Macnamara, B.N.; Barth, M.; Hambrick, D.Z. Further Muddying the Waters? A Comment on Bell et al.'s 2021 Definition of Youth Sport Specialization. *J. Athl. Train.* **2021**, *56*, 1252–1254. [CrossRef]
22. Bell, D.R.; Post, E.G.; Trigsted, S.M.; Hetzel, S.; McGuine, T.A.; Brooks, M.A. Prevalence of sport specialization in high school athletics: A 1-year observational study. *Am. J. Sports Med.* **2016**, *44*, 1469–1474. [CrossRef]
23. Sport for Life. Long-Term Development in Sport and Physical Activity. 2022. Available online: https://sportforlife.ca/wp-content/uploads/2019/06/Long-Term-Development-in-Sport-and-Physical-Activity-3.0.pdf (accessed on 1 January 2022).
24. Hopwood, M.J.; Farrow, D.; MacMahon, C.; Baker, J. Sibling dynamics and sport expertise. *Scand. J. Med. Sci. Sports* **2015**, *25*, 724–733. [CrossRef]
25. Wilson, S.G.; Wilson, M.J.; Baker, J. Parental sport achievement and the development of athlete expertise. *Eur. J. Sport Sci.* **2018**, *19*, 661–670. [CrossRef]
26. Hopwood, M. The Developmental History of Athletes Questionnaire: Towards a Comprehensive Understanding of the Development of Sport Expertise. Ph.D. Thesis, Victoria University, Footscray, VIC, Australia, 2013.
27. Hopwood, M.; Baker, J.; MacMahon, C.; Farrow, D. Reliability and validity of the Developmental History of Athletes Questionnaire (DHAQ). *J. Exerc. Mov. Sport* **2010**, *42*, 113.
28. Dehghansai, N.; Lemez, S.; Wattie, N.; Baker, J. Training and development of Canadian wheelchair basketball players. *Eur. J. Sport Sci.* **2017**, *17*, 511–518. [CrossRef]

29. Bridge, M.W.; Toms, M.R. The specialising or sampling debate: A retrospective analysis of adolescent sports participation in the UK. *J. Sports Sci.* **2013**, *31*, 87–96. [CrossRef]
30. Swindell, H.W.; Marcille, M.L.; Trofa, D.P.; Paulino, F.E.; Desai, N.N.; Lynch, T.S.; Ahmad, C.S.; Popkin, C.A. An Analysis of Sports Specialization in NCAA Division I Collegiate Athletics. *Orthop. J. Sports Med.* **2019**, *7*, 2325967118821179. [CrossRef]
31. Coakley, J. The "logic" of specialization: Using children for adult purposes. *J. Phys. Educ. Recreat. Danc.* **2010**, *81*, 16–25. [CrossRef]
32. DiFiori, J.P.; Benjamin, H.J.; Brenner, J.S.; Gregory, A.; Jayanthi, N.; Landry, G.L.; Luke, A. Overuse injuries and burnout in youth sports: A position statement from the American Medical Society for Sports Medicine. *Br. J. Sports Med.* **2014**, *48*, 287–288. [CrossRef]
33. Smucny, M.; Parikh, S.N.; Pandya, N.K. Consequences of Single Sport Specialization in the Pediatric and Adolescent Athlete. *Orthop. Clin. N. Am.* **2015**, *46*, 249–258. [CrossRef]
34. Gulbin, J.P.; Oldenziel, K.E.; Weissensteiner, J.R.; Gagné, F. A look through the rear view mirror: Developmental experiences and insights of high performance athletes. *Talent Dev. Excell.* **2010**, *2*, 149–164.
35. Wilhelm, A.; Choi, C.; Deitch, J. Early Sport Specialization: Effectiveness and Risk of Injury in Professional Baseball Players. *Orthop. J. Sports Med.* **2017**, *5*, 2325967117728922. [CrossRef] [PubMed]
36. Post, E.G.; Trigsted, S.M.; Riekena, J.W.; Hetzel, S.; McGuine, T.A.; Brooks, M.A.; Bell, D.R. The association of sport specialization and training volume with injury history in youth athletes. *Am. J. Sports Med.* **2017**, *45*, 1405–1412. [CrossRef] [PubMed]
37. Pasulka, J.; Jayanthi, N.; McCann, A.; Dugas, L.R.; LaBella, C. Specialization patterns across various youth sports and relationship to injury risk. *Phys. Sport Med.* **2017**, *45*, 344–352. [CrossRef] [PubMed]
38. Bell, D.R.; Snedden, T.; Biese, K.; Nelson, E.; Watson, A.; McGuine, T.; Brooks, M.A.; Brown, R.; Kliethermes, S.A. Consensus Definition of Sport Specialization in Youth Athletes Using a Delphi Approach. *J. Athl. Train.* **2021**, *56*, 1239–1251. [CrossRef]

sports MDPI

Article

In Pursuit of a Comprehensive Understanding of Expertise Development: A Comparison between Paths to World-Class Performance in Complex Technical vs. Endurance Demanding Sports

Martine Aalberg [1], Truls Valland Roaas [1], Morten Andreas Aune [1], Øyvind Bjerke [2] and Tore Kristian Aune [1,*]

[1] Department of Sport Science, Sport and Human Movement Science Research Group (SaHMS), Nord University, 7600 Levanger, Norway; martine_aalberg@hotmail.com (M.A.); truls.v.roaas@nord.no (T.V.R.); morten.a.aune@nord.no (M.A.A.)
[2] Department of Teacher Education, Norwegian University of Science and Technology, 7491 Trondheim, Norway; oyvind.bjerke@ntnu.no
* Correspondence: tore.k.aune@nord.no; Tel.: +47-740-22-774

Abstract: A comprehensive understanding of skill acquisition is important for different performance domains, and has practical implications for both sport sciences and public health. The study compared important constraints for expertise development in a physically demanding sport (cross-country skiing) versus a technically demanding sport (freeskiing). Eighteen world-class athletes reported the importance of different constraints for their developmental history subdivided into two age spans: (1) 7–15 years and (2) 16 years until present. The total amount of training did not differ between the groups, but from the age of 16, the cross-country skiers spend approximately 98% of their training specific to their main sport, compared to 75% for freeskiers. No differences were found between the distribution of organized versus non-organized training in main sport, but freeskiers reported a higher amount of unorganized training in other sports after the age of 16. No differences were found in perceived importance of facilities, enjoyment of performing their sport, or the need for early specialization of training. After the age of 16, the cross-country skiers reported a higher need for coach involvement compared to freeskiers. The two sports mainly share common paths to expertise but differ in the need for specific training and coach involvement.

Keywords: expertise; specialization; freeskiing; cross-country skiing; deliberate play; deliberate practice

Citation: Aalberg, M.; Roaas, T.V.; Aune, M.A.; Bjerke, Ø.; Aune, T.K. In Pursuit of a Comprehensive Understanding of Expertise Development: A Comparison between Paths to World-Class Performance in Complex Technical vs. Endurance Demanding Sports. *Sports* **2022**, *10*, 16. https://doi.org/10.3390/sports10020016

Academic Editors: Richard B. Kreider and Jarek Mäestu

Received: 14 September 2021
Accepted: 20 January 2022
Published: 28 January 2022

1. Introduction

A comprehensive understanding of the complexity of skill acquisition is important for different performance domains of all ages and tasks and has practical implications for both sport sciences and public health (e.g., lifespan skill development, physical education, rehabilitation, elite sport). Studies of paths to high performance in different domains have the potential to elucidate and rate knowledge about significant constraints for the development of specific expertise, and for the improvement in skills in general. Over the recent decades, there has been an increase in expert performance development, in general, and, in particular, within sports performance. Expertise development in sport is a result of an arduous process that emerges as a result of interactions between multiple sport-specific constraints. There seem to be different pathways concerning how this expertise is accomplished and how different athletes achieve gold medals in Olympic sports [1]. Researchers have mainly pointed out two different pathways to achieve expert performance in sports [2]. The pathways are usually described as either a coach-led and highly sport-specific way termed "deliberate practice" or as a non-organized (peer-led) activity termed "deliberate play" [1].

In sports where peak performance is attained before puberty (e.g., women's gymnastics, figure skating), early specialization is often required to reach elite performance; however, in sports where athletes reach peak performance in their late 20s or early 30s, early diversification is described as the best pathway to expertise [2]. While there are several definitions of expertise, sport performance expertise is defined as the ability to consistently demonstrate superior athletic performance [3].

Expertise in sports is a result of interacting constraints; these different constraints are described as the numerous variables that form the expert's developmental trajectory [4]. The literature about expert performance development describes several constraints that are considered to be crucial to reach a level of expertise within a field. A crucial constraint that has been studied to a large extent is the amount of time spent on one's main sport. Simon and Chase [5] estimated that expert chess players had spent approximately 10,000 to 50,000 h playing or studying chess. They also found that a minimum of 10 years of training is necessary to reach an expert level, which is also revealed in the power law of practice by Newell and Rosenbloom [6]. The "10-year rule" has been widely used in the sports domain [7,8]. The validity of the "10,000-h rule" has been investigated [9–11], and the main critique is the lack of measures for variation in the amount of training leading to expertise. Gobet and Campitelli [10] studied the variation in the number of training hours required to reach the level of expertise and found that chess players spent between 3000 and 23,000 h of training to reach expertise, with the average number of hours being 11,000. Gobet and Campitelli [10] concluded that the number of hours with domain-specific training could not be the only requirement for achieving expertise. In the context of Olympic gymnastics, research found that the athletes had accumulated 18,835 training hours by the age of 16 [12]. The key point is diversity in accumulated training hours to reach expertise in different sports and for individual athletes, as shown by Güllich [13].

Another interesting constraint for expertise development is whether the athletes are a result of early specialization or early diversification. The ongoing debate is whether athletes become experts through early specialization in their main sport or through a varied and diverse road of experiences before specialization at a later stage [13]. There is no consensus in the definition of specialization in sports [14]; but it refers to athletes who focus on one sport and on specific training, often defined as intensive year-round training in one sport to the exclusion of other sports, participation in the main sport at least eight months of a year, or the termination of practice of all other sports to focus on the main sport [15,16]. This corresponds to the widely used conceptual term *deliberate practice* [7], where frequent repetitions and coach corrections in training are made, often through a monotone repetition in a specific context. Ericsson et al. [7] claimed that practice should be domain specific, and that individuals who started specializing early would always perform better than those who started late. This is exemplified with expert musicians, figure skaters, and soccer players who began practice at five years of age [8,9,17]. Deliberate practice is described as structured and organized activity that requires cognitive or physical effort with the primary goal of improving an important aspect of current performance [8]. Through coach-led organized training, the intention is to improve previous performance and deliberate efforts to change particular aspects of performance without immediate reward. In sports where peak performance is attained before puberty (e.g., women´s gymnastics, figure skating), early specialization is often required to reach elite performance [12]. The study by Law et al. [12] on rhythmic gymnastics demonstrated that Olympic gymnasts participated in fewer than two additional activities from age 4 to 16, stating that early specialization in their main sport is necessary to reach a world-class level.

A contrasting pathway to expertise is early diversification. Early diversification refers to reduced early sport-specific training and, instead, experience throughout various sports experiences that is hypothesized to facilitate later performance (see Güllich, [1]) for a more detailed description). Early diversification is related to the conceptual framework of deliberate play, which builds on the idea that sports activities are non-organized and self-regulated and that the athletes themselves are interested in discovering and exploring activities that are peer-led and without coach involvement [18]. Several studies have argued that deliberate play is important to improving motor skills, such as skill acquisition in Brazilian soccer players [19], or tactical creativity and tactical intelligence for basketball players [20].

Some evidence from team sports suggests that a considerable number of training hours in both deliberate practice and deliberate play can contribute significantly to the development of expertise in sports [21,22]. Furthermore, Soberlak and Côté [22] found that elite ice hockey players spent the same amount of time in deliberate play activities as deliberate practice activities before the age of 20. Using an ecological approach [23,24], sports practitioners can be seen as landscape designers for learning, shaping an environment, and adopting tasks to explore learning. With an ecological approach, the athletes are wayfinders that individually learn to self-regulate through unfamiliar landscapes in a skillful way. However, independent of categorization, training specificity is crucial for performance enhancement [25]. Presumably, there are several contents that are valuable for athletes, as Güllich [13] proclaimed, there are many roads that lead to Rome.

Research has shown that expertise in sports develops through interactions between an individual and their performance environment [26,27]. To understand pathways towards expertise, it is interesting to evaluate how the athletes themselves perceive the importance of different constraints on their expertise development. Constraints such as (1) physical facilities (access to training facilities and equipment) [28], (2) coach involvement [29,30], (3) athletes needing to be part of an organized group [31,32], (4) athletes feeling the enjoyment of training and competition [28,33], and (5) the need for specialization in their main sport [12,34].

The purpose of the present study is to examine potential similarities and differences in pathways to skill development and world-class expertise in complex technically versus physically demanding sports: (1) Freeski, which is regarded as a high demand of coordinative and technical skills, accompanied by courage [35], compared to (2) Cross-country skiing which is a very physically demanding sport that mainly requires high endurance capacity, high peak oxygen uptake (VO_{2PEAK}), movement-specific strength, and only a few repetitive skiing techniques [36–38].

As freeski is a relatively new discipline in alpine sports; there is limited research on how expert performance is acquired. Freeskiing has two subgenres: freeriding, otherwise known as big mountain skiing, which refers to skiing extremely steep runs off-piste and on exposed terrain; and freestyling, which consists of performing aerial tricks using kickers, half-pipes, or other obstacles. Performance in competition is scored on dynamical criteria such as progression, amplitude, variety, execution, and difficulty. Sports that require complex coordinative skills and techniques (e.g., gymnastics) have firm traditions of early specialization and coach-led practice. As peak performance is reached at an early age, there might be reasons to consider that this is the case for freeskiers, as well. There are similarities in the requirements for complex coordinative skills and techniques between freeskiers and gymnasts. However, a common perception is that freeskiers, as the term implies, are less or not dependent on coach-led activity.

In comparison, cross-country skiing is regarded as a very physically demanding sport where high-performance athletes are associated with high VO_{2max} [36,39–41]. Cross-country skiing has normally focused on the periodization of different training methods [42], different physiological [39,43], and biomechanical constraints [44], or a combination of the above [45]. Due to the high physical demands of cross-country skiing, young cross-country skiers normally follow a traditional coach-led and specified training regime influenced by prescriptive "best practices" [40,46].

Study Purpose

Based on the presented considerations, the aim of the current study was to investigate the impact of significant constraints on expert developmental history for world-class freeride skiers compared to cross-country skiers. Potential similarities and differences in the two groups on the road to expertise are described throughout the following subcategories of constraints:

1. Comparison of the total number of hours of training.
2. Early specialization versus early diversification: Comparison of time spent on main sport versus other sports.
3. Comparison of the distribution of organized versus non-organized training.
4. Comparison of the athletes' perceived importance of different constraints on expertise development.

2. Materials and Methods

Two groups of world-class winter athletes participated in this study: (1) freeskiers and (2) cross-country skiers. The selection criteria were that athletes had to be competing regularly at the highest level in their sport (e.g., World Cup, World Championship, Olympic Games, Freeride World Tour). The sample included 8 freeskiers (three females and five males) and 10 cross-country skiers (eight females and two males), 18 participants in total. See Table 1.

Table 1. Descriptive statistic for biographic information of the athletes in each sport.

Group	Freeriders (N = 8)			Cross-Country Skiers (N = 10)		
	Minimun	Maximum	Mean (SD)	Minimun	Maximum	Mean (SD)
Age range of the participants	19	37	27.5 (7.1)	22	32	27.5 (3.4)
Age for first organized training in main sport	6	16	11.3 (4.2)	6	16	9.1 (3.1)
Age for first unorganized training in main sport	4	16	10 (5.1)	4	15	7.3 (4.2)
Start age for systematic training in main sport	13	16	15 (1.2)	12	16	14.6 (1.3)

The participants were recruited by personal communication, and the questionnaire was sent to the participants by e-mail or social media. All participants received written consent to sign if they wanted to participate in the study. The participants were informed of the procedures, including information about the voluntary nature of participation and the possibility to withdraw from the study at any time without giving any reasons or facing any consequences. All participants received a survey link to a digital questionnaire. All participants provided written consent to participate in the study. The study was conducted according to the guidelines of the Declaration of Helsinki, and approved by the Norwegian Social Science Data Services (NSD) with reference number 502539.

The questionnaire was adapted from previous research [9,47,48] with some additional questions and customized adjustments. The study by Hodges and Starkes [47] examined athletes' training history by asking the athletes to recall their practice and play throughout their career and, likewise, examined the relevance of different activities. In addition, some customized adjustments were made in the questionnaire by adding questions about the impact of other significant constraints. These new constraints were the athletes' perceived importance of facilities, coach involvement, being part of an organized group, enjoyment, and specialization, which were not investigated in Helsen et al. [9], Hodges and Starkes [47], or Hodges et al. [48]. Due to the lack of research on athletes' perceived importance, new information and knowledge will be vital for a better understanding of athletes' expert development. A 5-point Likert-type scale was used to assess the athletes' ratings of the importance of the respective constraints.

2.1. Questionnaire

The first section of the questionnaire asked for biographical information about the athlete: (a) level of competition, (b) the age they initiated both organized and unorganized training, and (c) the age they began systematic training in their main sport. The second section asked for information about the athletes' training history. The training history was subdivided into two age spans: (1) 7–15 years and (2) 16 until the present. In both age spans, the subjects were required to estimate the number of hours of training in: (a) organized and unorganized training in their main sport, (b) organized and unorganized training in other sports, (c) training alone in their main sport, and (d) time spent in other activities in a typical week at this age. In the third and last part of the questionnaire, the athletes had to rate their perceived importance of different constraints. The questionnaire took approximately 15 min to complete.

2.2. Data Analysis

Athletes' recall of training volume was reported as hours of training during a year from the age of seven until the date of the survey. Mean training time in organized and unorganized training, as well as hours of training alone in main sport and other sports in both age groups, was found by multiplying the mean weekly time in practice by weeks within a year. This was later converted to percentages and used to analyze differences between the two groups.

Welch's unequal variances *t*-tests were used to determine differences between groups (freeskiers versus cross-country skiers), while dependent t-tests were used to determine differences between age categories in the respective sports. For independent *t*-tests, Cohen's (d_z) was applied as a measure of effect size [49–51], in which the criteria to interpret the magnitude of the ES were: 0.0–0.2 trivial, 0.2–0.6 small, 0.6–1.2 moderate, 1.2–2.0 large, and >2.0 very large [50,52]. A 95% confidence interval (CI) on the effect size between groups was used. All statistical calculations were performed with Predictive Analytics Software (PASW, IBM, US; previously SPSS) Version 26.0 with an alpha level of significance set at $p = 0.05$ as the criterion for statistical significance.

3. Results

Figure 1 reports mean accumulated training hours for both sports and age groups from the age of seven until the present day. The overall results showed no significant difference in mean accumulated training hours between the freeskiers and the cross-country skiers, except at the age of seven [$t(16) = 2.79$, $p = 0.039$, $d_z = 1.32$ (95% CI [0.72, 2.34])]. In total, the mean accumulated hours of training from seven years of age until the time of the survey is 12,162 h for the freeskiers and 10,698 h for the cross-country skiers.

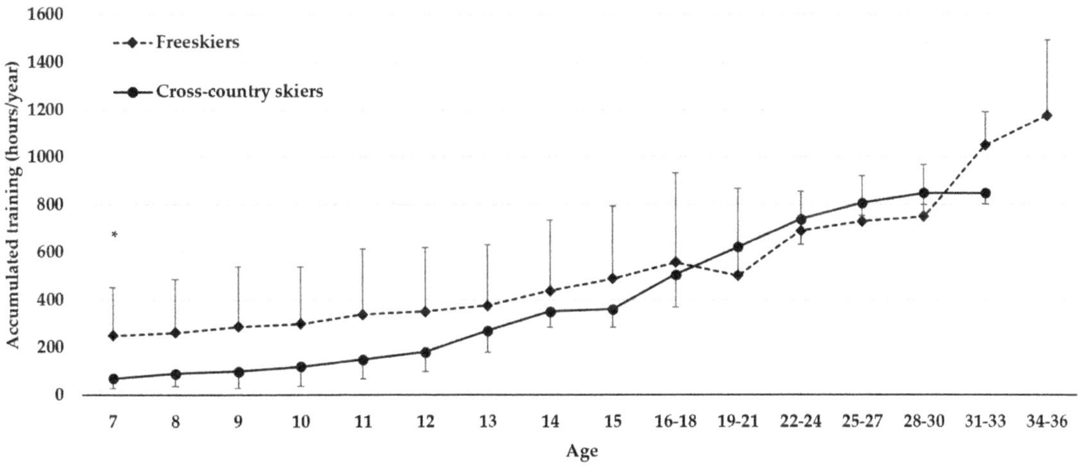

Figure 1. Total number (mean and standard deviation) of accumulated training hours from the age of seven until the time of the survey. * Indicates a significant difference between freeskiers and cross-country skiers.

3.1. Content and Organization of Training

Table 2 reports accumulated training for the freeskiers and the cross-country skiers for each age span. Between 7 and 15 years of age, the main sport accounted for approximately 63% of accumulated training hours for both freeskiers and cross-country skiers, while other sports accounted for the remaining 37% of the total training hours. From age 16 years until the time of the survey, there is a significant difference in the distribution total amount of training between the main sport and other sports for the freeskiers compared to the cross-country skiers. The freeskiers spend 75% of their total amount of training time in their main sport, while the cross-country skiers spend up to 98% of their total accumulated training time [t(16) = 3.94, $p = 0.009$, $d_z = 01.87$, (95% CI [0.72, 2.98])] in their main sport.

Table 2. Comparison of the distribution of total amount of training in main sport (specific training in main sport) vs. other sports for cross-country skiers and freeskiers. * Indicates a significant difference between freeskiers and cross-country skiers.

		Freeskiers 7–15 Years	Cross-Country Skiers 7–15 Years	Freeskiers 16+ Years	Cross-Country Skiers 16+ Years
		% (SD)	% (SD)	% (SD)	% (SD)
Main sport	% distribution of hours of training in main sport	63% (17.88)	63% (12.40)	75% (18.77)	98% * (2.85)
Other sports	% distribution of hours of training in other sport	37% (17.87)	37% (12.27)	25% (18.77)	2% * (2.85)

In Table 3, accumulated hours of training are divided between organized versus non-organized training. No significant differences were found between the freeskiers and the cross-country skiers in the distribution of organized versus non-organized training in the span of 7–15 years of age. In the age span from 16 years until the time of the survey, there is a significant difference in the distribution of organized versus non-organized training between the freeskiers and the cross-country skiers. The freeskiers spend 24% of their training in other non-organized sports, compared to only 1.8% for the cross-country skiers [t(16) = 4.05, $p = 0.008$, $d_z = 1.92$, (95% CI [0.76, 3.04])].

Table 3. Comparison of the distribution of total number of hours of organized vs. unorganized training between main sport and other sports for cross-country skiers and freeskiers. * Indicates a significant difference between freeskiers and cross-country skiers.

		Freeskiers 7–15 Years	Cross-Country Skiers 7–15 Years	Freeskiers 16+ Years	Cross-Country Skiers 16+ Years
		% (SD)	% (SD)	% (SD)	% (SD)
Organized training	% distribution of hours of organized training in main sport	21.9% (17.1)	29.1% (22.8)	9.8% (0) †	17.5% (49.0)
	% distribution of hours of organized training alone in main sport	24.4% (29.4)	16.3% (14.9)	23.4% (36.4)	30.1% (27.3)
Unorganized training	% distribution of hours of unorganized training in main sport	28.2% (31)	31.2% (37.2)	42.8% (15.2)	50.7% (14.4)
	% distribution of hours of involvement in other unorganized sports	25.5% (22.5)	23.3% (25.2)	24.0% * (48.4)	1.8% (9.3)

Note: † explaination of SD = 0; because only one athlete reported practicing the present type of training.

3.2. The Impact of Athletes' Perceived Importance of Different Constraints on Expertise Development

Evaluation of the athletes' perceived importance of various constraints is presented in Table 4. The results showed no significant difference in how the freeskiers and cross-country skiers rate the importance of the respective constraints in the age span of 7–15 years. Nevertheless, the results show that both coach involvement and being part of an organized group become more important for the cross-country skiers throughout their careers (from the age of 16 years until the present) compared to the freeskiers, [t(16) = 3.74, $p = 0.006$, $d_z = 1.79$, (95% CI [0.65, 2.87])] and [t(16) = 4.48, $p = 0.001$, $d_z = 2.13$, (95% CI [0.95, 3.29])], respectively.

Table 4. Freeskiers vs. cross-country skiers perceived importance of different constraints upon expertise development. * Indicates a significant difference between freeskiers compared to cross-country skiers.

Constraint	Freeskiers 7–15 Years	Cross-Country Skiers 7–15 Years	Freeskiers 16+ Years	Cross-Country Skiers 16+ Years
	Mean (SD)	Mean (SD)	Mean (SD)	Mean (SD)
Facilities	4.0 (1.31)	2.9 (1.10)	4 (1.31)	4.1 (0.57)
Coach involvement	3.25 (1.04)	3.9 (0.74)	2.25 (1.28)	4.0 * (0.67)
Being part of organized group	ND	ND	2.13 (1.13)	4.1 * (0.74)
Enjoyment	4.13 (0.64)	4 (1.05)	4.25 (0.46)	3.9 (0.88)
Specialization	2.25 (1.58)	2.2 (0.79)	3.75 (0.71)	4.1 (0.57)

Notes: 0 = low; 5 = high. No data (ND).

4. Discussion

The purpose of the present study was to examine potential similarities and differences in pathways to skill development of complex technically versus physically demanding sports in general, and the impact of significant constraints on expert development for world-class freeskiers compared to cross-country skiers in particular.

Essential constraints for expert development are the amount of training and the distribution of the type of training. While the results showed no difference in the total amount of training between the cross-country skiers and freeskiers, the distribution of the amount of training between the main sport and other sports differentiated. From the age of 16 onward, the cross-country skiers define approximately 98% of their training as specific to their main sport, compared to 75% for the freeskiers. No differences were found between the distribution of organized versus non-organized training for the cross-country skiers versus freeskiers in their main sport, but the freeskiers reported that they continued after the age of 16 with a significantly higher amount of unorganized training in other sports. In addition, the present study also evaluated how the athletes perceive the importance of significant environmental, organismic, and task constraints for their expert development. No differences were found for how the athletes perceived the importance of facilities, enjoyment of performing their sport, or the need for specialization, but interestingly, after the age of 16, the cross-country skiers reported a significantly higher need for coach involvement for their expertise development compared to the freeskiers.

The world-class cross-country skiers and freeskiers have engaged on average between 10,000 and 12,000 h of training, respectively. The present results are consistent with former research on the number of hours of training expert athletes invest towards their status as experts, suggesting a relationship between investment in training hours and expert performance [2,22,52]. Previous studies in cross-country skiing have shown an accumulation of 750–900 h of training per year [39,53], and, in total, an accumulation of 10,709 h from the age of 13 to 30. The most successful Olympic cross-country skier accumulated 14,300 h of training throughout her career [42]. While there are no earlier studies of freeskiers regarding the amount of training, the present results are in line with what is observed in adolescent alpine skiing and Olympic gymnasts [12,54].

While hours of training is a constraint for performance development towards expertise, the discussions about the variation in the amount of training leading to expertise could be more nuanced [13]. It is necessary to underline that several studies have shown a variation in accumulated training hours in the development of expert performance [10–12]. For example, Gobet and Campitelli [10] found large differences in the number of training hours required to reach the level of expertise, indicating that chess players spent between 3000 and 23,000 h of training to reach expertise. It is also necessary to emphasize that even relatively equal numbers of training hours do not ensure equal performance development, highlighting that the athletes' performance might unfold at individual and uneven speeds. There has been a general myth about differences between sport disciplines that are seemingly more-or-less organized and the amount of training necessary to reach expertise (i.e., that traditional organized endurance sports, such as cross-country skiing, are more demanding regarding training hours than unorganized sports, such as freeskiing). The present study challenges this point of view, and the results busted this myth. The key point must be individualized diversity in accumulated training hours to reach expertise, as shown by Güllich [13].

The present results show no difference in mean age between freeskiers and cross-country skiers regarding the age they started organized training in their main sport, respectively, 11.3 and 9.1 years of age. Both groups started remarkably later compared to what previous expertise research has reported, i.e., expert musicians, figure skaters and soccer players tend to start organized training at five years of age [7,9,17]. In addition, there is no difference between freeskiers and cross-country skiers in the distribution of training between their main sport and other sports in the age span of 7 and 15 years. Between 7 and 15 years of age, the main sport accounted for approximately 63% of the accumulated training hours for both freeskiers and cross-country skiers, while other sports accounted for the remaining 37% of total training hours. From the age of 16 onward, there is a significant difference in the distribution of the total amount of training between the main sport and other sports for freeskiers compared to cross-country skiers. The freeskiers spend 75% of their total amount of training in their main sport, while the cross-country skiers define as much as 98% of the total accumulated training in their main sport. Based on these findings, both sports consider specialization as a necessity for expertise development after the age of 16. However, Gobet and Campitelli [10] concluded that the number of hours with domain-specific training alone cannot be the only requirement for achieving expertise. Based on the current results, or former research as such, it is by no means possible to conclude whether expertise development in either freeskiing or cross-country skiing is dependent on early diversification or early specialization [14]. However, the results are supported by previous research, suggesting that training hours related to the conceptual terms deliberate practice and deliberate play contribute positively to expertise development [21,22].

The third aim of the present study was to compare differences in the organization of training as a constraint for expert development. The distribution of organized versus non-organized training for cross-country skiers versus freeskiers in the main sport was not significant, but the freeskiers reported that they continued after the age of 16 with a significantly higher amount of unorganized training in other sports. The results showed that freeskiers and cross-country skiers started participating in unorganized training in their main sport earlier than organized training. Freeskiers started participating in unorganized training at 10 years of age, while cross-country skiers started at 7.3 years of age. A plausible explanation for this is the sporting culture related to winter sports, with an unorganized entrance to training reinforced by family influence [31,55,56]. No significant differences were found between groups in their time spent in organized or unorganized training between 7 to 15 years of age, and both groups reported no significant difference in the number of hours in organized versus non-organized training during this age span. This is surprising, considering the nature of the two sports, i.e., cross-country skiing is seen as a highly organized sport and freeskiing is seen as an unorganized sport. However, a significant difference between freeskiers and cross-country skiers was found in the number of hours of involvement in non-organized sports for the age span of 16 and older. A plausible explanation for this may be found in the cultures and traditions of the two sports. Freeskiing is a relatively new discipline and is apparently open and less prescriptive regarding development. However, cross country skiing has strong traditions regarding training and best practices in the sport, as seen through the prescriptive developmental stages proposed by the Norwegian Ski Federation [31,53,57].

The fourth aim of the present study was to compare the athletes' perceived importance of the different constraints on expertise development. No significant differences were found between freeskiers and cross-country skiers in their perceived importance of facilities. Although the freeskiers rated the importance of facilities considerably higher than cross-country skiers between 7–15 years of age. It is interesting to note that freeskiers perceive the importance of facilities similarly in both age spans, while cross-country skiers viewed facilities to be of increasingly higher importance with age. The significant increase in training hours in the main sport for cross-country skiers might be an explanation of why facilities are perceived to be of greater importance above 16 years of age. The present results are in accordance with previous research and highlight the importance of natural accessibility to facilities for expert development [58,59].

There is no significant difference in rating between the perceived importance of coach involvement for freeskiers and cross-country skiers at the age of 7–15 years, but a significant difference was found between the two sports in the age span of 16 years and older. Freeskiers rate the importance of coach involvement below average, while cross-country skiers rate coach involvement as very important. These findings support the notion about freeskiers with a parallel to their involvement in non-organized activities (discovering their own pathway to expertise) and in contrast with the cross-country skiers' organized and prescriptive approach, making the need for coach involvement very different. The present results for cross-country skiers are in line with former research, showing that coach involvement is perceived to be necessary for expertise development as athletes grow older [60]. This could also be related to findings from Ericsson et al. [61] about deliberate practice, where coaches seemingly have the ability to maximize training time, make the "right" choices, prioritize what has to be improved, and create an optimal training environment for the athlete. The contextual differences can pinpoint an essential nuance to former research about coach involvement for expertise development.

It could be unnecessary to compare how freeskiers and cross-country skiers perceive the importance of being part of an organized group to develop expertise, since they are individual athletes. However, former research has shown that being part of a performance group with teammates is important for the development of individual athletes [31]. The present results show that cross-country skiers above 16 years of age perceived being part of an organized group as significantly more important compared to freeskiers. The cross-country skiers' ratings of being part of an organized group are supported by other studies [31,32], which argue that teammates are important to ensure quality in training, increase the effect of matching in training, and exchange experiences and knowledge. The findings are in line with previous studies in cross-country skiing, showing that former Olympic champions Thomas Alsgaard and Marit Bjørgen have emphasized the importance of discussions with teammates for their expertise development as cross-country skiers [42,62]. In contrast, freeskiers deviate from traditional rule-bound sport cultures because skill development is often conducted by individual discovery learning with subcultural similarities found in other action sports [63]. Due to these factors, these athletes place less importance on being part of an organized group for their expertise development. Freeskiing athletes have, from the historical beginning of the sport, manipulated task constraints (i.e., tricks, jumps, and other features in the environment) to create movement challenges and overload, resulting in the need for teammates, strict organization, coach instructions, and assessment to be unessential by design.

Both freeskiers and cross-country skiers have spent a substantial number of hours in organized training throughout their careers, and both groups rate the importance of organized training as average for their expertise development. Both groups consider organized training to be more important in early age (between 7–15 years of age) compared to older age (16 years and older), but no significant differences between groups were found. In addition, freeskiers also reported a high importance of extensive unorganized training, and confirmed their preference of independence, use of discovery learning, and deliberate play. This is in accordance with Ryan and Deci [64], who emphasized the athletes' desire to initiate their own actions, e.g., skiing individual lines off-piste or jumping over obstacles that are mostly prepared by themselves [65]. In addition, cross-country skiers perceive unorganized training as important, which might be interpreted as conflicting since they perceive coach involvement and being part of an organized group as important, as well. This might be explained by the governing policies in cross-country skiing that, on the one hand, recommend and encourage the development of independent athletes [57] but, on the other hand, have an indoctrinated system to emphasize the importance of best-organized practices and the significance of coaches. This will possibly lead athletes to find support and knowledge from coaches if they are seen to have a vital role in the planning and evaluation of training [42], but then execute the training without the coach present, making the appearance of training to be unorganized when training alone.

Acquisition of expertise requires athletes to invest time and effort, as well as have the passion and desire to improve performance. A collective term in this matter is enjoyment, which is seen as fundamental for prolonged activity, no matter the context. Both freeskiers and cross-country skiers rate enjoyment as highly important for their expertise development, and the results showed no significant differences in the rating of enjoyment between the two groups at either age span. High ratings for enjoyment at an early age can be an important factor for sport participation later in their career [56]. Nevertheless, nuances in the factors behind individual enjoyment must not be underestimated. In the original work of Ericsson et al. [7], deliberate practice towards expertise was not inherently enjoyable. In light of this alleged notion, a separation between practice that was high on relevance but low on enjoyment was put forward. A priori, this separation is speculative. Research has shown that when consequences of an activity are differentiated from its inherent enjoyment, enjoyment ratings are generally lower, yet still relatively high [48,66]. Studies have shown that demanding practice activities with high relevance can be enjoyable [9,47,66,67]. The collective term enjoyment could be viewed as polysemic because there is a range of causes behind athletes' personal preferences regarding enjoyment, and future research should nuance the multifaced nature of individual enjoyment.

Whether athletes become experts through early specialization in their main sport or through early diversification with various experiences before specialization at a later stage remains ambiguous. Neither freeskiers nor cross-country skiers reported that they perceive early specialization as particularly important between 7–15 years of age, but the results show an increase in the perceived importance of specialization in their main sport above the age of 16. No significant differences concerning the perceived importance of early specialization were found between freeskiers and cross-country skiers at either age span. The increase in the perceived importance of specialization combined with an increase in the number of training hours in their main sport is supported by previous studies [28,68], which found that athletes gradually specialized throughout their career. The ongoing debate of whether early specialization or early diversification is best for athletes is futile. Arguments are pervaded with agents' perspectives and confirmation biases concerning advantages. With certainty and through practical experience, it can be proclaimed that it is possible for athletes to reach an expert level in sports with either early specialization or diversification in the early stages of their sporting careers [69,70].

Practical Implications for Researchers, Coaches and Athletes

The results from the present study have to be interpreted according to the modest sample size. The total n is a consequence of examining the importance of different constraints in world-class freeski and cross-country skiing athletes, and when the inclusion criteria are restricted to top-ranked elite-level performers the total sample size has to be relatively low. Based upon the nature of expertise, the key practical implication emphasized to develop expertise is individualization, not generalization. Hence, there are some logical practical implications related to the present findings. Firstly, the problem when describing and comparing differences in the organization of training is the unnatural polarization of categorization. Researchers often attempt to place descriptive data in a distinct dichotomy manner, which leads to conclusive speculative "evidence" for athletes' pathways to expertise. Based upon the present results, the invented borders between play and practice are merely fulfilling a need to categorize training when it comes to expertise development, instead of providing nuanced directives based upon specific principles. Both coach-led and unorganized regimes could be helpful for expertise development, instead of biasing one side, researchers should recognize the individuality of an athlete's path to expertise. Observational, empirical, and longitudinal data can contribute to supporting the principle of individuality. It is perhaps time for researchers to change some of the terms used to describe experts' developmental pathways since the unsubtle conceptual frameworks of deliberate practice and play often act as an argument for "what to do". In order to reach world-class performance, there are numerous factors involved, but in general, everyone performing at the highest level has spent a tremendous number of hours training the sport they are experts in (specificity). Thus, for coaches, the principle of specificity in combination with a continuous increase in task and environmental complexity in an individualized setting is crucial in constructing representative learning conditions. The awareness for coaches when athletes make the transition into specializing and extensive training is also to "take the hand of" when the time is right in order to give the athlete self-awareness and allow them to be more self-regulated when exploring and discovering movement patterns and possibilities (individualization). Perhaps athletes should participate in more unfolding self-initiated training and positive competition with others—facilitating subjective enjoyment components. We proclaim that the multitude of different ways to organize training may not be captured in single terms. However, as world-class freeskiers reported, specificity is a crucial constraint to achieve expert performance, although organized training and coach involvement were less decisive constraints to attain expert performance from the age of 16. As a consequence, it is interesting to discuss whether the term unorganized training should be shelved or renamed with the more nuanced term self-organized training, and as a consequence, researchers, coaches, and athletes have to define what self-organize is in their particular context.

5. Conclusions

The purpose of the present study was to compare the similarities and differences of the important constraints on the development of expertise for world-class freeskiers versus cross-country skiers. In conclusion, freeskiers (complex technically demanding sport) and cross-country skiers (very physically demanding sport) in general share common elements and roads to expertise. However, interestingly, the observed higher need for specific training and coach involvement for cross-country skiers compared to freeskiers creates new questions for future research in the search for arbitrative constraints for specific expertise, and for skill development in general.

Author Contributions: Conceptualization, M.A., T.V.R., M.A.A. and T.K.A.; methodology, M.A., T.V.R., M.A.A., Ø.B. and T.K.A.; formal analysis, M.A., T.V.R., M.A.A., Ø.B. and T.K.A.; investigation, M.A. and T.K.A.; data curation, M.A. and T.V.R.; writing—original draft preparation, M.A., T.V.R., M.A.A., Ø.B. and T.K.A.; supervision, T.V.R. and T.K.A.; project administration, T.K.A. All authors have read and agreed to the published version of the manuscript.

Funding: This research received no external funding.

Institutional Review Board Statement: The study was conducted according to the guidelines of the Declaration of Helsinki, and approved by the Norwegian Social Science Data Services (NSD) with reference number 502539.

Informed Consent Statement: All participants received written consent to sign if they wanted to participate in the study. The participants were informed of the procedures, including information about the voluntary nature of participation and the possibility to withdraw from the study at any time without giving any reasons or facing any consequences. All participants received a survey link to a digital questionnaire. All participants provided written consent to participate in the study.

Data Availability Statement: The data presented are available on request from the corresponding author. The data are not publicly available due to rules of Norwegian Center for Research Data.

Conflicts of Interest: The authors declare no conflict of interest.

References

1. Güllich, A. International medallists' and non-medallists' developmental sport activities–a matched-pairs analysis. *J. Sports Sci.* **2017**, *35*, 2281–2288. [CrossRef] [PubMed]
2. Coutinho, P.; Mesquita, I.; Davids, K.; Fonseca, A.M.; Còtê, J. How structured and unstructured sport activities aid the development of expertise in volleyball players. *Psychol. Sport Exerc.* **2016**, *25*, 51–59. [CrossRef]
3. Janelle, C.M.; Hillman, C.H. Expert performance in sport: Current perspectives and critical issues. In *Expert Performance in Sports: Advances in Research on Sport Expertise*; Starks, J.L., Ericsson, K.A., Eds.; Human Kinetics: Champaign, IL, USA, 2003; pp. 17–47, 406–412. 38p.
4. Phillips, E.; Davids, K.; Renshaw, I.; Portus, M. Expert performance in sport and the dynamics of talent development. *Sports Med.* **2010**, *40*, 271–283. [CrossRef] [PubMed]
5. Simon, H.S.; Chase, W.G. Skill in chess. *Am. Sci.* **1973**, *61*, 394–403.
6. Newell, A.; Rosenbloom, P.S. Mechanisms of skill acquisition and the law of practice. *Cogn. Ski. Acquis.* **1981**, *1*, 1–55.
7. Ericsson, K.A.; Krampe, R.T.; Tesch-Römer, C. The role of deliberate practice in the acquisition of expert performance. *Psychol. Rev.* **1993**, *100*, 363–406. [CrossRef]
8. Baker, J.; Young, B. 20 years later: Deliberate practice and the development of expertise in sport. *Int. Review Sport Exerc. Psychol.* **2014**, *7*, 135–157. [CrossRef]
9. Helsen, W.F.; Starkes, J.L.; Hodges, N.J. Team sports and the theory of deliberate practice. *J. Sport Exerc. Psychol.* **1998**, *20*, 12–34. [CrossRef]
10. Gobet, F.; Campitelli, G. The role of domain-specific practice, handedness, and starting age in chess. *Dev. Psychol.* **2007**, *43*, 159–172. [CrossRef]
11. Tucker, R.; Collins, M. What makes champions? A review of the relative contribution of genes and training to sporting success. *Br. J. Sport Med.* **2012**, *46*, 555–561. [CrossRef]
12. Law, M.P.; Côté, J.; Ericsson, K.A. Characteristics of expert development in rhythmic gymnastics: A retrospective study. *Int. J. Sport Exerc. Psychol.* **2007**, *5*, 82–103. [CrossRef]
13. Güllich, A. Many roads lead to Rome–developmental paths to Olympic gold in men's field hockey. *Eur. J. Sport Sci.* **2014**, *14*, 763–771. [CrossRef] [PubMed]
14. Güllich, A.; Macnamara, B.N.; Hambrick, D.Z. What makes a champion? Early multidisciplinary practice, not early specialization, predicts world-class performance. *Perspect. Psychol. Sci.* **2021**, *17*, 1745691620974772. [CrossRef] [PubMed]
15. Hughes, J.N. Athletes from Football Bowl Subdivision Schools Perceptions of Sport Specialization and Its Relationship with Perfectionism. Ph.D. Thesis, Middle Tennessee State University, Murfreesboro, TN, USA, 2018.
16. Jayanthi, N.; Pinkham, C.; Dugas, L.; Patrick, B.; LaBella, C. Sport Specialization in Young Athletes: Evidence-Based Recommendations. *Sports Health Multidiscip. Approach* **2013**, *5*, 251–257. [CrossRef]
17. Starkes, J.L.; Deakin, J.M.; Allard, F.; Hodges, N.J.; Hayes, A. Deliberate practice in sports: What is it anyway? In *The Road to Excellence: The Acquisition of Expert Performance in the Arts, Sciences, Sports and Games*; Ericsson, K.A., Ed.; Erlbaum: Mahwah, NJ, USA, 1996; pp. 81–106.
18. Côté, J.; Baker, J.; Abernethy, B. Practice and play in the development of sport expertise. *Handb. Sport Psychol.* **2007**, *3*, 184–202.
19. Araújo, D.; Fonseca, C.; Davids, K.W.; Garganta, J.; Volossovitch, A.; Brandão, R.; Krebs, R. The role of ecological constraints on expertise development. *Talent Dev. Excell.* **2010**, *2*, 165–179.
20. Greco, P.; Memmert, D.; Morales, J.C. The effect of deliberate play on tactical performance in basketball. *Percept. Mot. Ski.* **2010**, *110*, 1–8. [CrossRef]
21. Baker, J.; Côté, J.; Abernethy, B. Learning from the experts: Practice activities of expert decision makers in sport. *Res. Q. Exerc. Sport* **2003**, *74*, 342–347. [CrossRef]
22. Soberlak, P.; Côté, J. The developmental activities of elite ice hockey players. *J. Appl. Sport Psychol.* **2003**, *15*, 41–49. [CrossRef]
23. Button, C.; Seifert, L.; Chow, J.-Y.; Araújo, D.; Davids, K. *Dynamics of Skill Acquisition: An Ecological Dynamics Rationale*; Human Kinetics: Champaign, IL, USA, 2020.
24. Woods, C.T.; Rudd, J.; Robertson, S.; Davids, K. Wayfinding: How ecological perspectives of navigating dynamic environments can enrich our understanding of the learner and the learning process in sport. *Sports Med. Open* **2020**, *6*, 51. [CrossRef]
25. Sigmundsson, H.; Trana, L.; Polman, R.; Haga, M. What is trained develops! theoretical perspective on skill learning. *Sports* **2017**, *5*, 38. [CrossRef] [PubMed]

26. Davids, K.; Baker, J. Genes, environment and sport performance. *Sports Med.* **2007**, *37*, 961–980. [CrossRef] [PubMed]
27. Farrow, D.; Baker, J.; MacMahon, C. *Developing Sport Expertise. Researchers and Coaches Put Theory into Practice*; Routledge: London, UK, 2008.
28. Baker, J.; Côté, J.; Abernethy, B. Sport-specific practice and the development of expert decision-making in team ball sports. *J. Appl. Sport Psychol.* **2003**, *15*, 12–25. [CrossRef]
29. Black, S.J.; Weiss, M.R. The relationship among perceived coaching behaviors, perceptions of ability, and motivation in competitive age-group swimmers. *J. Sport Exerc. Psychol.* **1992**, *14*, 309–325. [CrossRef]
30. Sinclair, D.A.; Vealey, R.S. Effects of coaches' expectations and feedback on the self-perceptions of athletes. *J. Sport Behav.* **1989**, *12*, 77.
31. Bloom, B. *Developing Talent in Young People*; Ballantine: New York, NY, USA, 1985.
32. Rogn, G.E. Lag i individuell idrett. In *Oppstart og Utvikling av Elitelag I Langrenn*; Norges idrettshøyskole: Oslo, Norway, 2008.
33. Scanlan, T.K.; Carpenter, P.J.; Lobel, M.; Simons, J.P. Sources of enjoyment for youth sport athletes. *Pediatric Exerc. Sci.* **1993**, *5*, 275–285. [CrossRef]
34. Côté, J.; Fraser-Thomas, J. Play, Practice, and athlete development. In *Developing Sport Expertise*; Farrow, D., Baker, J., MacMahon, C., Eds.; Routledge: London, UK, 2008; pp. 17–25.
35. Langran, M. Alpine skiing and snowboarding injuries. In *Adventure and Extreme Sports Injuries*; Springer: London, UK, 2013; pp. 37–67.
36. Karlsen, T.; Solli, G.S.; Samdal, S.T.; Sandbakk, Ø. Intensity control during block-periodized high-intensity training: Heart rate and lactate concentration during three annual seasons in world-class cross-country skiers. *Front. Sports Act. Living* **2020**, *2*, 153. [CrossRef]
37. Sandbakk, Ø; Holmberg, H.C. Physiological capacity and training routines of elite cross-country skiers: Approaching the upper limits of human endurance. *Int. J. Sports Physiol. Perform.* **2017**, *12*, 1003–1011. [CrossRef]
38. Solli, G.S.; Kocbach, J.; Seeberg, T.M.; Tjønnås, J.; Rindal OM, H.; Haugnes, P.; Sandbakk, Ø. Sex-based differences in speed, sub-technique selection, and kinematic patterns during low- and high-intensity training for classical cross-country skiing. *PLoS ONE* **2018**, *13*, e0207195. [CrossRef]
39. Sandbakk, Ø.; Holmberg, H.C. A reappraisal of success factors for Olympic cross-country skiing. *Int. J. Sports Physiol. Perform.* **2014**, *9*, 117–121. [CrossRef]
40. Solli, G.S.; Tønnessen, E.; Sandbakk, Ø. Block vs. traditional periodization of HIT: Two different paths to success for the world's best cross-country skier. *Front. Physiol.* **2019**, *10*, 375. [CrossRef] [PubMed]
41. Åstrand, P.O.; Rodahl, K.; Dahl, H.A.; Strømme, S.B. *Textbook of Work Physiology: Physiological Bases of Exercise*; Human Kinetics: Champaign, IL, USA, 2003.
42. Solli, G.S. The Development Process of the Most Successful Winter Olympian in History. Ph.D. Thesis, Norwegian University of Science and Technology, Trondheim, Norway, 2020.
43. Mahood, N.V.; Kenefick, R.W.; Kertzer, R.; Quinn, T.J. Physiological determinants of cross-country ski racing performance. *Med. Sci. Sports Exerc.* **2001**, *33*, 1379–1384. [CrossRef] [PubMed]
44. Zoppirolli, C.; Hébert-Losier, K.; Holmberg, H.C.; Pellegrini, B. Biomechanical determinants of cross-country skiing performance: A systematic review. *J. Sports Sci.* **2020**, *38*, 2127–2148. [CrossRef] [PubMed]
45. Lindinger, S.J.; Ishikawa, M.; Komi, P.V.; Ohtonoen, O.; Linnamo, V. *Methodological Aspects, Movement Control, and Performance in Cross-Country Skiing*; Second International Congress on Science and Nordic Skiing: Vuokatti, Finland, 2012.
46. Solli, G.S.; Tønnessen, E.; Sandbakk, Ø. The training characteristics of the world's most successful female cross-country skier. *Front. Physiol.* **2017**, *8*, 1069. [CrossRef] [PubMed]
47. Hodges, N.J.; Starkes, J.L. Wrestling with the nature of expertise: A sport specific test of Ericsson, Krampe, and Tesch-Römer's (1993) theory of deliberate practice. *Int. J. Sport Psychol.* **1996**, *27*, 400–424.
48. Hodges, N.J.; Kerr, T.; Starkes, J.L.; Weir, P.L.; Nananidou, A. Predicting performance times from deliberate practice hours for triathletes and swimmers: What, when, and where is practice important? *J. Exp. Psychol. Appl.* **2004**, *10*, 219. [CrossRef]
49. Cohen, J. Set correlation and contingency tables. *Appl. Psychol. Meas.* **1988**, *12*, 425–434. [CrossRef]
50. Lakens, D. Calculating and reporting effect sizes to facilitate cumulative science: A practical primer for t-tests and ANOVAs. *Front. Psychol.* **2013**, *4*, 863. [CrossRef]
51. Hopkins, W.; Marshall, S.; Batterham, A.; Hanin, J. Progressive statistics for studies in sports medicine and exercise science. *Med. + Sci. Sports+ Exerc.* **2009**, *41*, 3. [CrossRef]
52. Baker, J.; Côté, J.; Deakin, J. Expertise in ultra-endurance triathletes early sport involvement, training structure, and the theory of deliberate practice. *J. Appl. Sport Psychol.* **2005**, *17*, 64–78. [CrossRef]
53. Rusko, H. Physiology of cross-country skiing. In *Handbook of Sports Medicine and Science: Cross Country Skiing*; Blackwell Sceince: Malden, MA, USA, 2003; pp. 1–31.
54. DeCouto, B.S.; Fawver, B.; Taylor, T.; Williams, A.M. Physical fitness is associated with better technical performance in adolescent alpine ski racers after controlling for practice time: A retrospective regression analysis. *J. Sports Sci.* **2021**, *39*, 380–387. [CrossRef] [PubMed]
55. Appleton, P.R.; Hall, H.K.; Hill, A.P. Family patterns of perfectionism: An examination of elite junior athletes and their parents. *Psychol. Sport Exerc.* **2010**, *11*, 363–371. [CrossRef]

56. Côté, J. The influence of the family in the development of talent in sport. *Sport Psychol.* **1999**, *13*, 395–417. [CrossRef]
57. *Norges Skiforbund*; Utviklingstrappa: Oslo, Norway, 2013.
58. Gilberg, R.; Breivik, G. *Hvorfor ble de Beste Best? Barndom, Oppvekst og Idrettslig Utvikling hos 18 av Norges Mestvinnende Idrettsutøvere*; Norges idrettshøgskole: Oslo, Norway, 1998.
59. Tønnessen, E. *Hvorfor ble de beste best? En Casestudie av Kvinnelige Verdensenere I Orientering, Langrenn og Langdistanseløp*; Norges idrettshøgskole: Oslo, Norway, 2009.
60. Morgan, T.K.; Giacobbi, P.R., Jr. Toward two grounded theories of the talent development and social support process of highly successful collegiate athletes. *Sport Psychol.* **2006**, *20*, 295–313. [CrossRef]
61. Ericsson, K.A.; Hoffman, R.R.; Kozbelt, A.; Williams, A.M. (Eds.) *The Cambridge Handbook of Expertise and Expert Performance*; Cambridge University Press: Cambridge, UK, 2018.
62. Alsgaard, T. *Best på ski*; Tun Forlag: Oslo, Norway, 2008.
63. Thorpe, H.; Wheaton, B. 'Generation X Games', action sports and the Olympic movement: Understanding the cultural politics of incorporation. *Sociology* **2011**, *45*, 830–847. [CrossRef]
64. Ryan, R.M.; Deci, E.L. Self-determination theory and the facilitation of intrinsic motivation, social development, and well-being. *Am. Psychol.* **2000**, *55*, 68. [CrossRef]
65. Willmott, T.; Collins, D. A retrospective analysis of trick progression in elite freeskiing and snowboarding. *Int. Sport Coach. J.* **2017**, *4*, 1–12. [CrossRef]
66. Ward, P.; Hodges, N.J.; Williams, A.M.; Starkes, J.L. The road to excellence in soccer: A quasi-longitudinal approach to deliberate practice. *High Abil. Stud.* **2007**, *18*, 119–153. [CrossRef]
67. Helsen, W.F.; Hodges, N.J.; van Winckel, J.; Starkes, J.L. The roles of talent, physical precocity and practice in the development of soccer expertise. *J. Sports Sci.* **2000**, *18*, 727–736. [CrossRef]
68. Berry, J.; Abernethy, B. *Expert Game-Based Decision-Making in Australian Football: How Is It Developed and How Can It Be Trained?* University of Queensland, School of Human Movement Studies: Brisbane, Australia, 2003.
69. Güllich, A.; Faß, L.; Gies, C.; Wald, V. On the Empirical Substantiation of the Definition of "Deliberate Practice" (Ericsson et al., 1993) and "Deliberate Play" (Côté et al., 2007) in Youth Athletes. *J. Expertise/March* **2020**, *3*, 1.
70. Lidor, R.; Tenenbaum, G.; Ziv, G.; Issurin, V. Achieving expertise in sport: Deliberate practice, adaptation, and periodization of training. *Kinesiol. Rev.* **2016**, *5*, 129–141. [CrossRef]

sports

MDPI

Article

A Multidisciplinary Investigation into the Talent Development Processes in an English Premiership Rugby Union Academy: A Preliminary Study through an Ecological Lens

Francesco Dimundo [1,2,*], Matthew Cole [1], Richard C. Blagrove [3], Kevin Till [4] and Adam L. Kelly [1]

1 Research Centre for Life and Sport Sciences (CLaSS), Department of Sport and Exercise, School of Health Sciences, Birmingham City University, Birmingham B15 3TN, UK; matthew.cole@bcu.ac.uk (M.C.); Adam.Kelly@bcu.ac.uk (A.L.K.)
2 Worcester Warriors Rugby Football Club, Sixways Stadium, Worcester WR3 8ZE, UK
3 School of Sport, Exercise, and Health Sciences, Loughborough University, Loughborough LE11 3TU, UK; r.c.blagrove@lboro.ac.uk
4 Carnegie School of Sport, Leeds Beckett University, Leeds LS1 3HE, UK; K.Till@leedsbeckett.ac.uk
* Correspondence: francesco.dimundo@mail.bcu.ac.uk

Abstract: (1) Background: The progression of youth rugby union (RU) players towards senior professional levels can be the result of various different constraints. The aim of this study was to examine characteristics that differentiated playing positions and player rankings in an English Premiership RU academy. (2) Methods: Thirty players (mean age = 18.5 ± 2.8 years) were divided by playing positions (forwards = 18, backs = 12) and ranked (one to thirty) by coaches based on their potential to achieve senior professional status. Players were analysed across 32 characteristics from eight overreaching factors based on task, environmental, and performer constraints. MANOVA and ANOVA were used to calculate differences among variables in players' positions (i.e., forwards vs. backs) and ranks (i.e., top 10 vs. bottom 10), with a Welch's *t*-test applied to identify individual differences amongst groups and effect sizes calculated. (3) Results: Large effect sizes were found between groups for socioeconomic, sport activity, anthropometric, physical, and psychological factors. Moreover, environmental and performer constraints differentiated playing positions, whereas task and environmental constraints discriminated player ranks. (4) Conclusion: Present findings showed that playing positions and player ranks can be distinguished according to specific constraints.

Keywords: psychology; socioeconomic; social identity; physical; cognitive skills

Citation: Dimundo, F.; Cole, M.; Blagrove, R.C.; Till, K.; Kelly, A.L. A Multidisciplinary Investigation into the Talent Development Processes in an English Premiership Rugby Union Academy: A Preliminary Study through an Ecological Lens. *Sports* **2022**, *10*, 13. https://doi.org/10.3390/sports10020013

Academic Editor: Alun Williams

Received: 1 December 2021
Accepted: 17 January 2022
Published: 18 January 2022

Publisher's Note: MDPI stays neutral with regard to jurisdictional claims in published maps and institutional affiliations.

1. Introduction

One of the main challenges of youth development in rugby union (RU) is to predict future talents at both professional club and national governing body levels [1]. The talent development (TD) processes have been observed in sports' literature [2–7] and more recently applied to the context of RU [8]. There is currently an acceptance by clubs and organisations that the progression of RU players towards the top levels of competition is multifactorial, which can be underscored using an ecological dynamics theoretical approach [9]. This implies that developing expertise in RU cannot be the result of a single independent factor and is instead the result of a combination of *task* (i.e., participation history), *performer* (i.e., psychological, technical-tactical, anthropometric, physiological), and *environmental* (i.e., relative age, sociocultural) constraints [10,11].

The weaknesses of the TD processes in RU are represented by the limited multidimensional examinations of players within the current literature [8,12–14], which often do not take into account all the ecological constraints or analyse players' positions or playing levels [12]. This is despite evidence indicating that athletes' progression is largely affected by a range of factors, such as anthropometric [15], physiological [16,17], psychological [18,19], social identity [20], socioeconomic [21,22], and perceptual-cognitive expertise (PCE) [1,23] traits. Moreover, it has been found that the participation in adult-led practice and peer-led play in sport-specific and multisport activities [24,25], as well as the accumulation of hours of game-exposure at different ages [26], can impact the development of a young player. In addition, population density in the town of growth [27,28] and levels of deprivation [29] have been shown to have a significant impact on the TD opportunities and outcomes in RU. Indeed, researchers have recommended that future investigations in RU should consider these aspects in unison when studying professional academy contexts in order to better understand the holistic demands of the TD process [3,30].

A RU team is comprised of 15 playing athletes with a maximum of eight substitutes. The squad is generally split into forward (e.g., props, hooker, flankers) and back (e.g., inside centre, outside centre, full back) players. Forwards are those who normally engage first with opponents and are involved in set pieces and close high-force collisions. In contrast, backs are players who tackle the opposition at a later stage of the game, are engaged in rapid actions, and cover longer distances in high speed running [31,32]. Since the two main playing positions require the development of specific characteristics, a different TD path is often needed for these players [31]. Existing investigations on long-term athlete development (LTAD) pathways in RU have yet to elucidate the most suitable qualities to train forward and back players through an ecological lens [33]. In addition, there are only limited suggestions on how to differentiate playing levels based on player rankings in RU [34]. Indeed, questions remain surrounding the most appropriate processes that facilitate players' progression towards RU senior professional status, since sport organisations' *modus operandi* can often result in missing future professionals due to the pyramidal structure of the talent identification (TID) system, in which, at each stage of selection, the number of places for players to follow a development path decreases [35]. Unfortunately, existing research is yet to report a multidisciplinary investigation based on the aforementioned areas that are important for TD in RU. Thus, the present investigation aimed to examine a range of task, environmental, and performer constraints in an English Premiership RU academy. Specifically, both playing positions (i.e., forwards and backs) and player rankings (i.e., top-10 potentials vs. bottom-10 potentials) were analysed to: (a) offer a preliminary study to better understand the TD processes in RU, (b) provide professional RU academies a novel approach of assessing players, and (c) establish a methodological framework that may be useful for other researchers in the future.

2. Materials and Methods

2.1. Participants

Thirty players (under 16 (U16) = 11, U18 = 9, U21 = 10) from an English Premiership RU academy agreed to participate in this study. Table 1 reports the descriptive statistics of the participants. All participants were analysed based on playing position to compare possible differences (forwards = 18, backs = 12). They were also ranked on their potential to become a senior professional RU player, regardless of playing position and age, from one to thirty by three Level 4 academy coaches. Coaches ranked players using subjective criteria based on both their own vision of the game and personal philosophy of coaching. This produced a linear classification of higher-ranked players down to their lower-ranked peers, who were then split into thirds using tertiles. This created a cohort of 'top-10 potentials' ($n = 10$), who represent the top third, and a cohort of 'bottom-10 potentials' ($n = 10$), who represent the bottom third. This enabled a distinct comparison between the higher- and lower-ranked potentials across the group, with the middle third discarded from the player rank analysis ($n = 10$). Ethical approval was granted by the Faculty of Health, Education, and Life Sciences Research Ethics Committee at Birmingham City University.

Table 1. Descriptive statistics for forwards, backs, top-10 potentials, and bottom-10 potentials.

Factors	All Forwards (n = 18)	All Backs (n = 12)	Top-10 Potentials	Bottom-10 Potentials
	Mean ± SD (z-Score)	Mean ± SD (z-Score)	Mean ± SD (z-Score)	Mean ± SD (z-Score)
Age (year)	18.1 ± 3.1	18.4 ± 2.9	19.0 ± 2.9	18.7 ± 2.3
BQs	1.9 ± 1.1	2.0 ± 1.1	1.7 ± 1.0	2.0 ± 1.0
Task constraints				
Participation history				
Number of sports	2.9 ±1.8	3.7 ± 1.9	3.5 ± 2.1	3.5 ± 2.2
Sport activities				
Game exposure U8-U11 (h)	74.1 ± 47.5	99.0 ± 50.1	120.7 ± 52.3	59.8 ± 24.3
Coach-led U8-U11 (h)	300.8 ± 182.3	216.5 ± 131.3	296.4 ± 112.1	216.0 ± 193.8
Peer-led U8-U11 (h)	126.8 ± 159.0	81.0 ± 72.3	139.0 ± 209.1	82.7 ± 62.9
Game exposure U12-U15 (h)	226.1 ± 114.4	222.4 ± 93.0	234.8 ± 122.2	215.5 ± 71.7
Coach-led U12-U15 (h)	411.9 ± 274.1	343.6 ± 150.7	391.0 ± 175.5	368.4 ± 225.3
Peer-led U12-U15 (h)	255.2 ± 233.1	287.5 ± 316.2	311.5 ± 274.8	124.1 ± 48.3
Environmental constraints				
Socioeconomic				
Town population (AU)	4.7 ± 0.5	4.2 ± 1.0	4.3 ± 1.0	4.3 ± 0.8
IMD decile	7.7 ± 1.8	6.3 ± 1.2	6.7 ± 1.5	8.3 ± 1.2
Performer constraints				
Anthropometric				
Body mass (kg)	98.7 ± 11.6 (0.606 ± 0.745)	85.4 ± 7.5 (−0.908 ± 0.390)	96.0 ± 11.2 (−0.118 ± 0.906)	94.4 ± 11.9 (0.234 ± 1.017)
Height (cm)	180.4 ± 4.7 (0.025 ± 0.737)	171.9 ± 42.9 (−0.022 ± 1.114)	178.3 ± 6.2 (−0.304 ± 1.057)	163.6 ± 56.8 (0.001 ± 1.017)
Physical factors				
Hand grip (kg)	48.2 ± 5.7 (0.022 ± 0.966)	50.4 ± 5.0 (−0.025 ± 1.004)	52.6 ± 4.3 (0.072 ± 1.016)	46.1 ± 5.5 (−0.252 ± 1.102)
IHE (kg)	144.1 ± 16.7 (0.389 ± 0.650)	131.5 ± 20.2 (−0.583 ± 1.091)	147.2 ± 22.8 (−0.088 ± 1.035)	130.5 ± 13.5 (0.057 ± 1.067)
CMJ (cm)	35.6 ± 5.7 (−0.261 ± 0.862)	41.3 ± 3.5 (0.417 ± 1.014)	40.3 ± 4.8 (−0.225 ± 0.999)	35.4 ± 5.4 (−0.209 ± 0.734)
Peak power (W)	4585.53 ± 654.94 (0.539 ± 0.769)	4323.81 ± 476.07 (−0.927 ± 0.695)	4743.9 ± 644.5 (−0.218 ± 1.012)	4339.53 ± 586.3 (−0.084 ± 1.194)
Relative peak power (W/kg)	46.44 ± 3.62 (−0.366 ± 0.797)	50.55 ± 2.27 (0.291 ± 0.829)	49.36 ± 3.15 (−0.201 ± 0.930)	46.30 ± 3.56 (−0.356 ± 0.567)
RSI (m/m·s)	1.2 ± 0.3 (−0.256 ± 0.988)	1.7 ± 0.4 (0.383 ± 0.846)	1.8 ± 0.4 (0.219 ± 0.863)	1.2 ± 0.4 (−0.202 ± 0.845)
20 m sprint (s)	3.11 ± 0.19 (0.233 ± 0.949)	2.98 ± 0.13 (−0.333 ± 0.942)	2.97 ± 0.09 (−0.328 ± 0.537)	3.18 ± 0.19 (0.312 ± 0.906)
20 m momentum (m·s^{-1})	635.4 ± 76.7 (0.517 ± 0.745)	574.5 ± 57.1 (−0.792 ± 0.709)	647.6 ± 85.5 (−0.036 ± 0.952)	592.8 ± 68.8 (0.075 ± 1.176)
VO_2max (mL·kg^{-1}·min^{-1})	47.6 ± 5.0 (−0.333 ± 0.973)	52.7 ± 3.1 (0.517 ± 0.748)	54.2 ± 5.1 (0.299 ± 1.006)	46.7 ± 2.9 (−0.323 ± 0.844)
Psychological				
Factor 1—adverse response to failure (AU)	2.7 ± 0.6 (−0.083 ± 0.691)	3.2 ± 0.9 (0.117 ± 1.321)	3.1 ± 0.7 (−0.099 ± 1.003)	2.7 ± 0.7 (−0.285 ± 0.822)

Table 1. *Cont.*

Factors	All Forwards (n = 18)	All Backs (n = 12)	Top-10 Potentials	Bottom-10 Potentials
	Mean ± SD (z-Score)	Mean ± SD (z-Score)	Mean ± SD (z-Score)	Mean ± SD (z-Score)
Factor 2—imagery and active preparation (AU)	3.8 ± 0.8 (0.006 ± 0.978)	3.7 ± 0.9 (1.619 ± 0.990)	3.6 ± 0.7 (−0.262 ± 0.757)	3.7 ± 1.1 (0.050 ± 1.194)
Factor 3—self-directed control and management (AU)	4.4 ± 0.6 (−0.239 ± 0.996)	4.7 ± 0.5 (0.350 ± 0.847)	4.5 ± 0.8 (0.242 ± 0.971)	4.6 ± 0.6 (0.162 ± 1.072)
Factor 4—perfectionistic tendencies (AU)	3.1 ± 0.6 (−0.078 ± 0.869)	3.3 ± 0.6 (0.117 ± 1.128)	3.3 ± 0.4 (−0.025 ± 0.999)	3.0 ± 0.8 (−0.379 ± 0.997)
Factor 5—seeking and using social support (AU)	4.6 ± 0.6 (0.117 ± 0.875)	4.4 ± 0.7 (−0.158 ± 1.108)	4.5 ± 0.6 (−0.014 ± 0.938)	4.6 ± 0.6 (0.116 ± 0.836)
Factor 6—active coping (AU)	4.4 ± 0.5 (−0.289 ± 0.896)	4.7 ± 0.6 (0.442 ± 0.931)	4.3 ± 0.5 (−0.099 ± 0.962)	4.6 ± 0.6 (0.299 ± 1.005)
Factor 7—clinical indicators (AU)	2.0 ± 0.5 (0.167 ± 0.999)	2.0 ± 0.5 (−0.242 ± 0.866)	2.2 ± 0.4 (−0.012 ± 0.742)	1.8 ± 0.5 (−0.252 ± 0.987)
Perceptual-cognitive expertise				
PCE (AU)	3.1 ± 1.3 (0.033 ± 0.970)	2.8 ± 1.5 (−0.067 ± 1.013)	2.3 ± 1.5 (−0.431 ± 0.960)	2.7 ± 1.1 (−0.188 ± 0.907)
Social identity				
In group ties (AU)	6.1 ± 1.0 (0.128 ± 0.883)	5.7 ± 1.1 (−0.208 ± 1.052)	5.9 ± 0.8 (−0.047 ± 0.912)	5.8 ± 1.4 (−0.117 ± 1.212)
Cognitive centrality (AU)	4.9 ± 1.5 (0.083 ± 0.978)	5.2 ± 1.5 (−0.100 ± 0.989)	5.3 ± 1.7 (0.004 ± 1.034)	5.3 ± 1.6 (0.045 ± 1.160)
In group affect (AU)	6.6 ± 0.7 (0.061 ± 1.035)	6.5 ± 0.5 (−0.050 ± 0.923)	6.6 ± 0.5 (0.178 ± 0.871)	6.7 ± 0.4 (0.222 ± 0.818)
Total score SIQ (AU)	5.8 ± 0.9 (0.094 ± 0.967)	5.8 ± 0.9 (−0.158 ± 1.000)	5.9 ± 1.0 (0.030 ± 0.959)	5.9 ± 0.9 (0.037 ± 1.112)

Note: Shows descriptive difference between forwards and backs and top-10 and bottom-10 potentials. BQs = birth quartiles; IMD decile = index of multiple deprivation decile; SIQ = social identity questionnaire; IHE = isometric hip extension; CMJ = countermovement jump; RSI = reactive strength index; PCE = perceptual-cognitive expertise; VO_2max = maximal aerobic capacity; AU = arbitrary unit.

2.2. Procedure

Data were collected during the first 9 weeks of the 2019 pre-season where athletes were tested before afternoon training. Participants were instructed to follow a standardised training and recovery procedure in the 48 h before each physical testing session. All physical tests were proceeded by a familiarisation trial and were conducted during the same day. Each anthropometric and physical test was explained and demonstrated with physical assessment preceded by a standardised RAMP warm up, a type of activation similar to what players usually perform before training and competition (e.g., mobility, dynamic stretching, low level plyometrics, and running drills). The PCE video simulation test was performed in a room that comprised a setting similar to a classroom to enhance players' concentration and comfort at the club. Psychological, socioeconomic, social identity, and participation history were collected using validated questionnaires distributed via an online platform (Online surveys Jisc, Bristol, UK), which participants were asked to complete in their own time. In total, players were analysed over 32 characteristics from eight overreaching factors based on task (i.e., participation history and sport activities), environmental (i.e., socioeconomic), and performer (i.e., anthropometrical, physical, PCE, and social identity) constraints.

2.3. Task Constraints

Participation History

An adapted participation history questionnaire was used to gather the participants' engagement in activities throughout their youth [36]. Following the Developmental Model of Sport Participation (DMSP [24,37]), data were collected using estimated time (in hours) spent in RU competition, coach-led practice, and peer-led play between the ages of 8–11 and 12–15 years. The number of sports played until the age of 15 years was also recorded to provide information on the variety of players' motor ability and competency in basic and complex motor athletic skills. This study followed guidelines indicated previously [36].

2.4. Environmental Constraints

Socioeconomic Factors

The town where participants spent the most of their life during childhood and adolescence was recorded via an online questionnaire. The number of inhabitants and index of multiple deprivation decile was calculated using the UK government data available online [38]. The size of the town was ranked using the classification adopted by Cobley et al. [27], where the crescent number of inhabitant per town was labelled according to a number ranging from 1 to 5: 1 = 0–9999, 2 = 10,000–19,999, 3 = 20,000–49,999, 4 = 50,000–99,999, and 5 = 100,000–199,999. Moreover, according to the government norms, the index of multiple deprivation (IMD) decile reflected the players' socioeconomic situation from the most deprived (scored with '1') to the least deprived (scored with '10').

2.5. Performer Constraints

2.5.1. Anthropometric

Body mass and height were measured to the nearest 0.1 kg and 0.1 cm using calibrated Seca Alpha (model 220) scales and Seca Alpha stadiometer (Seca, Hamburg, Germany), respectively.

2.5.2. Physical

Isometric hip extension (IHE) strength was measured using a portable Takei Back and Leg Dynamometer (Takei Scientific Instruments Co., Ltd., Tokyo, Japan), whereby participants stood on a portable platform with knees fully extended, back in a neutral position, and hips flexed. The length of the handle's chain was set according to the participant's height by asking the subject to stand with extended knees. The handle was then positioned at the height of the intra-articular space of the knee joint. Subjects were instructed to lift vertically in order to generate isometric contractions of the extensors of the knees, hips, and lower back while pulling the handle as hard and as fast as possible for 5 s. Dominant handgrip strength was measured using the Takei 5401 Handgrip Dynamometer (Takei Scientific Instruments Co., Ltd., Tokyo, Japan). Participants performed the test sitting and holding the shoulder at 0° flexion, abduction and rotation, the elbow flexed at 90° and wrist positioned between 0° and 30° dorsiflexion and between 0° and 15° of ulnar deviation. Participants were instructed to 'squeeze' as hard as possible for 5 s and the best results of three attempts was recorded, with a 3-min rest between tests. Thus, the muscle strength primarily generated by the flexor muscles of the hand and the forearm could have been recorded. Strong verbal encouragement was provided during each repetition. These tests followed standardised validated procedure explained in previous literature [39,40]. All participants' positions for both the isometric hip extension and handgrip strength test were checked in a previous recording. The dominant hand was determined by asking the participants which hand they normally write with. Recorded measure from the two dynamometers consisted of the maximal force expressed in kg.

All participants were familiar with the CMJ as this was used frequently in testing and training at the club. Players performed three trials of a CMJ by jumping as high as possible while positioned between two parallel infrared beams (Microgate, OptoGait, Italy) and following a standard procedure already used in literature (e.g., [41]). After circa ninety seconds of recovery, players then completed three attempts for the reactive strength index (RSI) test whereby they performed ten consecutive jumps trying to reach maximal height for every bounce whilst spending as little time in contact with the ground between jumps as possible. RSI was calculated for each jump as the ratio between height (in metres) and contact time (in seconds). The best score of the three attempts on both tests was recorded. Peak power was calculated using Sayers equation [42]:

$$\text{Peak power (W)} = (60.7 \cdot \text{H}) + (45.3 \cdot \text{W}) - 2055$$

where 'H' refers to the CMJ height in cm; 'W' to body mass in kg.

Relative peak power ($W \cdot kg^{-1}$) was calculated by dividing peak power by player's body mass.

Sprint time over 20 m was recorded using timing gates (Brower Timing Systems, IR Emit. Draper, UT, USA). Timing gates were placed at the starting point and at 20-m distance. Following the warm-up, participants completed three maximal sprints from a staggered start with a 3 min passive rest between attempts. Each sprint started behind the initial timing gate (0.3 m), with players instructed to set off in their own time and run maximally through the final 20 m timing gate. Participants' starting point was checked before they were allowed to proceed. The best of the three attempts was taken for analysis with times measured to the nearest 0.01 s. Momentum was simply calculated by multiplying body mass and estimated final velocity over 20 m sprint, as previously used [16].

The 30–15IFT consisted of a 30 s shuttle run over a 40 m distance, interspersed with a 15-s recovery. The test began at 8 $km \cdot h^{-1}$ and increased by 0.5 $km \cdot h^{-1}$ at each successive running shuttle. All procedures were followed as reported in previous literature [43]. The test was terminated when participants were no longer able to maintain the imposed speed of the test or when they did not reach a 3 m tolerance zone on three consecutive occasions. The velocity from the last completed stage was noted and used to the estimate VO_2max ($mL \cdot kg^{-1} \cdot min^{-1}$) through the following formula [43]:

$$VO_2max\ (mL \cdot kg^{-1} \cdot min^{-1}) = 28.3 - (2.15 \cdot G) - (0.74 \cdot A) - (0.0357 \cdot W) + (0.0586 \cdot A \cdot V_{IFT}) + (1.03 \cdot V_{IFT})$$

where 'V_{IFT}' is the final running velocity; 'G' refers to gender (male = 1; female = 2); 'A' is age; 'W' is subject's body mass (kg).

2.5.3. Perceptual-Cognitive Expertise (PCE)

A perceptual-cognitive video simulation test was used to examine the participants' decision-making skills based on a combination of tactical situations, which have been used in RU literature [1] and demonstrated to produce valid and reliable measures for PCE research in several sport environments, e.g., [44]. Game situations of fifteen video clips were chosen from live rugby match footage, filmed from different elevated angles to provide a wide-range view of the pitch. Following moments of build-up play, the screen unexpectedly froze for 8 s prior to a critical decision-making moment. A game-related question with four possible options appeared and participants were required to select an answer on their response sheet before the next clip automatically began. As per examination conditions, participants were seated and were unable to engage with each other. Participants overall score was ranked using percentiles (i.e., 90th, 75th, 50th, and 25th) and then classified (i.e., 1 = excellent, 2 = good, 3 = average, 4 = low, and 5 = poor) for analysis. The total accuracy of the participants' responses was recorded for analysis.

2.5.4. Psychological Characteristics of Developing Excellence Questionnaire Version 2 (PCDEQ2)

To measure psychological characteristics, the seven factor (factor 1 = adverse response to failure, factor 2 = imagery and active preparation, factor 3 = self-directed control and management, factor 4 = self-directed control and management, factor 5 = seeking and using social support, factor 6 = active coping, and factor 7 = clinical indicators) and 88 item PCDEQ2 was used [19]. The answers were ranked with a Likert score ranging from 1 ('very unlike me') to 6 ('very like me') and then were converted into final scores on the seven factors. This conversion finally led to a score of one to ten for each of the seven items, as explained by Hill et al. [19].

2.5.5. Social Identity Questionnaire for Sport (SIQS)

The SIQS was used to evaluate players' social identity within their respective academy team. Nine items in a Likert score system (1 = 'strongly disagree' and 7 'strongly agree') reflected three underlying dimensions: (a) in-group ties (items 1–3), (b) cognitive centrality (items 4–6), and (c) in-group affect (items 7–9). SIQS total score was also calculated [20]. These data were collected via an online questionnaire that players were requested to complete in their own time.

3. Statistical Analysis

The Shapiro–Wilk test was used to check data normal distribution. Anthropometrical, physical, psychological, PCE, and SIQS scores were then normalised using *z*-scores ($z = (x - \mu)/\delta$), where x is the raw score, μ is the population (U16, U18, and U21) mean, and δ is the population standard deviation. A multivariate analysis of variance (MANOVA) was used to calculate the difference among the combined participation history, socioeconomic, social identity, psychological, anthropometric, and physical factors between both forwards and backs and top-10 and bottom-10 potential players. A one-way analysis of variance (ANOVA) was used to explore the differences for the cognitive test as it was comprised of one variable. A Welch's *t*-test was then conducted for all the variables to compare differences among players' positions and ranks. A Cohen's *d* was also used to calculate the effect size of these factors. Cohen's *d* effect size was calculated as reported in previous literature [45] with threshold values of 0.20 (small), 0.50 (medium), and 0.80 (large), with corresponding 95% confidence intervals (CIs). Significance was set for an α level of 0.05 with the statistical analysis conducted using IBM SPSS Statistics Version 24 (SPSS Inc., Chicago, IL, USA).

4. Results

The descriptive statistics are reported in Table 1. The MANOVA for training type factors, socioeconomic factors, social identity factor, psychological factors, anthropometric and physical factors, and the ANOVA for cognitive factors are reported in Table 2. The Welch's *t*-test analysis is reported in Table 3.

Table 2. MANOVA results for socioeconomic, social identity, anthropometric, physical, psychological, and sport activity factors, as well as ANOVA results for perceptual-cognitive expertise and participation history.

Factor	All Forwards vs. Backs		Top-10 vs. Bottom-10 Potentials	
	p	F	*p*	F
Socioeconomic	0.030 *	3.985	0.049 *	3.581
Social identity	0.918	0.231	0.963	0.144
Anthropometric	<0.001 **	22.135	0.604	0.519
Physical	0.004 *	4.340	0.784	0.548
Psychological	0.273	1.354	0.954	0.273
Perceptual-cognitive expertise	0.788	0.074	0.550	0.371
Sport activities	0.172	1.678	0.018 *	3.820
Participation history	0.270	1.268	0.963	0.002

Note: Significance set for $p = 0.05$; * denotes a statistical significance of ≤ 0.05; ** denotes a statistical significance of ≤ 0.001.

Table 3. Welch's t-tests for forwards and backs and top-10 and bottom-10 potentials.

Characteristic	Welch's t-Test (p)	Cohen's d
Number of sports		
Forwards vs. backs	0.275	−0.41 (−1.15; 0.32)
Top-10 potentials vs. bottom-10 potentials	0.963	−0.21 (−0.87; 0.83)
Game exposure U8-U11		
Forwards vs. backs	0.187	−0.51 (−1.25; 0.24)
Ranked top 10 vs. ranked bottom 10	0.003 *	0.80 (0.52; 2.48)
Coach-led U8-U11		
Forwards vs. backs	0.153	0.53 (−0.21; 1.26)
Top-10 potentials vs. bottom-10 potentials	0.266	0.50 (−0.37; 1.36)
Peer-led U8-U11		
Forwards vs. backs	0.296	0.37 (−0.37; 1.10)
Top-10 potentials vs. bottom-10 potentials	0.404	0.37 (−0.49; 1.23)
Game exposure U12-U15		
Forwards vs. backs	0.922	0.03 (−0.69; 0.76)
Top-10 potentials vs. bottom-10 potentials	0.661	0.19 (−0.66; 1.05)
Coach-led U12-U15		
Forwards vs. backs	0.439	0.29 (−0.44; 1.02)
Top-10 potentials vs. bottom-10 potentials	0.802	0.11 (−0.74; 0.96)
Peer-led U12-U15		
Forwards vs. backs	0.765	−0.11 (−0.84; 0.61)
Top-10 potentials vs. bottom-10 potentials	0.038 *	0.97 (0.05; 1.87)
Town population		
Forwards vs. backs	0.177	0.55 (−0.20; 1.30)
Top-10 potentials vs. bottom-10 potentials	0.880	−0.06 (−0.92; 0.79)
IMD decile		
Forwards vs. backs	0.020 *	0.88 (0.11; 1.64)
Top-10 potentials vs. bottom-10 potentials	0.015 *	−0.79 (−1.08; −0.22)
In group ties		
Forwards vs. backs	0.372	0.34 (−0.39; 1.08)
Top-10 potentials vs. bottom-10 potentials	0.870	0.07 (−0.78; 0.92)
Cognitive centrality		
Forwards vs. backs	0.622	0.18 (−0.54; 0.91)
Top-10 potentials vs. bottom-10 potentials	0.913	−0.04 (−0.90; 0.80)
In group affect		
Forwards vs. backs	0.761	0.11 (−0.61; 0.84)
Top-10 potentials vs. bottom-10 potentials	0.905	−0.05 (−0.90; 0.80)
Total score SIQ		
Forwards vs. backs	0.499	0.25 (−0.48; 0.98)
Top-10 potentials vs. bottom-10 potentials	0.939	−0.03 (−0.89; 0.82)
Body mass		
Forwards vs. backs	<0.001 **	0.82 (0.53; 1.53)
Top-10 potentials vs. bottom-10 potentials	0.427	−0.35 (−1.21; 0.51)

Table 3. *Cont.*

Characteristic	Welch's *t*-Test (p)	Cohen's *d*
Height		
Forwards vs. backs	0.890	−0.05 (−0.78; 0.68)
Top-10 potentials vs. bottom-10 potentials	0.504	0.29 (−0.56; 1.15)
Handgrip		
Forwards vs. backs	0.899	0.04 (−0.68; 0.77)
Top-10 potentials vs. bottom-10 potentials	0.505	0.29 (−0.56; 1.15)
IHE		
Forwards vs. backs	0.013 *	0.81 (0.25; 1.88)
Top-10 potentials vs. bottom-10 potentials	0.741	−0.14 (−1.00; 0.71)
CMJ		
Forwards vs. backs	0.050 *	−0.73 (−1.48; 0.02)
Top-10 potentials vs. bottom-10 potentials	0.959	−0.02 (−0.87; 0.83)
Peak power		
Forwards vs. backs	<0.001 **	0.81 (0.73; 0.90)
Top-10 potentials vs. bottom-10 potentials	0.718	−0.16 (−1.01; 0.70)
Relative peak power		
Forwards vs. backs	0.041 *	−0.80 (−1.56; −0.03)
Top-10 potentials vs. bottom-10 potentials	0.633	0.21 (−0.64; 1.07)
RSI		
Forwards vs. backs	0.070	−0.69 (−1.44; 0.06)
Top-10 potentials vs. bottom-10 potentials	0.280	0.48 (−0.39; 1.35)
20 m sprint		
Forwards vs. backs	0.121	0.59 (−0.15; 1.34)
Top-10 potentials vs. bottom-10 potentials	0.049 *	−0.83 (−1.72; 0.06)
20 m momentum		
Forwards vs. backs	<0.001 **	0.89 (0.90; 2.67)
Top-10 potentials vs. bottom-10 potentials	0.827	−0.09 (−0.95; 0.76)
VO_2max		
Forwards vs. backs	0.012 *	−0.98 (−1.74; −0.19)
Top-10 potentials vs. bottom-10 potentials	0.128	0.69 (−0.19; 1.57)
Factor 1		
Forwards vs. backs	0.637	−0.19 (−0.92; 0.54)
Top-10 potentials vs. bottom-10 potentials	0.624	0.21 (−0.64; 1.07)
Factor 2		
Forwards vs. backs	0.988	0.00 (−0.72; 0.73)
Top-10 potentials vs. bottom-10 potentials	0.459	−0.33 (−1.18; 0.53)
Factor 3		
Forwards vs. backs	0.094	−0.63 (−1.38; 0.11)
Top-10 potentials vs. bottom-10 potentials	0.851	0.08 (−0.77; 0.93)
Factor 4		
Forwards vs. backs	0.619	−0.19 (−0.92; 0.54)
Top-10 potentials vs. bottom-10 potentials	0.417	0.36 (−0.50; 1.22)

Table 3. *Cont.*

Characteristic	Welch's *t*-Test (p)	Cohen's *d*
Factor 5		
Forwards vs. backs	0.478	0.27 (−0.46; 1.00)
Top-10 potentials vs. bottom-10 potentials	0.744	−0.14 (−1.00; 0.71)
Factor 6		
Forwards vs. backs	0.043 *	−0.80 (−1.55; 0.02)
Top-10 potentials vs. bottom-10 potentials	0.351	−0.41 (−1.27; 0.45)
Factor 7		
Forwards vs. backs	0.245	0.43 (−0.30; 1.17)
Top-10 potentials vs. bottom-10 potentials	0.544	0.27 (−0.59; 1.12)
PCE		
Forwards vs. backs	0.790	0.86 (0.10; −0.63)
Top-10 potentials vs. bottom-10 potentials	0.550	−0.26 (−1.12; 0.59)

Note. Shows difference between forwards and backs and top-10 and bottom-10 potentials post-hoc and Cohen's d effect size (90% confidence interval). IMD decile = index of multiple deprivation decile; SIQ = social identity questionnaire; IHE = isometric hip extension; CMJ = countermovement jump; RSI = reactive strength index; PCE = perceptual-cognitive expertise; VO_2max = maximal aerobic capacity; * denotes a statistical significance of ≤0.05; ** denotes a statistical significance of ≤0.001.

5. Forwards vs. Backs

Results showed that there was a significant difference between playing positions for both anthropometric ($p < 0.001$) and physical ($p = 0.004$) factors. The Welch's *t*-tests reported large differences between players for IMD decile (forwards = 7.7 ± 1.8 vs. backs = 6.3 ± 1.2; $p = 0.020$, $d = 0.88$), body mass (forwards = 98.7 ± 11.6 kg vs. backs = 85.4 ± 7.5 kg; $p < 0.001$, $d = 0.82$), IHE (forwards = 144.1 ± 16.7 kg vs. backs = 131.5 ± 20.2 kg; $p = 0.013$, $d = 0.81$), CMJ (forwards = 35.6 ± 5.7 cm vs. backs = 41.3 ± 3.5 cm; $p = 0.050$, $d = 0.73$), peak power (forward = 4585 ± 654.9 W vs. backs = 4323 ± 476.0 W; $p < 0.001$, $d = 2.00$), relative peak power (forwards = 46.4 ± 3.6 W/kg, backs = 50.6 ± 2.3 W/kg; $p = 0.041$), 20 m momentum (forwards = 635.4 ± 76.7 $m \cdot s^{-1}$ vs. backs = 574.6 ± 57.2 $m \cdot s^{-1}$; $p < 0.001$, $d = 0.89$), VO_2max (forwards = 47.6 ± 5.0 $mL \cdot kg^{-1} \cdot min^{-1}$ vs. backs = 52.7 ± 3.1 $mL \cdot kg^{-1} \cdot min^{-1}$; $p = 0.012$, $d = 0.98$), and factor 6 (forwards = 4.4 ± 0.5 vs. backs = 4.7 ± 0.6; $p = 0.043$, $d = 0.80$). In addition, there was no significant differences between positions for the other variables.

6. Top 10 vs. Bottom 10

When examining groups based on coaches' rank, the analysis displayed statistical significance for socioeconomic ($p = 0.049$) and sport activities ($p = 0.018$) cumulative variables. The Welch's *t*-test showed significant differences among four different factors, whereby the top-10 players: (a) came from a more deprived area (6.7 ± 1.5 vs. 8.3 ± 1.2; $p = 0.015$, $d = 0.79$), (b) were more exposed to hours of rugby game when they were between 8 and 11 years old (120.7 ± 52.3 vs. 59.8 ± 24.3 h; $p = 0.003$, $d = 0.80$), (c) accumulated greater amount of time in training led by peers between ages 12 and 15 years (311.5 ± 274.8 vs. 124.1 ± 48.3 h; $p = 0.038$, $d = 0.97$), and (d) were faster over 20 m sprint (2.97 ± 0.09 s vs. 3.18 ± 0.19 s; $p = 0.049$, $d = 0.83$) compared to bottom-10 players. Moreover, despite small to moderate effect sizes among other variables, these were not statistically significant.

7. Discussion

Key findings revealed that environmental and performer constraints differentiated players based on positions. Academy forwards came from less deprived areas, were heavier, stronger, more powerful, and possessed greater momentum. Backs possessed greater relative peak power, RSI, VO_2max, and were characterised by superior active coping strategies (PCDEQ2 Factor 6) compared to forwards. Moreover, task and environmental constraints discriminated player ranks, whereby the top-10 potential players came from more deprived areas, were exposed to more RU competition between ages 8 to 11 years, accumulated a greater amount of engagement in peer-led play between ages 12 and 15 years, and were significantly faster over the 20 m sprint when compared to the bottom-10 potential players.

The IMD decile indicated that forwards originate from less deprived areas compared to backs (i.e., higher IMD score), possibly implying developmental differences in these players. Previous research from Winn et al. [29] found that more deprived young Welsh players engaged in less sports and accumulated less hours of rugby-specific training. In contrast to Winn et al. [29], however, although the present study revealed that backs originated from a *more deprived* areas, it does not reflect the fact that backs were excluded from sports (mean number of sports = 3.7 ± 1.9) and RU activities (e.g., games, coach-led practice, and peer-led play from U8 to U15), nor were *critically deprived* (e.g., IMD below 5). Several studies have attempted to analyse the influence of socioeconomic status on anthropometrical qualities in young RU players [21,22,46,47]. These investigations revealed that players with a lower socioeconomic status were physically smaller and lighter than those players from a higher status. According to present findings and the importance that some qualities have in characterising players in RU [31], the results on IMD decile provide an important indicator to consider when researching and developing young RU players in relation to their position, suggesting more investigation is needed on this aspect.

When analysing players according to their ranking, top-10 potentials came from *more deprived* areas compared to bottom-10 potentials (IMD decile = 6.7 ± 1.5 vs. 8.3 ± 1.2). Thus, it could be suspected that deprivation may help somehow in shaping characteristics useful for unlocking players' potential. As explained in the rocky road theory of Collins et al. [48], it is possible that the top-10 potentials had both the opportunity to challenge themselves and to have adequate social support to interpret adversities as positive growth experiences. Moreover, it could be speculated that deprivation reduces the engagement of young players with organised sport environments [29], whereas from another perspective it might increase vital opportunities of practice sport-related activities in deliberate play settings with parents, peers, and siblings [49]. In fact, a more enjoyable and peer-led environment has already been adopted from international professional RU teams to stimulate self-awareness, decision-making, tactical awareness, and in general, athlete functionality in adult players [50]. Therefore, this social discrepancy can lead to the possible theory that the IMD decile variable could help in forming attributes relevant to diverse playing positions (e.g., anthropometric, physical, psychological, social identity, PCE), as well as a higher ranking in RU academies. In this light, professional RU environments could add this parameter in a novel format of players' assessment.

From an anthropometric perspective, this investigation revealed that forwards were heavier than backs (98.7 ± 11.6 kg vs. 85.4 ± 7.5 kg, $p < 0.001$). This is in agreement with previous results across RU academies [31], senior squads [51], and clubs from different countries [12]. Due to players' positional requirements, a higher body mass in forwards aids in attenuating impacts during tackles and collisions [52]. The variation in anthropometric measures among playing positions consolidates how forwards and backs require diverse anthropometric characteristics in order to perform position-specific tasks during games [31]. From a ranking viewpoint, although not statistically significant, top-10 potentials were heavier than bottom-10 potentials, indicating this may be important for players to succeed in an academy. Recent studies demonstrated how body mass was pivotal to distinguish selected and non-selected academy players in England [53], predict players' progression in an Italian academy [54], as well as to discriminate positions in South African [52], Zimbabwean [55], and Argentinian [56] academy environments. Therefore, coaches should consider the importance of body mass in developing players and their progression across an academy. However, practitioners should be aware that players of the same chronological age can differ in their maturity status, and therefore caution should be placed when selecting players based on morphology parameters only.

Physical parameters have been shown to differentiate both playing positions [31] and age-grade players [56], as well as to distinguish levels [53] in RU academies. In the present investigation, forwards were significantly stronger than backs in the IHE test (144.1 ± 16.7 kg vs. 131.5 ± 20.2 kg, $p = 0.013$), demonstrating the importance of this physical characteristic for this playing position. One of the reasons why forwards are typically stronger than backs is because these players are required to produce higher maximal isometric force during games in holding scrums and competing for the ball in rucks and mauls when compared to backs [32,52]. Together, these findings indicate that different aspects of strength should be developed in RU academies according to players' individual needs.

Sprint momentum has been defined as a key parameter for performance in RU, as well as differentiating playing levels [56,57] and playing positions [31] in various academy settings across the globe. In the current study, forwards performed 20 m sprint momentum similar to results from U18 forwards in a previous investigation [33] (637.6 ± 91.9 $m \cdot s^{-1}$ vs. 635.4 ± 76.7 $m \cdot s^{-1}$). Present results suggest that forwards outperformed backs due to their heavier body mass. Specifically, when a heavier body reaches a higher velocity, it possesses a greater kinetic energy compared to a lighter body. For instance, maximising sprint momentum through increasing body mass while maintaining linear speed capabilities appears to be an important characteristic for forwards to possess, since such a position involves ball carrying in situations where contact is unavoidable [57]. From a ranking point of view, momentum did not statistically differentiate top-10 from bottom-10 potentials, however, top-10 potentials recorded a medium effect size difference compared to bottom-10 potentials, suggesting that this parameter should be trained in TD environments.

Findings from the CMJ and power-related measures reported that backs jumped significantly higher and possessed greater relative peak power than forwards, indicating that these players could had superior jumping technique and were able to express more power per kg of body mass when compared to forwards (CMJ = 35.6 ± 5.7 cm vs. 41.3 ± 3.5 cm, $p = 0.05$, $d = -0.73$; relative peak power = 46.44 ± 3.62 W/kg vs. 50.55 ± 2.27 W/kg, $p = 0.041$, $d = -0.80$). Similar results were found between positions in a LTAD study within RU academies on CMJ [58] and relative peak power analysis [33]. An explanation for backs' possessing greater jumping performance and relative peak power is that these factors contribute to optimising linear sprints, changes of direction, agility, and to achieving higher speed from different starting positions during games. Similar to the findings of Howard et al. [59], peak power was significantly greater in forwards than backs in this current study (4585 ± 654 W vs. 4323 ± 476 W), indicating that, in general, players from this playing position often rely on this physical parameter during powerful actions of a match (e.g., closer stance explosive tackles). From a rank perspective, although top-10 potentials recorded superior CMJ, peak power, and relative peak power compared to bottom-10 potentials players, it was not statistically significant. Together, these results indicate that the evaluation and development of power-related qualities should be included in the RU TD process.

In the present study, 20 m sprint was the only physical factor that distinguished the top-10 and bottom-10 potentials, whilst no significant differences were found among positions. Sprint time has recently been shown to be a key factor in TID and TD processes in RU [53]. Moreover, sprint ability was linked both to different levels of RU [17,57], as well as different age groups and positions in different countries [31,56,57,60]. Speed has been increasingly recognised as important by RU practitioners since RU games are becoming more dynamic and faster than previous years [61]. Another possible explanation is that, as per body mass characteristics, sprinting speed has been correlated to momentum, which is a key component in RU matches [57]. Therefore, practitioners are encouraged to focus on maximising the development of the different phases of sprint mechanics in academies.

Aerobic capacity was estimated using the 30–15IFT. The only statistically significant difference was found between positions, whereby backs had a greater VO_2max when compared to forwards (52.7 ± 3.1 $mL \cdot kg^{-1} \cdot min^{-1}$ vs. 47.6 ± 5.0 $mL \cdot kg^{-1} \cdot min^{-1}$, $p = 0.012$), which aligns with previous literature [62]. Indeed, backs are generally leaner and have a lower body fat percentage when compared to forwards, which may have facilitated a superior aerobic profile when expressed relative to body mass [31]. Moreover, the specific demands of forwards requires them to cover less distance in a game when compared to backs [63], which may be explained with the present findings. Although not statistically significant, the top-10 potentials possessed greater VO_2max when compared to the bottom-10 potentials ($d = 0.69$), suggesting that this may have a certain degree of importance to differentiate ranks in players. Therefore, aerobic capacity should be trained based on position during a LTAD pathway [35] and be part of an assessment battery in RU.

Previous studies attempted to distinguish psychological traits in different playing positions [34,64,65], ranking [34,64,66], and based on coaches' perspectives [67] across RU players. Specifically, existing literature shows that forwards generally possess greater psychological skills, such as relaxation, stress reaction, and fear control [64,65] when compared to backs. Indeed, only one study [34] has shown that both forwards and backs possess equally good psychological traits (i.e., determination, goal directedness, self-confidence, concentration, and mental preparation). On the contrary, however, the results from the present study showed how backs were characterised by superior perceived active coping strategies (PCDEQ2 Factor 6) when compared to forwards. It is plausible to suggest that backs may experience more pressurised situations during competitive match-play when compared to forwards, since their role includes critical moments, such as executing penalty kicks and kicking conversions which require quick decision-making skills. Moreover, since the current study showed how backs come from higher deprivation, it could be speculated that a greater perceived active coping was a result of an adaptation to a more challenging socioeconomic environment during their development. However, further research is required to substantiate these suggestions and explore the association between socioeconomic status and the development of psychological characteristics in talent pathways. No significant differences were reported in psychological variables between top-10 and bottom-10 potential players. Thus, the present findings could be used to help explain the role of the environment and psychological development in RU players and guide future research.

With regards to the engagement in sport activities (i.e., game exposure, coach-led practice, and peer-led play), there were no positional differences at both aged 8–11 and 12–15 years. In comparison, however, the top-10 potentials engaged in more hours of game exposure at a younger age (i.e., aged 8–11 years) and accumulated more time in peer-led play during late childhood and early adolescence (i.e., aged 12–15 years) when compared to the bottom-10 potentials. An early exposure to competition has been considered an important part of the athlete development process [26,68], which aligns with the understanding that young players should be exposed to various enjoyable games that gradually produce more demanding performance-specific situations with an older age [18]. Similar to the present results, in a recent meta-analysis from Güllich et al. [68] it was reported that although world champions started their main sport at a later stage in life, higher performing athletes accumulated significant early exposure of their main sport than lower performers ($p = 0.010$; $d = 0.20$). In handball, for instance, Bjørndal et al. [26] stated that an early exposure to the competitive experience represented a vital part for player development towards their high performer status. Thus, coaches should take into account the potential long-term benefits that high-quality game exposure could have on players' status. With regards to player rankings, the top-10 potentials accumulated a greater number of hours in peer-led play between aged 12–15 years when compared to the bottom-10 potentials. Although these findings report controversies with conclusions of a recent study on athletes' progression [69], they align with rugby league research that has shown the importance of peer-led activities in development of professional players [64,70]. Thus, more varied learning experiences during early-adolescence could facilitate a later rugby-specific skill learning and refinement [71]. From an ecological dynamic perspective [9,10], it is possible to explain present results through the variation in learning tasks and environments, which may facilitate a players' ability to adapt their actions in learning and to familiarise their movement across various unpredictable environments (i.e., enhanced athlete functionality, see Rothwell et al., [50]). As such, a players' later exposure to peer-led play may continue during the transition from childhood to adolescence, which is a crucial stage for young RU players since they are generally selected to be part of a professional academy for the first time (i.e., at U15).

Overall, these findings offer a preliminary study to better understand the TD processes in RU, provide professional RU academies a novel approach of assessing players, and establish a methodological framework that may be useful for other researchers in the future.

8. Limitations and Future Directions

One limitation of this study was the small number of participants. A larger sample may have altered the outcomes of the current findings, especially those in relation to ranked players [72]. Another limitation of this study was that no age-related differences were investigated (i.e., it could be possible that different ages influenced players' ranks). However, the novelty of this study also compares those who have already been selected into an academy environment through analysing potential to achieve senior professional status, rather than the traditional approach of comparing 'elite' vs. 'non-elite' or 'selected' vs. 'non-selected', thus further limiting the prospective pool of participants. Moreover, it is important to mention that present results only reflect the status of a single Premiership RU academy, and thus it is possible that this is not representative of other environments in RU. However, other studies surrounding TD in RU [16] and football [44] adopted similar methodological procedures when analysing academies of professional clubs. Furthermore, some data were collected retrospectively (e.g., game exposure, peer-led play, and coach-led training), and therefore recall bias may have influenced findings. Nevertheless, previous research has applied these tools and demonstrated a good level of reliability and validity (e.g., [69]). In addition, due to the large number of data collection methods required to be completed in order to be included in the current study, only those academy players who conducted all the measures were analysed. Therefore, it is important to recognise that this study may have not considered participants whose results may have changed the outcomes should they have completed all the protocols. However, due to these limitations, this study was denoted as a preliminary investigation to ensure the reader acknowledges the exploratory nature of the research being performed. Thus, the present investigation can be used to guide future research methodologies which are encouraged to maintain a multidisciplinary approach and use a longitudinal protocol with a greater and more diverse sample.

9. Conclusions

To the authors' knowledge, this is the first multidisciplinary study that has analysed 32 characteristics from eight overarching factors in an English Premiership RU academy through an ecological dynamic lens. Present findings showed how playing positions can be differentiated by environmental and performer constraints. Moreover, top-10 potential players were distinguished from bottom-10 potential players in task and environmental constraints. Rugby practitioners are encouraged to follow a similar multidisciplinary approach and use these findings as framework when assessing professional academy players. Researchers could also use the methodology employed in this investigation as the basis for future work in this area.

Author Contributions: Conceptualization, F.D. and A.L.K.; methodology, F.D. and A.L.K.; software, F.D.; validation, F.D.; formal analysis, F.D.; investigation, F.D. and A.L.K.; resources, F.D.; data curation, F.D.; writing—original draft preparation, F.D.; writing—review and editing, F.D., A.L.K., M.C., R.C.B., K.T.; visualization, A.L.K., M.C., R.C.B., K.T.; supervision, A.L.K.; project administration, F.D. and A.L.K. All authors have read and agreed to the published version of the manuscript.

Funding: This research received no external funding.

Institutional Review Board Statement: The study was conducted in accordance with the Declaration of Helsinki, and approved by the Institutional Review Board (or Ethics Committee) of Birmingham City University (protocol code 3334 /R(B) /HELS FAEC and date of approval 24 July 2019).

Informed Consent Statement: Informed consent was obtained from all subjects involved in the study.

Data Availability Statement: MDPI Research Data Policies.

Conflicts of Interest: The authors declare no conflict of interest.

References

1. Sherwood, S.; Smith, T.; Masters, R.S. Pattern Recall, Decision Making and Talent Identification in Rugby Union. *Eur. J. Sport Sci.* **2019**, *19*, 834–841. [CrossRef]
2. Clarke, N.J.; Cushion, C.J.; Harwood, C.G. Players' Understanding of Talent Identification in Early Specialization Youth Football. *Soccer Soc.* **2018**, *19*, 1151–1165. [CrossRef]
3. Cobley, S.; Schorer, J.; Baker, J. Identification and Development of Sport Talent: A Brief Introduction to a Growing Field of Research and Practice. In *Talent Identification and Development in Sport*; Routledge: Abingdon, UK, 2013; pp. 21–30.
4. Côté, J.; Baker, J.; Abernethy, B. Practice and Play in the Development of Sport Expertise. In *Handbook of Sport Psychology*; Tenenbaum, G., Eklund, R.C., Eds.; Wiley: New York, NY, USA, 2007; pp. 184–202.
5. McCarthy, N.; Collins, D. Initial Identification & Selection Bias versus the Eventual Confirmation of Talent: Evidence for the Benefits of a Rocky Road? *J. Sports Sci.* **2014**, *32*, 1604–1610. [PubMed]
6. Reilly, T.; Williams, A.M.; Nevill, A.; Franks, A. A Multidisciplinary Approach to Talent Identification in Soccer. *J. Sports Sci.* **2000**, *18*, 695–702. [CrossRef] [PubMed]
7. Vaeyens, R.; Lenoir, M.; Williams, A.M.; Philippaerts, R.M. Talent Identification and Development Programmes in Sport. *Sports Med.* **2008**, *38*, 703–714. [CrossRef] [PubMed]
8. Till, K.; Barrell, D.; Lawn, J.; Lazenby, B.; Rock, A.; Cobley, S. 'Wide and Emergent–Narrow and Focussed': A Dual-Pathway Approach to Talent Identification and Development in England Rugby Union. In *Talent Identification and Development in Sport*; Routledge: Abingdon, UK, 2020; pp. 170–183.
9. Davids, K.; Araújo, D.; Hristovski, R.; Passos, P.; Chow, J.Y. Ecological Dynamics and Motor Learning Design in Sport. In *Skill Acquisition in Sport: Research, Theory and Practice*; Routledge: Abingdon, UK, 2012; pp. 112–130.
10. Araújo, D.; Fonseca, C.; Davids, K.; Garganta, J.; Volossovitch, A.; Brandão, R.; Krebs, R. The Role of Ecological Constraints on Expertise Development. *Talent. Dev. Excell.* **2010**, *2*, 165–179.
11. Sarmento, H.; Anguera, M.T.; Pereira, A.; Araújo, D. Talent Identification and Development in Male Football: A Systematic Review. *Sports Med.* **2018**, *48*, 907–931. [CrossRef]
12. Dimundo, F.; Cole, M.; Blagrove, R.; Till, K.; McAuley, A.; Hall, M.; Gale, C.; Kelly, A. Talent Identification and Development in Male Rugby Union: A Systematic Review. *J. Expertise* **2021**, *4*, 33–55.
13. Till, K.; Emmonds, S.; Jones, B. Talent Identification. In *Strength and Conditioning for Young Athletes*; Taylor & Franics: Abingdon, UK, 2019; pp. 21–44.
14. Till, K.; Baker, J. Challenges and [Possible] Solutions to Optimizing Talent Identification and Development in Sport. *Front. Psychol.* **2020**, *11*, 664. [CrossRef]
15. Fontana, F.Y.; Colosio, A.; De Roia, G.F.; Da Lozzo, G.; Pogliaghi, S. Anthropometrics of Italian Senior Male Rugby Union Players: From Elite to Second Division. *Int. J. Sports Physiol. Perform.* **2015**, *10*, 674–680. [CrossRef] [PubMed]
16. Darrall-Jones, J.D.; Jones, B.; Till, K. Anthropometric and Physical Profiles of English Academy Rugby Union Players. *J. Strength Cond. Res.* **2015**, *29*, 2086–2096. [CrossRef] [PubMed]
17. Jones, B.; Weaving, D.; Tee, J.; Darrall-Jones, J.; Weakley, J.; Phibbs, P.; Read, D.; Roe, G.; Hendricks, S.; Till, K. Bigger, Stronger, Faster, Fitter: The Differences in Physical Qualities of School and Academy Rugby Union Players. *J. Sports Sci.* **2018**, *36*, 2399–2404. [CrossRef] [PubMed]
18. Doré, I.; Sabiston, C.M.; Sylvestre, M.-P.; Brunet, J.; O'Loughlin, J.; Abi Nader, P.; Gallant, F.; Bélanger, M. Years Participating in Sports during Childhood Predicts Mental Health in Adolescence: A 5-Year Longitudinal Study. *J. Adolesc. Health* **2019**, *64*, 790–796. [CrossRef] [PubMed]
19. Hill, A.; MacNamara, Á.; Collins, D. Development and Initial Validation of the Psychological Characteristics of Developing Excellence Questionnaire Version 2 (PCDEQ2). *Eur. J. Sport Sci.* **2019**, *19*, 517–528. [CrossRef] [PubMed]
20. Bruner, M.W.; Benson, A.J. Evaluating the Psychometric Properties of the Social Identity Questionnaire for Sport (SIQS). *Psychol. Sport Exerc.* **2018**, *35*, 181–188. [CrossRef]
21. Arkell, R. Body Size, Socioeconomic Status and Training Background of a Select Group of U16 South African Rugby Union Players (2010–2013): The Impact on National Selection. Master's Thesis, University of Cape Town, Cape Town, South Africa, 2016.
22. Lambert, M.I. Long-Term Player Development in Rugby-How Are We Doing in South Africa? *S. Afr. J. Sports Med.* **2010**, *22*, 67–68. [CrossRef]
23. Farrow, D.; Mccrae, J.; Gross, J.; Abernethy, B. Revisiting the Relationship between Pattern Recall and Anticipatory Skill. *Int. J. Sport Psychol.* **2010**, *41*, 91–106.

24. Côté, J.; Erickson, K.; Abernethy, B. Play and Practice during Childhood. In *Conditions of Children's Talent Development in Sport*; Côté, J., Lidor, R., Eds.; Fitness Information Technology: Morgantown, WV, USA, 2013; pp. 9–20.
25. Côté, J.; Lidor, R. Early Talent Development in Sport: A Multifaceted Approach. In *Conditions of Children's Talent Development in Sport*; Fitness Information Technology: Morgantown, WV, USA, 2013; pp. 1–8.
26. Bjørndal, C.T.; Luteberget, L.S.; Holm, S. The Relationship between Early and Senior Level Participation in International Women's and Men's Handball. *J. Hum. Kinet.* **2018**, *63*, 73. [CrossRef]
27. Cobley, S.; Hanratty, M.; O'Connor, D.; Cotton, W. First Club Location and Relative Age as Influences on Being a Professional Australian Rugby League Player. *Int. J. Sports Sci. Coach.* **2014**, *9*, 335–346. [CrossRef]
28. MacDonald, D.J.; Cheung, M.; Côté, J.; Abernethy, B. Place but Not Date of Birth Influences the Development and Emergence of Athletic Talent in American Football. *J. Appl. Sport Psychol.* **2009**, *21*, 80–90. [CrossRef]
29. Winn, C.O.; Ford, P.R.; McNarry, M.A.; Lewis, J.; Stratton, G. The Effect of Deprivation on the Developmental Activities of Adolescent Rugby Union Players in Wales. *J. Sports Sci.* **2017**, *35*, 2390–2396. [CrossRef]
30. Baker, J.; Cobley, S.; Schorer, J.; Wattie, N. Talent Identification and Development in Sport. In *Routledge Handbook of Talent Identification and Development in Sport*; Taylor & Francis: Abingdon, UK, 2013; pp. 1–7.
31. Owen, C.; Till, K.; Weakley, J.; Jones, B. Testing Methods and Physical Qualities of Male Age Grade Rugby Union Players: A Systematic Review. *PLoS ONE* **2020**, *15*, e0233796. [CrossRef]
32. Quarrie, L.K.; Wilson, B.D. Force Production in the Rugby Union Scrum. *J. Sports Sci.* **2000**, *18*, 237–246. [CrossRef]
33. Dimundo, F.; Cole, M.; Blagrove, R.C.; McAuley, A.B.; Till, K.; Hall, M.; Pacini, D.; Kelly, A.L. The Anthropometric, Physical, and Relative Age Characteristics of an English Premiership Rugby Union Academy. *Int. J. Strength Cond.* **2021**, *1*, 1–20. [CrossRef]
34. Vaz, L.; Martín, I.; Batista, M.; Almeida, L.; Fernandes, H.M. Differences in the Psychological Skills and Strategies Used by Elite Male Under-19 Rugby Union Players in Competition According to Playing Position. *Rev. Iberoam. Psicol. Ejerc. Deporte* **2017**, *12*, 225–230.
35. Till, K.; Weakley, J.; Read, D.B.; Phibbs, P.; Darrall-Jones, J.; Roe, G.; Chantler, S.; Mellalieu, S.; Hislop, M.; Stokes, K. Applied Sport Science for Male Age-Grade Rugby Union in England. *Sports Med.-Open* **2020**, *6*, 14. [CrossRef]
36. Ford, P.R.; Ward, P.; Hodges, N.J.; Williams, A.M. The Role of Deliberate Practice and Play in Career Progression in Sport: The Early Engagement Hypothesis. *High Abil. Stud.* **2009**, *20*, 65–75. [CrossRef]
37. Côté, J.; Strachan, L.; Fraser-Thomas, J. Participation, Personal Development, and Performance through Youth Sport. In *Positive Youth Development through Sport*; Routledge: Abingdon, UK, 2007; pp. 48–60.
38. English Indices of Deprivation. Available online: https://www.gov.uk/government/collections/english-indices-of-deprivation (accessed on 23 September 2021).
39. Coldwells, A.; Atkinson, G.; Reilly, T. Sources of Variation in Back and Leg Dynamometry. *Ergonomics* **1994**, *37*, 79–86. [CrossRef] [PubMed]
40. Massy-Westropp, N.M.; Gill, T.K.; Taylor, A.W.; Bohannon, R.W.; Hill, C.L. Hand Grip Strength: Age and Gender Stratified Normative Data in a Population-Based Study. *BMC Res. Notes* **2011**, *4*, 127. [CrossRef] [PubMed]
41. Román, P.Á.L.; Macias, F.J.V.; Pinillos, F.G. Effects of a Contrast Training Programme on Jumping, Sprinting and Agility Performance of Prepubertal Basketball Players. *J. Sports Sci.* **2018**, *36*, 802–808. [CrossRef] [PubMed]
42. Sayers, S.P.; Harackiewicz, D.V.; Harman, E.A.; Frykman, P.N.; Rosenstein, M.T. Cross-Validation of Three Jump Power Equations. *Med. Sci. Sports Exerc.* **1999**, *31*, 572–577. [CrossRef] [PubMed]
43. Buchheit, M.; Laursen, P.B.; Millet, G.P.; Pactat, F.; Ahmaidi, S. Predicting Intermittent Running Performance: Critical Velocity versus Endurance Index. *Int. J. Sports Med.* **2008**, *29*, 307–315. [CrossRef] [PubMed]
44. Kelly, A.L.; Wilson, M.R.; Jackson, D.T.; Turnnidge, J.; Williams, C.A. Speed of Thought and Speed of Feet: Examining Perceptual-Cognitive Expertise and Physical Performance in an English Football Academy. *J. Sci. Sport Exerc.* **2020**, *3*, 88–97. [CrossRef]
45. Cohen, J. *Statistical Power Analysis for the Behavioural Sciences*; Laurence Erlbaum Associates Inc.: Hillsdale, NJ, USA, 1988.
46. Armstrong, M.E.; Lambert, E.V.; Lambert, M.I. Physical Fitness of South African Primary School Children, 6 to 13 Years of Age: Discovery Vitality Health of the Nation Study. *Percept. Mot. Ski.* **2011**, *113*, 999–1016. [CrossRef] [PubMed]
47. Krause, L.M.; Naughton, G.A.; Denny, G.; Patton, D.; Hartwig, T.; Gabbett, T.J. Understanding Mismatches in Body Size, Speed and Power among Adolescent Rugby Union Players. *J. Sci. Med. Sport* **2015**, *18*, 358–363. [CrossRef]
48. Collins, D.J.; Macnamara, A.; McCarthy, N. Putting the Bumps in the Rocky Road: Optimizing the Pathway to Excellence. *Front. Psychol.* **2016**, *7*, 1482. [CrossRef]
49. Fraser-Thomas, J.; Beesley, T.; Dickler, L.; Harlow, M.; Mosher, A.; Preston, C.; Wolman, L. Developing Talent While Promoting Positive Youth Development: A Balancing Act. In *Routledge Handbook of Talent Identification and Development in Sport*; Routledge: Abingdon, UK, 2017; pp. 377–393.
50. Rothwell, M.; Davids, K.; Stone, J.; Araujo, D.; Shuttleworth, R. The Talent Development Process as Enhancing Athlete Functionality: Creating Forms of Life in an Ecological Niche. In *Routledge Handbook of Talent Identification and Development in Sport*; Routledge: Abingdon, UK, 2020.
51. Argus, C.K.; Gill, N.D.; Keogh, J.W. Characterization of the Differences in Strength and Power between Different Levels of Competition in Rugby Union Athletes. *J. Strength Cond. Res.* **2012**, *26*, 2698–2704. [CrossRef]
52. Durandt, J.; Du Toit, S.; Borresen, J.; Hew-Butler, T.; Masimla, H.; Jokoet, I.; Lambert, M. Fitness and Body Composition Profiling of Elite Junior South African Rugby Players. *S. Afr. J. Sports Med.* **2006**, *18*, 38–45. [CrossRef]

53. Dimundo, F.; Cole, M.; Blagrove, R.C.; McAuley, A.B.; Till, K.; Kelly, A.L. Talent Identification in an English Premiership Rugby Union Academy: Multidisciplinary Characteristics of Selected and Non-Selected Male Under-15 Players. *Front. Sports Act. Living* **2021**, *3*, 162. [CrossRef]
54. Fontana, F.Y.; Colosio, A.L.; Da Lozzo, G.; Pogliaghi, S. Player's Success Prediction in Rugby Union: From Youth Performance to Senior Level Placing. *J. Sci. Med. Sport* **2017**, *20*, 409–414. [CrossRef]
55. Chiwaridzo, M.; Ferguson, G.D.; Smits-Engelsman, B.C. Anthropometric, Physiological Characteristics and Rugby-Specific Game Skills Discriminating Zimbabwean under-16 Male Adolescent Rugby Players by Level of Competition. *BMJ Open Sport Exerc. Med.* **2019**, *5*, e000576. [CrossRef]
56. Zabaloy, S.; Alcaraz, P.E.; Pereira, L.A.; Giráldez, J.; González, J.G.; Loturco, I.; Freitas, T.T. Anthropometric and Physical Performance of Amateur Rugby Players within Specific Playing Positions. *Isokinet. Exerc. Sci.* **2021**, *29*, 429–441. [CrossRef]
57. Barr, M.J.; Sheppard, J.M.; Gabbett, T.J.; Newton, R.U. Long-Term Training-Induced Changes in Sprinting Speed and Sprint Momentum in Elite Rugby Union Players. *J. Strength Cond. Res.* **2014**, *28*, 2724–2731. [CrossRef]
58. Wood, D.J.; Coughlan, G.F.; Delahunt, E. Fitness Profiles of Elite Adolescent Irish Rugby Union Players. *J. Strength Cond. Res.* **2018**, *32*, 105–112. [CrossRef]
59. Howard, S.M.; Cumming, S.P.; Atkinson, M.; Malina, R.M. Biological Maturity-Associated Variance in Peak Power Output and Momentum in Academy Rugby Union Players. *Eur. J. Sport Sci.* **2016**, *16*, 972–980. [CrossRef]
60. Chiwaridzo, M.; Ferguson, G.D.; Smits-Engelsman, B.C.M. Anthropometric, Physiological Characteristics and Rugby-Specific Game Skills of Schoolboy Players of Different Age Categories and Playing Standards. *BMC Sports Sci. Med. Rehabil.* **2020**, *12*, 3. [CrossRef] [PubMed]
61. Vahed, Y.; Kraak, W.; Venter, R. The Effect of the Law Changes on Time Variables of the South African Currie Cup Tournament during 2007 and 2013. *Int. J. Perform. Anal. Sport* **2014**, *14*, 866–883. [CrossRef]
62. Scott, A.C.; Roe, N.; Coats, A.J.; Piepoli, M.F. Aerobic Exercise Physiology in a Professional Rugby Union Team. *Int. J. Cardiol.* **2003**, *87*, 173–177. [CrossRef]
63. Phibbs, P.J.; Jones, B.; Read, D.B.; Roe, G.A.; Darrall-Jones, J.; Weakley, J.J.; Rock, A.; Till, K. The Appropriateness of Training Exposures for Match-Play Preparation in Adolescent Schoolboy and Academy Rugby Union Players. *J. Sports Sci.* **2018**, *36*, 704–709. [CrossRef] [PubMed]
64. Andrew, M.; Grobbelaar, H.W.; Potgieter, J.C. Sport Psychological Skill Levels and Related Psychosocial Factors That Distinguish between Rugby Union Players of Different Participation Levels. *S. Afr. J. Res. Sport Phys. Educ. Recreat.* **2007**, *29*, 1–14. [CrossRef]
65. Batista, M.; Honório, S.; Catarino, J.; Vaz, L.; Fernandes, H.; Petrica, J.M. Psychological Profile of Rugby Players-Analysis between Athletes of the U18 and U20 National Elite Teams in the Positions of Forwards and Defenders. *Rev. Iberoam. Psicol. Ejerc. Deporte* **2019**, *14*, 108–111.
66. Potgieter, J.C.; Grobbelaar, H.W.; Andrew, M. Sport Psychological Skill Levels and Related Psychosocial Factors That Distinguish between Rugby Union Players of Different Participation Levels. *J. Soc. Psychol. Sci.* **2008**, *1*, 43–64.
67. Hill, A.; MacNamara, Á.; Collins, D. Psychobehaviorally Based Features of Effective Talent Development in Rugby Union: A Coach's Perspective. *Sport Psychol.* **2015**, *29*, 201–212. [CrossRef]
68. Güllich, A.; Macnamara, B.N.; Hambrick, D.Z. What Makes a Champion? Early Multidisciplinary Practice, Not Early Specialization, Predicts World-Class Performance. *Perspect. Psychol. Sci.* **2021**, *12*, 1745691620974772. [CrossRef]
69. Barth, M.; Güllich, A. Non-Linear Association of Efficiency of Practice of Adult Elite Athletes with Their Youth Multi-Sport Practice. *J. Sports Sci.* **2021**, *39*, 915–925. [CrossRef] [PubMed]
70. Cupples, B.; O'Connor, D.; Cobley, S. Distinct Trajectories of Athlete Development: A Retrospective Analysis of Professional Rugby League Players. *J. Sports Sci.* **2018**, *36*, 2558–2566. [CrossRef]
71. Bransford, J.D.; Schwartz, D.L. Chapter 3: Rethinking Transfer: A Simple Proposal with Multiple Implications. *Rev. Res. Educ.* **1999**, *24*, 61–100. [CrossRef]
72. Hecksteden, A.; Kellner, R.; Donath, L. Dealing with Small Samples in Football Research. *Sci. Med. Footb.* **2021**, *19*, 1978106. [CrossRef]

sports

MDPI

Article

Coaches' Criteria for Talent Identification of Youth Male Soccer Players

Jan Fuhre, Arild Øygard and Stig Arve Sæther *

Department of Sociology and Political Science, Norwegian University of Science and Technology (NTNU), 7491 Trondheim, Norway; jan_fuhre@hotmail.com (J.F.); arild-oygard@hotmail.com (A.Ø.)
* Correspondence: stigarve@ntnu.no; Tel.: +47-7355-1133

Abstract: Introduction: The main aim of this study was to examine which criteria coaches considered in the talent identification of youth male soccer players aged 13–16. The second aim was to describe how the coaches considered these criteria when identifying players for their club or regional teams and how these criteria take the impact of the relative age effect into consideration. Methods: We conducted qualitative, semi-structured interviews with six male coaches from a professional club academy or a regional team within the Norwegian Football Federation. Results: In line with earlier research, the results showed that the coaches considered the technical, tactical, and mental factors as the most important in talent identification. Further in line with earlier research, they considered that the physiological and sociological factors were of secondary importance, while anthropometric measures were considered the least important. Regarding the relative age effect, the coaches were aware of the effect and its consequences, while few of them had ways to reduce the effect and its impact on their talent identification process. Even so, the coaches highlighted the importance of considering a holistic approach to talent identification. Conclusion: The results show similarities with earlier research, but there is still a need for more longitudinal studies that investigate criteria for talent identification in youth football.

Keywords: talent development; TID; relative age effect

Citation: Fuhre, J.; Øygard, A.; Sæther, S.A. Coaches' Criteria for Talent Identification of Youth Male Soccer Players. *Sports* **2022**, *10*, 14. https://doi.org/10.3390/sports10020014

Academic Editors: Adam Leigh Kelly, Sergio L. Jiménez Sáiz, Sara Diana Leal dos Santos and Alberto Lorenzo Calvo

Received: 9 December 2021
Accepted: 13 January 2022
Published: 18 January 2022

Publisher's Note: MDPI stays neutral with regard to jurisdictional claims in published maps and institutional affiliations.

1. Introduction

Talent identification (TID) in soccer involves identifying talented players with the prerequisites and potential to become a professional player based on specific criteria [1]. It is considered to be a complex process due to the many factors and skills that affect the performance in soccer, often considered dynamic, impacting each other, and influenced by exercise, which increases the complexity even further [2,3]. It has also traditionally been based on coaches or scouts watching players in matches or training contexts over time who judge their performance and future potential to reach the elite level [4]. The concept of talent has also been related to TID and has usually been understood and used on players who have an above-average level of ability within a domain [5]. This traditional approach is a process seldom based on objective criteria, but on the coach's subjective perceptions of the ideal player, skillset, and/or potential, where previous experiences and intuition of the recruiter influence the assessment [6]. Such a subjective assessment practice has been demonstrated by earlier research to be the norm in professional soccer around the world [1,7–9] as well as a practice that can lead to repeated misjudgments and limited continuity in identifying talent [10].

Since the review paper by Williams and Reilly [11] and their model on potential predictors of talent in soccer, research has tried to elaborate on which of the predictors would be considered the most important. In 2000, they introduced the physical, physiological, sociological, and psychological predictors, while Williams et al. [1] reintroduced the potential predictors of adult high performance in soccer with four predictors: skills, physical, psychological, and social. In the meantime, research has highlighted several skills that have shown to be of high importance for later success at the senior level, on both the motorial and cognitive levels [1]. In general, technical performance characteristics, such as ball control, dribbling, and passing, have proven to be important predictors of future success in elite-level soccer [12,13]. Huijgen et al. [12] found that technique was a discriminating factor already in youth soccer and that it was a potential indicator of which players would later succeed at the elite level. The sport-specific perceptual-cognitive skills, such as tactical skills, game understanding, creativity, positioning, and orientation, have also proven to be valid predictors of future success in football [12–14]. When developing tactical skills and specifically perceptual–cognitive skills, the content of training and the learning environment that the players are exposed to have shown to have great influence [15–17]. When it comes to physiological factors, young elite athletes generally tend to score higher than their less skilled peers, which is in line with what one sees in senior football [18]. Physical characteristics and anthropometric proportions are factors that the practitioner has little or no influence on him/herself, such as height, growth, physique, and muscle composition [11]. Research has also identified psychological factors, such as self-confidence, resilience, concentration, commitment, discipline, adaptability, motivation, and the ability to cope with different challenges, as significant predictors of development and later success [19–24]. Furthermore, there seems to be an increasingly broad consensus that this is an area of importance and has gained more attention in both the identification and development of young talented players [25].

The identifiers in the process, the coaches, highlight technique as highly important among coaches and practitioners [6], especially related to ball control [26], dribbling, first touch, passing/shooting, and technique under pressure [6,27]. Perceptual–cognitive characteristics such as decision making, positioning, and game understanding have been highlighted as especially important by coaches and practitioners [6,26,28], while anthropometric, physiological, mental, and social factors have been deemed less important by coaches and practitioners [6,28]. However, in the recent study by Bergkamp et al. [26], the anthropometric and physiological factors were deemed as important as tactical skills by the participating scouts. The coach's role in relation to the identification of talent is also related to the coach's perception of skills that are important to succeed at the elite level and awareness of how they develop them most effectively [16,29]. Due to the complexity of soccer as a sport, there are many factors to consider [1]. Therefore, the research suggests that a holistic approach to identifying and developing talent, which includes technical factors, mental factors, physiological factors, and social factors, such as family relations, parental support, and training load, etc., may provide better insights and decisions when trying to identify talent [1,30,31].

An essential factor to consider when identifying and selecting players is the relative age effect (RAE), which is prominent when dividing players into chronological age groups with a span of one year [32], reinforcing the physical, cognitive, and emotional differences based on maturation [33]. However, there appears to be little or no difference in technical skill performance between early and late matured players [34]. This has been related to the compensation phenomenon [1], where players compensate for poor skills in one area with good skills in another to be able to compete with their lower maturation. This is reflected in the fact that academy players born in the last quarter have a higher chance than those born in the first quarter of achieving professional status in senior football in the end [35]. Studies have shown that football coaches usually do not take the RAE into account in the selection process, despite knowledge of the phenomenon [36].

The main aim of this study was to examine which criteria coaches considered in the TID of youth male soccer players aged 13–16. The second aim was to describe how the coaches considered these criteria when identifying players for their club or regional teams and how these criteria take the impact of the RAE into consideration.

2. Materials and Methods

In this study, we adopted a qualitative research design, using semi-structured interviews.

2.1. Participants

This study is based upon six semi-structured interviews with active male coaches in regional teams under the auspices of the Norwegian Football Federation and academy coaches connected to professional Norwegian soccer clubs. The coaches had a mean age of 34.16 (std 6.36), with an average of 6.66 (std 2.06) years' experience as coaches. Two coaches had a UEFA-A license and the other four a UEFA-B license. Two of the coaches had a master's degree in sport science, while one of the coaches had a bachelor's degree in sport science.

2.2. Data Collection

The semi-structured interviews were conducted to gain further insight into the coaches' perceptions of talent, TID, and the selection process. A formal interview guide was used to gain insight into the coaches' perspective of talent, TID, identification criteria, and selection mechanisms. The interview guide was also open to relevant follow-up questions in relation to the coaches' responses to the questions asked to gain relevant perspectives and insights from each of the participants. All interviews were individual and conducted digitally, using either Zoom or Microsoft Teams. The interviews were approximately one hour each and were transcribed in their full length by the interviewer.

2.3. Data Analysis

The interviews were audio-recorded and transcribed. To provide anonymity, the coaches were given identification codes, I1–I6 (Informant 1–Informant 6). Following the six steps suggested for qualitative analysis by Braun et al. [37], the following steps were used when analysing the data: (1) transcribing, reading, and re-reading the data; (2) generating initial codes, such as TID and perception of skills; (3) making use of deductive codes and identifying lower-order themes under the initial codes (e.g., ball control under technical skills), the data was coded by the second author and thereafter by author one and three with the intention to increase inter-rater reliabilities, and furthermore; (4) laying out the main topics from the data material with the use of ongoing member reflections; (5) reviewing the final categories and sub-themes; and (6) writing a report and presenting the data. Having processed all the data from the six interviews, we ended up with four central topics: perception of talent, talent development, talent identification, and relative age effect, reflecting the interview guide and presented in the results section. In the following presentation of the results from the six participants from the clubs and the local football federation (FA), I1, I5, and I6 represent region 1, and I2–I4 represent region 2.

3. Results

3.1. Coaches' Perception of Talent

To get an impression of how coaches in the end identify talents, the coaches were asked about how they view talented players in soccer in a broader perspective. Mostly, the coaches said they viewed talent as someone who possessed abilities out of the ordinary. In addition, they point to talent as the potential to be a top-level athlete later in their career:

I3: There is someone who has special abilities, and it can be in football or music (. . .) and it is that you as a player have something more than what the others have.

One of the coaches, however, had quite a different view on talent when linking it with interest in the sport, the willingness to train, and feeling joy when practicing.

I2: No, it is a comprehensive concept, which deals with a lot. But for me, the talent is about the one who likes to play football the most and who is willing to go to great lengths to play football. It is in a way the first thing I think of as a talent concept, and then I know for myself that it is much more comprehensive than just that.

Overall, the coaches view talent as something extraordinary, regarding either performance or skills or regarding an extraordinary interest for the sport, which again leads to many accumulated hours of practice activity and joy when practicing the sport.

3.2. Talent Development: The Skills That Count

Asked about which skills characterise a talent in the age group of 13–16 years, the coaches varied a bit in their perceptions, but all of them somehow mentioned technical and tactical skills as the main characteristics of talent in this age group. Three of the coaches highlighted the technical and tactical aspects, such as first touch and decision making, as the most important skills when judging players' potential and talent:

I4: Creativity and choices. I am interested in football players who have good close-up technique, good touch (. . .) I think it is very important to have a good deal of self-training with the ball (. . .) And it is now an experience I have had, that those who I have had, and whom I have seen are our best, creative players; they have spent a lot of time with the ball (. . .) so I look at skills, choice, execution, what you do with the ball rather than the physical.

The other three coaches also mention technical and tactical aspects as highly important, but they are also concerned with factors such as interest, motivation, drive, and the will to train as important characteristics in that specific age group, considering the ability to practice and, through practice, developing soccer-specific skills:

I3: You must have a presence, motivation, an inner drive. That is probably what is most important. And if you have it, then you usually have something else too, I think. And it must come from a young age; you must have played a lot with the ball since you were little. Technical and tactical, as I said, you can be very technical, but you lack everything else within the mental or the psychological aspect, like that you are not a nice person. Football is a team game, so it is clear to me that the psychological factors that we are talking about in a football context are very important.

Regarding how the coaches take their own skill valuation into account in their daily work with their players, all the coaches again point to the technical and tactical skills as the focal starting point for their coaching sessions. However, the two that get the most attention and the reasoning behind them differ among the coaches:

I3: (. . .) technique and tactics are a bit connected, but technique, call it basic skills, if you do not have it when you are 14 years old, then it limits the possibilities, I think. Tactical is like . . . you learn a lot through play and games at a younger age, and then it comes more and more into the 13–19 age group, and then it's about cracking the codes quite quickly.

One of the coaches (I3) considers technical skills as the most important at an early stage for young players and considers tactical skills as something that comes later in the development process. This points to a fragmented understanding of the developmental process, such as developing one skill at a time. I1 also considered the technical skills as more important than the tactical skills but still pointed out that, in his club, the tactical skills tended to have more focus in the daily work:

I1: It is difficult to say that we look more at one or the other, but at the same time I can add that the club I am in has been very concerned with the decisions they make, the understanding of the game, and making good choices. And I may have been a little soft for it myself, which players make good choices. Then you can, to a certain extent, imagine that it is easier to develop (decision making) than the technical execution, so a slightly subjective answer to that (. . .).

3.3. Talent Identification

All the coaches in this study pointed out that their assessment of players' talent and potential was based on subjective measures, and usually, a player's performance in match situations was an essential factor:

I1: (...) The assessments we make on who we think fits and does not become subjective. And we are concerned with getting them into our context, to come in and train with us, to see if they can adapt to our tactics (...).

The coaches' answers indicated some differences between the club and regional team coaches, where factors concluded as important by the Norwegian Football Federation were also highlighted by the regional teams' coaches. The coaches at the club level were guided more by their club's playing style and tried to identify players who would fit within that specific way of playing. Both, however, were still based on subjective measures and opinions.

Coach 2 coupled the definition of talent with interest and motivation towards the sport, which also influenced his perception of how early a talent could be identified:

I2: If you just take ... that of finding a player who is genuinely interested in playing football, then you can see it quite early. We have a couple of examples from our own club. We have two players who were born in 2010 and are in their eleventh year, and you see they are genuinely interested in playing football. Whatever opportunity they have, they play football. And that is a starting point for something that can help them in later years. (...) They will lay an insanely good foundation in relation to the others, who are there only on team training, for example.

The reasoning behind this perception is that an extreme interest and joy in playing the sport itself will cumulate in more hours of practice on their own.

The coaches considered it possible, but challenging, to identify talented players quite early, especially if you use the players' current abilities as the benchmark, even if they are 6, 10, or 13 years old. At the same time, they were clear about the complex nature of identifying talent regarding future potential as the following citations illustrate:

I1: (...) it is clear that a six-year-old can be much better than the other six-year-olds, and then he is better than the others at that time. And then the question is whether you can call it a talent or not. So, it's probably easy to mix talent with the one who is good there and then, but I think more that it is aimed at the one who has potential. So, you can see if players are good already when they are six years old, but if it is a talent is very difficult to say, because it is impossible to say who will succeed.

I6: Basically, in my opinion, you see the football technical and coordinative abilities when they are 12–14 years of age, but then it differs even more when they are 16 years old. When you see the choices they make regarding school, you start to make some life choices (...).

Even if the coaches state that it is possible to identify the talented players quite early based on their current abilities, they are also very clear about the difference between current abilities, which may be better than others, and the future potential to reach the top level and develop the already good skills even further. Some also point out that different factors such as motivation, personal traits, and choices in life when they are old enough to make their own choices play a big role in determining the level of success a young player will have at the senior level. Those factors, according to the coaches, are harder to identify early, and they make a point out of differing between abilities and potential.

3.4. Relative Age Effect: The Consequence?

Identifying talented footballers at an early age often involves an impact from the RAE. The coaches in this study all answered that they knew about the concept of RAE and the potential challenges that it leads to. They answered that they do not consider physical skills or characteristics as important as the technical and tactical skills, which again plays a role when considering the RAE in their decisions. Even so, as Coach 5 pointed out, there is still an advantage if you are physically well-developed from an early age, because players' abilities in real time often are higher, regarding obtaining results here and now. Coach 5 also made the point that the early-developed players may allow themselves bad habits technically and tactically because they can compensate with their physique.

I5: The advantage is that you are very efficient if you are well-developed physically. The disadvantage of this is that one then tends to acquire some bad habits, which they must unlearn when things level out more physically. In the same way, if one is very weak physically, then one can acquire some good habits to cope.

Another challenge related to selecting early developed players is that they will get more and better follow-up from coaches and clubs in their path, which, again, may lead to other talented players getting too little attention in their important development years.

When asked if they believe that talented players slip under the radar because of the RAE, the coaches partly disagreed, where some meant that they absolutely missed out on potential elite players, while others did not. Some of the arguments were that the deselected players potentially could give up too early or that the focus was on results, while some meant that the best players would always be identified:

I3: Yes, I think so. I think the players may not even reach their maximum level because they give up maybe a little early. And then we see that we may also be a little impatient not to give them the time they need to grow up then, as they call it.

I2: No, I do not think so. I think in a way that we can pick up those who can be the best or are the best; we will probably get them, I think. And then you can ask yourself the question of whether you might lose out on a broader group with potential to get to an OK Norwegian level. Maybe, but the best I think we will get.

4. Discussion

The main aim of this study was to examine which criteria coaches considered in the TID of youth male soccer players aged 13–16. The second aim was to describe how the coaches considered these criteria when identifying players for their club or regional teams and how these criteria take the impact of the RAE into consideration. The results indicated that the complexity of soccer makes it a very challenging task to identify the most talented players who will go on to succeed at the senior level from an early age [38]. These coaches, like earlier studies, show a high appreciation for the technical and tactical factors [11,30]. This also corresponds well with what has been seen in previous studies on coaches' criteria for selection [6,28]. This also highlights the tradition of identifying and selecting talented players based on subjective criteria, which vary across cultures and clubs [1,6,7,9].

The coaches' overall perspective is in line with earlier research's definition of talent as someone who possesses abilities above average [8]. In particular, technical and tactical skills, such as ball control, dribbling, passing, creativity, positioning, and decision making, have proven to be valid predictors of future success [12–14]. Several of the coaches in this study also emphasised game understanding and positioning as important tactical qualities, which is in line with previous research in both qualitative studies [27,39,40] and quantitative studies [6,26,28]. This also relates to the complexity of judging talent, which corresponds well to earlier research on players who later succeed at the senior level [4]. The coaches also mention the player's motivation, willingness to learn and train, and, maybe most importantly, the joy they get from playing the game as important factors. This differs a bit from the sport-specific focus of many coaches but nonetheless shows a broad understanding related to the complexity of identifying and nurturing talent. This also falls in line with research into traits and personalities who have reached the top level in their sports, which highlights the inner drive, discipline, and motivation of the succeeding athletes but also shows the differences between personalities, even if some traits are alike [19,21–23]. Earlier research has highlighted that interest, motivation, and joy lead to more self-organized training [41]. Part of the reasoning behind this perception is that an extreme interest and joy in playing the sport itself will cumulate into more hours of practice on their own, which research has shown to be a discriminating factor regarding later success at the elite level [41].

This study also found that the coaches, to a lesser degree, focused on the physiological attributes, which is also in line with earlier research and often related to the fact that these skills are trainable [42] and that differences may result from more systematic training over time [11]. Research has still shown that, for example, running velocity may be a useful predictor for later success at the elite level [12], but in general, one should still be careful about selecting talent based on physiological superiority. Another challenge related to selecting early developed players has been the access to better and more follow-up from coaches and clubs in their path, which, again, may lead to talented players receiving too little attention in their important development years, which studies have pointed to in the past [1,34,43]. This has also often been related to the RAE and the awareness of the effect, which have been found to be prominent in the TID process [4]. The coaches in this study, however, highlighted that the late-blooming players probably still could succeed later, even if they were not selected at the first crossroad. Even so, studies have pointed to the possible benefits one receives from early selection, which may reinforce the differences between players' abilities because of better coaches and closer monitoring of their development [1]. Club coaches have this impression of players breaking through even if they are not selected because their potential and talent might be expected since they mostly work on a short-term basis. A valid point that the coaches make is that the differences between players, based on the RAE, are smaller when it comes down to technical and tactical skills, which they emphasise as most important in their view [12,32], and this may even lead to compensational development from the less physically developed players to be able to cope in games.

It is interesting to note that all the coaches seem to think that they can identify talent quite early based on players' technical abilities here and now, even if they also point to the complexity of the development and especially the identification process. What supports the notion of being able to identify talents early on based on their technical abilities are the earlier findings that point to the technical skills as a potential early discriminator between elite and sub-elite players [12]. The technical skills are also less affected by the RAE, as we have seen earlier [12,32]. Even if the technical abilities might give an early indication, many coaches tend to mix the technical abilities and potential with early achievements, which, again, can stem from physical superiority and so forth.

4.1. Limitations and Methodological Issues

The small sample size must be considered as a limitation of this study. Furthermore, interviews with the use of teams have disadvantages because of the lack of closeness to the coaches interviewed. This might have impacted the results in the way that the participants might have misunderstood the questions or the reduced access to the interviewers' body language might have been misunderstood or vice versa.

4.2. Future Studies

There is still a need for more longitudinal studies that investigate criteria for TID in youth football. Furthermore, studies comparing coaches with different roles, such as club coaches both at grassroot and elite levels and football federation coaches representing the national TID system, are needed. The TID process will most likely be impacted by the context in which the coaches are coaching, such as the degree of follow up and responsibility in everyday training, where club coaches have a much bigger impact on the players' development compared to regional federation coaches in this study. Another vital question would be the background of the coaches related to both experience (years and level) and coach education. Knowledge and experiences from different levels (club, federation, school, etc.) might impact the coach's perception on both talent development and the TID process. Matin and Sæther [44] found in a study on male and female football players in a sport specialization program (SSP) and elite sport specialization program (ESSP) in upper secondary schools in Norway that the players from the SSP school considered the school coach to focus significantly (<0.01) more on technical skills than the club coach. The players from the ESSP, however, considered the club coach to focus on technical skills as compared to the school coach. Furthermore, the players from the ESSP regarded the club coach to be significantly (<0.01) more focused on physical skills as compared to the players from the SSP. As discussed in the study, there are obvious reasons as to why coaches in different contexts (club vs. school) have different focuses in their training sessions and periods since only club coaches are responsible for a team, while school coaches are more naturally focused on individual physical and technical development.

Even though the intention of the present study was not to address the coaches' background, which impacted the TID process, a vital question would be to address how their experience, education, and knowledge impacted the TID process. Earlier research has shown that coaches refer to "gut feeling" [40] and "practical sense" in the TID process. Peterson [45] has furthermore raised the question if football players succeed despite the talent selection system (the case of Sweden), both based on the prevalence of the RAE in the TID processes and the lack of precision in the TID criteria. He even questioned the idea of identifying talent at an early age based on the lack of success in most TID systems [2]. Future studies on perceptions of TID and RAE should also include the issue of bio-banding and how it impacts the coaches' TID process—an issue raised in the newly published book by Kelly et al. [46].

Another issue future research should also address is the impact of how the national TID system and football federation impacts the coaches' approach in the TID process. The coaches in this system may have an impact on each other's approach in the TID process, with the risk of group thinking when the coaches discuss the players in the TID process.

5. Conclusions

This study has provided qualitative insight into the process of TID among youth soccer coaches in Norway. The results implied that the coaches considered the technical, tactical, and mental factors as the most important in TID—more specifically, ball control, decision making, and motivation (inner drive). They furthermore considered, in line with earlier research, that the physiological and sociological factors were of secondary importance, while the anthropometric measures were considered the least important. Regarding the RAE, the coaches were aware of the effect and its potential consequences, while few of them had ways to reduce the effect and its impact on their TID process. Even so, the coaches highlighted the importance of considering a holistic approach to TID, with all four potentially predicting categories of Williams et al.'s [1] model. There is still a need for more longitudinal studies that investigate the criteria for TID in youth football and studies comparing coaches with different roles in the TID system.

Key findings:

- Coaches considered the technical, tactical, and mental factors as the most important TID criteria.
- Despite awareness of the relative age effect and its potential consequences, few of them had ways to reduce the effect and its impact on the TID process.

The results show similarities with earlier research, still indicating a need for longitudinal studies that investigate criteria for TID in youth football.

Author Contributions: Conceptualization, A.Ø. and S.A.S.; methodology, A.Ø.; formal analysis, A.Ø.; investigation, A.Ø.; writing—original draft preparation, J.F., A.Ø. and S.A.S.; writing—review and editing, J.F., A.Ø. and S.A.S.; supervision, S.A.S.; project administration, A.Ø. All authors have read and agreed to the published version of the manuscript.

Funding: This research received no external funding.

Institutional Review Board Statement: The research was undertaken according to the ethical guidelines of the Norwegian Centre for Research Data (NSD).

Informed Consent Statement: Informed consent was obtained from all subjects involved in the study.

Data Availability Statement: The data are not publicly available as data sharing was not included within the original study ethics submission or participant consent form.

Acknowledgments: The author would like to thank all coaches involved in the project for their participation.

Conflicts of Interest: The authors declare no conflict of interest.

References

1. Williams, A.M.; Ford, P.R.; Drust, B. Talent Identification and Development in Soccer Since the Millennium. *J. Sports Sci.* **2020**, *38*, 1199–1210. [CrossRef] [PubMed]
2. Güllich, A. Selection, de-selection and progression in German football talent promotion. *Eur. J. Sport Sci.* **2014**, *14*, 530–537. [CrossRef] [PubMed]
3. Suppiah, H.T.; Low, C.Y.; Chia, M. Detecting and developing youth athlete potential: Different strokes for different folks are warranted. *Br. J. Sports Med.* **2015**, *49*, 878–882. [CrossRef] [PubMed]
4. Baker, J.; Cobley, S.; Schorer, J.; Wattie, N. Talent identification and development in sport: An introduction. In *Routledge Handbook of Talent Identification and Development in Sport*; Routledge: New York, USA, 2017; pp. 1–8.
5. Tansley, C. What do we mean by the term "talent" in talent management? *Ind. Commer. Train.* **2011**, *43*, 266–274. [CrossRef]
6. Larkin, P.; O'Connor, D. Talent identification and recruitment in youth soccer: Recruiter's perceptions of the key attributes for player recruitment. *PLoS ONE* **2017**, *12*, e0175716. [CrossRef] [PubMed]
7. Dugdale, J.H.; Sanders, D.; Myers, T.; Williams, A.M.; Hunter, A.M. A case study comparison of objective and subjective evaluation methods of physical qualities in youth soccer players. *J. Sports Sci.* **2020**, *38*, 1304–1312. [CrossRef]
8. Nesse, M.; Moe, V.F.; Sæther, S.A. Unified approaches on talent development in football? Differences and similarities among representatives from a football association and two professional football clubs. *Cogent Soc. Sci.* **2020**, *6*, 1785216.
9. Sieghartsleitner, R.; Zuber, C.; Zibung, M.; Conzelmann, A. Science or coaches' eye?–Both! Beneficial collaboration of multi-dimensional measurements and coach assessments for efficient talent selection in elite youth football. *J. Sports Sci. Med.* **2019**, *18*, 32.
10. Meylan, C.; Cronin, J.; Oliver, J.; Hughes, M. Talent identification in soccer: The role of maturity status on physical, physiological and technical characteristics. *Int. J. Sports Sci. Coach.* **2010**, *5*, 571–592. [CrossRef]
11. Williams, A.M.; Reilly, T. Talent identification and development in soccer. *J. Sports Sci.* **2000**, *18*, 657–667. [CrossRef]
12. Huijgen, B.C.; Elferink-Gemser, M.T.; Lemmink, K.A.; Visscher, C. Multidimensional performance characteristics in selected and deselected talented soccer players. *Eur. J. Sport Sci.* **2014**, *14*, 2–10. [CrossRef] [PubMed]
13. Murr, D.; Feichtinger, P.; Larkin, P.; O'Connor, D.; Höner, O. Psychological talent predictors in youth soccer: A systematic review of the prognostic relevance of psychomotor, perceptual-cognitive and personality-related factors. *PLoS ONE* **2018**, *13*, e0205337.
14. Roca, A.; Ford, P.R.; Memmert, D. Creative decision making and visual search behavior in skilled soccer players. *PLoS ONE* **2018**, *13*, e0199381. [CrossRef] [PubMed]
15. Ford, P.R.; Yates, I.; Williams, A.M. An analysis of practice activities and instructional behaviours used by youth soccer coaches during practice: Exploring the link between science and application. *J. Sports Sci.* **2010**, *28*, 483–495. [CrossRef]
16. O'Connor, D.; Larkin, P.; Williams, A.M. What learning environments help improve decision-making? *Phys. Educ. Sport Pedagog.* **2017**, *22*, 647–660. [CrossRef]
17. O'Connor, D.; Larkin, P.; Williams, A.M. Observations of youth football training: How do coaches structure training sessions for player development? *J. Sports Sci.* **2018**, *36*, 39–47. [CrossRef]
18. Dodd, K.D.; Newans, T.J. Talent identification for soccer: Physiological aspects. *J. Sci. Med. Sport* **2018**, *21*, 1073–1078. [CrossRef]
19. Holt, N.L.; Dunn, J.G. Toward a grounded theory of the psychosocial competencies and environmental conditions associated with soccer success. *J. Appl. Sport Psychol.* **2004**, *16*, 199–219. [CrossRef]
20. Vaeyens, R.; Lenoir, M.; Williams, A.M.; Philippaerts, R.M. Talent identification and development programmes in sport. *Sports Med.* **2008**, *38*, 703–714. [CrossRef]
21. Van Yperen, N.W. Why some make it and others do not: Identifying psychological factors that predict career success in professional adult soccer. *Sport Psychol.* **2009**, *23*, 317–329. [CrossRef]
22. Zuber, C.; Zibung, M.; Conzelmann, A. Motivational patterns as an instrument for predicting success in promising young football players. *J. Sports Sci.* **2015**, *33*, 160–168. [CrossRef] [PubMed]
23. Toering, T.T.; Elferink-Gemser, M.T.; Jordet, G.; Visscher, C. Self-regulation and performance level of elite and non-elite youth soccer players. *J. Sports Sci.* **2009**, *27*, 1509–1517. [CrossRef]
24. MacNamara, Á.; Button, A.; Collins, D. The role of psychological characteristics in facilitating the pathway to elite performance part 1: Identifying mental skills and behaviors. *Sport Psychol.* **2010**, *24*, 52–73. [CrossRef]
25. Musculus, L.; Lobinger, B.H. Psychological characteristics in talented soccer players–recommendations on how to improve coaches' assessment. *Front. Psychol.* **2018**, *9*, 41. [CrossRef] [PubMed]
26. Bergkamp, T.L.; Frencken, W.G.; Niessen, A.S.M.; Meijer, R.R.; den Hartigh, R.J. How soccer scouts identify talented players. *Eur. J. Sport Sci.* **2021**, 1–11. [CrossRef] [PubMed]
27. Berber, E.; McLean, S.; Beanland, V.; Read, G.J.; Salmon, P.M. Defining the attributes for specific playing positions in football match-play: A complex systems approach. *J. Sports Sci.* **2020**, *38*, 1248–1258. [CrossRef]

28. Roberts, S.J.; McRobert, A.P.; Lewis, C.J.; Reeves, M.J. Establishing consensus of position-specific predictors for elite youth soccer in England. *Sci. Med. Footb.* **2019**, *3*, 205–213. [CrossRef]
29. Stoszkowski, J.; Collins, D. Sources, topics and use of knowledge by coaches. *J. Sports Sci.* **2016**, *34*, 794–802. [CrossRef]
30. Sarmento, H.; Anguera, M.T.; Pereira, A.; Araújo, D. Talent identification and development in male football: A systematic review. *Sports Med.* **2018**, *48*, 907–931. [CrossRef]
31. Unnithan, V.; White, J.; Georgiou, A.; Iga, J.; Drust, B. Talent identification in youth soccer. *J. Sports Sci.* **2012**, *30*, 1719–1726. [CrossRef]
32. Vaeyens, R.; Philippaerts, R.M.; Malina, R.M. The relative age effect in soccer: A match-related perspective. *J. Sports Sci.* **2005**, *23*, 747–756. [CrossRef]
33. Musch, J.; Grondin, S. Unequal competition as an impediment to personal development: A review of the relative age effect in sport. *Dev. Rev.* **2001**, *21*, 147–167. [CrossRef]
34. Cripps, A.J.; Hopper, L.; Joyce, C. Maturity, physical ability, technical skill and coaches' perception of semi-elite adolescent Australian footballers. *Pediatric Exerc. Sci.* **2016**, *28*, 535–541. [CrossRef]
35. Kelly, A.L.; Wilson, M.R.; Gough, L.A.; Knapman, H.; Morgan, P.; Cole, M.; Jackson, D.T.; Williams, C.A. A longitudinal investigation into the relative age effect in an English professional football club: Exploring the 'underdog hypothesis'. *Sci. Med. Footb.* **2020**, *4*, 111–118. [CrossRef]
36. Hill, B.; Sotiriadou, P. Coach decision-making and the relative age effect on talent selection in football. *Eur. Sport Manag. Q.* **2016**, *16*, 292–315. [CrossRef]
37. Braun, V.; Clarke, V.; Weate, P. Using thematic analysis in sport and exercise research. In *Routledge Handbook of Qualitative Research in Sport and Exercise*; Routledge: London, UK, 2016; pp. 191–205.
38. Skorski, S.; Skorski, S.; Faude, O.; Hammes, D.; Meyer, T. The relative age effect in elite German youth soccer: Implications for a successful career. *Int. J. Sports Physiol. Perform.* **2016**, *11*, 370–376. [PubMed]
39. Christensen, M.K. "An eye for talent": Talent identification and the "practical sense" of top-level soccer coaches. *Sociol. Sport J.* **2009**, *26*, 365–382. [CrossRef]
40. Lund, S.; Söderström, T. To see or not to see: Talent identification in the Swedish Football Association. *Sociol. Sport J.* **2017**, *34*, 248–258. [CrossRef]
41. Hornig, M.; Aust, F.; Güllich, A. Practice and play in the development of German top-level professional football players. *Eur. J. Sport Sci.* **2016**, *16*, 96–105.
42. Pearson, D.; Naughton, G.A.; Torode, M. Predictability of physiological testing and the role of maturation in talent identification for adolescent team sports. *J. Sci. Med. Sport* **2006**, *9*, 277–287. [CrossRef] [PubMed]
43. Ramos-Filho, L.; Ferreira, M.P. The reverse relative age effect in professional soccer: An analysis of the Brazilian National League of 2015. *Eur. Sport Manag. Q.* **2021**, *21*, 78–93. [CrossRef]
44. Matin, V.; Sæther, S.A. Talented high school football players' perception of talent identification criteria. *Sport Mont J.* **2017**, *2*, 3–7.
45. Peterson, T. *Talangutveckling eller Talangavveckling?* SISU Idrottsböcker: Malmö, Sweden, 2011.
46. Kelly, A.L.; Côté, J.; Turnnidge, J.; Hancock, D. (Eds.) *Birth Advantages and Relative Age Effects: Exploring Organisational Structures in Youth Sport*; Frontiers Media SA: Lausanne, Sweden, 2021.

sports

MDPI

Article

Maturity Has a Greater Association than Relative Age with Physical Performance in English Male Academy Soccer Players

John M. Radnor [1,*], Jacob Staines [2], James Bevan [2], Sean P. Cumming [3], Adam L. Kelly [4], Rhodri S. Lloyd [1,5,6] and Jon L. Oliver [1,5]

1 Youth Physical Development Centre, School of Sport and Health Sciences, Cardiff Metropolitan University, Cardiff CF23 6XD, UK; rlloyd@cardiffmet.ac.uk (R.S.L.); joliver@cardiffmet.ac.uk (J.L.O.)
2 Exeter City Football Club, Exeter EX4 6PX, UK; jacob.staines@ecfc.co.uk (J.S.); James.bevan@ecfc.co.uk (J.B.)
3 Department for Health, University of Bath, Bath BA2 7AY, UK; sc325@bath.ac.uk
4 Faculty of Health, Education and Life Sciences, Birmingham City University, Birmingham B15 3TN, UK; Adam.Kelly@bcu.ac.uk
5 Sport Performance Research Institute New Zealand, AUT University, Auckland 0632, New Zealand
6 Centre for Sport Science and Human Performance, Waikato Institute of Technology, Hamilton 3200, New Zealand
* Correspondence: jradnor@cardiffmet.ac.uk

Abstract: This study aimed to: (1) examine differences in physical performance across birth-quartiles and maturity-status, and (2) determine the relationships among relative age, maturation and physical performance in young male soccer players. The sample included 199 males aged between 8.1 and 18.9 years, from two professional soccer academies in the English Football League. Data were collected for height, weight, self-reported biological parent heights, 30 m sprint time and countermovement jump (CMJ) height. Relative age was conveyed as a decimal, while maturity status was determined as the percentage of predicted adult height (PAH). There were no significant differences in any measure between birth quartiles, however early maturers outperformed on-time and later maturers in most performance measures. Pearson-product-moment correlations revealed that maturation was inversely associated with 30 m sprint time in U12 to U16 ($r = -0.370$–0.738; $p < 0.05$), but only positively associated with CMJ performance in U12 ($r = 0.497$; $p < 0.05$). In contrast, relative age was unrelated to sprint performance and only significantly associated with superior CMJ performance in U16. This study indicates that maturity has a greater association with sprint performance than relative age in English male academy soccer players. Practitioners should monitor and assess biological maturation in young soccer players to attempt to control for the influence on physical performance, and avoid biasing selection on absolute performance rather than identifying the most talented player.

Keywords: sprint; countermovement jump; youth; maturation; predicted adult height; football

Citation: Radnor, J.M.; Staines, J.; Bevan, J.; Cumming, S.P.; Kelly, A.L.; Lloyd, R.S.; Oliver, J.L. Maturity Has a Greater Association than Relative Age with Physical Performance in English Male Academy Soccer Players. *Sports* **2021**, *9*, 171. https://doi.org/10.3390/sports9120171

Academic Editor: Pedro E. Alcaraz

Received: 8 November 2021
Accepted: 9 December 2021
Published: 20 December 2021

Publisher's Note: MDPI stays neutral with regard to jurisdictional claims in published maps and institutional affiliations.

1. Introduction

Soccer academies are a vital pathway in the long-term development of youth players, with the primary objective of identifying and developing talented individuals to compete at senior levels [1,2]. Two factors that have been shown to impact both player performance and selection in youth soccer are relative age and biological maturation [3–5]. Contrary to lay opinion, relative age and biological maturation are distinct constructs that exist and operate independent of one another [6].

Relative age is determined by date of birth and the selection cut-off date and refers to a player's chronological age within their specific age group. Due to the application of arbitrary and chronologically aged (bi)annual groupings for soccer academies (e.g., U9, U10, U11, etc.), players within the same age group can be by almost twelve months apart in chronological age. This results in the phenomenon known as the relative age effect (RAE), where players born earlier in their selection year (e.g., birth quartile [BQ] one and BQ2) have a greater likelihood of being selected into talent pathways (~38–40% and ~24–30%, respectively) when compared to those born later in the year (BQ3: ~15–21% and BQ4: ~13–16%, respectively) [4,7].

Biological maturation is the process of progressing toward a mature state and varies in magnitude (extent of change), timing (onset of change) and tempo (rate of change) between different systems in the body [8] and between individuals [9]. Variance in biological maturation is a result of genetic and environmental factors and players of the same chronological age can vary by as much as five to six years in skeletal age [10], an established index of maturation in youth. As such, it is entirely possible for a player to be both the youngest and most mature player within their age group, as well as vice versa. Youth can be classified as biologically "ahead of" (early maturer), "on-time with" (average maturer) or "behind" (late maturer) their chronological age [11].

Whereas chronological age is predictable and easily assessed, biological age is significantly more difficult to assess. The gold standard method of assessing maturation is using skeletal age, but due to the expense and requirement for specialised radiographers using this method [9], other methods are often utilised. Somatic age refers to the use of growth in stature or specific dimensions of the body for the estimation of maturity [9]. The most simple level of assessment involves longitudinal anthropometric assessments [9], and the repeated collection of height over a period of time would enable the analysis of growth curves that allow information related to the initiation of the growth spurt and peak-height-velocity (PHV) to be obtained. Considering the limitations associated with collecting longitudinal data to identify PHV, predictive equations can be used to predict the age at PHV from single measurements of anthropometric variables [12,13]. Mirwald and colleagues [12] proposed a predictive equation based on the theory of differential growth rates between the lower limbs and torso. Despite this method being a popular tool for measuring maturity, it does have potential limitations. In particular, the method has received criticism from researchers who suggest a bias is prevalent towards chronological age at the time of estimation, or low sensitivity to identify early and late maturing individuals [14]. The percentage of predicted adult height (%PAH) can be calculated at a given time point during childhood and adolescence and this can be used to determine the maturational status of a young athlete [15]. This approach may be useful to differentiate between those who are early-maturing and those who are naturally predisposed to being tall, especially as it is possible that two individuals in this situation could present with the same absolute stature at a given chronological age [9]. Khamis and Roche have proposed a prediction equation to calculate final adult height, using mid-parental height but also included the child's current stature and weight in addition to specific coefficients for each of these variables at 0.5-year intervals serve to improve the accuracy of the prediction model [16]. Recent longitudinal analysis to observe timing of PHV illustrated that %PAH was accurate 96% of the time, with maturity offset correct only 61% of the time [17]. The error of prediction in the %PAH equation has been estimated to be ~2 cm [16] and %PAH has been shown to correlate with skeletal age [18]. This has resulted in %PAH being used as a popular method of estimating maturity in youth [19] and has become increasingly popular within soccer [20] where it is used throughout the Premier League's management application.

Differences in the maturity status and timing of individuals has been shown to have implications on the physical, psychological and athletic development of adolescent males [3,21,22]. Boys who mature in advance of their peers experience the adolescent growth spurt at an earlier age and, thus, are invariably taller and heavier from late childhood and possess greater absolute and relative lean mass [3,5,22]. As a consequence of their advanced maturity, early maturing players also tend to tend to outperform their less mature counterparts on tests of speed, power, strength, momentum, and agility [5,21]. In addition to these physical advantages, early maturing boys also tend to perceive themselves as more athletic and competent in sport [22]. Given the inherent benefits associated with advanced maturation, it is therefore not surprising that early maturing males are more likely to be represented and selected for sports where greater size, strength and power are desirable attributes, such as in soccer [3]. The selection bias towards advanced maturity in males emerges from late childhood/early adolescence and increase in size and magnitude with age and level of competition [11].

While players born earlier in the selection year are heavily represented in youth soccer [4,23], there is limited evidence to suggest that relatively older players possess advantages in functional capacities. Relatively older players are often assumed to be biologically more mature and, thus, physically superior in comparison to their relatively younger peers [24]. Despite these assumptions, relative age does not necessarily imply more advanced maturity [23], with relative age shown to be weakly correlated with maturity status in young athletes [25,26]. There is some research to suggest that players born earlier in the selection year have greater anthropometric characteristics, in addition to greater physiological attributes, which are associated with successful performance in elite youth football, however the differences between players in BQ1 and BQ4 were often unclear and predominantly trivial or small [26].

Further evidence of the independent nature of relative age and maturation can be seen in their associations with both physical and psychological variables. In a recent study investigating predictors of physical fitness in male academy soccer players, maturation was found to have a significant association with a range of physical performance measures, whereas relative age was only weakly correlated with 20 m speed and CMJ performance [27]. However, U12 to U16 were pooled together for the analysis and the effects of maturity and relative age on physical performance were not established for individual age groups. Similarly, an investigation of the 'underdog hypothesis' revealed that delayed maturity, but not younger relative age, was associated with greater use of adaptive self-regulated learning strategies in academy football [23]. In light of this evidence, the primary aim of the current study was to investigate the relationships between relative age, maturity and physical performance in soccer players from U9 and U18 age groups. In accordance with previous research, it was hypothesised that advanced maturation, rather than greater relative age, would be associated with superior performance on tests of sprint and jumping ability in English male academy soccer players.

2. Materials and Methods

2.1. Participants

One hundred and ninety-nine elite male junior soccer players from two professional soccer academies in the United Kingdom, between the ages of U9 and U18, volunteered to participate in the cross-sectional study. In line with the Elite Player Performance Plan (EPPP), participants trained two to four days per week, depending on age group, and typically had one competitive match per week. All players participated in a structured strength and conditioning programme, delivered by qualified coaches within the academy. Data collection occurred within the academies during the 2018–2019 and 2019–2020 seasons. None of the players reported injuries at the time of testing, nor had a major injury six months prior to testing. Parental consent and participant assent were collected for all elements of the study, in addition to a standardised health questionnaire. Ethical approval was granted by the University Research Ethics Committee for all elements of the study.

2.2. Procedure

2.2.1. Anthropometrics

Standing height was measured using the nearest 0.1 cm with the use of a stadiometer (SECA, 321, Vogel & Halke, Hamburg, Germany). Body mass was measured to the nearest 0.1 kg on an electronic scale (SECA, 321, Vogel & Halke, Hamburg, Germany). During both anthropometric assessments, participants were instructed to stand in normal posture with weight equally distributed between feet [27].

2.2.2. Birth-Date Distribution

The selection year for youth soccer in the UK spans 1st September to 31st August and consistent with previous research [4], the year was split into four quartiles. September, October and November were classified as 'BQ1', December, January and February classified as 'BQ2', March, April and May classified as 'BQ3', and June, July and August as 'BQ4'. The measure of relative age was also expressed as a decimal, using the difference between a participant's birthdate and the selection cut-off date, divided by the number of days in a year [23].

2.2.3. Biological Age

To estimate biological maturation, the Khamis-Roche method was used, which requires chronological age, current height and weight of the child, and calculation of mid-parental height of the biological parents, to estimate final adult height (Equation (1)) [16]. When predicting final adult height of males between 4.0 and 17.5 years of age, the median error associated with the use of the Khamis-Roche method is 2.2 cm [16]. The standing height of participants' biological parents was collected by academy staff or, where collection was not possible, self-reported by the parents [27]. In instances where the heights were self-reported, these were adjusted for overestimation using sex-specific equations [28] (Equation (2)).

$$\text{Predicted adult height} = \beta 0 + \beta 1\ \text{height} + \beta 2\ \text{weight} + \beta 3\ \text{mid} - \text{parent height} \quad (1)$$

Equation (1). Equation for predicting final adult height [16].
$\beta 0$ is a sex- and age-specific intercept and $\beta 1$, $\beta 2$, and $\beta 3$ are sex- and age-specific coefficients, in which height, weight and mid-parent height should be multiplied [29].

$$\begin{array}{c}\text{Male adult height (inches)} = 2.316 + (0.955 \times \text{height [inches]})\\ \text{Female adult height (inches)} = 2.803 + (0.953 \times \text{height [inches]})\end{array} \quad (2)$$

Equation (2). Equation to adjust for self-reported heights in adults [29].

To estimate biological maturity, %PAH attained was calculated by dividing current height by PAH and multiplying by 100 [15]. Players with a greater %PAH can be expected to be more advanced in maturation compared to those further away from their PAH [27]. To estimate each participants timing of maturity, %PAH was calibrated with age- and sex-specific reference standards obtained from the UK 1990 growth reference data [30]. The age that the participants current %PAH aligned with was identified as participants biological age [31]. Maturity status was then determined using the discrepancy score between biological age (BA) and chronological age (CA). Using the traditional method of +1.0 and −1.0 for early and late maturers, respectively, fails to differentiate between individuals who differ markedly in maturity (e.g., BA-CA of +0.99 and −0.99 are both deemed on-time) [32]. Therefore, a less conservative set of criteria was applied (currently employed in the Premier League Player Management Application), and those participants with a BA-CA score of below −0.5 were classified as "late maturers", between −0.49 and 0.49 as "on-time", and those above +0.5 as "early maturers" [32].

2.2.4. Physical Performance Tests

Countermovement Jump (CMJ): Participants performed three trials of the CMJ on a mobile contact mat (Smart Jump; Fusion Sport, Queensland, Australia), with the best jump being used for further analysis. Participants were instructed to keep their hands on their hips, and lower themselves rapidly from an initial standing position to a self-selected squat position, followed immediately by an explosive vertical jump [33]. This protocol has been reported to be a valid and reliable assessment of neuromuscular performance in youth (Intraclass correlation coefficient [ICC] = 0.83 [34]).

30 m Sprint: Sprint times during three trials of maximal sprinting over 30 m were assessed using photo-electric timing gates (Smart Speed, Fusion Sport, Queensland, Australia) on an outdoor 3G pitch. The timing gates were placed at 0 m, 5 m, 10 m, 20 m and

30 m. Participants were instructed to begin their sprint in a split stance on a line 50 cm from the first gate, to avoid starting the timer early when in their set position. Participants were then instructed to "get ready" and "go", and were given verbal encouragement throughout each trial to ensure they were sprinting maximally through the final timing gate. A minimum of four minutes passive rest was given between trials to ensure sufficient recovery [35]. The best 30 m time was used for further analysis.

2.3. Statistical Analyses

The assumption of normality was assessed via the Shapiro-Wilk test, and descriptive statistics were calculated for all variables as mean and standard deviation (SD). Separate one-way analysis of covariance (ANCOVA) tests, with age as the covariate, were used to determine the differences in all measured variables between age groups, birth quartiles and maturity classifications, with a Bonferroni post-hoc analysis applied to identify any significant between-group differences.

Frequency counts were used to determine the number of players within each birth quartile (BQ1–4) and each maturity classification (early, on-time, late). Chi-square ($\chi 2$) analysis was then used to compare maturity distributions from within each birth quartile to what would be expected based on a normal distribution (30.3% as early and late maturers, and 38.3% as average maturers). Cramer's V was also calculated to determine the magnitude of difference in frequency counts and interpreted as a value of 0.06–0.16 as a small effect size, 0.17–0.28 as a medium effect size and >0.29 as a large effect size [36]. Furthermore, analysis of the adjusted standardized residuals was completed to identify frequencies that were greater than 1.96 or less than −1.96 z-scores ($p < 0.05$), highlighting a significant difference to the expected distribution for each age group.

Relationships between both relative age and percentage of PAH, and CMJ jump height and split times from the 30 m sprint (0–5 m, 0–10 m, 0–20 m and 0–30 m) were assessed via Pearson's correlation coefficients and interpreted as: <0.2 (no relationship), 0.2–0.45 (weak), 0.45–0.7 (moderate) and >0.7 (strong) based on previous recommendations [37].

3. Results

The descriptive statistics of each age group for height, weight, PAH, percentage of PAH (%PAH) and performance parameters including 5 m, 10 m, 20 m, 30 m speed and CMJ jump height are presented in Table 1.

Table 1. Frequency count of birth quartile (BQ) and maturity classification, and descriptive statistics for anthropometric characteristics for each age group (mean ± SD).

Age Group	BQ1 (n)	BQ2 (n)	BQ3 (n)	BQ4 (n)	Early (n)	On-Time (n)	Late (n)	Height (cm)	Body Mass (kg)	PAH (cm)	%PAH
U9	4	6	2	0	2	10	0	135.6 ± 4.7	31.2 ± 2.7	178.6 ± 5.7	0.75 ± 0.01
U10	9	2	5	2	2	16	0	137.8 ± 4.8	32.7 ± 3.8	175.3 ± 5.6	0.78 ± 0.01
U11	11	5	1	0	2	11	3	143.3 ± 6.3	38.8 ± 6.1	177.0 ± 6.9	0.81 ± 0.02 $^{\#a}$
U12	5	7	9	1	3	18	1	148.7 ± 6.8 $^{\#a}$	40.8 ± 6.2 $^{\#a}$	178.2 ± 4.9	0.84 ± 0.02 $^{\hat{}\#a}$
U13	14	6	7	9	8	23	5	154.1 ± 8.4 $^{\hat{}\#a}$	44.6 ± 7.9 $^{\#a}$	180.4 ± 6.9	0.86 ± 0.03 $^{\hat{}\#a}$
U14	13	9	7	5	8	21	5	161.6 ± 7.5 $^{¢§\hat{}\#a}$	51.4 ± 9.1 $^{¢§\hat{}\#a}$	181.2 ± 5.5 $^{\#}$	0.89 ± 0.03 $^{¢§\hat{}\#a}$
U16	15	19	6	4	14	22	8	170.7 ± 7.0 $^{\infty¢§\hat{}\#a}$	60.0 ± 8.9 $^{\infty¢§\hat{}\#a}$	180.3 ± 4.8 $^{\#}$	0.95 ± 0.03 $^{\infty¢§\hat{}\#a}$
U18	12	3	2	0	4	13	0	180.1 ± 5.3 *	75.3 ± 7.0 *	180.5 ± 5.2	1.00 ± 0.02 *

[a] significantly different to U9; # significantly different to U10; ^ significantly different to U11; § significantly different to U12; ¢ significantly different to U13; ∞ significantly different to U14; * significantly different to all groups.

Older age group players were significantly taller, heavier and more mature than the younger age groups ($p < 0.05$). However, there were no differences between the U11, U10 and U9 for height, weight, or PAH ($p > 0.05$), but the U11 were significantly more mature than the U10 and U9 ($p < 0.05$). From a physical performance aspect, older players significantly outperformed younger players across most sprint distances and in the CMJ ($p < 0.05$). However, there were no significant differences in sprint performance at any distance between U9 to U13, other than U12 being significantly faster than U10 at 20 m and 30 m. Specific differences in anthropometric and performance scores between age groups are shown in Tables 1 and 2.

Table 2. Descriptive statistics for sprint times and CMJ height for each age group (mean ± SD).

Age Group	5 m (s)	10 m (s)	20 m (s)	30 m (s)	CMJ (cm)
U9	1.12 ± 0.04	1.96 ± 0.05	3.55 ± 0.13	5.16 ± 0.22	24.0 ± 3.5
U10	1.15 ± 0.04	2.02 ± 0.07	3.64 ± 0.14	5.27 ± 0.26	22.8 ± 2.4
U11	1.13 ± 0.06	1.99 ± 0.10	3.60 ± 0.19	5.18 ± 0.30	24.5 ± 3.4
U12	1.09 ± 0.05	1.92 ± 0.08	3.42 ± 0.18 #	4.92 ± 0.27 #	27.2 ± 3.7
U13	1.15 ± 0.08	1.99 ± 0.12	3.50 ± 0.20	4.96 ± 0.29 #	30.6 ± 5.9 ^#a
U14	1.10 ± 0.12	1.90 ± 0.14 ¢#a	3.34 ± 0.21 ¢^#a	4.73 ± 0.28 ¢^#a	32.4 ± 5.4 §^#a
U16	1.08 ± 0.09 ¢#	1.85 ± 0.11 ¢^#a	3.19 ± 0.16 ∞¢§^#a	4.46 ± 0.22 ∞¢§^#a	36.9 ± 6.2 ∞¢§^#a
U18	0.99 ± 0.05 *	1.71 ± 0.06 *	2.96 ± 0.09 *	4.15 ± 0.12 *	41.9 ± 6.5 *

[a] significantly different to U9; # significantly different to U10; ^ significantly different to U11; § significantly different to U12; ¢ significantly different to U13; ∞ significantly different to U14; * significantly different to all groups.

The adjusted means of each birth quartile for height, weight, PAH, %PAH and performance parameters including 5 m, 10 m, 20 m, 30 m speed and CMJ jump height are presented in Table 3. There were no significant differences between any birth quartile for any of the measured variables.

Table 3. Descriptive statistics for all measured variables across birth quartiles (adjusted mean ± adjusted SD).

BQ	Height (cm)	Body Mass (kg)	PAH (cm)	%PAH	5 m (s)	10 m (s)	20 m (s)	30 m (s)	CMJ (cm)
1	156.4 ± 7.1	48.5 ± 7.4	178.7 ± 5.8	87.4 ± 2.7	1.10 ± 0.08	1.92 ± 0.11	3.40 ± 0.18	4.84 ± 0.26	30.3 ± 5.4
2	158.5 ± 7.1	50.0 ± 7.3	179.7 ± 5.8	88.4 ± 2.2	1.09 ± 0.08	1.89 ± 0.10	3.35 ± 0.18	4.75 ± 0.25	31.3 ± 5.3
3	156.4 ± 7.1	47.8 ± 7.4	179.5 ± 5.8	87.1 ± 2.5	1.10 ± 0.08	1.90 ± 0.11	3.34 ± 0.18	4.73 ± 0.26	32.8 ± 5.4
4	157.8 ± 7.1	50.7 ± 7.3	181.5 ± 5.8	87.2 ± 2.3	1.14 ± 0.08	1.96 ± 0.11	3.42 ± 0.18	4.84 ± 0.26	32.1 ± 5.3

The adjusted means of each maturity classification are presented in Table 4. Early maturers were significantly taller and heavier compared with both on-time and late maturers ($p < 0.05$). From a performance aspect, early and on-time maturers significantly outperformed late maturers in 5 m,10 m, 20 m and 30 m sprint times ($p < 0.05$), but there were no differences in CMJ height between groups ($p > 0.05$).

Table 4. Descriptive statistics for all measured variables across maturity classifications (adjusted mean ± adjusted SD).

Maturity Classification	Height (cm)	Body Mass (kg)	PAH (cm)	BA-CA (Years)	5 m (s)	10 m (s)	20 m (s)	30 m (s)	CMJ (cm)
Early	164.4 ± 6.4	56.5 ± 6.1	182.3 ± 5.9	0.89 ± 0.35	1.07 ± 0.08	1.87 ± 0.10	3.29 ± 0.18	4.69 ± 0.26	32.9 ± 5.5
On Time	156.1 ± 6.2 *	48.9 ± 5.9 *	178.7 ± 5.7 *	0.04 ± 0.29 *	1.09 ± 0.08	1.90 ± 0.10	3.37 ± 0.17	4.80 ± 0.26	30.7 ± 5.4
Late	153.3 ± 6.3 *	42.1 ± 6.0 *#	179.6 ± 5.8	−0.73 ± 0.21 *	1.16 ± 0.08 *#	2.00 ± 0.10 *#	3.49 ± 0.18 *#	4.92 ± 0.26 *#	31.7 ± 5.5

* significantly different to "early" maturers; # significantly different to "on time" maturers. BA: biological age; CA: chronological age.

The maturity distributions within each birth quartile were significantly skewed with a large effect size compared to normal distribution (χ^2 ($df = 2$) = 73.1, $p < 0.05$, $V = 0.429$) (see Figure 1). The adjusted residuals showed that there were significantly more on-time maturers and significantly less early and late maturers for the BQ1 and BQ3 than expected ($p < 0.05$).

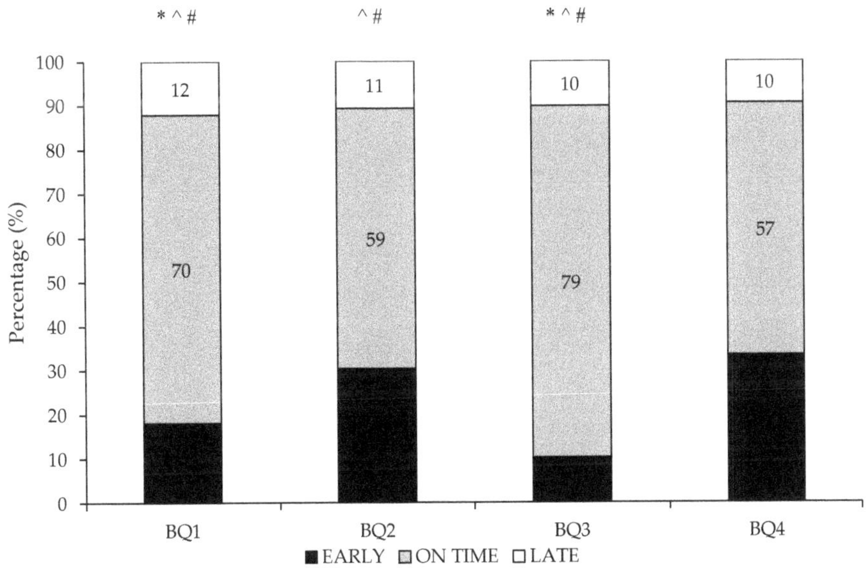

Figure 1. Frequency count for number of players in each maturity band from each birth quartile. * significantly fewer early maturers than expected based on normal distribution ($p < 0.05$). ^ significantly more on time maturers than expected based on normal distribution ($p < 0.05$). # significantly fewer late maturers than expected based on normal distribution ($p < 0.05$).

The relationships between relative age, maturity and sprint and jump performance are displayed in Table 5. There was a significant, weak relationship between relative age and CMJ height in U16 ($r = 0.416$; $p < 0.05$), however there were no other significant associations between relative age and physical performance in any age group. U12 to U16 showed weak to strong relationships between maturity and sprint performance ($r = 0.366$–0.711; $p < 0.05$), except for 30 m time in U13. There was also a moderate, significant relationship between maturity and CMJ height in U12 ($r = 0.497$; $p < 0.05$).

Table 5. Pearson correlations between relative age and biological age for each age group.

Age Group	Relative Age					Maturity Status				
	5 m (s)	10 m (s)	20 m (s)	30 m (s)	CMJ (cm)	5 m (s)	10 m (s)	20 m (s)	30 m (s)	CMJ (cm)
U9	0.152	−0.179	−0.109	−0.091	−0.012	−0.264	−0.552	−0.467	−0.423	−0.291
U10	0.295	0.293	0.321	0.322	−0.077	−0.034	0.087	0.275	0.286	−0.045
U11	−0.005	−0.118	−0.105	−0.025	0.136	−0.167	−0.153	−0.163	−0.102	0.150
U12	−0.073	−0.114	−0.148	−0.146	0.216	−0.738 *	−0.655 *	−0.686 *	−0.680 *	0.497 *
U13	−0.058	−0.096	−0.052	−0.071	−0.045	−0.477 *	−0.427 *	−0.366 *	−0.291	−0.305
U14	−0.062	−0.051	−0.008	0.013	−0.222	−0.706 *	−0.711 *	−0.652 *	−0.607 *	0.026
U16	0.190	0.179	0.141	0.057	0.416 *	−0.497 *	−0.654 *	−0.609 *	−0.616 *	0.200
U18	−0.236	−0.122	−0.289	−0.272	0.265	−0.257	−0.261	−0.296	−0.299	0.348

* significant correlation ($p < 0.05$).

4. Discussion

The main finding of the current study was that maturity status and relative age were differentially associated with sprint performance in young soccer players. Specifically, advanced maturity was associated with superior sprint performance in most age groups, whereas relative age was, in the majority of cases, unrelated to sprint performance. CMJ performance was significantly associated with more advanced maturity at U12, and older relative age at U16. Collectively, these findings generally support the conclusion that advanced maturity, but not older relative age, is associated with superior sprint speed in English male academy soccer players. Thus, the arguments that relatively older players possess superior speed are not supported in this context, and the initial hypothesis can be accepted.

There were a number of significant associations observed between maturity and sprint performance in U12 to U16, however, there were no significant relationships between relative age and sprint performance in any age groups. There was also a significant association between maturity and CMJ performance in U12, whereas U16 had a significant relationship between relative age and CMJ performance. Similar findings have recently been reported, where maturity status was shown to have a much greater influence on sprint, change of direction and CMJ performance in young soccer players [27]. The findings from the current study expands on this previous research, identifying that maturity influences sprint performance between 12 and 16 years, but has limited influence prior to and after these age groups. Considering that the onset of PHV is ~85% PAH [17], the majority of players U12 and below within the current study were yet to experience their growth spurt (66/68 players < 85% PAH) and therefore may explain why maturity has no influence on sprint speed prior to this age group. Additionally, the weakest significant association between sprint performance and maturation was within the U13 age group, and this group had an average PAH of 85.7%, suggesting they were at the onset of the adolescent growth spurt. It is possible that some of the challenges associated with adapting and adjusting to the growth spurt may mitigate some of the advantages associated with advanced maturity at this stage of development.

As expected, CMJ performance significantly increased with advancing age across the entire population. However, when considered within specific age groups, which become more homogenous, relationships between maturity and CMJ were mostly non-significant. The only group where a significantly relationship did exist was for the U12, an age which is associated with the start of the growth spurt and may represent a time of more variability in maturity and performance across players [38]. Furthermore, players within the same age group would have similar resistance training ages, due to starting at the academy at the same time. These similar training ages of players may have off-set any potential benefits of advanced maturity status on CMJ performance within individual age groups.

In accordance with previous research [39–41], the older age groups were significantly taller, heavier and closer to their predicted adult height compared to younger age groups, while the older groups also outperformed the younger groups in sprint and jump tests. Interestingly, there were no significant differences in anthropometric characteristics between each birth quartile. This may suggest that BQ4s need to be relatively taller and heavier to be selected into soccer academies, which supports previous findings where the mean height and weight of relatively younger soccer players lay above the normal development curve, whereas the means of relatively older players lay on or under that curve [42].

Although superior values were reported across the majority of the fitness variables in players born in the first three quarters of the year compared with the last quarter, the between-group differences were not significant. As with comparable studies, these findings may be limited by the small number of BQ4 compared to the other quartiles. However, similar outcomes have been reported in previous literature [1,43], where the only difference between young players from each birth quartile was in chronological age and %PAH, with no significant difference in physical performance across birth quartiles. One explanation of these findings may be that the BQ4s who are entering into academies are better physically than the average, school-aged BQ4, and one of the reasons why they are being selected in the first instance. This could explain the lack of differences between BQ's in performance and the lack of relationship between relative age and performance.

The current study found that early maturers were taller, heavier, faster and jumped higher than the on time and late maturers. Typically, research has reported that earlier maturing athletes have greater anthropometric characteristics (height and body mass) than later-maturing athletes [7], with previous research highlighting improvements in sprint performance with increasing maturation in young soccer players [43,44]. Cumulatively, the findings from the current study suggest that maturity status has a significant influence on sprint performance in English male academy soccer players, whereas relative age did not. As children mature, they will experience natural increases in strength and power [11], underpinned by structural and neural changes [45–47]. Recently, increases in muscle thickness throughout maturation were shown to be the underpinning factor in improvements in sprint speed in a cohort of school-aged boys [46]. Considering that relative maximal force is a strong predictor of sprint performance in boys [48], the increased force producing capabilities in boys as they mature may explain the influence of maturation on sprint performance.

There was a relative age bias present within the academies assessed within the current study, whereby ~70% of players were born in the first half of the year, with ~41% born in BQ1. Interestingly, although the percentage of late maturers was similar from each birth quartile (~10–12%), a greater percentage of BQ4′s were early maturers (33%) compared to the other birth quartiles (10–30%). It is often assumed that relatively older academy soccer players are further advanced in maturation and, thus, possess greater anthropometric qualities and superior performance characteristics [6]. However, the findings from the current study supports the notion that maturation and relative age are different constructs [6], and that being BQ4 does not mean that an academy soccer player will be a later maturer. However, these findings suggest that it may be important for players born in BQ4 to be early maturing to increase their likelihood of overcoming the relative age bias and being selected into an English male soccer academy. Previous research has also reported that early maturing soccer players were overrepresented in the last BQ, whereas late maturing athletes were overrepresented in the first BQ, suggesting that relatively younger soccer players may only have an opportunity of selection if they were early maturing, whereas relatively older athletes have an increased likelihood for selection independent of their biological maturity status [49,50].

Maturation influences physical performance, with early maturing boys outperforming on-time and late maturers, which has a subsequent impact on match-performance in soccer [51]. While advanced maturity offers an initial benefit in performance and selection [52], it may be detrimental in the long term, due to early maturing players neglecting their technical and tactical development in favour of using their physical prowess [52]. Research has suggested that 'elite' status in soccer gradually excludes early maturing boys and favour late maturing boys as age increased [53]. Those involved in the identification and development of academy players should be aware of, and accommodate for, individual differences in maturation. Bio-banding is the process of periodically grouping athletes on the basis of attributes associated with growth or maturation, rather than chronological age [38]. This approach has been used as a method to ensure holistic development of soccer players in academies and can theoretically benefit both early and late maturers, by levelling out physical requirements, ensuring that players develop technical and tactical abilities as well as using their physical qualities [20,38]. Bio-banding exists as an adjunct to, and not a replacement for, age group competition, meaning late maturing youth can also continue to experience the challenges of competing against their more mature peers in the traditional formats, which is important in the context of the underdog hypothesis [4]. Late maturing players have been found to possess superior technical skills [54] and more adaptive self-regulated learning strategies [23], and it may be important for these later maturers to compete against more mature peers in order to develop these traits that result in their success transitioning towards adulthood. One key use of bio-banding in soccer may be when comparing fitness testing data across age groups [55]. The current study has identified the influence of maturation on sprint speed, and therefore it seems prudent to identify and develop boys of the same maturational stage, as well as chronological age.

A limitation of the current study was that maturity was not assessed using the gold standard method of skeletal imaging [9]. This method requires access to specialist equipment and expertise, and is not accessible to most practitioners working in youth sport. Instead, maturity was estimated using %PAH, which is widely used in youth sport and particularly soccer, and has been shown to be a reliable method for estimating maturity [18,27]. While the current study has made a significant contribution to the literature surrounding the relationship between maturity status and performance, field-based methods were used to assess performance. Future research should attempt to collect more detailed metrics, such as force-time characteristics to better understand the influence of maturity on performance.

The RAE is well established within soccer academies, despite no clear benefit of being relatively older in terms of physical performance in those selected into an academy. Therefore, future research should aim to identify the processes and mechanisms that underpin the RAE in soccer, with a particular emphasis upon developmental attributes that afford a distinct advantage from early childhood. Moreover, the differences in physical performance outcomes between BQs who are selected into academies compared to those who are not should be explored to help better understand the role of sprint and power attributes as part of the selection process, as well as take the existing literature beyond the current academy soccer context.

5. Conclusions

The current study aimed to establish the relationship between maturation, relative age and physical performance. Sprint performance was associated with maturation, but not relative age, while there was no consistent relationship between relative age or maturation and CMJ performance. It is key for practitoners to understand that the RAE and maturity status are two distinct constructs, highlighted by the signifcant associaiton between sprint performance and maturation, but not relative age. Practitioners should be encouraged to monitor growth and maturation (frequent assessments of height and weight to establish predicted adult height and maturity status) to help interpret changes in physical performance of young English male academy soccer players. Furthermore, maturity status should be considered when comparing fitness scores in players to ensure practitioners are not comparing early and late maturers within the same age group, but rather are comparing boys of the same maturity status.

Author Contributions: Conceptualization, J.M.R. and J.L.O.; methodology, J.M.R., J.L.O., A.L.K. and S.P.C.; validation, J.M.R., J.S., S.P.C., A.L.K., R.S.L. and J.L.O.; data curation, J.S. and J.B.; formal analysis, J.M.R.; writing—original draft preparation, J.M.R.; writing—review and editing, A.L.K., S.P.C., J.S., J.L.O. and R.S.L.; visualization, J.M.R. and J.L.O.; project administration, J.M.R. All authors have read and agreed to the published version of the manuscript.

Funding: This research received no external funding.

Institutional Review Board Statement: The study was conducted according to the guidelines of the Declaration of Helsinki, and approved by the Ethics Committee of Cardiff Metropolitan University (Protocol Code: 17-4-06P and date of approval: 4 April 2017).

Informed Consent Statement: Informed consent was obtained from all subjects, or parents when participants were younger than 18 years old, involved in the study.

Data Availability Statement: The data is not yet publicly available.

Conflicts of Interest: The authors declare no conflict of interest.

References

1. Carling, C.; Le Gall, F.; Reilly, T.; Williams, A.M. Do Anthropometric and Fitness Characteristics Vary According to Birth Date Distribution in Elite Youth Academy Soccer Players?: Relative Age Effect in Elite Youth Soccer. *Scand. J. Med. Sci. Sports* **2008**, *19*, 3–9. [CrossRef]
2. Reilly, T.; Williams, A.M.; Nevill, A.; Franks, A. A Multidisciplinary Approach to Talent Identification in Soccer. *J. Sports Sci.* **2000**, *18*, 695–702. [CrossRef]
3. Clemente, F.M.; Clark, C.C.T.; Leão, C.; Silva, A.F.; Lima, R.; Sarmento, H.; Figueiredo, A.J.; Rosemann, T.; Knechtle, B. Exploring Relationships Between Anthropometry, Body Composition, Maturation, and Selection for Competition: A Study in Youth Soccer Players. *Front. Physiol.* **2021**, *12*, 651735. [CrossRef]
4. Kelly, A.L.; Wilson, M.R.; Gough, L.A.; Knapman, H.; Morgan, P.; Cole, M.; Jackson, D.T.; Williams, C.A. A Longitudinal Investigation into the Relative Age Effect in an English Professional Football Club: Exploring the 'Underdog Hypothesis'. *Sci. Med. Footb.* **2020**, *4*, 111–118. [CrossRef]
5. Meylan, C.; Cronin, J.; Oliver, J.; Hughes, M. Talent Identification in Soccer: The Role of Maturity Status on Physical, Physiological and Technical Characteristics. *Int. J. Sports Sci. Coach.* **2010**, *5*, 571–592. [CrossRef]
6. Towlson, C.; MacMaster, C.; Parr, J.; Cumming, S. One of These Things Is Not like the Other: Time to Differentiate between Relative Age and Biological Maturity Selection Biases in Soccer? *Sci. Med. Footb.* **2021**. [CrossRef]
7. Peña-González, I.; Fernández-Fernández, J.; Moya-Ramón, M.; Cervelló, E. Relative Age Effect, Biological Maturation, and Coaches' Efficacy Expectations in Young Male Soccer Players. *Res. Q. Exerc. Sport* **2018**, *89*, 373–379. [CrossRef]
8. Beunen, G.P.; Malina, R.M. Growth and Biologic Maturation: Relevance to Athletic Performance. In *The Child and Adolescent Athlete*; Bar-Or, O., Hebestreit, H., Eds.; Blackwell Publishing: Oxford, UK, 2005; pp. 3–17.
9. Lloyd, R.S.; Oliver, J.L.; Faigenbaum, A.D.; Myer, G.D.; De Ste Croix, M.B.A. A Chronological Age vs. Biological Maturation: Implications for Exercise Programming in Youth. *J. Strength Cond. Res.* **2014**, *28*, 1454–1464. [CrossRef]
10. Johnson, A.; Farooq, A.; Whiteley, R. Skeletal Maturation Status Is More Strongly Associated with Academy Selection than Birth Quarter. *Sci. Med. Footb.* **2017**, *1*, 157–163. [CrossRef]
11. Malina, R.M.; Bouchard, C.; Bar-Or, O. *Growth, Maturation and Physical Activity*, 2nd ed.; Human Kinetics: Champaign, IL, USA, 2004.
12. Mirwald, R.L.; Baxter-Jones, A.D.; Bailey, D.A.; Beunen, G.P. An Assessment of Maturity from Anthropometric Measurements. *Med. Sci. Sports Exerc.* **2002**, *34*, 689–694. [CrossRef] [PubMed]
13. Moore, S.A.; McKay, H.A.; Macdonald, H.; Nettlefold, L.; Baxter-Jones, A.D.; Cameron, N.; Brasher, P.M. Enhancing a Somatic Maturity Prediction Model. *Med. Sci. Sports Exerc.* **2015**. [CrossRef]
14. Malina, R.M.; Kozieł, S.M. Validation of Maturity Offset in a Longitudinal Sample of Polish Boys. *J. Sports Sci.* **2014**, *32*, 424–437. [CrossRef] [PubMed]
15. Roche, A.; Tyleshevski, F.; Rogers, E. Non-Invasive Measurements of Physical Maturity in Children. *Res. Q. Exerc. Sport* **1983**, *54*, 363–371. [CrossRef]
16. Khamis, H.; Roche, A. Predicting Adult Stature without Using Skeletal Age: The Khamis-Roche Method. *Pediatrics* **1994**, *94*, 504–507.
17. Parr, J.; Winwood, K.; Hodson-Tole, E.; Deconinck, F.; Parry, L.; Hill, J.; Malina, R.; Cumming, S. Predicting the Timing of the Peak of the Pubertal Growth Spurt in Elite Male Youth Soccer Players: Evaluation of Methods. *Ann. Hum. Biol.* **2020**, *47*, 400–408. [CrossRef] [PubMed]
18. Olivares, L.A.F.; De León, L.G.; Fragoso, M.I. Skeletal Age Prediction Model from Percentage of Adult Height in Children and Adolescents. *Sci. Rep.* **2020**, *10*, 15768. [CrossRef]
19. Malina, R.M.; Cumming, S.P.; Rogol, A.D.; Coelho-e-Silva, M.J.; Figueiredo, A.J.; Konarski, J.M.; Kozieł, S.M. Bio-Banding in Youth Sports: Background, Concept, and Application. *Sports Med.* **2019**, *49*, 1671–1685. [CrossRef] [PubMed]
20. Bradley, B.; Johnson, D.; Hill, M.; McGee, D.; Kana-ah, A.; Sharpin, C.; Sharp, P.; Kelly, A.; Cumming, S.P.; Malina, R.M. Bio-Banding in Academy Football: Player's Perceptions of a Maturity Matched Tournament. *Ann. Hum. Biol.* **2019**, *46*, 400–408. [CrossRef]
21. Guimarães, E.; Ramos, A.; Janeira, M.A.; Baxter-Jones, A.D.G.; Maia, J. How Does Biological Maturation and Training Experience Impact the Physical and Technical Performance of 11–14-Year-Old Male Basketball Players? *Sports* **2019**, *7*, 243. [CrossRef]
22. Lintunen, T.; Rahkila, P.; Silvennoinen, M.; Osterback, L. Psychological and Physical Correlates of Early and Late Biological Maturation in 9- to 11-Year-Old Girls and Boys. In *Young Athletes: Biological, Psychological, and Educational Perspectives*; Malina, R.M., Ed.; Human Kinetics: Champaign, IL, USA, 1988; pp. 85–91.
23. Cumming, S.P.; Searle, C.; Hemsley, J.K.; Haswell, F.; Edwards, H.; Scott, S.; Gross, A.; Ryan, D.; Lewis, J.; White, P.; et al. Biological Maturation, Relative Age and Self-Regulation in Male Professional Academy Soccer Players: A Test of the Underdog Hypothesis. *Psychol. Sport Exerc.* **2018**, *39*, 147–153. [CrossRef]
24. Wattie, N.; Cobley, S.; Baker, J. Towards a Unified Understanding of Relative Age Effects. *J. Sports Sci.* **2008**, *26*, 1403–1409. [CrossRef]
25. Figueiredo, A.J.; Gonçalves, C.E.; Coelho E Silva, M.J.; Malina, R.M. Youth Soccer Players, 11–14 Years: Maturity, Size, Function, Skill and Goal Orientation. *Ann. Hum. Biol.* **2009**, *36*, 60–73. [CrossRef] [PubMed]

26. Lovell, R.; Towlson, C.; Parkin, G.; Portas, M.; Vaeyens, R.; Cobley, S. Soccer Player Characteristics in English Lower-League Development Programmes: The Relationships between Relative Age, Maturation, Anthropometry and Physical Fitness. *PLoS ONE* **2015**, *10*, e0137238. [CrossRef]
27. Parr, J.; Winwood, K.; Hodson-Tole, E.; Deconinck, F.J.A.; Hill, J.P.; Teunissen, J.W.; Cumming, S.P. The Main and Interactive Effects of Biological Maturity and Relative Age on Physical Performance in Elite Youth Soccer Players. *J. Sports Med.* **2020**, *2020*, 1957636. [CrossRef]
28. Epstein, L.; Valoski, S.; Kalarchian, M.; McCurley, J. Do Children Lose and Maintain Weight Easier than Adults: A Comparison of Child and Parent Weight Changes from Six Months to Ten Years. *Obes. Res.* **1995**, *3*, 411–417. [CrossRef]
29. Cumming, S.P.; Sherar, L.; Esliger, D.; Riddoch, C.J.; Malina, R.M. Concurrent and prospective associations among biological maturation, and physical activity at 11 and 13 years of age. *Scand. J. Med. Sci. Sports* **2014**, *24*, e20–e28. [CrossRef]
30. Freeman, J.V.; Cole, T.J.; Chinn, S.; Jones, P.R.; White, E.M.; Preece, M.A. Cross Sectional Stature and Weight Reference Curves for the UK, 1990. *Arch. Dis. Child.* **1995**, *73*, 17–24. [CrossRef] [PubMed]
31. Gillison, F.; Cumming, S.; Standage, M.; Barnaby, C.; Katzmarzyk, P. Assessing the Impact of Adjusting for Maturity in Weight Status Classification in a Cross-Sectional Sample of UK Children. *BMJ Open* **2017**, *7*, e015769. [CrossRef]
32. Hill, M.; Scott, S.; Malina, R.M.; McGee, D.; Cumming, S.P. Relative age and maturation selection biases in academy football. *J. Sports Sci.* **2020**, *38*, 1359–1367. [CrossRef]
33. Lloyd, R.; Oliver, J.; Hughes, M.; Williams, C. Specificity of Test Selection for the Appropriate Assessment of Different Measures of Stretch-Shortening Cycle Function in Children. *J. Sports Med. Phys. Fit.* **2011**, *51*, 595–602.
34. Lloyd, R.S.; Oliver, J.L.; Hughes, M.G.; Williams, C.A. Reliability and Validity of Field-Based Measures of Leg Stiffness and Reactive Strength Index in Youths. *J. Sports Sci.* **2009**, *27*, 1565–1573. [CrossRef]
35. Lloyd, R.S.; Radnor, J.M.; De Ste Croix, M.; Cronin, J.B.; Oliver, J.L. Changes in Sprint and Jump Performances after Traditional, Plyometric, and Combined Resistance Training in Male Youth Pre- and Post-Peak Height Velocity. *J. Strength Cond. Res.* **2015**, *30*, 1239–1247. [CrossRef]
36. Cohen, J. *Statistical Power Analysis for the Behavioural Sciences*, 2nd ed.; Routledge: London, UK, 1988. [CrossRef]
37. O'Donoghue, P. *Statistics for Sport and Exercise Studies: An Introduction*; Routledge: London, UK, 2012.
38. Cumming, S.P.; Lloyd, R.S.; Oliver, J.L.; Eisenmann, J.C.; Malina, R.M. Bio-Banding in Sport: Applications to Competition, Talent Identification, and Strength and Conditioning of Youth Athletes. *Strength Cond. J.* **2017**, *39*, 34–47. [CrossRef]
39. Mendez-Villanueva, A.; Buchheit, M.; Kuitunen, S.; Douglas, A.; Peltola, E.; Bourdon, P. Age-Related Differences in Acceleration, Maximum Running Speed, and Repeated-Sprint Performance in Young Soccer Players. *J. Sports Sci.* **2011**, *29*, 477–484. [CrossRef] [PubMed]
40. Morris, R.; Emmonds, S.; Jones, B.; Myers, T.D.; Clarke, N.D.; Lake, J.; Ellis, M.; Singleton, D.; Roe, G.; Till, K. Seasonal Changes in Physical Qualities of Elite Youth Soccer Players According to Maturity Status: Comparisons with Aged Matched Controls. *Sci. Med. Footb.* **2018**, *2*, 272–280. [CrossRef]
41. Williams, C.A.; Oliver, J.L.; Faulkner, J. Seasonal Monitoring of Sprint and Jump Performance in a Soccer Youth Academy. *Int. J. Sports Physiol. Perform.* **2011**, *6*, 264–275. [CrossRef]
42. Votteler, A.; Höner, O. The Relative Age Effect in the German Football TID Programme: Biases in Motor Performance Diagnostics and Effects on Single Motor Abilities and Skills in Groups of Selected Players. *Eur. J. Sport Sci.* **2014**, *14*, 433–442. [CrossRef]
43. Figueiredo, A.J.; Coelho-e-Silva, M.J.; Cumming, S.P.; Malina, R.M. Relative Age Effect: Characteristics of Youth Soccer Players by Birth Quarter and Subsequent Playing Status. *J. Sports Sci.* **2019**, *37*, 677–684. [CrossRef]
44. Philippaerts, R.M.; Vaeyens, R.; Janssens, M.; Van Renterghem, B.; Matthys, D.; Craen, R.; Bourgois, J.; Vrijens, J.; Beunen, G.; Malina, R.M. The Relationship between Peak Height Velocity and Physical Performance in Youth Soccer Players. *J. Sports Sci.* **2006**, *24*, 221–230. [CrossRef]
45. Dotan, R.; Mitchell, C.; Cohen, R.; Klentrou, P.; Gabriel, D.; Bareket, F. Child-Adult Differences in Muscle Activation—A Review. *Clin. Neurophysiol.* **2012**, *123*, 106–116. [CrossRef] [PubMed]
46. Radnor, J.M.; Oliver, J.L.; Waugh, C.M.; Myer, G.D.; Lloyd, R.S. Muscle Architecture and Maturation Influences Sprint and Jump Ability in Young Boys: A Multi-Study Approach. *J. Strength Cond. Res.* **2021**. [CrossRef]
47. Tumkur Anil Kumar, N.; Oliver, J.L.; Lloyd, R.S.; Pedley, J.S.; Radnor, J.M. The Influence of Growth, Maturation and Resistance Training on Muscle-Tendon and Neuromuscular Adaptations: A Narrative Review. *Sports* **2021**, *9*, 59. [CrossRef]
48. Meyers, R.W.; Oliver, J.L.; Hughes, M.G.; Lloyd, R.S.; Cronin, J.B. The Influence of Maturation on Sprint Performance in Boys over a 21-Month Period. *Med. Sci. Sports Exerc.* **2016**, *48*, 2555–2562. [CrossRef] [PubMed]
49. Deprez, D.; Coutts, A.; Fransen, J.; Deconinck, F.; Lenoir, M.; Vaeyens, R.; Philippaerts, R. Relative Age, Biological Maturation and Anaerobic Characteristics in Elite Youth Soccer Players. *Int. J. Sports Med.* **2013**, *34*, 897–903. [CrossRef]
50. Müller, L.; Gehmaier, J.; Gonaus, C.; Raschner, C.; Müller, E. Maturity Status Strongly Influences the Relative Age Effect in International Elite Under-9 Soccer. *J. Sports Sci. Med.* **2018**, *17*, 216. [PubMed]
51. Buchheit, M.; Mendez-Villanueva, A. Effects of Age, Maturity and Body Dimensions on Match Running Performance in Highly Trained under-15 Soccer Players. *J. Sports Sci.* **2014**, *32*, 1271–1278. [CrossRef]

52. Malina, R.M.; Rogol, A.D.; Cumming, S.P.; Coelho e Silva, M.J.; Figueiredo, A.J. Biological Maturation of Youth Athletes: Assessment and Implications. *Br. J. Sports Med.* **2015**, *49*, 852–859. [CrossRef] [PubMed]
53. Ostojic, S.M.; Castagna, C.; Calleja-González, J.; Jukic, I.; Idrizovic, K.; Stojanovic, M. The Biological Age of 14-Year-Old Boys and Success in Adult Soccer: Do Early Maturers Predominate in the Top-Level Game? *Res. Sports Med.* **2014**, *22*, 398–407. [CrossRef]
54. Zibung, M.; Zuber, C.; Conzelmann, A. The Motor Subsystem as a Predictor of Success in Young Football Talents: A Person-Oriented Study. *PLoS ONE* **2016**, *11*, e0161049. [CrossRef]
55. Thomas, C.; Oliver, J.L.; Kelly, A.L. Bio-Banding in Youth Soccer: Considerations for Researchers and Practitioners. In *Birth Advantages and Relative Age Effects in Sport: Exploring Organizational Structures and Creating Appropriate Settings*; Routledge: London, UK, 2021; pp. 125–156.

sports

MDPI

Review

Training Management of the Elite Adolescent Soccer Player throughout Maturation

Alistair J. McBurnie [1,2,*], Thomas Dos'Santos [3,4], David Johnson [5] and Edward Leng [1]

1 Football Medicine & Sports Science, Manchester United F.C., AON Training Complex, Manchester M31 4BH, UK; ed.leng@manutd.co.uk
2 Department of Sport Science, School of Science and Technology, Nottingham Trent University, Nottingham NG1 4FQ, UK
3 Musculoskeletal Science and Sports Medicine Research Centre, Department of Sport and Exercise Sciences, Manchester Metropolitan University, Manchester M15 6BH, UK; T.DosSantos@mmu.ac.uk
4 Manchester Institute of Sport, Manchester Metropolitan University, Manchester M1 7EL, UK
5 Department for Health, University of Bath, Bath BA2 7AY, UK; dmj32@bath.ac.uk
* Correspondence: alistair.mcburnie@manutd.co.uk

Abstract: Professional soccer clubs invest significantly into the development of their academy prospects with the hopes of producing elite players. Talented youngsters in elite development systems are exposed to high amounts of sports-specific practise with the aims of developing the foundational skills underpinning the capabilities needed to excel in the game. Yet large disparities in maturation status, growth-related issues, and highly-specialised sport practise predisposes these elite youth soccer players to an increased injury risk. However, practitioners may scaffold a performance monitoring and injury surveillance framework over an academy to facilitate data-informed training decisions that may not only mitigate this inherent injury risk, but also enhance athletic performance. Constant communication between members of the multi-disciplinary team enables context to build around an individual's training status and risk profile, and ensures that a progressive, varied, and bespoke training programme is provided at all stages of development to maximise athletic potential.

Keywords: long-term athlete development; soccer; growth and maturation; performance monitoring; injury surveillance

Citation: McBurnie, A.J.; Dos'Santos, T.; Johnson, D.; Leng, E. Training Management of the Elite Adolescent Soccer Player throughout Maturation. *Sports* **2021**, *9*, 170. https://doi.org/10.3390/sports9120170

Academic Editor: Jon Oliver

Received: 31 October 2021
Accepted: 14 December 2021
Published: 17 December 2021
Corrected: 3 November 2022

1. Introduction

Professional soccer clubs invest significant amounts of time and money into the development of academy prospects with the hopes of producing elite players (e.g., individuals formally registered to a professional soccer club) [1,2]. Popularity, competition, and new codes of conduct have contributed to a new scientific approach to the development of the academy soccer player, delivered by multi-disciplinary teams focussed on optimising training strategy. This approach aims to improve training methods and monitoring systems within developmental programmes to enhance long-term athletic development (LTAD), mitigate the relative risk of injury, and subsequently, improve chances of progression into the professional game to compete in their first team or at another club [3–6]. Increased investment into evidence-based talent identification and development models often apply systematic training in a single sport from a young age in an attempt to accumulate training exposure and practise [1,7–9]. Talent development models in soccer often create pathways of 'early specialisation' for young athletes, whereby early age involvement in one chosen sport (typically early- to middle-childhood), is partaken at the expense of participation in other sports or activities. In the United Kingdom (UK), the adoption of the Elite Player Performance Plan (EPPP) by elite soccer organisations is an example of such a model [10]. The EPPP was introduced by the English Premier League in 2011, aiming to provide an increased volume of on-pitch weekly training hours for talented youth soccer players [10].

The number of required contact hours for coaching has increased to 8500 h for clubs in the highest academy classification category, in contrast to the original 3760 h required in the original UK academy system set out in 1998. Furthermore, the linear approach adopted by the EPPP will mean that training exposure for individuals participating in these programmes may systematically increase by 20–50% as they progress through the age groups (from the age of 5 through to 16) [10,11].

There still remains much debate as to whether early-specialisation versus early-diversification practises are more effective training methods [12,13]. Researchers have provided evidence for the positive effects a high exposure to technical and tactical skill development from an early age has on fitness and motor control [13], which may thus provide a physical and technical foundation for the sports-specific capabilities needed in the game [14,15]. However, these types of talent development structures may expose the youth soccer player to an increased risk of injury [9,16–18]. This can not only be catastrophic in the sense of withdrawal from training and competition, but also means the end of participation in physical activity and sport for the youth athlete that may lead to long-term detrimental outcomes [19]. This increase in injury risk may be in part attributed to an increase in musculoskeletal system demands without allowing sufficient time for recovery and adaptation [16]. Chronic overuse injuries can account for up to 40% of all injuries in youth athletes and 20% of which can typically be classified as severe (i.e., an injury absence from sport for 4 weeks or more) [20]. Resultantly, a conflict may exist whereby the accumulated training hours necessary to exceed at the highest level (i.e., skill development) may come 'hand-in-hand' with an increased injury risk. Indeed, it has been shown that youth soccer players may be up to 3 times more likely to sustain an injury since the onset of the EPPP [21], which in turn will reduce a player's availability to train, improve their sports-specific skill, and potential chances of progression [22]. Opposingly, other work has found a decrease in injury incidence when comparing pre-EPPP to post-EPPP (3.0/1000 h vs. 2.1/1000 h, respectively) [23], highlighting the role different management strategies play in these outcomes. Consequently, sports science and medicine practitioners working in highly-specialised elite development programmes should strive to ensure effective management systems are in place that maximise player availability and potential, reduce the likelihood of injury, and fundamentally ensure a long sporting career and healthy life [19].

Youth athletes (categorised as children $\leq$ 13 years and adolescents between 14–18 years [24]) face unique challenges with regards to the inconsistency of the timing and tempo of their physical growth and maturity [25,26]. Pertinently, individuals of the same chronological age can vary in maturity status by as much as 5–6 years in biological age [27]. Importantly, individuals who possess advanced maturity in comparison to their age-matched peers can possess advantages in sporting performance through improvements in aerobic and anaerobic capacities, muscular strength, power, and sprint speed, as a product of pubertal development [25,28]. Growth and maturation are not synonymous, as growth rate is defined as changes in an individual's body size or parts of the body (e.g., lower limb growth) over time, whereas biological maturation refers to the status, timing, and tempo of progress towards a fully 'adult' state (Figure 1) [28,29]. Additionally, 'maturity status' can be defined as the state of maturation at the time of observation and 'maturity timing' is defined as the age at which a specific maturational event occurs [28,29]. During the pubertal growth spurt, academy soccer players can have growth rates of up to 7.5 to 9.7 cm/year between 10.7 and 15.2 years of age [30,31]. These aforementioned factors are important elements to monitor in youth athlete settings, whereby growth [32–34] and maturation [35–37] have both been identified as dynamic moderators for potential injury risk.

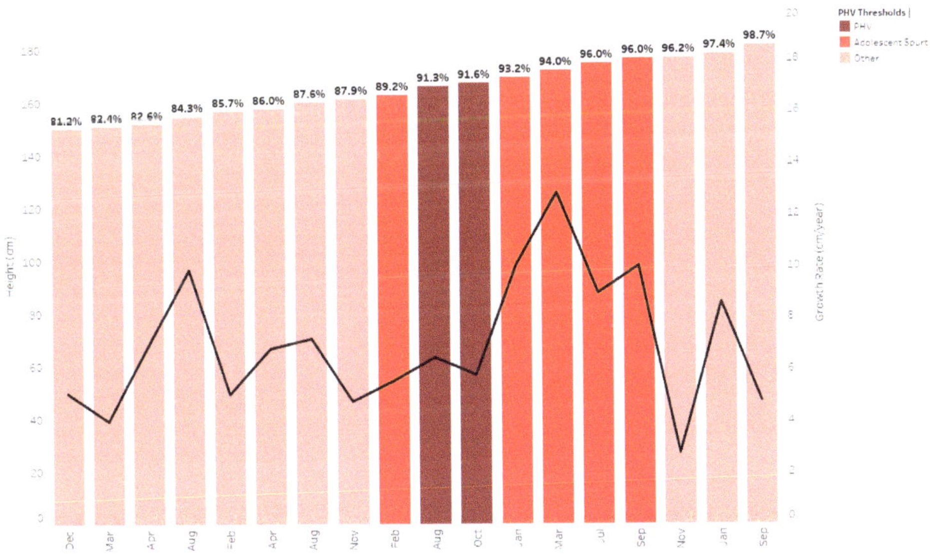

Figure 1. Longitudinal overview of an individual's growth curves throughout their academy journey. Key: PHV = 90–92% PAH; Adolescent Spurt = 88–96% PAH. PHV: Peak height velocity; PAH: Predicted adult stature. Percentage figures on bars represent % of predicted adult stature; the black line indicates growth rate. Dashboard produced in Tableau Desktop Software (version 2021.1, Salesforce Company, Seattle, WA, USA).

It is important to have an understanding of the sport's physical requirements to make informed decisions on training prescription at the individual level [8,38–40]. Undoubtedly, the most realistic approach to reveal the game's demands is via analysis of actual match play [39]. At the senior level, advanced automated analysis systems [38,41–45] allow for the tracking of work rate and activity patterns during match play, through external volume metrics—such as distance covered at various intensities [40,46–48], exercise intensity distribution [41,42], indications of fatigue [41,45], and position-specific differences in running performance [42,49]. The tracking of external measures can outline the physical output completed by a player, which can help identify an individual's physical capacities; however, the utility of internal measures as a means of identifying the relative physiological and psychological stress imposed on an individual is also vital in determining the game requirements and subsequent adaptation imposed by these demands [50]. Thus, the combination of both external and internal volume, intensity, and frequency parameters can highlight the relative physiological and biomechanical demands of soccer for an individual. This will enable the identification of strengths and deficits within a player's fitness capabilities, which can provide practitioners with a 'data-informed' framework from which to develop optimal conditioning strategies for the preparation of the team and the individual [51]. These methods have become commonplace within elite senior soccer team settings and have allowed for the monitoring and periodisation of players' fitness and fatigue throughout a competitive season to optimise performance levels and mitigate injury risk to improve player availability [52–54]. However, the financial, time, and sometimes ethical constraints (e.g., philosophical differences) faced by scientists and coaches working at the junior level can create barriers in their application [55]. Thus, those working with youth athletes require practical, yet scientifically driven, monitoring systems that can be implemented which ensures their athletes are supported during years of rapid growth, maturation, and development.

This review will therefore aim to provide a theoretical and practical reference for sports science and medicine practitioners to help them navigate through the complexities of managing the exercise demands of youth soccer players during their academy journey.

It is acknowledged that barriers may exist in the application of best practice; however, through an understanding of growth, maturation, and how these elements interact with the youth soccer player's relative injury risk, performance capabilities, and welfare, these obstacles can be overcome. Those working in academy soccer clubs should be able to implement foundational, age-appropriate, and progressive training methods to optimise athletic performance, reduce injury risk, and ensure the attainment of future potential.

2. Holistic Approach within the Multi-Disciplinary Team

A coordinated and systematic approach to training development and exercise monitoring in youth soccer as a means of guiding decisions on training and match exposure can serve to maximise positive training outcomes (e.g., athletic and technical skill development), player welfare, and minimize negative effects (e.g., illness, injury, overtraining, and drop-out) [56]. With that said, the ability to consistently capture and aggregate all aspects of a developing athlete's training demands remains extremely difficult in youth sport, which may stifle the effectiveness of the intended monitoring practises [57]. As such, at the forefront of any youth athlete management strategy should be the communication feedback loop between the athlete, coach, and members of the multi-disciplinary team (MDT) [58]. Although there is now a breadth of craft knowledge available to us within members of the MDT (i.e., technical coaching, sports science, physiotherapy, psychology, nutrition, player social care and education, etc.), it is crucial that contributions do not become siloed points of view and instead synergistically contribute to a holistic picture of the youth athlete in question. Furthermore, the MDT should avoid a reactive 'red-flag' culture with a risk-averse approach to athlete management (i.e., prevention rather than cure). Instead, it should be the goal of the MDT to strive towards a forward-planning, progression-oriented culture, where appropriate support structures are already in place for identified 'at risk' individuals. Subsequently, adjustments to the training programme for these individuals can be planned prospectively, which can reduce their relative risk of injury and subsequently maximise the attainment of athletic potential. A crucial tenet of this approach should be to maximise the exposure and subsequent development of each youth soccer player through a progressive, varied, and bespoke stage-appropriate training programme.

Adolescent soccer players have the capacity to train and improve physical performance in response to training demands imposed in the soccer training environment [59]. They may also be exposed to a heightened injury risk through greater frequencies, volumes, and intensities of training, particularly when coupled with the complexities of growth, maturation, sub-optimal physical conditioning, and reduced movement quality associated with adolescence [25,60–62]. It remains a major challenge for those working with athletes to determine the nature and magnitude of specific training stressors required to induce positive responses and balance these stressors with adequate recovery to avoid maladaptation and promote supercompensation [59]. Although increasingly complex models are attempting to expand on our understanding of these causal mechanisms underpinning performance and injury risk [63–66], the ability to predict such outcomes can still aptly be described as the "Quest for the Holy Grail" within sports science and sports medicine [67]. Thus, the use of and regular monitoring of growth, maturation, performance, and training data within a youth setting should not be used as definitive predictive tools for these aspects, but rather, guide our ongoing decision-making processes that are based on scientific rationale, fundamental training principles, and framed through a performance and training progression mindset [58]. The following sections will detail how these elements can be used to inform decisions on youth athlete training management.

3. Physiological Considerations for Injury Risk

As mentioned previously, growth [32–34] and maturation [35–37] have been highlighted as potential injury risk factors in the adolescent athlete population. Research findings have shown that episodes of rapid growth and the period around peak-height velocity

(PHV) are associated with increased injury risk in elite adolescent sports [32,37,68–70]. The rapid growth of bone length and mass contrast with the 'lag' in the adaptation of the muscles, tendons, and apophyses, which can increase the stress on muscle-tendon junctions [32,37,68–70]. This potentially explains the development of traction apophyseal growth-plate injuries, such as Osgood-Schlatter's disease, in youth athletes [71]. These types of injuries, often classified as 'overuse' injuries, are prevalent in adolescent soccer populations [72]. Overuse injuries (e.g., tendinopathies and stress fractures) are defined by the concept of an injury occurring in the absence of a singular, identifiable traumatic cause [73], and can result from the failure of the musculoskeletal system to withstand repetitive, submaximal forces over much longer time frames [74]. In theory, the structural tolerance of growth plates and developing bone may be exceeded if a rapidly growing youth athlete is exposed to excessive or repetitive stresses on their musculoskeletal structures [75]. Furthermore, it has been shown that sports specialisation is an independent risk factor for injury in youth athletic populations [16,17] and the prevalence of Osgood-Schlatter's disease is 4 times greater in highly-specialised athletes [76]. The high exposures to repetitive patterning of soccer activities (e.g., kicking) typically seen in highly specialised academy soccer players can also lead to morphological maladaptations (i.e., Cam-type deformities and femoral acetabular impingement) [75,77,78]. The sensitivity to these issues can be exacerbated during the stage of skeletal maturation (i.e., typically in boys between 12 and 14 years) when the growth plates are open [77–79], coincident with the increases in circulating growth hormone and insulin-like growth factor-1, which heightens the bone's osteogenic responsiveness to joint loading [80].

The lags in musculoskeletal growth and subsequent changes in body segment inertial parameters (i.e., mass, position of COM, moments of inertia, and radii of gyration) apparent in the growing athlete can lead to compromised neuromuscular control during dynamic activities (e.g., running, cutting, and landing) and is suggested to be a key mechanism for lower-limb ligament injuries [81,82]. A high proportion (20%) of injuries in male youth soccer players are acute traumatic ligament sprains at the ankle and knee [72,83,84]. This may be explained by deficits in active muscular protective mechanisms that are unable to adequately support joint torques during dynamic movements involving deceleration and high forces [85], although a causative link is yet to be substantiated. The compromised neuromuscular control, coupled with the vulnerability of bodily tissues—including musculotendinous junctions, ligament structures, growth cartilage, and bone mineral density—during this period may subject the highly specialised elite youth soccer player to increased injury risk; particularly if movement competency is not maintained, loading variability is limited, and tissue-specific weaknesses are not addressed.

4. Elite Training Programme

The training practises of highly-specialised elite youth soccer players are a key component in ensuring players receive the appropriate coaching and training exposure necessary to develop their skillset. Yet it should also be acknowledged that early-specialisation practises have been shown to predispose these individuals to an increased injury risk [70,72]. Therefore, a technical training programme that fails to consider the aspects of growth, maturation, and fundamental training principles may exacerbate these risks, if not addressed through appropriate compensatory programming [15,16]. A highly popular training method in soccer is the use of small-sided games (SSGs), which can be manipulated using various constraints to elicit an intensified physiological stimulus that typically serves as a form of sports-specific high-intensity interval training [86]. Further, pedagogical principles can be integrated to help the youth athlete improve specific technical skills or tactical behaviours aligned to a club's coaching philosophy [87]. As such, SSGs can provide simultaneous development of physical, technical, and tactical performance qualities [87,88]. However, academy systems may fall guilty of an over-emphasis on small-sided games in order to enhance technical skill development [89]. This may increase the risk of over-use injury through the performance of high and repeated exposure to mechanically demanding activity, repetitive movement patterns, and sports-specific skills [75], as well as under-expose players to the high-speed running demands of match play [89].

Elite academies often partake in seasonal domestic or international tournament matches, whereby condensed fixture schedules require players to perform multiple matches within a few days (typically between 2–4 days). Youth athletes may in fact be more fatigue-resistant than senior athletes [90]; however, the periods of intense competitive match play during these tournaments are reflective of the physical demands placed their elite senior counterparts, who are required to repeat performance in over-intensified competitive schedules with limited recovery opportunities [91]. These experiences certainly represent a challenge and are valuable development opportunities from technical, tactical, physical, and psychological perspectives. Practitioners should also be wary of the reduced loading capabilities of the rapidly growing athlete, which needs to be managed during dense periods of match play to allow sufficient time for rest and recovery for tissue homeostasis [16]. Moreover, disparities in growth and maturity status between individuals performing within the same chronological age groups are often apparent and have been identified as factors that may affect the injury risk [62,92]. This may be explained by the large inter-individual variation in physical and neuro-developmental pathways between players, which brings with them a contrast in morphological structures and neuromuscular control [62,92]. This relationship remains unclear; however, the same way an individual may demonstrate a physical competitive 'advantage' by being more mature, their underdeveloped counterpart may demonstrate a physical 'disadvantage' which predisposes them to increased risk of sustaining an injury [32,69].

5. Developing a Framework for Performance Monitoring and Injury Surveillance

Elite soccer clubs aim to offer an advanced soccer development and education programme, with age groups starting potentially starting from under-5 s through to under-23 s, supported by expert and dedicated staff. A soccer academy high-performance training programme should strive to facilitate an environment where success is inevitable, using every available resource and knowledge base to maximise appropriate player exposure at any age, whilst still mitigating injury risk [93] and promoting player welfare. As such, the goal of the MDT should be to push athletic boundaries and inspire athletes to unlock their true potential [93]. Furthermore, a performance-oriented approach may in fact present as a dual benefit in progressing athletic performance while simultaneously reducing injury risk [94–101]. Resultantly, practitioners should aim to evaluate the effectiveness of their programmes by establishing global and specific key performance indicators (KPIs) [93]. This approach is common in business strategy and facilitates the objective assessment and monitoring of organisational performance relative to its objectives [102], which in this instance would be to improve athletic performance, reduce injuries, and maximise player availability. The use of the following global statistics are therefore suggested for these means and can be presented at both the squad- and individual-level:

1. Match exposure: The total number of soccer match hours or minutes.
2. Training exposure: The total number of training hours or minutes.
3. Match availability: Percentage of the total available team matches against those missed due to injury or illness.
4. Training availability: Percentage of the total available training sessions against those missed due to injury or illness.
5. Injury incidence: Number of injuries per 1000 h of exposure (training, match play, or combined) [103].
6. Injury burden: The total number of days lost to injury per 1000 h of exposure (training, match play, or combined) [103].
7. Bradford factor: Time absent multiplied by the square of the number of injuries [19].

As such, benchmarks can be established for specifically identified KPIs (e.g., strength, power, speed, endurance, movement quality, growth, and maturation) which may be assessed, trained, and monitored cross-sectionally and longitudinally (Tables 1 and 2) [93].

In addition, with recognition of maturity- and population-specific injury risk patterns, and how these may interact with the elite youth soccer player's training programme, the development of 'injury-risk profiles' can be established as a key initiative for effective performance monitoring and surveillance frameworks for youth athletes [104]. This information will inform more bespoke training strategies that are relative to the individual's needs. Recent work has shown that youth soccer players may sustain specific injuries at different percentages of their final adult stature [105]. In this instance, growth injuries may appear to follow a chronological distal to proximal pattern, with muscle and joint injuries occurring more frequently in mature players [104,105]. Importantly, although the period of PHV may represent an increased incidence of growth-related injuries [32,37,68–70], the high growth rates of the lower extremities pre-PHV—and the trunk post-PHV—increase the susceptibility of these areas to injury without necessarily being in a period of rapid increases in stature [105]. Consequently, estimating and longitudinally monitoring the percentage of player's predicted adult stature (PAH) may represent a viable basis for which effective maturity-specific injury mitigation programmes can be implemented (Table 2) [105–107].

Table 1. Injury risk profile and potential mitigation strategies for the elite adolescent soccer player with reference to the technical training programme. A periodised, multi-component athletic training programme is advised at all stages of the programme in addition to these specific strategies.

Injury Risk Mechanism	Rationale	Assessment and Monitoring Considerations	Intervention Strategy and Exercise Example (s)
Highly specialised sports-specific technical practise [70,72]	Increase movement diversification and variability to reduce repetitive soccer-specific movement patterns	• Monitor soccer-specific, gym-based, multi-sports and extra-curricular training exposure [19,108].	• Supplement or replace soccer drill exposure with multi-sports activity (e.g., rugby, basketball, hockey, American football, gymnastics, free running, swimming, tennis, combat sports) [15].
Over-exposure to SSG format technical practise with reduced pitch dimensions [75,89]	Reduce over-exposure to mechanically demanding activity and ensure players are prepared for HSR demands of match play	• Monitoring acute and chronic acceleration, deceleration, COD *, and HSR volumes [109]. • Quantitative and qualitative assessment of sprinting intensity [110] and technique [96].	• Supplement technical training with HSR exposure in athletic development training [89,111]. • Ensure weekly attainment of >95% MSS intensities [110].
Large disparities in maturation within chronological age groups [32,62,92]	Reduce variation in maturity status between individuals to balance physicality	• Regular growth and maturation assessment (e.g., 2–3 months) [29].	• Bio-banded training and match play [112–114]. • Individualised and stage-appropriate training targets and standards [14,15,115].

Key: SSG = small-sided games; LSG = large-sided games; HSR = high-speed running; COD = change of direction; MSS = maximum sprinting speed. * COD count and distance should be monitored during athletic development training due to current limitations of tracking technologies.

Table 2. Injury risk profile and potential mitigation strategies for the elite adolescent soccer player with reference to maturity-based injury risk factors. A periodised, multi-component athletic training programme is advised at all stages of maturation in addition to these specific strategies.

Injury Risk Mechanism	Rationale	Assessment and Monitoring Considerations	Intervention Strategy and Exercise Example (s)
	Pre-PHV (<88% PAH)		
Higher incidence of growth-related injuries in extremities (e.g., Sever's) and ankle joint/ligament injuries [104,105]	Develop foot/ankle strength, neuromuscular control, and localised tissue robustness and mobility.	• Movement quality, neuromuscular control (e.g., QASLS [116], LESS [117], TJA [118]). • Symptomology [119].	• Linear and curvilinear sprint exercises (e.g., sprint races and sprint drills) [61,98]. • Jumping/landing competency and impact attenuation training (e.g., plyometrics in various directions, intensities, and surfaces) [120,121]. • GSA complex training (e.g., pogo hop variations, end-range isometrics) [121].
	Circa-PHV (88–96% PAH)		
Higher proportion of knee joint/ligament injuries [81,105]	Improve movement competency, frontal plane neuromuscular control, and localised tissue robustness.	• Movement quality during COD, jump-landing and squatting patterns (e.g., CMAS [122], QASLS [116], LESS [117], TJA [118]). • Multi-joint lower-limb maximal and rapid force production (e.g., IMTP, CMJ, RSI, horizontal jumps/hops) [123–127]. • Symptomology [119].	• MDS competency training (e.g., deceleration and cutting technique) [61,100,101]. • Jumping/landing competency and capacity training (e.g., plyometrics in various directions, intensities and surfaces) [120,121].
Higher incidence of growth-related injuries more proximally (e.g., Osgood's and pelvic avulsions) [104,105]	Improve knee flexor and extensor strength.	• Isolated strength qualities during single- and multi-joint actions (e.g., isokinetics, Nordics) [127–129]. • Symptomology [119].	• Compound and isolated RT (e.g., multi-planar SL compound lifts, Nordics) [130,131].
	Post-PHV (>96% PAH)		
Higher proportion of groin/spine injuries [104,105]	Develop glute, groin, hip and core strength, and stability.	• Multi-joint lower-limb maximal and rapid force production (e.g., IMTP, CMJ, RSI, horizontal jumps/hops) [123–127]. • Isolated strength qualities during single- and multi-joint actions (e.g., isokinetics, adductor squeeze, abductor pull) [127–129]. • Symptomology [119].	• Multi-component RT (e.g., heavy DL compound lifts *, glute bridge progressions, Copenhagen exercises). • MDS capacity training (e.g., high-intensity deceleration, pivoting and maximal velocity sprinting) [61,98,100,101].
Higher prevalence of muscular injuries [105]	Increase physical capabilities to tolerate the systematic increase in training exposure and physical demands	• Quantification of training volume, intensity, frequency, type and response (e.g., GPS, RPE, wellness) [132–134].	• Establish age, maturity and positional benchmarks for match and training demands [11,108,135–137].
Soccer-specific overuse injuries	Avoid repetitive actions and provide variation in soccer-specific activity	• Quantification of kicking volume, intensity, frequency and type (e.g., wearable monitoring devices).	• Gradual and progressive overload of kicking volumes.

Key: PAH = predicted adult height; LESS = landing error score system; TJA = tuck jump assessment; QASLS = qualitative analysis of single leg squat; IMTP = isometric mid-thigh pull; CMJ = countermovement jump; RSI = reactive strength index; GPS = global positioning system; RPE = rating of perceived exertion; GSA = gastrocnemius-soleus-Achilles; MDS = multi-directional speed; RT = resistance training; SL = single leg; DL = double leg. * Strength training intensities at >85% 1 RM.

6. Longitudinal Training Monitoring

Youth soccer match play is an essential part of a player's physical development across all stages of development. It is frequently underlined that young athletes should not be treated as 'miniature adults' [38,138,139] as they possess markedly different physical and physiological processes that attribute to their soccer performance [139]. These differences are observed in match play, where recent studies have suggested that age and maturation impact match running performance in soccer [137,140–145]. As such, the relative physiological demands for young soccer players playing the game within the same age category may be hugely variable within different contexts [8]. These more nuanced elements should be considered when evaluating match performance capabilities at the individual level and efforts should be made to ensure comparisons within chronological age bands are coupled with standardised scores relative to their maturity [115,145]. Importantly, such data can provide age-appropriate insights into the athletic and technical requirements of the game, allowing for bespoke training methods [146] and improvement of long-term training management of an academy player [140], thus avoiding the replication of methods utilised in senior players [8]. Moreover, when considered in absolute terms, this data can also be used to inform when talented young players are physically capable of demonstrating performance outputs that are sufficient to compete when moving up to play in older age brackets [8]. Such information facilitates the effective 'scaffolding' of age group requirements around a LTAD framework, allowing practitioners to reverse-engineer their weekly training programmes at the micro- and meso-level to guide developmentally appropriate training prescription.

In contrast to the senior game, those working in academy settings are afforded the opportunity to evaluate a training programme with a long-term vision in mind. Therefore, the use of periodisation strategies should permit the structure of a training programme in logical and appropriately sequenced phases and cycles, following specificity and progressive overload principles. With the correct balance of training (i.e., stress) and recovery (i.e., adaptation), training strategies can be used to physically prepare the youth soccer player for the increasing physical demands as they progress towards the senior game [147]. For this management to be effective, an understanding of the longitudinal structure of the adolescent player's programme is important to define suitable training doses and risk thresholds which can optimise performance and recovery (Figure 2). Limited published information is available with regards to the longitudinal training demands in youth soccer academies (i.e., $\leq$16 years). To date, studies have been reduced to quantifying youth player match and training demands over one-to-two weeks [11,148], or have solely reported the training durations [108], internal [11], or external demands of training [135]. Collectively, a general trend is reported towards a progressive increase in overall physical loading (i.e., exercise volume) as chronological age increases. More detailed investigations are certainly warranted that examine the chronic physical demands and exercise volumes of academy soccer players from different academy environments with different training philosophies. This detailed insight will enable practitioners to suitably design contextualised training methods which considers the interaction between training, growth, and maturation over longer time frames. It is, therefore, advised that data specific to the individuals in question be collected to inform practitioners in this regard, where possible. This will allow more appropriate physical trajectories to be established, risk thresholds to be identified, and enable long-term training decisions to be informed through a progressive standpoint.

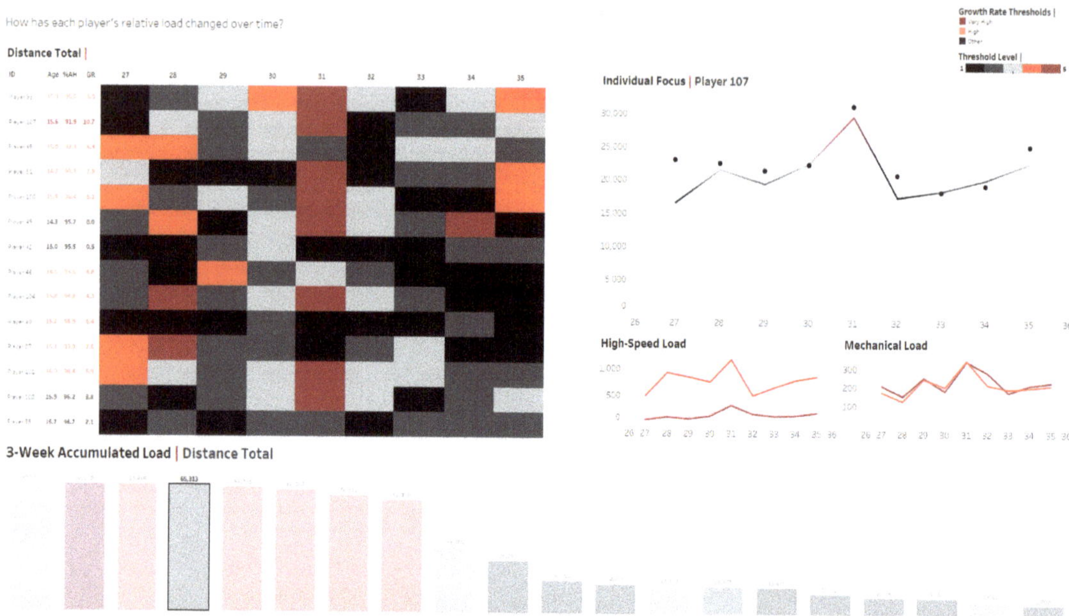

Figure 2. An interactive dashboard which provides a macroscopic overview of external training volume over time—alongside age, maturity status, and growth rate—in a group of academy players. Key: %AH = Percentage of predicted adult stature; GR = Growth rate; 1 to 5 colour thresholds are based on z-scores that represent a players' normal distribution curve for each external metric (i.e., 15/20/30/20/15th percentiles). Dashboard produced in Tableau Desktop Software (version 2021.1, Salesforce Company, Seattle, WA, USA).

The need to monitor the short- and long-term implications of the internal and external training stressors remain key in longitudinally managing young soccer players. Injury patterns in youth athletes follow a specific aetiology according to their stage of maturation [105], which will have implications for how these stressors interact with the youth athlete at both the acute and chronic level [149]. In addition, the intermittent and multi-directional nature of soccer means the variation in stimuli and response will impart varying degrees of physiological and musculoskeletal demands [61,98,99,150]. For example, neuromuscular fatigue driven by high-speed activity may cycle more transiently within the physiological system [151] and numerous investigations have found associations between acute weekly rapid increases in high-speed running volumes and increased risk of soft tissue injury [109,110,152–154]. Conversely, high cumulative totals (i.e., 2- to 4-week accumulative totals) of distance covered [152,155], number of decelerations [152], as well as 'relative velocity change' (e.g., algorithmically derived total accelerations, decelerations and change of directions) [155] have been linked to overuse injury. This may be explained by a 'mechanical fatigue failure' phenomenon, in which the mechanical fatigue of tissue is perpetuated by accumulated damage as a result of summative and repetitive loading events, subsequently surpassing the remodelling rate of the biological tissue [156]. The heightened responsiveness of the musculoskeletal structures to joint loading in the rapidly growing athlete [80] points towards the importance of monitoring chronic external training volume (Figure 2), alongside individual growth and maturation data (Figure 1), to support informed decisions on youth soccer players during their intensive training programme.

Equally, the evaluation of internal stress should remain a central component within monitoring practises, and the utility of practical, subjective measures allow this data to be readily collected on the youth soccer player [52,157–160]. Indeed, the psychophysiological response experienced from a specific exercise bout may vary depending on a host of modifiable and non-modifiable contextual factors [161]. The complexities of rapid and varied growth, maturation, and development within a group of adolescent athletes may exacerbate these differences in the response to the prescribed the same training [28]. Furthermore, it should be acknowledged that common indices of training intensity (e.g., session-RPE) may still only provide a 'global' evaluation of the internal response and will not detect the mechanical stresses and strains experienced within the muscles, tendons, ligaments, bones, and cartilage (i.e., musculoskeletal tissue). As such, being able to differentiate between the physiological and biomechanical internal response to various external stressors may assist in developing more appropriate and specific loading paradigms in youth soccer [150]. For example, using differential-RPE, a player may be specifically asked to rate their degree of 'breathlessness' and 'leg muscle exertion' during a session or activity [157,162–164], or—through a wellness questionnaire—gage their degree of 'muscle or joint soreness' and 'fatigue' in the days following a session, and also sleep quality [52,158]. In addition, monitoring symptomology and encouraging the youth athlete to delineate specifically the location of their pain or soreness may deliver further insights [119,165]. This data can be used on the daily level to adjust training or analysed longitudinally to evaluate trends with respect to the external versus internal demands of training, and subsequently, inform periodization strategies. More research is certainly warranted to validate these novel methods, particularly in their applicability for use in youth athletes [163,166]. The harmony of the MDT is pivotal in situations such as these, where roles and responsibilities are clear, expertise is channelled suitably, and athletes are continually educated to maximise the outcome of these measures [93].

7. Applying Theory to Practise

Practitioners working in soccer academies should be able to readily apply the tools and theoretical considerations discussed in this article to support their ongoing management processes of the elite soccer player. Regular assessment (e.g., 2–3 months) of anthropometrics (e.g., standing stature, sitting stature, body mass), in addition to parental stature, can support the development of growth and maturity calculations which can be tracked longitudinally [29,149]. Although a detailed discussion on maturity assessment methods is beyond the scope of this current discussion, the authors advocate the use of the Khamis-Roche predicted adult stature method [106,107], and readers are referred to the following texts for more detailed discussions on this topic [28,29,149]. As illustrated in Figure 1, some of an individual's highest growth rates can occur within the generic PHV thresholds (e.g., 88% to 95% of predicted adult stature) that have been substantiated by previous findings [107] and utilising these growth and maturation indicators can provide valuable insights to inform training. Indeed, recent work [167] has demonstrated a positive linear relationship between smoothed week-to-week changes in total exposure and injury incidence ($p = 0.001$), resulting in a 168% increase in injury likelihood with a 2 SD change in training duration. Furthermore, a positive linear relationship between growth rate and injury incidence ($p = 0.031$), as well as a non-linear relationship between the percentage of predicted adult stature (peak risk occurring at 92%) with injury incidence. In follow-up work, Johnson [168] found that identifying these aforementioned risk factors, and subsequently providing a modified programme, could significantly reduce injury incidence and burden in adolescent soccer players. This adapted training programme involved small adjustments of team-based soccer training volume, which was replaced with low intensity individual sport-specific skills, balance, coordination, impact attenuation, and individualised strength sessions [15,169,170].

As such, by no means do these situations have to become reactive 'red flag' scenarios. Instead, when planned for proactively and with appropriate rationale, instances such as these should be seen as an opportunity to provide an individualised stage-specific training programme and progress their athleticism, while reducing the likelihood of injury. Regular assessment of an individual's athletic profile (e.g., strength, power, speed, flexibility, and movement quality) can determine an individual's strengths and weaknesses which can inform training prescription. This can be further supported by the numerous performance models [14,15,61,171] and training guidelines that exist within the LTAD area, which can be used to guide athletic development training (e.g., resistance training [24,172–174] and multi-directional speed training [61,175–177]).

Using evidence-based training programmes which conform to the scientific principles of training, not only can these individuals' relative injury risk be reduced, but they may also concurrently enhance their biomotor and physical performance capabilities which are fundamental for a successful transition into the senior game [178].

A Case Study Example

Using the working example presented in Figure 2, a deeper look into 'Player 107' was warranted due to their high growth rate (i.e., 7.2 cm/year; "very high") combined with their current maturity status (i.e., 91.6% AH; "PHV territory"). Notably, the heightened training volume experienced during 'Week 31' due to training with an older age group caused the player to reach a high relative loading threshold for several external training volume measures. With acknowledgement of this, alongside the utility of the global (e.g., training availability, Bradford factor, etc.) and specific (e.g., strength, power, and speed profile) KPIs discussed in earlier sections, objective information around the youth athlete can be aggregated, enabling a 'decision tree' process to formulate and context to build around an individual's training status and risk profile. Consequently, conversations between members of the MDT can be implemented, centring around both preparation for the approaching weeks, as well as a potential reduction in soccer-specific training volume during the following weeks, as demonstrated by the individual's relative '3-Week Accumulated Load' for total distance at 'Week 33' [109,167,179]. As a result, a balance may be reached between mitigating the risk of overuse injury while appropriately exposing the player to 'challenging' situations, which is key for their continued development.

The resource hierarchy often observed in academy settings means that more advanced monitoring systems (e.g., player tracking technology) tend to be incorporated with older age brackets (e.g., >U15), or an academy with limited resources may not have this facility at all. However, getting the 'basics' right, and obtaining simple, accurate, and standardised information on a few research-validated measurement tools can still readily be scaffolded over an entire academy as an effective management system. As mentioned previously, the use of such data within an academy environment should not be solely relied upon, but rather, used to build context around the player and allow subsequent conversations between the MDT to be informed through an objective perspective.

8. Conclusions

The information we now have on players is becoming increasingly more applicable to the end-user with the development of new technologies and data visualisation software. For example, there is now the availability for athletes to remotely self-report psychophysiological measures on their devices away from the training facility increasing digital accessibility [159,160]. However, the art of face-to-face communication becomes important as ever in distilling the swathes of information into meaningful messages, in particular to the 'non-experts' who are typically at the centre of which all decisions are impacting (i.e., the athlete and technical coach) [58]. Furthermore, the role of the coach and sports scientist in educating their youth athletes on understanding the "why" of these abovementioned strategies is important in achieving their long-term goals [93]. The young athlete needs to build trust in the process, and eventually possess the tools and understanding to become a self-sufficient, resilient individual who thrives in the senior game on their own.

The large disparities in maturation status within academy settings and the growth-related issues, particularly for individuals going through accelerated periods of growth, necessitates an individualised approach to the management of the adolescent soccer player. Furthermore, elite youth soccer players are inherently predisposed to greater injury risk due to their large involvement in highly-specialised sports-specific practice, and so practitioners have a duty of care to provide safe and effective sports science practise for their athletes. Developing a performance monitoring and injury surveillance framework within an academy system can be supported by an understanding of growth, maturation, youth soccer performance, and epidemiology. Subsequently, bespoke programming can be provided which may offset these inherent injury risks elite soccer players are predisposed to, while developing athletic performance. Practitioners should possess a sound theoretical understanding of the training process, and utilise data, where available, to guide ongoing decision-making. Those with limited resources, however, can still use practical methods discussed in this review as evidence-based alternatives when managing their youth athletes. To conclude, a summary list of practical strategies, ordered from simple to more advanced, is provided for consideration:

1. Communication between the MDT and the player—Utilising the extensive resource of the MDT to establish context around an individual while continuing to have ongoing discussions with players to understand their needs.
2. Monitoring exercise activity—Using exercise duration to track exposure and further categorising into specific activity types (e.g., match play, sports-specific, athletic development and extra-curricular activity). Establishing normative values and thresholds for an individual's activity to determine relative increases or decreases in exposure (e.g., 'time on feet').
3. Collection of growth and maturity data—Regular anthropometric assessment (e.g., 2–3 months) to allow for cross-sectional (e.g., maturity status) and longitudinal (e.g., growth curves) evaluation of individuals.
4. Establish athlete profiles, plan and progress athlete training—Evaluate key athletic qualities and establish benchmarks specific to age, maturity, and position. Tracking of specific athletic development exercises (e.g., gym-based and field-based activity) and using published literature as well as previous experience as a guide for age-appropriate training prescription. Using fundamental scientific training principles while using the above points to inform progression or regression of training.
5. Monitoring subjective indices of load—Daily collection of RPE and wellness (e.g., sleep, fatigue, soreness, and stress) to establish values and thresholds for individuals to detect changes. Further classifying scores relative to the physiological (e.g., "breathlessness") and biomechanical (e.g., "leg exertion") characteristics to establish more specific load-response profiles.
6. Monitoring of training activity using tracking technology—Establish age-, position-, and maturity-specific match activity profiles and extending this longitudinally to map out a progressive training framework over a LTAD pathway. Monitor chronic exposure to training stressors, understanding the unique physiological and biomechanical load-adaptation pathways underpinning KPIs, and how these may interact with growth and maturation.

Author Contributions: Conceptualization, A.J.M.; writing—original draft preparation, A.J.M.; writing—review and editing, A.J.M., T.D., D.J., E.L.; visualization, A.J.M. All authors have read and agreed to the published version of the manuscript.

Funding: This research received no external funding.

Institutional Review Board Statement: Not applicable.

Informed Consent Statement: Not applicable.

Conflicts of Interest: The authors declare no conflict of interest.

References

1. Unnithan, V.; White, J.; Georgiou, A.; Iga, J.; Drust, B. Talent Identification in Youth Soccer. *J. Sports Sci.* **2012**, *30*, 1719–1726. [CrossRef] [PubMed]
2. Williams, A.M.; Reilly, T. Talent Identification and Development in Soccer. *J. Sports Sci.* **2000**, *18*, 657–667. [CrossRef] [PubMed]
3. Goto, H.; Morris, J.G.; Nevill, M.E. Match Analysis of U9 and U10 English Premier League Academy Soccer Players Ising a Global Positioning System. *J. Strength Cond. Res.* **2015**, *29*, 954–963. [CrossRef]
4. Reilly, T.; Bangsbo, J.; Franks, A. Anthropometric and Physiological Predispositions for Elite Soccer. *J. Sports Sci.* **2000**, *18*, 669–683. [CrossRef]
5. Carling, C.; Le Gall, F.; Reilly, T.; Williams, A.M. Do Anthropometric and Fitness Characteristics Vary According to Birth Date Distribution in Elite Youth Academy Soccer Players? *Scand. J. Med. Sci. Sports* **2008**, *19*, 3–9. [CrossRef]
6. Lloyd, R.S.; Cronin, J.B.; Faigenbaum, A.D.; Haff, G.G.; Howard, R.; Kraemer, W.J.; Micheli, L.J.; Myer, G.D.; Oliver, J.L. National Strength and Conditioning Association Position Statement on Long-Term Athletic Development. *J. Strength Cond. Res.* **2016**, *30*, 1491–1509. [CrossRef] [PubMed]
7. Burgess, D.; Geraldine, A.N. Talent Development in Adolescent Team Sports: A Review. *Int. J. Sports Physiol. Perform.* **2010**, *5*, 103–116. [CrossRef] [PubMed]
8. Palucci Vieira, L.H.; Carling, C.; Barbieri, F.A.; Aquino, R.; Santiago, P.R.P. Match Running Performance in Young Soccer Players: A Systematic Review. *Sports Med.* **2019**, *49*, 289–318. [CrossRef] [PubMed]
9. Malina, R.M. Early Sport Specialization. *Curr. Sports Med. Rep.* **2010**, *9*, 364–371. [CrossRef] [PubMed]
10. Elite Player Performance Plan. Document Prepared by the English Premier League. 2011. Available online: https://www.premierleague.com/youth/EPPP (accessed on 13 December 2021).
11. Wrigley, R.; Drust, B.; Stratton, G.; Scott, M.; Gregson, W. Quantification of the Typical Weekly In-Season Training Load in Elite Junior Soccer Players. *J. Sports Sci.* **2012**, *30*, 1573–1580. [CrossRef] [PubMed]
12. Bridge, M.W.; Toms, M.R. The Specialising or Sampling Debate: A Retrospective Analysis of Adolescent Sports Participation in the UK. *J. Sports Sci.* **2013**, *31*, 87–96. [CrossRef] [PubMed]
13. Fransen, J.; Pion, J.; Vandendriessche, J.; Vandorpe, B.; Vaeyens, R.; Lenoir, M.; Philippaerts, R.M. Differences in Physical Fitness and Gross Motor Coordination in Boys Aged 6–12 Years Specializing in One versus Sampling More than One Sport. *J. Sports Sci.* **2012**, *30*, 379–386. [CrossRef] [PubMed]
14. Lloyd, R.S.; Oliver, J.L. The Youth Physical Development Model: A New Approach to Long-Term Athletic Development. *Strength Cond. J.* **2012**, *34*, 61–72. [CrossRef]
15. Wormhoudt, R.; Savelsbergh, G.J.P.; Teunissen, A.J.W.; Davids, K. *The Athletic Skills Model: Optimizing Talent Development through Movement Education*; Routledge: Oxfordshire, UK, 2017. [CrossRef]
16. DiFiori, J.P.; Benjamin, H.J.; Brenner, J.S.; Gregory, A.; Jayanthi, N.; Landry, G.L.; Luke, A. Overuse Injuries and Burnout in Youth Sports: A Position Statement from the American Medical Society for Sports Medicine. *Br. J. Sports Med.* **2014**, *48*, 287–288. [CrossRef]
17. Jayanthi, N.A.; LaBella, C.R.; Fischer, D.; Pasulka, J.; Dugas, L.R. Sports-Specialized Intensive Training and the Risk of Injury in Young Athletes. *Am. J. Sports Med.* **2015**, *43*, 794–801. [CrossRef]
18. Lloyd, R.S.; Oliver, J.L.; Faigenbaum, A.D.; Myer, G.D.; De Ste Croix, M.B.A. Chronological Age vs. Biological Maturation. *J. Strength Cond. Res.* **2014**, *28*, 1454–1464. [CrossRef] [PubMed]
19. Murray, A. Managing the Training Load in Adolescent Athletes. *Int. J. Sports Physiol. Perform.* **2017**, *12*, S2-42–S2-49. [CrossRef] [PubMed]
20. Theisen, D.; Malisoux, L.; Seil, R.; Urhausen, A. Injuries in Youth Sports: Epidemiology, Risk Factors and Prevention. *Dtsch. Z. Sportmed.* **2014**, *2014*, 248. [CrossRef]
21. Read, P.J.; Oliver, J.L.; De Ste Croix, M.B.A.; Myer, G.D.; Lloyd, R.S. An Audit of Injuries in Six English Professional Soccer Academies. *J. Sports Sci.* **2018**, *36*, 1542–1548. [CrossRef]
22. Larruskain, J.; Lekue, J.A.; Martin-Garetxana, I.; Barrio, I.; McCall, A.; Gil, S.M. Injuries Are Negatively Associated with Player Progression in an Elite Football Academy. *Sci. Med. Footb.* **2021**, 1–10. [CrossRef]
23. Tears, C.; Chesterton, P.; Wijnbergen, M. The Elite Player Performance Plan: The Impact of a New National Youth Development Strategy on Injury Characteristics in a Premier League Football Academy. *J. Sports Sci.* **2018**, *36*, 2181–2188. [CrossRef]
24. Lloyd, R.S.; Faigenbaum, A.D.; Stone, M.H.; Oliver, J.L.; Jeffreys, I.; Moody, J.A.; Brewer, C.; Pierce, K.C.; McCambridge, T.M.; Howard, R.; et al. Position Statement on Youth Resistance Training: The 2014 International Consensus. *Br. J. Sports Med.* **2014**, *48*, 498–505. [CrossRef]
25. Naughton, G.; Farpour-Lambert, N.J.; Carlson, J.; Bradney, M.; Van Praagh, E. Physiological Issues Surrounding the Performance of Adolescent Athletes. *Sports Med.* **2000**, *30*, 309–325. [CrossRef] [PubMed]
26. Malina, R.M.; Reyes, M.E.P.; Eisenmann, J.C.; Horta, L.; Rodrigues, J.; Miller, R. Height, Mass and Skeletal Maturity of Elite Portuguese Soccer Players Aged 11–16 Years. *J. Sports Sci.* **2000**, *18*, 685–693. [CrossRef] [PubMed]
27. Johnson, A. Monitoring the Immature Athlete. *Aspetar Sports Med. J.* **2015**, *1*, 114–118.

28. Malina, R.M.; Bouchard, C.; Bar-Or, O. *Growth, Maturation, and Physical Activity*, 2nd ed.; Human Kinetics: Champaign, IL, USA, 2004.
29. Malina, R.M.; Rogol, A.D.; Cumming, S.P.; Coelho, E.; Silva, M.J.; Figueiredo, A.J. Biological Maturation of Youth Athletes: Assessment and Implications. *Br. J. Sports Med.* **2015**, *49*, 852–859. [CrossRef]
30. Towlson, C.; Cobley, S.; Parkin, G.; Lovell, R. When Does the Influence of Maturation on Anthropometric and Physical Fitness Characteristics Increase and Subside? *Scand. J. Med. Sci. Sports* **2018**, *28*, 1946–1955. [CrossRef]
31. Philippaerts, R.M.; Vaeyens, R.; Janssens, M.; Van Renterghem, B.; Matthys, D.; Craen, R.; Bourgois, J.; Vrijens, J.; Beunen, G.; Malina, R.M. The Relationship between Peak Height Velocity and Physical Performance in Youth Soccer Players. *J. Sports Sci.* **2006**, *24*, 221–230. [CrossRef]
32. Kemper, G.L.J.; Van Der Sluis, A.; Brink, M.S.; Visscher, C.; Frencken, W.G.P.; Elferink-Gemser, M.T. Anthropometric Injury Risk Factors in Elite-Standard Youth Soccer. *Int. J. Sports Med.* **2015**, *36*, 1112–1117. [CrossRef] [PubMed]
33. Rommers, N.; Rössler, R.; Goossens, L.; Vaeyens, R.; Lenoir, M.; Witvrouw, E.; D'Hondt, E. Risk of Acute and Overuse Injuries in Youth Elite Soccer Players: Body Size and Growth Matter. *J. Sci. Med. Sport* **2020**, *23*, 246–251. [CrossRef]
34. Wik, E.H.; Martínez-Silván, D.; Farooq, A.; Cardinale, M.; Johnson, A.; Bahr, R. Skeletal Maturation and Growth Rates Are Related to Bone and Growth Plate Injuries in Adolescent Athletics. *Scand. J. Med. Sci. Sports* **2020**, *30*, 894–903. [CrossRef] [PubMed]
35. Rejeb, A.; Johnson, A.; Farooq, A.; Verrelst, R.; Pullinger, S.; Vaeyens, R.; Witvrouw, E. Sports Injuries Aligned to Predicted Mature Height in Highly Trained Middle-Eastern Youth Athletes: A Cohort Study. *BMJ Open* **2019**, *9*, e023284. [CrossRef] [PubMed]
36. Bult, H.J.; Barendrecht, M.; Tak, I.J.R. Injury Risk and Injury Burden Are Related to Age Group and Peak Height Velocity among Talented Male Youth Soccer Players. *Orthop. J. Sports Med.* **2018**, *6*, 2325967118811042. [CrossRef]
37. Van-der-Sluis, A.; Elferink-Gemser, M.; Coelho-e-Silva, M.; Nijboer, J.; Brink, M.; Visscher, C. Sport Injuries Aligned to Peak Height Velocity in Talented Pubertal Soccer Players. *Int. J. Sports Med.* **2013**, *35*, 351–355. [CrossRef] [PubMed]
38. Castagna, C.; D'Ottavio, S.; Abt, G. Activity Profile of Young Soccer Players during Actual Match Play. *J. strength Cond. Res.* **2003**, *17*, 775–780. [CrossRef] [PubMed]
39. Drust, B.; Atkinson, G.; Reilly, T. Future Perspectives in the Evaluation of the Physiological Demands of Soccer. *Sports Med.* **2007**, *37*, 783–805. [CrossRef]
40. Carling, C.; Bloomfield, J.; Nelsen, L.; Reilly, T. The Role of Motion Analysis in Elite Soccer. *Sports Med.* **2008**, *38*, 839–862. [CrossRef] [PubMed]
41. Mohr, M.; Krustrup, P.; Bangsbo, J. Match Performance of High-Standard Soccer Players with Special Reference to Development of Fatigue. *J. Sports Sci.* **2003**, *21*, 519–528. [CrossRef]
42. Di-Salvo, V.; Baron, R.; Tschan, H.; Calderon Montero, F.; Bachl, N.; Pigozzi, F. Performance Characteristics According to Playing Position in Elite Soccer. *Int. J. Sports Med.* **2007**, *28*, 222–227. [CrossRef]
43. Dellal, A.; Wong, D.P.; Moalla, W.; Chamari, K. Physical and Technical Activity of Soccer Players in the French First League-with Special Reference to Their Playing Position. *Int. Sports Med J.* **2010**, *11*, 278–290.
44. Barros, R.M.L.; Misuta, M.S.; Menezes, R.P.; Figueroa, P.J.; Moura, F.A.; Cunha, S.A.; Anido, R.; Leite, N.J. Analysis of the Distances Covered by First Division Brazilian Soccer Players Obtained with an Automatic Tracking Method. *J. Sports Sci. Med.* **2007**, *6*, 233–243.
45. Mohr, M.; Krustrup, P.; Bangsbo, J. Fatigue in Soccer: A Brief Review. *J. Sports Sci.* **2005**, *23*, 593–599. [CrossRef] [PubMed]
46. Bradley, P.S.; Sheldon, W.; Wooster, B.; Olsen, P.; Boanas, P.; Krustrup, P. High-Intensity Running in English FA Premier League Soccer Matches. *J. Sports Sci.* **2009**, *27*, 159–168. [CrossRef]
47. Di-Salvo, V.; Gregson, W.; Atkinson, G.; Tordoff, P.; Drust, B. Analysis of High Intensity Activity in Premier League Soccer. *Int. J. Sports Med.* **2009**, *30*, 205–212. [CrossRef]
48. Reilly, T.; Thomas, V. A Motion Analysis of Work-Rate in Different Positional Roles in Professional Football Match-Play. *J. Hum. Mov. Stud.* **1976**, *2*, 87–97.
49. Di-Salvo, V.; Baron, R.; González-Haro, C.; Gormasz, C.; Pigozzi, F.; Bachl, N. Sprinting Analysis of Elite Soccer Players during European Champions League and UEFA Cup Matches. *J. Sports Sci.* **2010**, *28*, 1489–1494. [CrossRef] [PubMed]
50. Halson, S.L. Monitoring Training Load to Understand Fatigue in Athletes. *Sports Med.* **2014**, *44*, 139–147. [CrossRef]
51. Atan, S.A.; Foskett, A.; Ali, A. Special Populations: Issues and Considerations in Youth Soccer Match Analysis. *Int. J. Sports Sci.* **2014**, *4*, 103–114. [CrossRef]
52. Thorpe, R.T.; Atkinson, G.; Drust, B.; Gregson, W. Monitoring Fatigue Status in Elite Team-Sport Athletes: Implications for Practice. *Int. J. Sports Physiol. Perform.* **2017**, *12*, S2–S27. [CrossRef]
53. Hägglund, M.; Waldén, M.; Magnusson, H.; Kristenson, K.; Bengtsson, H.; Ekstrand, J. Injuries Affect Team Performance Negatively in Professional Football: An 11-Year Follow-up of the UEFA Champions League Injury Study. *Br. J. Sports Med.* **2013**, *47*, 738–742. [CrossRef]
54. Nimmo, M.A.; Ekblom, B. Fatigue and Illness in Athletes. *J. Sports Sci.* **2007**, *25*, S93–S102. [CrossRef]
55. Salter, J.; Croix, M.B.A.D.S.; Hughes, J.D.; Weston, M.; Towlson, C. Monitoring Practices of Training Load and Biological Maturity in UK Soccer Academies. *Int. J. Sports Physiol. Perform.* **2021**, *16*, 395–406. [CrossRef] [PubMed]
56. Booth, M.; Orr, R.; Cobley, S. Call for Coordinated and Systematic Training Load Measurement (and Progression) in Athlete Development: A Conceptual Model with Practical Steps. *Br. J. Sports Med.* **2017**, *51*, 559–560. [CrossRef] [PubMed]

57. Buchheit, M. Applying the Acute:Chronic Workload Ratio in Elite Football: Worth the Effort? *Br. J. Sports Med.* **2017**, *51*, 1325–1327. [CrossRef]
58. West, S.W.; Clubb, J.; Torres-Ronda, L.; Howells, D.; Leng, E.; Vescovi, J.D.; Carmody, S.; Posthumus, M.; Dalen-Lorentsen, T.; Windt, J. More than a Metric: How Training Load Is Used in Elite Sport for Athlete Management. *Int. J. Sports Med.* **2021**, *42*, 300–306. [CrossRef]
59. Gabbett, T.J.; Whyte, D.G.; Hartwig, T.B.; Wescombe, H.; Naughton, G.A. The Relationship between Workloads, Physical Performance, Injury and Illness in Adolescent Male Football Players. *Sports Med.* **2014**, *44*, 989–1003. [CrossRef]
60. Brenner, J.S.; Small, E.W.; Bernhardt, D.T.; Congeni, J.A.; Gomez, J.E.; Gregory, A.J.M.; Gregory, D.B.; McCambridge, T.M.; Reed, F.E.; Rice, S.G.; et al. Overuse Injuries, Overtraining, and Burnout in Child and Adolescent Athletes. *Pediatrics* **2007**, *119*, 1242–1245. [CrossRef] [PubMed]
61. McBurnie, A.J.; Parr, J.; Kelly, D.M.; Dos'Santos, T. Multi-Directional Speed in Youth Soccer Players: Practical Applications and Programming Considerations. *Strength Cond. J.* **2021**, publish ahead of print.
62. Quatman-Yates, C.C.; Quatman, C.E.; Meszaros, A.J.; Paterno, M.V.; Hewett, T.E. A Systematic Review of Sensorimotor Function during Adolescence: A Developmental Stage of Increased Motor Awkwardness? *Br. J. Sports Med.* **2012**, *46*, 649–655. [CrossRef]
63. Ruddy, J.D.; Cormack, S.J.; Whiteley, R.; Williams, M.D.; Timmins, R.G.; Opar, D.A. Modeling the Risk of Team Sport Injuries: A Narrative Review of Different Statistical Approaches. *Front. Physiol.* **2019**, *10*, 829. [CrossRef]
64. Bahr, R.; Krosshaug, T. Understanding Injury Mechanisms: A Key Component of Preventing Injuries in Sport. *Br. J. Sports Med.* **2005**, *39*, 324–329. [CrossRef]
65. Windt, J.; Gabbett, T.J. How Do Training and Competition Workloads Relate to Injury? The Workload-Injury Aetiology Model. *Br. J. Sports Med.* **2017**, *51*, 428–435. [CrossRef]
66. Kalkhoven, J.T.; Watsford, M.L.; Impellizzeri, F.M. A Conceptual Model and Detailed Framework for Stress-Related, Strain-Related, and Overuse Athletic Injury. *J. Sci. Med. Sport* **2020**, *23*, 726–734. [CrossRef]
67. McCall, A.; Fanchini, M.; Coutts, A.J. Prediction: The Modern-Day Sport-Science and Sports-Medicine "Quest for the Holy Grail". *Int. J. Sports Physiol. Perform.* **2017**, *12*, 704–706. [CrossRef]
68. Materne, O.; Farooq, A.; Johnson, A.; Greig, M.; McNaughton, L. Relationship between Injuries and Somatic Maturation in Highly Trained Youth Soccer Players. In Proceedings of the World Congress on Science and Soccer, Copenhagen, Denmark, 20–23 May 2015.
69. Johnson, D.M.; Williams, S.; Bradley, B.; Sayer, S.; Murray Fisher, J.; Cumming, S. Growing Pains: Maturity Associated Variation in Injury Risk in Academy Football. *Eur. J. Sport Sci.* **2019**, *20*, 544–552. [CrossRef] [PubMed]
70. Hall, E.C.R.; Larruskain, J.; Gil, S.M.; Lekue, J.A.; Baumert, P.; Rienzi, E.; Moreno, S.; Tannure, M.; Murtagh, C.F.; Ade, J.D.; et al. An Injury Audit in High-Level Male Youth Soccer Players from English, Spanish, Uruguayan and Brazilian Academies. *Phys. Ther. Sport* **2020**, *44*, 53–60. [CrossRef] [PubMed]
71. Adirim, T.A.; Cheng, T.L. Overview of Injuries in the Young Athlete. *Sports Med.* **2003**, *33*, 75–81. [CrossRef]
72. Price, R.; Hawkins, R.; Hulse, M.; Hodson, A.; Gibson, M. The Football Association Medical Research Programme: An Audit of Injuries in Academy Youth Football. *Br. J. Sports Med.* **2004**, *38*, 466–471. [CrossRef]
73. Chéron, C.; Le Scanff, C.; Leboeuf-Yde, C. Association between Sports Type and Overuse Injuries of Extremities in Adults: A Systematic Review. *Chiropr. Man. Ther.* **2017**, *25*, 4. [CrossRef]
74. Tenforde, A.S.; Kraus, E.; Fredericson, M. Bone Stress Injuries in Runners. *Phys. Med. Rehabil. Clin. N. Am.* **2016**, *27*, 139–149. [CrossRef] [PubMed]
75. Read, P.J.; Oliver, J.L.; De Ste Croix, M.B.A.; Myer, G.D.; Lloyd, R.S. The Scientific Foundations and Associated Injury Risks of Early Soccer Specialisation. *J. Sports Sci.* **2016**, *34*, 2295–2302. [CrossRef]
76. Hall, R.; Foss, K.B.; Hewett, T.E.; Myer, G.D. Sport Specialization's Association with an Increased Risk of Developing Anterior Knee Pain in Adolescent Female Athletes. *J. Sport Rehabil.* **2015**, *24*, 31–35. [CrossRef] [PubMed]
77. Agricola, R.; Bessems, J.H.J.M.; Ginai, A.Z.; Heijboer, M.P.; Van Der Heijden, R.A.; Verhaar, J.A.N.; Weinans, H.; Waarsing, J.H. The Development of Cam-Type Deformity in Adolescent and Young Male Soccer Players. *Am. J. Sports Med.* **2012**, *40*, 1099–1106. [CrossRef]
78. Tak, I.; Weir, A.; Langhout, R.; Waarsing, J.H.; Stubbe, J.; Kerkhoffs, G.; Agricola, R. The Relationship between the Frequency of Football Practice during Skeletal Growth and the Presence of a Cam Deformity in Adult Elite Football Players. *Br. J. Sports Med.* **2015**, *49*, 630–634. [CrossRef] [PubMed]
79. Agricola, R.; Waarsing, J.H.; Thomas, G.E.; Carr, A.J.; Reijman, M.; Bierma-Zeinstra, S.M.A.; Glyn-Jones, S.; Weinans, H.; Arden, N.K. Cam Impingement: Defining the Presence of a Cam Deformity by the Alpha Angle. Data from the CHECK Cohort and Chingford Cohort. *Osteoarthr. Cartil.* **2014**, *22*, 218–225. [CrossRef] [PubMed]
80. Ferguson, M.; Patricios, J. What Is the Relationship between Groin Pain in Athletes and Femoroacetabular Impingement? *Br. J. Sports Med.* **2014**, *48*, 1074–1075. [CrossRef]
81. Hewett, T.E.; Myer, G.D.; Ford, K.R. Decrease in Neuromuscular Control about the Knee with Maturation in Female Athletes. *J. Bone Jt. Surg.* **2004**, *86*, 1601–1608. [CrossRef]
82. Hewett, T.E.; Myer, G.D.; Ford, K.R.; Heidt, R.S.; Colosimo, A.J.; McLean, S.G.; Van Den Bogert, A.J.; Paterno, M.V.; Succop, P. Biomechanical Measures of Neuromuscular Control and Valgus Loading of the Knee Predict Anterior Cruciate Ligament Injury Risk in Female Athletes: A Prospective Study. *Am. J. Sports Med.* **2005**, *33*, 492–501. [CrossRef]

83. Le Gall, F.; Carling, C.; Reilly, T.; Vandewalle, H.; Church, J.; Rochcongar, P. Incidence of Injuries in Elite French Youth Soccer Players: A 10-Season Study. *Am. J. Sports Med.* **2006**, *34*, 928–938. [CrossRef]
84. Rumpf, M.C.; Cronin, J. Injury Incidence, Body Site, and Severity in Soccer Players Aged 6-18 Years: Implications for Injury Prevention. *Strength Cond. J.* **2012**, *34*, 20–31. [CrossRef]
85. Powell, J.W.; Barber-Foss, K.D. Sex-Related Injury Patterns among Selected High School Sports. *Am. J. Sports Med.* **2000**, *28*, 385–391. [CrossRef] [PubMed]
86. Buchheit, M.; Laursen, P.B. High-Intensity Interval Training, Solutions to the Programming Puzzle. *Sports Med.* **2013**, *43*, 313–338. [CrossRef]
87. Hammami, A.; Gabbett, T.J.; Slimani, M.; Bouhlel, E. Does Small-Sided Games Training Improve Physical Fitness and Team-Sport-Specific Skills? A Systematic Review and Meta-Analysis. *J. Sports Med. Phys. Fit.* **2018**, *58*, 1446–1455. [CrossRef]
88. Clemente, F.M.; Lourenço Martins, F.M.; Mendes, R.S. Developing Aerobic and Anaerobic Fitness Using Small-Sided Soccer Games: Methodological Proposals. *Strength Cond. J.* **2014**, *36*, 76–87. [CrossRef]
89. Clemente, F. The Threats of Small-Sided Soccer Games: A Discussion about Their Differences with the Match External Load Demands and Their Variability Levels. *Strength Cond. J.* **2020**, *42*, 100–105. [CrossRef]
90. Ratel, S.; Duché, P.; Williams, C.A. Muscle Fatigue during High-Intensity Exercise in Children. *Sports Med.* **2006**, *36*, 1031–1065. [CrossRef]
91. Carling, C.; Gregson, W.; McCall, A.; Moreira, A.; Wong, D.P.; Bradley, P.S. Match Running Performance during Fixture Congestion in Elite Soccer: Research Issues and Future Directions. *Sports Med.* **2015**, *45*, 605–613. [CrossRef] [PubMed]
92. McKay, D.; Broderick, C.; Steinbeck, K. The Adolescent Athlete: A Developmental Approach to Injury Risk. *Pediatr. Exerc. Sci.* **2016**, *28*, 488–500. [CrossRef] [PubMed]
93. Turner, A.; Bishop, C.; Cree, J.A.; Carr, P.B.; McCann, A.; Bartholomew, B.; Halsted, L. Building a High-Performance Model for Sport: A Human Development-Centered Approach. *Strength Cond. J.* **2019**, *41*, 100–107. [CrossRef]
94. Mendiguchia, J.; Conceição, F.; Edouard, P.; Fonseca, M.; Pereira, R.; Lopes, H.; Morin, J.B.; Jiménez-Reyes, P. Sprint versus Isolated Eccentric Training: Comparative Effects on Hamstring Architecture and Performance in Soccer Players. *PLoS ONE* **2020**, *15*, e0228283. [CrossRef]
95. Mendiguchia, J.; Edouard, P.; Guex, K.; Lahti, J.; Samozino, P.; Morin, J.B. Sprinting: A Potential Vaccine for Hamstring Injury? *Sport Perform. Sci. Rep.* **2019**, *1*, 1–2.
96. Mendiguchia, J.; Castaño-Zambudio, A.; Jiménez-Reyes, P.; Morin, J.B.; Edouard, P.; Conceição, F.; Colyer, S.L. Can We Modify Maximal Speed Running Posture? Implications for Performance and Hamstring Injuries Management. *Int. J. Sports Physiol. Perform.* **2021**, *1*, 1–10. [CrossRef]
97. McCall, A.; Pruna, R.; Van der Horst, N.; Dupont, G.; Buchheit, M.; Coutts, A.J.; Impellizzeri, F.M.; Fanchini, M. Exercise-Based Strategies to Prevent Muscle Injury in Male Elite Footballers: An Expert-Led Delphi Survey of 21 Practitioners Belonging to 18 Teams from the Big-5 European Leagues. *Sports Med.* **2020**, *50*, 1667–1681. [CrossRef]
98. McBurnie, A.J.; Dos'Santos, T. Multi-Directional Speed in Youth Soccer Players: Theoretical Underpinnings. *Strength Cond. J.* **2021**, publish ahead of print. [CrossRef]
99. McBurnie, A.J.; Harper, D.J.; Jones, P.A.; Dos'Santos, T. Deceleration Training: Another Potential Vaccine for Sports-Related Injury? *Sports Med.* **2021**, in press. [CrossRef]
100. Dos'Santos, T.; McBurnie, A.; Comfort, P.; Jones, P.A. The Effects of Six-Weeks Change of Direction Speed and Technique Modification Training on Cutting Performance and Movement Quality in Male Youth Soccer Players. *Sports* **2019**, *7*, 205. [CrossRef] [PubMed]
101. Dos'Santos, T.; Thomas, C.; McBurnie, A.; Comfort, P.; Jones, P.A. Change of Direction Speed and Technique Modification Training Improves 180° Turning Performance, Kinetics, and Kinematics. *Sports* **2021**, *9*, 73. [CrossRef]
102. Maté, A.; Trujillo, J.; Mylopoulos, J. Conceptualizing and Specifying Key Performance Indicators in Business Strategy Models. In *Conceptual Modeling*; Atzeni, P., Cheung, D., Ram, S., Eds.; Springer: Berlin/Heidelberg, Germany, 2012; pp. 282–291.
103. Fuller, C.W.; Ekstrand, J.; Junge, A.; Andersen, T.E.; Bahr, R.; Dvorak, J.; Hägglund, M.; McCrory, P.; Meeuwisse, W.H. Consensus Statement on Injury Definitions and Data Collection Procedures in Studies of Football (Soccer) Injuries. *Scand. J. Med. Sci. Sports* **2006**, *16*, 83–92. [CrossRef] [PubMed]
104. Materne, O.; Chamari, K.; Farooq, A.; Tabben, M.; Weir, A.; Holmich, P.; Bahr, R.; Greig, M.; McNaughton, L.R. Shedding Light on Incidence and Burden of Physeal Injuries in a Youth Elite Football Academy: A 4-Season Prospective Study. *Scand. J. Med. Sci. Sports* **2022**, *32*, 165–176. [CrossRef]
105. Monasterio, X.; Gil, S.M.; Bidaurrazaga-Letona, I.; Lekue, J.A.; Santisteban, J.; Diaz-Beitia, G.; Martin-Garetxana, I.; Bikandi, E.; Larruskain, J. Injuries According to the Percentage of Adult Height in an Elite Soccer Academy. *J. Sci. Med. Sport* **2021**, *24*, 218–223. [CrossRef] [PubMed]
106. Khamis, H.J.; Roche, A.F. Predicting Adult Stature without Using Skeletal Age: The Khamis-Roche Method. *Pediatrics* **1994**, *94*, 504–507. [PubMed]
107. Parr, J.; Winwood, K.; Hodson-Tole, E.; Deconinck, F.J.A.; Parry, L.; Hill, J.P.; Malina, R.M.; Cumming, S.P. Predicting the Timing of the Peak of the Pubertal Growth Spurt in Elite Youth Soccer Players: Evaluation of Methods. *Ann. Hum. Biol.* **2020**, publish ahead of print. [CrossRef] [PubMed]

108. Brownlee, T.E.; O'Boyle, A.; Morgans, R.; Morton, J.P.; Erskine, R.M.; Drust, B. Training Duration May Not Be a Predisposing Factor in Potential Maladaptations in Talent Development Programmes That Promote Early Specialisation in Elite Youth Soccer. *Int. J. Sports Sci. Coach.* **2018**, *13*, 674–678. [CrossRef]
109. Bowen, L.; Gross, A.S.; Gimpel, M.; Li, F.X. Accumulated Workloads and the Acute: Chronic Workload Ratio Relate to Injury Risk in Elite Youth Football Players. *Br. J. Sports Med.* **2017**, *51*, 452–459. [CrossRef]
110. Malone, S.; Roe, M.; Doran, D.A.; Gabbett, T.J.; Collins, K. High Chronic Training Loads and Exposure to Bouts of Maximal Velocity Running Reduce Injury Risk in Elite Gaelic Football. *J. Sci. Med. Sport* **2017**, *20*, 250–254. [CrossRef] [PubMed]
111. Beato, M.; Drust, B.; Iacono, A. Implementing High-Speed Running and Sprinting Training in Professional Soccer. *Int. J. Sports Med.* **2021**, *42*, 295–299.
112. Cumming, S.P.; Lloyd, R.S.; Oliver, J.L.; Eisenmann, J.C.; Malina, R.M. Bio-Banding in Sport: Applications to Competition, Talent Identification, and Strength and Conditioning of Youth Athletes. *Strength Cond. J.* **2017**, *39*, 34–47. [CrossRef]
113. Cumming, S.P.; Brown, D.J.; Mitchell, S.; Bunce, J.; Hunt, D.; Hedges, C.; Crane, G.; Gross, A.; Scott, S.; Franklin, E.; et al. Premier League Academy Soccer Players' Experiences of Competing in a Tournament Bio-Banded for Biological Maturation. *J. Sports Sci.* **2018**, *36*, 757–765. [CrossRef]
114. Bradley, B.; Johnson, D.; Hill, M.; McGee, D.; Kana-ah, A.; Sharpin, C.; Sharp, P.; Kelly, A.; Cumming, S.P.; Malina, R.M. Bio-Banding in Academy Football: Player's Perceptions of a Maturity Matched Tournament. *Ann. Hum. Biol.* **2019**, *46*, 400–408. [CrossRef]
115. Lovell, R.; Fransen, J.; Ryan, R.; Massard, T.; Cross, R.; Eggers, T.; Duffield, R. Biological Maturation and Match Running Performance: A National Football (Soccer) Federation Perspective. *J. Sci. Med. Sport* **2019**, *22*, 1139–1145. [CrossRef]
116. Herrington, L.; Myer, G.; Horsley, I. Task Based Rehabilitation Protocol for Elite Athletes Following Anterior Cruciate Ligament Reconstruction: A Clinical Commentary. *Phys. Ther. Sport* **2013**, *14*, 188–198. [CrossRef]
117. Padua, D.A.; Boling, M.C.; DiStefano, L.J.; Onate, J.A.; Beutler, A.I.; Marshall, S.W. Reliability of the Landing Error Scoring System-Real Time, a Cinical Assessment Tool of Jump-Landing Biomechanics. *J. Sport Rehabil.* **2011**, *20*, 145–156. [CrossRef]
118. Herrington, L.; Myer, G.D.; Munro, A. Intra and Inter-Tester Reliability of the Tuck Jump Assessment. *Phys. Ther. Sport* **2013**, *14*, 152–155. [CrossRef]
119. Horobeanu, C.; Jones, T.; Johnson, A. Can We Limit Training Days Lost Due to Osgood Schlatters Disease in Junior Squash Athletes? *Br. J. Sports Med.* **2017**, *51*, 331–332. [CrossRef]
120. Root, H.; Trojian, T.; Martinez, J.; Kraemer, W.; DiStefano, L.J. Landing Technique and Performance in Youth Athletes After a Single Injury-Prevention Program Session. *J. Athl. Train.* **2015**, *50*, 1149–1157. [CrossRef] [PubMed]
121. Lloyd, R.S.; Meyers, R.W.; Oliver, J.L. The Natural Development and Trainability of Plyometric Ability during Childhood. *Strength Cond. J.* **2011**, *33*, 23–32. [CrossRef]
122. Dos'Santos, T.; McBurnie, A.; Donelon, T.; Thomas, C.; Comfort, P.; Jones, P.A. A Qualitative Screening Tool to Identify Athletes with 'High-Risk' Movement Mechanics during Cutting: The Cutting Movement Assessment Score (CMAS). *Phys. Ther. Sport* **2019**, *38*, 152–161. [CrossRef] [PubMed]
123. Pedley, J.S.; DiCesare, C.A.; Lloyd, R.S.; Oliver, J.L.; Ford, K.R.; Hewett, T.E.; Myer, G.D. Maturity Alters Drop Vertical Jump Landing Force-Time Profiles but Not Performance Outcomes in Adolescent Females. *Scand. J. Med. Sci. Sports* **2021**, *31*, 2055–2063. [CrossRef] [PubMed]
124. Harper, D.J.; Cohen, D.D.; Carling, C.; Kiely, J. Can Countermovement Jump Neuromuscular Performance Qualities Differentiate Maximal Horizontal Deceleration Ability in Team Sport Athletes? *Sports* **2020**, *8*, 76. [CrossRef] [PubMed]
125. Harper, D.J.; Cohen, D.D.; Rhodes, D.; Carling, C.; Kiely, J. Drop Jump Neuromuscular Performance Qualities Associated with Maximal Horizontal Deceleration Ability in Team Sport Athletes. *Eur. J. Sport Sci.* **2021**, 1–12. [CrossRef]
126. Thomas, C.; Comfort, P.; Jones, P.A.; Dos'Santos, T. A Comparison of Isometric Midthigh-Pull Strength, Vertical Jump, Sprint Speed, and Change-of-Direction Speed in Academy Netball Players. *Int. J. Sports Physiol. Perform.* **2017**, *12*, 916–921. [CrossRef] [PubMed]
127. Thomas, C.; Dos'Santos, T.; Comfort, P.; Jones, P.A. Relationships between Unilateral Muscle Strength Qualities and Change of Direction in Adolescent Team-Sport Athletes. *Sports* **2018**, *6*, 83. [CrossRef]
128. Jones, S.; Mullen, R.; Clair, Z.; Wrigley, R.; Andersen, T.E.; Williams, M. Field Based Lower Limb Strength Tests Provide Insight into Sprint and Change of Direction Ability in Academy Footballers. *Scand. J. Med. Sci. Sports* **2021**, *31*, 2178–2186. [CrossRef]
129. Lehance, C.; Binet, J.; Bury, T.; Croisier, J.L. Muscular Strength, Functional Performances and Injury Risk in Professional and Junior Elite Soccer Players. *Scand. J. Med. Sci. Sports* **2009**, *19*, 243–251. [CrossRef] [PubMed]
130. Dedinsky, R.; Baker, L.; Imbus, S.; Bowman, M.; Murray, L. Exercises That Facilitate Optimal Hamstring and Quadriceps Co-Activation to Help Decrease Acl Injury Risk in Healthy Females: A Systematic Review of the Literature. *Int. J. Sports Phys. Ther.* **2017**, *12*, 3–15. [PubMed]
131. Ribeiro-Alvares, J.B.; Marques, V.B.; Vaz, M.A.; Baroni, B.M. Four Weeks of Nordic Hamstring Exercise Reduce Muscle Injury Risk Factors in Young Adults. *J. Strength Cond. Res.* **2018**, *32*, 1254–1262. [CrossRef] [PubMed]
132. Buchheit, M.; Simpson, B.M. Player-Tracking Technology: Half-Full or Half-Empty Glass? *Int. J. Sports Physiol. Perform.* **2017**, *12*, S2-35–S2-41. [CrossRef] [PubMed]
133. Haddad, M.; Stylianides, G.; Djaoui, L.; Dellal, A.; Chamari, K. Session-RPE Method for Training Load Monitoring: Validity, Ecological Usefulness, and Influencing Factors. *Front. Neurosci.* **2017**, *11*, 612. [CrossRef]

134. Saw, A.E.; Main, L.C.; Gastin, P.B. Monitoring the Athlete Training Response: Subjective Self-Reported Measures Trump Commonly Used Objective Measures: A Systematic Review. *Br. J. Sports Med.* **2016**, *50*, 281–291. [CrossRef]
135. Hannon, M.P.; Coleman, N.M.; Parker, L.J.F.; McKeown, J.; Unnithan, V.B.; Close, G.L.; Drust, B.; Morton, J.P. Seasonal Training and Match Load and Micro-Cycle Periodization in Male Premier League Academy Soccer Players. *J. Sports Sci.* **2021**, *39*, 1838–1849. [CrossRef]
136. Gastin, P.B.; Bennett, G.; Cook, J. Biological Maturity Influences Running Performance in Junior Australian Football. *J. Sci. Med. Sport* **2013**, *16*, 140–145. [CrossRef]
137. Harley, J.A.; Barnes, C.A.; Portas, M.; Lovell, R.; Barrett, S.; Paul, D.; Weston, M. Motion Analysis of Match-Play in Elite U12 to U16 Age-Group Soccer Players. *J. Sports Sci.* **2010**, *28*, 1391–1397. [CrossRef]
138. Welsman, J.R.; Armstrong, N. The Measurement and Interpretation of Aerobic Fitness in Children: Current Issues. *J. R. Soc. Med.* **1996**, *89*, 281P–285P. [CrossRef] [PubMed]
139. Rosenbloom, C.A.; Loucks, A.B.; Ekblom, B. Special Populations: The Female Player and the Youth Player. *J. Sports Sci.* **2006**, *24*, 783–793. [CrossRef] [PubMed]
140. Buchheit, M.; Mendez-Villanueva, A.; Simpson, B.M.; Bourdon, P.C. Match Running Performance and Fitness in Youth Soccer. *Int. J. Sports Med.* **2010**, *31*, 818–825. [CrossRef] [PubMed]
141. Atan, S.A.; Foskett, A.; Ali, A. Motion Analysis of Match Play in New Zealand U13 to U15 Age-Group Soccer Players. *J. Strength Cond. Res.* **2016**, *30*, 2416–2423. [CrossRef]
142. Al Haddad, H.; Simpson, B.M.; Buchheit, M.; Di Salvo, V.; Mendez-Villanueva, A. Peak Match Speed and Maximal Sprinting Speed in Young Soccer Players: Effect of Age and Playing Position. *Int. J. Sports Physiol. Perform.* **2015**, *10*, 888–896. [CrossRef]
143. Bellistri, G.; Marzorati, M.; Sodero, L.; Sforza, C.; Bradley, P.S.; Porcelli, S. Match Running Performance and Physical Capacity Profiles of U8 and U10 Soccer Players. *Sport Sci. Health* **2017**, *13*, 273–280. [CrossRef]
144. Buchheit, M.; Mendez-Villanueva, A. Effects of Age, Maturity and Body Dimensions on Match Running Performance in Highly Trained under-15 Soccer Players. *J. Sports Sci.* **2014**, *32*, 1271–1278. [CrossRef]
145. Parr, J.; Winwood, K.; Hodson-Tole, E.; Deconinck, F.J.A.; Hill, J.P.; Cumming, S.P. Maturity-Associated Differences in Match Running Performance in Wlite Male Youth Soccer Players. *Int. J. Sports Physiol. Perform.* **2021**, *1*, 1–9. [CrossRef]
146. Vieira, L.H.P.; Aquino, R.; Moura, F.A.; de Barros, R.M.; Arpini, V.M.; Oliveira, L.P.; Bedo, B.L.; Santiago, P.R. Team Dynamics, Running, and Skill-Related Performances of Brazilian U11 to Professional Soccer Players During Official Matches. *J. Strength Cond. Res.* **2019**, *33*, 1179–1462. [CrossRef]
147. Reynolds, J.; Connor, M.; Jamil, M.; Beato, M. Quantifying and Comparing the Match Demands of U18, U23, and 1st Team English Professional Soccer Players. *Front. Physiol.* **2021**, *12*, 948. [CrossRef]
148. Coutinho, D.; Gonçalves, B.; Figueira, B.; Abade, E.; Marcelino, R.; Sampaio, J. Typical Weekly Workload of under 15, under 17, and under 19 Elite Portuguese Football Players. *J. Sports Sci.* **2015**, *33*, 1229–1237. [CrossRef]
149. Towlson, C.; Salter, J.; Ade, J.D.; Enright, K.; Harper, L.D.; Page, R.M.; Malone, J.J. Maturity-Associated Considerations for Training Load, Injury Risk, and Physical Performance in Youth Soccer: One Size Does Not Fit All. *J. Sport Health Sci.* **2020**, *10*, 403–412. [CrossRef]
150. Vanrenterghem, J.; Nedergaard, N.J.; Robinson, M.A.; Drust, B. Training Load Monitoring in Team Sports: A Novel Framework Separating Physiological and Biomechanical Load-Adaptation Pathways. *Sports Med.* **2017**, *47*, 2135–2142. [CrossRef]
151. Seiler, S.; Haugen, O.; Kuffel, E. Autonomic Recovery after Exercise in Trained Athletes: Intensity and Duration Effects. *Med. Sci. Sports Exerc.* **2007**, *39*, 1366–1373. [CrossRef] [PubMed]
152. Jaspers, A.; Kuyvenhoven, J.P.; Staes, F.; Frencken, W.G.P.; Helsen, W.F.; Brink, M.S. Examination of the External and Internal Load Indicators' Association with Overuse Injuries in Professional Soccer Players. *J. Sci. Med. Sport* **2018**, *21*, 579–585. [CrossRef] [PubMed]
153. Malone, S.; Owen, A.; Mendes, B.; Hughes, B.; Collins, K.; Gabbett, T.J. High-Speed Running and Sprinting as an Injury Risk Factor in Soccer: Can Well-Developed Physical Qualities Reduce the Risk? *J. Sci. Med. Sport* **2018**, *21*, 257–262. [CrossRef] [PubMed]
154. Duhig, S.; Shield, A.J.; Opar, D.; Gabbett, T.J.; Ferguson, C.; Williams, M. Effect of High-Speed Running on Hamstring Strain Injury Risk. *Br. J. Sports Med.* **2016**, *50*, 1536–1540. [CrossRef] [PubMed]
155. Colby, M.J.; Dawson, B.; Heasman, J.; Rogalski, B.; Gabbett, T.J. Accelerometer and GPS-Derived Running Loads and Injury Risk in Elite Australian Footballers. *J. Strength Cond. Res.* **2014**, *28*, 2244–2252. [CrossRef]
156. Edwards, W.B. Modeling Overuse Injuries in Sport as a Mechanical Fatigue Phenomenon. *Exerc. Sport Sci. Rev.* **2018**, *46*, 224–231. [CrossRef]
157. Weston, M.; Siegler, J.; Bahnert, A.; McBrien, J.; Lovell, R. The Application of Differential Ratings of Perceived Exertion to Australian Football League Matches. *J. Sci. Med. Sport* **2015**, *18*, 704–708. [CrossRef]
158. Hooper, S.L.; Mackinnon, L.T.; Howard, A.; Gordon, R.D.; Bachmann, A.W. Markers for Monitoring Overtraining and Recovery. *Med. Sci. Sports Exerc.* **1995**, *27*, 106–112. [CrossRef]
159. Phibbs, P.J.; Roe, G.; Jones, B.; Read, D.B.; Weakley, J.; Darrall-Jones, J.; Till, K. Validity of Daily and Weekly Self-Reported Training Load Measures in Adolescent Athletes. *J. Strength Cond. Res.* **2017**, *31*, 1121–1126. [CrossRef] [PubMed]
160. Scantlebury, S.; Till, K.; Sawczuk, T.; Phibbs, P.; Jones, B. Validity of Retrospective Session Rating of Perceived Exertion to Quantify Training Load in Youth Athletes. *J. Strength Cond. Res.* **2018**, *32*, 1975–1980. [CrossRef]

161. Impellizzeri, F.M.; Marcora, S.M.; Coutts, A.J. Internal and External Training Load: 15 Years On. *Int. J. Sports Physiol. Perform.* **2019**, *14*, 270–273. [CrossRef]
162. Barrett, S.; McLaren, S.; Spears, I.; Ward, P.; Weston, M. The Influence of Playing Position and Contextual Factors on Soccer Players' Match Differential Ratings of Perceived Exertion: A Preliminary Investigation. *Sports* **2018**, *6*, 13. [CrossRef] [PubMed]
163. Wright, M.D.; Songane, F.F.; Emmonds, S.; Chesterton, P.; Weston, M.; McLaren, S.J. Differential Ratings of Perceived Match and Training Exertion in Girls' Soccer. *Int. J. Sports Physiol. Perform.* **2020**, *15*, 1315–1323. [CrossRef] [PubMed]
164. McLaren, S.J.; Smith, A.; Spears, I.R.; Weston, M. A Detailed Quantification of Differential Ratings of Perceived Exertion during Team-Sport Training. *J. Sci. Med. Sport* **2016**, *20*, 290–295. [CrossRef]
165. Shields, B.J.; Cohen, D.M.; Harbeck-Weber, C.; Powers, J.D.; Smith, G.A. Pediatric Pain Measurement Using a Visual Analogue Scale: A Comparison of Two Teaching Methods. *Clin. Pediatrics* **2003**, *42*, 227–234. [CrossRef] [PubMed]
166. Wylde, M.; Baranee, K.; Young, L.; Callaway, A. Axis Specific Player Load to Quantify Lower Limb Biomechanical Loading in Adolescent Badminton Players. *Int. J. Racket Sports Sci.* **2019**, *1*, 37–44.
167. Johnson, D.M.; Williams, S.; Bradley, B.; Cumming, S. The Interaction between Training Volume, Growth, Maturation, and Injury Risk in Adolescent Football Players. Paper Presented at: Sport Kongress 2020; 30 January 2020. Available online: https://www.researchgate.net/publication/340175460_David_Johnson_Sports_Kongress_Poster_2020 (accessed on 1 September 2021).
168. Johnson, D.M.; Williams, S.; Bradley, B.; Cumming, S. Growth, Maturation and Bio-Banding in Football. Paper Presented at: Webinarium Elitprojektet Online; 16 April 2021. Available online: https://www.youtube.com/watch?t=1291&v=wDE8lyYfZ64&feature=youtu.be (accessed on 1 September 2021).
169. Moran, J.J.; Sandercock, G.R.H.; Ramírez-Campillo, R.; Meylan, C.M.P.; Collison, J.A.; Parry, D.A. Age-Related Variation in Male Youth Athletes' Countermovement Jump after Plyometric Training: A Meta-Analysis of Controlled Trials. *J. Strength Cond. Res.* **2017**, *31*, 552–565. [CrossRef]
170. Rathleff, M.S.; Winiarski, L.; Krommes, K.; Graven-Nielsen, T.; Hölmich, P.; Olesen, J.L.; Holden, S.; Thorborg, K. Activity Modification and Knee Strengthening for Osgood-Schlatter Disease: A Prospective Cohort Study. *Orthop. J. Sports Med.* **2020**, *8*, 2325967120911106. [CrossRef] [PubMed]
171. Van-Hooren, B.; De, M.; Croix, S. Sensitive Periods to Train General Motor Abilities in Children and Adolescents: Do They Exist? A Critical Appraisal. *Strength Cond. J.* **2020**, *46*, 7–14. [CrossRef]
172. Myer, G.D.; Faigenbaum, A.D.; Ford, K.R.; Best, T.M.; Bergeron, M.F.; Hewett, T.E. When to Initiate Integrative Neuromuscular Training to Reduce Sports-Related Injuries and Enhance Health in Youth? *Curr. Sports Med. Rep.* **2011**, *10*, 155–166. [CrossRef]
173. Zwolski, C.; Quatman-Yates, C.; Paterno, M.V. Resistance Training in Youth: Laying the Foundation for Injury Prevention and Physical Literacy. *Sports Health* **2017**, *9*, 436–443. [CrossRef]
174. Faigenbaum, A.D.; Myer, G.D. Pediatric Resistance Training: Benefits, Concerns, and Program Design Considerations. *Curr. Sports Med. Rep.* **2010**, *9*, 161–168. [CrossRef] [PubMed]
175. Oliver, J.L.; Lloyd, R.S.; Rumpf, M.C. Developing Speed throughout Childhood and Adolescence: The Role of Growth, Maturation and Training. *Strength Cond. J.* **2013**, *35*, 42–48. [CrossRef]
176. Cahill, M.J.; Oliver, J.L.; Cronin, J.B.; Clark, K.P.; Cross, M.R.; Lloyd, R.S. Sled-Push Load-Velocity Profiling and Implications for Sprint Training Prescription in Young Athletes. *J. Strength Cond. Res.* **2020**, *34*, 2751–2759. [CrossRef] [PubMed]
177. Cahill, M.J.; Oliver, J.L.; Cronin, J.B.; Clark, K.P.; Cross, M.R.; Lloyd, R.S. Sled-Pull Load–Velocity Profiling and Implications for Sprint Training Prescription in Young Male Athletes. *Sports* **2019**, *7*, 119. [CrossRef] [PubMed]
178. Saward, C.; Hulse, M.; Morris, J.G.; Goto, H.; Sunderland, C.; Nevill, M.E. Longitudinal Physical Development of Future Professional Male Soccer Players: Implications for Talent Identification and Development? *Front. Sports Act. Living* **2020**, *2*, 578203. [CrossRef]
179. Soligard, T.; Schwellnus, M.; Alonso, J.M.; Bahr, R.; Clarsen, B.; Dijkstra, H.P.; Gabbett, T.; Gleeson, M.; Hägglund, M.; Hutchinson, M.R.; et al. How Much Is Too Much? (Part 1) International Olympic Committee Consensus Statement on Load in Sport and Risk of Injury. *Br. J. Sports Med.* **2016**, *50*, 1030–1041. [CrossRef]

sports MDPI

Article

Differences in Squat Jump, Linear Sprint, and Change-of-Direction Performance among Youth Soccer Players According to Competitive Level

Michael Keiner [1,*], Andreas Kapsecker [1], Tobias Stefer [1], Björn Kadlubowski [1,2] and Klaus Wirth [3]

1 Department of Sports Science, German University of Health & Sport, 85737 Ismaning, Germany; Andreas.Kapsecker@edu.my-campus-berlin.com (A.K.); tobias.stefer@tsv1860.org (T.S.); bjoernkadlubowski@gmail.com (B.K.)
2 DSC Arminia Bielefeld e. V., 33615 Bielefeld, Germany
3 Department of Sports Science, University of Applied Sciences Wiener Neustadt, 2700 Vienna, Austria; klaus.wirth@fhwn.ac.at
* Correspondence: michaelkeiner@gmx.de

Abstract: The aim of this investigation was to analyze significant differences in performance depending on the level of play (elite vs. amateur) in youth soccer players (under 17 years. old (U17) and U19). A cross-sectional study was conducted, and 45 elite and amateur male youth soccer players (16.56 $\pm$ 0.9 years old) were evaluated in their performances in squat jump (SJ), 10 m linear sprint (LS), 20 m LS, 505 agility test (505) and Illinois agility test (IAT). Differences in performances were analyzed with a 2 $\times$ 2 MANOVA, post-hoc ANOVAs, and Hedges' g (g) for pairwise comparisons of subgroups (level of play and age group). This investigation showed that the elite player performance was significantly ($p < 0.05$) better in all performance tests than amateur players in both age groups. Interestingly, this investigation showed that the more complex the target exercise, the larger the effect sizes for group differences (SJ: g = 0.64–1.18, LS: g = 0.05–2.23, change-of-direction (COD): g = 3.01–6.84). The SJ, LS, 505, and IAT may prove useful in talent selection test batteries to separate between competitive levels in youth soccer players.

Keywords: talent identification; speed; football; performance diagnosis

Citation: Keiner, M.; Kapsecker, A.; Stefer, T.; Kadlubowski, B.; Wirth, K. Differences in Squat Jump, Linear Sprint, and Change-of-Direction Performance among Youth Soccer Players According to Competitive Level. *Sports* **2021**, *9*, 149. https://doi.org/10.3390/sports9110149

Academic Editors: Adam Leigh Kelly, Sergio L. Jiménez Sáiz, Sara Diana Leal dos Santos, Alberto Lorenzo Calvo and Aaron Scanlan

Received: 28 July 2021
Accepted: 25 October 2021
Published: 27 October 2021

1. Introduction

For match-play demands in soccer, players have to perform several high-intensity tasks during a game [1]. Because of changes in technical, tactical, and physical requirements, soccer matches have become more dynamic and fast paced [2,3]. It is becoming increasingly essential to consider physiological factors, such as the capacity of soccer players to produce various forceful and explosive actions, for players to exhibit optimal performance [4]. In addition to jumps, shots, and linear sprints (LS), athletes must also perform sprints with directional changes [5]. The capacity to perform change-of-direction (COD, excluding the decision-making process) is an essential physical fitness factor needed to perform effective and efficient COD maneuvers in multiple sports [6]. In modern soccer, which is speed-oriented, dynamic, and fundamentally more demanding, the physical demands of speed, especially COD speed, are of central importance for the development of the best possible performance and corresponding sports success [7–9]. Therefore, the ability to accelerate, change direction, and rapidly decelerate could increase the chance of players winning one-on-one duels or performing effective defending maneuvers in the match [10]. Soccer players perform approximately 727 $\pm$ 203 turns during a match [1]. A soccer player changes direction every 2–4 s and makes 1200–1400 changes of direction during a match [11]. LS, COD, and jump performances are therefore determinants of performance in soccer matches. The assessment of field-based physical performance can help coaches in evaluating the level of players, thus positively impacting the talent identification processes and training implications [12]. Since LS, COD, and jump performances also have a significant influence on the outcome of the match, it is to be expected that these performances differ between different performance levels. The literature shows performance differences between elite and sub-elite players [13] and elite and elite youth soccer players [14]. However, the literature is not exhaustive and is concordant on performance differences between elite and amateur soccer players. The literature reports small to large differences in squat jump (SJ) performance in favor of elite compared to amateur soccer players (Hedges' g (g) = 0.3–0.97) [13,15] and 16 years old elite and youth amateur soccer players (g = 0.3) [16]. For LS (10 meters (m)), the literature shows quite large differences in favor of elite compared with amateur soccer players (g = 0.79) [13], and also for elite compared to amateur youth soccer players (12–16 years old), small to moderate differences for LS 10 m (g = 0.22–0.65) [16–18], large differences for 20 m (g = 0.89–0.94) [18,19], moderate to large differences for LS 30 m (g = 0.5–1.2) [20] and moderate differences for LS 40 m (g = 0.5) [21]. Comparisons between elite and amateur soccer players for distances over 30 m in LS showed moderate differences as well [13]. Only two studies analyzed different COD tests in elite youth soccer players and youth amateur players. Trajković et al. [18] found large differences in COD tests with one 45° turn (left and right direction in separated run) on a 10 m course (g = 0.8–0.95) and large differences for the Illinois test (IAT) (63–65 m, 9 turns of 90°–180°, g = 1.5) and Trecroci et al. [17] found significant differences (g = 0.69) for a COD test with six 90° turns on a 15 m course in favor of elite and regional youth soccer players (15–16 years old).

In light of the considerations above, it would be important to assess whether elite and amateur players between 16 and 19 years old are distinguishable by means of field-based physical tests or whether this difference is even more pronounced than in the studies with younger subjects [16–21]. The outcomes could provide coaches additional data on the physical performances of their under-19-year-old (U19) and under-17-year-old (U17) players to make appropriate choices in terms of talent identification at this particular age stage. From this, the research question is whether there are statistically significant differences in LS, COD, and jumping performance as a function of playing level (elite vs. amateur) in youth soccer players. It was hypothesized that performance would differ by playing level in favor of elite soccer players.

2. Materials and Methods

In order to answer the research question, a cross-sectional study was conducted. Forty-five elite and amateur male youth soccer players were evaluated in their performances in SJ, 10 m LS, 20 m LS, 505 agility test (505), and IAT and checked for differences in performances (elite vs. amateur players). Tests were carried out on 2 test days within a 1-week period. On test day 1, the LS was performed first, followed by the SJ. Two days later, on test day 2, the COD tests were determined. Based on the total distance in ascending order, to avoid fatiguing effects, first the 505, then the IAT test, was performed. The elite players were familiar with the IAT, 505, LS, and SJ tests because these tests were part of their semi-annual performance diagnostics routine. The amateur players do not conduct regular performance tests; therefore, one week before test day 1, the subjects completed a familiarization session on two separate days (day 1: LS, SJ; day 2: 505, IAT).

2.1. Subjects

Forty-five male youth soccer players (16.56 ± 0.9 (range: 15–18) years old; body mass: 70.7 ± 9.4 kg; height: 1.78 ± 0.08 m, body mass index: 22.3 ± 2.6) were recruited from U17 and U19 teams of two youth training centers. The teams of the two training centers played in different competitive levels. Within the two age groups, a comparison was made between a soccer team in the highest German junior divisions (youth Bundesliga, elite players (U17: n = 14; U19: n = 12)) and a youth soccer team in the lowest German youth league (club level, amateur players (U17: n = 8; U19: n = 11)). The soccer teams were classified elite in reference to the definition used by Lorenz et al. [22], who considered elite athletes as those who played at a higher level than peers within a sport (national (junior Bundesliga, highest league in Germany) vs. club level (district league, lowest league in Germany)). The elite players of both age groups performed 4 to 5 soccer sessions per week (1.5–2 training sessions/day), whereas the amateur soccer players of both age groups performed 1 to 2 soccer sessions per week. Both groups regularly competed in their league on weekends during the season. The subjects did not participate in fatiguing training sessions for a minimum of 3 days before testing. None of the subjects reported any injuries at the time of testing.

Each subject and their parents (if the subject was not yet 18 years old) were informed about the aims of the study and the experimental risks involved with the research. All subjects and their parents (if the subject was not yet 18 years old) provided written informed consent to participate in the present study. Furthermore, this study was approved by the Ethics Committee of the German University of Health & Sport (DHGS-EK-2021-002). The study was performed with human subjects in accordance with the Helsinki Declaration.

2.2. Procedures

Body mass was analyzed using a personal scale (PSD, Neuss, Germany). Body height was determined by means of a meter stick and spirit level attached to a wall. The spirit level was placed on the head of the subjects in order to be able to read the body height. The warm-up for the jump and sprint tests consisted of nonspecific running at low-to-medium intensity for approximately 5 min. Then, coordination exercises, such as running with lifted knees, heeling, and sidestepping, were performed for approximately 5 min. Subsequently, 3 acceleration runs over approximately 30 m were performed with short intervening walking breaks. Jumping performance was measured using a contact mat (Refitronic, Schmitten, Germany) that operates as a switch. This system sent information to the computer regarding whether the mat was loaded. From this information, the flight time and the jump height were determined for all jumps. The jump height was calculated from the flight time ($gt^2/8$; g = the gravitational acceleration (9.81 $m{\cdot}s^{-2}$) and t = flight time). The squat jump was initiated at a knee angle of 90° without counter-movement. The subjects had 5 trials in which to achieve their best result. Between every jump, the athletes received a 1 min break. The test-retest reliability is reported to have an ICC = 0.97 [23]. The subjects performed three attempts per COD test, which were separated by a 3 min break. The best trial was used for the statistical analysis. The description of the test setups can be found in the literature [24,25]. As different COD tests have different requirements (e.g., length, turns), these tests seem to evaluate task-specific requirements [26]. Therefore, two tests that have a heterogeneous requirement profile (IAT: 63–65 m, 9 turns of 90°–180° vs. 505: 10 m, 1 turn of 180°) were selected. IAT and 505 tests were frequently used in studies with soccer players [17,18]; the choice of the same tests should lead to better comparability of own results and other studies. If the pylons or hurdle bars were knocked down or touched during COD testing, a follow-up attempt was completed. The tests were separated by a break of 15 min. LS performance was measured for a distance of 10 m and 20 m. Each athlete also had three attempts. Between each completed sprint, the athletes received a 3 min break. The time was measured for all COD and LS tests with a double-timing gate system (wk7 time watch, Ditzingen, Germany). The starting point was marked with a small cap 0.75 m away from the starting gate to avoid early triggering, e.g., by a hand movement or a bent body position. The subjects independently chose when the measurement began according to the activation of the barriers. Thus, the reaction time was excluded from the measurement. The test-retest reliability is reported to have an ICC = 0.85 for 505, ICC = 0.97 for IAT, and ICC = 0.89 for LS [23].

2.3. Statistical Analyses

Statistical analysis was performed using SPSS software version 27 (IBM, Ehningen, DE, Germany). Data were tested for normality using the Shapiro–Wilk test to determine whether parametric or non-parametric statistical methods were appropriate. Reliability analyses were performed using intraclass correlation coefficients (ICCs) and a 95% confidence interval (95% CI). For the reliability analyses, the ICCs and 95% CIs were calculated from the 3 trials of the testing day. ICCs greater than 0.70 indicate suitable reliability [27]. Additionally, variability was determined using mean coefficients of variation (CV %). Acceptable thresholds were determined using a CV of <10%. In addition, the variance of homogeneity or variance of heterogeneity was determined by the Levene test of equal variance. The homogeneity of covariances was assessed by Box's test. A 2 × 2 MANOVA was calculated to show a statistically significant difference between the age group and playing level on the combined 5 dependent variables (LS 10 m, LS 20 m, SJ, IAT, 505). A second 2 × 2 MANOVA was calculated to show a statistically significant difference between the age group and playing level on combined anthropometric data (body mass, height, BMI). Furthermore, partial eta-squared values (η2) were computed for all the analyses as an indicator of the effect size. A partial eta-squared value between 0.01 and 0.06 indicates a small effect size, and a partial eta-squared between 0.07 and 0.13 indicates a medium effect size, whereas a value equal or higher than 0.14 indicates a large effect [28]. Post-hoc univariate ANOVAs were performed for each dependent variable when a significant result was observed in the MANOVA. Significant effects of ANOVA were tested by post-hoc Scheffé test for effects in the subgroups. Effect sizes Hedges'g were calculated and were defined as trivial effects between 0.01 and 0.19, as small between 0.2 and 0.49, as moderate between 0.5 and 0.79, and as large above 0.8.

3. Results

The Shapiro–Wilk test showed normally distributed data for variables in all subgroups and the total group. The results of the reliability analyses are displayed in Table 1. All ICCs are clearly above the limit value of 0.7 and under the limit of CV (< 10%) and can therefore be classified as suitable reliability [27].

Table 1. Mean coefficients of variation, intraclass correlation coefficients, ninety-five percent confidence intervals of the performance tests.

Tests	ICC (95% CI)	CV (95% CI)
Linear sprint 10 m	0.97 (0.95–0.98)	1.4% (1.1%–1.6%)
Linear sprint 20 m	0.98 (0.97–0.99)	1.2% (0.1%–1.4%)
Squat jump	0.97 (0.95–0.98)	4.4% (3.6%–5.2%)
IAT	0.98 (0.97–0.99)	1.5% (1.0%–2.0%)
505	0.96 (0.94–0.98)	2.9% (2.3%–3.5%)

CV = coefficients of variation; ICC = intraclass correlation coefficients; 95% CI = ninety-five percent confidence intervals.

Levene's test calculated homogeneity of the error variances ($p > 0.05$), except for the variables body mass and BMI ($p = 0.031$, 0.005). There was heterogeneity of covariances, as assessed by Box's test ($p < 0.001$). MANOVA showed no statistically significant difference between age groups and playing level for the combined anthropometric variables ($F_{(3,39)} = 0.122$, 1.786, $p = 0.166$, 0.953, partial $\eta^2 = 0.009$, 0.121, Roy's largest root $\varphi = 0.009$–0.137). MANOVA showed no statistically significant difference between age groups ($F_{(5,37)} = 2.233$, $p = 0.071$, partial $\eta^2 = 0.232$, Roy's largest root $\varphi = 0.302$), but between playing level ($F_{(5,37)} = 81.782$, $p < 0.001$, partial $\eta^2 = 0.917$, Roy's largest root $\varphi = 11.052$) on the combined dependent performance variables. Post-hoc univariate ANOVAs showed for every depending performance variable statistically significant results between level of play, $F_{(1,41)} = 10.225$–365.256, $p < 0.003$, partial $\eta^2 = 0.200$–0.899). Tables 2–4 provide an overview of the mean values ± SD and effect sizes for subgroups of both anthropometric data (height, body mass, and BMI) and performance tests (SJ, LS and COD) in age groups and the total group. Elite player performance was significantly better in SJ, LS, and COD than amateur players. The results show that the more complex the target exercise, the larger the effect sizes for group differences (SJ: $g = 0.64$–1.18, LS: $g = 0.05$–2.23, COD: $g = 3.01$–6.84).

Table 2. Mean, standard deviation, absolute and relative difference, t-value, *p*-value, and effect size within the age group of under 19 years old.

Tests	Elite (n = 12) Mean ± SD	Amateur (n = 11) Mean ± SD	abs. Mean Δ (%)	Effect
Height (cm)	179.2 ± 6.7	176.6 ± 10.9	2.6 (1.5%)	0.27
Body mass (kg)	74.0 ± 7.7	67.0 ± 14.6	7.0 (9.5%)	0.59
BMI (%)	23.0 ± 1.3	21.4 ± 4.3	1.6 (7.0%)	0.48
Linear sprint 10 m (s)	1.67 ± 0.06	1.82 ± 0.08	0.15 (9.0%) *	2.03
Linear sprint 20 m (s)	2.89 ± 0.08	3.19 ± 0.17	0.30 (10.4%) *	2.24
Squat jump (cm)	40.45 ± 2.91	34.61 ± 6.21	5.84 (14.4%) *	1.18
IAT (s)	13.66 ± 0.36	17.19 ± 0.61	3.53 (25.8%) *	6.84
505 (s)	2.15 ± 0.06	2.64 ± 0.16	0.49 (22.8%) *	4.18

505 = 505 agility test, IAT = Illinois agility test, cm = centimeter, s = seconds; * = significant ($p < 0.05$), effect size = Hedges' g.

Table 3. Mean, standard deviation, absolute and relative difference, t-value, *p*-value, and effect size within the age group of under 17 years old.

Tests	Elite (n = 14) Mean ± SD	Amateur (n = 8) Mean ± SD	abs. Mean Δ (%)	Effect
Height (cm)	175.8 ± 5.0	182.0 ± 7.4	6.20 (3.5%)	1.00
Body mass (kg)	70.8 ± 7.2	70.7 ± 8.4	0.10 (0.1%)	0.02
BMI (%)	22.9 ± 1.8	21.3 ± 1.5	1.60 (7.0%)	0.93
Linear sprint 10 m (s)	1.80 ± 0.05	1.80 ± 0.12	0.00 (0.0%)	0.05
Linear sprint 20 m (s)	3.06 ± 0.09	3.22 ± 0.14	0.16 (5.2%) *	1.40
Squat jump (cm)	34.87 ± 2.64	32.51 ± 4.74	2.36 (6.8%)	0.64
IAT (s)	14.52 ± 0.22	16.23 ± 0.60	1.71 (−11.8%) *	4.12
505 (s)	2.25 ± 0.09	2.53 ± 0.08	0.28 (−12.4%) *	3.01

505 = 505 agility test, IAT = Illinois agility test, cm = centimeter, s = seconds; * = significant ($p < 0.05$), effect = Hedges' g.

Table 4. Mean, standard deviation, absolute and relative difference, t-value, *p*-value, and effect size within the total group.

Tests	Elite (n = 26) Mean ± SD	Amateur (n = 19) Mean ± SD	abs. Mean Δ (%)	Effect
Height (cm)	177.3 ± 6.0	178.9 ± 9.7	1.6 (0.9%)	0.20
Body mass (kg)	72.3 ± 7.4	68.5 ± 12.2	3.8 (5.3%)	0.38
BMI (%)	23.0 ± 1.5	21.4 ± 3.4	1.6 (7.0%)	0.63
Linear sprint 10 m (s)	1.74 ± 0.08	1.81 ± 0.10	0.07 (4.0%) *	0.80
Linear sprint 20 m (s)	3.98 ± 0.12	3.20 ± 0.16	0.22 (7.4%) *	1.62
Squat jump (cm)	37.45 ± 3.92	33.73 ± 5.60	3.72 (9.9%) *	0.78
IAT (s)	14.13 ± 0.52	16.79 ± 0.77	2.77 (18.8%) *	4.11
505 (s)	2.20 ± 0.09	2.60 ± 0.14	0.40 (18.2%) *	3.40

505 = 505 agility test, IAT = Illinois agility test, cm = centimeter, s = seconds; * = significant ($p < 0.05$), effect = Hedges' g.

4. Discussion

Elite player performance was significantly better in SJ, LS, and COD than amateur players. Interestingly this investigation showed that the more complex the target exercise, the larger the effect sizes for group differences (SJ: g = 0.64–1.18, LS: g = 0.05–2.23, COD: g = 3.01–6.84).

Performance differences in favor of elite populations are basically consistent with the literature [16–20]. Playing soccer (and, therefore, sprinting, changing direction, and jumping) may be an effective training stimulus for improving SJ, LS, and COD [29]. The higher performance in the elite subjects might be attributed to the higher volume of playing soccer. The analysis of data in this study cannot clarify why the level of significance was not reached for LS 10 m and SJ (g = 0.64) in subgroup U17. It is possible that from a statistical perspective, the non-significant differences in the SJ of both subgroups (elite vs. amateur) in U17 can be explained by the heterogeneous performance of the subgroup amateur players. Although, the variation (CV = 4.4%) of SJ, which could be explained by the generally unusual concentric-only movement of soccer players, may have influenced the calculation. It should be noted that the two groups in the U17 age group differ with a moderate effect in SJ (g = 0.64). The non-significant results in the LS 10 m variable are contrary to the findings of performance differences depending on the playing levels of LS 10 m in U19, as well as LS 20 m in both age groups and contrary to other findings in the literature [16,18,19]. Only Trajković et al. [18] also found non-significant differences between elite and amateur soccer players in LS 10 and argued that this might be related to the maturity stage of players, which can affect sprint performance. This effect may also mask group differences in LS 10 m and SJ between the level of play in U17 in this study, as effect sizes in body height were calculated to be large (g = 1.0) in favor of the amateur players. However, a clear justification of the non-significant differences in LS 10 m and SJ between the level of play in subgroup U17 in this study cannot be found in the anthropometric data, as this study did not calculate significant differences between the subgroups (amateur vs. elite) in accordance with the literature [17].

The effect sizes of performance differences in COD tests of this study (g = 3.01–6.84) far exceeded the effect sizes found by Trecroci et al. [17] and Trajković et al. [18] (g = 0.69–1.5). It is fundamentally difficult to compare the studies, and thus also the results, with each other because different studies may have used different definitions of the subject categorization into elite, sub-elite, and amateur [17]. Here, the demand for a uniform definition of the performance level for team sports and its consistent use in the literature must be established [22]. Nevertheless, even in comparison with studies that used an identical definition of playing status and comparable or identical COD tests in terms of the requirement profile, significantly higher performance differences between elite and amateur players were confirmed in this study (g = 0.95 vs. g = 3.01–6.84) [18]. Trajković et al. [18] recruited elite, sub-elite players from national championships without naming the country of the championship. It is possible that a difference in performance between the national leagues explains the difference in effect sizes. It is also possible that the difference in the magnitude of the effects is due to the fact that the subjects measured in this study were older compared with other studies (16.56 $\pm$ 0.9 vs. 15.7 $\pm$ 0.6 years old) [18]. Here, selection factors could be used again as an explanation for larger training volumes and intensity of older elite players. Nevertheless, on the basis of the available data, it is unclear why the effect sizes differ between the studies mentioned and thus points to a further need for scientific research on this topic. COD test performance represents the performance parameters that showed the highest effect sizes in this study (SJ: g = 0.64–1.18, LS: g = 0.05–2.23, COD: g = 3.01–6.84). Cometti et al. [13] also found higher effect sizes for performance differences depending on the level of play in linear sprint compared to jump performance; however, Trajković et al. [18] did not. It may be possible that the more complex target performances (COD, LS) benefit more from a higher training volume and intensity of the elite players than simpler exercises (e.g., SJ).

A limitation of this study is the difference in sample size between the groups (elite vs. amateur), which is due to the limited scheduling availability of the amateur athletes. Unbalanced sample sizes may increase the error in difference calculations, but this is mitigated by appropriate corrections to the calculations (e.g., Roy's largest root or Hedges' g instead of Cohen's d). An additional limitation is the ad hoc sample, which can generally increase the sampling error. However, the data set is valuable because only a few studies have investigated and compared young male soccer players [16–21].

5. Conclusions

The data from this study show performance differences in performance tests depending on the level of play. The SJ, LS, 505, and IAT may prove useful in talent selection test batteries to separate between competitive levels in youth soccer players. The more complex the performance tests were, the larger the effect sizes of the performance differences were. Therefore, the LS and COD tests should be integrated into test batteries. On the contrary, anthropomorphic factors do not significantly differentiate between the level of play and should therefore not be used for selection.

Author Contributions: Conceptualization, M.K.; data collection, A.K., M.K., B.K., and T.S.; writing—original draft preparation, M.K. and K.W.; writing—review and editing, B.K., K.W., and T.S. All authors have read and agreed to the published version of the manuscript.

Funding: This research received no external funding.

Institutional Review Board Statement: The study was conducted according to the guidelines of the Declaration of Helsinki and approved by the Ethics Committee of the German University of Health & Sport (DHGS-EK-2021-002).

Informed Consent Statement: Informed consent was obtained from all subjects, or parents when younger than 18 years old, involved in the study.

Data Availability Statement: The data is not yet publicly available.

Acknowledgments: The authors would like to thank the athletes for participating in the study.

Conflicts of Interest: The authors declare no conflict of interest.

References

1. Bloomfield, J.; Polman, R.; O'Donoghue, P. Physical Demands of Different Positions in FA Premier League Soccer. *J. Sports Sci. Med.* **2007**, *6*, 63–70.
2. Barnes, C.; Archer, D.T.; Hogg, B.; Bush, M.; Bradley, P.S. The Evolution of Physical and Technical Performance Parameters in the English Premier League. *Endoscopy* **2014**, *35*, 1095–1100. [CrossRef]
3. Wallace, J.L.; Norton, K.I. Evolution of World Cup soccer final games 1966–2010: Game structure, speed and play patterns. *J. Sci. Med. Sport* **2014**, *17*, 223–228. [CrossRef]
4. Reilly, T.; Bangsbo, J.; Franks, A.L. Anthropometric and physiological predispositions for elite soccer. *J. Sports Sci.* **2000**, *18*, 669–683. [CrossRef] [PubMed]
5. Jones, P.; Bampouras, T.M.; Marrin, K. An investigation into the physical determinants of change of direction speed. *J. Sports Med. Phys. Fit.* **2009**, *49*, 97–104.
6. Sheppard, J.M.; Young, W.B. Agility literature review: Classifications, training and testing. *J. Sports Sci.* **2006**, *24*, 919–932. [CrossRef] [PubMed]
7. Haugen, T.A.; Tønnessen, E.; Hisdal, J.; Seiler, S. The role and development of sprinting speed in soccer. *Int. J. Sports Physiol. Perform.* **2013**, *9*, 432–441. [CrossRef] [PubMed]
8. Sheppard, J.; Dawes, J.J.; Jeffreys, I.; Spiteri, T.; Nimphius, S. Broadening the view of agility: A scientific review of the literature. *J. Aust. Strength Cond.* **2014**, *22*, 6–30.
9. Stølen, T.; Chimari, K.; Castagna, C.; Wisløff, U. Physiology of soccer—An update. *Sports Med.* **2005**, *35*, 501–536. [CrossRef]
10. Trecroci, A.; Milanović, Z.; Rossi, A.; Broggi, M.; Formenti, D.; Alberti, G. Agility profile in sub-elite under-11 soccer players: Is SAQ training adequate to improve sprint, change of direction speed and reactive agility performance? *Res. Sports Med.* **2016**, *24*, 331–340. [CrossRef]
11. Bangsbo, J.; Mohr, M.; Krustrup, P. Physical and metabolic demands of training and match-play in the elite football player. *J. Sports Sci.* **2006**, *24*, 665–674. [CrossRef]
12. Young, W.B.; Dawson, B.; Henry, G.J. Agility and Change-of-Direction Speed are Independent Skills: Implications for Training for Agility in Invasion Sports. *Int. J. Sports Sci. Coach.* **2015**, *10*, 159–169. [CrossRef]
13. Cometti, G.; Maffiuletti, N.A.; Pousson, M.; Chatard, J.-C.; Maffulli, N. Isokinetic Strength and Anaerobic Power of Elite, Subelite and Amateur French Soccer Players. *Endoscopy* **2001**, *22*, 45–51. [CrossRef] [PubMed]
14. Lehance, C.; Binet, J.; Bury, T.; Croisier, J.L. Muscular strength, functional performances and injury risk in professional and junior elite soccer players. *Scand. J. Med. Sci. Sports* **2008**, *19*, 243–251. [CrossRef]
15. Keiner, M.; Sander, A.; Wirth, K.; Hartmann, H. Differences in the performance tests of the fast and slow stretch and shortening cycle among professional, amateur and elite youth soccer players. *J. Hum. Sport Exerc.* **2015**, *10*. [CrossRef]
16. Gissis, I.; Papadopoulos, C.; Kalapotharakos, V.I.; Sotiropoulos, A.; Komsis, G.; Manolopoulos, E. Strength and Speed Characteristics of Elite, Subelite, and Recreational Young Soccer Players. *Res. Sports Med.* **2006**, *14*, 205–214. [CrossRef] [PubMed]
17. Trecroci, A.; Longo, S.; Perri, E.; Iaia, F.M.; Alberti, G. Field based physical performance of elite and sub-elite middle-adolescent soccer players. *Res. Sports Med.* **2018**, *27*, 60–71. [CrossRef]

18. Trajković, N.; Sporiš, G.; Krističević, T.; Madić, D.M.; Bogataj, Š. The Importance of Reactive Agility Tests in Differentiating Adolescent Soccer Players. *Int. J. Environ. Res. Public Health* **2020**, *17*, 3839. [CrossRef]
19. Stewart, A.; Bloom, L.; Clarkson, B.; Comfort, P. A Comparison of Field Test Performances be-tween Elite and Sub-Elite Academy Soccer Players. *J. Atheletic Enhauncement* **2014**, *3*, 4.
20. Vaeyens, R.; Malina, R.M.; Janssens, M.; Van Renterghem, B.; Bourgois, J.; Vrijens, J.; Philippaerts, R.M. A multidisciplinary selection model for youth soccer: The Ghent Youth Soccer Project Commentary. *Br. J. Sports Med.* **2006**, *40*, 928–934. [CrossRef] [PubMed]
21. le Gall, F.; Carling, C.; Williams, M.; Reilly, T. Anthropometric and fitness characteristics of international, professional and amateur male graduate soccer players from an elite youth academy. *J. Sci. Med. Sport* **2010**, *13*, 90–95. [CrossRef] [PubMed]
22. Lorenz, D.S.; Reiman, M.P.; Lehecka, B.J.; Naylor, A. What Performance Characteristics Determine Elite Versus Nonelite Athletes in the Same Sport? *Sports Health* **2013**, *5*, 542–547. [CrossRef]
23. Kadlubowski, B.; Keiner, M.; Stefer, T.; Kapsecker, A.; Hartmann, H.; Wirth, K. Influence of linear-sprint performance, concentric power and maximum strength on change of direction performance in elite youth soccer players. *Ger. J. Exerc. Sport Res.* **2020**, *51*, 116–121. [CrossRef]
24. Nimphius, S.; Callaghan, S.J.; Bezodis, N.E.; Lockie, R.G. Change of Direction and Agility Tests: Challenging Our Current Measures of Performance. *Strength Cond. J.* **2018**, *40*, 26–38. [CrossRef]
25. Guarav, V.; Singh, S.; Singh, M.; Rathi, B. A comparative study of arm and shoulder gridle strength and agility of college-lavel baseball pitchers and non-pothers. *J. Phys. Educ. Sports Manag.* **2011**, *2*, 17–20.
26. Kadlubowski, B.; Keiner, M.; Hartmann, H.; Wirth, K.; Frick, U. The Relationship between Change of Direction Tests in Elite Youth Soccer Players. *Sports* **2019**, *7*, 111. [CrossRef] [PubMed]
27. Shrout, P.E.; Fleiss, J.L. Intraclass correlations: Uses in assessing rater reliability. *Psychol. Bull.* **1979**, *86*, 420–428. [CrossRef]
28. Cohen, J. *Statistical Power Analysis for the Behavioral Sciences*, 2nd ed.; Erlbaum Associates: Hillsdale, MI, USA, 1988; p. 567.
29. Ramirez-Campillo, R.; Castillo, D.; Raya-González, J.; Moran, J.; de Villarreal, E.S.; Lloyd, R.S. Effects of Plyometric Jump Training on Jump and Sprint Performance in Young Male Soccer Players: A Systematic Review and Meta-analysis. *Sports Med.* **2020**, *50*, 2125–2143. [CrossRef]

sports

Article

Elite Youth Soccer Players' Sources and Types of Soccer Confidence

Iain Greenlees [1,*], Aimee Parr [1], Sarah Murray [1] and Esther Burkitt [2]

1 Institute of Sport, Nursing and Allied Health, University of Chichester, Chichester PO19 6PE, UK; aparr1@chialumni.ac.uk (A.P.); s.murray@chi.ac.uk (S.M.)
2 Department of Psychology and Counselling, University of Chichester, Chichester PO19 6PE, UK; e.burkitt@chi.ac.uk
* Correspondence: i.greenlees@chi.ac.uk

Abstract: Sport confidence is a psychological characteristic considered vital for youth soccer players to possess. However, only limited research has explored the types and sources of sport confidence important to elite youth performers in professional soccer academies. Semi-structured interviews were conducted with 11 academy footballers (aged 10 or 11). Abductive hierarchical content analysis identified types of confidence to include achievement, skill execution, psychological factors, superiority to opposition and tactical awareness. Key sources of confidence identified by players were performance accomplishments, coaching, social support, and preparation. Even though the dimensions reported were similar to previous research, a number of unique sub-themes of confidence sources emerged, including pre-training/competition emotions, coach and team-mate feedback. The results demonstrate the importance of considering maturation levels and context when seeking to understand and develop confidence in youth performers.

Keywords: self-confidence sources; self-efficacy; elite youth athletes; soccer

Citation: Greenlees, I.; Parr, A.; Murray, S.; Burkitt, E. Elite Youth Soccer Players' Sources and Types of Soccer Confidence. *Sports* **2021**, *9*, 146. https://doi.org/10.3390/sports9110146

Academic Editor: Pedro E. Alcaraz

Received: 20 August 2021
Accepted: 20 October 2021
Published: 22 October 2021

1. Introduction

Numerous factors (physiological, technical, tactical, sociological and psychological) have been proposed to be important predictors of talent development and long-term playing success in soccer [1]. One psychological factor that appears pivotal is self-confidence [1]. Within high level youth soccer, confidence has been implicated as a central requirement not only for development and the transition through youth to elite performance [2,3] but also for the maintenance of positive psychological well-being [4]. Despite this, we still know relatively little about the determinants of confidence in high-level youth sport, yet alone in soccer.

In sport, understanding of confidence has been largely informed by the work of Bandura's [5] concept of self-efficacy and by the work of Vealey and colleagues [6–8]. These approaches conceptualise confidence as a dynamic psychological state influenced by a range of personal, demographic and environmental factors. In their general model of sport confidence, Vealey and Chase [8] propose that sport-confidence (one's belief that one can be successful in a chosen sport) is a multidimensional construct made up of confidence in executing skills (SC—physical skills and training), decision-making and concentration (SC—cognitive efficiency) and dealing with setbacks (SC—resiliency). Vealey and Chase [8] further propose nine sources of confidence nested within 3 domains (Achievement (including mastery or improvement of personal skills and demonstration of ability by showing others or demonstrating superior ability to opponents), Self-regulation (including physical and mental preparation and physical self-presentation) and Climate (including social support, vicarious experiences, coach's leadership, environmental comfort and situational favourableness). Importantly, Vealey and Chase [8] propose that, rather than being universally important, the importance of each of the sources of confidence will differ across individuals and contexts.

The principles of the model have received broad support in the literature. Hays, Maynard, Thomas and Bawden [9], in a study of world-class, adult performers, provided support for the multi-dimensionality of confidence types and sources but differed from Vealey and Chase's [8] model in two ways. First, performers reported six, rather than three, types of confidence (skill execution, achievement, physical factors, psychological factors, superiority to opposition and tactical awareness). Second, whilst preparation, performance accomplishments, coaching, and social support all emerged as sources of confidence from the data, little evidence was found to support the role of vicarious experiences, physical self-presentation, environmental comfort or situational favourableness. In addition, Hays et al. [9] found that trust, competitive advantage and self-awareness emerged as sources of confidence information.

Even though this model and research provide some insight into sources of confidence in sport, extrapolating these findings to elite youth sport is problematic for a number of reasons. First, Vealey and Chase [8] stress that age influences the sources of confidence that are most relevant to an individual. Vealey et al. [7] argued that schemas concerning sources of confidence are developed during childhood and adolescence and so this age period may witness fundamental changes in sources of confidence. Research in youth sport settings (for reviews see [4,10]) has shown that, compared to adult and older adolescents, children utilise fewer sources of competence information, show less ability to integrate and differentiate between different sources of information, and rely more on concrete and external information sources (in the form of actual performance outcomes and coach/parent/peer feedback). Second, Vealey and Chase [8] also propose that organisational and cultural characteristics (such as expectation levels, the quantity and quality of feedback provided and the nature of the social relations) will also influence the types and sources of confidence that are most relevant to performers. Thus, whilst findings from adult performers [9] and from children engaged in youth sport [10] may not truly reflect the types and sources of confidence important to performers in the unique and rarefied environment of elite youth soccer.

In acknowledgement of the potential lack of generalisability from elite, adult performers to youth athletes, Thomas, Thrower, Lane and Thomas [11] explored types, sources and debilitators of confidence in elite (signed to the academy system of professional football clubs), adolescent soccer players. The ages of these players ranged from 12–15, with a mean of 14.28 years. Even though their results were broadly consistent with Vealey and Chase's [8] and previous research [9] in terms of the types and sources of confidence reported, they also showed that there were differences in the number of sources of confidence reported and the importance of different sources of confidence. Specifically, whilst Thomas et al. [11] did find that performance accomplishments, preparation and social support were important sources of confidence, their results did not match the work of Hays et al. [9] or Vealey and Chase's [8] model in demonstrating the importance of sources such as environmental comfort, situational favourableness, self-awareness or experience. This research demonstrates the need for more research to explore the distinct profiles of confidence types and sources across a range of populations and settings.

In light of this, the aim of the current research was to explore the types and sources of confidence of elite youth soccer players at the younger age range (10–11 year olds) than those used in Thomas et al.'s [11] research (early adolescents aged between 12 and 15). This age was chosen for three reasons. First, our sample represents an earlier stage of development than players included in Thomas et al.'s research. Whilst Thomas et al. sampled from the early adolescent stage, our players were in middle to late childhood according to established criteria [4]. This, theoretically, will entail different sources of confidence being used due to different self-perceptions and complexity of self-perceptions at this age [4]. Specifically, it could be predicted that our sample would rely on fewer sources of confidence, would rely on tangible achievements to gain confidence (e.g., scoring goals, being made captain) and would have a greater tendency to rely on others to interpret and evaluate their competence [12]. Second, research has highlighted the importance of confidence at this age for the development of youth soccer players [3]. Thus, it seems a particularly important age range to explore sources of confidence within. Third, the origins of this project were rooted in informal discussions with coaches at the soccer club involved which identified confidence in this age group as being particularly important to understand.

2. Materials and Methods

2.1. Philosophical Underpinning

A critical realist [13] philosophical perspective informed the current project. This perspective proposes that there is a reality that exists independently of subjective perceptions (ontological realism) but that access to this depends on fallible perceptions and subjective interpretations that are limited, incomplete and theory-dependent (epistemological relativism).

2.2. Context

Within English soccer, academies are run by professional clubs with the aim of producing professional players from the age of nine. Academies are classified from Category One to Four, with one being the highest level. Categorisation is determined by factors such as training facilities, productivity of the academy, welfare and development provisions. Places are highly prized and competitive but difficult to retain with regular rounds of re-selection [14] based on individual progression. The academy that this research was situated in considered itself a progressive academy with player development and welfare firmly at its heart. However, competition for places was high and, as a result, players (and their parents) were acutely aware of the need to impress coaches to retain their place in the academy.

2.3. Participants

In this case, 11 male academy players (aged between 10 and 12) from a Category One academy of an English Premiership club participated in the study. Participants were either in the under 11's (9 players) or 12's squad (two players) and all player positions (goalkeeper, defender, midfielder and forward) were represented. The average age of the participants was 11.57 years. Time spent at the academy ranged from one to four years. All participants self-reported as White-British.

2.4. Procedure

Ethical approval was provided by the Research Ethics Committee of the authors' university. Following approval for the study being provided by the football club and relevant coaches, informed written consent was provided by parents. Then, all players in the targeted squads were provided with written information and a verbal summary of the project from the academy's lead sport psychologist and asked if they would want to volunteer. We used a stratified purposeful sampling strategy [15] to select participants from those who volunteered. Participants were selected to represent a range of confidence levels, playing positions and years of experience. Written consent was provided by all participants prior to the data collection commencing. All interviews were conducted on a one-to-one basis in a quiet area of the football club's training facilities and lasted between 31 and 51 min (M = 42 min).

Interview Guide

A semi-structured interview guide, based on information provided by Hays et al. [9], was developed. This guide was divided into five sections. First, introductory comments reminded participants of the study's purpose and reinforced confidentiality and anonymity. Second, participants were asked general questions about self-confidence (e.g., "what does the term self-confidence mean to you?"). Similar to Hays et al., we did not provide a standardised definition of self-confidence as the aim was to develop greater knowledge of the participants' perspectives. Next, participants were asked questions related to their perceptions of the types and sources of confidence within youth football. The fourth section of the interview asked about times of high and low confidence and situations in which they had experienced changes in confidence. The interview concluded with an invitation for the participants to provide further information concerning their confidence. Throughout the interview, clarification and elaboration probes were used to ensure that participants provided a rich and detailed response to the questions asked [16].

2.5. Data Analysis

In line with similar approaches used in research examining youth sport experiences (e.g., [2,11]), qualitative data were analysed using an abductive hierarchical content analysis. Here, the transcripts were examined and any meaningful units of information (such as key words or phrases) relating to the types and sources of confidence were identified. The first and second author independently listened to the recordings and read and re-read the transcripts to ensure familiarity with each script before any analysis took place. The second author then completed initial coding which was developed in discussion with the first author, who acted as a critical friend. After the inductive stage, we used a more deductive phase whereby the existing literature (e.g., [7,9,11]) was compared with our findings. The deductive phase was also used to prompt the use of terminology to increase consistency with previous research findings. Themes which did not fit any of the underlying dimensions were grouped together, sorted into common themes and subsequently labelled according to the meaning of the data. This approach permits the emergence of new raw-data themes (units of information) whilst ensuring that the theoretical frameworks underpinning the research retains a central position in the enquiry [2].

2.6. Enhancing Ontological Plausibility and Validity

In line with our critical realist perspective, and the writings of Maxwell [17], we accept the inherent fallibility of the research methods and analytical procedures that we have used but also took a number of steps to address threats to descriptive (the factual accuracy of the research account), interpretive (the extent to which our interpretations accurately reflect the participants' experience) and theoretical (the extent to which our findings explain the experiences of participants) validity. In doing this, we were guided by the suggestions of Ronkainen and Wiltshire [18]. Specifically, to enhance descriptive validity the interviewer (second author) made notes following each interview and transcripts were checked by two authors. To enhance interpretive validity, we considered competing explanations of the evidence through the lead and second author independently coding the data before comparing interpretations, the use of critical friends, discussions with the participants' coaches and sport psychologists to consider emerging themes and to offer reflections on our findings and, ultimately, via peer-review. The discussions with coaches and sport psychologists involved the presentation of research findings to coaches in a workshop that sought to discuss the findings and to consider ways in which the information could be used to inform practice. This confirmed our interpretations and also provided further support for the powerful role of moods in determining self-confidence. In addition, the second author kept a reflexive diary [19]. Diary entries were made after each interview allowing a focus on internal responses to being a researcher and emotional feelings in the process of data collection to be noted [20]. Finally, we argue that theoretical validity was enhanced as the research project was developed with the coaches in light of a perceived need to enhance their knowledge of self-confidence in academy soccer players and through the practical suggestions that were developed in light of our research findings.

3. Results

We present the results in two parts. First, we cover types of confidence and, second, we present sources of confidence. The data are presented in Tables 1 and 2 and through the use of representative quotations (with pseudonyms used throughout). A fuller outline of the content analysis can also be found in Supplementary Materials Tables S1 and S2. Even though we recognise the limitations of response frequencies, we present them to facilitate comparison with previous findings [9,11].

3.1. Types of Confidence

In this case, 22 raw-data themes were categorized into seven global dimensions. These related to skill execution, achievement, psychological factors, superiority to opposition, tactical awareness and athlete specific factors. These are summarised in Table 1 below.

Table 1. Numbers of participants referring to specific confidence types.

Type of Confidence	Number of Athletes Citing Type. *N* = 11	Total % of Athletes	Total % Citing in Hays et al. (2007) *N* = 14	Total % Citing in Thomas et al. (2021) *N* = 28
Skill Execution	9	82%	71%	100%
Achievement	6	64%	71%	–
Outcome	–	–	64%	–
Performance	–	–	29%	–
Physical Factors	–	–	64%	61%
Psychological Factors	3	27%	57%	46%
Superiority to Opposition	3	27%	50%	–
Tactical Awareness	2	18%	14%	–
Athlete Specific Factors	2	18%	50%	–

3.1.1. Skill Execution

Nine of the 11 athletes found skill execution to be a type of confidence important to them, referring to their belief in their own ability to execute skills in soccer. This often included specific skills during a match. As Bailey put it "I think I'm quite good at passing. I'm not, I'm not the best at shooting, so I just, I don't shoot, I just assist people".

3.1.2. Achievement

Six athletes identified achievement. This concerned belief that certain outcomes or performance-based skills could be achieved. Athletes talked about the need to be confident in their own ability: "You have to be confident in yourself and your ability to play football and like confident that you can actually achieve things." (Will). Only Felix mentioned achievement in outcomes, highlighting his need to be confident in "scoring a good goal and winning".

3.1.3. Psychological Factors

Three athletes spoke of their need to be confident of psychological factors, including having a positive mind set for soccer. "You have to have a good mentality, like if you have a bad mentality then your confidence won't be like, it will just drop, it will keep dropping." (Freddie).

3.1.4. Superiority to Opposition

Three athletes identified superiority to an opponent as a type of confidence, with one player needing to feel confident in "beating players with skills and dribbles." (Will).

3.1.5. Tactical Awareness

Tactical awareness was identified as a type of confidence by only two athletes. This referred to being confident about their own tactical ability, such as ball placement or decision making.

3.1.6. Athlete Specific Factors

Two athletes identified unique types of confidence which were individual to them. This included having confidence in their communication with team-mates and having confidence in team-mates and their soccer ability.

3.2. Sources of Confidence

Inductive analysis of confidence sources yielded 117 raw-data themes. We conceptualised these into 24 lower order and sub-themes and 15 higher order themes. These are summarised in Table 2 below.

Table 2. Numbers of participants referring to specific confidence sources.

Source of Confidence	Number of Athletes Citing Source (*N* = 11)	Total % of Athletes	Total % Citing in Hays et al. (2007)	Total % Citing in Thomas et al. (2021)
Preparation	9	82%	100%	82%
Physical preparation	7	64%	100%	75%
Mental preparation	1	9%	76%	50%
Holistic preparation	7	64%	57%	14%
Performance Accomplishments	11	100%	100%	100%
Competition accomplishments	11	100%	100%	96%
Training accomplishments	9	82%	50%	96%
Other accomplishments	4	36%	–	64%
Coaching	11	100%	93%	–
Individual interactions with coach	11	100%	–	–
Interactions with coach in group	3	27%	–	–
Social Support	10	91%	57%	100%
Family	7	64%	–	–
Team-mates	9	82%	–	–
School/Friends	3	27%	–	–
Other	2	18%	–	–
Innate Factors	4	36%	50%	29%
Experience	4	36%	43%	–
Athlete Specific Factors	4	36%	50%	–
Competitive Advantage	–	–	36%	–
Trust	–	–	14%	–
Self-Awareness	–	–	14%	–

3.2.1. Performance Accomplishments

This consisted of three higher order themes, competition accomplishments, training accomplishments and other accomplishments. All 11 players highlighted performing successfully and mastering specific skills, both in matches and training. Competition accomplishments included in-game performance elements, being selected, gaining awards and general wins and losses. For example, Dan reported increased confidence when he "scored a great goal from like the half-way line then that gave me confidence and I done this good run, beat 2 players and scored again and I set up another goal" (competition performance accomplishment) whilst Luke stated:

"I got an assist about 30 s as soon as I came on cos someone got injured and then I kept doing good passes and keeping the ball and then it made me feel happy so then I could like play better and like forget anything bad that happened in that game."

A range of training accomplishment sub-themes were apparent. These included skill development and mastery, general performance levels, displaying superiority to others and being given the captaincy. The importance of training performance accomplishments was encapsulated by Bradley when he commented:

"I did well in training so like I kept trying it and then whenever I try I get better at it and then you get more confident doing it. If I keep practicing it and get really good at it then I can like do it in matches and then it will like work all the time and then I will just do it all the time."

Players also identified how achieving success in school level football and in other sports could enhance their feelings of soccer-specific confidence.

3.2.2. Coaching

All players identified the coach as a confidence source. Recognition, praise/criticism, attention/feedback and athlete-handling were identified as important. Praise and criticism appeared to be a big contributor, with Bradley stating, "when they encourage you to do stuff and like praise you it boosts it so much cos you know you've achieved something" and a second stating:

"If like you get more praise, if you keep getting praise in training and you are doing everything right they'll give you more praise and it will make you play better and then your confidence will rise but if you're playing bad and they're like moaning at you it can affect some players."

Recognition from coaches played a role in increasing confidence levels. Bailey spoke of recognition within the academy when he said "Getting a two-year contract I think, cos then you know that you've done it, like then you know that you're good" whilst others spoke of recognition in terms of selection to their preferred position or being selected to be the captain. As Freddie put it:

"We were playing at the xxxxx [home stadium] and whoever got trainer of the night that Friday would be captain and I go through the training session and get trainer of the night and I'm, I walk out onto the xxxxxx with the captain's armband on and I just felt really good. Wearing the captain's arm band on the actual xxxxxxxx [home stadium] makes you feel really good!"

Whilst performance accomplishments were important, a number of the performers relied on their coaches to inform them when a performance was good and to recognise, beyond obvious markers of good play such as goals scored, when they had performed well. As Felix commented "Yeah, cos if I think that I'm not playing well at all then someone (coach) comes over to me and say what I have done well and then I feel more confident".

3.2.3. Social Support

Ten players discussed social support. This consisted of the higher order themes of family, team-mates, school/friends and external others (other coaches and spectators) support. Within the higher order family support theme, lower-order themes of atmosphere, feedback and support were seen. Having family support was seen as important, as was gaining positive feedback from family members with Max stating "if I've played well then my dad gives me more confidence and he says 'oh you've done well.'". The participants reported both positive, negative and neutral impacts of parental feedback on their confidence, with Ashley explaining "If I've played well then like my dad gives me more confidence and he says oh you've done well and if my mums says I played well and I have played well then it sort of stays the same". For some, the negative impact of parental feedback and criticism appeared to be exceptionally strong. For example, Max said "Like sometimes my dad will moan at me [pause], a lot. He won't stop until I'm like almost in tears and that will affect my confidence" whilst Jordan commented:

"They [referring to parents] say it in kind of a bad way and they should just say "that wasn't very good today, try and do better tomorrow" but they say "that was terrible, if you're gonna carry on doing that you're gonna get kicked, there's better boys out there who want to be there and it looks like when you're on the pitch you don't even wanna be there."

Further social support was found from teammates and team support. The dangers of a negative, critical team atmosphere was highlighted frequently, with Bailey stating:

"If you didn't have team work then you would just be like running with the ball, no one's passing, everyone's moaning at each other and everyone's confidence will be down."

Furthermore, Sam commented:

"Sometimes my teammates don't give me as much confidence because sometimes they put me down. Especially coming to the end of the season it's really hard because they, they're fighting for the same position that you are so they are trying to knock you down."

Finally, participants also mentioned the important role of people outside of the immediate family and academy in developing and maintaining their confidence levels. References were made to school friends, teachers and opposition and neutral spectators in developing and reducing confidence.

3.2.4. Preparation

This consisted of two higher-order themes, physical/mental preparation and holistic preparation. Whilst physical/mental preparation encompassed a range of sources (e.g., having warmed up well), we were struck by the seeming importance attached to what we referred to as holistic preparation. Specifically, the players commonly reported that general life-events and hassles had an important influence on their confidence (both negatively and positively) that seemed to reflect the effect of general mood on confidence. The following quotes illustrate this:

"Home, so say somethings going on at home, it would make it [referring to confidence] lower. So say if something is going on in school maybe it'll make it lower or something's ... like you're doing well in school, you're doing well in something it will make it higher (Will)."

"Well I want to do well most of the time but sometimes like I wake up and I'm a bit like I don't really feel like playing football and I don't really have a good game. Yeah, it makes me ... cos I don't really feel like playing, my confidence just drops (Freddie)."

"If you've just had like a row with someone or your family you wouldn't really want to be there because you know you're not going to play as well if you're, like, confident (Ashley)."

3.2.5. Innate Factors

Innate factors mostly comprised of players having a good mentality for soccer, with a good mind-set giving them confidence to train and compete. Dan stated:

"You have to like start again because you might be a bit disappointed but if you stay disappointed you will just keep on losing, you gotta try and like boost so you can have confidence on the field."

3.2.6. Experience

Four players identified experience to be a source of self-confidence. This included playing experience and knowledge of teammates and the academy. It became apparent that becoming acquainted with aspects of the club was a determinant in confidence level for some performers, with Felix stating, "starting to know everyone and just starting playing up has just like increased my confidence a lot".

3.2.7. Athlete Specific Factors

These were factors that were unique to individuals and their confidence levels. This included role models performing well, seeing others fail, impression of opponents, crowd criticism, referee decisions, decision making in game, physical inferiority, situational favourableness, opposition comments and environment comfort.

4. Discussion

The present study represents the first attempt to examine the sources and types of confidence salient to academy football players, aged 11–12. We predicted that whilst Vealey and Chase's [8] model and previous research [9,11] would provide a useful base from which to view our results, we would see differences in the types and sources of confidence deemed important to the performers due to maturational and environmental factors. Overall, this prediction was broadly supported. Our interpretations of the data do fit reasonably well into the dimensions generated in previous research but notable differences emerged in the lower- and higher order themes that contributed to these global dimensions.

In terms of the types of confidence, we found support for Vealey and Chase's [8] contention of the multi-dimensionality of sport confidence and also found support for some of the dimensions seen in previous research. What is striking about our data is the relatively low quantity of raw data themes that emerged in the discussions with the participants. Whilst the need to be confident in their psychological factors (3 participants), superiority to the opposition (3 participants), tactical awareness (2 participants) and athlete specific factors (2 participants) was evident, the majority of the participants focused on either skill execution or achievement (confidence in their ability to gain selection/a continued contract). This is in contrast to the findings of Hays et al. [9] but is in line with the limited types of confidence identified by Thomas et al. [11]. In addition, a number of athletes found difficulty identifying their types of confidence. There are a number of reasons why this finding may have been seen. First, as Vealey et al. [7] noted, youth athletes are developing their own self-confidence and so the rather narrow types of confidence deemed important to the participants may reflect developing perception of competence and the determinants of success. Second, theory and research [12] within the achievement goal literature suggests that children may not develop a mature concept of ability and the ability to differentiate the determinants of success (e.g., the relative roles of luck and ability) until they are at least 11. Thus, the current findings may reflect a relative lack of knowledge of, or the ability to differentiate between, different determinants of performance.

In terms of confidence sources, we were able to categorise the responses of the participants into a number of the dimensions that have been identified in previous research. However, in contrast to the sources of confidence reported by elite, adult athletes [9], participants did not mention competitive advantage, trust, or self-awareness. Further, although our findings were in line with Thomas et al.'s [11] findings showing that the youth soccer players did seem to rely heavily on accomplishments, preparation and social support, the players in our research seemed to feel that the coach was a particularly powerful source of confidence. However, differences between the lower- and higher-order themes that emerged in our and previous research were more notable. First, whilst performance accomplishments were mentioned by all athletes in the Hays et al. paper, and all athletes in the current research, the range of responses in our study seemed to be less wide-ranging. From our analysis, participants relied on tangible indicators of accomplishment, such as goals scored/assisted and man of the match awards. This is consistent with the broader developmental literature which would suggest that the primary sources of competence information used by children in the childhood years (7–12 years old) are concrete in nature [4].

Such reliance on concrete sources of information is also consistent with the importance attached to social support and significant others. In our study, the information and feedback provided by coaches was identified by all participants as being influential in determining their confidence. Interestingly, our findings concerning the specific impact of coaching seems to be qualitatively different to the way in which coaches informed the confidence of participants in Hays et al. [9]. In Hays et al.'s research, lower order themes were more concerned with general support and belief in the coaches' ability than with the provision of praise and criticism as seen in this study. From a development perspective, coach feedback represents a concrete source of information (Horn, 2004). Equally, the environment of the elite youth soccer academy is one in which selection is an ever-present stressor [21] and coaches are the people who make such decisions. As Will commented, "at the end of the season, they're the ones who are going to decide whether I stay or go".

A further difference from our results and both Hays et al.'s [9] and Thomas et al.'s [11] research is the role of holistic preparation and general mood as a determinant of confidence. In Hays et al.'s research, holistic preparation seems to refer predominantly to the perception that training had gone well whilst Thomas et al. focused on the practical impact of good preparation (e.g., tactical awareness, sound nutrition). Whilst we saw this within the current data, confidence was also influenced by general mood prior to competition or training. This can be seen through references to the moods and through references to the impact of external events such as events in the family or at school. Equally, the perception that family and team "atmospheres" could be interpreted as evidence of how an individual's confidence may be influenced by their general feelings of contentment and happiness. The importance of mood resonates with Bandura's [5] theorising concerning the role of emotional states as a source of efficacy beliefs. Specifically, Bandura argues that mood may influence perceptions of self-efficacy in a number of ways. First, negative life events and depressed mood may activate feelings of personal inadequacy and low self-worth, which may contribute to lower perceptions of one's own competence. Second, negative moods activate memories of past failings, thus creating more negative self-efficacy, whilst a positive mood will activate memories of past successes, thus creating more positive self-efficacy. Finally, Bandura proposes that moods can influence the way in which events are evaluated so that mood may influence perceptions of the demands of the task at hand as well as perceptions of the resources available to oneself. Whilst our findings do not shed light on the ways in which mood may influence perceptions of confidence, they do suggest that the potential impact of mood on self-confidence should not be underestimated when working with children.

4.1. Applied Implications

There are two broad directions to take when considering the applied implications of this research. First, we can consider how best to enhance confidence by working through the sources of confidence known to be most salient for the client population [8]. In this instance, coaches and parents should be mindful of the role of concrete performance accomplishments, coaching feedback, peer and parental support and general mood/affect in determining confidence. That is, our findings first suggest that structuring practices and drills to provide the opportunity for concrete performance accomplishments and being mindful that feedback, particularly the beneficial effects of positive feedback, could be one way for coaches to enhance the confidence of the children they coach. A second suggestion that stems from our findings is to consider ways in which positive moods can be induced in youth performers before, during and after competition and training. This may be achieved through the actions of coaches and/or parents to provide activities that are enjoyed by the participants and are designed to lift moods or through the systematic use of positive psychology interventions such as the practice of grateful thinking or visualising best possible selves [22].

The second intervention option is to change the range of sources of confidence that are salient for the individual, so that (a) performers draw confidence from a range of sources and (b) performers become more reliant on controllable sources of confidence, such as mental preparation [8]. The results of this study suggest that young performers may be most reliant on the information provided by others and on the achievement of concrete performance accomplishments. Self-awareness and the ability to self-reflect and self-evaluate have been proposed [23,24] to be essential to the development of strong and stable perceptions of personal efficacy/confidence and have also been implicated in the development of elite level soccer players [25]. Thus, any strategy that develops self-reflection and self-evaluation, such as performance evaluation sheets [26], may be worthwhile.

4.2. Limitations

Even though this study provides an insight into the types and sources of confidence of elite level youth athletes, it does have limitations. First, the sample was, due to access issues, limited to one soccer academy. Given that Vealey and Chase [8] acknowledge that context and culture will influence sources of confidence then it is unclear the extent to which the current findings reflect the broader population of young soccer players or the context of this specific academy. Further research is needed to confirm these findings in a broader sample. Second, again mainly due to access and ethical reasons, we were not able to mitigate against some threats to validity as identified by critical realist researchers [18]. Specifically, although the lead researcher had completed a period of supervised experience within the academy, they had not prolonged engagement within the academy setting to collect triangulating data (e.g., observations, interviews with coaches/parents) or to build rapport and trust with the performers that may have led to richer data.

5. Conclusions

The sources of confidence identified by the athletes in this study broadly supported previous literature [8,9,11], but some notable differences were observed. Specifically, the young, male, elite-level, soccer performers interviewed provided evidence to suggest a relatively narrow range of types of confidence used (possibly indicative of a less mature understanding of the determinants of performance) and a reliance on more concrete sources of information such as tangible performance results (goals scored, goals saved, assists) and feedback from others (coaches predominantly but also family members and team-mates). We also saw the potential importance of general mood/affect in determining confidence levels. These results support the idea that sources of confidence are not universal and that they may be influenced by internal factors such as maturation levels and external factors such as coaching climate.

Supplementary Materials: The following are available online at https://www.mdpi.com/article/10.3390/sports9110146/s1, Table S1: Global dimensions and themes for types of confidence, Table S2: Themes and categories for sources of confidence identified by the academy youth athletes.

Author Contributions: Conceptualization, I.G., S.M. and A.P.; methodology, I.G. and E.B.; formal analysis, A.P. and I.G.; investigation, S.M. and A.P.; writing—original draft preparation, A.P. and I.G.; writing—review and editing, S.M. and E.B. All authors have read and agreed to the published version of the manuscript.

Funding: This research received no external funding.

Institutional Review Board Statement: The study was conducted according to the guidelines of the Declaration of Helsinki and approved by the Institutional Review Board (or Ethics Committee) of The University of Chichester (1516_40).

Informed Consent Statement: Informed consent was obtained from all subjects involved in the study.

Data Availability Statement: Data not available due to ethical restrictions—Due to the nature of this research, participants of this study did not agree for their data to be shared publicly, so supporting data is not available.

Conflicts of Interest: The authors declare no conflict of interest.

References

1. Forsman, H.; Blomqvist, M.; Davids, K.; Liukkonen, J.; Kontinnen, N. Identifying technical, physiological, tactical and psychological characteristics that contribute to career progression in soccer. *Int. J. Sports Sci. Coach.* **2016**, *11*, 505–513. [CrossRef]
2. Mills, A.; Butt, J.; Maynard, I.; Harwood, C. Identifying factors perceived to influence the development of elite youth football academy platers. *J. Sport Sci.* **2012**, *30*, 1593–1604. [CrossRef] [PubMed]
3. Vaeyens, R.; Coehlo e Silva, M.; Visscher, C.; Philippaerts, R.M.; Williams, A.M. Identifying young performers. In *Science and Soccer: Developing Elite Performers*, 3rd ed.; Williams, A.M., Ed.; Routledge: London, UK, 2013; pp. 289–306.
4. Horn, T.S. Developmental perspectives on self-perceptions in children and adolescents. In *Developmental Sport and Exercise Psychology: A Developmental Perspective*; Weiss, M.R., Ed.; Fitness Information Technology: Morgantown, WI, USA, 2004; pp. 101–143.
5. Bandura, A. *Self-Efficacy: The Exercise of Control*; W.H. Freeman: New York, NY, USA, 1997.
6. Vealey, R.S. Conceptualization of sport-confidence and competitive orientation: Preliminary investigation and instrument development. *J. Sport Psychol.* **1986**, *8*, 221–246. [CrossRef]
7. Vealey, R.S.; Hayashi, S.W.; Garner-Holman, M.; Giacobbi, P. Sources of sport-confidence: Conceptualization and instrument development. *J. Sport Exerc. Psychol.* **1998**, *20*, 54–80. [CrossRef]
8. Vealey, R.S.; Chase, M.A. Self-Confidence in Sport. In *Advances in Sport Psychology*, 2nd ed.; Horn, T.S., Ed.; Human Kinetics: Champaign, IL, USA, 2008; pp. 65–98.

9. Hays, K.; Maynard, I.; Thomas, O.; Bawden, M. Sources and types of confidence identified by world class sport performers. *J. Appl. Sport Psychol.* **2007**, *19*, 434–456. [CrossRef]
10. Weiss, M.R.; Amorose, A.J. Motivational Orientations and Sport Behaviour. In *Advances in Sport Psychology*, 2nd ed.; Horn, T.S., Ed.; Human Kinetics: Champaign, IL, USA, 2008; pp. 115–155.
11. Thomas, O.; Thrower, S.N.; Lane, A.; Thomas, J. Types, sources and debilitating factors in sport confidence in elite early adolescent academy soccer players. *J. Appl. Sport Psychol.* **2021**, *33*, 192–217. [CrossRef]
12. Weiss, M.R.; Raedeke, T.D. Developmental Sport and Exercise Psychology: Research Status on Youth and Directions toward a Lifespan Perspective. In *Developmental Sport and Exercise Psychology: A Developmental Perspective*; Weiss, M.R., Ed.; Fitness Information Technology: Morgantown, WI, USA, 2004; pp. 1–26.
13. Bhaskar, R. *Reclaiming Realism: A Critical Introduction to Contemporary Philosophy*; Verso: London, UK, 1989.
14. Cushion, C.; Jones, R.L. Power, discourse, and symbolic violence in professional youth soccer: The case of Albion football club. *Sociol. Sport J.* **2006**, *23*, 142–161. [CrossRef]
15. Palinkas, L.A.; Horwitz, S.M.; Green, C.A.; Wisdom, J.P.; Duan, N.; Hoagwood, K. Purposeful sampling for qualitative data collection and analysis in mixed method implementation research. *Adm. Pol. Ment. Health* **2015**, *42*, 533–544. [CrossRef] [PubMed]
16. Patton, M.Q. *Qualitative Evaluation and Research Methods*; Sage: Newbury Park, CA, USA, 2002.
17. Maxwell, J.A. *A Realist Approach for Qualitative Research*; Sage: London, UK, 2012.
18. Ronkainen, N.J.; Wiltshire, G. Rethinking validity in qualitative sport and exercise psychology research: A realist perspective. *Int. J. Sport Exerc. Psychol.* **2021**, *19*, 13–28. [CrossRef]
19. Fade, S.A. Communicating and judging the quality of qualitative research: The need for a new language. *J. Hum. Nutr. Diet.* **2003**, *16*, 139–149. [CrossRef] [PubMed]
20. Culver, D.M.; Gilbert, W.; Sparkes, A. Qualitative research in sport psychology journals: The next decade 2000–2009 and beyond. *Sport Psychol.* **2012**, *26*, 261–281. [CrossRef]
21. Taylor, I.M.; Bruner, M.W. The social environment and developmental experiences in elite youth soccer. *Psychol. Sport Exerc.* **2012**, *13*, 390–396. [CrossRef]
22. Sheldon KLyubomirsky, S. How to increase and sustain positive emotion: The effects of expressing gratitude and visualising best possible selves. *J. Posit. Psychol.* **2006**, *1*, 73–82. [CrossRef]
23. Petchlikoff, L.M. Self-Regulation Skills for Children and Adolescents. In *Developmental Sport and Exercise Psychology: A Developmental Perspective*; Weiss, M.R., Ed.; Fitness Information Technology: Morgantown, WI, USA, 2004; pp. 269–288.
24. Zimmerman, B.J.; Cleary, T.J. Adolescents' Development of Personal Efficacy: The Role of Self-Efficacy Beliefs and Self-Regulatory Skill. In *Self-Efficacy Beliefs of Adolescents*; Parajes, F., Urdan, T., Eds.; Information Age Publishing: Charlotte, NC, USA, 2006; pp. 45–70.
25. Toering, T.T.; Elferink-Gemser, M.T.; Jordet, G.; Visscher, C. Self-regulation and performance level of elite and non-elite youth soccer players. *J. Sport Sci.* **2009**, *27*, 1509–1517. [CrossRef] [PubMed]
26. Allen, M.S.; Jones, M.V.; Sheffield, D. The influence of positive reflection on attributions, emotions and self-efficacy. *Sport Psychol.* **2010**, *24*, 211–226. [CrossRef]

sports

MDPI

Article

Comparison of High and Low Responders to a Cross-Country Skiing Talent Transfer Program: A Coach's Perspective

Stig Arve Sæther [1,*], Mats Iversen [1], Rune Kjøsen Talsnes [2,3] and Øyvind Sandbakk [4]

1 Department of Sociology and Political Science, Norwegian University of Science and Technology, 7491 Trondheim, Norway; mats.iversen@hotmail.com
2 Meråker High School, Trøndelag County Council, 7735 Steinkjer, Norway; runta@trondelagfylke.no
3 Department of Sports Science and Physical Education, Nord University, 8049 Bodø, Norway
4 Centre for Elite Sports Research, Department of Neuromedicine and Movement Science, Norwegian, University of Science and Technology, 7491 Trondheim, Norway; oyvind.sandbakk@ntnu.no
* Correspondence: stigarve@ntnu.no; Tel.: +47-73551133

Abstract: Purpose: To examine how coaches differentiate athletes with successful and non-successful development during a cross-country (XC) skiing talent transfer (TT) program. Methods: We conducted qualitative, semi-structured interviews with seven Norwegian coaches working with a group of 23 Chinese summer endurance athletes transferring from running, rowing, and kayaking to the winter endurance sport XC skiing over a six-month training period. The athletes were grouped as either high (n = 9), moderate (n = 3), or low responders (n = 11) based on objective performance development, quantified using laboratory testing. The interview guide contained six sections: physiological development, technical development, psychological characteristics, training and recovery routines, athlete background, and considerations about the effectiveness of TT initiatives in general. Results: The assessments of the coaches revealed that greater development of both physiological and technical capacities among the high-responding TT athletes were associated with higher motivation, as well as superior ability to deal with adversity in the development process. Conclusion: The coaches considered the TT program to be effective; however, successful transfer of athletes to a world class level in a complex sport such as XC skiing requires a multidisciplinary selection process and a longer time frame than the six-month period used in the current project.

Keywords: endurance sport; performance development; cross-country skiing; talent transfer; training responses; Winter Olympic Games

Citation: Sæther, S.A.; Iversen, M.; Talsnes, R.K.; Sandbakk, Ø. Comparison of High and Low Responders to a Cross-Country Skiing Talent Transfer Program: A Coach's Perspective. *Sports* **2021**, *9*, 138. https://doi.org/10.3390/sports9100138

Academic Editors: Adam Leigh Kelly, Sergio L. Jiménez Sáiz, Sara Diana Leal dos Santos and Alberto Lorenzo Calvo

Received: 12 August 2021
Accepted: 30 September 2021
Published: 2 October 2021

1. Introduction

For several decades, coaches, athletes, and researchers have searched for keys to optimize performance development among elite athletes [1]. Most of the previous literature has focused on the development process of athletes within a specific sport [2]. However, an alternative approach is talent transfer (TT, also referred to as athlete transfer). TT can be defined as the process in which athletes make a change from their original sport (i.e., donor sport) to a new sport (i.e., transfer sport). TT has been adopted by various sport government bodies as a means of capitalizing on the developmental investment made in previously identified athletes ('recycling') and fast-tracking athletes to new sports where they might have a greater chance to succeed [3]. The overall idea of TT initiatives is to recruit athletes who have shown a high level of performance in a sport, through investment in both time and effort, without taking the last step to an elite level, and they are therefore provided the opportunity to transfer to a new sport where they might have better potential for success. These programs have been conducted in various sports during the last several decades, despite claims of low success rates and a high degree of serendipity, such as in Great Britain prior to the London Olympics in 2012 [4]. The TT approach has mainly been used in summer sports, but the latest approach by China is directed towards select winter sports, aiming to develop athletes for the upcoming Olympic Games in Beijing 2022 (https://www.chinadaily.com.cn/china/2017-08/08/content_30364992.htm, accessed on 11 August 2021). Research has found a close relationship between the host country and the number of medals won in the competitions [5]. The Olympics has become a showcase, making the host country want to invest in their own games to win the overall medal count in their host Olympics. To be able to be among the most successful nations in the medal count, the host nations are dependent on developing athletes that can compete at an international level in most sports held during the Olympics. Investing in a TT program could increase the possibility to perform in sports that the nations, historically, have been less successful in. Still, a recent systematic review showed that the available evidence on TT is scarce and included only five studies, and it concluded that similarities between donor and transfer sports are associated with success and that psychological factors play an important role for individual athletes' development in the TT program [6]. Furthermore, most studies, both on talent development and TT, have been conducted on western athletes, even though some recent examples of studies are done among Chinese athletes [7].

One vital aspect of TT programs is to understand the connection between 'donor' sports (the original sport an athlete used to practice in) and 'recipient' sports (the destination sport to which an athlete has transferred) [8]. Currently, the connection between 'donor' sports and 'recipient' sports is less examined; although, Talsnes et al. [9] found runners to improve more than kayakers and rowers when Chinese athletes were transferred to cross-country (XC) skiing over a six-month training period. Another important aspect is to find the appropriate selection criteria, in which previous studies have criticized TT programs for adopting inappropriately narrow criteria, focusing mainly on physical and anthropometric factors [10]. Accordingly, studies have emphasized the importance of assessing athletes' abilities to learn in new, knowledge-rich environments. Furthermore, the role of psycho-behavioral factors and self-regulation to foster the use of knowledge and skills that should facilitate effective transfer has been emphasized [11,12]. In other words, the authors argue that a greater effort should be placed on identifying an athlete's capacity to learn instead of measuring what has already been learned.

In an approach by Talsnes et al. [9], a group of 24 Chinese runners, rowers, and kayakers were transferred to XC skiing, one of the most technically and physiologically demanding endurance sports, over a six-month training period using state of-the-art training methodology [13] and high-level coaches. Here, large improvements in the development of sport-specific capacities (i.e., roller skiing and double-poling ergometry) were measured in the laboratory, with fewer changes in more general capacities (i.e., running) [9]. In a follow-up study by Talsnes et al. [14], they compared high and low responders in the same group of TT athletes, and better physiological adaptations were found among high responders, whereas technical capacities were largely improved in both groups. To separate the athletes that best responded to the six-month training period, a quantitative performance index was used to classify high- and low-responders. Relative changes from pre to post peak velocity during an incremental treadmill running and roller-ski skating test, in addition to average power output during both 30 sec and 5 min double-poling ergometry tests, were used to determine this index. Superior physiological improvements coincided with a higher training load, higher training volume, and less injury and illness during the training period in the group of high responders [14]. However, the underlying mechanisms related to the ability to concurrently respond both physiologically and technically, as well as the factors associated with high training tolerance and low injury rates, are not known. Here, the study by Talsnes et al. [14] indicated that high motivation and strong coach–athlete alliances played a significant role, which are aspects that require more attention in future studies.

While the selection and identification process related to talent development is widely investigated (for a review, see Till et al. [1]), the TT process could be considered an exploration of the rest of the potential from athletes that have achieved a high-performance level without reaching a world-class level within their original sport. The idea of doing a TT is to use the athlete's potential in a donor sport and invest training in a 'recipient' sport, where the athlete is regarded to have the potential to reach a high international level. The athlete's level of progress (response) is therefore vital in the TT process. Building upon the previous study by Talsnes et al. [14], the primary aim of the present study was, therefore, to examine how coaches differentiated high- from low-responding athletes in a six-month XC skiing TT program. Since both the TT process and identification of athletes have been studied to a lesser degree, a secondary aim was to examine how coaches perceived the value of such TT initiatives based on their experience as elite coaches.

2. Materials and Methods

In this study, a qualitative approach was chosen to gather detailed information from Norwegian XC-skiing coaches working with TT athletes about their reflections on differences between successful and less successful development in their respective TT program, used to develop XC skiers aiming for success in the Beijing 2022 Winter Olympics. Laboratory tests were performed to determine the TT athletes' development in performance, physiological, and technical capacities as described in detail by Talsnes et al. [9]. These data further defined the groups of responders and non-responders and thereby formed the basis for the qualitative interviews presented in this study. The coaches were aware of these results when they were interviewed regarding the TT athletes.

2.1. Participants

The seven interviewed coaches (six men, one woman) were all experienced Norwegian XC-skiing coaches who worked on the project during the entire six-month period (see Table 1). The age and years of coaching experience for the group was (mean $\pm$ standard deviation [SD]) 29 $\pm$ 4 and 6 $\pm$ 5 years, respectively. The group consisted of some younger coaches with 1–2 years of coaching experience and some older coaches with 7–15 years of experience, including one previous national team head coach in XC skiing for Norway and one previous head coach for the development team in the biathlon for Norway. The project was conducted in Meråker, Norway from November 2018 to May 2019. All participants volunteered for this study and signed informed consent forms prior to their participation. The research was undertaken according to the ethical guidelines of the Norwegian Centre for Research Data (NSD).

2.2. Data Collection

Prior to the interview, participants were informed both verbally and in writing about the study and interview. During the interviews, a semi-structured approach was adopted where an identical set of questions was employed in a similar manner. Although this procedure resulted in a certain element of structure to each interview, the ordering of questions varied depending on the responses of each participant, where some of the issues raised were explored further by the interviewer. Although the discussions varied in their content due to the participant responses, a variety of probe (e.g., 'Why do you think this is?') and elaboration (e.g., 'What do you mean by that?') questions were employed to ensure that all issues were investigated in depth (see Table 2). At the end of all interviews, participants were asked if all appropriate factors had been discussed and if they wanted to add anything regarding some of the aspects, and none of them did. All of the interviews were conducted face-to-face in an environment comfortable for the participant, and the interviews were also tape-recorded in their entirety (30–45 min duration) and transcribed verbatim.

Table 1. Coach demographic information.

Coach	Gender of Coach	Sport	Gender of Athletes	Coach Experience
Coach 1	Male	Cross-country skiing	Male/Female	Medium 3–6 years, regional/national
Coach 2	Male	Cross-country skiing	Male/Female	Highly 6+ years, elite/international
Coach 3	Male	Cross-country skiing	Male/Female	Little 1–3 years, local/club
Coach 4	Male	Cross-country skiing	Male/Female	Little 1–3 years, local/club
Coach 5	Male	Cross-country skiing	Male/Female	Medium 3–6 years, regional/national
Coach 6	Male	Cross-country skiing	Male/Female	Little 1–3 years, local/club
Coach 7	Female	Cross-country skiing	Male/Female	Little 1–3 years, local/club

Table 2. Examples of interview questions.

1. Physiological development: How is the correspondence between the development in cross-country skiing performance and the physiological development of the athletes?
2. Technical development: How crucial is the technical development in cross-country skiing for performance?
3. Psychological characteristics: What mental qualities characterize the skiers with best performance-development?
4. Training and recovery routines: How do athletes balance training and recovery?
5. Athlete background: What would you say are the sports backgrounds of those with the greatest performance-progress?
6. Considerations about the effectiveness of TT initiatives: What do you consider as the success factors for a successful TT?

2.3. Data Analysis

The interviews were audio recorded and transcribed. All players were given an identification code, C1–C7 (Coach 1–Coach 7), to provide anonymity. According to the six steps suggested by Braun et al. [15] for qualitative analysis, the data were analyzed using the following process: (1) transcribing, reading, and re-reading the data; (2) generating initial codes, such as physiological and technical development; (3) applying deductive codes and identifying lower-order themes, such as motivation within psychological characteristics; (4) outlining of the overall topics from the data, such as training and recovery routines; (5) reviewing and refining the subthemes and final categories; (6) writing a report and presenting the data (p. 191). Having gone through all the data from the interviews, we ended up with our six topics presented in the results section.

In this manuscript, rigor is considered through the meaningful coherence between the purpose of the study, the procedure, and the findings [16]. Building on Tracy [16], we have sought to ensure transparency by making a detailed description of the research process and role by creating distance when reflecting and interpretating our findings. Moreover, we have continually sought to verify and validate the analysis and provide critical interpretations of the data. During the data collection, the research team discussed various results thereby ensuring peer validity [17]. Additionally, ongoing member reflections took place during the individual studies [16], with the intention of making sure that the descriptions and explanations were rich, bountifully supplied, generous, and unstinting.

2.4. Performance Index

The athletes' development of performance, physiological, and technical capacities were measured in the laboratory as described in detail by Talsnes et al. [9]. A performance index was developed to classify high and low responders to the six-month training period [14]. Pre- to post-relative changes in performance measures during incremental running and roller-ski skating tests, in addition to average power output in a 30 s and 5 min double-poling ergometry test, were used. These four performance measures were further summated, resulting in the performance index with cutoffs set to classify high ($n = 9$), moderate ($n = 3$), and low responders ($n = 11$). Moderate responders were excluded from further analyses to establish distinct group differences. High responders consisted of eight men and one woman, whereas low responders consisted of six men and five women, respectively. Furthermore, eight athletes in the high responder group and five athletes in the low responder group had a previous sports background in running (i.e., long or middle distance), whereas the remaining athletes in both groups had a sports background in kayaking or rowing.

3. Results

The current investigation examined how coaches differentiated athletes with successful and non-successful development in a XC skiing TT program. These are presented for each of the interview topics below (see Table 3).

Table 3. Emerging themes and sub-categories following the process of thematic analysis.

Physiological Development	Technical Development	Psychological Characteristics	Training and Recovery Routines	Athlete Background	Effectiveness of Talent Transfer Programs
-Sport-specific physiological capacities (5)	-Most crucial factor for performance development (3)	-Motivation (7)	-Balance optimizing (7)	-Runners positively responders (7)	-Short-term progress (7)
-Lack of development in general physiological capacities (5)	-Positive response to feedback (2)	-Curiosity and desire (3)	-Training load adjustment (3)	-Younger athletes positively responders (7)	-Positive impressions (7)
-Training developed both sport-specific and general physiological capacities (1)	-Combination of physiological and technical capacities (7)	-Dealing with adversity (2)	-Lack of ability to adjust training load and training response (2)	-Rowers and kayakers' disadvantage (2)	-Positive development, less in XC skiing (3)
-Variation in development of physiological capacities (1)		-Self-criticisms, reflections, and wellbeing (5)	-Adjustment of training load challenge during the TT program (5)	-Men positive responders (7)	-More specific selection criteria (3)
-Maximal oxygen uptake (5)			-'Extra sessions' characterized highest responders (1)		-More play and less seriousness in the beginning (2)
-Initial aerobic endurance capacity (3)			-Catch up training not the way to go (1)		-'Donor athletes' with experience with XC skiing (5)
-Endurance capacity of young male runners (2)			-Injuries and illness (4)		
			-Self-challenging athletes more injury days (1)		

Numbers in parentheses represent number of coaches highlighting each themes/sub-category.

3.1. Physiological Development

In general, most coaches (5) stated that the greatest development was seen in sport-specific physiological capacities (i.e., roller-ski skating and double-poling ergometry), whereas more general physiological capacities (i.e., running) were less changed. Regarding the development of general capacities, there were different opinions between the coaches. Although, most of the coaches (4) that perceived those general physiological capacities seemed unchanged in the entire group, one of the coaches reported that the athletes who responded positively to the training developed both sport-specific and general physiological capacities:

C2: 'The ones who have a positive development, have development in all capacities I would say. Even general physiological capacities.'

One coach points out that the development of physiological capacities for the athletes varied, and some of them had good development also in this aspect:

C4: 'For some, the development of physiological capacities has been good. Others had almost no development, while some had negative development of their physiological capacities.'

Most of the coaches (5) experienced that the highest responders had good development in maximal oxygen uptake (VO_{2max}), while the lowest responders had little or even negative changes.

In addition, a high initial aerobic endurance capacity is highlighted by several coaches (3) as one of the most important characteristics of high responders. In particular, the high endurance capacity of the young runners, especially the men, seems to be a determining factor for a successful transfer to XC skiing, and several of the coaches (2) point this out in the following quotes:

C6: 'It seems that athletes transferring from running improved more than athletes transferring from kayaking and rowing. Also, the boys revealed better development than the girls, together with athletes with high initial endurance capacities. The youngest athletes had also better physiological and technical development than the older ones.'

C7: 'It's the former runners who developed the most ... They had a good endurance capacity when entering the project, and you may think that running is more comparable to the technical patterns in XC skiing than rowing or kayaking. Thus, runners had endurance capacities and muscular characteristics more relevant for the demands of XC skiing.'

While all of the coaches agreed that strength and endurance capacity is important in XC skiing, they also highlight the importance of a good perception of the technical aspects of XC skiing as crucial:

C4: 'Several of the athletes had better physiological capacities than their Norwegian peers, but this does not reflect their actual performance in competitions. It's the ski perception of skiing that matters the most and how to efficiently go on skis. They still miss this feeling.'

C3: 'You have to obtain a minimum level of VO_{2max} but still, it is not necessarily the one with highest VO_{2max} who is the best performing skier.'

3.2. Technical Development

Three of the coaches emphasized technical development as the most crucial factor for performance development, as outlined in the following quote from a coach:

C5: 'It's clear that technical development is the most important factor. The ones who perform best in competitions are the ones in the program who have had largest technical development.'

The coaches seem to have different opinions on the main factors associated with technical development. Some coaches (2) consider the ability to positively respond to feedback as the most important factor of technical development, and they highlight the importance of a strong coach–athlete relationship in this context. They state that the ones with the largest technical development are the ones who are constantly seeking technical feedback from their coaches:

C6: 'The ability to positively respond to feedback, so they can work with the things they are instructed to the highest responders understand they have to work with the technical tasks provided, not only for that specific session, but over a long period of time, even the times when there is no one watching them'.

The coaches seem convinced that concurrent development of both physiological and technical capacities is crucial for performance development in XC skiing. However, due to the large demand for rapid technical development in athletes transferring from summer sports to XC skiing, the coaches argue that more emphasis should be placed on the athletes' technical capacity, to learn technique, enjoy skiing, respond well to feedback, and develop motivation to work with technical tasks over time.

3.3. Psychological Characteristics

All of the coaches point out motivation as crucial for performance development in the process of transferring to XC skiing, and many coaches (3) also highlight the athlete's curiosity and desire to develop as important for inducing positive training responses. As one of the coaches says:

C1: 'Motivation, enjoyment and passion. Curiosity and the desire to develop and become better. These are clearly characteristics of the highest responders if you are not happy and motivated with what you are doing in life, then it doesn't matter what you do. If you have the best coaches, the best training program, it doesn't matter If they don't like what they are doing, then it doesn't work.'

Others (2) emphasize the ability to deal with adversity, as quoted below:

C7: 'Maybe the lowest responders give up easier due to lack of inner passion to improve.'

C6: 'The best athletes give it several tries before they give up; they react constructively to feedback, and don't give up and this motivation can be hard to maintain in this project. That's why the athlete's wellbeing is highly important. To keep up their motivation.'

When asked about characteristics of the highest responding athletes, the coaches showed many similarities in their opinions, with high motivation as a main point described from the following statements:

C6: 'It's the ones who have been most motivated. The most motivated athletes do as they are told, and you see they are more focused and have more discipline. They are ready for training sessions early, are concentrated during the sessions, and you see that they want to become better athletes.'

C4: 'It's the ones who have maintained the highest motivation, have been most satisfied with living here in Norway, and have established a good environment and training group.'

In addition to a high motivation, other mental capacities associated with a large development in XC skiing were outlined, such as self-criticisms, reflections, and wellbeing:

C6: 'The ones who are critical to themselves and to the training program we carry out develop well. The best have always asked a lot of questions, reflect if they do things good enough, and are constantly looking after better ways to develop the ones who doesn't [sic] try and fail hard, have stagnated.'

C1: 'They see the joy in training, like the new life situation here. And their wellbeing.'

3.4. Training and Recovery Routines

The coaches also highlighted the importance of optimizing the balance between training and recovery, of individual adjustments of training load, and of the absence of injuries and sickness. Some of the coaches (3) perceived athletes with the greatest performance development as the ones who were more able to adjust their training load. Especially early in the project, one coach felt that the best responders were the ones who made more adjustments in their training:

C5: 'Some of the athletes were better than others at making small adjustments in their training load. These ones probably perceived the balance between load and recovery better, and therefore had a more positive performance development.'

Other coaches (2) perceived that there was no clear relationship between the athlete's ability to adjust their training load and their subsequent training response. Here, most athletes completed the same training sessions and thus had a similar overall training load; although, each athlete's training response differed extensively. As coach 3 mentions:

C3: 'I see no clear relationships between making adjustments in load and performance development among these athletes. I feel that some of the ones who have responded negatively to the training also have tried to do everything right, including adjustments of load, but just without having the same training response and performance development as others.'

Most of the coaches (5) point out that this aspect has been a challenge during the TT program since most athletes were not familiar with making individual adjustments in their daily training. One coach also felt this had been a challenging aspect of the entire TT project:

C5: 'Either it's us [sic] who hasn't [sic] communicated this well enough, or it's cultural differences between Norway and China in the training and development of endurance athletes which explains this challenge.'

One of the coaches felt that it was the athletes who performed 'extra sessions' who are the highest responders, as pointed out in the following quote:

C2: 'It's like the ones who become the best are the ones you often see out on an extra training session, the ones who dare to train a little more than the others.'

However, coach 7 has a different opinion on this:

C7: 'On rest days, some of the athletes performed individual training sessions. These ones are not the highest responding athletes, but the ones to strive to catch up with the rest the best ones are normally better at recovery so it's often the ones who are a little behind and think they have to do some extra training between the sessions. But these ones are not the best.'

One other factor that is pointed out as important for the athletes' development is continuity in their training process (4). Injuries and illness can negatively influence adaptations and performance development, and the highest responders seem to have more continuity in their training and less injuries:

C3: 'It's the athletes who have continuity in their training, stay healthy and free of injury in addition to finding the optimal balance between load and recovery which had the largest performance development . . . '

C1: 'Yes, you can see that it's the ones in the lowest responding group who have most cases of sick days and days with injuries, that is clearly.'

One of the coaches had another take on this and claimed that the athletes who challenged themselves the most also had more injury days due to falling on the slopes:

C2: 'but the ones with best development are also the ones that dare to challenge themselves, and due to this, the highest responding group has some more days with injuries as a result of falling on roller-skis/skis.'

3.5. Athlete Background

When asked about the athletes' sports background and how this may have influenced differences in performance development between athletes, all coaches shared the same opinion that runners most positively responded to a XC skiing TT initiative:

C3: 'The rowers seem to have "heavy legs" and poor development; the kayakers seem to fit a little better, and have somewhat better development, but it's definitely the runners who developed the most during the six-month training period.'

All of the coaches seemed to agree that many of the oldest athletes struggled to develop, especially technically, and seemed to have a harder time adapting new movement patterns, as coach 7 pointed out:

C7: 'It's like the older ones are harder to adapt to a new movement pattern. So, it seems to be preferable with the younger athletes. And also, mentally, the younger ones seem to be more curious and unexperienced'.

In addition, some of the coaches (2) point out that athletes from rowing and kayaking had more muscles and higher body mass than the runners, which could be a disadvantage for developing as skiers. As coach 5 says:

C5: 'You have to carry your own body weight when you are skiing, and some of them seemed untrained they were a little bit too big, strong in the upper body, but weak in the core and the lower parts of the body.'

Furthermore, it seems to be an agreement among the coaches that the men displayed better development than the women. As the coaches highlight:

C2: 'It can be the differences in gender, in how we are built. It can be the amount of testosterone in the boys, it can be the toughness within the boys because of the testosterone, and it can be that the girls perform better at lower intensity but struggle to develop high top speed.'

C5: 'It seemed like the boys liked it better here. They were like a boy's club who were at a training camp here and had an awesome time together with each other both during and between the training.'

The last quote is related to the athlete's motivation, which was regarded as one of the most important characteristics of high responders.

3.6. Considerations about the Effectiveness of Talent Transfer Programs

All coaches pointed out that the athletes' development during the six-month period had been impressive. However, the development was not linear and leveled out during the last weeks of the project, as coach 4 points out:

C4: 'Isolated, they have developed a lot. In the beginning, the progression curve was steep, while now, we are finding us in a period where we have somewhat less progress in performance.'

Coach 1 points out that even though he is satisfied with the development they have seen, his satisfaction is limited because of the ambitious goal of the project:

C1: 'On one side, I am very satisfied with the development the athletes have had on skis. In a 10 km race, they came 30 min behind their Norwegian peers in the beginning, while they are now only 2–5 min behind the same athletes today. But on the other side, when the project has the 2022 Olympics as a goal, the athletes' performance is far from a level good enough for World Cup and Olympic races.'

The coaches' impression of TT initiatives as a development model seemed to differ, but all of the coaches pointed out many positive aspects of the TT model, and it seemed to work pretty well, with some precautions. One coach pointed this out in the following quote:

C2: 'It works well if you start with 2000 athletes you select from, and always narrow it in I'm sure that when you come down to 10–20 highly motivated athletes, who work hard and like to train, and if you work with these ones over a long period, maybe 6 to 10 years, then I am 100% sure you would get results in XC skiing.'

Some of the coaches (3) pointed out that they think it is a good model for development of sporting talents, but it can be a challenging approach in XC skiing because of the complex demands of the sport. Coach 1 outlines this in the following quote:

C1: 'The project has worked well but regarding TT and XC skiing, it's something I wouldn't recommend. I think it is an advantage to have more "ski feeling" from early stages in life.'

This is something another coach also points out:

C6: 'They have absolutely no history with XC skiing from before, and they started pretty late I think the transfer model would work better in some other endurance sports, because XC skiing has so many technical aspects, so to perform well is hard I think it would be easier to transfer their physical capacities to for example cycling or rowing.'

When the coaches were asked about what changes they would have made if they could have started the project over again, they have different answers. Three of the coaches pointed out that the selection process was not optimally handled, as highlighted in the following:

C7: 'I would have spent more time with the selection process, had more specific tests and conducted personal interviews with each athlete, to hear their thoughts about the project and how their motivation was.'

Two of the coaches also highlighted more playing and less seriousness in the beginning of the project as something they would have changed in retrospect. This would have made the skiing-specific technical aspects easier to manage for the athletes. One of the coaches pointed this out in the following quote:

C4: 'A little more playing on skis in the beginning. Even more than we did. In the beginning this was forgotten, and things got a little too serious in retrospect this would have helped them become better in downhills.'

Another aspect pointed out is that the 'donor athletes' should have had more experience with XC skiing when arriving. This would have made the technical development easier and would have made the progression faster. One coach said it as:

C6: 'If you had taken a group that had tried skis before, maybe a group of runners for example with some kind of experience with skiing then I would assume they would have had better development.'

This section may be divided by subheadings. It should provide a concise and precise description of the experimental results, their interpretation, as well as the experimental conclusions that can be drawn.

4. Discussion

The present study examined how coaches differentiated athletes with successful and non-successful development during a six-month XC skiing TT initiative. The main findings were as follows: (1) The coaches experienced that a greater development of both physiological and technical abilities in high responders was coupled with a superior motivation for learning a new sport and the development process, which led to higher performance development in XC skiing compared to low responders. (2) Athletes transferring from running, especially men, adapted more rapidly to XC skiing and tolerated XC skiing-specific training better. (3) The coaches were more ambiguous about the importance of the ability to perceive and adjust training load or to deal with adversity in the development process for differentiating high responders from lower-responding peers. (4) The coaches considered the TT program to have potential for athlete development, although the TT program seemed less appropriate for such a complex and technically demanding sport as XC skiing.

As expected, all athletes improved their sport-specific performance following six months of XC ski-specific training. However, as previously described by Talsnes et al. [14], the coaches perceived that the highest-responding athletes concurrently improved physiological capacities in both general (i.e., running) and sport-specific (i.e., roller-ski skating and double-poling ergometry) exercise modes. Here, the coaches emphasized that high responders especially developed their XC skiing technique more rapidly due to a higher motivation for learning a new sport and for the development process in general leading to a larger performance development in XC skiing, in comparison to the low responders. This consequently influenced the ability to work hard with given technical assignments, that involved searching for individual solutions in addition to a superior ability to react to feedback from coaches. A vital challenge for most athletes was, however, the lack of perception for the technical aspects related to skiing, which, according to the coaches, limited their ability to develop as XC skiers.

In this context, the athletes' age and gender seemed to have influenced their technical development, as some coaches stated that the youngest male athletes in the project developed their technical capacities the most. The fact that the youngest athletes demonstrated better technical development than the older ones corresponds with previous literature on TT initiatives [8]. Furthermore, the fact that athletes with backgrounds in running were most represented in the high-responder group is further supported by the findings of Talsnes et al. [9], which found a significantly larger sport-specific performance development among runners than kayakers and rowers. Furthermore, this also agrees with Vaeynes et al. [8], who pointed out that some sports are more likely to act as donor sports in TT initiatives, and additionally, they outlined rowing and kayaking as sports that may be more suitable as recipient sports than donor sports. Furthermore, anthropometric factors, such as body height and weight, were highlighted by the coaches as important factors, as runners with lower body weight developed more than rowers and kayakers with higher body weights, which also agrees with the findings of Rees et al. [18]. Overall, the most successful athletes of the project were able to concurrently develop their physiological and technical capacities, which coincided with a high level of motivation for the development process, and thereby an improved working alliance with their coach.

These findings are supported by previous literature, emphasizing psychological factors as a main determinant of successful performance development in sports [18,19]. As mentioned above, the coaches particularly pointed out motivation as the most important factor associated with a successful transfer from summer sports to XC skiing over a six-month training period. Here, the coaches related high motivation to have an influence on several aspects among the high responders, such as a superior ability to reflect upon their own training process and to deal with adversity. The coaches were, however, more ambiguous on how the ability to reflect upon the training process and to deal with adversity was different between the two groups of athletes. This might be explained by the normally high training load expected by XC skiers [9]. Altogether, these findings are supported by the existing literature, which highlights motivation as a key factor with influence on several underlying mechanisms related to sports development [12,19,20].

When we asked the coaches how they considered this TT initiative as a model for developing world-class XC skiers, the coaches had two-sided opinions on its effectiveness. On one hand, the coaches were satisfied with the large progress of the athletes during the six-month training period in Norway. On the other hand, all of them pointed out several challenges with the ambitious goal of the project, developing XC skiers in such a short period. Thus, XC skiing takes, most likely, longer time to learn and may therefore not be the optimal sport for TT initiatives. In addition, some of the large differences in development between athletes may have been caused by the selection process of the current TT program. Thus, this process should have been performed in a more sophisticated manner, involving experts in XC skiing, which could have resulted in fewer low responders, due to a selection of younger athletes with a larger physical (including sports background) and motivational potential for a transfer to XC skiing.

Seen in retrospect, the coaches stated that it would have been preferable if the athletes had more experience with XC skiing before they came to Norway and started the TT process. Lastly, the coaches would have put more focus on play and less on seriousness in the initial phase of the project. Such aspects were suggested to have a positive influence on the technical development in the later phases of the project, which was considered the most crucial skill to develop within the project. Here, the coaches' assumptions corresponded well with previous findings on TT initiatives [8,19], as a more thorough selection process by recruiting athletes from sports with similar physical characteristics, such as the coaches highlighted, could have made the program more efficient in the goal of developing high-level XC skiers.

4.1. Methodological Considerations

The present study extended upon previous findings by Talsnes et al. [14], using a qualitative research design to provide a more holistic understanding of the underlying mechanisms explaining successful versus less successful TT development to XC skiing. We believe that the main strength of the study is the comprehensive amount of data. This is supported by Martindale et al. [20] who argued that an athletic development process should be considered in a holistic manner to gain insight into what an effective talent development environment (TDE) is. The main limitation is related to cultural differences and the language barrier since these aspects could lead to wrong interpretations by the coaches. Another obvious limitation is the lack of athletes' perspective on the TT process in our design, which most likely would have given complementary insights into the mechanisms explaining successful versus less successful TT to XC skiing.

4.2. Practical Implications

Conducting a TT initiative has obvious challenges for an athlete, which is particularly related to changing from a well-known sport to a less known sport and to trying to reach an elite level over a relatively short time. As earlier research has highlighted [6], such transfer initiatives would be easier when there are large similarities between the donor and the transfer sport, indicating that both athletes and coaches should highlight the importance of choosing a relevant transfer sport. Still, the same research has shown that psychological factors play an important role in TT with the high quality required associated with effective talent development. Our study confirmed the earlier studies on both these topics, indicating that there are donor sports that seemed more physiologically suitable for recruitment to XC skiing, even though the athletes' psychological skills were considered additionally essential by the coaches. Overall, the coaches in this study believed in the general concept of TT, but they highlighted that the identification process must be performed more thoroughly and should be based on experts in the transfer sport. These results would indicate that TT is an aspect to consider for most sport governing bodies to supplement their talent development process. Still, the coaching TT athletes would probably require different approaches compared to their "normally recruited" athletes.

5. Conclusions

The assessments of coaches working with a group of summer endurance athletes transferring to XC skiing revealed that greater development of both physiological and technical capacities among the high-responding TT athletes were associated with higher motivation, greater ability to perceive and adjust their training, as well as superior ability to deal with adversity in the development process. More specifically, superior motivation for learning a new sport and the development process leading to a higher performance in XC skiing seemed to be key differences between high and low responders. Most of these high responders, who also adapted faster to XC skiing and tolerated the training better than low responders, had a background in running and were younger men. Furthermore, according to some of the coaches, the better ability to perceive and adjust training load appropriately might have induced higher compliance to training, and thereby larger training loads as well as fewer injuries and less sickness, even though other coaches did not share this opinion. The coaches agreed that superior ability to reflect upon their own training process and deal with adversity differentiated high responders from their lower-responding peers. In general, the coaches considered the TT program to be positive; however, successfully transferring athletes to a world class level in a complex sport such as XC skiing requires a multidisciplinary selection process and a longer time frame than the six-month period used in the current project.

Author Contributions: Conceptualization, M.I. and S.A.S.; methodology, M.I.; formal analysis, M.I.; investigation, M.I.; writing—original draft preparation, S.A.S., M.I., R.K.T. and Ø.S.; writing—review and editing, S.A.S., M.I., R.K.T. and Ø.S.; supervision, S.A.S.; project administration, M.I. All authors have read and agreed to the published version of the manuscript.

Funding: The study is funded by Meråker High School and the Research Council of Norway (RCN), project number 298645.

Institutional Review Board Statement: The research was undertaken according to the ethical guidelines of the Norwegian Centre for Research Data (NSD).

Informed Consent Statement: Informed consent was obtained from all subjects involved in the study.

Data Availability Statement: The data are not publicly available as data sharing was not included within the original study ethics submission or participant consent form.

Acknowledgments: The author would like to thank all coaches and athletes involved in the project for their participation.

Conflicts of Interest: The authors declare no conflict of interest.

References

1. Till, K.; Baker, J. Challenges and [Possible] Solutions to optimizing talent identification and development in sport. *Front. Psychol.* **2020**, *11*, 664. [CrossRef] [PubMed]
2. Bridge, M.W.; Toms, M.R. The specialising or sampling debate: A retrospective analysis of adolescent sports participation in the UK. *J. Sports Sci.* **2013**, *31*, 87–96. [CrossRef] [PubMed]

3. Halson, S.; Martin, D.T.; Gardner, A.S.; Fallon, K.; Gulbin, J. Persistent fatigue in a female sprint cyclist after a talent-transfer initiative. *Int. J. Sports Physiol. Perform.* **2006**, *1*, 65–69. [CrossRef] [PubMed]
4. Collins, R.; Collins, D.; MacNamara, Á.; Jones, M.I. Change of plans: An evaluation of the effectiveness and underlying mechanisms of successful talent transfer. *J. Sports Sci.* **2014**, *32*, 1621–1630. [CrossRef] [PubMed]
5. Weber, A.C.; De Bosscher, V.; Kempf, H. Positioning in Olympic Winter sports: Analysing national prioritisation of funding and success in eight nations. *Eur. Sport Manag. Q.* **2018**, *18*, 8–24. [CrossRef]
6. van Harten, K.; Bool, K.; van Vlijmen, J.; Elferink-Gemser, M. Talent transfer: A systematic review. *Curr. Issues Sport Sci.* **2021**, *6*. [CrossRef]
7. Bonal, J.; Jiménez, S.L.; Lorenzo, A. The Talent Development Pathway for Elite Basketball Players in China. *Int. J. Environ. Res. Public Health* **2020**, *17*, 5110. [CrossRef] [PubMed]
8. Vaeyens, R.; Güllich, A.; Warr, C.R.; Philippaerts, R. Talent identification and promotion programmes of Olympic athletes. *J. Sports Sci.* **2009**, *27*, 1367–1380. [CrossRef] [PubMed]
9. Talsnes, R.K.; Hetland, T.-A.; Cai, X.; Sandbakk, Ø. Development of performance, physiological and technical capacities during a six-month cross-country skiing talent transfer program in endurance athletes. *Front. Sports Act. Living* **2020**, *2*, 103. [CrossRef] [PubMed]
10. MacNamara, Á.; Button, A.; Collins, D. The role of psychological characteristics in facilitating the pathway to elite performance part 2: Examining environmental and stage-related differences in skills and behaviors. *Sport Psychol.* **2010**, *24*, 74–96. [CrossRef]
11. Bransford, J.; Schwartz, D. *Rethinking Transfer: A Simple Proposal with Multiple Implications*; American Educational Research Association: Washington, DC, USA, 1999; Volume 24.
12. MacNamara, Á.; Button, A.; Collins, D. The role of psychological characteristics in facilitating the pathway to elite performance part 1: Identifying mental skills and behaviors. *Sport Psychol.* **2010**, *24*, 52–73. [CrossRef]
13. Sandbakk, Ø.; Holmberg, H.C. Physiological capacity and training routines of elite cross-country skiers: Approaching the upper limits of human endurance. *Int. J. Sports Physiol. Perform.* **2017**, *12*, 1003–1011. [CrossRef] [PubMed]
14. Talsnes, R.K.; van den Tillaar, R.; Cai, X.; Sandbakk, Ø. Comparison of high- vs. low-responders following a 6-month XC ski-specific training period: A multidisciplinary approach. *Front. Sports Act. Living* **2020**, *2*, 114. [CrossRef] [PubMed]
15. Braun, V.; Clarke, V.; Wheate, P. Using thematic analysis in sport and exercise research. In *Routledge Handbook of Qualitative Research in Sport and Exercise*; Smith, B., Sparkes, A., Eds.; Routledge: London, UK, 2016; pp. 191–205.
16. Tracy, S.J. Qualitative quality: Eight 'big-tent' criteria for excellent qualitative research. *Qual. Inq.* **2010**, *16*, 837–851. [CrossRef]
17. Kvale, S.; Brinkmann, S. *Interviews: Learning the Craft of Qualitative Research Interviewing*, 2nd ed.; Sage Publications: Thousand Oaks, CA, USA, 2009.
18. Rees, T.; Hardy, L.; Güllich, A.; Abernethy, B.; Côté, J.; Woodman, T.; Montgomery, H.; Laing, S.; Warr, C. The great British medalists project: A review of current knowledge on the development of the world's best sporting talent. *Sports Med.* **2016**, *46*, 1041–1058. [CrossRef] [PubMed]
19. McCormick, A.; Meijen, C.; Marcora, S. Psychological determinants of whole-body endurance performance. *Sports Med.* **2015**, *45*, 997–1015. [CrossRef] [PubMed]
20. Martindale, R.J.; Collins, D.; Daubney, J. Talent development: A guide for practice and research within sport. *Quest* **2005**, *57*, 353–375. [CrossRef]

sports

Article

Physical and Anthropometric Characteristics Do Not Differ According to Birth Year Quartile in High-Level Junior Australian Football Players

Paul Larkin [1,2,*], Carl T. Woods [1], Jade Haycraft [1] and David B. Pyne [3]

1 Institute for Health and Sport, Victoria University, Melbourne, VIC 8001, Australia; carl.woods@vu.edu.au (C.T.W.); jade.haycraft@vu.edu.au (J.H.)
2 Maribrynong Sports Academy, Melbourne, VIC 3032, Australia
3 Research Institute for Sport and Exercise, University of Canberra, Canberra, ACT 2617, Australia; David.Pyne@canberra.edu.au
* Correspondence: paul.larkin@vu.edu.au

Abstract: The aim of this study was to explore differences in the physical fitness and anthropometric profiles between birth year quartiles of players attending the Australian Football League (AFL) National Draft Combine. Date of birth, anthropometric, 20 m sprint, vertical and running vertical jump, AFL planned agility, and 20 m Multi-Stage Fitness Test (MSFT) data were obtained for players selected to attend the Combine between 1999 and 2019 (n = 1549; M_{age} = 18.1; SD_{age} = 0.3). The underlying density distributions of the data were visually explored using violin plots overlaid with box and whisker plots. A multivariate analysis of variance (MANOVA) was then used to model the main effect of birth quartile (four levels) on the physical and anthropometric scores. Results showed that physical and anthropometric test scores did not significantly differ according to birth quartile ($V = 0.008$, $F = 0.880$, $p = 0.631$). We conclude that the physical and anthropometric profiles of high-level junior Australian Football players were similar according to birth year quartile across the modeled period. Therefore, how players utilize their physical and anthropometric attributes during game-play via contextualized, representative assessments, such as small-sided games, should be considered when examining potential causes of a RAE.

Keywords: talent identification; talent development; performance assessment; relative age effect

Citation: Larkin, P.; Woods, C.T.; Haycraft, J.; Pyne, D.B. Physical and Anthropometric Characteristics Do Not Differ According to Birth Year Quartile in High-Level Junior Australian Football Players. *Sports* **2021**, *9*, 111. https://doi.org/10.3390/sports9080111

Academic Editors: Adam Leigh Kelly, Sergio L. Jiménez Sáiz, Sara Diana Leal dos Santos and Alberto Lorenzo Calvo

Received: 22 June 2021
Accepted: 10 August 2021
Published: 12 August 2021

1. Introduction

Annual age-grouping policies are common across most team sports and involve organization of athletes into defined chronological age groups. The Australian Football League (AFL) participation pathway has two competition streams: the local participation pathway and the talent pathway that flows into the elite competition. The local participation pathway consists of age-grouped levels from Under (U) 10 years old to open age competition, with the talent pathway encompassing regional and development squads, state squads at the U14–U16 levels, and national squads at the U16–U18 levels, with players then potentially selected for a professional senior club. Players identified as talented within the local participation pathway are selected by coaches and talent managers to join their regional team to compete in state-level competitions (see Woods [1]). The relative age effect (RAE) is a demographic phenomenon involving a bias towards the selection of athletes born earlier in an age group year compared to those born later in the same year [2–4]. Further, the RAE is commonly observed in male invasion sports that require physical precocity, such as Australian football [3,5,6], basketball [7–9], ice hockey [10–12], rugby [13–15], and soccer [4,16–22].

Talent in sport is not dependent on birth month, though, but rather on a complex interaction of multifactorial performance attributes including physiological, technical, tactical, psychological, and sociological influences [23,24]. The RAE has been reported to occur early in the development pathways of junior athletes in invasion sports, implying a bias towards the selection of early maturing athletes over late-maturing talented athletes [5,25–28]. Physiological growth and maturation have been proposed as the underlying mechanism for the RAE in invasion sport development pathways [25], with the superior physicality of early maturing players often confused as talent by coaches and talent scouts [6]. However, the chronological age of players may have a greater impact on an athlete's playing experience, perceptual and motor skills, and social and psychological development, more than their physiological maturation [29].

Additionally, in popular team invasion sports, such as Australian football, which have a large grass-root participation base, the RAE may be amplified across key stages of development, given the high participation rates, creating selection pressure at both local participation and talent levels [30]. In Australian football, specifically, the RAE has been reported within the U10-U12 competitions [6], amplifying as players are selected into the talent pathway [5,6,28]. To mitigate the RAE within the AFL talent pathway, two rule changes to the age-policy of National U18 draft attendees in 2003 and 2009 were implemented. While policy changes removed birth month bias in the first half of the selection year, the RAE was still evident in the first and last quarter of the selection calendar [28]. Selection bias also occurred at the State U16 level and likely carried through to the National Draft level (i.e., U18), making age-policy changes at the National U18 level irrelevant [28]. In mature-aged AFL draftees (>20 years-old), birth month bias was reversed, with 63% of mature-aged players drafted to elite AFL teams having been born in the latter half of the year [3]. Therefore, given this known RAE in Australian football [28] it is possible that the physical fitness and anthropometric attributes of junior (U18) talent selected Australian football players is influenced by their birth quartile, manifesting in the noted selection bias. This distribution is, however, yet to be examined.

Despite increased interest and proposed solutions to address the issue, researchers acknowledge there is limited impact of research outcomes on talent selection practices [31]. Current talent selection practices within the AFL talent pathway involve coaches and selectors inviting talent identified players to attend State and National Draft Combines. AFL clubs then select players at the annual National Draft to join their club playing lists [32]. The annual National Draft Combine places a heavy focus on selected anthropometric measures (i.e., height and body mass) and physical tests including 20 m sprint, vertical and running vertical jump, AFL planned agility, and aerobic endurance, previously the 20 m Multi-Stage Fitness Test (MSFT), and now the YOYO Intermittent Recovery (IR) 2 [33,34]. Birth quartile and birth half of the selection year did not seem to have a major impact on physical and anthropometric profiles of players drafted to the AFL following their attendance at the National Draft Combine between 2010 and 2013 [5]. However, birth quartile may have a greater impact on physical and anthropometric profiles of players selected to attend the National Combine. While anthropometric and physical fitness attributes have a role in talent selection, training, and management, there is limited understanding of the implications of the physical profile of Australian football players born in different birth quartiles on the RAE over an extended period within the AFL talent pathway.

As Haycraft and colleagues [28] identified a RAE within a high-level junior sample (U18 age group) attending the National Combine, the aim of this study was to conduct a 20-year retrospective cross-sectional analysis of the AFL National Draft Combine. As the National Draft Combine places an emphasis on physical performance, this analysis would determine whether the RAE identified by Haycraft and colleagues [28] is influenced by physical and anthropometric differences between the birth year quartiles. Given the reported RAE within this U18 age group sample [28], we expected that physical and anthropometric profiles would differ relative to birth year quartile.

2. Materials and Methods

This study used a 20-year retrospective cross-sectional design to examine differences in the physical and anthropometric profiles between male Australian footballers born in differing age group quartiles. Date of birth (DOB), physical fitness, and anthropometric data were obtained for U18 players who were selected to attend the AFL National Draft Combine (n = 1549; M_{age} = 18.1; SD_{age} = 0.3; $M_{attendees/year}$ = 79; $SD_{attendees/year}$ = 16) between 1999 and 2019. National player data were available for all years between 1999 and 2019. Players were classified into the following birth quartiles: Q1: January–March, N = 1396, 34%; Q2: April–June, N = 1068, 26%; Q3: July–September, N = 913, 23%; Q4: October–December N = 694, 17%. In 2003 and 2008, changes to age eligibility policies were imposed at the National U18 level to target the RAE of players invited to attend the AFL National Draft Combine. Specifically, between 1999 and 2003 players were required to turn 17 years of age by June 30, with this cut-off date shifting to April 30 in 2008. Player birth quartiles of those attending National U18 drafts in these years were classified based on the first month of the new age eligibility date (e.g., 2004–2008 Q1: May–July). All players 17 years of age were excluded from analysis between 2010 and 2013, as during these years the acquisition of these players was limited to trades, given the introduction of two new AFL teams [28]. The study was approved by the university's human research ethics committee.

The following test data were provided by the AFL: anthropometric measures (standing height (cm), Typical Error (TE) = 1.0 cm; body mass (kg), TE = 1.0 kg); 20-m sprint (s), TE = 0.03 s; stationary vertical and running vertical jump (cm), TE = 1.4 cm; AFL planned agility (s), TE = 0.04 s; and aerobic endurance (20 m MSFT, measured via total distance reached, TE = 3%); with the best of three efforts recorded for all tests except the 20 m MSFT [34,35]. The 20 m sprint was performed from a standing stationary start, with time recorded using electronic timing gates. The vertical jump test was also performed from a stationary start, with the running vertical jump test requiring players to take a 5 m run up and perform three jumps off both their left and right legs. The AFL agility course was measured via a timed trial through a course of cones positioned 5 m apart and included three left and two right 90° angle turns. Aerobic endurance was measured using the 20 m MSFT, which is a repeated 20 m interval test that progressively increases speed until players are no longer capable of keeping pace [36,37]. All testing was conducted over a 3–4-day period at the conclusion of the junior competition season (October) and preceded the AFL National Draft by approximately 5–6 weeks.

Statistical Analysis

First, the underlying distribution of each assessment relative to each year was visually inspected using overlaid density distribution plots in the ggplot2 package (Wickham, 2016) in the R computing language (R Core Team, 2019). This step enabled us to confidently pool the dataset. Next, to explore the underlying density distributions of the data, violin plots with an overlaid box and whisker plot (showing the median and interquartile range, with minimum and maximum test values as jittered points) were developed for all physical and anthropometric tests relative to birth quartile. This analysis was completed using the ggplot2 package (Wickham, 2016), with the plots arranged and annotated using the cowplot package (Wilke, 2017) in the R computing language (R Core Team, 2019, Vienna, Austria). A multivariate analysis of variance (MANOVA) was then used to model the main effect of birth quartile (four levels) on the physical and anthropometric test scores. This modeling was completed using the "manova()" function, with significance set $\alpha < 0.05$. Pending the outcomes of the MANOVA (noted via a significant Pillai's Trace), univariate analysis of variance was planned to identify which (if any) of the physical and/or anthropometric tests differed significantly according to birth quartile.

3. Results

Visual inspection of the density distributions of the data using the violin plots (Figure 1) indicates that physical and anthropometric test scores do not appear to meaningfully differ relative to birth quartile. In support of this observation, the MANOVA did not yield a significant Pillai's Trace ($V = 0.008$, $F = 0.880$, $p = 0.631$), indicating the physical and anthropometric profiles of players were similar across all four birth quartiles. Given this outcome, follow up univariate analysis was not performed on the data. Mean data for the physical and anthropometric characteristics relative to birth quartile are provided in the online Supplementary Materials.

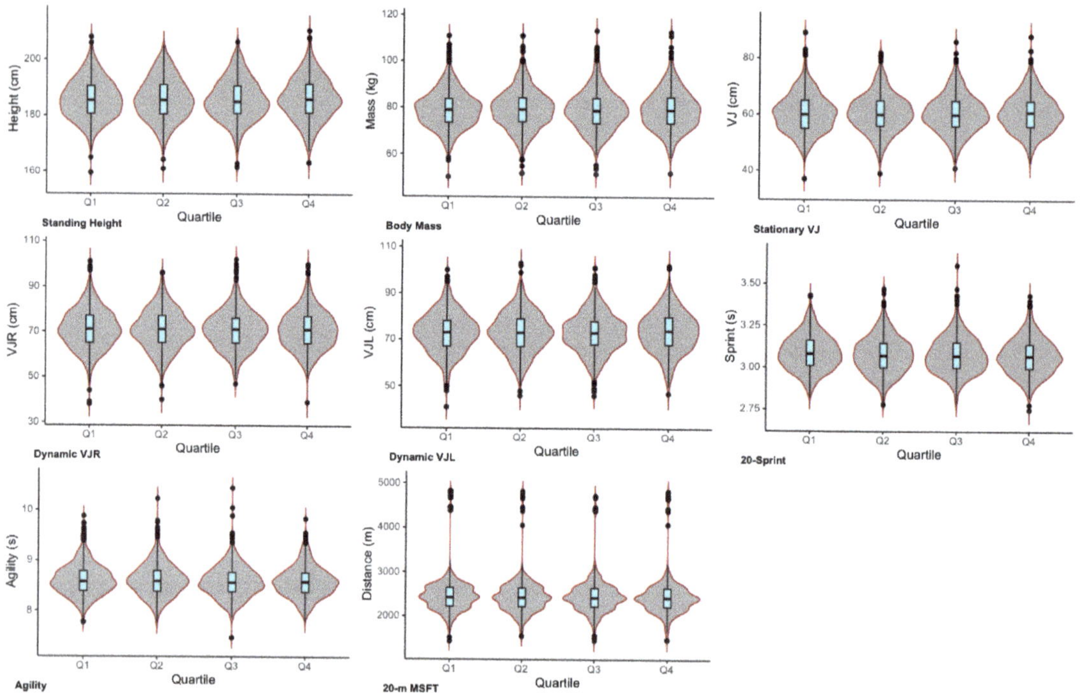

Figure 1. Violin plots with overlaid box and whisker plots showing the underlying density distributions of the data. Note: the 'dots' reflect the maximum and minimum test scores, "VJR" denotes vertical jump right leg take-off, and "VJL" denotes vertical jump left leg take-off.

4. Discussion

The aim of this study was to explore if the physical and anthropometric attributes of players attending the AFL National Draft Combine differed according to birth quartile. Results did not yield a significant effect of birth quartile, with all players at this U18 level possessing a relatively homogenous physical and anthropometric profile. Therefore, it appears that players are largely selected for the AFL National Draft Combine based on relatively homogenized physical and anthropometric profiles, supporting previous studies which have suggested this occurs in professional adult populations [38,39]. Therefore, the RAE in this population may manifest in technical and tactical skills, not necessarily physical precocity.

Several studies have investigated the prevalence of RAE within talent development programs and elite senior competitions of invasion sports [3–5,19,40]. However, there has been limited investigation into the physical profile differences relative to birth quartile in Australian football. Our results indicate that, over the 20-year period, there was a relatively uniform physical and anthropometric profile for high-level junior Australian football players attending the National Draft Combine. This outcome supports the earlier work of Woods, Robertson [5] who, on a smaller sample, demonstrated that physical and anthropometric assessments did not differ according to birth year quartile or half year in drafted National U18 Australian football players. The RAE reported by Haycraft et al. [28] may, therefore, be the result of other factors that influence player performances, such as the technical and tactical skills needed to compete within Australian football [29], rather than physical capabilities. Taken together these data imply that, in late adolescence, early maturing athletes may not be able to maintain their physical advantages–an observation seen in soccer [37] and rugby league [13,14]. This outcome is in contradiction to several studies showing that athletes born early in a selection year are more biologically mature than those born later in the same year, with substantial differences in body size, height, strength, and power, and who may, therefore, possess an advantage during physical testing [5,28,33]. Physical and anthropometric profiles may only assist players in gaining initial access into the talent development pathway [26,28,41], with the RAE manifested by other domain-specific attributes at later stages in development. Future work in Australian football should ascertain more contextual information on how players utilize their physical and anthropometric attributes during game-play (e.g., winning the ball in a tight position and then breaking clear from the pack) to explain the observed RAE.

While the AFL National Draft Combine provides a detailed assessment of the physical and anthropometric qualities of talented U18 players [34,42], they are limited in their contextualization. This shortcoming has been highlighted by Bonney, Berry [43] who proposed that isolated, decontextualized laboratory assessments provide information on movement patterns or physical capabilities, but do not identify the proficiency of a player's skill under match conditions within Australian football. To better understand the impact of RAE on talent development and selection in Australian football, contextualized and more representative tests should be considered, such as small-sided game assessments [44,45]. The likely benefit of incorporating such assessments is the integration of both physical and technical components in game-like contexts [36,43]. Future work should investigate whether incorporating a small-sided game assessment into the Australian football and other sporting talent pathways mitigates the RAE on talent development and selection [28].

While this study examined the influence of physical and anthropometric profiles on RAE at the National Draft Combine over a substantial time period, the outcomes should be considered with respect to limitations. A key limitation of the current study is the representativeness of the cohort analyzed. We acknowledge that the athletes assessed at the National Combine are selected to attend this event by coaches and selectors. Therefore, there is potential for an inherent bias within the sample based on how the players were invited to participate in the testing Combine. Further, while not a specific aim of this study, future research should examine whether the current findings are replicated between the local competition levels (i.e., U12′s to U18′s), and the entry levels of the talent development pathway (i.e., State U15′s and U16′s). This work will establish whether players who do not exhibit a physical profile within the associated bandwidths are discriminated against in relation to selection for more high-level talent development programs.

5. Conclusions

This study examined the physical and anthropometric profiles of high-level junior Australian football players attending the AFL National Draft Combine according to birth year quartile over a 20-year period. There were no significant differences in the physical and anthropometric profiles across the birth year quartiles, despite there being a known RAE across the same cohort. Therefore, it appears that players are selected for the AFL National Draft Combine based on relatively homogenized physical and anthropometric profiles, and that the RAE is influenced through other aspects of performance, such as technical and tactical skill. We recommend that future work examines the performance of players attending the National Combine via contextualized, representative assessments, such as small-sided games, when examining potential causes of a RAE.

Supplementary Materials: The following are available online at https://www.mdpi.com/article/10.3390/sports9080111/s1, Table S1: Physical and anthropometric characteristics of players relative to birth quartile.

Author Contributions: Conceptualization, P.L., C.T.W., J.H. and D.B.P.; methodology, P.L., C.T.W., J.H. and D.B.P.; formal analysis, C.T.W.; writing—original draft preparation, P.L, C.T.W. and J.H.; writing—review and editing, D.B.P.; visualization, C.T.W. All authors have read and agreed to the published version of the manuscript.

Funding: This research received no external funding.

Institutional Review Board Statement: The study was conducted according to the guidelines of the Declaration of Helsinki and approved by the Victoria University Human Ethics Research Committee (HRE 16-064 and 30 May 2016).

Informed Consent Statement: Informed consent was waived due to analysis being conducted on a retrospective data set which was provided to the researchers for analysis.

Data Availability Statement: Due to institutional ethics requirements the data could not be shared publicly.

Acknowledgments: The authors would like to thank the Australian Football League for allowing access to the data for analysis.

Conflicts of Interest: The authors declare no conflict of interest.

References

1. Woods, C. The development of an objective multidimensional approach to talent identification in junior Australian football. In *School of Exercise and Health Sciences*; Edith Cowan University: Perth, Australia, 2015; p. 181.
2. Andronikos, G.; Elumaro, A.I.; Westbury, T.; Martindale, R.J.J. Relative age effect: Implications for effective practice. *J. Sports Sci.* **2015**, *34*, 1124–1131. [CrossRef]
3. Sullivan, C.; Bilsborough, J.C.; Cianciosi, M.; Hocking, J.; Cordy, J.; Coutts, A.J. Match score affects activity profile and skill performance in professional Australian Football players. *J. Sci. Med. Sport* **2014**, *17*, 326–331. [CrossRef]
4. Simmons, C.; Paull, G.C. Season-of-birth bias in association football. *J. Sports Sci.* **2001**, *19*, 677–686. [CrossRef]
5. Woods, C.T.; Robertson, S.J.; Gastin, P.B. Does relative age distribution influence the physical and anthropometric profiles of drafted under 18 Australian footballers? An investigation between the 2010 and 2013 seasons. *Talent. Dev. Excell.* **2015**, *7*, 83–90.
6. Tribolet, R.; Watsford, M.L.; Coutts, A.J.; Smith, C.; Fransen, J. From entry to elite: The relative age effect in the Australian football talent pathway. *J. Sci. Med. Sport* **2019**, *22*, 741–745. [CrossRef]
7. Delorme, N.; Raspaud, M. The relative age effect in young French basketball players: A study on the whole population. *Scand. J. Med. Sci. Sports* **2008**, *19*, 235–242. [CrossRef]
8. Arrieta, H.; Torres-Unda, J.; GIL Orozko, S.M.; Irazusta, J. Relative age effect and performance in the U16, U18 and U20 European Basketball Championships. *J. Sports Sci.* **2015**, *34*, 1530–1534. [CrossRef] [PubMed]
9. Riaza, A.D.L.R.; Calvo, J.L.; Mon-López, D.; Lorenzo, A. Impact of the Relative Age Effect on Competition Performance in Basketball: A Qualitative Systematic Review. *Int. J. Environ. Res. Public Health* **2020**, *17*, 8596. [CrossRef]
10. Wattie, N. A historical examination of relative age effects in Canadian hockey players. *Int. J. Sport Psychol.* **2007**, *38*, 178–186.
11. Boucher, J.L.; Mutimer, B.T.P. The Relative Age Phenomenon in Sport: A Replication and Extension with Ice-Hockey Players. *Res. Q. Exerc. Sport* **1994**, *65*, 377–381. [CrossRef]
12. Lemoyne, J.; Pelletier, V.H.; Trudeau, F.; Grondin, S. Relative Age Effect in Canadian Hockey: Prevalence, Perceived Competence and Performance. *Front. Sports Act. Living* **2021**, *3*, 14. [CrossRef] [PubMed]
13. Till, K.; Cobley, S.; Hara, J.O.; Cooke, C.; Chapman, C. Considering maturation status and relative age in the longitudinal evaluation of junior rugby league players. *Scand. J. Med. Sci. Sports* **2014**, *24*, 569–576. [CrossRef]
14. Till, K.; Cobley, S.; Wattie, N.; O'Hara, J.; Cooke, C.; Chapman, C. The prevalence, influential factors and mechanisms of relative age effects in UK Rugby League. *Scand. J. Med. Sci. Sports* **2009**, *20*, 320–329. [CrossRef]
15. Kelly, A.L.; Till, K.; Jackson, D.; Barrell, D.; Burke, K.; Turnnidge, J. Talent Identification and Relative Age Effects in English Male Rugby Union Pathways: From Entry to Expertise. *Front. Sports Act. Living* **2021**, *3*, 12. [CrossRef]
16. Craig, T.P.; Swinton, P. Anthropometric and physical performance profiling does not predict professional contracts awarded in an elite Scottish soccer academy over a ten-year period. *Eur. J. Sport Sci.* **2020**, *21*, 1101–1110. [CrossRef]
17. Höner, O.; Leyhr, D.; Kelava, A. The influence of speed abilities and technical skills in early adolescence on adult success in soccer: A long-term prospective analysis using ANOVA and SEM approaches. *PLoS ONE* **2017**, *12*, e0182211. [CrossRef] [PubMed]

18. Cumming, S.P.; Searle, C.; Hemsley, J.K.; Haswell, F.; Edwards, H.; Scott, S.; Gross, A.; Ryan, D.; Lewis, J.; White, P.; et al. Biological maturation, relative age and self-regulation in male professional academy soccer players: A test of the underdog hypothesis. *Psychol. Sport Exerc.* **2018**, *39*, 147–153. [CrossRef]
19. Van den Honert, R. Evidence of the relative age effect in football in Australia. *J. Sport Sci.* **2012**, *30*, 1365–1374. [CrossRef]
20. Kelly, A.L.; Wilson, M.R.; Gough, L.A.; Knapman, H.; Morgan, P.; Cole, M.; Jackson, D.T., Williams, C.A. A longitudinal investigation into the relative age effect in an English professional football club: Exploring the 'underdog hypothesis'. *Sci. Med. Footb.* **2019**, *4*, 111–118. [CrossRef]
21. Leyhr, D.; Bergmann, F.; Schreiner, R.; Mann, D.; Dugandzic, D.; Höner, O. Relative Age-Related Biases in Objective and Subjective Assessments of Performance in Talented Youth Soccer Players. *Front. Sports Act. Living* **2021**, *3*, 94. [CrossRef]
22. Dugdale, J.H.; McRobert, A.P.; Unnithan, V.B. "He's just a wee laddie": The relative age effect in male Scottish soccer. *Front. Psychol.* **2021**, *12*, 103. [CrossRef]
23. Larkin, P.; Marchant, D.; Syder, A.; Farrow, D. An eye for talent: The recruiters' role in the Australian Football talent pathway. *PLoS ONE* **2020**, *15*, e0241307. [CrossRef] [PubMed]
24. Reeves, M.J. Factors affecting the identification of talented junior-elite footballers: A case study. *Soccer Soc.* **2018**, *19*, 1106–1121. [CrossRef]
25. Cobley, S.; Baker, J.; Wattie, N.; McKenna, J. Annual Age-Grouping and Athlete Development. *Sports Med.* **2009**, *39*, 235–256. [CrossRef]
26. Toum, M.; Tribolet, R.; Watsford, M.L.; Fransen, J. The confounding effect of biological maturity on talent identification and selection within youth Australian football. *Sci. Med. Footb.* **2020**, 1–9. [CrossRef]
27. Schorer, J. Relative age, talent identification and youth skill development: Do relatively younger athletes have superior technical skills. *Talent Dev. Excell.* **2009**, *1*, 45–56.
28. Haycraft, J.A.; Kovalchik, S.; Pyne, D.B.; Larkin, P.; Robertson, S. The influence of age-policy changes on the relative age effect across the Australian Rules football talent pathway. *J. Sci. Med. Sport* **2018**, *21*, 1106–1111. [CrossRef]
29. Romann, M.; Rüeger, E.; Hintermann, M.; Kern, R.; Faude, O. Origins of Relative Age Effects in Youth Football—A Nationwide Analysis. *Front. Sports Act. Living* **2020**, *2*, 2. [CrossRef] [PubMed]
30. Musch, J.; Grondin, S. Unequal Competition as an Impediment to Personal Development: A Review of the Relative Age Effect in Sport. *Dev. Rev.* **2001**, *21*, 147–167. [CrossRef]
31. Steidl-Müller, L.; Hildebrandt, C.; Raschner, C.; Müller, E. Challenges of talent development in alpine ski racing: A narrative review. *J. Sports Sci.* **2018**, *37*, 601–612. [CrossRef] [PubMed]
32. Coutts, A.J.; Kempton, T.; Vaeyens, R. Relative age effects in Australian Football League National Draftees. *J. Sports Sci.* **2013**, *32*, 623–628. [CrossRef]
33. Haycraft, J.; Kovalchik, S.; Pyne, D.; Robertson, S. Classification of Players Across the Australian Rules Football Participation Pathway Based on Physical Characteristics. *J. Strength Cond. Res.* **2020**. [CrossRef]
34. Woods, C.T.; Raynor, A.; Bruce, L.; McDonald, Z.; Collier, N. Predicting playing status in junior Australian Football using physical and anthropometric parameters. *J. Sci. Med. Sport* **2015**, *18*, 225–229. [CrossRef]
35. Pyne, D.; Gardner, A.; Sheehan, K.; Hopkins, W. Fitness testing and career progression in AFL football. *J. Sci. Med. Sport* **2005**, *8*, 321–332. [CrossRef]
36. Bonney, N.; Berry, J.; Ball, K.; Larkin, P. Validity and reliability of an Australian football small-sided game to assess kicking proficiency. *J. Sports Sci.* **2020**, *38*, 79–85. [CrossRef]
37. Malina, R.M.; Eisenmann, J.C.; Cumming, S.P.; Ribeiro, B. Maturity-associated variation in the growth and functional capacities of youth football (soccer) players 13–15 years. *Graefe's Arch. Clin. Exp. Ophthalmol.* **2004**, *91*, 555–562. [CrossRef]
38. Franks, A. Talent identification in elite youth soccer players: Physical and physiological characteristics. *J. Sport Sci.* **1999**, *17*, 812.
39. Ford, P.R.; Williams, M.A. No Relative Age Effect in the Birth Dates of Award-Winning Athletes in Male Professional Team Sports. *Res. Q. Exerc. Sport* **2011**, *82*, 570–573. [CrossRef] [PubMed]
40. Finnegan, L.; Richardson, D.; Littlewood, M.; McArdle, J. The influence of date and place of birth on youth player selection to a National Football Association elite development programme. *Sci. Med. Footb.* **2017**, *1*, 30–39. [CrossRef]
41. Ackland, T.; Bloomfield, J. Stability of human proportions through adolescent growth. *Aust. J. Sci. Med. Sport* **1996**, *28*, 57–60.
42. Robertson, S.; Woods, C.; Gastin, P. Predicting higher selection in elite junior Australian Rules football: The influence of physical performance and anthropometric attributes. *J. Sci. Med. Sport* **2015**, *18*, 601–606. [CrossRef] [PubMed]
43. Bonney, N.; Berry, J.; Ball, K.; Larkin, P. Australian Football Skill-Based Assessments: A Proposed Model for Future Research. *Front. Psychol.* **2019**, *10*, 429. [CrossRef]
44. Fenner, J.S.J.; Iga, J.; Unnithan, V. The evaluation of small-sided games as a talent identification tool in highly trained prepubertal soccer players. *J. Sports Sci.* **2016**, *34*, 1983–1990. [CrossRef] [PubMed]
45. Bennett, K.J.M.; Novak, A.R.; Pluss, M.A.; Stevens, C.J.; Coutts, A.J.; Fransen, J. The use of small-sided games to assess skill proficiency in youth soccer players: A talent identification tool. *Sci. Med. Footb.* **2018**, *2*, 231–236. [CrossRef]

sports MDPI

Article

Relative Age Effects in Basketball: Exploring the Selection into and Successful Transition Out of a National Talent Pathway

Adam L. Kelly [1,*], Sergio Lorenzo Jiménez Sáiz [2], Alberto Lorenzo Calvo [3], Alfonso de la Rubia [3], Daniel T. Jackson [1], Mark A. Jeffreys [1], Charlie Ford [4], Dave Owen [4] and Sara Diana Leal dos Santos [5]

1 Research Centre for Life and Sport Sciences (CLaSS), Department of Sport and Exercise, School of Health Sciences, Birmingham City University, Birmingham B15 3TN, West Midlands, UK; Daniel.Jackson6@mail.bcu.ac.uk (D.T.J.); Mark.Jeffreys@bcu.ac.uk (M.A.J.)
2 Centre for Sport Studies, Universidad Rey Juan Carlos, Fuenlabrada, 28933 Madrid, Spain; sergio.jimenez.saiz@urjc.es
3 Departamento de Deportes, Facultad de Ciencias de la Actividad Física y del Deporte—INEF, Universidad Politécnica de Madrid, 28040 Madrid, Spain; alberto.lorenzo@upm.es (A.L.C.); alfonso.delarubia@upm.es (A.d.l.R.)
4 Basketball England, NCS, Gate 13, Etihad Campus, Rowsley Street, Manchester M11 3FF, Lancashire, UK; charlie.ford@englandboxing.org (C.F.); dave.owen@basketballengland.co.uk (D.O.)
5 Research Center in Sports Sciences, Health Sciences and Human Development (CIDESD), University of Trás-os-Montes and Alto Douro, 5000-801 Vila Real, Portugal; sarasantos_8@hotmail.com
* Correspondence: Adam.Kelly@bcu.ac.uk

Abstract: Relative age effects (RAEs) appear consistently prevalent throughout the youth basketball literature. However, the selection into and successful transition out of a national talent pathway in basketball is yet to be explored. Thus, the primary aim of this study was to explore the influence of relative age, gender, and playing time based on the selection into the Regional Talent Hubs and Basketball England youth teams (U16, U18, and U20) and the successful transition into the England National Senior Teams. Participants who were selected into the male ($n = 450$) and female ($n = 314$) Basketball England Talent Pathway were allocated into one of three cohorts: (a) Regional Talent Hubs (U12 to U15; $n = 183$), (b) England National Youth Teams (U16, U18, and U20; $n = 537$), and (c) England National Senior Teams ($n = 44$). A chi-square test was used to compare the birth quarter (BQ) distributions of each cohort against the expected distributions, with a Cramer's V (Vc) used to interpret effect sizes. Odds ratios (OR) and 95% confidence intervals were also calculated to compare the likelihood of each BQ being represented. Males revealed significant RAEs across both the Regional Talent Hubs ($p < 0.001$, Vc > 0.29, OR = 10) and England National Youth Teams ($p < 0.001$, Vc > 0.17, OR = 3.1). In comparison, females only had significant RAEs in the Regional Talent Hubs ($p < 0.001$, Vc > 0.29, OR = 2.3). Despite RAEs being prevalent throughout youth levels, there were no significant differences in the BQ distribution based on playing time and those who made the successful transition to the England National Senior Teams. These findings demonstrate the potential mechanisms of RAEs in basketball, as well as the impetus to explore more equitable competition structures within the England Basketball Talent Pathway.

Keywords: talent identification; talent development; athlete development; expertise; sports coaching; growth and maturation

Citation: Kelly, A.L.; Jiménez Sáiz, S.L.; Lorenzo Calvo, A.; de la Rubia, A.; Jackson, D.T.; Jeffreys, M.A.; Ford, C.; Owen, D.; Santos, S.D.L.d. Relative Age Effects in Basketball: Exploring the Selection into and Successful Transition Out of a National Talent Pathway. *Sports* **2021**, 9, 101. https://doi.org/10.3390/sports9070101

Academic Editor: Corrado Lupo

Received: 21 June 2021
Accepted: 7 July 2021
Published: 12 July 2021

1. Introduction

Achieving expertise in basketball is a complex and multidimensional process [1]. For instance, a diverse sporting background during childhood [2], high jump and fast sprint capabilities [3], and advanced achievement and competitiveness motivation [4] have all been revealed as contributing factors towards greater long-term player development in basketball [5]. However, despite these multiple factors, senior professional and international basketball players must often acquire one common characteristic if they are to be selected: being tall [6]. For example, although the national average height for an American adult male is 5 ft 9 in (180 cm; [7]), the average height for a male professional basketball player competing in the National Basketball Association (NBA) is 6 ft 6 in (201 cm; [8]). Similarly, females competing in the Women's NBA (WNBA) are on average taller (6 ft; 183 cm; [9]) when compared with the average American adult female national norm (5 ft 4 in; 165 cm; [7]). Taller players can have an advantage in basketball due to their shots travelling less distance to the basket, they start closer to the rebound, and their ability to reach higher into the air offers a greater opportunity of blocking shorter players' passes and shots [10]. Since being taller can offer advantages in senior professional and international basketball competitions, it may also have important implications on talent identification and development processes in youth basketball [6].

Common selection and participation biases in youth sport are based on relative age effects—RAEs [11]. Relative age effects are based on a concept that demonstrates how youth athletes who are born at the beginning of a (bi)annual age group (e.g., under-16, under-18, and under-20) are more likely to be selected and/or participate in youth sport compared with their relatively younger peers [12]. For example, in their systematic review of 57 studies across a range of sports, Smith et al. [13] revealed 25% more female athletes born in the first birth quarter (BQ1; January, February, and March) of an annual age group were selected and/or participated in youth sport compared with those born in the last BQ (BQ4; October, November, and December). A possible explanation for this overrepresentation of relatively older athletes is due to the physiological and psychosocial advantages of being born earlier [14]. In basketball, for instance, height, body mass, running speed, and explosive power are important factors for greater performance [15], and these can be greatly affected by relative age [16]. In a systematic review of nine studies, de la Rubia et al. [17] demonstrated that RAEs had a significant impact on basketball performance, particularly within males and at formative ages (14–18 years). Since there appears to be a reliance on advanced physiological and anthropometrical characteristics for greater selection opportunities and performance levels in basketball, these factors may fortuitously exacerbate RAEs in youth settings [18].

Pronounced RAEs in youth basketball have been previously reported in various mixed-gender case studies from Brazil [19,20], France [21], Germany [22], Japan [23], North America [24], Poland [25], Portugal [26], and Spain [27]. The prevalence of RAEs has also been revealed in international competitions, such as the Olympic Games [28], World Championships [29], European Championships [30], and the Adidas Next Generation Tournament (the top European competition for under-18s; [31]). As an example, Arrieta et al. [32] found a significant overrepresentation of those born earlier in the selection year, as well as an association between relatively older age and performance outcomes (e.g., increased playing time), during the under-16, under-18, and under-20 International Basketball Federation (FIBA) European Championships. Further research has suggested competition level, age, and gender are important considerations when examining who is at risk of RAEs. For instance, García et al. [33] found weaker RAEs in the female youth FIBA World Championships compared with their male equivalents. Furthermore, they illustrated how RAEs decreased with older age, whereby they were highest in the under-17 competition, slightly less but also significant in the under-19 competition, and insignificant in the under-21 competition. Together, these findings highlight the importance of exploring a nation-specific context, as well as considering the influence of competition level, age, gender, and performance outcomes as contributing sources of RAEs in youth basketball.

Although RAEs are common throughout youth sport and especially in team sports, findings at senior levels appear mixed. On the one hand, some research in basketball has reported *knock-on effects* of RAEs during adulthood [34]. For instance, López de Subijana and Lorenzo [27] revealed long-term success at the senior professional level in Spanish basketball (e.g., BQ1 30% vs. BQ4 19%) was due to a relative age bias in selection at youth level (e.g., BQ1 38% vs. 9%). On the other hand, however, some literature has highlighted no RAEs at the senior professional level when compared with their youth cohorts (e.g., [35]). In the context of basketball, *no RAEs* were demonstrated at the senior international level in athletes who participated in the Olympic Games of London 2012 [36], and Rio de Janeiro 2016 [28]. In fact, *reversal effects* of relative age (e.g., [37]) or an *underdog hypothesis* [38] have been reported during the transition from youth levels to senior status in some team sports. For example, Kelly et al., [39] demonstrated potential late birthday benefits in soccer through the lens of the underdog hypothesis, whereby those born in BQ4 were approximately four times more likely to achieve a professional contract once selected into a youth academy compared with those born in BQ1. In sum, the variabilities in the outcomes at the senior level suggest RAEs may be associated with a combination of socio-environmental factors and sport-specific performance demands [40]. Interestingly, however, it appears the transition from youth level to senior representation in international basketball is yet to be explored. In doing so, it may offer further evidence of RAEs, while also depicting the effectiveness of existing talent identification and development processes.

Basketball England has participated in male and female international youth competitions at under-16, under-18, and under-20 age groups for over two decades. The main aim of these national youth teams is to develop and prepare young players for the senior Basketball England teams [41]. More recently, Basketball England created a pool of ten Regional Talent Hubs as an entry level into their Talent Pathway. The main purpose of these Talent Hubs is to identify talented young players aged 11 to 15 years, and offer them the greatest opportunity to develop towards the national teams in the future [42]. Despite the range of literature exploring the talent identification and development processes in youth basketball, the selection into and successful transition out of a national talent pathway in youth basketball is yet to be explored across both male and female cohorts. As such, the first aim of this current study was to explore the BQ distribution of the male and female Regional Talent Hubs (i.e., under-12 to under-15) as the entry level to the Basketball England Talent Pathway. Next, the BQ distribution of the male and female England National Youth Teams (i.e., under-16, under-18, and under-20) was analysed. Further analysis explored the BQ distribution of those who successfully transitioned from these England National Youth Teams to the male and female England National Senior Teams. Lastly, to test the competition opportunities once participants were selected for the England National Youth Teams and England National Senior Teams, differences between the average number of minutes played per game between BQs were compared. Based on the pre-existing literature, it was hypothesised that there would be an overrepresentation of relatively older players who played more minutes at youth levels. In comparison, it was hypothesised that there would be no RAEs at senior levels.

2. Methods

2.1. Sample and Procedures

The sample comprised of 764 participants who were selected into the male (n = 450) and female (n = 314) Basketball England Talent Pathway. Participants were allocated into one of three mixed-gender cohorts based on their playing level: (a) Regional Talent Hubs (under-12 to under-15; n = 183), (b) England National Youth Teams (under-16, under-18, and under-20; n = 537), and (c) England National Senior Teams (n = 44). Prior to 2016, the home nations of England, Scotland, and Wales held FIBA licences individually and competed in European competitions as England, Scotland, or Wales. As a result of an Olympic legacy agreement, the three home nations gave up their individual licences with Great Britain Basketball holding the licence for senior competition, which accordingly meant that the under-16, under-18, and under-20 teams came together and competed internationally as Great Britain rather than individual nations (i.e., England). Thus, those selected after 2016 may have represented Great Britain rather than England at youth and senior levels in certain competitions (e.g., Olympic Games), although Basketball England still competes in other competitions (e.g., Commonwealth Games). Male and female cohorts were also analysed independently in order to investigate gender-specific contexts. Participants from the Regional Talent Hubs were registered during the last three seasons since their inception

(2017/18, 2018/19, and 2019/20). Participants from the England National Youth Teams were selected for the under-16, under-18, and/or under-20 age groups during the last 20 years (2000 to 2020) since data collection began. Participants from the England National Senior Teams only included those who were previously selected for the England National Youth Teams in order to explore the selection into and successful transition out of a national youth development programme. Data were provided to the research team by Basketball England in an attempt to better understand their existing organisational structures, which is part of an ongoing collaboration.

This methodology divided the year into four three-month BQs in accordance with each respective cohort's selection cut-off date. Thus, international regulation age group cut-off dates were applied, with the 1st of January as *month 1* and 31st December as *month 12* (e.g., [43]). Each participant was subsequently assigned a BQ corresponding to their birthdate to create an observed BQ distribution within each of the cohorts. The observed BQ distributions from each cohort were compared against the expected BQ distribution calculated from average national live births (i.e., National Norms applied from the Office for National Statistics [ONS], 2015). In addition, the average number of minutes played across the England National Youth Teams and England National Senior Teams were analysed to test the competition opportunities between BQs. This study was ethically approved at both organisational (Basketball England) and institutional (Birmingham City University) levels

2.2. Data Analysis

A chi-square (χ^2) goodness of fit test was used to compare the BQ distributions of each cohort against the expected BQ distributions (ONS, 2015), following procedures outlined by McHugh [44]. Since this test does not reveal the magnitude of difference between the BQ distributions for significant χ^2 outputs, a Cramer's V (Vc) was also used. The Vc was interpreted as per conventional thresholds for correlation, whereby a value of 0.06 or more indicated a small effect size, 0.17 or more indicated a medium effect size, and 0.29 or more indicated a large effect size [45]. Odds ratios (ORs) and 95% confidence intervals (CIs) were calculated in order to compare the likelihood of each BQ being represented (CIs including one marked no association). In addition, the differences between the average number of minutes played per game across each BQ were analysed using a one-way ANOVA. Results were considered significant for $p < 0.05$.

3. Results

There was a significant difference between the BQ distributions of the Regional Talent Hubs compared with the National Norms for the male, female, and combined cohorts ($p < 0.001$), with large effect sizes ($Vc > 0.29$). The ORs showed an increased likelihood of relatively older players being selected, with the highest OR being BQ1 vs. BQ4 (ranging from 2.27 to 9.98). Similarly, there was a significant difference between the BQ distributions of the England National Youth Teams compared with the National Norms for the male and combined cohorts ($p < 0.001$), with medium effect sizes ($Vc > 0.17$). The ORs showed an increased likelihood of relatively older players being selected, with the highest OR being BQ1 vs. BQ4 (ranging from 2.11 to 3.07). However, there were no significant differences between the BQ distributions of England National Youth Teams compared with the National Norms for the female cohort ($p = 0.153$). Moreover, there were no significant differences between the BQ distributions of the England National Senior Teams compared with the National Norms for the male, female, and combined cohorts ($p = 0.35$). Table 1 and Figure 1 show the χ^2 analysis and BQ distributions, respectively. Lastly, the ANOVAs for the England National Youth Teams ($F(_{1.529}) = 0.071$, $p = 0.79$) and the England National Senior Teams ($F(_{1.42}) = 0.095$, $p = 0.76$) revealed no significant differences between the average number of minutes played across BQs.

Table 1. BQ distributions compared with National Norms (ONS, 2015) with chi-square analysis and BQ1 vs. BQ4 analysis.

Pathway	Cohort	BQ1	BQ2	BQ3	BQ4	Total	χ^2 (df = 3)	*p*	Cramer's V	BQ1 vs. BQ4 OR (95% CI)
Regional Talent Hubs	Male	40	28	16	4	88	32.338	**<0.001**	0.43	**9.98 (3.05; 32.59)**
	Female	38	29	14	14	95	17.365	**<0.001**	0.3	**2.27 (1.18; 6.23)**
	Combined	78	57	30	18	183	47.052	**<0.001**	0.36	**4.33 (2.25; 8.32)**
England National Youth Teams	Male	129	101	65	42	337	52.105	**<0.001**	0.28	**3.07 (1.94; 4.85)**
	Female	51	62	44	43	200	5.264	0.153	0.11	1.18 (0.64; 2.08)
	Combined	180	163	109	85	537	44.858	**<0.001**	0.2	**2.11 (1.49; 3.00)**
England National Senior Teams	Male	6	9	7	3	25	3.264	0.35	0.26	1.99 (0.34; 11.71)
	Female	5	5	6	3	19	1.095	0.778	0.17	1.66 (0.14; 4.59)
	Combined	11	14	13	6	44	3.806	0.283	0.21	1.83 (0.92; 6.68)

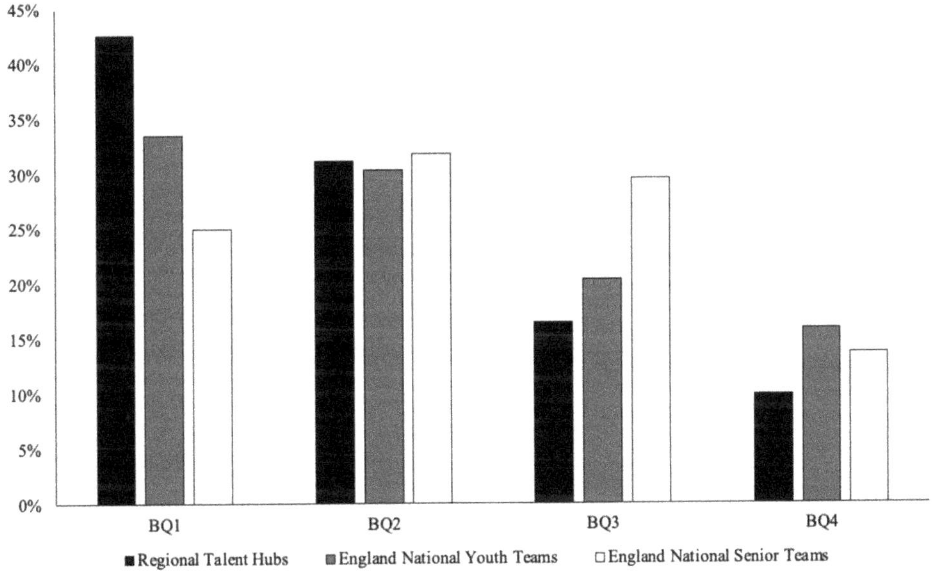

Figure 1. Regional Talent Hubs, England National Youth Teams, and England National Senior Teams BQ distributions.

4. Discussion

The main purpose of the current study was to explore the influence of RAEs throughout the Basketball England Talent Pathway according to gender. Key aims included the analysis of: (a) the BQ distributions of the Regional Talent Hubs (i.e., under-12 to under-U15) as an entry level into the Basketball England Talent Pathway, (b) the BQ distributions of the England National Youth Teams (i.e., under-16, under-18, and under-20), (c) the BQ distributions of those players who successfully transitioned from England National Youth Teams to the England National Senior Teams, and (d) the BQ distribution differences of the England National Youth Teams and England National Senior Teams based on the average number of minutes played. Key findings revealed RAEs were prevalent across the Regional Talent Hubs in both males and females, whereby BQ1s were up to ten times more likely to be selected compared with BQ4s. However, the BQ distributions across the England National Youth Teams were only significant in males. When exploring the youth- to senior-level transitions and the average number of minutes, the results show no differences between the BQ distributions in both males and females. In sum, the results of this study help to better understand the potential mechanisms of RAEs, as well as providing the impetus to explore more equitable selection and competition structures within the England Basketball Talent Pathway. As such, this discussion will attempt to offer considerations to reflect upon when designing implementing, and evaluating organisational structures in youth basketball, including: (a) gender, (b) age and competition levels, (c) youth- to senior-level transitions and playing time, and (d) proposing potential relative solutions and alternative group banding strategies.

According to gender, the impact of RAEs was greater in males compared to females, particularly in the England National Youth Teams and England National Senior Teams. In this regard, several studies revealed stronger RAEs in male youth sports compared to females [33,46]. The latest Sport England data show 26% of people aged 16+ years who play basketball regularly are female. Thus, similar to other sport contexts, English basketball comprises of lower participation in females compared to males [41]. As a result, the number of English teams and leagues is also lower (males = 99 teams, 9 leagues; females = 26 teams, 2 leagues), while the male top league is professional (i.e., players are salaried full-time) whereas the female top league is not. Therefore, a lower number of active players [13], and a lower depth of competition [47] could initially explain why these results showed a weaker prevalence of RAEs in females. Moreover, diverse developmental dynamics at puberty [48], an accelerated stabilisation of conditional-biological differences [32], and a variation in game demands [49] may have also contributed to weaker RAEs in females. Notwithstanding, RAEs in females are present in the Regional Talent Hubs. Accordingly, Smith et al. [13] revealed a higher relative age magnitude in female sports contexts among players aged 12 to 14 years. Furthermore, Delorme and Raspaud [21] observed significant RAEs in all youth categories of French female basketball players aged from 7 to 17 years, which appears more pronounced during puberty. These findings demonstrate that the current understanding of the mechanisms that explain RAEs in female players still needs to be improved.

The presence of RAEs decreases along the England Basketball Talent Pathway, which coincides with an increase in age and competition level. As in other talent identification and development systems (e.g., [50]), the physical and anthropometric advantages of relatively older players tend to reduce as the chronological age of the player increases. Thus, the selection processes may gradually become less affected by RAEs to the detriment of other factors, such as sport-specific skill level [51]. Nevertheless, and especially in basketball, factors such as height [25] or years at peak high velocity [52] are considered determinants in participation and competition performance, favouring relatively older players due to possible greater maturational development. On the other hand, however, it seems that a larger pool of potentially eligible basketball players at the senior levels would allow the recruitment methodology to solve itself [53]. Furthermore, the selection and re-selection processes at high-performance levels (e.g., talent pathways) have undergone significant changes in recent years, balancing the presence of relatively older and young players (e.g., [20,54]). In addition, Kalén et al. [30] demonstrated that relatively young players were 20–25% more likely to be reselected until age 20 than relatively older players due to initial selection age and long-term performance. Overall, it appears RAEs reduce with age and competition levels in high-performance basketball settings across both genders. Moving forward, selection processes should aim to recruit players based on their potential to achieve expertise in the long-term rather than focusing on individual and collective performance levels in the short-term.

Given the number of studies analysing the relative age phenomena in basketball, the information focused on the consequences of (bi) annual age group selection and performance is limited. Key findings showed the overrepresentation of relatively older players at youth levels did not translate into RAEs at senior levels or more minutes played for these players. More specialised sport contexts [55] and higher training levels [56] would imply an adjustment in the performance of players participating in talent pathways. Thus, players born at the end of the selection year would have the opportunity to overcome initial disadvantages (e.g., poorer access to training facilities, less expert coach support, and lower playing time among quality players), which may help them to acquire superior sport-specific skills compared with those born at the beginning of the selection year [39]. This reversal effect of relative age [37] or underdog hypothesis [38] may subsequently lead to higher performance levels for relatively younger players in the long-term. For instance, psychological characteristics, such as resilience, would help to reduce the differences caused by RAEs. Moreover, experiencing more stressful [57] or traumatic situations [58], greater effort in the learning process [59], training in pressurised conditions [60], and playing matches with higher levels of achievement [61] are factors that are often present in the sport transition processes of relatively younger players. Indeed, these experiences could enhance their performance capabilities in the long-term and, thus, increase the increased likelihood of successfully transitioning to senior levels or the number of minutes played. Furthermore, relatively younger players could suffer fewer injuries due to a low exposure to competition at formative ages, which could reduce their dropout rate and allow them to reach high-performance levels in the long-term [62]. On the other hand, however, those relatively younger players who are selected may be advanced in their maturation status and, thus, may have the physical skill set required in order to compete with their relatively older peers (e.g., [63]). Indeed, this may be why minutes played across the BQs were equally distributed. Overall, this study provides a relevant contribution in guiding the actions of coaches, clubs, and governing bodies to mitigate age unbalances in order to reduce the influence of RAEs. Further research is required to better understand the impact of RAEs on long-term performance, participation, and personal development outcomes to substantiate these suggestions.

4.1. Practical Implications and Future Research

Since there appears to be pronounced RAEs throughout the Basketball England Talent Pathway, it is important to consider possible relative age solutions and offer directions for future research. In the context of soccer, Mann and van Ginneken [64] designed an *age-ordered shirt numbering system* by providing stakeholders with the knowledge that the numbers on individuals' playing shirts corresponded with relative age. Bennett et al. [65] recommended a *selection quota*, whereby stakeholders endorse policies that ensure clubs select a minimum number of players from each BQ. Tribolet et al. [66] proposed *avoiding early deselection* by encouraging stakeholders to avoid releasing players at young ages to ensure they have continued exposure to practice, competition, and resources without the option of being deselected. Romann et al. [67] suggested clubs and governing bodies should *delay the selection process* preferentially until post-maturation in order to make more fair and accurate decisions based on potential and negate the possible drawbacks of early specialisation. Grossmann and Lames [68] advised governing bodies to include the relative age phenomena into *coach education*, with the purpose of enhancing knowledge and understanding of RAEs as part of coaches' formal coaching courses and/or training. Kelly et al., [39] conceptualised a *flexible chronological approach*, whereby early birth quartiles (i.e., BQ1s) and late birth quartiles (i.e., BQ4s) should be offered the opportunity to "play-up" (e.g., [69,70]) and "play-down" annual age groups, respectively. Kelly et al. [54] introduced birthday-banding, in which young athletes move up to their next birthdate group on their birthday with the aim to remove particular selection time points and specific chronological age groups. However, despite this range of possible relative age solutions, they are yet to be empirically evaluated in a basketball setting. As such, future research is required to explore the practical implications of these strategies within a youth basketball context.

The literature on alternative grouping strategies to moderate RAEs is limited when compared with the body of research demonstrating its prevalence. Moreover, where proposed grouping strategies have been suggested, little evidence has documented their effectiveness or directly implemented those strategies [71]. A useful strategy that may be utilised in basketball could be from the organisational policies incorporated in youth American football. More specifically, contrary to many other youth team sports, previous research has identified no RAEs in American football, which may lend credibility to the *age and anthropometric bandings* that are often employed to group their players (e.g., [72,73]). Another recent grouping approach that has produced promising results in youth soccer is *bio-banding*, which groups young players based on anthropometric and maturational status (see [74]). For instance, during their maturity-matched soccer competition, Bradley et al. [75] demonstrated how later-maturing players believed the bio-banded games offered greater opportunities to express themselves, adopt positions of leadership, and have a more important influence on gameplay. In comparison, early maturing players believed the bio-banded games were more physically and technically challenging. Both age and anthropometric and biological bands appear to systematically address one of the key mechanisms of RAEs, whereby relatively older athletes may have an advanced maturity status [14]. However, it is important to consider how these may look in the context of youth basketball, while the interaction effect of relative age and maturational status requires further study. Indeed, current studies have primarily focused their attention on academy soccer, thus, it is difficult to fully interpret how it will be conveyed within a basketball setting that is comprised of diverse talent development systems. Overall, although these banding approaches remain unproven in their impact on RAEs, an introduction to grouping players by height, weight, and/or some maturational variables may prove beneficial in moderating RAEs in youth basketball and, thus, warrants further research.

4.2. Limitations

It is important to consider the limitations of this study when interpreting its findings. First, playing position was not included because these data were not available. Previous research into playing position in basketball has shown a greater prevalence of RAEs in the guard position, whereas the centre position appears less affected [31]. Thus, it is important to consider playing position in future research to better understand who is more vulnerable to RAEs throughout respective talent pathways. Second, only one appearance for the England National Youth Teams and the England Senior National Teams were required to be included in this study. Since some players may have competed in considerably more games at both these levels, performance outcomes and career duration should be considered in future studies to examine the implications of RAEs on long-term development outcomes. Lastly, only within-1-year effects were explored based on annual competition cycles (i.e., BQ1 to BQ4), whereas constituent year effects based on biannual cycles (i.e., BQ1 to BQ8) as the national youth competition is organised were not measured. Previous research has shown how constituent year effects can impact opportunities to be selected into talent development pathways in basketball [22], thus, further study examining the impact of biannual age grouping is warranted.

5. Conclusions

It is evident that there is a complicated relationship between the BQ a player is born in, their opportunities to be selected into the Basketball England Talent Pathway, and their likelihood of successfully transitioning to senior levels. Key findings showed how RAEs were prevalent across the Regional Talent Hubs in both males and females, although they were only significant across the England National Youth Teams in males. When exploring the youth- to senior-level transitions and the average number of minutes played, the results show no differences between the BQ distributions in both males and females, suggesting possible reversal or underdog effects. Since there appears to be RAEs throughout the Basketball England Talent Pathway, researchers and practitioners are encouraged to work collaboratively to design, implement, and evaluate relative age solutions and alternative grouping strategies to create more equitable opportunities in youth basketball.

Author Contributions: Conceptualization, A.L.K. and M.A.J.; methodology, A.L.K.; formal analysis, D.T.J.; data curation, C.F. and D.O.; writing—original draft preparation, A.L.K.; writing—review and editing, S.L.J.S., A.L.C., A.d.l.R., and S.D.L.d.S. All authors have read and agreed to the published version of the manuscript.

Funding: This research received no external funding.

Institutional Review Board Statement: The study was conducted according to the guidelines of the Declaration of Helsinki, and approved by the Health, Education, and Life Sciences Faculty Academic Ethics Committee at Birmingham City University (protocol code 7383 dated 30/04/2020).

Informed Consent Statement: Participant consent was waived because the dataset was unidentifiable and considered very-low risk by the Health, Education, and Life Sciences Faculty Academic Ethics Committee at Birmingham City University.

Data Availability Statement: Descriptive data is presented in Table 1.

Conflicts of Interest: The authors declare no conflict of interest.

References

1. Carvalho, H.M.; Gonçalves, C.E.; Collins, D.; Paes, R.R. Growth, functional capacities and motivation for achievement and competitiveness in youth basketball: An interdisciplinary approach. *J. Sport Sci.* **2018**, *36*, 742–748. [CrossRef]
2. Leite, N.; Sampaio, J. Long-term athletic development across different age groups and gender from portuguese basketball players. *Int. J. Sports Sci. Coach.* **2012**, *7*, 285–300. [CrossRef]
3. Arede, J.; Esteves, P.; Ferreira, A.P.; Sampaio, J.; Leite, N. Jump higher, run faster: Effects of diversified sport participation on talent identification and selection in youth basketball. *J. Sport Sci.* **2019**, *37*, 2220–2227. [CrossRef]
4. Soares, A.L.A.; Kós, L.D.; Paes, R.R.; Nascimento, J.V.; Collins, D.; Gonçalves, C.E.; Carvalho, H.M. Determinants of drop-out in youth basketball: An interdisciplinary approach. *Res. Sport. Med.* **2019**, *28*, 84–98. [CrossRef]
5. Sáenz-López, P.; Ibáñez, S.J.; Giménez, J.; Sierra, A.; Sánchez, M. Multifactor characteristics in the process of development of the male expert basketball player in Spain. *Int. J. Sport Psychol.* **2005**, *36*, 151–171.
6. Zarić, I.; Kukić, F.; Jovićević, N.; Zarić, M.; Marković, M.; Toskić, L.; Dopsaj, M. Body height of elite basketball players: Do taller basketball teams rank better at the FIBA world cup? *Int. J. Environ. Res. Public Health* **2020**, *17*, 3141. [CrossRef]
7. US Census. Statistical Abstract of the United States: 2012. Available online: http://www.census.gov/compendia/statab/2012/tables/12s0209.pdf (accessed on 11 January 2021).
8. NBA. 13 Facts to Know from 2019–20 NBA Roster Survey. 2019. Available online: https://www.nba.com/article/2019/11/01/2019-20-nba-roster-survey (accessed on 11 January 2021).
9. WNBA. 2004 WNBA Player Survey Results: Height. 2004. Available online: https://www.wnba.com/archive/wnba/statistics/survey_height_2004.html (accessed on 11 January 2021).
10. Gryko, K.; Bodasinski, S.; Bodasinska, A.; Zielinski, J. Offensive and defensive play in handball in a 2-year world championship cycle: Characteristics and tendencies. *Pol. J. Sport Tour.* **2018**, *25*, 10–16. [CrossRef]
11. Barnsley, R.; Thompson, A.; Barnsley, P. Hockey success and birthdate: The relative age effect. *J. Can. Assoc. Health Phys. Educ. Recreat.* **1985**, *51*, 23–28.
12. Kelly, A.L.; Williams, C.A. Physical characteristics and the talent identification and development processes in male youth soccer: A narrative review. *Strength Cond. J.* **2020**, *42*, 15–34. [CrossRef]
13. Smith, K.L.; Weir, P.L.; Till, K.; Romann, M.; Cobley, S.P. Relative age effects across and within female sport contexts: A systematic review and meta-analysis. *Sport. Med.* **2018**, *48*, 1451–1478. [CrossRef]
14. Cobley, S.; Baker, J.; Wattie, N.; Mckenna, J. Annual Age-grouping and Athlete Develop- ment. A Metas-analytical Review of Relative Age Effects in Sport. *Sport. Med* **2009**, *39*, 235–256. [CrossRef]
15. Karalejic, M.; Jakovljevic, S.; Macura, M. Anthropometric characteristics and technical skills of 12 and 14 year old basketball players. *J. Sports Med. Phys. Fitness* **2011**, *51*, 103–110.
16. Gil, S.M.; Badiola, A.; Bidaurrazaga-Letona, I.; Zabala-Lili, J.; Gravina, L.; Santos-Concejero, J.; Lekue, J.A.; Granados, C. Relationship between the relative age effect and anthropometry, maturity and performance in young soccer players. *J. Sport Sci.* **2014**, *32*, 479–486. [CrossRef]
17. De la Rubia, A.; Lorenzo-Calvo, J.; Mon-López, D.; Lorenzo, A. Impact of the relative age effect on competition performance in basketball: A qualitative systematic review. *Int. J. Environ. Res. Public Health* **2020**, *17*, 8596. [CrossRef]
18. Torres-Unda, J.; Zarrazquin, I.; Gravina, L.; Zubero, J.; Seco, J.; Gil, S.M.; Gil, J.; Irazusta, J. Basketball Performance Is Related to Maturity and Relative Age in Elite Adolescent Players. *J. Strength Cond. Res.* **2016**, *30*, 1325–1332. [CrossRef]
19. De Oliveira, H.Z.; Ribeiro Junior, D.B.; Vianna, J.M.; Werneck, F.Z. Efeito da idade relativa no Campeonato Brasileiro de Basquete: Categoria sub-15. *Braz. J. Kineanthropometry Hum. Perform.* **2017**, *19*, 526. [CrossRef]
20. Ribeiro Junior, D.B.; Werneck, F.Z.; de Oliveira, H.Z.; Panza, P.S.; Ibáñez, S.J.; Vianna, J.M. From talent identification to Novo Basquete Brasil (NBB): Multifactorial analysis of the career progression in youth Brazilian elite basketball. *Front. Psychol.* **2021**, *12*, 1–12. [CrossRef] [PubMed]

21. Delorme, N.; Raspaud, M. The relative age effect in young French basketball players: A study on the whole population. *Scand. J. Med. Sci. Sport.* **2009**, *19*, 235–242. [CrossRef] [PubMed]
22. Steingröver, C.; Wattie, N.; Baker, J.; Helsen, W.F.; Schorer, J. The interaction between constituent year and within-1-year effects in elite German youth basketball. *Scand. J. Med. Sci. Sport.* **2017**, *27*, 627–633. [CrossRef] [PubMed]
23. Nakata, H.; Sakamoto, K. Relative age effect in Japanese male athletes. *Percept. Mot. Skills* **2011**, *113*, 570–574. [CrossRef]
24. Chittle, L.; Horton, S.; Dixon, J. Time out or fast break? The relativea age effect in NCAA division 1 basketball. *J. Sport Behav.* **2016**, *39*, 107–125.
25. Rubajczyk, K.; Świerzko, K.; Rokita, A. Doubly disadvantaged? The relative age effect in Poland's basketball players. *J. Sport. Sci. Med.* **2017**, *16*, 280–285.
26. Leite, N.; Borges, J.; Santos, S.; Sampaio, J. The relative age effect in school and federative sport in basketball. *Rev. Psicol. Deport.* **2013**, *22*, 219–223.
27. De Subijana, C.L.; Lorenzo, J. Relative age effect and long-term success in the Spanish soccer and basketball national teams. *J. Hum. Kinet.* **2018**, *65*, 197–204. [CrossRef]
28. Vegara-Ferri, J.M.; García-Mayor, J.; Pérez, A.M.; Cabezos, H. Effect of relative age in the basketball world championships sub-17, sub-19 and Olympic Games of Brazil 2016. *J. Sport Health Res.* **2019**, *11*, 33–42.
29. Saavedra-García, M.; Gutierrez-Aguilar, O.; Galatti, L.; Romero, J.J.F. Relative age effect in lower categories of FIBA Basketball World Championships (1979–2011). *Cuad. Psicol. Deport.* **2015**, *15*, 237–242.
30. Kalén, A.; Lundkvist, E.; Ivarsson, A.; Rey, E.; Pérez-Ferreirós, A. The influence of initial selection age, relative age effect and country long-term performance on the re-selection process in European basketball youth national teams teams. *J. Sport Sci.* **2020**, 1–7. [CrossRef]
31. Ibañez, S.J.; Mazo, A.; Nascimento, J.; Garcıa-Rubio, J. The relative age effect in under-18 basketball: Effects on performance according to playing position. *PLoS ONE* **2018**, *13*, e0200408. [CrossRef]
32. Arrieta, H.; Torres-Unda, J.; Gil, S.M.; Irazusta, J. Relative age effect and performance in the U16, U18 and U20 european basketball championships. *J. Sport Sci.* **2016**, *34*, 1530–1534. [CrossRef]
33. García, M.S.; Aguilar, Ó.G.; Romero, J.J.F.; Lastra, D.F.; Oliveira, G.E. Relative age effect in lower categories of international basketball. *Int. Rev. Sociol. Sport* **2014**, *49*, 526–535. [CrossRef]
34. Schorer, J.; Neumann, J.; Cobley, S.P.; Tietjens, M.; Baker, J. Lingering effects of relative age in basketball players ' post athletic career. *Int. J. Sports Sci. Coach.* **2011**, *6*, 143–148. [CrossRef]
35. González-Víllora, S.; Pastor-Vicedo, J.C.; Cordente, D. Relative age effect in UEFA championship soccer players. *J. Hum. Kinet.* **2015**, *47*, 237–248. [CrossRef] [PubMed]
36. Werneck, F.Z.; Coelho, E.F.; de Oliveira, H.Z.; Ribeiro Júnior, D.B.; Almas, S.P.; de Lima, J.R.P.; Matta, M.O.; Figueiredo, A.J. Relative age effect in Olympic basketball athletes. *Sci. Sports* **2016**, *31*, 158–161. [CrossRef]
37. McCarthy, N.; Collins, D.; Court, D. Start hard, finish better: Further evidence for the reversal of the RAE advantage. *J. Sport Sci.* **2016**, *34*, 1461–1465. [CrossRef]
38. Gibbs, B.G.; Jarvis, J.A.; Dufur, M.J. The rise of the underdog? The relative age effect reversal among Canadian-born NHL hockey players: A reply to Nolan and Howell. *Int. Rev. Sociol. Sport* **2012**, *47*, 644–649. [CrossRef]
39. Kelly, A.L.; Wilson, M.R.; Gough, L.A.; Knapman, H.; Morgan, P.; Cole, M.; Jackson, D.T.; Williams, C.A. A longitudinal investigation into the relative age effect in an English professional football club: Exploring the 'underdog hypothesis'. *Sci. Med. Footb.* **2020**, *4*, 111–118. [CrossRef]
40. Wattie, N.; Schorer, J.; Baker, J. The relative age effect in sport: A developmental systems model. *Sport. Med.* **2015**, *45*, 83–94. [CrossRef]
41. Basketball England. Facts and Figures: Basketball in England in 2018. 2018. Available online: https://www.basketballengland.co.uk/about/research/facts-and-figures/ (accessed on 11 January 2021).
42. Basketball England. Talent Pathway. 2020. Available online: https://www.basketballengland.co.uk/talent/talent-pathway/ (accessed on 11 January 2021).
43. Kelly, A.L.; Till, K.; Jackson, D.T.; Barrell, D.; Burke, K.; Turnnidge, J. Talent identification and relative age effects in English male rugby union pathways: From entry to expertise. *Front. Sport. Act. Living* **2021**, *3*, 1–9. [CrossRef]
44. Mchugh, M.L. The Chi-square test of independence. *Biochem. Medica* **2013**, *23*, 143–149. [CrossRef] [PubMed]
45. Cohen, J. *Statistical Power Analysis for the Behavioral Sciences*, 2nd ed.; Routledge: London, UK, 1998.
46. Vincent, J.; Glamser, F. Gender differences in the relative age effect among US olympic development program youth soccer players. *J. Sport Sci.* **2006**, *24*, 405–413. [CrossRef]
47. Baker, J.; Cobley, S.P.; Winckel, V. Gender, depth of competition and relative age effects in team sports. *Asian J. Exerc. Sport. Sci.* **2009**, *6*, 1–8.
48. Malina, R.M.; Rogol, A.D.; Cumming, S.P.; Coelho-e-Silva, M.J.; Figueiredo, A.J. Biological maturation of youth athletes: Assessment and implications. *Br. J. Sports Med.* **2015**, *49*, 852–859. [CrossRef]
49. Leicht, A.; Gomez, M.A.; Woods, C.T. Team performance indicators explain outcome during women's basketball matches at the Olympic Games. *Sports* **2017**, *5*, 96. [CrossRef]
50. Bjørndal, C.T.; Luteberget, L.; Till, K.; Holm, S. The relative age effect in selection to international team matches in Norwegian handball. *PLoS ONE* **2018**, *13*, e0209288. [CrossRef]

51. Karcher, C.; Ahmaidi, S.; Buchheit, M. Effect of birth date on playing time during international handball competitions with respect to playing positions. *Kinesiology* **2014**, *46*, 23–32.
52. Torres-Unda, J.; Zarrazquin, I.; Gil, J.; Ruiz, F.; Irazusta, A.; Kortajarena, M.; Seco, J.; Irazusta, J. Anthropometric, physiological and maturational characteristics in selected elite and non-elite male adolescent basketball players. *J. Sport Sci.* **2013**, *31*, 196–203. [CrossRef]
53. Steingröver, C.; Wattie, N.; Baker, J.; Helsen, W.F.; Schorer, J. Geographical variations in the interaction of relative age effects in youth and adult elite soccer. *Front. Psychol.* **2017**, *8*, 1–13. [CrossRef]
54. Kelly, A.L.; Jackson, D.T.; Taylor, J.J.; Jeffreys, M.A.; Turnnidge, J. " Birthday-Banding " as a strategy to moderate the relative age effect: A case study into the England squash talent pathway. *Front. Psychol.* **2020**, *2*, 1–9. [CrossRef]
55. Rees, T.; Hardy, L.; Güllich, A.; Abernethy, B.; Côté, J.; Woodman, T.; Montgomery, H.; Laing, S.; Warr, C. The Great British medalists project: A review of current knowledge on the development of the world's best sporting talent. *Sport. Med.* **2016**, *46*, 1041–1058. [CrossRef]
56. Hancock, D.J.; Adler, A.L.; Côté, J. A proposed theoretical model to explain relative age effects in sport. *Eur. J. Sport Sci.* **2013**, *13*, 630–637. [CrossRef] [PubMed]
57. Andronikos, G.; Elumaro, A.I.; Westbury, T.; Martindale, R.J.J. Relative age effect: Implications for effective practice. *J. Sport Sci.* **2016**, *34*, 1124–1131. [CrossRef]
58. Collins, D.; MacNamara, A. The rocky road to the top. Why talent needs trauma. *Sport. Med.* **2012**, *42*, 907–914. [CrossRef]
59. Roberts, S.J.; Stott, T. A new factor in UK students' university attainment: The relative age effect reversal? *Qual. Assur. Educ.* **2015**, *23*, 295–305. [CrossRef]
60. Skorski, S.; Skorski, S.; Faude, O.; Hammes, D.; Meyer, T. The relative age effect in elite German youth soccer: Implications for a successful career. *Int. J. Sports Physiol. Perform.* **2016**, *11*, 370–376. [CrossRef]
61. Romann, M.; Fuchslocher, J. Influence of the selection level, age and playing position on relative age effects in Swiss women's soccer. *Talent Dev. Excell.* **2011**, *3*, 239–247.
62. Wattie, N.; Cobley, S.; Macpherson, A.; Howard, A.; Montelpare, W.J.; Baker, J. Injuries in Canadian youth ice hockey: The influence of relative age. *Pediatrics* **2007**, *120*, 1–7. [CrossRef] [PubMed]
63. Parr, J.; Winwood, K.; Hodson-Tole, E.; Deconinck, F.J.A.; Hill, J.P.; Teunissen, J.W.; Cumming, S.P. The main and interactive effects of biological maturity and relative age on physical performance in elite youth soccer players. *J. Sports Med.* **2020**, *2020*, 1–11. [CrossRef]
64. Mann, D.L.; van Ginneken, P.J.M.A. Age-ordered shirt numbering reduces the selection bias associated with the relative age effect. *J. Sport Sci.* **2017**, *35*, 784–790. [CrossRef] [PubMed]
65. Bennett, K.J.M.; Vaeyens, R.; Fransen, J. Creating a framework for talent identification and development in emerging football nations. *Sci. Med. Footb.* **2019**, *3*, 36–42. [CrossRef]
66. Tribolet, R.; Watsford, M.L.; Coutts, A.J.; Smith, C.; Fransen, J. From entry to elite: The relative age effect in the Australian football talent pathway. *J. Sci. Med. Sport* **2019**, *22*, 741–745. [CrossRef]
67. Romann, M.; Rüeger, E.; Hintermann, M.; Kern, R.; Faude, O. Origins of relative age effects in youth football—A nationwide analysis. *Front. Sport. Act. Living* **2020**, *2*, 1–7. [CrossRef]
68. Grossmann, B.; Lames, M. Relative age effect (RAE) in football talents—The role of youth academies in transition to professional status in Germany. *Int. J. Perform. Anal. Sport* **2013**, *13*, 120–134. [CrossRef]
69. Goldman, D.E.; Turnnidge, J.; Kelly, A.L.; Devos, J.; Cote, J. Ahtlete perceptions of 'playing up' in youth soccer. *J. Appl. Sport Psychol.* **2021**, 1–24. [CrossRef]
70. Kelly, A.L.; Wilson, M.; Jackson, D.T.; Goldman, D.; Turnnidge, J.; Côté, J.; Williams, C.A. A multidisciplinary investigation into "playing-up" in academy football according to age phase. *J. Sport Sci.* **2021**, *39*, 854–864. [CrossRef]
71. Webdale, K.; Baker, J.; Schorer, J.; Wattie, N. Solving sport's 'relative age' problem: A systematic review of proposed solutions. *Int. Rev. Sport Exerc. Psychol.* **2019**, 1–18. [CrossRef]
72. Jones, C.; Visek, A.J.; Barron, M.J.; Hyman, M.; Chandran, A. Association between relative age effect and organisational practices of American youth football. *J. Sport Sci.* **2019**, *37*, 1146–1153. [CrossRef]
73. Macdonald, D.J.; Cheung, M.; Cote, J.; Abernethy, B. Place but not date of birth influences the development and emergence of athletic talent in American football. *J. Appl. Sport Psychol.* **2009**, *21*, 80–90. [CrossRef]
74. Thomas, C.; Oliver, J.; Kelly, A.L. Bio-banding in youth soccer: Considerations for researchers and practitioners. In *Birth Advantages and Relative Age Effects in Sport*; Routledge: London, UK, 2021; pp. 107–124.
75. Bradley, B.; Johnson, D.; Hill, M.; Mcgee, D.; Kana-ah, A.; Sharpin, C.; Sharp, P.; Kelly, A.; Cumming, S.P.; Malina, R.M. Bio-banding in academy football player's perceptions of a maturity matched tournament. *Ann. Hum. Biol.* **2019**, *46*, 400–408. [CrossRef]

sports

MDPI

Article

How Relative Age Effects Associate with Football Players' Market Values: Indicators of Losing Talent and Wasting Money

Michael Romann [1,*], Marie Javet [1], Stephen Cobley [2] and Dennis-Peter Born [1,3]

1 Department for Elite Sport, Swiss Federal Institute of Sport Magglingen, 2532 Magglingen, Switzerland; Marie.Javet@baspo.admin.ch (M.J.); dennis.born@swiss-aquatics.ch (D.-P.B.)
2 Faculty of Health Science, University of Sydney, Lidcombe, NSW 2141, Australia; stephen.cobley@sydney.edu.au
3 Swiss Swimming Federation, 3063 Berne, Switzerland
* Correspondence: michael.romann@baspo.admin.ch

Abstract: Background: In football, annual age-group categorization leads to relative age effects (RAEs) in talent development. Given such trends, relative age may also associate with market values. This study analyzed the relationship between RAEs and market values of youth players. Methods: Age category, birthdate, and market values of 11,738 youth male football players were obtained from the "transfermarkt.de" database, which delivers a good proxy for real market values. RAEs were calculated using odds ratios (OR) with 95% confidence intervals (95%CI). Results: Significant RAEs were found across all age-groups ($p < 0.05$). The largest RAEs occurred in U18 players (Q1 [relatively older] v Q4 [relatively younger] OR = 3.1) ORs decreased with age category, i.e., U19 (2.7), U20 (2.6), U21 (2.4), U22 (2.2), and U23 (1.8). At U19s, Q1 players were associated with significantly higher market values than Q4 players. However, by U21, U22, and U23 RAEs were inversed, with correspondingly higher market values for Q4 players apparent. While large typical RAEs for all playing positions was observed in younger age categories (U18–U20), inversed RAEs were only evident for defenders (small-medium) and for strikers (medium-large) in U21–U23 (not goalkeepers and midfielders). Conclusions: Assuming an equal distribution of football talent exists across annual cohorts, results indicate the selection and market value of young professional players is dynamic. Findings suggest a potential biased selection, and undervaluing of Q4 players in younger age groups, as their representation and market value increased over time. By contrast, the changing representations and market values of Q1 players suggest initial overvaluing in performance and monetary terms. Therefore, this inefficient talent selection and the accompanying waste of money should be improved.

Keywords: talent selection; talent development; return of investment; market value; drafts

Citation: Romann, M.; Javet, M.; Cobley, S.; Born, D.-P. How Relative Age Effects Associate with Football Players' Market Values: Indicators of Losing Talent and Wasting Money. *Sports* **2021**, *9*, 99. https://doi.org/10.3390/sports9070099

Academic Editors: Adam Leigh Kelly, Sergio L. Jiménez Sáiz, Sara Diana Leal dos Santos and Alberto Lorenzo Calvo

Received: 28 May 2021
Accepted: 7 July 2021
Published: 9 July 2021

1. Introduction

During childhood and adolescence, young football players are categorized by annual age groups. However, the chronological age gap of up to 12 months between players born in early (January) and late (December) in the year leads to substantial differences in performance and biased talent selection decisions [1,2]. The result of participation or selection bias, specifically the overrepresentation of chronologically older soccer players within one age category, is called relative age effects (RAEs). RAEs have been shown to affect talent development systems in a wide range of team and individual sports, e.g., ice hockey, football, swimming, tennis, in both females and males from 4 years of age to adulthood [2,3]. Relatively older children within annual cohorts are more likely to be selected in talent development teams, with selection commensurate with additional training, and access to higher quality coaching likely leading to accumulated performance advantages [4,5]. By contrast, the relatively younger children are underrepresented, are less likely to be selected to talent development systems, and are more likely to withdraw from the sport [6–8]. Interestingly, research has subsequently shown how relatively younger players, who are selected for a talent development system, actually have a greater chance of becoming a professional player than their relatively older counterparts [9]. Such observations have become synonymous with the proposition of a "underdog hypothesis" [10]. In talent development contexts, late-born players have been shown to be more likely to achieve senior professional status, as they may benefit more from competitive play with their older counterparts [10–12]; that said multiple factors and processes may contribute to the outcome. Furthermore, a study of German professional soccer players has shown that players born late (Q4) had systematically higher wages than their fellow Q1 players [13].

For football clubs, the capability to accurately identify athletic potential, and recruit potential, in the early stages of development has several organisational benefits [14]. Given how athletic talent can influence team achievements, being able to secure athletic potential can have performance benefit [15]. That said, research which examines the hiring decisions in professional sports, recognized the difficulty of being able to accurately identify youthful talent, which may lead to future performance productivity [16].

In addition to the traditional assessment methods of talent scouts, fans and football experts have established a large online community called "transfermarkt.de". Transfermarkt.de assesses the market value of professional footballer players at an age range from U15 to retirement. The community has become the main source for reporting on market values [17,18]. From an economic perspective, the aim of many professional football clubs is to buy undervalued players to achieve both higher performance and higher returns on investment [18]. Moreover, a rapidly growing body of literature emphasizes the importance of collective judgements for assessing actual and future values [17,19]. Recent studies showed that the variance of actual transfer fees paid (for players) in the German Bundesliga can almost entirely be explained (R^2 = 0.90) by the market values reported on transfermarkt.de [17]. Current literature suggests that player market values on transfermarkt.de are good proxy estimate indicators of current as well as future players' real market values and will, therefore, play an increasing role in talent recruitment, sports economics and talent development [17,19].

Given the relevance of RAEs and market values for professional soccer clubs this study had two objectives. The first objective was to identify the presence of (changing) RAEs in professionally contracted players across the developmental to professional years (e.g., 18–23 years of age). The second objective was to assess the relationship between RAEs and player market values (as indicated on Transfermarkt) and whether age-group and playing position moderated the relationship.

2. Materials and Methods

2.1. Participants

Participants were n = 11,738 players included in this study. Inclusion criteria were 2000 players with the highest market valuein every age categories from U18 to U23. In the U18 category all 1738 listed players were analysed (Table 1).

Table 1. Subject characteristics per age category.

under (U)	n	Age (Years)	Height (cm)	Market Value (€)	n (Clubs)	n (Countries)
18	1738	17.4 ± 0.5	180.0 ± 7.0	326,252 ± 1,878,569	941	98
19	2000	18.6 ± 0.3	180.6 ± 6.8	399,588 ± 1,957,796	1149	105
20	2000	19.6 ± 0.3	180.5 ± 7.0	853,200 ± 4,360,673	1140	110
21	2000	20.6 ± 0.3	180.6 ± 6.9	1,255,337 ± 4,761,941	1118	118
22	2000	21.6 ± 0.3	180.6 ± 6.8	1,367,525 ± 6,070,609	1119	119
23	2000	22.6 ± 0.3	180.8 ± 6.9	1,968,675 ± 5,863,952	1077	117
Total	11,738	20.1 ± 1.7	180.6 ± 6.9	1,043,561 ± 4,552,652	2861	153

Note: Data presented as mean ± the standard deviation or frequency (n).

Data were provided by the owner of the open-source football database transfermarkt.de, with permission to anonymously analyse and publish the results. All data were extracted on 17 July 2020 and included current data of players age, height, market value, club and nationality. The website provides independent estimates of players' market value and is regularly updated (last update in March 2020) by more than 190,000 professional and non-professional individuals with the approval of Transfermarkt.de experts [17,20]. Transfermarkt.de has been used in several previous studies [20–23], and has been shown to be a valid and useful database for game performance indicators and market values [17]. Data exported for this study included birthdate, market value, nationality, club and playing position. The study was pre-approved by the institutional review board of the Swiss Federal Institute of Sport Magglingen (Reg.-Nr. HLP-2021-131) and is in accordance with the Declaration of Helsinki.

2.2. Procedures and Data Analysis

The cut-off date for age group selections in international football in all countries and according to FIFA rules is 1 January. Players were categorized into four relative age quarters (Q) and two relative age semesters (S) according to their birth month, independently of birth year (i.e., S1 = January to June; S2 = July to December and Q1 = January to March; Q2 = April to June; Q3 = July to September; and Q4 = October to December). Due to the multi-nation sample (n = 152) within the current investigation, potential national differences in birth rates per month could not be taken into consideration which has to be considered as a limitation. Therefore, equal distribution of births across all months and years was assumed for the expected birth distribution of the general population [1,5]. The following age categories were analysed for their relative age distributions: U18 to U23. RAEs were calculated using odds ratios (OR; Q1 vs. Q4) with 95% confidence intervals (95%CI). The OR was interpreted as an effect size as follows: we assumed a significant RAE if the CI did not include 1 and interpreted $1.00 \leq OR < 1.22$, $1.22 \leq OR < 1.86$, $1.86 \leq OR < 3.00$, and $OR \geq 3.00$, as negligible, small, medium and large, respectively [24]. If the OR was <1 and the CI did not include 1, this finding was interpreted as a significant inverse RAE. Inverse ORs < 0.33 (1/3), $0.33 \leq OR < 0.53$ (1/1.86), $0.53 \leq OR < 0.81$, $0.81 \leq OR < 1.0$ were, respectively, interpreted as large, medium, small, and negligible. Market values were extracted in €, playing positions were categorized as goalkeepers, defenders (central and outside), midfielders (central and outside) and strikers. Using these data mean market values per age category and Q were calculated using crosstabulations. In a second step the difference of observed and expected market values (Δ) were calculated. Observed market values were the sum of the market values of all players per age category and per Q. Estimated market values were calculated in the same way, but with the assumption of an equal distribution of players per Q. For instance, if the expected number of players in each Q is 500, the observed number in Q4 is 400 and the mean market value of the Q4players of the age category is 1,000,000€, the calculated Δ is $400 - 500 = -100 \times 1{,}000{,}000€ = -100{,}000{,}000€$.

3. Results

Distribution of players per Q with 95% CI are illustrated in Table 2. There were medium RAEs in the U18 to U22 and small RAEs in the U23. With a large OR of 3.1, RAEs were highest in the youngest age category (U18) and consistently/continuously decreases to small RAEs demonstrated by an OR of 1.8 in the U23 (Table 2).

Table 3 shows market values across each age group and all playing positions separated by birth quartile. In the U19 a small effect with an OR of 1.2 was found. The RAEs in the U18, U20, and U21 were negligible. A medium inverse effect (OR 0.5 [95%CI 0.4, 0.5]), where Q4 players had a higher market value, were found in the U22 and a small effect in the U23 (OR 0.7 [95%CI 0.6, 0.8]).

Table 4 shows the difference of observed and expected market values across each age group and Q. In Q1 and Q2 observed values were constantly higher than expected values. In contrast, in Q3 and Q4 observed values were constantly lower than expected.

Within the age categories, there was a constant decrease in values from Q1 to Q4. In the overall group, this leads to a deviation/overestimation of €1.2 billion in Q1 and a deviation/underestimation of €1.4 billion in Q4.

Table 2. Distribution of players per age category and quarter (Q).

under (U)	n	Q1	Q2	Q3	Q4	OR Q1/Q4	95% CI	Effect Size
18	1738	705 (40.6%)	462 (26.6%)	340 (19.6%)	231 (13.3%)	3.1 *	(2.6, 3.6)	large
19	2000	746 (37.3%)	574 (28.7%)	402 (20.1%)	278 (13.9%)	2.7 *	(2.3, 3.1)	medium
20	2000	783 (39.2%)	509 (25.5%)	410 (20.5%)	298 (14.9%)	2.6 *	(2.3, 3.0)	medium
21	2000	722 (36.1%)	537 (26.9%)	439 (22%)	302 (15.1%)	2.4 *	(2.1, 2.8)	medium
22	2000	700 (35%)	560 (28%)	417 (20.9%)	323 (16.2%)	2.2 *	(1.9, 2.5)	medium
23	2000	659 (33%)	531 (26.6%)	452 (22.6%)	358 (17.9%)	1.8 *	(1.6, 2.1)	small
Total	11,738	4315 (36.8%)	3173 (27%)	2460 (21%)	1790 (15.2%)	2.4 *	(2.2, 2.6)	medium

Note: RAEs of players listed in Tranfermarkt.de. Q1 to Q4 = Quartile 1 to 4; OR = Odds ratio; CI = Confidence Interval; * $p < 0.05$; OR < 1.22, 1.22 ≤ OR < 1.86, 1.86 ≤ OR < 3.00, and OR ≥ 3.00, was interpreted as negligible, small, medium and large.

Table 3. Mean market values per age category and relative age quartile (Q).

under (U)	n	Q1 (€)	Q2 (€)	Q3 (€)	Q4 (€)	OR Q1/Q4	95% CI	Effect Size
18	1738	318,950	320,963	328,971	327,597	1.0	(0.9, 1.0)	no
19	2000	469,437	373,563	317,910	383,993	1.2 *	(1.1, 1.3)	small
20	2000	942,593	711,690	809,939	919,547	1.0	(1.0, 1.0)	non
21	2000	1,183,587	1,311,778	1,252,790	1,330,215	0.9 *	(0.9, 0.9)	non
22	2000	1,136,464	1,285,982	1,052,338	2,416,563	0.5 *	(0.5, 0.5)	medium
23	2000	1,789,416	2,112,712	1,519,967	2,651,536	0.7 *	(0.7, 0.7)	small
Total	11,738	960,002	1,031,007	913,638	1,445,796	0.7 *	(0.6, 0.7)	small

Note: Q1 to Q4 = Quartile 1 to 4; OR = Odds Ratio; CI = Confidence Interval; * $p < 0.05$; 1.00 ≤ OR < 1.22, 1.22 ≤ OR < 1.86, 1.86 ≤ OR < 3.00, and OR ≥ 3.00, was interpreted as negligible, small, medium and large. Inverse ORs < 0.33 (1/3), 0.33 ≤ OR < 0.53 (1/1.86), 0.53 ≤ OR < 0.81, 0.81 ≤ OR < 1.0 were interpreted as large, medium, small and negligible.

Table 4. Δ of total market values per age category and relative age quartile (Q).

under (U)	n	Δ Q1 (€)	Δ Q2 (€)	Δ Q3 (€)	Δ Q4 (€)
18	1738	86,276,071	8,826,488	−31,087,721	−66,666,071
19	2000	115,481,501	27,643,641	−31,155,224	−85,246,403
20	2000	266,753,704	6,405,206	−72,894,512	−185,748,490
21	2000	262,756,371	48,535,801	−76,420,216	−263,382,616
22	2000	227,292,857	77,158,929	−87,344,065	−427,731,734
23	2000	284,517,109	65,494,068	−72,958,407	−376,518,156
Total	11,738	1,243,077,614	234,064,133	−371,860,145	−1,405,293,470

Note: Difference of observed and expected market values (Δ). Q1 to Q4 = Quartile 1 to 4.

Distribution of player positions per Q with 95% CI are illustrated in Table 5. There were medium to large RAEs in all positions from U18 to U22. The highest ORs were found in the U18 age category, except for goalkeepers. There were no significant differences between the different playing positions.

Table 5. Distribution of player positions per age category and quarter (Q).

Position	under (U)	n	Q1 (%)	Q2 (%)	Q3 (%)	Q4 (%)	OR Q1/Q4	95% CI	Effect Size
Goalkeeper	18	240	89 (37.1)	65 (27.1)	49 (20.4)	37 (15.4)	2.4 *	(2.3, 2.5)	medium
	19	191	67 (35.1)	52 (27.2)	44 (23.0)	28 (14.7)	2.4 *	(2.3, 2.5)	medium
	20	149	62 (41.6)	38 (25.5)	30 (20.1)	19 (12.8)	3.3 *	(3.1, 3.4)	large
	21	144	55 (38.2)	34 (23.6)	27 (18.8)	28 (19.4)	2.0 *	(1.9, 2.1)	medium
	22	109	43 (39.4)	31 (28.4)	23 (21.1)	12 (11.0)	3.6 *	(3.4, 3.7)	large
	23	114	39 (34.2)	35 (30.7)	21 (18.4)	19 (16.7)	2.1 *	(2.0, 2.1)	medium
Defender	18	431	212 (49.2)	114 (26.5)	89 (20.6)	50 (11.6)	4.2 *	(4.0, 4.4)	large
	19	558	226 (40.5)	161 (28.9)	101 (18.1)	70 (12.5)	3.2 *	(3.1, 3.4)	large
	20	560	219 (39.1)	143 (25.5)	116 (20.7)	82 (14.6)	2.7 *	(2.6, 2.8)	medium
	21	609	230 (37.8)	165 (27.1)	120 (19.7)	94 (15.4)	2.4 *	(2.4, 2.6)	medium
	22	639	224 (35.1)	169 (26.4)	136 (21.3)	110 (17.2)	2.0 *	(2.0, 2.1)	medium
	23	676	238 (35.2)	172 (25.4)	144 (21.3)	122 (18)	2.0 *	(1.9, 2.0)	medium
Midfielder	18	606	236 (38.9)	164 (27.1)	119 (19.6)	87 (14.4)	2.7 *	(2.6, 2.8)	medium
	19	711	255 (35.9)	218 (30.7)	141 (19.8)	97 (13.6)	2.6 *	(2.5, 2.7)	medium
	20	713	272 (38.1)	183 (25.7)	144 (20.2)	114 (16.0)	2.4 *	(2.3, 2.5)	medium
	21	681	243 (35.7)	173 (25.4)	171 (25.1)	94 (13.8)	2.6 *	(2.5, 2.7)	medium
	22	669	250 (37.4)	204 (30.5)	120 (17.9)	95 (14.2)	2.6 *	(2.5, 2.7)	medium
	23	627	179 (28.5)	179 (28.5)	155 (24.7)	114 (18.2)	1.6 *	(1.5, 1.6)	small
Striker	18	461	202 (43.8)	119 (25.8)	83 (18)	57 (12.4)	3.5 *	(3.4, 3.7)	large
	19	540	198 (36.7)	143 (26.5)	116 (21.5)	83 (15.4)	2.4 *	(2.3, 2.5)	medium
	20	578	230 (39.8)	145 (25.1)	120 (20.8)	83 (14.4)	2.8 *	(2.7, 2.9)	medium
	21	566	194 (34.3)	165 (29.2)	121 (21.4)	86 (15.2)	2.3 *	(2.2, 2.4)	medium
	22	583	183 (31.4)	156 (26.8)	138 (23.7)	106 (18.2)	1.7 *	(1.6, 1.8)	small
	23	583	203 (34.8)	145 (24.9)	132 (22.6)	103 (17.7)	2.0 *	(1.9, 2.1)	medium

Note: Q1 to Q4 = Quartile 1 to 4; OR = Odds Ratio; CI = Confidence Interval; * $p < 0.05$; 1.00 ≤ OR < 1.22, 1.22 ≤ OR < 1.86, 1.86 ≤ OR < 3.00, and OR ≥ 3.00, was interpreted as negligible, small, medium and large.

Table 6 displays position specific RAEs between Q1 and Q4 players for each age group based on market values. Market value was greater for relatively older goalkeepers (Q)1 compared to Q4, with a small to large effect depending on age group. The market values of defenders, midfielders and strikers were significantly higher for Q4 compared to Q1 players in the U21, U22 and U23 with small to large effects. Over- and undervaluing due to RAEs were highest for strikers, followed by defenders, midfielders, and goalkeepers.

Table 6. Market values per playing position, age category and relative age quartile (Q).

Position	under (U)	n	Q1 (€)	Q2 (€)	Q3 (€)	Q4 (€)	OR Q1/Q4	95% CI	Effect Size
Goalkeeper	18	240	159,238	140,860	33,724	24,024	6.6 *	(6.4, 6.9)	large
	19	191	107,090	90,385	101,705	83,929	1.3 *	(1.2, 1.3)	small
	20	149	318,548	265,789	251,667	236,842	1.3 *	(1.3, 1.4)	small
	21	144	1,750,909	217,647	432,407	413,393	4.2 *	(4.0, 4.4)	large
	22	109	786,628	1,112,903	370,652	581,250	1.4 *	(1.3, 1.4)	small
	23	114	2,008,333	1,289,286	1,257,143	707,895	2.8 *	(2.7, 3.0)	medium
Defender	18	431	192,191	265,570	114,326	74,500	2.6 *	(2.5, 2.7)	medium
	19	558	473,341	385,093	135,891	261,786	1.8 *	(1.7, 1.9)	medium
	20	560	689,954	455,944	560,129	1,484,146	0.5 *	(0.5, 0.5)	medium
	21	609	1,016,739	1,233,636	1,392,292	1,347,340	0.8 *	(0.7, 0.8)	small
	22	639	1,221,205	999,260	767,463	2,872,045	0.4 *	(0.4, 0.4)	medium
	23	676	1,457,248	1,458,285	1,083,854	2,216,189	0.7 *	(0.6, 0.7)	small
Midfielder	18	606	501,907	334,146	271,639	245,690	2.0 *	(2.0, 2.1)	medium
	19	711	288,431	329,128	324,823	177,835	1.6 *	(1.6, 1.7)	small
	20	713	694,210	877,869	479,688	664,035	1.0 *	(1,0, 1.1)	non
	21	681	1,331,173	1,533,671	1,350,585	615,160	2.2 *	(2.08, 2.3)	medium
	22	669	1,311,600	1,616,299	1,373,958	1,462,895	0.9 *	(0.9, 0.9)	non
	23	627	2,199,581	2,425,279	1,407,903	2,472,368	0.9 *	(0.9, 0.9)	non
Striker	18	461	318,688	492,731	808,735	863,158	0.4 *	(0.4, 0.4)	medium
	19	540	820,707	531,294	550,000	829,217	1.0	(1.0, 1.0)	non
	20	578	1,645,109	871,034	1,587,292	868,976	1.9 *	(1.8, 2.0)	medium
	21	566	1,035,696	1,382,727	1,159,298	2,391,570	0.4 *	(0.4, 0.5)	medium
	22	583	875,683	1,199,038	1,167,029	3,006,368	0.3 *	(0.3, 0.3)	large
	23	583	1,775,123	2,701,897	2,169,129	3,724,029	0.5 *	(0.5, 0.5)	medium

Note: Difference of observed and expected market values (Δ). Q1 to Q4 = Quartile 1 to 4; OR = Odds Ratio; CI = Confidence Interval; * $p < 0.05$; 1.00 ≤ OR < 1.22, 1.22 ≤ OR < 1.86, 1.86 ≤ OR < 3.00, and OR ≥ 3.00, was interpreted as negligible, small, medium and large. Inverse ORs < 0.33 (1/3), 0.33 ≤ OR < 0.53 (1/1.86), 0.53 ≤ OR < 0.81, 0.81 ≤ OR < 1.0 were interpreted as large, medium, small and negligible.

4. Discussion

Results from the present study, illustrate the following main findings: (i) the analysis of relative age distribution illustrated significant overrepresentations of Q1 players in all age categories. Effect sizes diminished progressively from the U18 (large) to the U23 (small). This trend only existed when analyzing the whole sample, not when separated by playing positions. (ii) Relative age was also associated with biased market values. Initially, higher market values were apparent for Q1 players at U19. Thereafter, the effect was inversed, with Q4 players showing a significantly higher market value across U21, U22, and U23. (iii) Playing positions analysis revealed higher market values for Q4 defenders, midfielders, and strikers at U23, compared to Q1. By contrast, relatively older goalkeepers (Q1) had a higher market value than Q4 goalkeepers in all age categories.

Present findings align with previous studies, where RAEs biases were evident in the sample [2]. Biased selection during youth talent development programs may reduce a relatively younger athlete's chances of succeeding later in their career. The relatively younger are disadvantaged by lower selection quotas, which in turn may lead to less competition experience, lower motivation, as well as a lower opportunities of accessing high-quality training [2]. However, particular RAE studies identify inverse RAEs in talent development programs post-puberty [9,12], suggesting delayed benefits if the relatively young can remain within the sporting development system. For instance, Deaner (2013) showed how compared to those born in Q1, Q3 and Q4 players were twice as likely to reach professional career benchmarks. Similarly, Fumarco (2017) identified how Q4 players scored more often, and receive higher salaries, than Q1 players. When considered alongside present findings, the underdog effect is supported, reflected by the increased likelihood of being drafted, career length, performance productivity, and now market value at the professional level [25].

The phenomenon that Q4 athletes are over-represented among those who successfully transition from youth systems to senior professional status has been called the 'underdog hypothesis'. Being younger essentially facilitates long-term development by necessitating them to overcome the relative age disadvantage, through being challenged by their older and more advanced peers [10–12]. A previous study by Doyle and Bottomley (28), who analyzed the market values of the top 1000 players on transfermarkt.de in the season of 2013–2014, noted that relatively older players had greater opportunities due to assessment selection bias, but were valued equally to players born later in the year. Although the current study confirms these results, the market values of players do represent the underdog effect. As such, selected Q4 players are often initially undervalued, but later are valued higher than Q1 players [9]. Additionally, a recent study of Perez-Gonzalez et al. [26] analyzed the market value of 2577 adult professional players of the biggest European football leagues. Small to medium RAEs were shown in all leagues ($p < 0.05$). However, this bias did not affect the market value of the professional elite soccer players examined. The authors concluded that identification and promotion of talent at young ages are often biased by RAEs, however once players have reached the professional stage, their market value is independent of RAEs [26]. In our study, from a return of investment point of view, market value of Q1 players increases by 560% from U18 to U23, whereas market value of Q4 players increases by 810%. This phenomenon is even more pronounced when differentiated by playing position. The value of Q4 goalkeepers and defenders increases by approximately 3000%, while the value of Q1 players "only" increases by 1260% and 760%, respectively. In the U23, the highest mean values in terms of playing positions were found for defenders, midfielders and strikers born in Q4, except for goalkeepers. This leads to the assumption that the underdog effect exists as well if the sample is subdivided by playing positions. To sum up, RAEs and biased market values likely lead to inefficient selection and return of investment of football talent. To gain further insight into this issue, longitudinal studies analyzing the evolution of market values of players throughout talent development should be conducted.

Limitations: while the present analysis was performed using a cross sectional dataset, future studies should use a longitudinal design to analyze the evolution of market values and their interrelationships with RAEs. Furthermore, as financial loss due to over- and undervaluing was calculated on a theoretical estimate assuming an equal distribution of players between birth quarters, future studies which particularly focus on this aspect, should also include factors such as the evolution of market values in the long run, differences between female and male sports and the optimal talent development from a sports-scientific and economic point of view.

5. Conclusions

The analysis of relative age distribution illustrated significant overrepresentations of Q1 players in all age categories. This trend only existed when analyzing the whole sample, not when separated by playing positions. Relative age was also associated with biased market values. Initially, higher market values were apparent for Q1 players at U19. Thereafter, the effect was inversed, with Q4 players showing a significantly higher market value across U21, U22, and U23. Playing positions analysis revealed higher market values for Q4 defenders, midfielders, and strikers at U23, compared to Q1. By contrast, relatively older goalkeepers (Q1) had a higher market value than Q4 goalkeepers in all age categories. Assuming an equal distribution of football talent exists across annual cohorts, findings suggest the selection and market value of young professional players is dynamic. Findings suggest a potential biased selection, and undervaluing of Q4 players in younger age groups, as their representation and market value increased over time. By contrast, the changing representations and market values of Q1 players suggest initial overvaluing in performance and monetary terms.

Author Contributions: Conceptualization, M.R., M.J., S.C., D.-P.B.; methodology, M.R., M.J., D.-P.B.; validation, M.R., M.J., S.C., D.-P.B.; formal analysis, M.R., D.-P.B.; investigation, M.R., M.J., S.C., D.-P.B.; data curation, M.R., D.-P.B.; writing—original draft preparation, M.R.; writing—review and editing, M.R., M.J., S.C., D.-P.B.; visualization, M.R., D.-P.B.; supervision, D.-P.B.; project administration, M.R. All authors have read and agreed to the published version of the manuscript.

Funding: This research received no external funding.

Institutional Review Board Statement: The study was conducted according to the guidelines of the Declaration of Helsinki and approved by the Institutional Review Board (or Ethics Committee) of the Swiss Federal Institute of Sport Magglingen (Reg.-Nr. HLP-2021-131).

Informed Consent Statement: Patient consent was waived due to the analysis of secondary data which is available on the website www.transfermarkt.de (accessed on 17 July 2020).

Data Availability Statement: All data used in this study are available on the website www.transfermarkt.de (accessed on 17 July 2020).

Acknowledgments: We want to thank www.transfermarkt.de (accessed on 17 July 2020) for providing the data.

Conflicts of Interest: The authors declare no conflict of interest.

References

1. Smith, K.L.; Weir, P.L.; Till, K.; Romann, M.; Cobley, S. Relative Age Effects across and within Female Sport Contexts: A Systematic Review and Meta-Analysis. *Sports Med.* **2018**, *48*, 1451–1478. [CrossRef] [PubMed]
2. Cobley, S.; Baker, J.; Wattie, N.; McKenna, J. Annual Age-Grouping and Athlete Development. *Sports Med.* **2009**, *39*, 235–256. [CrossRef]
3. Romann, M.; Rössler, R.; Javet, M.; Faude, O. Relative Age Effects in Swiss Talent Development—A Nationwide Analysis of All Sports. *J. Sports Sci.* **2018**, *36*, 2025–2031. [CrossRef] [PubMed]
4. Wattie, N.; Schorer, J.; Baker, J. The Relative Age Effect in Sport: A Developmental Systems Model. *Sports Med.* **2015**, *45*, 83–94. [CrossRef] [PubMed]
5. Helsen, W.F.; Van Winckel, J.; Williams, A.M. The Relative Age Effect in Youth Soccer across Europe. *J. Sports Sci.* **2005**, *23*, 629–636. [CrossRef] [PubMed]

6. Cobley, S.P.; Till, K. Participation Trends According to Relative Age across Youth UK Rugby League. *Int. J. Sports Sci. Coach.* **2017**, *12*, 339–343. [CrossRef]
7. Helsen, W.F.; Starkes, J.L.; Van Winckel, J. The Influence of Relative Age on Success and Dropout in Male Soccer Players. *Am. J. Hum. Biol. Off. J. Hum. Biol. Assoc.* **1998**, *10*, 791–798. [CrossRef]
8. Romann, M.; Rüeger, E.; Hintermann, M.; Kern, R.; Faude, O. Origins of Relative Age Effects in Youth Football—A Nationwide Analysis. *Front. Sports Act. Living* **2020**, *2*, 591072. [CrossRef]
9. Deaner, R.O.; Lowen, A.; Cobley, S. Born at the Wrong Time: Selection Bias in the NHL Draft. *PLoS ONE* **2013**, *8*, e57753. [CrossRef]
10. Gibbs, B.G.; Jarvis, J.A.; Dufur, M.J. The Rise of the Underdog? The Relative Age Effect Reversal among Canadian-Born NHL Hockey Players: A Reply to Nolan and Howell. *Int. Rev. Sociol. Sport* **2012**, *47*, 644–649. [CrossRef]
11. Kelly, A.L.; Wilson, M.R.; Gough, L.A.; Knapman, H.; Morgan, P.; Cole, M.; Jackson, D.T.; Williams, C.A. A Longitudinal Investigation into the Relative Age Effect in an English Professional Football Club: Exploring the 'Underdog Hypothesis'. *Sci Med. Footb.* **2020**, *4*, 111–118. [CrossRef]
12. Fumarco, L.; Gibbs, B.G.; Jarvis, J.A.; Rossi, G. The Relative Age Effect Reversal among the National Hockey League Elite. *PLoS ONE* **2017**, *12*, e0182827. [CrossRef]
13. Ashworth, J.; Heyndels, B. Selection Bias and Peer Effects in Team Sports: The Effect of Age Grouping on Earnings of German Soccer Players. *J. Sports Econ.* **2007**, *8*, 355–377. [CrossRef]
14. Treadway, D.C.; Adams, G.; Hanes, T.J.; Perrewé, P.L.; Magnusen, M.J.; Ferris, G.R. The Roles of Recruiter Political Skill and Performance Resource Leveraging in NCAA Football Recruitment Effectiveness. *J. Manag.* **2014**, *40*, 1607–1626. [CrossRef]
15. Pawlowski, T.; Breuer, C.; Hovemann, A. Top Clubs' Performance and the Competitive Situation in European Domestic Football Competitions. *J. Sports Econ.* **2010**, *11*, 186–202. [CrossRef]
16. Fuchslocher, J.; Romann, M.; Gulbin, J. Strategies to Support Developing Talent. *Schweiz. Z. Sportmed. Sporttraumatol.* **2013**, *61*, 10–14.
17. Herm, S.; Callsen-Bracker, H.-M.; Kreis, H. When the Crowd Evaluates Soccer Players' Market Values: Accuracy and Evaluation Attributes of an Online Community. *Sport Manag. Rev.* **2014**, *17*, 484–492. [CrossRef]
18. Gavião, L.O.; Sant'Anna, A.P.; Alves Lima, G.B.; de Almada Garcia, P.A. Evaluation of Soccer Players under the Moneyball Concept. *J. Sports Sci.* **2020**, *38*, 1221–1247. [CrossRef]
19. Atanasov, P.; Rescober, P.; Stone, E.; Swift, S.A.; Servan-Schreiber, E.; Tetlock, P.; Ungar, L.; Mellers, B. Distilling the Wisdom of Crowds: Prediction Markets vs. Prediction Polls. *Manag. Sci.* **2017**, *63*, 691–706. [CrossRef]
20. Bryson, A.; Gomez, R.; Zhang, T. All-star or benchwarmer? Relative age, cohort size and career success in the NHL. In *Breaking the Ice*; Springer: Berlin/Heidelberg, Germany, 2017; pp. 57–91.
21. Gyimesi, A.; Kehl, D. Relative Age Effect on the Market Value of Elite European Football Players: A Balanced Sample Approach. *Eur. Sport Manag. Q.* **2021**, 1–17. [CrossRef]
22. Schorer, J.; Deja, J.; Steingröver, C.; Baker, J.; Loffing, F.; Helsen, W.; Wattie, N. The Association of Two Relative Age Effects and Estimated Money Value in Elite Soccer. *J. Exerc. Mov. Sport* **2017**, *49*, 128.
23. Herberger, T.A.; Wedlich, F. Does Selection Bias Matter in Football Players' Valuation? A Crowdsourced Valuation Approach on Players' Athletic Characteristics. *J. Glob. Sport Manag.* **2017**, *2*, 196–214. [CrossRef]
24. Olivier, J.; Bell, M.L. Effect Sizes for 2 × 2 Contingency Tables. *PLoS ONE* **2013**, *8*, e58777. [CrossRef] [PubMed]
25. Ramos-Filho, L.; Ferreira, M.P. The Reverse Relative Age Effect in Professional Soccer: An Analysis of the Brazilian National League of 2015. *Eur. Sport Manag. Q.* **2021**, *21*, 78–93. [CrossRef]
26. Perez-Gonzalez, B.; Fernandez-Luna, A.; Castillo, D.; Burillo, P. Are European Soccer Players Worth More If They Are Born Early in the Year? Relative Age Effect on Player Market Value. *Int. J. Environ. Res. Public Health* **2020**, *17*, 3301. [CrossRef] [PubMed]

sports

MDPI

Article

Early Drop-Out from Sports and Strategic Learning Skills: A Cross-Country Study in Italian and Spanish Students

Carla Consoni [1,*], Caterina Pesce [2] and Domenico Cherubini [1]

1 Sports Faculty, UCAM Catholic University of Murcia, 30107 Murcia, Spain; dcherubini@ucam.edu
2 Department of Movement, Human and Health Sciences, University of Rome "Foro Italico", 00135 Rome, Italy; caterina.pesce@uniroma4.it
* Correspondence: carla.consoni@gmail.com

Abstract: The search for overarching factors involved in both sport and broader lifestyle and achievement domains may help to understand the early drop-out phenomenon. This study aimed to analyze the association between early sport drop-out and strategic learning skills, checking for the individual and joint role of nationality, school type, gender, age and sport habits. Six hundred and fourteen Italian and Spanish students aged 14–18 years completed two self-assessment questionnaires concerning physical activity, sports habits and learning strategies. Outcomes were analyzed with frequency analysis. Higher affective–motivational strategic learning skills were associated with lower drop-out rates in Italian but not Spanish students. In high schools with an enhanced sports curriculum, drop-out rates were negligible compared to other Italian and Spanish curricula. A lack of persistence in the same sport type was significantly associated with a higher drop-out rate in males but not in female students, who had overall higher drop-out rates. This study suggests that overarching personal skills, cultural characteristics and sports habits may independently and jointly contribute to sport drop-out. Specifically, affective–motivational learning skills may play a key role in sport persistence and in strategies tailored to drop-out prevention.

Keywords: physical activity; strategic skills; affective; motivational; sport habits

Citation: Consoni, C.; Pesce, C.; Cherubini, D. Early Drop-Out from Sports and Strategic Learning Skills: A Cross-Country Study in Italian and Spanish Students. *Sports* **2021**, *9*, 96. https://doi.org/10.3390/sports9070096

Academic Editors: Adam Leigh Kelly, Sergio L. Jiménez Sáiz, Sara Diana Leal dos Santos and Alberto Lorenzo Calvo

Received: 18 May 2021
Accepted: 23 June 2021
Published: 30 June 2021

1. Introduction

Physical and sports activities provide benefits to physical, mental and socio-emotional health [1]. All people who exercise regularly improve their quality of life at any age and without gender differences [2,3]. Moreover, in young people, sports training promotes good health, an optimal psycho-physical development and the acquisition of healthy lifestyles that will persist even in later ages [4]. Generally, children start playing sports to socialize, improve motor skills and have fun [5]. The more the approach is enjoyable and gratifying, the stronger the motivation to continue doing sports over time and the probability to ensure greater investment in sport and talent identification [6,7].

Nevertheless, young athletes, especially the talented ones, may quit the sport prematurely during their school education before reaching their potential peak performance due to personal, social and contextual factors [8,9]. The combination of socioeconomic [10] and parental/coaches/peers support factors can predict sports participation/abandonment during childhood [11]. Therefore, sports drop-out has been identified as a multifactorial and complex phenomenon, strongly influenced by different cultural backgrounds [10] and behavioral factors, as well as by personal characteristics, types of sports, attitudes and motivations [12]. The drop-out rates from sports gradually increase across adolescence [13]; this is associated with physical inactivity later in life, thus contributing to unhealthy lifestyles [14]. Conversely, young people who exercise keep healthy lifestyle habits, such as continued physical activity and healthy nutrition [15]. Findings in Sport and Physical Activity surveys [16,17] showed that the percentage of young people aged 15–24 years who exercise regularly decreased in ten years from 14% to 9%. In this group, 15% were males and 33% females (expressed in percentage of the whole sample interviewed: 15–55 years). In addition, several authors confirmed the drop-out rates from sports being higher in females than in males [18,19].

Reasons for dropping-out of sports have been classified into performance and training factors, education and work obligations [20,21], motivational factors, social environment and other interests [22,23]. Most recently, Crane & Temple [24] and Witt & Dangi [25] discussed the *leisure constraints theory* [26] and categorized the variables associated with drop-out from sports as intrapersonal, interpersonal and structural constraints. The findings related to this theoretical model showed drop-out rates being associated with sport-specific attrition factors at intra-personal and interpersonal levels. According to the cognitive–affective model [27], as well as Sorkkila et al. [28], demands, critical situations and personal resource deficit are associated with the development of sport and school burnout. On the other hand, Cosh and Tully [29] observed that playing sport could upskill young athletes by enhancing time management, self-care, self-efficacy and specific strategies for coping with stress in different life areas in young people. Moreover, transferring these skills to academic commitment may help young people successfully achieve their goals in other life domains. Adolescent sports participation [30] and, more broadly, physical activity are linked to academic achievement [31,32]. This is consistent with the evidence that systematic exercise programs may actually enhance the development of cognitive and metacognitive skills known to be important in addressing challenges of achievements both academically and beyond [33].

In summary, these studies confirm an important relationship linking sport and physical activity to functional personal resources. Furthermore, these resources have a cross-boundary influence on every learning context and are relevant to dealing with challenges, changes and critical situations. The linkage of cross-functional personal skills in more domains encourages the investigation of unspecific dimensions and factors, which could play a role in drop-out from, or persistence in, sports activity but are still underexplored.

In the framework of the soft skills that have to be promoted in the educational field, Bay, Grządziel and Pellerey [34] define specific learning skills as strategic. These skills are cognitive, metacognitive, affective, motivational and volitional in nature, and refer to the capacity for self-determination and self-regulation. Self-determination is related to motivation, choice and intention of the action in different salient and emotionally laden life conditions. Self-regulation refers to the monitoring, evaluation and control system of an action to monitor its consistency, stability and orientation, and to regulate its functionality [35]. These capacities enable positive behaviour towards challenges and changes; they are functional in order to carry out personal objectives. The strategic learning skills are general and transferable, and seem to constitute stable internal dispositions indispensable to dealing with study, work, sport and any relevant tasks in life autonomously and successfully. From an early age, good management of these skills allows the person to play a self-orienting role in different learning and life domains [36]. In other life contexts, the sport experiences may provide a favorable environment to promote positive youth development through sport participation [37].

While the search for factors associated with sport drop-out rates is ongoing, the role played by skills that may affect performance in broader life domains, different from sport, is still under-investigated. Thus, the main aim of this study was to explore the association between early drop-out from sport in adolescence and strategic learning skills. We hypothesized that these skills functional to learning processes, mainly studied in academic achievement research [38], may also be associated with sport participation and, vice versa, an insufficient level/lack of such skills could be associated to early drop-out from sport. The hypothesis is grounded on indirect evidence of piecemeal associations between strategic learning skills, sport participation and academic achievement [39]. These associations have been mainly studied in a unidirectional manner, focusing on the beneficial effect of the sports activity on other life domains and academic achievement [40,41]. However, it is possible that these relations are bidirectional and the co-variation of sport participation and academic achievement is due to commonalities in the underlying domain-general of functional skills that rely on high-level cognitive function [42,43]. Thus, we expected that strategic learning skills considered functional to academic achievement might also be supportive of sport participation adherence. We tested this hypothesis in an exploratory manner, looking for associations with a cross-sectional design. Since early drop-out seems influenced by demographic variables such as gender and age [23], and by the quantitative and qualitative characteristics of sports activity [44,45], we further explored the hypothesized association of sport drop-out rates with strategic learning skills as a function of these variables. This association was investigated also as a function of nationality in Italian and Spanish adolescents to perform a cross-country comparison and extend the generalizability of results.

2. Materials and Methods

2.1. Participants

Participants were sampled in a stratified manner, purposefully identifying different types of high schools and urban areas of the cities of Rome (Italy) and Murcia (Spain) before sampling. Both are large cities located in a central region of their respective countries, close to the Mediterranean coast. To reduce possible influences determined by the socio-economic status of students participating in the research, school structures operating in districts characterized by a prevailing middle social status were included in the study. All participants were screened and selected based on three criteria: age (14 to 18 years); high school students; eligible senior high school curriculum. The Spanish participants attended the compulsory secondary school (ESO) that is the unique eligible curriculum until compulsory school. In contrast, Italian students have the option to choose among different school curricula. To the aim of this study, we recruited Italian students who attended upper secondary schools with an enhanced sports curriculum (sports senior high school), and other non-sport enhanced (scientific, applied sciences, linguistic) senior high schools. In this way, we could perform a cross-country comparison between Spanish and Italian students attending non-sport enhanced senior high schools, as well as a comparison of students attending senior high school with an enhanced sport curriculum with the other traditional Italian and Spanish school curricula. Since the Italian students could attend many different high-school curricula, it was deemed appropriate to recruit a larger number of Italian rather than Spanish students to have a representative sample for each type of eligible Italian high school. Thus, the sample consisted of 188 Spanish (109 male, 79 female) and 426 Italian (257 male, 169 female) students aged 14–18 years. The students consented to fill in the questionnaires and their parents/guardians (if younger than 18 years) signed an informed written consent before participating in the study.

2.2. Instruments

The students, in their own native language, completed two self-assessment questionnaires after forward–backward translation of the Italian version of the CAPAFD (originally developed in Spanish) and of the Spanish version of the QSA-R (originally developed in Italian).

The first questionnaire asked about the motivation and training of physical-sporting activity: Questionnaire for the Analysis of the Practice of Physical-Sports Activities, CAPAFD [46]. This questionnaire was developed to evaluate physical and sport habits associated with different socio-demographic variables. It has been proven suitable to collect descriptive data on objectively reportable demographics and behavioral habits [46]. The original questionnaire is made of eleven blocks comprising twenty-six questions targeted at people aged 15–64 years. In order to collect information on sport habits and drop-out relevant to the aim of the present study, two sections comprising seventeen questions were selected. These questions included information concerning sport drop-out, sport participation and background variables relevant to the study. Specifically, the selected items allowed collecting information about: age, gender, type of attended school, onset age of sports activity, current/past registration to a sports Federation, sport practice/drop-out current and past sports practiced (years of practice for each type of sport practiced, if any) workout frequency (hours/days and days/week), competitive sports activity performed (if any) and sport practiced in organized club/association. This information was collected with three types of questions. Five of the 17 items were open-ended questions regarding demographic information (e.g., School attended); 5 items were close-ended questions with dichotomous answer (yes or no; e.g., Have you dropped-out of sports?); the remaining 7 items were multiple choice with three to six answers (e.g., How much time do you spend working out daily on average?).

The second questionnaire, the Questionnaire on Learning Strategies (reduced version) QSA-R validated by Margottini [47], was about students' strategic competences for learning to self-evaluate their studying habits and/or the critical situations encountered in schoolwork. The QSA-R estimates the perceived mastery of eight strategic skills, grouped in two dimensions of four scales each: cognitive/metacognitive (C1, C2, C3, C4) and affective/motivational (A1, A2, A3, A4). Each scale refers to a specific strategic competence (Table 1). The questionnaire is composed of 46 questions about school and homework. Twenty-one items are grouped in the four cognitive–metacognitive scales as follows: C1 (n = 6; e.g., I try to find relationships between what I learn and what I already know) C2 (n = 7; e.g., I organize my study based on the time I have available), C3 (n = 5; e.g., I build diagrams, graphs or summary tables to summarize what I study) and C4 (n = 3; e.g. While the teacher explains, I get distracted). Twenty-five items are grouped in the four affective–motivational scales as follows: A1 (n = 6; e.g., I quickly get nervous for a question or problem that I don't understand immediately), A2 (n = 6; e.g., Even if I don't like the subject, I equally work hard to succeed), A3 (n = 8; e.g., When an oral exam goes well, I think I have done well studying hard) and A4 (n = 5; e.g., I feel capable of successfully completing my study commitments). The answers were given on a four-point Likert-type scale from 1 (never or almost never) to 4 (always or almost always).

Table 1. QSA-R: Scales and Dimensions.

Dimensions		Scales	C. α *
Cognitive Metacognitive	C1	Processing strategies for understanding and remembering	0.74
	C2	Self-regulation strategies	0.64
	C3	Graphic strategies to understand, summarize, and memorize	0.71
	C4	Attention control strategies	0.63
Affective Motivational	A1	Emotion management strategies	0.79
	A2	Volition	0.74
	A3	Causal attribution-Locus of control	0.52
	A4	Perception of competence	0.72

Note: * Cronbach's alpha of 1840 Italian and Spanish students.

From raw scores for each of the eight subscales, the formula [(sum raw score-mean)/standard deviation × 2 + 5] provided the final profile of students on a nine-point standard Stanine-scale [47]. For the majority of subscales (C1, C2, C3 and A2, A3, A4), the higher the scoring, the more positive the outcome. For the remaining subscales (C4 and A1), conversely, the lower the scoring, the more positive the outcome. Thus, the scores for C4 and A1 subscales were reversed to obtain higher scores corresponding to higher functional skills consistently across scales. According to normative data [47], 1 to 3 are below average skill scores, 4 to 6 points are the average range and 7 to 9 are above average scores. All stanine-transformed scores of each subscale (C1–4 and A1–4) were further transformed in dichotomous variables for data analysis. It was attributed "0" (=no-criticality, i.e., skills above critical level) to all factors which showed a value from 4 to 9 and "1" (=criticality, i.e., skills below critical level) for value from 1 to 3. Summing up the critical values across the four affective/motivational (A1–4) subscales and across the four (C1–4) cognitive/metacognitive subscales [47], each participant could have from 0 (no critical values) to 4 (critical values in all four subscales). Based on this sum of critical values, participants were divided in five (0, 1, 2, 3, 4) categories of criticality in cognitive/metacognitive and affective/motivational skills.

As reported in the validation study [47] performed with 1.182 students, Cronbach's alpha ranges between 0.59–0.78 across the eight scales, with corrected item–scale correlations being mostly moderate. This Cronbach's alpha range is deemed acceptable for tools measuring cognitive and affective constructs in learning [48]. A recent Italian study with a larger sample of 3.091 upper secondary school students [49] confirmed a similar Cronbach's alpha range for the eight scales (0.62–0.78). In the present study, the internal consistency was computed on a larger sample of 1840 Italian and Spanish students (n = 1451 and 389, respectively) involved in a wider research project that the present study on youth drop-out is a part of. For the Spanish version of the QSA-R, Cronbach's alpha was 0.76 for all four cognitive/metacognitive scales (C1–4) and 0.71 for all four affective/motivational (A1–4). Overall, Cronbach's alpha was 0.76. The values for each scale are reported in Table 1.

2.3. Procedure

Data were collected between March and May 2019 by a paper-and-pencil version of the tools. Questionnaires were administered during school time and the compilation of the two questionnaires was performed in a single session (about 20 min) to facilitate the correct matching of tools (and data) for the same student. All students received information for filling out the two questionnaires. Data anonymity was ensured by assigning a unique numerical code to the two questionnaires and an acronym to identify the academic course. Data collection was performed in compliance with school time constraints in order to not overload the students in the self-assessment task and not exceed the maximum time teachers of participant classes agreed on. Students who affirmed that they had never played sports were excluded from the sample, as their data could not contribute to the study of the drop-out phenomenon.

2.4. Data Analysis

Descriptive statistics were computed for all variables separately for students who declared drop-out or no drop-out from sports (Table 2).

Frequency analyses were performed to analyze the distribution of sport drop-out rates as a function of: gender, age, nationality, school curriculum, strategic learning skills (affective/motivational and cognitive/metacognitive) and persistence in sport type or change therein. The association of sport drop-out with strategic learning skills (primary focus of the present study) and with persistence in/change of sport (secondary focus) was also analyzed as a function of demographic variables (gender, age and nationality). To this aim, the frequency distribution of drop-out rates was compared among categories of strategic skills deficits (0, 1, 2, 3, 4), and among categories of persistence in/change of sport as a function of gender, age or nationality. After verifying the non-normality of the variables analyzed through the Kolmogorov–Smirnov test ($p < 0.001$), to compute both main and associated effects, bivariate analyses with contingency tables and the chi-square test (χ^2) were conducted. Effect sizes were also computed (phi-coefficient, φ, for 2×2 contingency tables and Cramer's V for contingency tables other than 2×2). Data were processed by SPSS Statistics software (25.0; IBM, Armonk, NY, USA).

Table 2. Frequency of students who dropped out/did not dropped out of sport as a function of age, gender, nationality, school curriculum, sport type and strategic learning skills. Absolute frequencies and percentages (*N*/total of each corresponding category) and χ^2 test results are reported.

Variables	Drop-Out *N* (%)		No Drop-Out *N* (%)		χ^2	df	*p*	Effect Size
Age					4.18	2	0.123	0.083 [a]
14 years	26 (18.4)		115 (81.6)					
15 years	31 (11.9)		230 (88.1)					
16+ years	37 (17.4)		175 (82.6)					
Gender					6.35	1	0.012	−0.102 [b]
Male	45 (12.3)		321 (87.7)					
Female	49 (19.8)		199 (80.2)					
Nationality					13.69	1	<0.001	−0.149 [b]
Italian	50 (11.7)		376 (88.3)					
Spanish	44 (23.4)		144 (76.6)					
School curriculum		1 vs 2			45.29	1	<0.001	−0.326 [b]
		1 vs 3			47.38	1	<0.001	−0.342 [b]
1. Italian: sports high schools	3 (1.4)		213 (98.6)					
2. Italian: other high schools *	47 (22.4)		163 (77.6)					
3. Spanish: compulsory secondary school	44 (23.4)		144 (76.6)					
Type (individual/team) and amount of sports practiced								
Males					13.87	2	0.001	0.224 [a]
Only one sport	10 (8.3)		110 (91.7)					
Multiple sports of one type	13 (14.4)		77 (85.6)					
Multiple sports of different type	19 (28.8)		47 (71.2)					
Females					1.59	2	0.450	0.089 [a]
Only one sport	15 (28.3)		38 (71.7)					
Multiple sports of one type	25 (21.9)		89 (78.1)					
Multiple sports of different type	6 (17.1)		29 (82.9)					
Strategic Learning Skills								
Cognitive/Metacognitive								
Scales	No CL **	CL **	No CL	CL				
C1	63 (67.0)	31 (33.0)	361 (69.4)	159 (30.6)	0.21	1	0.643	−0.019 [a]
C2	66 (70.2)	28 (29.8)	379 (72.9)	141 (27.1)	0.28	1	0.594	−0.022 [a]
C3	68 (72.3)	26 (27.7)	339 (65.2)	181 (34.8)	1.82	1	0.177	0.054 [a]
C4	38 (40.4)	56 (59.6)	247 (47.5)	273 (52.5)	1.60	1	0.206	−0.051 [a]
Affective/Motivational								
A1	69 (73.4)	25 (26.6)	428 (82.3)	92 (17.7)	4.09	1	0.043	−0.082 [a]
A2	34 (36.2)	60 (63.8)	236 (45.4)	284 (54.6)	2.74	1	0.098	−0.067 [a]
A3	56 (59.6)	38 (40.4)	371 (71.3)	149 (28.7)	5.20	1	0.022	−0.092 [a]
A4	62 (66.0)	32 (34.0)	417 (80.2)	103 (19.8)	9.40	1	0.002	−0.124 [a]

a = Cramer's V; b = φ-coefficient; * Scientific, applied sciences, linguistic senior high school; ** No CL: no criticality (below critical level) = "0"; CL: criticality (above critical level) = "1" (dichotomous variables).

3. Results

As reported in Table 2, students dropped-out of sport at an average age of 13.4 ± 1.6 years. A significantly greater dropout of females compared to male students was found, whereas no effect for age emerged. Findings also revealed a significant effect for nationality, with greater drop-out from sports for Spanish as compared to Italian students. However, further analyses distinguishing between sport and non-sport high school curricula in the Italian sample revealed that the cross-country effect depended on the sample composition of school type, which is different in Italy and Spain. The drop-out rate was negligible in Italian students attending sports high schools with enhanced curricular physical education/sport as compared to both Italian students attending high schools with a focus different from sport (such as scientific, applied sciences and linguistic schools) and Spanish ones (Table 2).

Most importantly, an association of drop-out rates with affective–motivational strategic learning skills was found. These skills are associated with drop-out rates independently of gender and age, but interactively with nationality (Table 3). The overall association model showed an increase in the frequency of students who dropped out of sport with the number of critical values in affective–motivational strategic learning skills. As shown in Table 3, this pattern of differences was more pronounced and significant in Italian students, with a progressive increment in drop-out rates from none to all critical values, and less pronounced and non-significant in Spanish students. Cognitive–metacognitive skills were non-discriminant for sport drop-out rates both in Italian ($p = 0.102$) and in Spanish students ($p = 0.872$).

Table 3. Frequency of students who dropped out of sport as a function of the amount of critical values in affective-motivational strategic skills. Absolute frequencies and percentages (N/total of each corresponding category of critical values) and χ^2 test results are reported.

Sample	Amount of Critical Values						χ^2	df	p	Effect Size
	0 N (%)	1 N (%)	2 N (%)	3 N (%)	4 N (%)	Total N (%)				
Italian	8 (6.2)	16 (10.7)	12 (14.1)	11 (22.9)	3 (21.4)	50 (11.7)	11.50	4	0.021	0.164 [a]
Spanish	6 (15.3)	15 (23.1)	16 (28.1)	5 (20.8)	2 (66.7)	44 (23.4)	5.31	4	0.256	0.168 [a]
Total	14 (8.3)	31 (14.4)	28 (19.7)	16 (22.2)	5 (29.4)	94 (15.3)	13.83	4	0.008	0.150 [a]

a = Cramer's V.

A further analysis also showed that a lack of persistence in the same (individual/team) sport type was predictive of higher drop-out rates in males (Table 2). Highest drop-out rates emerged when the change was not between similar sport types (same-individual or same-team sport), but between individual and team sports. A non-significant reverse trend was observed in females, who more often showed drop-out from sport when they persisted in the same sport or individual/team skill sport type (Table 2).

4. Discussion

The main purpose of this study was to evaluate the association between early drop-out from sport and strategic learning skills relevant to both sport and academic career in senior high school students. It was expected that a critical value in and, therefore, lack of strategic learning skills functional to academic performance could be a factor related to the early drop-out from sport and, conversely, a high value of such skills could be related to adherence to sport. The presence or absence of an enhanced sport focus in the senior high school curriculum, as well as behavioral habits in extra-curricular sport, as the persistence of training in a given sport, or switching to another sport—of similar or different type—were also analyzed as potential correlates of different drop-out rates. Moreover, we evaluated the individual and joint role of gender, age and nationality with the strategic learning skills and the persistence in/change of sport.

The results of this study are mostly consistent with the literature and add novel information to the relevance of domain-general strategic skills in relation to the dropout phenomenon in the sport domain. In line with Eime et al. [44] and Molinero et al. [8], we found that young people dropped-out of sport during their teenage age (13.4 ± 1.6 years). Regarding gender, a larger drop-out emerged in females than in males. This is in line with evidence reported by some authors [20,50,51], showing that males are more active than females who have higher sport drop-out rates. Additionally, a female's tendency to be less active than a male is confirmed as a global gender difference [52]. Further research is warranted to analyze the role of gender stereotypes in sport drop-out with relation to sport self-perceptions of value and match/mismatch of the gender-identity with the gendered-nature of a sport [53].

We did not find comparable studies in literature on sport drop-out between Italian and Spanish students; to our knowledge, this is the first comparative study of adolescent populations from these countries. Our results highlighted a less pronounced drop-out in sports high school vs. other Italian and Spanish high schools. The sports high school is currently a school curriculum that is started up in Italy with no comparable curriculum in Spain. This may be one of the reasons explaining why, among the Italian students of the present sample, which included sports high school students, the sports drop-out is less pronounced than among their Spanish counterparts. Speculatively, the sports high school curriculum may have a favorable impact on sport persistence. Future research is needed to address the hypothesis that the targeted methodological strategies in this school curriculum could help student-athletes pursue their dual career [54].

Beyond the descriptive analysis of the drop-out phenomenon in sport, the most important finding of this study has to do with the association between sport drop-out and domain-general strategic learning skills. Our findings showed an association only for the affective–motivational dimension, but not for the cognitive–metacognitive dimension. This suggests that, in adolescence, an insufficient/critical level of affective skills (e.g., anxiety and emotions control), motivational skills (e.g., perception of competence and causal attribution) and volitional skills (e.g., perseverance) has a direct link to the abandonment of sport. Since sport and academic performance are interrelated [39], quitting sport activity means that students cannot capitalize on the beneficial influence of sport on academic achievement [30,41]. We speculatively suggest that the strategic learning skills may be a linking element within this association. This hypothesis deserves future mediation research.

Although there is a lack of literature on strategic learning skills associated with sport drop-out, we found some agreement with evidence on similar skills and maladaptive sport processes (burnout profiles) that allow us to frame a reference theory. Burnout profiles have been defined as a state of sport/school exhaustion related to chronic stress and have been studied in sport and at school jointly [55]. This burnout process involves a lack/insufficient level of skills that are common to the two areas (sport and school), similar to the strategic learning skills, objects of the present investigation. Thus, our suggestions are based on evidence that highlights the relevance of skills as resilience, similar to the concept of volition addressed in the present study. This evidence comes into play in a phenomenon such as burnout in sport that has commonalities with the drop-out from sport investigated here. A stronger risk of sport drop-out is related to a low level of resilience-related skills [55], high level of anxiety [25] and low perception of competence [24]. The critical level of these competences reported in the literature may represent a risk factor comparable to the critical value of affective–motivational strategic learning skills that emerged in our study. Conversely, athletes with an internal locus of control are more likely to exercise regularly [56], feel that their skill level is appropriate and tend to be more autonomously motivated towards their sport than athletes who drop-out [57].

Although the cross-sectional nature of the present study does not allow us to address causality, we hypothesize that bettering the affective–motivational strategic learning skills may enable us to select the most effective learning objectives and strategies, improving decision making to monitor and evaluate the progresses achieved, engage and persist over time to achieve goals successfully [36]. Being unable to face life transitions—in sport as well as in school and work settings—may foster maladaptive behaviors and negatively impact sport habits [23]. In the sport domain, improving learning processes fosters positive sport experiences; this may, in turn, support long-term sport participation and oppose the drop-out phenomenon. Our findings are in line with the literature showing that self-management of social-cognitive strategies (e.g., goals, plans and acts), similar to skills addressed in this study, may operate as mediators in the association between self-efficacy and physical activity/sports training in order to achieve goals and foster sport persistence [58]. Moreover, personal characteristics, such as personality traits, goal orientation and volitional skills, may contribute in the same way to face critical situations, such as dropping-out or pursuing a career in sport [59,60].

These findings call for studies that investigate the association of strategic skills with sport drop-out longitudinally. Based on the present cross-sectional evidence, it may be hypothesized that developing competences and personal skills common to the sporting and school context may be crucial to achieve excellence in both domains, and to prevent maladaptive consequences such as burnout and drop-out as a behavioral aspect of a broader maladaptive profile. According to Camiré, Trudel and Forneris [61], we believe that a common functional context may be favorable to acquire skills useful both in sport and in school settings. Since school and sport domains are interrelated [55], and strategic learning skills are transferable and general [36], integrating them in learning processes [62] from early schooling age may help capitalize on these skills later in life—such as, but not limited to, sport context—to achieve performance goals.

The outcomes highlighted that the association between strategic learning skills and drop-out rates could be influenced by nationality, but not by gender and age. This nationality effect leads us to suppose that factors other than strategic learning skills may intervene in Spanish students' sport drop-out processes. This latter aspect deserves an in-depth analysis of the potential intrapersonal, interpersonal, cultural and societal factors that may underlie the observed cross-country differences. Alternatively, the imbalance in sample size between Italian and Spanish students may have caused insufficient power in the case of the Spanish sample, thus not allowing us to draw definitive conclusions of cross-country differences in sport drop-out.

A second aim of the study was the role played by the persistence in/change to another sport in the students' sport history. A lack of persistence in the same (individual/team) sport type was predictive of higher drop-out rates, but only in males. Sampling different sports until about 12 years (the so-called 'sampling years') develops fun and participation [63]. Successively, in adolescence (specialization years), persisting in the same sport seems appropriate for males to pursue personal achievements, maintain individual interest, re-engage and persist in sport [63]. Playing the same sport type probably motivates them to persist in sport activity to excel and win in competition supported by a performance orientation [64]. Following a cascade process, the greater the fun in sports competitions, the greater the increase in fun of sports training and the persistence in sports over time [65]. Females show an opposite tendency, suggesting that a more process-oriented approach to sport and a stronger motivation to learn different skills without a primary performance goal might render females more persistent in sports if they play different sport types. Reasons for the higher drop-out rates of females may include a weak perception of their own competence associated with sport demand and pressure [23]. Indeed, those females who perceive themselves as athletic are more likely to maintain participation in sport [8]. To understand this difference between males and females in sport persistence/abandonment, other determinants or moderators, such as motivational aspects, must be considered in the future.

This study is not without limitations. Although our large sample size ($N = 614$) and stratified sampling ensured representativeness, the homogeneity of the prevalent socio-economic status of the urban districts and the selected schools were located in limits of the generalizability to youths living in different socio-economic contexts, typically influencing their physical activity and sport habits. This similarity allowed us to dampen the influence of potential major socio-economic covariates and to compare cross-country and cross-school system differences and their impact on sports habits. Moreover, a more balanced sample in the two countries could allow for a stronger interpretation in terms of cross-country differences in the association between sport drop-out and strategic skills, excluding power issues that may have undermined the possibility to detect this association in the smaller sample of Spanish students.

This study is cross-sectional and therefore does not allow any causal inference. However, its results pave the way for an intervention study aimed to clarify the causal relationship between strategic learning skills and sports abandonment, sport participation and academic achievement and the role played by gender, nationality and sport type. Finally, this study was conducted only in two cultural settings—Spain and Italy—within their specific schooling contexts. It would be important to implement this study in wider multi-national cultural settings.

5. Conclusions

This study suggests that the affective–motivational strategic learning skills relevant to sport and non-sport life domains may play a role in relation to youths' drop-out. Moreover, the cultural and personal characteristics, and sports habits, may independently and jointly contribute to the phenomenon of drop-out. We believe that these interrelations must be considered to develop targeted strategies for prevention of drop-out, starting as early as schooling. We assume that, similarly to any life skills, strategic learning skills are not learned automatically, but can and should be taught in appropriate contexts, such as school and sport [66], through meaningful experiences [40]. Sport programs tailored to include this aim might be more successful in promoting youth sport participation, optimizing performance and preventing drop-out with positive consequences for talent identification and promotion. Specifically, in adolescents, reflective processes on their strategic learning skills should be encouraged. This may help the student-athlete develop self-determination and self-regulation, in order to be able to monitor and evaluate the learning progression, orient themselves in study and sport and build a successful and long-term continuous developmental path.

Author Contributions: All authors contributed equally to the manuscript. All authors have read and agreed to the published version of the manuscript.

Funding: This research received no external funding.

Institutional Review Board Statement: The ethics committee of the UCAM Catholic University of Murcia in Spain approved this study (CE102012).

Informed Consent Statement: Informed consent was obtained from all subjects involved in the study.

Data Availability Statement: The data presented in this study are available on request from the corresponding author. The data are not publicly available, as they are part of a larger research project not yet completed.

Acknowledgments: We would like to thank M. Margottini (Roma Tre University) and his collaborators, who gave us precious advice for the development of this study and access to their www.competenzestrategiche.it (accessed on 7 November 2019) platform. This acknowledgment of our gratefulness does not imply endorsement, as only the authors have responsibility for this work. For Spanish translation, we would like to thank M. Aulicino. Furthermore, we would like to thank all the students, professors and school managers of secondary schools for their willingness to participate, which has allowed the realization of this study.

Conflicts of Interest: The authors declare no conflict of interest.

References

1. Fraser-Thomas, J.; Côté, J. Youth Sports: Implementing Findings and Moving Forward with Research. *Athl. Insight* **2006**, *8*, 12–27.
2. Garber, C.E.; Blissmer, B.; Deschenes, M.R.; Franklin, B.A.; Lamonte, M.J.; Lee, I.; Nieman, D.C.; Swain, D.P. Quantity and Quality of Exercise for Developing and Maintaining Cardiorespiratory, Musculoskeletal, and Neuromotor Fitness in Apparently Healthy Adults: Guidance for Prescribing Exercise. *Msse* **2011**, *43*, 1334–1359. [CrossRef] [PubMed]
3. Gill, D.L.; Hammond, C.C.; Reifsteck, E.; Jehu, C.M.; Williams, R.A.; Adams, M.M.; Lange, E.H.; Becofsky, K.; Rodriguez, E.; Shang, Y.T. Physical Activity and Quality of Life. *J. Prev. Med. Public Health* **2013**, *46*, S28–S34. [CrossRef]
4. Bailey, R.; Armour, K.; Kirk, D.; Jess, M.; Pickup, I.; Sandford, R.; BERA Physical Education and Sport Pedagogy Special Interest Group. The Educational Benefits Claimed for Physical Education and School Sport: An Academic Review. *Res. Pap. Educ.* **2009**, *24*, 1–27. [CrossRef]

5. Côté, J.; Hay, J. Children's involvement in sport: A developmental perspective. In *Psychological Foundations of Sport*; Silva, J.M., Stevens, D.E., Eds.; Allyn and Bacon: Boston, MA, USA, 2002; pp. 484–502.
6. Kirk, D. Physical Education, Youth Sport and Lifelong Participation: The Importance of Early Learning Experiences. *Eur. Phys. Educ. Rev.* **2005**, *11*, 239–255. [CrossRef]
7. Wall, M.; Côté, J. Developmental Activities that Lead to Dropout and Investment in Sport. *Phys. Educ. Sport Pedagog.* **2007**, *12*, 77–87. [CrossRef]
8. Molinero, O.; Salguero, A.; Tuero, C.; Alvarez, E.; Márquez, S. Dropout Reasons in Young Spanish Athletes: Relationship to Gender, Type of Sport and Level of Competition. *J. Sport Behav.* **2006**, *29*, 255–269.
9. Fraser-Thomas, J.; Côté, J.; Deakin, J. Examining Adolescent Sport Dropout and Prolonged Engagement from a Developmental Perspective. *J. Appl. Sport Psychol.* **2008**, *20*, 318–333. [CrossRef]
10. Strandbu, A.; Bakken, A.; Sletten, M.A. Exploring the minority–majority gap in sport participation: Different patterns for boys and girls? *Sport Soc.* **2019**, *22*, 606–624. [CrossRef]
11. Vella, S.A.; Cliff, D.P.; Okely, A.D. Socio-ecological predictors of participation and dropout in organised sports during childhood. *Int. J. Behav. Nutr. Phys. Act.* **2014**, *11*, 1–10. [CrossRef]
12. Carlman, P.; Wagnsson, S.; Patriksson, G. Causes and Consequences of Dropping Out from Organized Youth Sports. *Swed. J. Sports Res.* **2013**, *2*, 26–54. Available online: https://www.diva-portal.org/smash/record.jsf?pid=diva2:750473 (accessed on 25 April 2021).
13. Lagestad, P.; Van Den Tillaar, R.; Mamen, A. Longitudinal Changes in Physical Activity Level, Body Mass Index, and Oxygen Uptake Among Norwegian Adolescents. *Front. Public Health* **2018**, *6*, 97. [CrossRef] [PubMed]
14. Palomäki, S.; Hirvensalo, M.; Smith, K.; Raitakari, O.; Männistö, S.; Hutri-Kähönen, N.; Tammelin, T. Does Organized Sport Participation during Youth Predict Healthy Habits in Adulthood? A 28-Year Longitudinal Study. *Scand. J. Med. Sci. Sports* **2018**, *28*, 1908–1915. [CrossRef]
15. Moa, I.F.; Berntsen, S.; Lagestad, P. Cardiorespiratory Fitness Is Associated with Drop Out from Sport in Norwegian Adolescents. A Longitudinal Study. *Front. Public Health* **2020**, *8*, 502307. [CrossRef]
16. European Commission. *Special Eurobarometer 334: Sport and Physical Activity*; Directorate-General for Communication: Brussels, Belgium, 2010; Available online: https://data.europa.eu/data/datasets/s776_72_3_ebs334?locale=en (accessed on 25 April 2021).
17. European Commission. *Special Eurobarometer 472: Sport and Physical Activity*; Directorate-General for Communication: Brussels, Belgium, 2018. [CrossRef]
18. Craike, M.J.; Symons, C.; Zimmermann, J. Why do Young Women Drop Out of Sport and Physical Activity? A Social Ecological Approach. *Ann. Leis. Res.* **2009**, *12*, 148–172. [CrossRef]
19. Moreno, J.M.; Cerezo, C.R.; Guerrero, J.T. Motivos de Abandono de la Práctica de Actividad Físico-Deportiva en los Estudiantes de Bachillerato de la Provincia de Granada. *Rev. Educ.* **2010**, *353*, 495–519.
20. Enoksen, E. Dropout rates and drop-out reasons among promising Norwegian track and field athletes. *Scand. Sport Stud. Forum* **2011**, *2*, 19–43.
21. Molinero, O.; Salguero, A.; Álvarez, E. Reasons for Dropout in Youth Soccer: A Comparison with Other Team Sports. *Mot. Eur. J. Hum. Mov.* **2009**, *22*, 21–30. Available online: https://www.redalyc.org/articulo.oa?id=274219861002 (accessed on 25 April 2021).
22. Monteiro, D.M.; Cid, L.S.; Marinho, D.A.; Moutão, J.M.; Vitorino, A.P.; Bento, T. Determinants and Reasons for Dropout in Swimming. Systematic Review. *Sports* **2017**, *5*, 50. [CrossRef] [PubMed]
23. Monteiro, D.M.; Marinho, D.A.; Moutão, J.M.; Vitorino, A.P.; Antunes, R.N.; Cid, L.S. Reasons for Dropout in Swimmers, differences between Gender and Age and Intentions to Return to Competition. *J. Sports Med. Phys. Fit.* **2018**, *57*, 1–13. [CrossRef]
24. Crane, J.; Temple, V. A Systematic Review of Dropout from Organized Sport among Children and Youth. *Eur. Phys. Educ. Rev.* **2015**, *21*, 114–131. [CrossRef]
25. Witt, P.A.; Dangi, T. Why Children/Youth Drop out of Sports. *J. Park Recreat. Adm.* **2018**, *36*, 191–199. [CrossRef]
26. Crawford, D.W.; Jackson, E.L.; Godbey, G. A Hierarchical Model of Leisure Constraints. *Leis. Sci.* **1991**, *13*, 309–320. [CrossRef]
27. Smith, R.E. Toward a cognitive-affective model of athletic burnout. *J. Sport Psychol* **1986**, *8*, 36–50. [CrossRef]
28. Sorkkila, M.; Ryba, T.V.; Selanne, H.; Aunola, K. Development of school and sport burnout in adolescent student-athletes: A longitudinal mixed-methods study. *J. Res. Adolesc.* **2020**, *30*, 115–133. [CrossRef]
29. Cosh, S.; Tully, P.J. Stressors, coping, and support mechanisms for student athletes combining elite sport and tertiary education: Implications for practice. *Sport Psychol.* **2014**, *29*, 120–133. [CrossRef]
30. Ishihara, T.; Nakajima, T.; Yamatsu, K.; Okita, K.; Sagawa, M.; Morita, N. Relationship of participation in specific sports to academic performance in adolescents: A 2-year longitudinal study. *Scand. J. Med. Sci. Sports* **2020**, *30*, 1471–1482. [CrossRef]
31. Fraile-García, J.; Tejero-González, C.M.; Esteban-Cornejo, I.; Veiga, Ó.L. Asociación entre disfrute, autoeficacia motriz, actividad física y rendimiento académico en educación física. *Retos* **2019**, *36*, 58–63. [CrossRef]
32. Haverkamp, B.F.; Wiersma, R.; Vertessen, K.; Van Ewijk, H.; Oosterlaan, J.; Hartman, E. Effects of physical activity interventions on cognitive outcomes and academic performance in adolescents and young adults: A meta-analysis. *Sports Sci.* **2020**, *38*, 2637–2660. [CrossRef]
33. Álvarez-Bueno, C.; Pesce, C.; Cavero-Redondo, I.; Sánchez-López, M.; Garrido-Miguel, M.; Martínez-Vizcaíno, V. Academic achievement and physical activity: A meta-analysis. *Pediatrics* **2017**, *140*, 40–55. [CrossRef]

34. Bay, M.; Grządziel, D.; Pellerey, M. Competenze strategiche e autoregolazione. In *Promuovere la Crescita Nelle Competenze Strategiche Che Hanno le Loro Radici Nelle Dimensioni Morali e Spirituali della Persona. Rapporto di Ricerca [Strategic Skills and Self-Regulation In Promoting Growth in Strategic Skills that Have Their Roots in the Moral and Spiritual Dimensions of the Person. Research Report]* Bay, M., Grządziel, D., Pellerey, M., Eds.; CNOS-FAP: Roma, Italy, 2010; pp. 67–75. Available online: https://www.cnos-fap.it/sites/default/files/pubblicazioni/promuovere%20la%20crescita%20nelle%20competenze%20strategiche.PDF (accessed on 25 April 2021).
35. Pellerey, M.; GrząDziel, D. Il quadro di riferimento teorico e gli strumenti valutativi ed auto-valutativi valorizzati. In *Imparare a Dirigere Se Stessi. Progettazione e Realizzazione di una Guida e di uno Strumento Informatico per Favorire L'autovalutazione e lo Sviluppo delle Proprie Competenze Strategiche nello Studio e nel Lavoro [Theoretical Framework and Evaluation/Self-Evaluation Tools. In Learning Self-Direction. Planning and Creation of a Guide and Computer-Based Tool to Facilitate the Self-Evaluation and Development of Strategic Competences for Study and Work]*; Pellerey, M., GrząDziel, D., Margottini, M., Epifani, F., Ottone, E., Eds.; Cnos-Fap: Roma, Italy 2013; pp. 13–100. Available online: https://www.cnos-fap.it/sites/default/files/pubblicazioni/Imparare%20a%20dirigere%20se%20stessi.pdf (accessed on 25 April 2021).
36. Pellerey, M. *Soft Skill e Orientamento Professionale [Soft Skill and Professional Orientation]*; Cnos-Fap: Roma, Italy, 2017; pp. 61–76. Available online: https://www.cnos-fap.it/sites/default/files/pubblicazioni/soft_skill.pdf (accessed on 25 April 2021).
37. Côté, J.; Hancock, D.J. Evidence-based policies for youth sport programmes. *Int. J. Sport Policy Politics* **2016**, *8*, 51–65. [CrossRef]
38. Joseph, N. Metacognition needed: Teaching middle and high school students to develop strategic learning skills. *Prev. Sch. Fail Altern. Educ. Child. Youth* **2009**, *54*, 99–103. [CrossRef]
39. Muñoz-Bullón, F.; Sanchez-Bueno, M.J.; Vos-Saz, A. The Influence of Sports Participation on Academic Performance among Students in Higher Education. *Sport Manag. Rev.* **2017**, *20*, 365–378. [CrossRef]
40. Holt, N.L.; Neely, K.C.; Slater, L.G.; Camirè, M.; Côté, J.; Fraser-Thomas, J.; McDonald, D.; Strachan, L.; Tamminen, K.A. A grounded theory of positive youth development through sport based on results from a qualitative meta-study. *Int. Rev. Sport Exerc. Psychol.* **2017**, *10*, 1–49. [CrossRef]
41. Dyer, A.M.; Kristjansson, A.L.; Mann, M.J.; Smith, M.L.; Allegrante, J.P. Sport participation and academic achievement: A longitudinal study. *Am. J. Health Behav.* **2017**, *41*, 179–185. [CrossRef] [PubMed]
42. Jacobson, J.; Matthaeus, L. Athletics and executive functioning: How athletic participation and sport type correlate with cognitive performance. *Psychol. Sport Exerc.* **2014**, *15*, 521–527. [CrossRef]
43. Ahmed, S.F.; Tang, S.; Waters, N.E.; Davis-Kean, P. Executive Function and Academic Achievement: Longitudinal Relations from Early Childhood to Adolescence. *J. Educ. Psychol.* **2019**, *111*, 446–458. [CrossRef]
44. Eime, R.M.; Harvey, J.T.; Charity, M.J. Sport drop-out during adolescence: Is it real, or an artefact of sampling behaviour? *Int. J. Sport Policy Politics* **2019**, *11*, 715–726. [CrossRef]
45. Keathley, K.; Himelein, M.J.; Srigley, G. Youth Soccer Participation and Withdrawal: Gender Similarities and Differences. *J. Sport Behav.* **2013**, *36*, 171–188.
46. Hellín, P.; Moreno, J.A.; Rodríguez, P.L. Motivos de Práctica Físico-Deportiva en la Región de Murcia. *Cuad. Psicol. Deporte* **2004**, *4*, 101–115.
47. Margottini, M. Validazione del QSA Ridotto. In *Strumenti e Metodologie di Orientamento Formativo e Professionale nel Quadro dei Processi di Apprendimento Permanente [Educational and Professional Guidance Tools and Methodologies in the Framework of Lifelong Learning Processes.]*; Pellerey, M., Ed.; Cnos-Fap: Roma, Italy, 2018; pp. 257–304. Available online: https://www.cnos-fap.it/sites/default/files/pubblicazioni/strumenti_e_metodologie.pdf (accessed on 25 April 2021).
48. Taber, K.S. The Use of Cronbach's Alpha When Developing and Reporting Research Instruments in Science Education. *Res. Sci. Educ.* **2018**, *48*, 1273–1296. [CrossRef]
49. Giacomantonio, A. Soft skills assessment: Factorial structure of QSA-R. *QTimes J. Educ.* **2019**, *9*, 56–70.
50. Montero, M.M.; Juan, F.R. Estudio longitudinal de los comportamientos y el nivel de actividad físico-deportiva en el tiempo libre en estudiantes de Costa Rica, México y España. *Retos* **2017**, *31*, 219–226.
51. Plaza, M.; Boiché, J.; Brunel, L.; Ruchaud, F. Sport = male ... But not all sports: Investigating the gender stereotypes of sport activities at the explicit and implicit levels. *Sex Roles* **2017**, *76*, 202–217. [CrossRef]
52. Guthold, R.; Stevens, G.A.; Riley, L.M.; Bull, F.C. Global trends in insufficient physical activity among adolescents: A pooled analysis of 298 population-based surveys with 1·6 million participants. *Lancet Child Adolesc. Health* **2020**, *4*, 23–35. [CrossRef]
53. Plaza, M.; Boiché, J. Gender stereotypes, self, and sport dropout: A one-year prospective study in adolescents. *Mov. Sport Sci. Sci. Mot.* **2017**, *96*, 75–84. [CrossRef]
54. Pato, A.S.; Isıdori, E.; Calderón, A.; Brunton, J. *An Innovative European Sports Tutorship Model of the Dual Career of Student-Athletes*; UCAM Catholic University of Murcia: Murcia, Spain, 2017. [CrossRef]
55. Sorkkila, M.; Tolvanen, A.; Aunola, K.; Ryba, T.V. The Role of Resilience in Student-Athletes' Sport and School Burnout and Dropout: A Longitudinal Person-Oriented Study. *Scand. J. Med. Sci. Sports* **2019**, *29*, 1059–1067. [CrossRef] [PubMed]
56. Cobb-Clark, D.A.; Kassenboehmer, S.C.; Schurer, S. Healthy habits: The connection between diet, exercise, and locus of control. *J. Econ. Behav. Organ.* **2014**, *98*, 1–28. [CrossRef]
57. Rottensteiner, C.; Tolvanen, A.; Laakso, L.; Konttinen, N. Youth athletes' motivation, perceived competence, and persistence in organized team sports. *J. Sport Behav.* **2015**, *38*, 1–18.

58. Dishman, R.K.; Motl, R.W.; Sallis, J.F.; Dunn, A.L.; Birnbaum, A.S.; Welk, G.J.; Bedimo-Rung, A.L.; Voorhees, C.C.; Jobe, J.B. Self-Management Strategies Mediate Self-Efficacy and Physical Activity. *Am. J. Prev. Med.* **2005**, *29*, 10–18. [CrossRef]
59. Baron-Thiene, A.; Alfermann, D. Personal characteristics as predictors for dual career dropout versus continuation—A prospective study of adolescent athletes from German elite sport schools. *Psychol. Sport Exerc.* **2015**, *21*, 42–49. [CrossRef]
60. Stambulova, N.B. Crisis-transitions in athletes: Current emphases on cognitive and contextual factors. *Curr. Opin. Psychol.* **2017**, *16*, 62–66. [CrossRef] [PubMed]
61. Camiré, M.; Trudel, P.; Forneris, T. Coaching and transferring life skills: Philosophies and strategies used by model high school coaches. *Sport Psychol.* **2012**, *26*, 243–260. [CrossRef]
62. Bean, C.; Kramers, S.; Forneris, T.; Camiré, M. The Implicit/Explicit Continuum of Life Skills Development and Transfer. *Quest* **2018**, *70*, 456–470. [CrossRef]
63. Côté, J.; Allan, V.; Turnnidge, J.; Erickson, K. Early Sport Specialization and Sampling. In *Handbook of Sport Psychology*; Eklung, R., Tenenbaum, G., Eds.; Wiley: Hoboken, NJ, USA, 2020.
64. Duda, J.L. Relationship between task and ego orientation and the perceived purpose of sport among high school athletes. *J. Sport Exerc. Psychol.* **1989**, *11*, 318–335. [CrossRef]
65. Lagestad, P.; Sorensen, A. Longitudinal changes in sports enjoyment among adolescents. *J. Phys. Educ. Sport* **2018**, *18*, 89–97.
66. Turnnidge, J.; Côté, J.; Hancock, D.J. Positive youth development from sport to life: Explicit or implicit transfer? *Quest* **2014**, *66*, 203–217. [CrossRef]

sports

Review

Motor Performance in Male Youth Soccer Players: A Systematic Review of Longitudinal Studies

Maryam Abarghoueinejad [1,*], Adam D. G. Baxter-Jones [2], Thayse Natacha Gomes [3], Daniel Barreira [1] and José Maia [1]

1 Centre of Research, Education, Innovation, and Intervention in Sport (CIFI2D), Faculty of Sport, University of Porto, 4200-450 Porto, Portugal; dbarreira@fade.up.pt (D.B.); jmaia@fade.up.pt (J.M.)
2 College of Kinesiology, University of Saskatchewan, Saskatoon, SK S7N 5B2, Canada; baxter.jones@usask.ca
3 Department of Physical Education, Federal University of Sergipe, São Cristóvão 49100-000, SE, Brazil; thayse_natacha@hotmail.com
* Correspondence: m.abarghouei.n@gmail.com; Tel.: +351-220425200

Abstract: The aim of this systematic review was to identify and synthesize the available information regarding longitudinal data addressing young soccer players' motor performance changes. Following the Preferred Reporting Items for Systematic and Meta-analyses (PRISMA) statement, literature searches were performed in three databases: PubMed, ISI Web of Science and SCOPUS. The following descriptors were used: football, soccer, youth, young, player, athlete, physical performance, motor performance, longitudinal. The inclusion criteria were original articles in English with longitudinal data of young males (aged 10–18 years), with the aim to investigate motor performance serial changes. The initial search returned 211 records, and the final sample comprised 32 papers. These papers covered the European continent, and used mixed and pure longitudinal design with variation in sample size and age range. The reviewed studies tended to use different tests to assess the motor performance and aimed to identify changes in motor performance in several ways. In general, they indicated motor performance improvements with age, with a marked influence of biological maturity, body composition, and training stimuli. This review highlights the need for coaches and stakeholders to consider players' motor performance over time whilst considering biological maturation, biological characteristics, and training stimuli.

Keywords: longitudinal; young; soccer players; motor performance

Citation: Abarghoueinejad, M.; Baxter-Jones, A.D.G.; Gomes, T.N.; Barreira, D.; Maia, J. Motor Performance in Male Youth Soccer Players: A Systematic Review of Longitudinal Studies. *Sports* **2021**, *9*, 53. https://doi.org/10.3390/sports9040053

Academic Editor: Toby Mündel

Received: 30 March 2021
Accepted: 12 April 2021
Published: 19 April 2021

1. Introduction

Soccer is the world's most popular sport and participants represent ~4.1% of the total sporting population [1]. With such large numbers of participants, governing bodies and other stakeholders invest significant amounts of money in soccer players' talent identification. The identification and development of the next generation of young soccer players is a key goal for these organizations [2,3]. Thus, the design and implementation of appropriate programs to uncover youth soccer players' potentials are common practice within soccer academies. These academies support the early development [4] and then the transition of young players into the senior professional world [5,6].

In a cross-sectional study, data are collected from many different individuals at a single time point and comparisons are made between different populations. In contrast, in a longitudinal study, the same data are collected in the same individuals over short or long periods of time. Therefore, whilst a cross-sectional study considers a snapshot in time, the longitudinal study design considers what happens before or after the snapshot is taken. The benefits of the cross-sectional design are that it allows researchers to compare many different variables at the same time. However, the disadvantage is that cross-sectional studies are not able to provide definitive information about cause-and-effect relationships. A longitudinal study can detect development or changes in population characteristics at both the group and individual level. Thus, longitudinal studies can establish sequences of events and enable the researcher to address cause-and-effect relationships. In youth soccer studies, the longitudinal design allows the researcher to distinguish the effects of training and competition from those associated with normal growth and development. There is much research devoted to describing and interpreting the manifold expressions of soccer players' characteristics and their response to training and competition. Unfortunately, most available evidence is based on cross-sectional data [2], with few longitudinal reports or well-controlled experimental studies. This limits the current knowledge concerning youth soccer players' development [7,8].

Recent systematic reviews of young soccer players have dealt with match running performance [9], talent identification [10], and anthropometric-physiological profiling [11]. These reviews identified a series of inconsistencies and gaps in the literature which have hampered practitioners' abilities to make evidence-informed decision making [2,12,13]. Furthermore, there is an absence of research in young soccer players' development processes such as the interactions of their physical growth and biological maturation with systematic training stimuli, estimation of velocities and spurts in their motor performance and specific skills' levels, as well as players' systematic responses to training and competition [8].

To the best of our knowledge, there apparently is no available systematic review dealing with young male soccer players' longitudinal development of motor performance. Therefore, our goal is to provide a summary of existing longitudinal data dealing with male soccer players' motor performance changes during adolescence, which is a very important time-window for the nurturing of soccer players' careers.

2. Materials and Methods

2.1. Protocol

This review used the "Preferable Reporting Items for Systematic Reviews and Meta-Analyses Protocols" (PRISMA-P) [14,15] to probe the literature of longitudinal studies into young soccer players' motor performance. We also complied with the Cochrane Handbook for Systematic Reviews of Interventions (version 5.1.0; http://handbook-5-1.cochrane.org/, accessed on 10 October 2020).

2.2. Information Sources and Search Strategy

The search strategy comprised two stages. First, the electronic databases MEDLINE (PubMed/PubMed Central interface), Web of Science™ Core Collection and SCOPUS were searched up to January 2021. The online search was performed based on the following strategy: (football OR soccer) AND (youth OR young OR player OR athlete) AND (physical performance OR motor performance) AND (longitudinal). Second, the reference list of the selected papers was searched for possible studies to be included in the review. A full description of the input arguments used in each database is also provided (Electronic Supplementary Material Table S1). EndNote software (version X9.0, X7.0.1, Clarivate Analytics, Philadelphia, PA, USA) was used as the citation manager during the processes of searching, deduplication, selection, and management of the studies.

2.3. Eligibility Criteria

To be included in the review, original studies had to: (i) have a longitudinal design following players over time, i.e., with at least two repeated observations; (ii) have a sample of young male soccer players, i.e., athletes aged between 10 and 18 years; (iii) aim to investigate physical fitness/physical/motor performance and/or functional capacity (expressed by muscular strength and/or power, aerobic/anaerobic power, agility, flexibility, movement coordination and speed, as well as specific soccer technical performance such as dribbling and shooting, for example) serial changes; and (iv) be published in English and in peer-reviewed journals. Studies were excluded if: (i) psychological facets were mainly assessed, (ii) they used impaired players, and (iii) they concentrated on match-analysis.

2.4. Study Selection

Two researchers (MA, TNG) independently conducted the online search. Grounded in the eligibility criteria, papers were firstly selected based on their title and abstract, and those selected had their full text screened. To be included in the present review, eligible papers had to be selected by the two researchers, and if any discrepancies were observed at this stage, reviewers discussed and resolved inclusion and/or asked for the judgement of a senior researcher (JM). The senior researcher examined each situation on a case-by-case basis and determined the inclusion or exclusion of a given article using his experience in the field. After the selection of the manuscript to be included, one of the researchers screened the reference lists of the selected papers to identify any other potential paper to be included in the review. Those studies selected in this stage were re-checked by the second researcher, and only those approved by both were considered for inclusion in the present study.

2.5. Methodological Quality Assessment

The quality of the included articles was assessed with the modified version of current established scale used in sport science, health care and rehabilitation (i.e., Cochrane, Coleman, Delphi, and Physiotherapy Evidence Database (PEDro)). The current scale (Table 1) was adapted from a recent review by Sarmento et al. [16]. Articles were assessed based on their purpose (Q1), participants' characteristics (Q2), sample justification (Q3), motor performance assessments (Q4), statistical procedures used (Q5), results and outcome (Q6), study method conclusion (Q7), practical implications (Q8), limitations (Q9), and future direction (Q10). All ten quality criteria were scored on three levels (2-point per item), i.e., a score of zero (no), one (maybe), and two (yes) given for each item. The total scores ranged between zero and twenty. A sum of scores from all questions was subsequently computed. To make a fair comparison between studies with different designs, the decision was made to calculate a percentage score as a final measure of methodological quality. For this, the total score was converted into percentages, ranging from 0 to 100%, to ensure that the quality assessment was equitable across all the included articles. Studies were categorized into 3 levels; high (≥75%), moderate (50–74%) and low (<50%) methodological quality scores [16]. Methodological quality was not evaluated for the purpose of including/excluding studies. Two researchers (MA, TNG) performed independent assessments. If discrepancies occurred, these were resolved in a consensus discussion with third senior researcher (JM).

Table 1. Methodological quality scoring system (adapted from Sarmento et al., 2018 [16]).

	Question	Answer	Score
Q1	Was(were) the aim(s) of study clearly set out?	Yes = 2; Maybe = 1; No = 0	0–2
Q2	Were characteristics of participants presented in detail in methods section? (number of subjects, sex, age, country/city)	Yes = 2; Maybe = 1; No = 0	0–2
Q3	Was sample size justified?	Yes = 2; Maybe = 1; No = 0	0–2
Q4	Are the motor performance to be measured clearly described in the methods section?	Yes = 2; Maybe = 1; No = 0	0–2
Q5	Were statistics clearly presented?	Yes = 2; Maybe = 1; No = 0	0–2
Q6	Results' details (means and standard deviations and/or change/difference, effect size/mechanistic magnitude-based inference)	Yes = 2; Maybe = 1; No = 0	0–2
Q7	Were conclusions appropriate given the study methods and the objectives?	Yes = 2; Maybe = 1; No = 0	0–2
Q8	Are there any implications for practice given the results of the study?	Yes = 2; Maybe = 1; No = 0	0–2
Q9	Were limitations of the study acknowledged and described by the authors?	Yes = 2; Maybe = 1; No = 0	0–2
Q10	Are there any future direction described by the authors?	Yes = 2; Maybe = 1; No = 0	0–2
Total			0–20

Strict rules applied (No information = 0 point; 1–2 items described = 1 point; all items described = 2 points).

2.6. Strategy for Data Synthesis

A descriptive synthesis of the findings from the included studies is presented in Table 2, where summaries with reference to authors and years of publication were provided. Then, the terminologies used in motor performance variable definition and assessment were examined. Demographic details were extracted, including sample size, age/age group of participations, and the geographical location of the players. Design aspects (mixed-longitudinal, longitudinal), configuration (duration), and measurement techniques/equipment were also included. Finally, general results regarding changes in motor performance were extracted and main findings were organized and described.

Table 2. Characteristics of studies included in the review.

Author/Country	Study Design Duration	Participants		Motor Performance Assessments (Tests)	Main Results	Quality Score
		Age	Number			
(Philippaerts et al., 2006) [17]/Belgium	Mixed-longitudinal 5 y	11–13 y at baseline	33	Physical performance: Plate tapping, sit and reach (SR), sit-ups, bent arm hang, standing long jump (SLJ), vertical jump (VJ), endurance shuttle run (ESHR). Soccer specific physical performance: 30 m repeated sprint (RSA), agility shuttle run 5 × 10 m (SHR).	Physical performance improved non-linearly and reached its peak around peak height velocity (PHV), yet with different timing and tempo.	65%
(Huijgen et al., 2010) [18]/Netherlands	Mixed-longitudinal 5 y	12–19 y at baseline	267	Physical performance: shuttle sprint and dribble test. Soccer specific skills: slalom sprint and dribble test.	Speed and dribbling improved with age mainly at 12–14 y, but with different tempo. Dribbling improved after 16 y and sprinting from 14 to 16 y. Additionally, fat free mass, weekly hours of practice and playing position were positively associated with dribbling changes.	70%
(Mirkov et al., 2010) [48]/Serbia	Longitudinal 4 y	11 y at baseline	S_g = 26 C_g = 63	Physical performance: SR, SLJ, countermovement jumps (CMJ) with and without arm swing, slalom running with obstacles, SHR.	Physical performance improved with age in both groups, yet soccer players performed better in agility and motor coordination.	85%
(Roescher et al., 2010) [40]/Netherlands	Mixed-longitudinal 5 y	12–19 y at baseline	Pro = 53 N_pro = 77	Soccer specific physical performance: intermittent endurance (ISRT).	Aerobic capacity increased non-linearly with age but differences between groups occurred from 17 y onwards. Training was positively associated with performance.	75%
(Williams et al., 2011) [39]/United Kingdom	Longitudinal 3 y	U12–U16 at baseline	200	Physical performance: sprints 10 m (S10 m), 30 m (S30 m), VJ.	Physical performance improved linearly but with different rates for 10 m speed, 30 m sprint and vertical jump.	80%
(Gonaus and Muller, 2012) [45]/Austria	Mixed-longitudinal 9 y	14–17 y at baseline	1642	Physical performance: S20 m, hurdles agility run, CMJ, drop jump, foot tapping reaction medicine ball throw 2 kg, SR, 20 m multi-stage endurance run (MSER). Soccer specific physical performance: (SHR).	Speed, power, flexibility, and endurance improved with age. Power and flexibility as well as endurance effect sizes decreased with age; however, in speed results were stable from 14 to 17 y.	80%
(Valente-dos-Santos et al., 2012) [25]/Portugal	Mixed-longitudinal 5 y	11–13 y at baseline	83	Physical performance: MSER.	Aerobic performance unfolding was related to chronological and skeletal ages, and training stimuli.	90%
(Valente-dos-Santos et al., 2012) [19]/Portugal	Mixed-longitudinal 5 y	11–13 y at baseline	83	Physical performance: MSER, CMJ. Soccer specific physical performance: RSA.	Repeated sprint performance changes were related to chronological and skeletal ages, as well as fat free mass, aerobic endurance, and lower limb explosive strength.	55%
(Valente-dos-Santos et al., 2012) [26]/Portugal	Mixed-longitudinal 5 y	11–13 y at baseline	83	Physical performance: MSER, CMJ. Soccer specific physical performance: RSA.	Repeated sprint performance development was related to chronological age, maturity status, fat free mass, body size, aerobic endurance, and lower limb explosive strength and annual training.	80%

Table 2. *Cont.*

Author/Country	Study Design Duration	Participants Age	Number	Motor Performance Assessments (Tests)	Main Results	Quality Score
(Valente-dos-Santos et al., 2012) [27]/Portugal	Mixed-Longitudinal 5 y	11–13 y at baseline	83	Physical performance: SHR, MSER, CMJ. Soccer specific physical performance: RSA. Soccer specific skills: Ball control, dribbling speed, Shooting accuracy, wall pass.	Overall physical performance development was related to chronological age, maturation status, fat mass, dribbling speed and training stimuli. In general, soccer skills unfolding was related to chronological age, playing position, fat and fat-free mass, repeated sprint and aerobic endurance and training stimuli.	90%
(Huijgen et al., 2013) [41]/Netherland	Mixed-longitudinal 3 y	10–18 y at baseline	270	Soccer specific skills: Loughborough Soccer Passing (LSPT).	Soccer skills improved non-linearly: 18% in speed pass, and 32% in speed and accuracy pass with age.	85%
(Carvalho et al., 2014) [42]/Spain	Mixed-Longitudinal 4 y	U11 age category at baseline	33	Soccer specific physical performance: The Yo-Yo Intermittent Recovery Test (YYIR1).	Aerobic performance increased non-linearly with chronological age; yet, between 12–13 y decreased. Additionally, aerobic performance was related to training stimuli but not with body size and maturity status.	80%
(Deprez et al., 2014) [31]/Belgium	Longitudinal 5 y	11–14 y at baseline	162	Soccer specific physical performance: YYIR1.	Aerobic performance improved non-linearly with age and was related to stature, fat-free mass, and motor coordination.	85%
(Valente-dos-Santos et al., 2014) [28]/Portugal	Mixed-longitudinal 5 y	10–14 y at baseline	83	Physical performance: SHR. Soccer specific skills: Dribbling.	Agility development was related to chronological and skeletal age, stature, fat-free mass and playing position. Dribbling changes were related to chronological and skeletal age, stature, playing position and training stimuli.	85%
(Valente-dos-Santos et al., 2014) [29]/Portugal	Mixed-longitudinal 5 y	11–13 y at baseline	83	Physical performance: SHR, MSSE, CMJ. Soccer specific skills: Dribbling.	Agility changes were related to skeletal age, maturity status, fat-free mass, aerobic endurance, and explosive strength. Dribbling changes were associated with skeletal age, maturity states, fat-free mass, aerobic endurance, explosive strength, and training stimuli.	75%
(Wrigley et al., 2014) [36]/United Kingdom	Longitudinal 3 y	U12–U16 age category at baseline	S_g = 27 C_g = 18	Physical performance: S10 m, S20 m, CMJ, agility (505 test). Soccer specific physical performance: RSA, YY IRT2.	Systematic soccer specific training stimuli had significant effects on physical performance changes in young male players independently from baseline levels of performance and biological maturation.	90%
(Bidaurrazaga-Letona et al., 2015) [24]/Spain	Mixed-Longitudinal 4 y	U11 age category at baseline	38	Physical performance: CMJ, agility (barrow zigzag run), S15 m.	Non-linear improvement in explosive strength and agility performance with higher development rates for early matures. However, late matures had better linear improvements in speed performance.	75%
(Deprez et al., 2015) [32]/Belgian	Longitudinal 2 y and 4 y	4 y: ~12 y at baseline 2 y: ~13 y at baseline	21 21	Soccer specific physical performance: YYIR1.	Aerobic performance stability was moderate in 4 y and high over 2 y.	85%
(Deprez et al., 2015) [33]/Belgian	Mixed-Longitudinal 7 y	7–17 y at baseline	555	Physical performance: CMJ, standing broad jump (SBJ).	Explosive strength development was related to chronological age and motor coordination. However, in 11–15 y was positively influenced by stature and negatively by fat mass, but in 16–20 y positively influenced by fat free mass.	75%

Table 2. *Cont.*

Author/Country	Study Design Duration	Participants Age	Number	Motor Performance Assessments (Tests)	Main Results	Quality Score
(Deprez et al., 2015) [34]/Belgian	Mixed-Longitudinal 7 y	11–14 y at baseline	356	Physical performance: CMJ, SBJ.	Explosive strength performance improved non-linearly with age in CMJ test and linearly is SBJ. Additionally, explosive strength performance was related to leg length, fat free mass, flexibility, and motor coordination also maturity status except in SBJ test.	90%
(Forsman et al., 2016) [46]/Finland	Longitudinal 1 y	12–14 y at baseline	288	Physical performance: S30 m, agility (8-figure). Soccer specific skills: Dribbling, passing.	Physical performance and soccer skills remained relatively high and stable across the period of one year.	85%
(Francioni et al., 2016) [20]/Italy	Longitudinal one season	U8-U12 age category at baseline	103	Physical performance: CMJ with and without free arm, S15 m. Soccer specific skills: Touch of the ball with the body and the head, passing, shooting, dribbling, dribbling with pass.	Physical performance and soccer specific skills increased with age in one season.	70%
(Zuber et al., 2016) [21]/Swiss	Longitudinal 3 y	U13 age category at baseline	119	Physical performance: S40 m, CMJ. Soccer specific physical performance: YY IRT1. Soccer specific skills: Dribbling, passing, Juggling.	Change pattern showed to be partial structural with high individual motor performance stability.	70%
(Carvalho et al., 2017) [43]/Spain	Mixed-Longitudinal 6 y	U11 age category at baseline	33	Physical performance: Agility (barrow zigzag run), S15 m, CMJ. Soccer specific physical performance: YYIR1.	Agility and aerobic performance improved non-linearly and reach a steady rate around 3–4 y after PHV. Sprint and explosive strength maximum velocity occurred around 2 y after PHV.	80%
(Fransen et al., 2017) [35]/Belgian	Mixed-Longitudinal 6 y	5–20 y at baseline	2228	Physical performance: Agility (T-Test), S10 m, S20 m, S30 m, SR, hand grip. Soccer specific physical performance: YYIR1.	Linear increases of all physical performance tests. Yet, there is a suggestion of reaching a plateau around 15–17 years of age.	75%
(Rebelo-Goncalves et al., 2017) [22]/Portugal	Mixed-Longitudinal 5 y	11–13 y at baseline	16	Physical performance: MSSE. Soccer specific skills: Dribbling speed, wall pass.	Aerobic capacity and passing skills improved linearly in goalkeepers yet dribbling speed development was non-linear. Soccer skills improvement were mostly explained by training stimuli not by fat-free mass increases.	60%
(Francioni et al., 2018) [44]/Italy	Longitudinal one season	U14 age category at baseline	33	Physical performance: CMJ with and without free arm, S15 m. Soccer specific skills: Touch of the ball with the body and the head, passing, shooting, dribbling, dribbling with pass.	Motor performance improved in U14 age categories during one soccer season independent of training exposure.	80%
(Coutinho et al., 2018) [30]/Portugal	Longitudinal 10 weeks	U15, U17 age category at baseline	E_g = 15 C_g = 15	Physical performance: CMJ, S30 m, agility (repeated change in direction).	Physical performance of U15E improved in 10 weeks. Training had a moderate effect in U15E agility and in U17E CMJ improvements.	95%
(Leyhr et al., 2018) [47]/Germany	Longitudinal 3 y	U12 age category at baseline	1134	Physical performance: S20m, agility (slalom course). Soccer specific skills: Dribbling, ball control, shooting.	Motor performance improved non-linearly with time. Future professional players performed better at baseline and maintained their superiority across time.	80%
(Bennett et al., 2019) [23]/Belgium	Mixed-longitudinal	6–20 y at baseline	2201	Physical performance: CMJ, SBJ.	Explosive strength improved non-linearly with age. The length of the time between assessments did not show a strong impact on player's future performance.	70%

Table 2. *Cont.*

Author/Country	Study Design Duration	Participants		Motor Performance Assessments (Tests)	Main Results	Quality Score
		Age	Number			
(Moran et al., 2020) [37]/United Kingdom	Longitudinal 6 y	U10 age category at baseline	6	Physical performance: S10 m, S20 m, CMJ.	Straight speed and lower limb explosive strength performance can arise rapidly and in radical fashions.	90%
(Saward et al., 2020) [38]/United Kingdom	Longitudinal 11 y	U9–U19 age category at baseline	2875	Physical performance: S20 m, agility (slalom test), CMJ, the multistage fitness tests/ 20 m multi (MSER) (MSFT).	Agility, explosive strength, and speed improved non-linearly except aerobic capacity which improved linearly with age. Differences in playing position occurred in physical performance development. Future professional players had a faster rate as they get older, with different development patterns in explosive strength and agility.	90%

y = years, g = group, Pro = professional, S = soccer, C = control, E = experimental.

3. Results

3.1. Included Studies

Study collection database searches retrieved 267 citations. Figure 1 shows the number of articles found in each electronic database and the literature search/selection processes, including all the steps performed. After exclusion of duplicates, two hundred and five articles remained, and eight additional articles identified through other sources were included in the selection process. The remaining 213 articles were screened based on their title and abstract, and one hundred and seventeen articles were excluded at this stage. The remaining 35 studies were screened for full text assessment. One study did not have its full text available, and two other articles were excluded since they did not precisely examine the development of motor performance variables. Thirty-two articles fulfilled all the inclusion criteria and were chosen at the end of the screening procedure for the in-depth analysis (i.e., qualitative analysis) and review.

3.2. Methodological Quality

Quality scores attributed to studies are found in Table 2 and in the Electronic Supplementary Materials (Table S2). The quality of indicators was as follows: (i) the mean $\pm$ standard deviation score of the 32 articles was 79 $\pm$ 10 percent; (ii) none of the studies achieved the maximum score of 100% or scored below 50% (low quality); (iii) eight articles were classified as of moderate quality (ranging between 51 and 75%) [17–24], but (iv) twenty-four received high methodological quality scores (>75%). Putative deficiencies were mostly related to question 3 (justification of the study sample size), and question 9 (limitations of the study acknowledged).

3.3. Studies' Characteristics

3.3.1. Location

All studies were from the European continent: eight were conducted in Portugal (25.8%) [19,22,25–30], seven in Belgium (22.6%) [17,23,31–35], four in the United Kingdom (12.9%) [36–39], three in the Netherlands (9.7%) [18,40,41], three in Spain (9.7%) [24,42,43], two in Italy (6.5%) [20,44]. Single studies were conducted in Austria [45], Finland [46], Germany [47], Switzerland [21], and Serbia [48] (Figure 2).

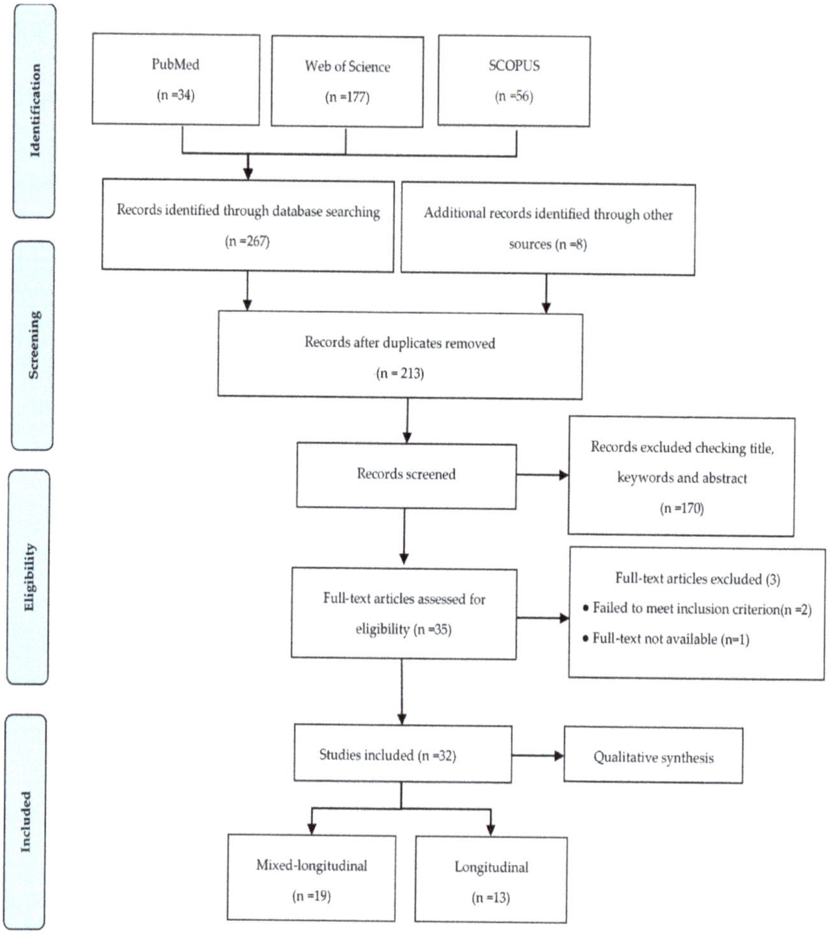

Figure 1. Flow chart including literature search and selection steps following the Preferred Reporting Items for Systematic and Meta-analyses (PRISMA) statement.

3.3.2. Sample Size and Design

Motor performance was investigated in a total of 12,190 youth soccer players, representing an average of ~380 players per study. Nineteen studies used a mixed-longitudinal design, with sample sizes ranging between 16 [22] and 2228 [35], and age ranging from 5 to 20 years. Time durations (serial data collection) ranged from three [41] to nine years [45]. Thirteen studies used a pure longitudinal design lasting from ten weeks [30] to eleven years of a prognostic period [38]. Sample sizes varied from 6 [37] to 2875 subjects [38], and player age ranged from 7 [20] to 19 years [38] (Figure 3).

Figure 2. Number of studies by country.

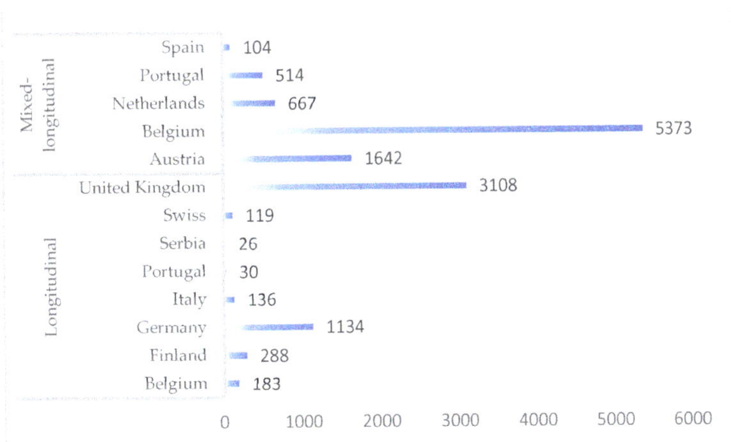

Figure 3. Total number of subjects across studies by country.

3.3.3. Motor Performance Assessments (Tests)

Motor performance, soccer-specific motor performance, and soccer-specific skills were distinctively assessed. Twenty-five studies [17–27,29–31,33–39,43–47] used tests such as plate tapping, sit and reach, sit-ups, bent arm hang, standing long jump, vertical jump with and without free arm, endurance shuttle run, sprints (10, 15, 20, and 30 m), medicine ball throw 2 kg, multi-stage endurance run, agility (505 test, barrow zigzag run, 8-figure T-Test hurdles run, slalom course, slalom running with obstacles) and the multistage fitness tests. Seventeen studies [17–19,21,26–29,31,32,35,36,40,42,43,45,48] assessed soccer-specific motor performance, namely: 30 m repeated sprint (RSA), agility shuttle run 5 × 10 m (SHR), intermittent endurance (ISRT), the Yo-Yo Intermittent Recovery Tests, slalom sprint and shuttle sprint. Additionally, a dozen [18,20–22,27–29,38,41,44,46,47] of them also assessed soccer-specific skills including dribbling, dribbling with a pass, shooting, shooting accuracy, ball control, touch of the ball with the body and the head, juggling, passing, and wall pass.

3.3.4. Changes in Motor Performance

Overall, the reviewed studies aimed at identifying changes in motor performance in different ways: (i) modelling mean trends as well as their covariates [18,19,22,23,25–29,31,33,34,38,40–43,47]; (ii) describing mean changes over time [20,30,32,35,36,39,44–46,48]; (iii) aligning changes by age-at-peak height velocity [37,43]; (iv) identifying timings of spurts in different motor performance tests [17]; and (v) describing patterns of change [21].

Most multilevel/mixed modelling with polynomial age trends (age, age^2, and age^3) showed systematic increases in soccer players' aerobic capacity [25,31,40,42]; however, two did not [22,38]. Training stimuli [25,40] and playing position [38] were linked to aerobic capacity differences, except for goal-keepers [22]; maturity status was not related to these trends [31,42]. There is evidence [24,33,34,38] for lower limb explosive strength (countermovement jump) increasing non-linearly with increasing age, while the increase is linear in standing broad jump test [23]. These increases were related to maturity status [24,34], fat-free mass [34], playing position [38], and previous performance [23]. There are also reports [24,28,29,38,43,47] showing non-linear improvements in change in direction, which were explained by training stimuli [24], fat-free mass [28,29], and playing position [28,38].

Most straight speed [24,38,43,47] and repeated sprint ability [19,26] showed non-linear trends, although one showed a linear trend [24]. Maturity status [24], fat-free mass, and playing position [38] were associated with these changes. Additionally, training stimuli, lower limb explosive strength, and fat-free mass were identified as additional repeated sprint covariates [19,26]. Furthermore, there was also evidence that future professional players had systematic higher physical performance levels than future non-professionals [38,47]. Additionally, non-linear trends were observed in soccer technical skills [22,28,29,41,47]. Players with more training stimuli [22,28,29] and more lower limb explosive strength [29] and midfielders [28] were better regarding dribbling speed.

During soccer seasons, significant differences were evident in motor performance [20,35,39,45]. However, one study did not identify such changes in different age groups [39]. There is also evidence that motor performance remained relatively high and stable during the period of one year [46], and in particular, aerobic capacity showed high stability over two years and moderate stability over four years [32]. Three years of training was associated with changes in physical performance independent of baseline levels and maturational change [36]. Ten weeks of physical and tactical training in small-sided games produced a moderate effect on U15 change in direction, moderate improvements in U17 lower limb explosive strength, and a positive effect on attackers' physical performance [30]. In contrast, one study reported that a season follow-up improved U14 players' motor performance independent of training stimuli [44].

Two studies aligned motor performance changes with age-at-peak height velocity (PHV) [37,43]. A case study reported systematic fluctuations in players' straight speed and lower limb explosive strength performance [37]. On the other hand, the maximum velocity of lower limb explosive strength occurred 2 years after PHV, straight speed was coincident with PHV, whereas change in direction and aerobic capacity started levelling off their increases 3–4 years after PHV [43]. Contrarily, one study showed that almost all physical performance peak spurts occurred at PHV and that a plateau in straight speed, lower limb explosive strength and upper-body endurance development occurred after

PHV [17]. Finally, one study used a person-centered approach aiming to identify players' patterns of change and showed partial structural stable clustering as well as high individual stability [21].

4. Discussion

In this systematic review, our aim was to provide a comprehensive overview of the current body of evidence of existing longitudinal research concerning young soccer players' motor performance. Across studies, there is evidence of motor performance improvements with chronological age, as well as marked influences of biological maturity, body composition and training stimuli. Further, researchers based their analyses and conclusions on data from pure longitudinal and mixed-longitudinal designs. Notably, all studies sampled European adolescent players.

4.1. Study Quality

Overall, studies tended to adhere to high quality standards. Yet, a less favorable point is related to the apparent absence of sample size justification and a putative insufficiency of this aspect is evident when discussing results' generalization. This, in turn, may weaken to a certain degree their external validity [49]. In any case, it is also important to consider pragmatic factors and/or research design requirements when sampling players and have their regular assessments which are often conditioned by their training schedules and academic obligations. This is a viable argument when there is a need for systematic and highly regular assessments [50]. As such, we suggest that future studies should discuss potential flaws of their designs, especially sample size, as well as ways of adequately dealing with missing data [51] before drawing conclusions about the results' transferability to other settings, namely coaches' decisions when planning their training sessions as well as their expectations.

4.2. Location

Although one important aim of the grassroots FIFA program focuses on "Develop the game" for all [52], there apparently is no doubt that appropriate nurturing of young soccer players is time- and money-consuming, as well as being a challenging process [53]. The studies retrieved in this systematic review are from European countries that received some form of funding from their governmental agencies. Furthermore, not only did their progressive governments' sport policies incorporate elements of soccer grassroots programs [54,55], but this is also in the interest of coaches and managers from private soccer clubs. We suggest that future longitudinal research with young soccer players should also be conducted worldwide. This requires, of course, the presence of collaborative research teams from different countries and continents, linking soccer producer countries with those apparently less advanced in terms of research, team building, and soccer education.

4.3. Motor Performance Assessments

Physical performance tests offer objective assessments of young soccer players that can generally be used for different purposes—description of systematic changes and their covariates, selection and placement, assess individual progress, i.e., diagnostics, prediction, and evaluation of training intervention programs [56]. Most reviewed papers dealt with the description of mean changes in important physiological markers such as aerobic capacity, lower limb explosive strength, and speed, by the use of different tests. A similar trend was observed for soccer-specific physical performance and technical skills. In general, technical skills improved with chronological age, as expected from players' regular training schedules. Even though tests were different for measuring the same construct across studies, similar results were identified and may be linked (i) to the fact that tests were appropriate for the age range and sample characteristics, and (ii) to expected changes during adolescence as part of their natural developmental course plus the systematic and cumulative effects of training and competition.

It was found that soccer's physiological demands and technical skills are different for goalkeepers and outfield players [57], and this is probably the main reason why most studies [18,25,28,41] decided to not include goalkeepers during data sampling and/or their analysis. However, Rebelo-Gonçalves et al. [22] sampled 16 goalkeepers that were similarly tested (aerobic capacity, dribbling and passing skills) as their outfield players peers. However, in a previous study [28], using the sample from the same research project from Rebelo-Gonçalves et al. [22], the authors decided to exclude goalkeepers during sample selection/data analysis because the sample size was very small. Hence, we emphasize the need for future research to direct its goal to goalkeepers' physical performance characteristics and as well as their specific technical skills.

4.4. Statistical Procedures and Changes in Motor Performance

There apparently is no specific trend in statistical procedures used to analyse motor performance changes across studies' publication years. Most studies [18,19,22,23,25–29,31,33,34,39–43,47] used multilevel/mixed modelling independent of study design, duration, and sample size. In general, they relied on polynomials of age (age, age^2, age^3) to estimate motor performance mean trends (linear and non-linear), as well as adding different predictors of such trends, namely training stimuli and maturity status [22,24,25,28,29,34,40], and reported their different effect sizes.

When focusing on mean changes across time, there apparently is no parallel trend across studies. For example, when using independent factors as group—control versus experimental [48]—or players' levels [30,45], training effects on motor performance as well as its stability vary by using different statistical methods such as analysis of variance [30,37] or the general linear model [35]. Yet, we were not able to localize a study that investigated the tracking of players' motor performance, notwithstanding the fact that stability of changes was mentioned [32,46]. One study [17] used a non-smoothed polynomial method to identify spurts in several physical performance markers aligned by age-at-PHV, and showed that in spite of their different intensities, they tend to peak around PHV. Another study used a person-centered approach with a cluster analysis to describe players' patterns of change and obtained partial structural stable clustering along with high individual stability [21]. One case study [37] showed that physical performance trajectories are irregular, occurring quickly and in a radical fashion, suggesting that individual differences between soccer players tend to be temporary. We concur with the authors that there is a novel need to longitudinally investigate young players as single cases, aiming to gain a better understanding of their erratic and systematic changes in order to assist coaches when structuring their training program as well as when making selection decisions.

There is a strong suggestion that motor performance changes are related to biological maturation differences, between and within players, as well as their training stimuli. Yet, there apparently is no unequivocal proof of the effects of different training interventions and bio-banding [58] on players' motor performance. Therefore, we recommend additional research for a deeper understanding of the impact of training interventions on motor performance during puberty, especially their hormonal and physiological mechanisms. Additionally, we could not find a theoretical basis for conducting research apart from using ANOVAs or the multilevel/mixed model. We contend that future research should also consider players' contexts, i.e., their families, coaches, and clubs. Young players' development occurs within these contexts and they should be acknowledged. In sum, there is a need to also use multidimensional and/or ecological approaches to enhance our understanding of the complexities of young players' development [8].

4.5. Limitations of the Current Review

This is most probably the first systematic review on young male soccer players' motor performance development based on serial data (pure longitudinal and mixed-longitudinal), and it is not without limitations. First, it is possible that the retrieved publications are not free from bias towards positive results. As such, we suggest future studies to combine available data for meta-analyses with proper statistical evaluation of publication bias. Second, we restricted our criteria to only include male players. We urge future research to also consider female players' motor performance serial data. Third, it is also possible that the review criteria and search strategy may have limited our scan. Fourth, although no study used in this systematic review reported injuries or orthopedic problems, care must be taken when interpreting data because of a putative equinus condition [59].

In spite of these limitations, we tried to present a comprehensive description of available longitudinal data during players' puberty given that it is considered a very important time window that may likely benefit soccer stakeholders to employ better developmental sporting strategies in their organizations to maximize young soccer players' potentials and smooth their career transitions.

5. Conclusions

The present study compiled current empirical evidence on longitudinal data dealing with male soccer players' motor performance changes during puberty. Puberty has been found to be a crucial time for nourishing soccer players' future quality vocations. Amongst studies, it was observed that motor performance improved with chronological age, which was linked to biological maturity, body composition changes and training stimuli. Coaches and stakeholder of young soccer players should be aware of the positive influence of physical and biological maturation, training stimuli and systematic fluctuations on players' immediate motor performance. This suggests that selection and deselection decisions should be made based on longitudinal rather than cross-sectional information. We propose that future longitudinal studies with young soccer players should also be global, with a focus on playing position, cases study, tracking methods, and deeper understanding of the impact of training interventions on motor performance during puberty, especially their hormonal and physiological mechanisms. Finally, there is a need for more research on the contextual and environmental aspects impacting motor performance development.

Supplementary Materials: The following are available online at https://www.mdpi.com/article/10.3390/sports9040053/s1, Table S1: Full search strategy for each database with arguments presented as they were used, Table S2: Scores attributed to each study according to twelve criteria used in evaluating methodological quality.

Author Contributions: Conceptualization, J.M. and M.A.; methodology, M.A. and T.N.G.; writing—original draft preparation, J.M., D.B., A.D.G.B.-J. and M.A.; writing—review and editing, J.M., M.A., D.B., A.D.G.B.-J. and T.N.G.; supervision, J.M., A.D.G.B.-J. and D.B. All authors have read and agreed to the published version of the manuscript.

Funding: This research was funded by the Portuguese Institute of Sports and Youth (IPDJ), grant number CP/594/DD/2018; and the Portuguese Olympic Committee (COP).

Institutional Review Board Statement: Not applicable.

Informed Consent Statement: Not applicable.

Data Availability Statement: Not applicable.

Conflicts of Interest: The authors declare no conflict of interest.

References

1. Haugaasen, M.; Jordet, G. Developing football expertise: A football-specific research review. *Int. Rev. Sport Exerc. Psychol.* **2012**, *5*, 177–201. [CrossRef]
2. Williams, A.M.; Ford, P.R.; Drust, B. Talent identification and development in soccer since the millennium. *J. Sports Sci.* **2020**, *38*, 1199–1210. [CrossRef]
3. Drust, B. Applied science and soccer: A personal perspective on the past, present and future of a discipline. *Sport. Perf Sci. Rep.* **2019**, *56*, 1–7.
4. Abarghoueinejad, M.; Barreira, D.; Dias, C.; Guimarães, E.; Baxter-Jones, A.D.G.; Maia, J. Body Physique, Body Composition, Physical Performance, Technical and Tactical Skills, Psychological Development, and Club Characteristics of Young Male Portuguese Soccer Players: The INEX Study. *Int. J. Environ. Res. Public Health* **2021**, *18*, 3560. [CrossRef]

5. Reeves, M.J.; Roberts, S.J. A bioecological perspective on talent identification in junior-elite soccer: A Pan-European perspective. *J. Sports Sci.* **2020**, *38*, 1259–1268. [CrossRef] [PubMed]
6. Ford, P.R.; Bordonau, J.L.D.; Bonanno, D.; Tavares, J.; Groenendijk, C.; Fink, C.; Gualtieri, D.; Gregson, W.; Varley, M.C.; Weston, M.; et al. A survey of talent identification and development processes in the youth academies of professional soccer clubs from around the world. *J. Sports Sci.* **2020**, *38*, 1269–1278. [CrossRef]
7. Bergeron, M.F.; Mountjoy, M.; Armstrong, N.; Chia, M.; Côté, J.; Emery, C.A.; Faigenbaum, A.; Hall, G.; Kriemler, S.; Léglise, M. International Olympic Committee consensus statement on youth athletic development. *Br. J. Sports Med.* **2015**, *49*, 843–851. [CrossRef]
8. Lloyd, R.S.; Cronin, J.B.; Faigenbaum, A.D.; Haff, G.G.; Howard, R.; Kraemer, W.J.; Micheli, L.J.; Myer, G.D.; Oliver, J.L. National Strength and Conditioning Association Position Statement on Long-Term Athletic Development. *J. Strength Cond. Res.* **2016**, *30*, 1491–1509. [CrossRef] [PubMed]
9. Palucci Vieira, L.H.; Carling, C.; Barbieri, F.A.; Aquino, R.; Santiago, P.R.P. Match Running Performance in Young Soccer Players: A Systematic Review. *Sports Med.* **2019**, *49*, 289–318. [CrossRef] [PubMed]
10. Sarmento, H.; Anguera, M.T.; Pereira, A.; Araujo, D. Talent Identification and Development in Male Football: A Systematic Review. *Sports Med.* **2018**, *48*, 907–931. [CrossRef] [PubMed]
11. Slimani, M.; Nikolaidis, P.T. Anthropometric and physiological characteristics of male soccer players according to their competitive level, playing position and age group: A systematic review. *J. Sports Med. Phys. Fit.* **2019**, *59*, 141–163. [CrossRef]
12. Johnston, K.; Wattie, N.; Schorer, J.; Baker, J. Talent Identification in Sport: A Systematic Review. *Sports Med.* **2018**, *48*, 97–109. [CrossRef]
13. Bergkamp, T.L.G.; Niessen, A.S.M.; den Hartigh, R.J.R.; Frencken, W.G.P.; Meijer, R.R. Methodological Issues in Soccer Talent Identification Research. *Sports Med.* **2019**, *49*, 1317–1335. [CrossRef] [PubMed]
14. Shamseer, L.; Moher, D.; Clarke, M.; Ghersi, D.; Liberati, A.; Petticrew, M.; Shekelle, P.; Stewart, L.A.; Group, P.-P. Preferred reporting items for systematic review and meta-analysis protocols (PRISMA-P) 2015: Elaboration and explanation. *BMJ* **2015**, *350*, g7647. [CrossRef] [PubMed]
15. Shamseer, L.; Sampson, M.; Bukutu, C.; Schmid, C.H.; Nikles, J.; Tate, R.; Johnston, B.C.; Zucker, D.; Shadish, W.R.; Kravitz, R.; et al. CONSORT extension for reporting N-of-1 trials (CENT) 2015: Explanation and elaboration. *J. Clin. Epidemiol.* **2016**, *76*, 18–46. [CrossRef] [PubMed]
16. Sarmento, H.; Clemente, F.M.; Araújo, D.; Davids, K.; McRobert, A.; Figueiredo, A. What Performance Analysts Need to Know About Research Trends in Association Football (2012–2016): A Systematic Review. *Sports Med.* **2018**, *48*, 799–836. [CrossRef]
17. Philippaerts, R.M.; Vaeyens, R.; Janssens, M.; Van Renterghem, B.; Matthys, D.; Craen, R.; Bourgois, J.; Vrijens, J.; Beunen, G.; Malina, R.M. The relationship between peak height velocity and physical performance in youth soccer players. *J. Sports Sci.* **2006**, *24*, 221–230. [CrossRef] [PubMed]
18. Huijgen, B.C.; Elferink-Gemser, M.T.; Post, W.; Visscher, C. Development of dribbling in talented youth soccer players aged 12–19 years: A longitudinal study. *J. Sports Sci.* **2010**, *28*, 689–698. [CrossRef]
19. Valente-dos-Santos, J.; Coelho-e-Silva, M.J.; Martins, R.A.; Figueiredo, A.J.; Cyrino, E.S.; Sherar, L.B.; Vaeyens, R.; Huijgen, B.C.H.; Elferink-Gemser, M.T.; Malina, R.M. Modelling Developmental Changes in Repeated-Sprint Ability by Chronological and Skeletal Ages in Young Soccer Players. *Int. J. Sports Med.* **2012**, *33*, 773–780. [CrossRef] [PubMed]
20. Francioni, F.M.; Figueiredo, A.J.; Terribili, M.; Tessitore, A. Analysis of the intraseasonal stability of field test performances in young academy soccer players. *J. Sports Sci.* **2016**, *34*, 966–972. [CrossRef]
21. Zuber, C.; Zibung, M.; Conzelmann, A. Holistic Patterns as an Instrument for Predicting the Performance of Promising Young Soccer Players—A 3-Years Longitudinal Study. *Front. Psychol.* **2016**, *7*, 1088. [CrossRef]
22. Rebelo-Goncalves, R.; Coelho-e-Silva, M.J.; Valente-dos-Santos, J.; Tessitore, A.; Figueiredo, A.J. Longitudinal study of aerobic performance and soccer-specific skills in male goalkeepers aged 11–18 years. *Sci. Med. Footb.* **2017**, *1*, 40–47. [CrossRef]
23. Bennett, N.; Woodcock, S.; Pluss, M.A.; Bennett, K.J.M.; Deprez, D.; Vaeyens, R.; Lenoir, M.; Fransen, J. Forecasting the development of explosive leg power in youth soccer players. *Sci. Med. Footb.* **2019**, *3*, 131–137. [CrossRef]
24. Bidaurrazaga-Letona, I.; Carvalho, H.M.; Lekue, J.A.; Santos-Concejero, J.; Figueiredo, A.J.; Gil, S.M. Longitudinal Field Test Assessment in a Basque Soccer Youth Academy: A Multilevel Modeling Framework to Partition Effects of Maturation. *Int. J. Sports Med.* **2015**, *36*, 234–240. [CrossRef] [PubMed]
25. Valente-dos-Santos, J.; Coelho-e-Silva, M.J.; Duarte, J.; Figueiredo, A.J.; Liparotti, J.R.; Sherar, L.B.; Elferink-Gemser, M.T.; Malina, R.M. Longitudinal Predictors of Aerobic Performance in Adolescent Soccer Players. *Medicina (Kaunas)* **2012**, *48*, 410–416. [CrossRef]
26. Valente-dos-Santos, J.; Coelho-e-Silva, M.J.; Severino, V.; Duarte, J.; Martins, R.S.; Figueiredo, A.J.; Seabra, A.T.; Philippaerts, R.M.; Cumming, S.P.; Elferink-Gemser, M.; et al. Longitudinal study of repeated sprint performance in youth soccer players of contrasting skeletal maturity status. *Int. J. Sports Med.* **2012**, *11*, 371–379.
27. Valente-dos-Santos, J.; Coelho-e-Silva, M.J.; Simoes, F.; Figueiredo, A.J.; Leite, N.; Elferink-Gemser, M.T.; Malina, R.M.; Sherar, L. Modeling Developmental Changes in Functional Capacities and Soccer-Specific Skills in Male Players Aged 11–17 Years. *Pediatr. Exerc. Sci.* **2012**, *24*, 603–621. [CrossRef]

28. Valente-dos-Santos, J.; Coelho-e-Silva, M.J.; Duarte, J.; Pereira, J.; Rebelo-Gonçalves, R.; Figueiredo, A.; Mazzuco, M.A.; Sherar, L.B.; Elferink-Gemser, M.T.; Malina, R.M. Allometric multilevel modelling of agility and dribbling speed by skeletal age and playing position in youth soccer players. *Int. J. Sports Med.* **2014**, *35*, 762–771. [CrossRef] [PubMed]
29. Valente-dos-Santos, J.; Coelho-e-Silva, M.J.; Vaz, V.; Figueiredo, A.J.; Capranica, L.; Sherar, L.B.; Elferink-Gemser, M.T.; Malina, R.M. Maturity-associated variation in change of direction and dribbling speed in early pubertal years and 5-year developmental changes in young soccer players. *J. Sports Med. Phys. Fit.* **2014**, *54*, 307–316.
30. Coutinho, D.; Santos, S.; Gonçalves, B.; Travassos, B.; Wong, D.P.; Schöllhorn, W.; Sampaio, J. The effects of an enrichment training program for youth football attackers. *PLoS ONE* **2018**, *13*, e0199008. [CrossRef] [PubMed]
31. Deprez, D.; Valente-dos-Santos, J.; Silva, M.C.; Lenoir, M.; Philippaerts, R.M.; Vaeyens, R. Modeling Developmental Changes in the Yo-Yo Intermittent Recovery Test Level 1 in Elite Pubertal Soccer Players. *Int. J. Sports Physiol.* **2014**, *9*, 1006–1012. [CrossRef]
32. Deprez, D.; Buchheit, M.; Fransen, J.; Pion, J.; Lenoir, M.; Philippaerts, R.M.; Vaeyens, R. A longitudinal study investigating the stability of anthropometry and soccer-specific endurance in pubertal high-level youth soccer players. *J. Sports Sci. Med.* **2015**, *14*, 418–426.
33. Deprez, D.; Valente-Dos-Santos, J.; Coelho-e-Silva, M.J.; Lenoir, M.; Philippaerts, R.; Vaeyens, R. Longitudinal Development of Explosive Leg Power from Childhood to Adulthood in Soccer Players. *Int. J. Sports Med.* **2015**, *36*, 672–679. [CrossRef] [PubMed]
34. Deprez, D.; Valente-Dos-Santos, J.; Coelho-e-Silva, M.J.; Lenoir, M.; Philippaerts, R.; Vaeyens, R. Multilevel Development Models of Explosive Leg Power in High-Level Soccer Players. *Med. Sci. Sports Exerc.* **2015**, *47*, 1408–1415. [CrossRef]
35. Fransen, J.; Bennett, K.J.M.; Woods, C.T.; French-Collier, N.; Deprez, D.; Vaeyens, R.; Lenoir, M. Modelling age-related changes in motor competence and physical fitness in high-level youth soccer players: Implications for talent identification and development. *Sci. Med. Footb.* **2017**, *1*, 203–208. [CrossRef]
36. Wrigley, R.D.; Drust, B.; Stratton, G.; Atkinson, G.; Gregson, W. Long-term Soccer-specific Training Enhances the Rate of Physical Development of Academy Soccer Players Independent of Maturation Status. *Int. J. Sports Med.* **2014**, *35*, 1090–1094. [CrossRef]
37. Moran, J.; Paxton, K.; Jones, B.; Granacher, U.; Sandercock, G.R.; Hope, E.; Ramirez-Campillo, R. Variable long-term developmental trajectories of short sprint speed and jumping height in English Premier League academy soccer players: An applied case study. *J. Sports Sci.* **2020**, *38*, 2525–2531. [CrossRef] [PubMed]
38. Saward, C.; Hulse, M.; Morris, J.G.; Goto, H.; Sunderland, C.; Nevill, M.E. Longitudinal Physical Development of Future Professional Male Soccer Players: Implications for Talent Identification and Development? *Front. Sports Act. Living* **2020**, *2*, 578203. [CrossRef] [PubMed]
39. Williams, C.A.; Oliver, J.L.; Faulkner, J. Seasonal monitoring of sprint and jump performance in a soccer youth academy. *Int. J. Sports Physiol. Perform.* **2011**, *6*, 264–275. [CrossRef]
40. Roescher, C.R.; Elferink-Gemser, M.T.; Huijgen, B.C.; Visscher, C. Soccer endurance development in professionals. *Int. J. Sports Med.* **2010**, *31*, 174–179. [CrossRef]
41. Huijgen, B.C.; Elferink-Gemser, M.T.; Ali, A.; Visscher, C. Soccer skill development in talented players. *Int. J. Sports Med.* **2013**, *34*, 720–726. [CrossRef]
42. Carvalho, H.M.; Bidaurrazaga-Letona, I.; Lekue, J.A.; Amado, M.; Figueiredo, A.J.; Gil, S.M. Physical Growth and Changes in Intermittent Endurance Run Performance in Young Male Basque Soccer Players. *Res. Sports Med.* **2014**, *22*, 408–424. [CrossRef]
43. Carvalho, H.M.; Lekue, J.A.; Gil, S.M.; Bidaurrazaga-Letona, I. Pubertal development of body size and soccer-specific functional capacities in adolescent players. *Res. Sports Med.* **2017**, *25*, 421–436. [CrossRef] [PubMed]
44. Francioni, F.M.; Figueiredo, A.J.; Lupo, C.; Terribili, M.; Condello, G.; Tessitore, A. Intra-seasonal variation of anthropometrical, conditional, and technical tests in U14 soccer players. *Rev. Int. Cienc. Deporte.* **2018**, *14*, 219–232. [CrossRef]
45. Gonaus, C.; Muller, E. Using physiological data to predict future career progression in 14-to 17-year-old Austrian soccer academy players. *J. Sports Sci.* **2012**, *30*, 1673–1682. [CrossRef]
46. Forsman, H.; Gråstén, A.; Blomqvist, M.; Davids, K.; Liukkonen, J.; Konttinen, N. Development of perceived competence, tactical skills, motivation, technical skills, and speed and agility in young soccer players. *J. Sports Sci.* **2016**, *34*, 1311–1318. [CrossRef]
47. Leyhr, D.; Kelava, A.; Raabe, J.; Honer, O. Longitudinal motor performance development in early adolescence and its relationship to adult success: An 8-year prospective study of highly talented soccer players. *PLoS ONE* **2018**, *13*, e0196324. [CrossRef]
48. Mirkov, D.M.; Kukolj, M.; Ugarkovic, D.; Koprivica, V.J.; Jaric, S. Development of anthropometric and physical performance profiles of young elite male soccer players: A longitudinal study. *J. Strength Cond. Res.* **2010**, *24*, 2677–2682. [CrossRef]
49. Hedeker, D.; Gibbons, R.D.; Waternaux, C. Sample size estimation for longitudinal designs with attrition: Comparing time-related contrasts between two groups. *J. Educ. Behav. Stat.* **1999**, *24*, 70–93. [CrossRef]
50. Vasileiou, K.; Barnett, J.; Thorpe, S.; Young, T. Characterising and justifying sample size sufficiency in interview-based studies: Systematic analysis of qualitative health research over a 15-year period. *BMC Med. Res. Methodol.* **2018**, *18*, 148. [CrossRef]
51. Hedeker, D.; Gibbons, R.D. *Longitudinal Data Analysis*; John Wiley & Sons: Hoboken, NJ, USA, 2006; p. 337.
52. Avry, Y.; Bernet, M.; Cornea, A.; Malouche, B.; Thondoo, G.; FIFA Education and Technical Development Department. Grassroots. (N/A). Available online: https://www.soccercoach.eu/fifa/grassroots_en.pdf (accessed on 22 March 2021).
53. Baker, J.; Schorer, J.; Wattie, N. Compromising Talent: Issues in Identifying and Selecting Talent in Sport. *Quest* **2018**, *70*, 48–63. [CrossRef]
54. UK Parliament. Funding and Supporting Grassroots Sports. 2012. Available online: https://publications.parliament.uk/pa/ld201011/ldselect/ldeucom/201130/213007.htm (accessed on 22 March 2021).

55. European Commission. Report to Commissioner Tibor Navracsics, Grassroots Sport—Shaping Europe. 2016. Available online: http://hu.eusport.org/files/242-nc0418026enn.en.pdf (accessed on 22 March 2021).
56. Safrit, M.J. *Introduction to Measurement in Physical Education and Exercise Science*; Mosby College Publishing: St. Louis, MO, USA, 1990.
57. Bangsbo, J.; Mohr, M.; Krustrup, P. Physical and metabolic demands of training and match-play in the elite football player. *J. Sports Sci.* **2006**, *24*, 665–674. [CrossRef] [PubMed]
58. Cumming, S.P.; Lloyd, R.S.; Oliver, J.L.; Eisenmann, J.C.; Malina, R.M. Bio-banding in Sport: Applications to Competition, Talent Identification, and Strength and Conditioning of Youth Athletes. *Strength Cond. J.* **2017**, *39*, 34–47. [CrossRef]
59. Rodriguez-Sanz, D.; Losa-Iglesias, M.E.; Becerro de Bengoa-Vallejo, R.; Palomo-Lopez, P.; Beltran-Alacreu, H.; Calvo-Lobo, C.; Navarro-Flores, E.; Lopez-Lopez, D. Skin temperature in youth soccer players with functional equinus and non-equinus condition after running. *J. Eur. Acad. Dermatol. Venereol.* **2018**, *32*, 2020–2024. [CrossRef] [PubMed]

MDPI
St. Alban-Anlage 66
4052 Basel
Switzerland
Tel. +41 61 683 77 34
Fax +41 61 302 89 18
www.mdpi.com

Sports Editorial Office
E-mail: sports@mdpi.com
www.mdpi.com/journal/sports

www.ingramcontent.com/pod-product-compliance
Lightning Source LLC
Chambersburg PA
CBHW061502220926
51850CB00004B/153